Toxicology of the Human Environment

Toxicology of the Human Environment

The critical role of free radicals

Edited by

Christopher J. Rhodes

School of Pharmacy and Chemistry,
Liverpool John Moores University, Liverpool, UK

CRC Press
Taylor & Francis Group
Boca Raton London New York

CRC Press is an imprint of the
Taylor & Francis Group, an **informa** business
A TAYLOR & FRANCIS BOOK

First published 2000 by Taylor & Francis

Published 2018 by CRC Press
Taylor & Francis Group
6000 Broken Sound Parkway NW, Suite 300
Boca Raton, FL 33487-2742

ISBN 13: 978-0-7484-0916-7 (hbk)

Visit the Taylor & Francis Web site at
http://www.taylorandfrancis.com

and the CRC Press Web site at
http://www.crcpress.com

Typeset in Sabon by Graphicraft Limited, Hong Kong

British Library Cataloguing in Publication Data
A catalogue record for this book is available from the British Library

Library of Congress Cataloging in Publication Data
Toxicology of the human environment : the critical role of free
radicals / edited by Christopher J. Rhodes.
p. cm.
Includes bibliographical references and index.
(alk. paper)
1. Environmental toxicology. 2. Free radicals (Chemistry) – Toxicology. 3. Nitroxides – Toxicology. 4. Electron paramagnetic resonance spectroscopy. I. Rhodes, Christopher J.
RA1226.T69 2000 99-31651
615.9′02–dc21

Contents

Contributors

Emanuele Albano, Dept of Medical Sciences, University of East Piedmont 'A. Avogadro', Novara, Italy.

Manuela Aragno, Department of Experimental Medicine and Oncology, Section of General Pathology, University of Torino, Torino, Italy.

Carmen M. Arroyo, Drug Assessment Division, US Army Medical Research Institute of Chemical Defense, Aberdeen Proving Ground, Maryland, USA.

Steven D. Aust, Biotechnology Center, Utah State University, Logan, Utah, USA.

Rodney Bilton, School of Biomolecular Sciences, Liverpool John Moores University, Liverpool, UK.

John Butler, Department of Biological Sciences, Salford University, UK and CRC Section of Drug Development and Imaging, Paterson Institute for Cancer Research, Christie Hospital NHS Trust, Manchester, UK.

Simonetta Camandola, Department of Experimental Medicine and Oncology, Section of General Pathology, University of Torino, Torino, Italy.

Elena Chiarpotto, Department of Experimental Medicine and Oncology, Section of General Pathology, University of Torino, Torino, Italy.

Colin F. Chignell, Laboratory of Pharmacology and Chemistry, National Institute of Environmental Health Services, National Institutes of Health, Research Triangle Park, North Carolina, USA.

Andrew W.D. Claxson, Bone and Joint Research Unit, St Bartholomew's and the Royal London Hospitals School of Medicine and Dentistry, London, UK.

Mark T.D. Cronin, School of Pharmacy and Chemistry, Liverpool John Moores University, UK.

Oliviero Danni, Department of Experimental Medicine and Oncology, Section of General Pathology, University of Torino, Torino, Italy.

Lennart Eberson, Department of Chemistry, Lund University, Lund, Sweden.

Peter Evans, Department of Public Health, University of Glasgow, UK.

Marco Ferrari, Department of Biomedical Sciences and Technology, University of L'Aquila, Italy.

Henry Jay Forman, Department of Environmental Health Sciences, University of Alabama, Birmingham, Alabama, USA.

Jürgen Fuchs, Department of Dermatology, Medical School, J.W. Goethe University, Frankfurt, Germany.

Martin Grootveld, Bone and Joint Research Unit, St Bartholomew's and the Royal London Hospitals School of Medicine and Dentistry, London, UK.

Stephen M. Hahn, Department of Radiation Oncology, University of Pennsylvania, Philadelphia, USA.

Dan Harris, Molecular Research Institute, Palo Alto, California, USA.

Anna Iannone, Department of Biomedical Sciences, University of Modena, Italy.

Nalini Kaul, Department of Pathology, University of Texas Southwestern, Dallas, Texas, USA.

Frank J. Kelly, Centre for Cardio+Vascular Biology and Medicine, King's College, and Rayne Institute, St Thomas' Hospital, London, England.

Murali C. Krishna, Radiation Biology Branch, National Cancer Institute, National Institutes of Health, Bethesda, Maryland, USA.

Periannan Kuppusamy, EPR Center, Johns Hopkins University, School of Medicine, Baltimore, Maryland, USA.

Joseph R. Landolph, School of Medicine and Pharmacy, University of Southern California, Los Angeles, California, USA.

Ronald P. Mason, Laboratory of Pharmacology and Chemistry, National Institute of Environmental Health Services, National Institutes of Health, Research Triangle Park, North Carolina, USA.

James B. Mitchell, Radiation Biology Branch, National Cancer Institute, National Institutes of Health, Bethesda, Maryland, USA.

Giuseppe Poli, Department of Experimental Medicine and Oncology, Section of General Pathology, University of Torino, Torino, Italy.

Valentina Quaresima, Department of Biomedical Sciences and Technology, University of L'Aquila, Italy.

Christopher A. Reilly, Biotechnology Center, Utah State University, Logan, Utah, USA.

Christopher J. Rhodes, School of Pharmacy and Chemistry, Liverpool John Moores University, Byrom Street, Liverpool, UK.

Angelo Russo, Radiation Biology Branch, National Cancer Institute, National Institutes of Health, Bethesda, Maryland, USA.

Amram Samuni, Molecular Biology, School of Medicine, Hebrew University, Jerusalem 91010, Israel.

Christopher J.L. Silwood, Bone and Joint Research Unit, St Bartholomew's and the Royal London Hospitals School of Medicine and Dentistry, London, UK.

Harold M. Swartz, EPR Center for the Study of Viable Systems, Department of Radiology, Dartmouth Medical School, Hanover, New Hampshire, USA.

Graham S. Timmins, Departments of Microbiology and Cardiology, University of Wales College of Medicine, Heath Hospital, Cardiff.

Aldo Tomasi, Department of Biomedical Sciences, University of Modena, Italy.

Rheal A. Towner, The Department of Physiology & Pharmacology, and the North Queensland Magnetic Resonance Centre, James Cook University, Townsville, Queensland, Australia.

Thuy T. Tran, School of Pharmacy and Chemistry, Liverpool John Moores University, UK.

Abbreviations

α-1PI	α-1 protease inhibitor
β_e	the Bohr magneton
ΔE	difference in energy of spin states
ν	frequency
π	pi
a_H a_H^β	hydrogen hyperfine splitting constant for a nitroxide spin adduct
AM	alveolar macrophage
ARDS	adult respiratory distress syndrome
ARE	antioxidant responsive elements
BAL	broncoalveolar lavage
BAPN	*N-tert*-butyl-α-phenylnitrone
BIM	biologically important molecules
BM_2PO	2-*tert*-butyl-5,5-dimethyl-1-pyrroline-*N*-oxide
CARET	carotene and retinol efficiency trial
Cat_1	4-trimethylammonium-2,2,6,6-tetramethylpiperidine-1-oxyl
$^{\bullet}CCl_3$	trichloromethyl radical
CCl_4	carbon tetrachloride (tetrachloromethane)
CF	cystic fibrosis
CHD	coronary heart disease
$^{\bullet}CO_2^-$	carbon dioxide radical anion
CRC	colorectal cancer
DBNBS	3,5-dibromo-4-nitrosobenzenesulphonate
DCT 1	divalent cation transporter
DDE	*trans*-2,4-decadienal
DEPMPO	5-diethoxyphosphoryl-5-methyl-1-pyrroline-*N*-oxide
DFT	density functional theory
DMPC	dimyristoyl phosphatidylcholine
DMPO	5,5-dimethyl-1-pyrroline-*N*-oxide
DMPO–OH	5,5-dimethyl-1-pyrroline-*N*-oxide–hydroxyl radical adduct
DMPOX	5,5-dimethyl-1-pyrroline-*N*-oxide product with hydroxamic acid
DNCB	2,4-dinitro-1-chlorobenzene
4-DOPBN	4-(dodecyloxy)-phenyl-*tert*-butyl nitrone
Doxo	2,2,5,5-tetramethyl-3-oxazolidinoxyl
EDTA	ethylene diamine tetra acetic acid
EGF	epidermal growth factor
ELISA	enzyme-linked immunosorbent assay

EPR	electron paramagnetic resonance
ESR	electron spin resonance
FEV1	forced expiratory volume in 1 second
G	gauss (SI unit)
g	gram (SI unit)
GC–MS	gas chromatography–mass spectrometry
g	g-factor
GHz	gigahertz
H	magnetic field
h	Planck's constant
HD	sulphur mustard
HH	hereditary haemochromatosis
H-MG	half-sulfur mustard
HNE	4-hydroxynonenal
HOMO	highest occupied molecular orbital
HPLC–ESR	high performance liquid chromatography–ESR spectrometry
I	nuclear spin quantum number
IBD	irritable bowel disorder
IgG	immunoglobulins
IL-1	interleukin-1
Imidazo	2,2,3,4,5,5-hexamethyl-imidazoline-1-yloxyl
I/O	iron overload
IP_3	inositol-1,4,5-trisphosphate
IRE	iron response element
LC–MS	liquid chromatography–mass spectrometry
LDL	low density lipoprotein
L-NOARG	*N*-nitro-L-arginine
LUMO	lowest occupied molecular orbital
M	molar concentration
MDA	malondialdehyde
MDL 101,002	3,4-dihydro-3,3-dimethylisoquinoline-2-oxide (cyclic analog of PBN)
MEP	molecular electrostatic potential
mg	milligram
MHz	megahertz
M_I	spin state
mM	millimolar concentration
MMC	mitomycin C
MMMF	man-made mineral fibres
MNP	2-methyl-2-nitrosopropane
M_3PO	2-methyl-5,5-dimethyl-1-pyrroline-*N*-oxide
M_s	electron spin quantum number
mT	millitesla
n	number of magnetically equivalent nuclei
NBT	nitroblue-tetrazolium
NF-κB	nuclear factor-kappa B
NHEK	human epidermal keratinocytes
NO	nitric oxide

NOS	nitric oxide synthase
$O_2^{-•}$	superoxide radical anion
•OH	hydroxyl radical
PAH	polycyclic aromatic hydrocarbons
PBN	phenyl-*tert*-butyl nitrone
PCA	(proxylcarboxylic acid) 2,2,5,5-tetramethylpyrroline-1-oxyl-3-carboxylic acid
PDGF	platelet derived growth factor
PDT	(perdeuterated tempone) 4-oxo-2,2,6,6-tetramethylpiperidine-1-oxyl
PhM_2PO	2-phenyl-5,5-dimethyl-1-pyrroline-*N*-oxide
PLD	phospholipase D
PMA	phorbol myristate acetate
PMN	polymorphonuclear leucocyte
POBN	α(4-pyridyl-1-oxide)-*N*-*tert*-butyl nitrone
Proxo	2,2,5,5-tetramethyl-1-dihydro-pyrrolinoxyl
PROXYL	2,2,5,5-tetramethylpyrroline-*N*-oxyl
PUFA	polyunsaturated fatty acid
PVPNO	polyvinylpyridine-*N*-oxide
Q	1,4-benzoquinone
QSAR	quantitative structure–activity relationships
R	functional group
R•	free radical
ROS	reactive oxygen species
SCF	self-consistent field
SCMPO	5-carboxyl-5-methyl-1-pyrroline-*N*-oxide
SDS	sodium dodecyl sulfate (identical to SLS)
SLS	sodium lauryl sulfate (identical to SDS)
SOD	superoxide dismutase
SQ	semiquinone
ST	spin-trap
T	tesla
TBARS	thiobarbituric acid reactive substances
TEAC	trolux equivalent antioxidant capacity
TEMPO	2,2,6,6-tetramethylpiperidine-1-oxyl
TEMPOL	4-hydroxy-2,2,6,6-tetramethylpiperidine-1-oxyl
tert	tertiary
TEWL	transepidermal water loss
TLCK	N^{α}*p*-tosyl-L-lysine chloromethyl ketone
TMPO	3,3,5,5-tetramethyl-1-pyrroline-*N*-oxide
TNF	tumour necrosis factor
TOLH	2,2,6,6-tetramethyl-1-hydroxypiperadine

1 Introduction

Christopher J. Rhodes

The environment in which humans live is complex, and we may encounter many different potentially toxic chemical substances during the course of our lives. It is becoming widely recognised that free radicals are involved in this toxicity, and that these chemicals (xenobiotics) may invade the living system in the form of environmental pollutants, in the diet, as pharmacologically administered compounds or even as chemical weapons. Organic radicals produced from a given material may be involved directly in its toxicity, or oxygen radicals may be produced from redox cycling. Free radicals from many xenobiotics are formed during their metabolism by enzymes, particularly cytochrome P450 or peroxidases; in other cases, the xenobiotic can redox cycle using reductases, such as cytochrome P450 reductase which can perform single-electron reductions. Superoxide, formed by one-electron reduction of molecular oxygen, can result in toxic effects, but is deactivated by superoxide dismutase (SOD); in the presence of transition metals, particularly iron, superoxide can generate the highly dangerous hydroxyl radical – for this reason, iron is controlled by transport and storage proteins, and so chemicals which can release iron are extremely toxic, causing oxidation of lipids, proteins and nucleic acids.

It should be stressed that free radicals are generally highly reactive, produced in low concentrations and are correspondingly very difficult to detect and identify; quantitative work poses particular challenges.

In this book, a theme is developed starting from the principal method for the detection of free radicals, namely spin-trapping combined with electron spin resonance (ESR) spectroscopy. Having outlined the essential chemistry involved in spin-trapping (Towner), and some of the problems encountered in that basic chemistry (Eberson), the specific application of spin-trapping to real biological systems is discussed: firstly, so called '*in-vivo*' methods are covered (Mason), in which radicals are trapped in body fluids such as blood, bile and urine; this is followed by a chapter entitled '*Ex vivo* detection of free radicals' (Tomasi *et al.*), which examines the detection of free radical intermediates occurring and trapped *in-vivo*, but detected in isolated, and in some cases purified, samples. In both the latter chapters, particular attention is given to the experimental techniques which are required to avoid artefacts of the biological media, and to maximise sensitivity in detection.

Then follows a section dealing with biological aspects of nitroxide free radicals, which can be thought of as relatively stable ESR detectable probe molecules. In the first chapter (Swartz and Timmins) it is shown that nitroxides are not completely stable in cells, but are metabolised by various cellular redox processes in which they can participate; the rate of nitroxide metabolism is determined from the change in its

ESR signal, and provides great insight into the nature of the cell metabolic mechanism; additionally, changes in the oxygen tension in particular regions of the cell can be monitored, overall providing a very complete view.

This leads on to a chapter covering some clinical uses of nitroxides (Mitchell *et al.*). In this context, the nitroxides are referred to as 'superoxide dismutase mimics', meaning that they can ameliorate damage from reactive oxygen species, effectively by radical scavenging and redox cycling as in the SOD–superoxide chain. One very important clinical application is the use of TEMPOL as a radiation modifier, to reduce hair-loss and skin-surface reactions in radiation therapy. The potential toxicological effect of nitroxides in skin is evaluated in the following chapter (Fuchs).

The next section of the book details the free radical toxicity of particular agents, chosen for their topical and timely concern. The first two chapters (chosen so as not to overlap significantly) deal with the role of iron in living systems: in particular, how iron may be released from its storage protein (ferritin) and thus catalyse a range of biological oxidation; and the role of antioxidants in ameliorating this (free) iron-dependent damage is developed.

There is much current angst over the effect of airborne pollutants on health, as is demonstrated by the next chapter (Evans) which discusses the toxicity of respirable particles: dusts, including silica, asbestos, synthetic mineral fibres, zeolites, coal, diesel particulates and cigarette smoke; therapeutic interventions are proposed.

There is a chapter on ethanol toxicity (Albano), which is included here rather than in the section dealing with dietary considerations. Evidence is given that the major free radicals produced from ethanol (hydroxyethyl) and its oxidation product, acetaldehyde (acetyl/methyl), are implicit in the initiation and propagation of various degenerative liver diseases.

With the memory of the 'Gulf War' with Iraq still poignant, it is probably timely to include coverage of the toxic effects of a chemical weapon, Sulphur Mustard (HD), for which there is evidence that nitric oxide (NO) participates in the initial inflammatory response (Arroyo).

The duplicity of thiols is considered in relation to their protective role against radical damage, but the destructive potential of the thiyl radicals which result from this 'repair' event (Rhodes). A new technique is outlined, involving muons, which can both detect and measure reaction kinetics of thiyl radicals in non-aqueous (membrane-type) environments; this is an advance over ESR to which thiyl radicals are undetectable in solution, and pulse-radiolysis which uses aqueous media. Thiyl radicals are shown to react fairly rapidly with lipids, but far more rapidly with glutathione and β-carotene, which, as membrane constituents, might deactivate them.

The section concludes with chapters on reactive oxygen species in physiology (Kaul and Forman) and oxidative stress and lipid peroxidation, and their role in disease, and the metabolism of halogenated organic compounds to free radicals and their role in liver cytotoxicity (Poli *et al.*).

While implicit in all the above, namely free radical involvement in the cause of disease, we expand this topic with two chapters which discuss the diseases of greatest concern in the latter part of the twentieth century: cancer (Landolph), and cardiovascular and respiratory disease (Kelly). The role of chemical free radical formation in these, and the protective role of antioxidants is enumerated.

There follows a chapter dealing with the toxicity of radicals in the diet, which describes very recent work on the toxicity of thermally stressed polyunsaturated oils:

in contrast to what we have been told for several years, it appears that if the oils have been heated (especially repeatedly so, as in a 'chip-pan') they can actually cause heart disease! The proposed mechanism for this effect involves production of acyl radicals from the aldehydes which they are shown to contain, by NMR, leading to membrane lipid peroxidation (Grootveld *et al.*).

It seems highly appropriate to include coverage of cytochromes P450, since it is largely these enzymes which are involved in the metabolism of xenobiotics and which sometimes convert them to free radicals. The chapter (Harris) describes very compelling computational work which enumerates the iron-centred free radical intermediates involved in the P450 enzymatic cycle.

The next two chapters are collected in a section on drug sensitization. First is the chapter on photosensitization of drugs, especially ketoprofen, and the evaluation of the mechanism for this. In the establishment of a triplet radical pair as the initial step in the photosensitization, magnetic field effects on the yield of free radicals which have escaped into the medium and on the rate of photohemolysis of human erythrocytes are used; it is proposed that the technique should be of value in determining the involvement of triplet radical pairs in photochemical processes occurring in organised biological systems such as membranes.

On a different tack, follows the chapter describing the thermodynamic dependence of free radical reactions: specifically concerned with the reduction potentials of antitumour drugs, and the energetics of their bioactivation (Butler).

Finally, it seems very appropriate in a book dealing with toxicology to mention structure–activity relationships: so, we end with a chapter (Cronin and Tran) which considers how such (QSAR) approaches may be used to predict the toxicity of free radicals in biological systems.

I thank my co-authors for their lucid and timely expositions in this book, and the editorial staff of Taylor & Francis, particularly Dilys Alam and Goober Fox, all of whom played their own fine role in the project.

Liverpool, 1999

Part I
Spin-trapping

2 Chemistry of spin-trapping

Rheal A. Towner

2.1 Introduction

A number of physiological and pathophysiological metabolic processes in animal tissues and organs seem to involve reactive transient free radicals that can only be observed effectively by the spin-trapping technique. The spin-trap reacts with reactive free radicals to form more stable radicals which can be observed by electron spin resonance (ESR) spectroscopy. In this chapter an overview of the basic principles of ESR spectroscopy and spin-trapping chemistry, the general types of spin-traps used in biological investigations, suggested selection criteria, and structural assignment considerations are discussed.

2.1.1 Basic principles of ESR spectroscopy

ESR, also referred to as electron paramagmetic resonance (EPR) spectroscopy, is considered the least ambiguous method for the detection and identification of chemical species containing one or more unpaired electrons, such as free radicals. ESR spectroscopy depends on the interaction of an external homogeneous magnetic field with the magnetic moment of unpaired electron(s) within free radical species.[1] An electron with its spin has an associated magnetic moment. When put in a strong magnetic field the electron will precess with a characteristic precessional frequency. For example, an electron in a 3400 G (0.34 T) magnetic field will precess at 9400 MHz.

A single unpaired electron, which has a spin quantum number (M_s) of $\pm\frac{1}{2}$, assumes two orientations in a strong magnetic field, i.e. either in the direction of the field or in the opposite direction. The two orientations of the unpaired electron exist at different energy levels, where the difference in energy (ΔE) is referred to as the Zeeman or fine splitting. The difference in energy of spin states is expressed in the equation:

$$\Delta E = h\nu = g_e \beta_e H$$

where h is Plank's constant, ν is the frequency associated with a transition from one energy level to the other (in the microwave region of 10^9–10^{11} Hz), g_e is a dimensionless proportionality constant characteristic for a given unpaired electron system (also called the g-factor; ~2.00), β_e is the Bohr magneton or the magnetic moment of the electron (-9.2732×10^{21} ergG^{-1}), and H is the external magnetic field (~3500 Gauss) (Poole, 1983). As the magnetic field (H) increases, the magnitude of the Zeeman splitting also increases. If radiation at the correct frequency is applied, the electrons

Table 2.1 Biological nuclei frequently observed by ESR spectroscopy

Nuclei	I	M_I
^{14}N	1	−1,0,1
^{13}C	$^1/_2$	$\pm^1/_2$
^{1}H	$^1/_2$	$\pm^1/_2$
^{19}F	$^1/_2$	$\pm^1/_2$
^{2}H	1	−1,0,1
^{17}O	$^5/_2$	$\pm^5/_2,\pm^3/_2,\pm^1/_2$

will undergo transitions between spin states. The frequency associated with the transition between spin states is expressed as follows:

$$\nu = \frac{g_e \beta_e H}{h}$$

In a magnetic field, an unpaired electron has an orbital angular momentum in addition to a spin angular momentum. The interaction between the spin and orbital angular momentums, called the spin–orbital coupling, results in a slightly different magnetic moment of the unpaired electron in a radical species compared with that of a free electron. Therefore, for a given frequency, radicals with different g-factors resonate at slightly different field strengths. The difference in the g-factor for a radical and that of a free electron is analogous to chemical shift in nuclear magnetic resonance (NMR) spectroscopy.

The unpaired electron can also interact with neighbouring nuclei with spin, which further modifies the energy of a given level by a given amount. This modification in energy is defined as the hyperfine coupling constant or hyperfine splitting constant (a_n; measured in mT or gauss, where 1 mT = 10 gauss). The hyperfine splitting constant, which is dependent on the spin quantum number of the nucleus (e.g. $I = {}^1/_2$, 1, $^3/_2$, etc), is characteristic of the interaction between the unpaired electron and the nucleus of a given molecule. Examples of biological nuclei with characteristic spin quantum numbers (I) and spin states (M_I) that are frequently observed in ESR spectroscopy are listed in Table 2.1.

The separation of lines, known as hyperfine splitting, is the most useful characteristic analytical feature of an ESR spectrum, which can be used for structural determination of a free radical. The number of resonant lines observed in the ESR spectrum can be predicted by the following relationship:

$$2nI + 1$$

where I is the spin of the nucleus and n is the number of magnetically equivalent nuclei with spin I. If for example, the nucleus interacting with the unpaired electron is nitrogen (^{14}N), which has a spin number $I = 1$, then the three transitions of nitrogen will appear as three spectral lines. The interaction of an unpaired electron with either ^{1}H or ^{13}C, which both have a spin of $^1/_2$, will result in two spectral lines. For multiple nuclei with a spin of $^1/_2$, such as hydrogens, the relative intensities and number of lines can be predicted using Pascal's triangle shown in Table 2.2.

Table 2.2 Prediction of ESR spectral lines using Pascal's triangle

Number of nuclei (n)	*Spectral intensities*
1	1 1
2	1 2 1
3	1 3 3 1
4	1 4 6 4 1

The spectrometer is designed to observe the ESR absorption by using a fixed frequency provided by a microwave source (~9.5 GHz for X band ESR spectroscopy) and varying or scanning the magnetic field. ESR spectrometers usually record the first derivative of the absorption curve, rather than the actual absorption curve itself (as in NMR spectroscopy). The first derivative curve gives a better resolution as well as a greater sensitivity. The area under the absorption curve is proportional to the number of spins in the measured sample. Integration of the first derivative gives the absorption curve, and further integration of the absorption curve is used to determine the area. By comparison of the area with those obtained from radicals of known concentrations, it is possible to quantitate the radical concentrations of the measured sample.

The sensitivity or detection limit of ESR spectrometers is in the order of $\sim 10^{-7}$ to 10^{-8} M (Wertz and Bolton, 1972). Direct detection of free radicals is possible only if the radicals are present or formed in relatively high concentrations. However, very few free radicals, especially those involved in biological systems, have sufficiently long lifetimes to maintain a steady state concentration of radicals above the threshold of $\sim 10^{-7}$ M. The direct observation of free radicals in complex biological systems is often hampered by their high reactivity, short lifetimes, and low concentrations. The spin-trapping technique provides a means of facilitating the visualisation of biological free radicals by ESR spectroscopy.

2.1.2 Basic spin-trapping chemistry

The spin-trapping method, a term coined by Janzen almost thirty years ago, was developed by various researchers independently (Janzen and Blackburn, 1968, 1969; Forshult *et al.*, 1969; Chalfont *et al.*, 1968; Leaver and Ramsay, 1969; and Terabe and Konaka, 1969) to extend the limits of ESR spectroscopy so that lower concentrations of free radicals could be detected indirectly.[2] This analytical technique involves the trapping of a reactive short-lived free radical ($R^\bullet$) by a diamagnetic nitroso or nitroxide spin-trap compound via an addition reaction to produce relatively persistent radical products or spin adducts. The detection of a free radical is only possible if the spin-trap and free radical concentrations are high enough and the rate of spin-trapping occurs rapidly. Persistent spin adducts accumulate until the detection limit of the ESR spectrometer is exceeded and a signal is recorded.

Reaction of a free radical ($R^\bullet$) with either a nitroxide or a nitroso compound results in the formation of an aminoxyl radical (also referred to as a nitroxyl or nitroxide radical) spin adduct. The spin-trapping reaction takes advantage of the well known stability of the aminoxyl free radical functional group (see Figure 2.1).

Figure 2.1 Nitroxide functional group

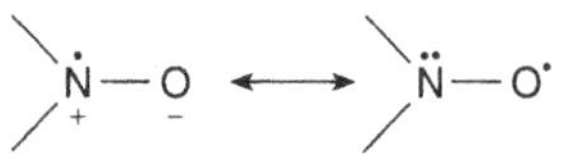

Figure 2.2 Stability of the aminoxyl (nitroxide) functional group

Figure 2.3 Reduction reaction of nitroxides

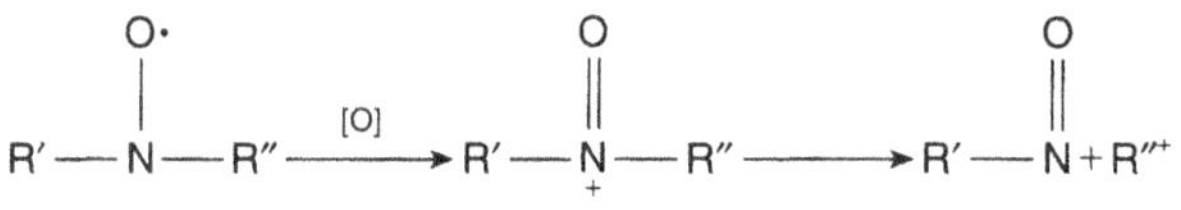

Figure 2.4 Oxidation reaction of nitroxides

Nitroxides are relatively stable free radicals due to the resonance stabilisation of the unpaired electron between the nitrogen and the oxygen of the aminoxyl functional group in the monomer form. The unpaired electron in nitroxides is located in a π orbital on the nitrogen and oxygen atoms of the aminoxyl functional group, resulting in interaction of the spin of the unpaired electron with both nitrogen and oxygen (see Figure 2.2).

The most stable nitroxides have inert functional groups attached to the nitrogen atom, such as a methylated carbon atom (*tert*-butyl group) or a phenyl group. The most common reactions of unstable nitroxides include dimerisation, disproportionation to a hydroxylamine, and dissociation. The bulky *tert*-butyl group reduces the possibilities of dimerisation and disproportionation to the hydroxylamine (Janzen *et al.*, 1985a).

A common reaction of stable nitroxides in aqueous solutions involves reduction (see Figure 2.3). For example, either ascorbic acid or sulfhydryl groups in the presence of transition metal ions are capable of reducing nitroxides rapidly to the corresponding hydroxylamine (Janzen *et al.*, 1985a).

In the presence of strong oxidising agents, oxidation may occur and result in the formation of the nitroxonium ion (see Figure 2.4). However, since the nitroxonium ion is not stable in aqueous solutions, it rapidly dissociates to a nitroso compound and a carbonium ion (see Figure 2.4; Janzen *et al.*, 1985a).

At temperatures above room temperature, it is also possible for nitroxides to dissociate to a nitroso compound (see Figure 2.5; Janzen *et al.*, 1985a).

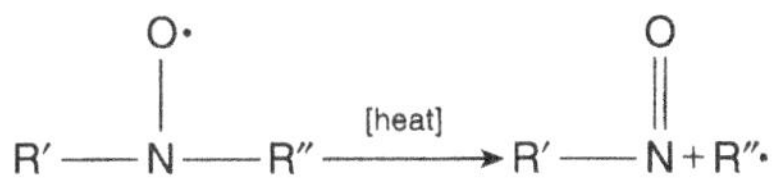

Figure 2.5 Dissociation reaction of nitroxides at temperatures above room temperature

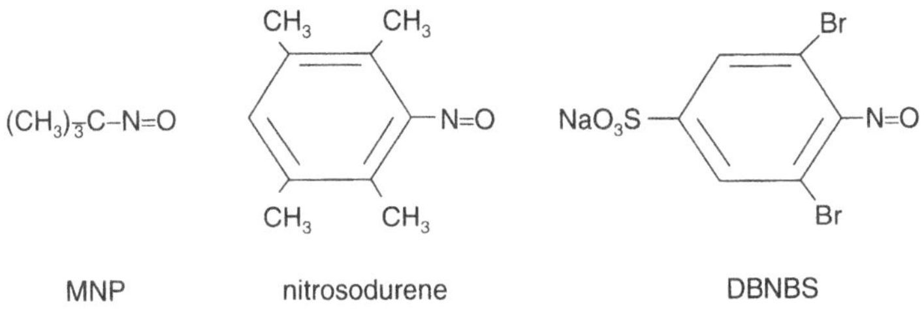

Figure 2.6 Commonly used nitroso spin-traps

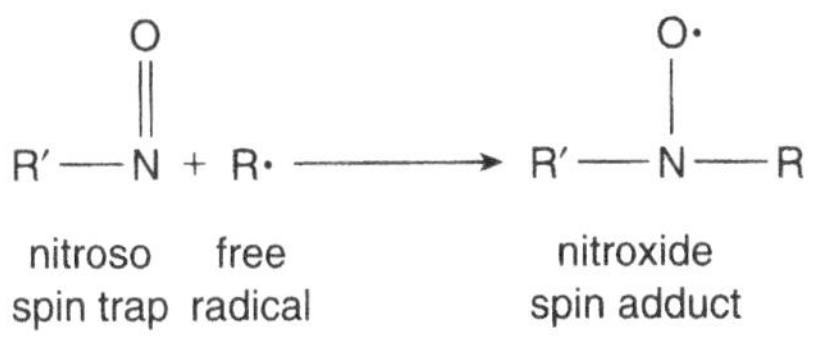

Figure 2.7 Spin-trapping reaction of nitroso spin-traps

2.2 Types of spin-trap

The type of spin-trap used is an important factor in determining how informative and sensitive the spin-trapping technique may be for a given free radical. Careful selection of spin-traps with particular solubilities, steric properties, adduct stability, and reaction specificity should be considered. The two categories of spin-trapping compounds that are most commonly used are nitroso and nitrone compounds.

2.2.1 Nitroso compounds

Examples of commonly used nitroso compounds are 2-methyl-2-nitrosopropane (MNP), 3,5-dibromo-4-nitrosobenzenesulfonate (DBNBS), and 2,3,5,6-tetramethylnitrosobenzene (or nitrosodurene) (see Figure 2.6). 2,4,6-tri-*tert*-butyl nitrosobenzene is also occasionally used.

The distinctive advantage of using nitroso spin-traps is that the nitroso nitrogen atom reacts directly with the free radical compound (see Figure 2.7).

Since ^{16}O has no spin and ^{14}N has a spin of 1, then an ESR spectrum of three lines, that are equally spaced and of equal intensity, is observed. If the attached free radical compound has atoms with nuclear spin, there will be additional hyperfine splitting. Structural information in the ESR spectrum of the detected spin adducts of a trapped free radical is obtained mainly from analysis of the multiplicity and magnitude of the hyperfine splitting constants. The additional hyperfine splitting information obtained from the direct coupling of the free radical to the nitroso nitrogen is more definitive in the identification and assignment of the trapped radical (Janzen, 1980).

The major disadvantages of the nitroso spin-traps are their thermal and photochemical instability, and a tendency to dimerize (Janzen, 1980; Makino *et al.*, 1981; Thornalley, 1985; Tomasi and Iannone, 1993). The monomer, which is involved in trapping, only exists in small amounts in solution at room temperature (Kirino *et al.*, 1981; Tomasi and Iannone, 1993). Therefore, quantitative analysis of the trapped radical with nitroso spin-traps is not possible due to the complex relation between the monomer and dimer forms. In addition, the possibility of a dissociation reaction of the nitroxide is also favourable due to the direct bonding of the radical to the nitroxide functional group. As an example, hydroxyl radicals can not be detected with nitroso spin-traps since the spin adducts are too short-lived as a result of a disproportionation reaction (Janzen, 1980; Tomasi and Iannone, 1993). Nitroso compounds can also undergo a nonradical -ene addition, which can readily form a hydroxylamine (Mason *et al.*, 1980; Tomasi and Iannone, 1993). Subsequently the hydroxylamine can undergo oxidation to form a nitroxide. This reaction is facilitated via organic solvent extractions, due to the relatively high concentration of oxygen in organic solvents in comparison to aqueous environments (Tomasi and Iannone, 1993). The use of nitrosobenzene derivative spin-traps (e.g. nitrosodurene) in biological systems is also restricted due to their limited solubility in water (Kalyanaraman, 1982).

The nitroso spin-trap most commonly used in biological studies is DBNBS. DBNBS, an aromatic C-nitroso compound with a sulfonate and bromine in the 3–5 position, is more hydrophilic and less sensitive to light and temperature variation in comparison to other nitroso spin-traps (Tomasi and Iannone, 1993). Although the dimeric form of DBNBS also exists, the concentration of the monomer is sufficient to form spin adducts in the presence of free radicals (Tomasi and Iannone, 1993). However, a major drawback of this spin-trap is that in the presence of cell cultures and red blood cells, DBNBS adducts have been found to rapidly decay and form artefactual radicals (Tomasi and Iannone, 1993; Ichimori *et al.*, 1993).

2.2.2 Nitrone compounds

The three most commonly used nitrone spin-traps include: phenyl-*tert*-butyl nitrone (PBN), α(4-pyridyl-1-oxide)-*N-tert*-butyl nitrone (POBN), and 5,5-dimethyl-1-pyrroline-*N*-oxide (DMPO) (see Figure 2.8). The nitrone function is the *N*-oxide of an imine.

Nitrones have become the most popular spin-traps in biological systems owing to their relative hydrophilicity, light insensitivity, low toxicity, solubility and stability leading to reproducible results (Janzen, 1971; Janzen, 1980; Tomasi and Iannone, 1993). Nitrones are also less sensitive to oxygen or water vapour in comparison to nitroso spin-traps (Janzen, 1971). In addition, nitrones are soluble in a large number of solvents at fairly high concentrations (> 0.1 M) (Janzen, 1980).

The nitrone spin-traps react with the free radical species via a carbon located in a β position relative to the nitrogen. Nitrone-derived spin adducts are considerably

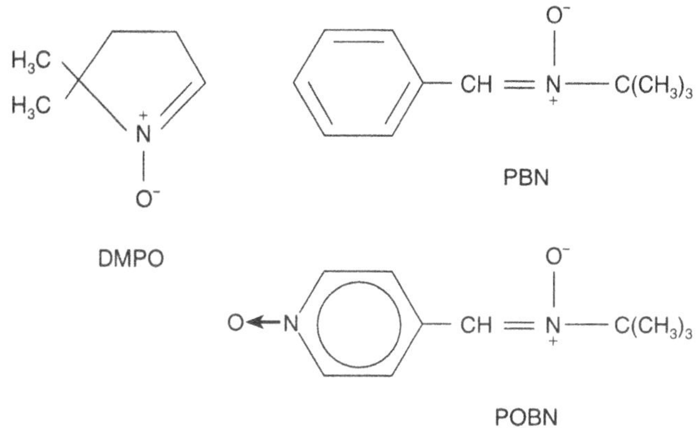

Figure 2.8 Commonly used biological nitrone spin-traps

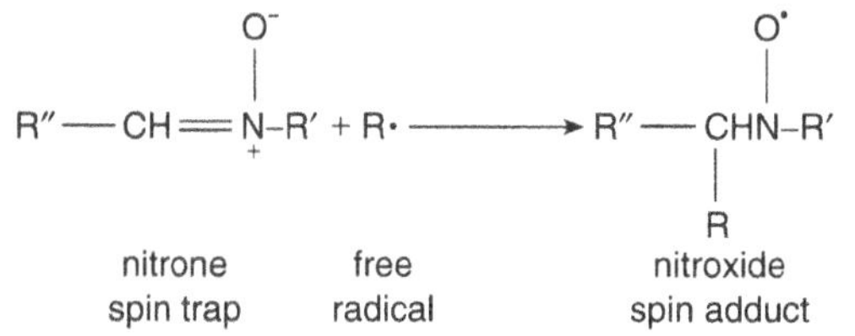

Figure 2.9 Spin-trapping reaction of nitrone spin-traps

more stable than nitroso-derived spin adducts, since a carbon atom separates the nitroxide functional group from the trapped radical species (see Figure 2.9).

The hydrogen ($I = \frac{1}{2}$) that is two bonds away from the nitroxyl functional group is referred to as the β-hydrogen. The appearance of the pattern in the ESR spectrum is dependent on the magnitudes of, and the degree of overlap between, the nitroxyl nitrogen (^{14}N; $I = 1$) and the β-hydrogen (^{1}H; $I = \frac{1}{2}$) hyperfine splitting constants. For example, if the nitrogen hyperfine splitting constant is much larger than the hydrogen splitting constant (as in the case of most PBN adducts) then the ESR spectral pattern will be six equally intense lines in a triplet of doublets pattern. If, on the other hand, the nitrogen hyperfine splitting constant is equal to the hydrogen splitting constant (as in the case of the DMPO–OH adduct) then the ESR spectral pattern will be four lines with an intensity ratio of 1:2:2:1.

The magnitude of the β-hydrogen splitting (a_H) in the nitroxide spin adduct is influenced by the bulk of the attached free radical (Janzen *et al.*, 1985a). In general, small R groups have a large a_H value. The bulk of the spin adduct R group also influences the rotation of all functional groups attached to the nitroxyl group, including the β-hydrogen. The magnitude of a_H depends on the dihedral angle between the C—H bond and the p-orbital on the nitrogen (Janzen *et al.*, 1985a). If the dihedral angle is large, then the resulting a_H value is expected to be small. On the other hand, if the dihedral angle is small then the splitting is expected to be large (Janzen *et al.*, 1985a).

Solvents can have a major effect on the hyperfine splitting observed for a spin adduct. In general, increases in solvent polarity produce an increase in the nitrogen-hyperfine splitting as the spin density on the nitrogen increases (Janzen *et al.*, 1982). For some spin adducts, the nitrogen- and hydrogen-hyperfine splitting constants can be linearly correlated in different solvents with excellent correlation coefficients (Janzen *et al.*, 1982). In addition, the hyperfine splittings can often be linearly correlated with physical-chemical properties of the solvent (Janzen *et al.*, 1982; Buettner, 1987).

DMPO is the most widely used spin-trap for the study of oxygen-centred free radicals, due to its solubility in water, its rapid diffusion in biological systems, and the formation of stable adducts (Janzen and Liu, 1973; Janzen *et al.*, 1973; Tomasi and Iannone, 1993). For example, DMPO reacts rapidly with the hydroxyl radical (rate of $K \sim 10^9 M^{-1}s^{-1}$) forming a relatively stable adduct (Janzen and Liu, 1973; Finkelstein *et al.*, 1980; Tomasi and Iannone, 1993). In contrast, PBN-oxygen centred radical adducts are relatively unstable in aqueous environments. The more lipophilic character of PBN limits its solubility in aqueous environments, but allows it to readily cross membranes and diffuse into cells (Tomasi and Iannone, 1993). As a result of its lipophilicity, PBN has been extensively used to trap lipid-derived radicals, as well as biotransformation products of lipophilic compounds, such as the trichloromethyl radical, a metabolite of carbon tetrachloride (McCay *et al.*, 1984; Janzen *et al.*, 1987). The more water-soluble PBN analogue, POBN, due to its elevated hydrophilicity and its ability to permeate cell membranes (Albano *et al.*, 1986; Tomasi and Iannone, 1993), can also effectively be used as an alternative to DMPO to trap and identify oxygen-centred radicals.

The use of nitrone spin-traps also has some disadvantages. In comparison to nitroso-derived spin adducts, the additional distance of the trapped species from the nitroxyl nitrogen in nitrone-derived spin adducts may result in a lack of spectral information. For example, ESR spectra of PBN spin adducts have characteristically small a_H^β (β-hyperfine splitting constant) values with a minor variation in magnitude when different free radical species are bound (Janzen, 1980). Therefore, the spectral features of PBN adducts can be inadequately sensitive, making it difficult to assign the trapped radical species from coupling constant measurements alone. In some cases, minor differences in the hyperfine splitting constants may be misinterpreted. Positive characterisation of most PBN- and POBN-spin adducts often requires isotopic substitution or chemical synthesis of the adduct (Janzen *et al.*, 1985a; Tomasi and Iannone, 1993). Alternatively, the magnitude of the β-hyperfine splitting is increased with the cyclic nitrones, such as DMPO (Janzen, 1980), resulting in enhanced spectral interpretation.

There are also some artefacts that may occur as a result of using nitrone spin-traps.[3] Photolysis of PBN in certain solvents (e.g. tetrahydrofuran) can result in the formation of spin adducts derived from solvent radicals (Janzen, 1980). The PBN–hydroxyl radical adduct decays with increasing pH forming a *tert*-butyl-hydroaminoxyl degradation product (Janzen *et al.*, 1991). Commercial preparations often do not require further purification, however the formation of a benzaldehyde breakdown product, favoured at low pH (Albano *et al.*, 1986; Tomasi and Iannone, 1993), must be cautioned.

DMPO is relatively sensitive to light and oxygen and therefore commercial preparations often contain contaminants and decomposition products, requiring further purification by redistillation under vacuum (Janzen, 1980). Methaemoglobin and ferric ions can oxidise DMPO to form a corresponding hydroxamic acid (Tomasi and Iannone, 1993). Subsequent oxidation of the hydroxamic acid results in the formation of the nitroxide, DMPOX (Thornalley *et al.*, 1983). The DMPO–superoxide adduct has been found to be very unstable, with a half-life of ~50s at room temperature and physiological pH (Buettner and Oberlay, 1978). The DMPO—OH adduct can, however, also arise as an artefact via the interaction of the superoxide adduct (DMPO—OOH) with redox active metal ions (Burkitt, 1993; Tomasi and Iannone, 1993). In the presence of cells, the DMPO—OH adduct, although relatively stable, can also be rapidly reduced to the ESR silent hydroxylamine by a single electron reduction (Samuni *et al.*, 1986).

2.3 Selection criteria

The choice of a particular spin-trap should be based on several criteria. Some of the criteria that should be considered include: (1) the ease of identification of the trapped radical from the ESR spectrum of the spin adduct; (2) the persistence of the spin adduct; (3) the stability of the spin-trap in an oxidising and reducing medium; (4) the rate of the radical trapping reactions; (5) the solubility of the spin-trap in biological environments; (6) the concentration of the spin-trap; and (7) the possible toxicity of the spin-trap, the spin adduct, and/or metabolites of both (McCay *et al.*, 1980).

2.3.1 Biological environment

Spin-trapping compounds are commonly used to probe biological systems for evidence of reactive free radical intermediates in specific physiological or pathophysiological processes. A particular problem encountered with the use of a spin-trap results in determining the biophysical environmental conditions within which the radicals are being formed and trapped, such as the lipophilic region of the membrane, or in the membrane–aqueous phase interface, which may have multiple layers of water molecules (McCay *et al.*, 1980). If the spin-trap is unable to penetrate a biological membrane, short-lived radicals which form and react within the lipophilic phase would not be trapped and therefore not be visible by ESR spectroscopy. On the other hand, if a radical is trapped in a system involving a membrane-bound enzyme system, it would then be necessary to determine whether the radical is formed by the enzymatic reaction or by subsequent reactions between the products of the enzymatic system and other system components.

The location of spin-traps can be evaluated using ^{13}C-NMR spectroscopy. The location of various nitrone spin-traps (PBN, 4-(dodecyloxy)-phenyl-*N*-*tert*-butylnitrone (4-DOPBN), DMPO, and sodium 5-carboxyl-5-methyl-1-pyrroline-*N*-oxide (SCMPO)) were determined in different heterogeneous biological media (sodium dodecyl sulfate (SDS) micelles, dimyristoyl phosphatidylcholine (DMPC) vesicles, and rat liver microsomal preparations). The nitronyl ^{13}C chemical shift was found to linearly correlate with solvent polarity parameters (e.g. Reichart's $E_{T(30)}$) (Janzen *et al.*, 1989). PBN was found to be amphiphilic in nature, partitioning in the membrane–aqueous phase interface, whereas DMPO was found to be more hydrophilic than PBN, partitioning into the bulk of the aqueous phase in membrane-containing systems (Janzen *et al.*, 1989).

2.3.2 *Biological effects and toxicity*

The concentration of the spin-traps may have a perturbing effect of their own in systems containing membranous organelles, which may influence the nature of the process being probed. Due to their relatively higher lipophilicity, both POBN and PBN have been found to alter membrane enzyme activities, such as cytochrome P450 (aminopyrine demethylase, ethoxycoumarin dethylase) (Albano *et al.*, 1986) and alter physiological processes, such as cardiac function (loss of systolic wall thickening, increased coronary blood flow) (Li *et al.*, 1993). POBN was also found to exert a potent vasodilatory effect in the isolated perfused rat heart (Konorev *et al.*, 1995). Concentrations of PBN greater than 25 mM were found to moderately affect hepatocyte integrity, which resulted in a slightly reduced cell viability (Albano *et al.*, 1986). With the use of radiolabelled PBN (^{14}C-PBN; ^{14}C label on the α-carbon), the highest amount of PBN and a metabolite was found in the liver (metabolic site) and adipose tissue (storage site) (Chen *et al.*, 1990a). The estimated biological half-life of PBN, in most tissues, was determined to be 134 min (Chen *et al.*, 1990b). The lethal dose of PBN was recently determined to be ~100 mg/100 g body weight (0.564 mmol/100 g) (Janzen *et al.*, 1995).

DMPO appears to be the least toxic of the biological spin-traps when used in concentrations of 25 mM or less (Albano *et al.*, 1986). Unlike PBN, DMPO was not found to inhibit cytochrome P450 or cyt. P450 reductase (Tomasi and Iannone, 1993). However, at concentrations of 50 mM DMPO can cause significant cell viability loss (~50%) (Pou and Rosen, 1990). In addition, both DMPO and PBN were found to inhibit the hexose monophosphate pathway in red blood cells (Thornalley and Stern, 1985).

To date, commonly used spin-traps have not been found to be mutagenic at the concentration levels used in biological systems (Hampton *et al.*, 1981).

2.3.3 *Metabolism of nitroxides*

The metabolism of nitroxides by biological systems can be of considerable importance in the evaluation of their usefulness as probes and in the interpretation of data obtained with them. A number of principles, as suggested by Swartz (1990), should be considered when evaluating metabolic interactions of nitroxides with cells, such as: (1) the reduction of nitroxides by cells occurs mainly intracellularly; (2) the reduction by cells occurs usually via enzyme associated reactions; (3) the rate of reduction of nitroxides by cells is oxygen dependent; (4) the principle site of reduction of nitroxides is in the mitochondria; (5) the microsomal fractions of cells can metabolise nitroxides; (6) the structural characteristics (ring structure, substituents, hydrophilicity/lipophilicity) of the nitroxide affect the rate of reduction; (7) the principle products of the metabolism of nitroxides are the corresponding hydroxylamines; (8) the relative rates of reduction of nitroxides in model systems may not reflect what is occurring in functional biological systems; (9) the rate of oxidation of hydroxylamines must be taken into consideration; (10) the oxidation of hydroxylamines mainly involves enzymes (cytochrome oxidase); (11) the enzymatic oxidation of hydroxylamines occurs mainly in membranes; (12) the rate of oxidation of hydroxylamines is dependent on the concentration of oxygen; and (13) the metabolism of nitroxides/hydroxylamines varies with the physiological state and environment of the cell (Swartz, 1990; Rosen *et al.*, 1990; Iannone *et al.*, 1990; Janzen *et al.*, 1993).

2.4 Structural assignment considerations

The major challenge of spin-trapping is assigning the correct structure of the spin adduct to ascertain the structural characteristics of the trapped free radical. In an ideal case, the resulting ESR spectrum will result in the complete characterisation of the original radical. However, in most cases only partial structural information is obtained. For instance, nuclei, with spin, three or more bonds away from the nitroxyl functional group do not normally contribute additional hyperfine splitting in the ESR spectrum.

Various cases of incorrect interpretation have resulted since the spin-trapping technique has been utilised in biological systems. In general, many of the reported artefacts may be attributed to: (1) changes in the spin-trap (nonradical, chemical, photochemical, enzymatic reactions), (2) pertubation of the biological system by the probe, and (3) artefactual reporting associated with intrinsic properties of the probe (Tomasi and Iannone, 1993). As an example, it has been recently determined that some disposable plastic syringes can contaminate organic extract solutions with a nitroxide free radical (Buettner *et al.*, 1991). In addition, the syringe nitroxide radical may serve as an antioxidant in the biological solution, which may alter natural free radical reactions (Buettner and Sharma, 1993).

It is important that as many of the possible reactions leading to the formation of the spin adducts are at least recognised if not understood. The identification of the radical adduct should not be done solely from the measurement of coupling constants. A number of methods that can be used to aid in the assignment of ESR spectra from spin adducts include: (1) oxygenation of an appropriate secondary amine precursor; (2) oxidation of the hydroxylamine of the spin adduct nitroxide; (3) spin-trapping with different types of spin-traps; (4) isotopic replacement of a nucleus with a different spin in the radical precursor; (5) isotopic replacement of the nitrogen, oxygen and/or hydrogen atoms in the spin-trap; (6) trapping of the same radical from different sources; (7) determination of g-value differences in similar adducts; (8) extensive simulation of all possible contributing hyperfine coupling constants (particularly if the spectrum consists of multiple components); and (9) using chromatographic techniques, such as gas chromatography–mass spectrometry (GC–MS) or liquid chromatography–mass spectrometry (LC–MS), to obtain further structural information (Janzen *et al.*, 1985a, 1985b).

2.4.1 Substituted spin-traps

Recent developments in spin-trapping involve the substitution of some of the functional groups of the common spin-traps, to improve their performance. For instance, the spin-trap 5-diethoxyphosphoryl-5-methyl-1-pyrroline-*N*-oxide (DEPMPO) was recently used to detect the superoxide radical anion. The DEPMPO–superoxide adduct, which is fifteen times more stable than the DMPO–superoxide adduct, does not spontaneously decay to the DEPMPO–hydroxyl radical adduct (Vásquez-Vivar *et al.*, 1997).

Thiyl radicals have been shown to be more readily trapped with the modified DMPO spin-trap, 3,3,5,5-tetramethyl-1-pyrroline-*N*-oxide (TMPO), than with DMPO (Davies *et al.*, 1987). TMPO was found to be more advantageous since the formed spin adducts are longer lived and have a more characteristic hyperfine coupling pattern. In contrast, reaction of certain thiyl radicals with DMPO produces adducts which are superficially similar to the DMPO–hydroxyl radical adduct.

Other DMPO derivatives have been recently synthesised for evaluation as better spin-traps for biological systems. Modifications to DMPO included replacing the hydrogen atom in the 2 position with a variety of inert groups, e.g. M_3PO (2-methyl-DMPO), BM_2PO (2-*tert*-butyl-DMPO), and PhM_2PO (2-phenyl-DMPO) (Janzen and Zhang, 1993). The 2-substituted DMPO spin-trap derived spin adducts were found to be less vunerable to oxidative degradation or disproportionation (Janzen and Zhang, 1993). In addition, the 2-substituted DMPOs were found to trap alkoxyl radicals adequately enough to be extracted from a biological system and be recorded by EPR spectroscopy (Janzen and Zhang, 1993).

Recently, a conformationally constrained cyclic analog of PBN, 3,4-dihydro-3,3-dimethylisoquinoline-2-oxide (MDL 101,002) was evaluated as a biological spin-trap (Thomas *et al.*, 1994, 1996; Dage *et al.*, 1997). MDL 101,002 was found to be a more effective radical trap than PBN in a membrane system, and was also able to trap hydroxyl radicals ($^{\bullet}OH$) and the superoxide radical anion ($O_2^{\bullet-}$) (Thomas *et al.*, 1996).

2.4.2 Isotopic labelling

Isotopic labelling using ^{13}C, ^{15}N, ^{2}H or ^{17}O has been of great value in the identification of spin adducts. These isotopic labels provide a different degree of multiplicity in the ESR spectrum, than that observed with the more prominent isotopes ^{12}C, ^{14}N, ^{1}H or ^{16}O.

Better characterisation of spin adducts can be obtained by isotopic substitution of appropriate nuclei in either the substrate and/or the spin-trap. For example, isotopic substitution of the carbon in $^{12}CCl_4$ with ^{13}C was used to confirm the assignment of the trichloromethyl radical ($^{\bullet}CCl_3$) (Poyer *et al.*, 1980; Janzen *et al.*, 1987) and the carbon dioxide radical anion ($^{\bullet}CO_2^-$) (Janzen *et al.*, 1988a). The interaction of the nitroxyl unpaired electron with ^{13}C ($I = ½$) resulted in an additional hyperfine splitting in the ESR spectrum. Alternatively, the spin-trap may be isotopically substituted to obtain additional hyperfine splitting information. For instance, a slight increase in linewidth is obtained from the *tert*-butyl hydrogens of PBN. If however, the hydrogens are substituted with deuteriums, then it is possible to obtain hyperfine splitting information from nuclei with spin that are three or four bonds away from the nitroxyl functional group. With the use of PBN-d_{14} (deuteration of both phenyl and *tert*-butyl groups) it was possible to determine that a 'carbon-centred' lipid-derived radical, generated from the metabolism of CCl_4, was a methylene carbon adduct of PBN (d_{14}-PBN-CH_2R) (Janzen *et al.*, 1987).

Use of d_2-^{15}N isotopically-labelled 3,5-dibromo-4-nitrosobenzenesulphonic acid (DBNBS-d_2-^{15}N, as its sodium salt) was also recently used to increase the sensitivity, and simplify the structural assignment, of complex spectra of *N*-methylacetamide-derived radical (produced from the hydrogen abstraction of *N*-methylacetamide by hydroxyl radicals) adducts of DBNBS-d_2-^{15}N compared with those of DBNBS (Timmins *et al.*, 1996).

2.4.3 Chromatography of spin adducts

In cases when several trapped free radicals are in the reaction mixture, such as in biological solutions, positive structural identification requires not only the use of ESR spectroscopy but also the use of various chromatographic techniques in conjunction with mass spectrometric analysis. The free radical chromatographic/mass spectrometric method allows separation of the mixture and the ability to study each trapped radical unambiguously. Recent studies on improving the structure determination of spin-trapped radicals include: gas chromatography–mass spectrometry (GC–MS) (Janzen *et al.*, 1988b; Janzen *et al.*, 1989; and Krygsman *et al.*, 1989), trimethylsilylated spin adduct GC–MS (Janzen *et al.*, 1985b; Abe *et al.*, 1984), HPLC (Hiraoka *et al.*, 1991), HPLC–ESR (Iwahashi *et al.*, 1991a), and HPLC–ESR–MS (Pan *et al.*, 1993).

Due to the stability of biological 'carbon-centred' PBN adducts, their structure can be further characterised by GC–MS, LC–MS or MS techniques. Further characterisation of free radicals generated from the metabolism of CCl_4 (Janzen *et al.*, 1990a) or halothane (Janzen *et al.*, 1990b) was obtained by using the isotopically labelled spin-trap, d_{14}-PBN, in combination with GC–MS. Recently, electron impact (EI) tandem MS (MS–MS) was used to identify free radical metabolites of a number of halocarbons (e.g. CCl_4, halothane) in rat liver microsomal preparations with perdeuterated *tert*-butyl PBN (d_9-PBN) (Sang *et al.*, 1997). Mass spectral markers of isotopically labelled spin-traps allows exclusive identification of specific spin-traps and resultant spin adducts.

A number of nitroso spin-trap adducts have also been structurally assigned using MS techniques. Mass spectrometry was used to identify as many as four radical adducts derived from the reaction of horse heart cytochrome *c* with hydrogen peroxide in the presence of the nitroso spin-trap DBNBS (Barr *et al.*, 1996). LC–thermospray–MS was used to structurally characterise three nitrosobenzene linoleic acid-derived radical adducts (Iwahashi *et al.*, 1991b).

It should also be noted that the quantitation of the concentrations of biological free radicals with the use of spin-traps requires that all reactions of the free radicals with the spin-trap or spin adduct (double spin adducts) must be taken into account. Many of the products formed during and after spin-trapping are silent to ESR spectroscopy.

2.5 Conclusions

Spin-trapping continues to be the only approach available to detect short-lived reactive free radicals at low concentrations in biological systems. The use of spin-trapping compounds for the study of radicals having extremely high rates of reactivity in biological reactions appears to have the potential for the detection and quantitation of radicals that would not be possible by other methods. The advantage of spin-trapping is to convert unstable or short-lived free radicals into more stable aminoxyl radicals for which ESR spectra can be easily recorded.

It is important to emphasise that no single spin-trap is optimal for the trapping of different types of reactive free radicals. The use of spin-trapping *in vivo* must also take into consideration the lifetime, stability, distribution, metabolism, and toxicity of both the spin-trap and the spin adduct, and their metabolites. Finally, caution must be exercised, and the spin-trapping literature consulted, before structural assignments of free radicals are made and conclusions are reached.

Notes

1 For a more in-depth review of ESR spectroscopy, please refer to Wertz and Bolton (1972), Nonhebel *et al.* (1979), Poole (1983), and Leffler (1993).
2 There are numerous reviews on spin-trapping, some of which include: Perkins (1970, 1980), Lagercrantz (1971), Janzen (1971, 1976, 1980, 1984, 1990), Evans (1979), Finkelstein *et al.* (1980), Kalyanaraman (1982), Buettner (1982, 1987), Mason (1984), Janzen *et al.* (1978), Janzen and Davis (1982), Janzen *et al.* (1985a), Rosen and Finkelstein (1985), Thornalley (1985), McCay (1987), Rosen *et al.* (1990), Tomasi and Iannone (1993), and Mason *et al.* (1994).
3 For a more comprehensive review on spin-trapping artefacts, see Janzen (1980), and Tomasi and Iannone (1993).

References

Abe, K., Suezawa, H., Hirota M. and Ishii, T. (1984). Mass spectrometric determination of spin adducts of hydroxyl and aryl free radicals, *J. Chem. Soc. Perkin Trans. II*, 29–34.

Albano, E., Cheeseman, K.H., Tomasi, A., Carini, R., Dianzani, M.U. and Slater, T.F. (1986). Effect of spin-traps in isolated rat hepatocytes and liver microsomes, *Biochem. Pharmacol.*, **35**, 3955–3960.

Barr, D.P., Gunther, M.R., Deterding, L.J., Tomer, K.B. and Mason, R.P. (1996). ESR spin-trapping of a protein-derived tyrosyl radical from the reaction of cytochrome c with hydrogen peroxide, *J. Biol. Chem.*, **271**, 15498–15503.

Buettner, G.R. (1982). The spin-trapping of superoxide and hydroxyl radicals, in Oberley, L.W. (ed.), *Superoxide Dismutase*, Vol. II, Boca Raton, Fl, USA: CRC Press, p. 63.

Buettner, G.R. (1987). Spin-trapping: ESR parameters of spin adducts, *Free Rad. Biol. Med.*, **3**, 259–303.

Buettner, G.R. and Oberley, L.W. (1978). Considerations in the spin-trapping of superoxide and hydroxyl radical in aqueous systems using 5,5-dimethylpyrroline-1-oxide, *Biochem. Biophys. Res. Comms.*, **83**, 69–74.

Buettner, G.R. and Sharma, M.K. (1993). The syringe nitroxide free radical: Part II, *Free Rad. Res. Comms.*, **19**, S227–S230.

Buettner, G.R., Scott, B.D., Kerber, R.E. and Mugge, A. (1991). Free radicals from plastic syringes, *Free Rad. Biol. Med.*, **11**, 69–70.

Burkitt, M.J. (1993). ESR spin-trapping studies into the nature of oxidising species formed in the Fenton reaction: Pitfalls associated with the use of 5,5-dimethyl-1-pyrroline-N-oxide in the detection of the hydroxyl radical, *Free Rad. Res. Comms.*, **18**, 43–57.

Chalfont, G.R., Perkins, M.J. and Horsfield, A. (1968). A probe for homolytic reactions in solution – II. The polymerisation of styrene, *J. Am. Chem. Soc.*, **90**, 7141–7142.

Chen, G., Bray, T.M., Janzen, E.G. and McCay, B.P. (1990a). Excretion, metabolism and tissue distribution of a spin-trapping agent, α-phenyl-*N-tert*-butyl nitrone (PBN) in rats, *Free Rad. Res. Comms.*, **9**, 317–323.

Chen, G., Griffin, M., Poyer, J.L. and McCay, P.B. (1990b). HPLC procedure for the pharmacokinetic study of the spin-trapping agent, α-phenyl-*N-tert*-butyl nitrone (PBN), *Free Rad. Biol. Med.*, **9**, 93–98.

Dage, J.L., Ackermann, B.L., Barbuch, R.L., Bernotas, R.C., Ohlweiler, D.F., Haegele, K.D. *et al.*, (1997). Evidence for a novel pentyl radical adduct of the cyclic nitrone spin-trap MDL 101,002, *Free Rad. Biol. Med.*, **22**, 807–812.

Davies, M.J., Forni, L.G. and Shuter, S.L. (1987). Electron spin resonance and pulse radiolysis studies on the spin-trapping of sulphur-centred radicals, *Chem.-Biol. Interactions*, **61**, 177–188.

Evans, C.A. (1979). Spin-trapping, *Aldrichimica Acta*, **12**, 23.

Finkelstein, E., Rosen, G.M. and Rauckman, E.J. (1980). Spin-trapping of superoxide and hydroxyl radical: practical aspects, *Arch. Biochem. Biophys.*, **200**, 1–16.
Forshult, S., Lagercrantz, C. and Torssell, K. (1969). Use of nitroso compounds as scavengers for the study of short-lived free radicals in organic reactions, *Acta Chem. Scand.*, **23**, 522–530.
Hampton, M.J., Floyd, R.A., Janzen, E.G. and Shetty, R.V. (1981). Mutagenicity of free radical spin-trapping compounds, *Mutation Res.*, **91**, 279–283.
Hiraoka, W., Kuwabara, M. and Sato, F. (1991). Characterisation of free radicals in γ-irradiated polycrystalline uridine 5′-monophosphate: a study combining ESR, spin-trapping and HPLC, *Int. J. Radiat. Biol.*, **59**, 875–883.
Iannone, A., Tomasi, A., Vianni, V. and Swartz, H.M. (1990). Metabolism of nitroxide spin labels in subcellular fraction of rat liver – I. Reduction by microsomes, *Biochim. Biophys. Acta*, **1034**, 285–289.
Ichimori, K., Arroyo, C.M., Prónai, L., Fukahori, M. and Nakazawa, H. (1993). The reactions of 3,5-dibromo-4-nitrosobenzenesulfonate and its biological applications, *Free Rad. Res. Comms.*, **19 Suppl. 1**, S129–S139.
Iwahashi, H., Albro, P.W., McGown, S.R., Tomer, K.B. and Mason, R.P. (1991a). Isolation and identification of α-(4-pyridyl-1-oxide)-*N-tert*-butylnitrone radical adducts formed by the decomposition of the hydroperoxides of linoleic acid, linolenic acid, and arachidic acid by soybean lipoxygenase, *Arch. Biochem. Biophys.*, **285**, 172–180.
Iwahashi, H., Parker, C.E., Mason, R.P. and Tomer, K.B. (1991b). Radical adducts of nitrosobenzene and 2-methyl-2-nitrosopropane with 12,13-epoxylinoleic acid radical, 12,13-epoxylinolenic acid radical and 14,15-epoxyarachidonic acid radical. Identification by h.p.l.c.–e.p.r. and liquid chromatography–thermospray–m.s., *Biochem. J.*, **276**, 447–453.
Janzen, E.G. (1971). Spin-trapping, *Accts. Chem. Res.*, **4**, 31–40.
Janzen, E.G. (1976). The application of ESR spin-trapping techniques in the detection of gas phase free radicals produced from photolysis of gas phase organic molecules, in Ware, W.R. (ed.), *Creation and Detection of the Excited State*, Vol. 4, New York: Marcel Dekker, p. 83.
Janzen, E.G. (1980). A critical review of spin-trapping in biological systems, in Pryor, W.A. (ed.), *Free Radicals in Biology*, Vol. IV, New York: Academic Press, p. 115.
Janzen, E.G. (1984). Spin-trapping, in Packer, L. (ed.), *Oxygen Radicals in Biological Systems*, Methods in Enzymology, Vol. 105, New York: Academic Press, pp. 188–189.
Janzen, E.G. (1990). Spin-trapping and associated vocabulary, *Free Rad. Res. Comms.*, **9**, 163–167.
Janzen, E.G. and Blackburn, B.J. (1968). Detection and identification of short-lived free radicals by an electron spin resonance trapping technique, *J. Am. Chem. Soc.*, **90**, 5909–5910.
Janzen, E.G. and Blackburn, B.J. (1969). Detection and identification of short-lived free radicals by electron spin resonance trappings techniques (spin-trapping): photolysis of organolead, -tin, and -mercury compounds, *J. Am. Chem. Soc.*, **91**, 4481.
Janzen, E.G. and Davis, E.R. (1982). The role of free radicals in arylamine carcinogenesis, in Floyd, R.A. (ed.), *Free Radicals in Cancer*, Ch. 10, New York: Marcel Dekker, pp. 361–394.
Janzen, E.G. and Liu, J.I.-P. (1973). Radical addition reactions of 5,5-dimethyl-1-pyrroline-1-oxide. ESR spin-trapping with a cyclic nitrone, *J. Mag. Res.*, **9**, 510–512.
Janzen, E.G. and Zhang, Y.-K. (1993). EPR spin-trapping alkoxyl radicals with 2-substituted 5,5-dimethylpyrroline-N-oxides (2-XM_2PO's), *J. Mag. Res. Series B*, **101**, 91–93.
Janzen, E.G., Evans, C.A. and Liu, J.I.-P. (1973). Factors influencing hyperfine splitting in the ESR spectra of five-membered ring nitroxides, *J. Mag. Resonance*, **9**, 513–516.
Janzen, E.G., Evans, C.A. and Davis, E.R. (1978). The spin-trapping reaction, in Pryor, W.A. (ed.), *Organic Free Radicals*, ACS Symp. Ser. 69, Washington DC: American Chemical Society, pp. 433–446.
Janzen, E.G., Coulter, G.A., Oehler, U.M. and Bergsma, J. (1982). Solvent effects on the nitrogen and β-hydrogen hyperfine splitting constants of aminoxyl radicals obtained in spin-trapping experiments, *Can. J. Chem.*, **60**, 2725–2733.

Janzen, E.G., Stonks, H.J., Dubose, C.M., Poyer, J.L. and McCay, P.B. (1985a). Chemistry and biology of spin-trapping radicals associated with halocarbon metabolism *in vitro* and *in vivo*, *Environ. Health Perpects.*, **64**, 151–170.

Janzen, E.G., Weber, J.R., Haire, D.L. and Fung, D.M. (1985b). Gas chromatography–mass spectrometry (GC–MS) of single and double spin adducts of PBN and the hydroxylamines of corresponding structure, *Anal. Letts*, **18**, 1749–1757.

Janzen, E.G., Towner, R.A. and Haire, D.L. (1987). Detection of free radicals generated from the *in vitro* metabolism of carbon tetrachloride using improved ESR spin-trapping techniques, *Free Rad. Res. Comms.*, **3**, 357–364.

Janzen, E.G., Towner, R.A. and Brauer, M. (1988a). Factors influencing the formation of the carbon dioxide anion ($^{\bullet}CO_2^-$) spin adduct of PBN in the rat liver metabolism of halocarbons, *Free Rad. Res. Comms.*, **4**, 359–369.

Janzen, E.G., Krygsman, P.H. and Haire, D.L. (1988b). The application of gas chromatographic/mass spectrometric techniques to spin-trapping. Conversion of α-phenyl-*N-tert*-butyl nitrone (PBN) spin adducts to stable trimethylsilylated derivatives, *Biomed. Environ. Mass Spectrometry*, **15**, 111–116.

Janzen, E.G., Haire, D.L., Coulter, G.A., Stronks, H.J., Krygsman, P.H., Towner, R.A. *et al.* (1989). Locating spin-traps in heterogeneous by ^{13}C NMR spectroscopy. Investigations in SDS micelles, DMPC vesicles, and rat liver microsomes, *J. Organic Chemistry*, **54**, 2915–2920.

Janzen, E.G., Towner, R.A., Krygsman, P.H., Lai, E.K., Poyer, J.L., Brueggemann, G. *et al.* (1990a). Mass spectroscopy and chromatography of the trichloromethyl radical adduct of phenyl-*tert*-butyl nitrone, *Free Rad. Res. Comms.*, **9**, 353–360.

Janzen, E.G., Towner, R.A., Krygsman, P.H., Haire, D.L. and Poyer, J.L. (1990b). Structures identification of free radicals by ESR and GC/MS of PBN spin adducts from the *in vitro* and *in vivo* rat liver metabolism of halothane, *Free Rad. Res. Comms.*, **9**, 343–351.

Janzen, E.G., Kotake, Y. and Hinton, R.D. (1991). Stabilities of hydroxyl radical spin adducts of PBN-type spin-traps, in Makino, K., Janzen, E.G. and Yoshikawa, T. (eds), *Proceedings of the 3rd International Symposium on Spin-Trapping and Aminoxyl Radical Chemistry*, Kyoto, Japan, Nov. 22–24, p. 28.

Janzen, E.G., Zhdanov, R.I. and Reinke, L.A. (1993). Metabolism of phenyl and alkyl spin adducts of PBN in rat hepatocytes. Rate dependence on size and type of addend group, *Free Rad. Res. Comms.*, **19**, S157–S162.

Janzen, E.G., Poyer, J.L., Schaefer, C.F., Downs, P.E. and DuBose, C.M. (1995). Biological spin-trapping – II. Toxicity of nitrone spin-traps: dose-ranging in the rat, *J. Biochem. Biophys. Methods*, **30**, 239–247.

Kalyanaraman, B. (1982). Detection of toxic free radicals in biology and medicine, in Hodgson, E., Bend, J.R. and Philpot, R.M. (eds), *Reviews in Biochemical Technology*, Vol. IV, New York: Elsevier, pp. 73–139.

Kirino, Y., Ohkuma, T. and Kwan, T. (1981). Spin-trapping with 5,5-dimethylpyrroline-*N*-oxide in aqueous solution, *Chem. Pharm. Bull.*, **29**, 29–34.

Konorev, E.A., Tarpey, M.M., Joseph, J., Baker, J.E. and Kalyanaraman, B. (1995). Nitronyl nitroxides as probes to study the mechanism of vasodilatory action of nitrovasodilators, nitrone spin-traps, and nitroxides: role of nitric oxide, *Free Rad. Biol. Med.*, **18**, 169–177.

Krygsman, P.H., Janzen, E.G., Towner, R.A. and Haire, D.L. (1989). Enhanced recognition of spin-trapped radicals in complex mixtures: deuterated nitronyl adducts provide a gas chromatographic/mass spectrometric marker, *Anal. Letts.*, **22**, 1009–1020.

Lagercrantz, C.J. (1971). Spin-trapping of some short-lived radicals by the nitroxide method, *J. Phys. Chem.*, **75**, 3466–3475.

Leaver, I.H. and Ramsey, G.C. (1969). Trapping of radical intermediates in the photoreduction of benzophenone, *Tetrahedron*, **25**, 5669–5675.

Leffler, J.P. (1993). Electron spin resonance, in *An Introduction to Free Radicals*, New York: Wiley–Interscience Publication, pp. 11–33.

Li, X.Y., Sun, J.Z., Bradamante, S., Piccinini, F. and Bolli, R. (1993). Effects of the spin-trap α-phenyl-*N-tert*-butyl nitrone on myocardial function and flow: a dose-response study in the open-chest dog and in the isolated rat heart, *Free Rad. Biol. Med.*, **14**, 277–285.

McCay, P.B. (1987). Application of ESR spectroscopy in toxicology, *Arch. Toxicol.*, **60**, 133–137.

McCay, P.B., Noguchi, T., Fong, K.-L., Lai, E.K. and Poyer, J.L. (1980). Production of radicals from enzyme systems and use of spin-traps, in *Free Radicals in Biology*, Vol. IV, London: Academic Press, pp. 155–186.

McCay, P.B., Lai, E.K., Poyer, J.L., DuBose, C.M. and Janzen, E.G. (1984). Oxygen- and carbon-centered free radical formation during carbon tetrachloride metabolism. Observation of lipid radicals *in vivo* and *in vitro*, *J. Biol. Chem.*, **259**, 2135–2143.

Makino, K., Suzuki, N., Moriya, F., Rokushika, S. and Hatano, H. (1981). A fundamental study on aqueous solutions of 2-methyl-2-nitrosopropane as a spin-trap, *Radiation Res.*, **86**, 294–310.

Mason, R.P. (1984). Spin-trapping free radical metabolites of toxic chemicals, in Holtzman, J.L. (ed.), *Spin Labelling in Pharmacology*, New York: Academic Press Inc., pp. 87–129.

Mason, R.P., Kalyanaraman, B., Trainer, B.E. and Eling, T.E. (1980). A carbon-centered free radical intermediate in the prostaglandin synthetase oxidation of arachidonic acid, *J. Biol. Chem.*, **255**, 5019–5022.

Mason, R.P., Hanna, P.M., Burkitt, M.K. and Kadiiska, M.B. (1994). Detection of oxygen-derived radicals in biological systems using electron spin resonance, *Environ. Health Perpect.*, **102**, Suppl. 10, 33–36.

Nonhebel, D.C., Tedder, J.M. and Walton, J.C. (1979). Electron spin resonance spectroscopy, in *Radicals*, London: Cambridge University Press, pp. 19–29.

Pan, K., Lin, C.-R. and Ho, T.-I. (1993). Substituent effects on ESR parameters of α-phenyl-*N-tert*-butylnitrone spin adducts. Resolution enhancement and mass spectrometry, *Mag. Res. Chemistry*, **31**, 632–638.

Perkins, M.J. (1970). The trapping of free radicals by diamagnetic scavengers, in *Essays on Free Radical Chemistry*, Chap. 5, London: The Chemical Society, Special Publication, pp. 97–115.

Perkins, M.J. (1980). Spin-trapping, *Adv. Phys. Org. Chem.*, **17**, 1–64.

Poole, C.P. (1983) in *Electron Spin Resonance*, New York: Wiley–Interscience.

Pou, S. and Rosen, G.M. (1990). Spin-trapping of superoxide by 5,5-dimethyl-1-pyrroline-*N*-oxide: application to isolated perfused organs, *Anal. Biochem.*, **190**, 321–325.

Poyer, J.L., McCay, P.B., Lai, E.K., Janzen, E.G. and Davis, E.R. (1980). Confirmation of assigment of the trichloromethyl radical spin adduct detected by spin-trapping during ^{13}C-carbon tetrachloride metabolism *in vitro* and *in vivo*, *Biochem. Biophys. Res. Comms.*, **94**, 1154–1160.

Rosen, G.M. and Finkelstein, E. (1985). Use of spin-traps in biological systems, *Adv. Free Rad. Biol. Medicine*, **1**, 345–375.

Rosen, G.M., Cohen, M.S., Britigan, B.E. and Pou, S. (1990). Application of spin-traps to biological systems, *Free Rad. Res. Comms.*, **9**, 187–195.

Samuni, A., Carmichael, A.J., Russo, A., Mitchell, J.B. and Riesz, P. (1986). On the spin-trapping and ESR detection of oxygen-derived radicals generated inside cells, *Proc. Natl. Acad. Sci. USA*, **83**, 7593–7597.

Sang, H., Janzen, E.G., Poyer, J.L. and McCay, P.B. (1997). The structure of free radical metabolites detected by ESR spin-trapping and mass spectroscopy from halocarbons in rat liver microsomes, *Free Rad. Biol. Med.*, **22**, 843–852.

Swartz, H.S. (1990). Principles of the metabolism of nitroxides and their implications for spin-trapping, *Free Rad. Res. Comms.*, **9**, 399–405.

Terabe, S. and Konaka, R. (1969). Electron spin resonance studies on oxidation with nickel peroxide: spin-trapping of free-radical intermediates, *J. Am. Chem. Soc.*, **91**, 5655–5657.

Thomas, C.E., Carney, J.M., Bernotas, R.C., Hay, D.A. and Carr, A.A. (1994). *In vitro* and *in vivo* activity of a novel series of radical trapping agents in model systems of CNS oxidative damage, *Am. N.Y. Acad. Sci.*, **738**, 243–249.

Thomas, C.E., Ohlweiler, D.F., Carr, A.A., Nieduzak, T.R., Hay, D.A., Adams, G. *et al.* (1996). Characterisation of the radical trapping activity of a novel series of cyclic nitrone spin-traps, *J. Biol. Chem.*, **271**, 3097–3104.

Thornalley, P.J. (1985). Theory and biological application of the electron spin resonance technique of spin-trapping, in *Life Chemistry Reports*, Vol. 1, UK: Harwood Academic Publications, pp. 1–56.

Thornalley, P.J. and Stern, A. (1985). The effect of nitrone spin-trapping agents on red cell glucose metabolism, *Free Rad. Res. Comms.*, **1**, 111–117.

Thornalley, P.J., Trotta, R.J. and Stern, A. (1983). Free radical involvement in the oxidative phenomena induced by *tert*-butyl hydroperoxide in erythrocytes, *Biochem. Biophys. Acta*, **759**, 16–22.

Timmins, G.S., Wei, X., Hawkins, C.L., Taylor, R.J.K. and Davies, M.J. (1996). The synthesis and use of a ^{15}N and ^{2}H isotopically-labelled derivative of the spin-trap 3,5-dibromo-4-nitrosobenzenesulfonic acid, *Redox Report*, **2**, 407–410.

Tomasi, A. and Iannone, A. (1993). ESR spin-trapping artefacts in biological model systems, in Berliner, L.J. and Reuben, J. (eds), *Biological Magnetic Resonance*, Vol. 13, New York: Plenum Press, pp. 353–384.

Vásquez-Vivar, J., Hogg, N., Pritchard Jr, K.A., Martasek, P. and Kalyanaraman, B. (1997). Superoxide anion formation from lucigenin: an electron spin resonance spin-trapping study, *FEBS Letts.*, **403**, 127–130.

Wertz, J.E. and Bolton, J.R. (1972) in *Electron Spin Resonance: Theory and Practical Applications*, New York: McGraw-Hill.

3 Spin-trapping: problems and artefacts

Lennart Eberson

3.1 Introduction

A spin-trap (ST) is ideally an inert additive to a chemical system under study, capable of reacting only with transient radicals $R^{\bullet}$ to form persistent diagnostic spin adducts R—ST$^{\bullet}$ (equation 3.1). However, such situations are rare, and one must always acknowledge the fact that spin-traps possess other reactivities which might lead to the formation of R—ST$^{\bullet}$ without the intervention of $R^{\bullet}$. Unfortunately, this problem pertains particularly to nitrones and nitroso compounds, exactly those compound classes among which the most popular spin-traps are included, as exemplified by *N*-phenyl-α-*tert*-butylnitrone (PBN), 4,5-dihydro-5,5-dimethylpyrrole-1-oxide (DMPO) and 2-methyl-2-nitrosopropane (MNP, *t*-BuNO), shown in Figure 3.1. Most often such artefacts originate from the redox (equation 3.2; R^{-} is the nucleophile obtained by addition of one electron to $R^{\bullet}$) or nucleophilic reactivity of ST (equation 3.3) (Eberson, 1998; Tomasi and Iannone, 1993), but recently it also has appeared that the possible participation of an important class of ST, namely nitrones, in cycloaddition reactions can cause problems.

$$\text{Radical reactivity:}\quad R^{\bullet} + ST \rightarrow R\text{—}ST^{\bullet} \tag{3.1}$$

$$\text{Redox reactivity:}\quad ST - e^{-} \rightarrow ST^{\bullet +} \xrightarrow{R^{-}} R\text{—}ST^{\bullet} \tag{3.2}$$

$$\text{Nucleophilic reactivity:}\quad ST + R^{-} \rightarrow R\text{—}ST^{-} \xrightarrow{-e^{-}} R\text{—}ST^{\bullet} \tag{3.3}$$

We shall only deal briefly below with the redox reactivity of the ST itself, since the reaction mode of equation 3.2 is dependent on the use of strong oxidants, typically with redox potentials above 1 V (*vs.* the saturated calomel electrode, SCE; all potentials

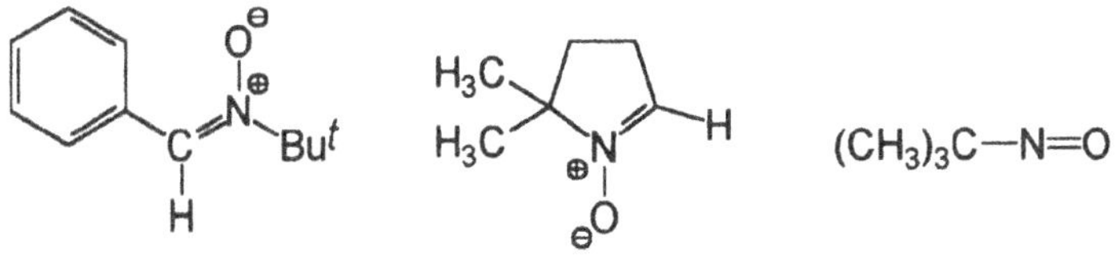

Figure 3.1 Formulas of the most often used spin-traps

in the following will be referred to this reference). Thus it is not likely that the sequence of equation 3.2 will be of any serious concern in applications of spin-trapping in biological systems where redox reagents generally are mild, unless, and this is an important exception, photochemically induced phenomena are under study. On the other hand, the nucleophilic reaction of equation 3.3, sometimes denoted the Forrester–Hepburn mechanism (Forrester and Hepburn, 1971), always carries the risk of being a source of misinterpretation since the redox step can be achieved by an oxidant as weak as molecular oxygen. It will therefore be discussed in detail. Finally some cycloaddition reactions of PBN and their possible relevance for the interpretation of spin-trapping results will be treated.

3.2 The redox reactivity of spin-traps and reagents involved in the radical cation mechanism; inverted spin-trapping

Much of the discussion to follow will involve redox chemistry, generally of the one-electron oxidation variety. Reductive processes, although possible, are much less likely to be encountered. It will therefore be necessary to provide a fair amount of data for judging redox reactivity, both in thermal and photochemical reactions. These come in the form of one-electron redox potentials, which means standard or reversible potentials of redox couples $B/B^{\bullet-}$ or, if such are not available, anodic peak potentials. The latter are numerically larger than the reversible potentials by an unknown increment, but the difference should seldom be greater than 0.1 V.

One might wonder why redox potentials, which are thermodynamic quantities, can be of any use for the discussion of reactivity which is a kinetic concept. Actually, redox potentials are used in the conventional way to estimate the free energy change $\Delta G°$ of the electron transfer reaction. As a crude approximation, this quantity can be used as a measure of its free energy of activation $\Delta G^{\ddagger}$ from which a *maximum* electron transfer rate constant can be estimated. For a more refined estimate, the Marcus theory of electron transfer (Marcus and Sutin, 1985) also includes a kinetic parameter, the reorganization energy λ, apart from $\Delta G°$, for the calculation of the electron transfer rate constant. Figure 3.2 illustrates the differences between the two approaches in a typical case; accounts of the detailed application of the Marcus theory to organic systems have been published (Astruc, 1995; Eberson, 1982, 1987). As a rule, the more endergonic an electron transfer reaction is, the less significant is the difference between the two approaches. For exergonic intermolecular electron transfer reactions, the difference is practically negligible, since such reactions are diffusion-controlled or nearly so. Only in an interval of $-15 < \Delta G° < 15$ kcal mol^{-1} is it necessary to make a Marcus analysis.

Equation 3.2 involves one-electron oxidation of the spin-trap to give its radical cation, a species with high reactivity toward nucleophiles (Bhattacherjee *et al.*, 1996; Zubarev and Brede, 1994, 1995), interchangeably denoted R^- or Nu^- in the following. The need for strong oxidants in this transformation is evident from the peak potentials E_{pa} of common spin-traps listed in Table 3.1. As an approximate rule of thumb, the oxidant should possess a redox potential which is less than 0.5 V lower than the appropriate E_{pa} for the electron transfer reaction to be fast enough.

For unambiguous operation of equation 3.2 it is also required that the nucleophile is difficult to oxidize in order to avoid competing or predominant oxidation of Nu^-

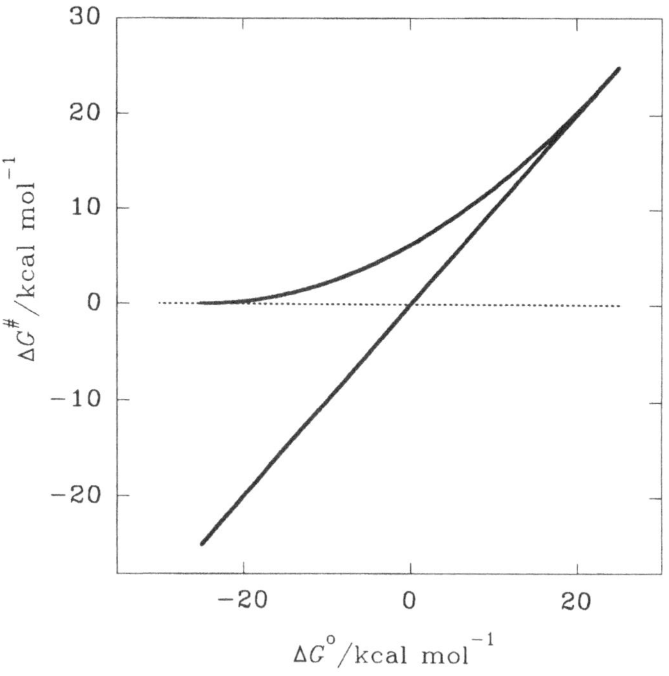

Figure 3.2 The curve is part of a Marcus parabola, represented by the equation $\Delta G^{\ddagger} = (1 + \Delta G^{\circ}/\lambda)^2$, where λ has been put at 25 kcal mol^{-1}. The straight line corresponds to the expression $\Delta G^{\ddagger} = \Delta G^{\circ}$

Table 3.1 Anodic peak potentials of common spin-traps, determined by cyclic voltammetry in acetonitrile, unless otherwise stated (McIntire *et al.*, 1980; Sosonkin *et al.*, 1982; Gronchi *et al.*, 1983; Eberson, 1994; Eberson *et al.*, 1996)

ST	E_{pa}/*V* vs. *SCE*
t-BuNO, monomeric	1.82
$(t\text{-BuNO})_2$	1.40
2,4,6-$(t\text{-Bu})_3C_6H_2NO$	1.38
CH_2=N(O)Bu^t	2.08
PhCH=N(O)Bu^t	1.47
4-PyCH=N(O)Bu^t [a]	1.93
4-PyOCH=N(O)Bu^t [b]	1.37
4-$MePy^+$CH=N(O)Bu^t [c]	2.32
4-$NO_2C_6H_4$CH=N(O)Bu^t	1.87[d]
4-$MeOC_6H_4$CH=N(O)Bu^t	1.25[d,e]
4-$Me_2NC_6H_4$CH=N(O)Bu^t	0.67[d,e]
DMPO	1.68
3,3,5,5-Tetramethyl-1-pyrroline-1-oxide (TMPO)	1.78

a Py = pyridino.
b PyO = 1-oxidopyridino.
c $MePy^+$ = 1-methylpyridino.
d In dichloromethane.
e Reversible potential.

Table 3.2 E°(Nu•/Nu⁻) of interest in inverted spin-trapping under oxidizing conditions (Pearson, 1986; Eberson, 1987; Carloni *et al.*, 1996; Alberti *et al.*, 1997)

Nu^-	E°($Nu^•/Nu^-$)/V vs. SCE *in water*	E°($Nu^•/Nu^-$)/V vs. SCE *in acetonitrile*
H_2O	3.4	–
F^-	3.4	2.7
Cl^-	2.4	1.9
Br^-	1.8	1.5
I^-	1.2	1.0
CN^-	2.3	2.0
SCN^-	1.4	1.3
N_3^-	1.1	0.8
NO_3^-	2.1	–
NO_2^-	0.8	0.7
HO^-	1.9	1.3
CH_3COO^-	2.2	1.6
$PhCOO^-$	1.5	0.9
CF_3COO^-	2.0	1.4
Succinimide anions	1.8–2.3	1.3–1.8
Benzotriazolate ion	–	1.1
Benzotriazole	–	2.0
Pyridine	2.2	–
Triethyl phosphite	2.1	–

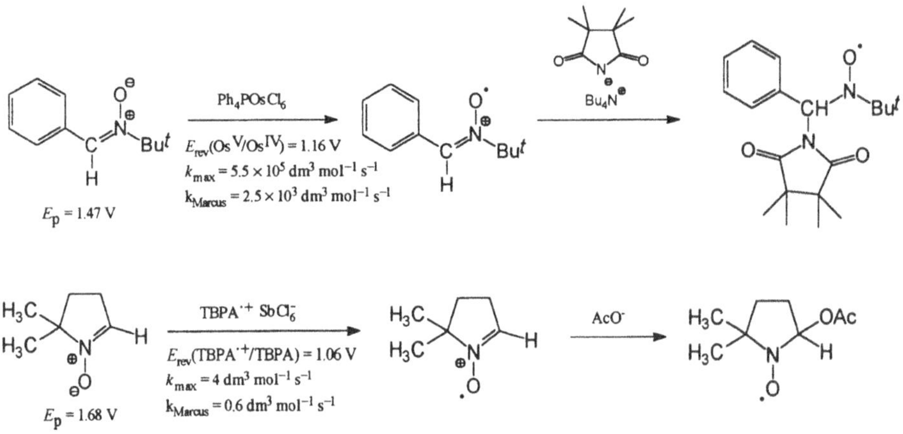

Figure 3.3 Showing two cases of inverted spin trapping, succinimide anion (upper formula) and acetate ion. The rate constants were evaluated according to the $\Delta G^{\ddagger} = \Delta G^{\circ}$ approximation (k_{max}) or the Marcus theory (k_{Marcus}), assuming a reorganization energy of 25 kcal mol^{-1}. The solvent was assumed to be acetonitrile and the electrostatic terms were neglected

to Nu•. Table 3.2 shows values of $E°(Nu^•/Nu^-)$ for a range of nucleophiles. Figure 3.3 shows some applications of equation 3.2, also denoted 'inverted spin-trapping' (Eberson, 1992) because of the inverted electron demand in relation to equation 3.1. Note again the need for strong oxidants, which makes this mechanism easily avoidable

Table 3.3 Excitation wavelengths λ_{exc} and excited state redox potentials of spin-traps (Eberson *et al.*, 1994)

ST	λ_{exc}/nm	E°$(ST^*/ST^{\bullet -})/V$ vs. *SCE*	E°$(ST^{\bullet +}/ST^*)/V$ vs. *SCE*
t-BuNO, monomeric	676	0.12	0.03
2,4,6-$(t\text{-Bu})_3C_6H_2NO$	338	2.4	−2.2
PBN	298	1.8	−2.2
$CH_2{=}N(O)Bu^t$	*ca.* 250	3.0	−2.8
DMPO	242	2.7	−3.4

in thermal reactions. The rate constants calculated for the electron transfer steps are based on both the $\Delta G^\circ = \Delta G^\ddagger$ approximation and on the Marcus treatment.

Photochemical reactions produce excited states which often are strong redox reagents (Kavarnos, 1993). This leads to not trivially recognizable inverted spin-trapping situations in spite of the seemingly mild reaction conditions. An excited state B* of a compound B can be regarded as a combination of a radical cation (half-filled HOMO) and a radical anion (half-filled LUMO), the redox potentials of which can be calculated from the thermal redox potentials and excitation energies $\Delta E_{0,0}$ according to equations 3.4 and 3.5. Calculations can be made for both singlet and triplet states, and as a general rule singlet states are more redox reactive than triplet states.

$$E^\circ(B^{\bullet +}/B^*) = E^\circ(B^{\bullet +}/B^*) - \Delta E_{0,0} \qquad (3.4)$$

$$E^\circ(B^*/B^{\bullet -}) = E^\circ(B/B^{\bullet -}) + \Delta E_{0,0} \qquad (3.5)$$

Table 3.3 shows excited state redox potentials of some common spin-traps, where it can be noted that excited singlet states of nitrones possess high redox potentials, 1.8–3.0 V. Photolysis of a nitrone in the presence of an electron acceptor will therefore lead to efficient production of its radical cation which will react with nucleophilic species present. Equation 3.6 shows an example of the photochemical excitation of PBN, reaction of the excited state with the weak acceptor 12-tungstocobalt(III)ate, and reaction of the radical cation with various nucleophiles (Eberson, 1994). It should be noted that PBN and similar nitrones (λ_{exc} around 300 nm) are easily accessible for excitation by UV light from Hg lamps whereas DMPO absorbs at about 240 nm, somewhat below the mercury line of shortest wavelength. From this point of view, DMPO and its congeneers should be the spin-traps of choice for photochemical studies.

$$\text{PBN} \xrightarrow{h\nu} \text{PBN}^* \xrightarrow{\text{Co}^{\text{III}}\text{W}_{12}\text{O}_{40}^{5-}} \text{PBN}^{\bullet +} \xrightarrow[\text{F}^-,\ \text{Cl}^-,\ \text{AcO}^-,\ \text{CN}^-]{\text{Nu}^-,\ \text{such as}} \text{Nu—PBN}^\bullet \qquad (3.6)$$

A different and very common situation in spin-trapping is photochemical excitation of an added sensitizer, the excited state of which is a strong oxidant. Table 3.4 includes ground and excited state potentials for a number of sensitizers used in photooxidation reactions. Recent studies of particular interest centre around the sensitized photooxidation of DMPO in aqueous media, using quinones as sensitizers. It was originally suggested (Ononye *et al.*, 1986) that the quinone triplet would

Table 3.4 Ground and excited state potentials of compounds B in organic media, commonly used as oxidants or photosensitizers in spin adduct formation (Mann and Barnes, 1970; Buettner, 1993; Murov *et al.*, 1993)

Ox form B^a *of redox couple* $B/B^{\bullet -}$	$E(B/B^{\bullet -})/V$ vs. SCE	E_S/eV	E_T/eV	$E°(^SB^*/B^{\bullet -})/V$ vs. SCE	$E°(^TB^*/B^{\bullet -})/V$ vs. SCE
Cl_4Q	0.02	2.76	2.14	2.78	2.16
F_4Q	−0.04		2.23		2.19
$(CN)_2Cl_2Q$	0.51		(2.14)		2.65
Q	−0.45	2.72	2.32	2.27	1.87
Me_4Q	−0.76		2.39		1.63
2,5-Me_2Q	−0.54		(2.39)		1.85
2,3-Cl_2Q	−0.19		(2.14)		1.95
Br_4Q	0.00		1.96		1.96
1,4-Naphthoquinone	−0.63		2.50		1.87
2-Me-1,4-naphthoquinone	−0.69		$(2.50)^x$		1.81
Anthraquinone	−0.94	2.94	2.71	2.00	1.77
TPP^+	−0.37	2.82	2.66	2.45	2.29
TAP^+	−0.60		2.34		1.74
1,2,4,5-$(CN)_4$-benzene	−0.66	4.10	2.81	3.44	2.15
9,10-$(CN)_2$-anthracene	−0.98	2.94	1.81	1.96	0.83
TBPA	1.06				
$OsCl_6^-$	1.18				
Anthraquinone-2-SO_3^-	−0.88		2.68		1.80
O_2	−0.57				
FcH^+	0.35				
Fe(III)	0.53				
Br_2	0.53				

a FcH^+ = ferricinium ion, Q = 1,4-benzoquinone, TAP^+ = 2,4,6-tri(4-methoxyphenyl)pyrylium ion, TPP^+ = 2,4,6-triphenylpyrylium ion, TBPA = tris(4-bromophenyl) aminium ion.

abstract hydrogen atom from a water molecule to give $HO^\bullet$ which was then trapped by DMPO. In view of the high H—O bond energy of water, 119 kcal mol^{-1}, this does not appear to be a likely pathway. Later (Monroe and Eaton, 1996; Eberson, 1999) it was concluded that the radical cation mechanism should be feasible, as suggested by the redox potentials of the triplet states involved (from benzoquinone, 1.87 V and 2-methyl-1,4-naphthoquinone, 1.81 V) in relation to that of DMPO, 1.68 V. However, the mechanism involving generation of $HO^\bullet$ from water still has its proponents (Alegría *et al.*, 1997), now in the form of direct one-electron transfer from water to the quinone triplet. The high value of $E°(H_2O/H_2O^{\bullet +})$, 3.4 V (Table 3.2), would seem to preclude this step in competition with electron transfer from DMPO.

Concluding this section on inverted spin-trapping, it can be stated that this mechanism can be easily controlled by avoiding strong thermal oxidants. This condition is most often fulfilled in biological systems (see below). However, if photochemistry is involved, as for example in spin-trapping studies of light-harvesting systems, there is need for a judicious choice of reaction components in order to avoid formation of spin-trap radical cations. A way to distinguish between true spin-trapping and the radical cation mechanism is to run the reaction in the presence of 1,1,1,3,3,3-hexafluoropropan-2-ol, a solvent which drastically attenuates nucleophilic reactivity by strong solvation of nucleophiles, particularly negatively charged ones (Eberson *et al.*, 1996).

Table 3.5 Anodic half-wave or peak potentials (*vs.* SCE) of hydroxylamines in various forms

Compound	$E_{1/2}/V^{a}$	E_{pa}/V^{b}	E_{pa1}/V^{c}	E_{pa2}/V^{d}
Hydroxylamine	−0.35		0.80	
N-Ethylhydroxylamine	−0.49		0.65	1.45
N-Isopropylhydroxylamine	−0.48			
N-*tert*-Butylhydroxylamine	−0.47		0.80	1.55
N-Cyclohexylhydroxylamine	−0.47		0.60	1.45
N-Phenylhydroxylamine	−0.48	−0.75	0.45	0.80
N-(4-Bromophenyl)hydroxylamine		−0.64	0.55	0.90
N-Benzylhydroxylamine	−0.52			
2-Phenyl-2-hydroxylaminopropane	−0.47			
3-Methyl-3-hydroxylamino-2-butanol	−0.51			
N,*N*-Dimethylhydroxylamine			0.50	1.25
N,*N*-Dibenzylhydroxylamine	−0.38			
N-Hydroxypiperidine		−1.10		

a $RNHO^-$ oxidation in aqueous solution at pH 13 (Iversen and Lund, 1965).
b $RNHO^-$ oxidation in dimethyl sulfoxide–Et_4NBF_4 (Bordwell and Liu, 1996).
c RNHOH oxidation in acetonitrile–$NaClO_4$ (Sayo *et al.*, 1973; Ozaki and Masui, 1978).
d RNH_2^+OH oxidation in acetonitrile–$NaClO_4$ (Sayo *et al.*, 1973; Ozaki and Masui, 1978).

3.3 The Forrester–Hepburn mechanism; nucleophilic addition–oxidation

This mechanism (equation 3.3) was in fact developed in parallel with the spin-trapping concept, in that one of the pioneering studies used the reaction between an alkyllithium or a Grignard compound and a spin-trap, followed by air oxidation, to make authentic spin adducts for comparison with those obtained by alkyl radical trapping (Janzen and Blackburn, 1968). Somewhat later, Forrester and Hepburn published a cautionary note regarding this mechanism (Forrester and Hepburn, 1971), showing that several types of common nucleophiles, such as acetate ion, cyanide ion or carbanions (for a review on nitrone chemistry, see Breuer, 1989), readily added to PBN and gave spin adducts upon oxidation with as weak an oxidant as dioxygen. It is interesting to speculate why the Forrester–Hepburn mechanism was not much discussed in the first 10–15 years of the development of the spin-trapping method. One reason might be its dependence on a redox process. Knowledge about electron transfer and organic redox reactivity always was scarce in the community of organic chemists, presumably due to its heavy addiction to 'two-electron' concepts and mechanisms, and it took a long time before electron transfer theory was firmly established in the curriculum of physical organic chemistry (Lowry and Richardson, 1987).

3.3.1 Redox properties of hydroxylamines and oxidants

As a prelude to the discussion of nucleophilic addition–oxidation, it is necessary to have an idea about the redox reactivity of hydroxylamines, the intermediate stage of equation 3.3. Recent studies of the mechanism (Hassan *et al.*, 1998) indicate that the oxidation of hydroxylamines proceeds by a one-electron transfer mechanism, and redox potentials are then needed to judge their reactivity. Such data are listed in Table 3.5. Essentially, two groups of data are available, namely from polarographic measurements of $E_{1/2}$ in aqueous medium at pH 13 and voltammetric determination of E_{pa1} and E_{pa2} in acetonitrile. The first group of potentials presumably pertains to oxidation of anions $RNHO^-$ and/or RN^-OH whereas the E_{pa1} represents oxidation of neutral RNHOH. A few determinations have also been carried out with the preformed anion of PhNHOH in non-aqueous medium, showing that $PhNHO^-$ is oxidized to the aminoxyl radical at as low a potential as −0.7 V. The anodic oxidation of neutral RNHOH leads to the simultaneous formation of protons, and therefore a second anodic peak at E_{pa2} due to oxidation of RNH_2^+OH could be recorded.

For the discussion of the Forrester–Hepburn mechanism, we will further need redox potentials of a number of oxidants B which might be used for the second step of equation 3.3. Table 3.4 includes redox potentials of a number of oxidants which are commonly used in spin-trapping situations. In appropriate cases, the excited state potentials are also given, demonstrating that a weak ground state oxidant often corresponds to a strong excited state oxidant.

3.3.2 Acid–base properties of RNHOH

The pK_a of RNHOH in water is not known, but from the pK_a = 24.2 in dimethyl sulfoxide it was deduced that the acidity of PhNHOH was approximately the same as that of isopropyl alcohol (Bordwell and Liu, 1996), meaning that the $RNHO^-$ is a strong base. The same study indicated that PhNHOH had both NH and OH acidity. The pK_a of $PhNH_2^+OH$ is 3.2, and if the relation between $PhNH_2^+OH$ and $AlkNH_2^+OH$ is the same as between $PhNH_3^+$ and $AlkNH_3^+$, the pK_a of $AlkNH_2^+OH$ should be around 9.

3.3.3 Addition of NuH to nitrones; thermodynamics

The addition of a negatively charged nucleophile to a nitrone, as exemplified by PBN, will initially involve the equilibrium of equation 3.7. With a proton source available, a second equilibrium (equation 3.8) will be established. The position of the overall equilibrium (equation 3.9) will be approximately governed by the difference between the bond energy of H—Nu and the C—Nu bond in X, given that the OH bond energy and the difference in C═N and C—N bond energy can be considered to be the same in all situations.

$$\text{PhCH═N(O)Bu}^t + \text{Nu}^- \rightleftharpoons \text{PhCH(Nu)N(O}^-) \qquad (3.7)$$

$$\text{PhCH(Nu)N(O}^-) + \text{H}^+ \rightleftharpoons \text{PhCH(Nu)N(OH)} \qquad (3.8)$$

$$\text{PhCH═N(O)Bu}^t + \text{Nu—H} \rightleftharpoons \text{PhCH(Nu)N(OH)} \qquad (3.9)$$

Table 3.6 gives bond energy data for a number of Nu—H and Nu—CH_3 bonds. In order to obtain an approximate value of $\Delta H°$ for equation 3.9, the C═N, C—N and O—H bond energies were taken to be 154, 79 and 100 kcal mol^{-1}. Despite the uncertainties in the estimates of $\Delta H°$ one can nevertheless calibrate one end of the scale by noting that HCN and RH (as *C*-centered anions, followed by protonation) are known to add irreversibly to nitrones and give isolatable hydroxylamine derivatives (Bonnett *et al.*, 1959). These values accordingly cannot be far off the mark. At the other end, it is not possible to judge how far to the right equation 3.9 is placed

Table 3.6 Bond dissociation energies of Nu—H and Nu—CH_3 and estimated enthalpy change for equation 3.9 (Lowry and Richardson, 1987; Benson and Cohen, 1997)

Nu—H	*pK of Nu—H*	*Nu—H bond strength/ kcal mol^{-1}*	*Nu—CH_3 bond strength/ kcal mol^{-1}*	*ΔH° of equation (3.9)/ kcal mol^{-1}*
HS—H	7.0	90	88	−23
NC—H	9.2	130	122	−17
CH_3CH_2—H	~50	98	85	−12
CH_3COO—H	4.8	112	97	−10
HO—H	15.7	119	102	−8
HOO—H	11.8	90	69	−4
H_2N—H	33	103	79	−1
F—H	3.2	136	108	4

for fluoride ion but at least one can say that *some* adduct must be formed, since the corresponding spin adduct has been detected after air oxidation by EPR spectroscopy both for PBN and DMPO (Eberson, 1994). With regard to amines, it is known that piperidine adds to the nitrone system, but the product undergoes further reactions. It would then appear that both water and hydrogen peroxide should add to a nitrone with the adduct predominating at equilibrium.

3.3.4 Kinetic factors in the Forrester–Hepburn mechanism

A second important matter of the reaction of equation 3.9 is its kinetics. Generally, the concentration of a spin adduct is dependent on both its rate of formation and its rate of decay where especially the latter is subject to factors seemingly outside the experimenter's control. At higher concentrations, the second-order reaction between two aminoxyls becomes important, leading to disproportionation to give a hydroxylamine and a nitrone. At lower concentrations (< 10 μmol dm^{-3}) first-order reactions prevail. In the present context we are dealing with reactions taking place under slightly oxidizing conditions, and hence the important decay process is likely to be further one-electron oxidation of the spin adduct to give a diamagnetic nitrosonium ion (equation 3.10). This reversible redox process takes place at potentials in the same range as those required for the formation of the spin adduct from the neutral hydroxylamines (compare Table 3.7 with Table 3.5).

$$\text{Ph—CH(R)N(O}^{\bullet}\text{)Bu}^t \xrightarrow{-e^-} \text{Ph—CH(R)N(O}^{+}\text{)Bu}^t \qquad (3.10)$$

For the special case of hydroxyl spin adducts one can add an efficient decay process caused by acidic conditions, namely protonation of the hydroxyl group and solvolysis of the intermediate to give the radical cation of the spin-trap (equation 3.11) (Davies *et al.*, 1992) which has several fast decomposition modes (Zubarev and Brede, 1994, 1995).

$$\text{HO—ST}^{\bullet} \rightarrow \text{H}_2\text{O}^{+}\text{—ST}^{\bullet} \rightarrow \text{H}_2\text{O} + \text{ST}^{\bullet+} \qquad (3.11)$$

Table 3.7 Reversible potentials for nitroxyl/nitrosonium (R_2N—$O^•$/R_2N=O^+) couples, as determined by cyclic voltammetry in acetonitrile (Sümmerman and Deffner, 1975; Bard *et al.*, 1974)

Compound R_2N—$O^•$	E°(R_2N—$O^•$/R_2N=O^+)/*V* vs. *SCE*
(t-Bu)$_2$N—$O^•$	0.57
2,2,6,6-Tetramethylpiperidin-1-oxyl	0.63
4-Hydroxy-2,2,6,6-tetramethylpiperidin-1-oxyl	0.69
4-Oxo–2,2,6,6-tetramethylpiperidin-1-oxyl	0.80
Ph—PBN$^•$	0.7

Table 3.8 First-order rate constants for the disappearance of hydroxyl spin adducts under various conditions (Haire and Janzen, 1982; Kotake and Janzen, 1991; Janzen *et al.*, 1992; Tuccio *et al.*, 1995, Eberson, 1999)

Spin adduct	*Conditions*	k/*min*$^{-1}$	*Half-life/s*
HO—PBN$^•$	CH_3CN–2% water	3.8	11
HO—PBN$^•$	Aqueous phosphate buffer, pH 6.0	0.46	90
HO—PBN$^•$	Aqueous phosphate buffer, pH 8.0	3.8	11
HO—PBN$^•$	Aqueous phosphate buffer, pH > 9	> 40	< 1
HO—DMPO$^•$	Aqueous buffer, pH 7.4	0.0044	9360
HO—DMPO$^•$	CH_3CN–12.5% water (v/v)	3.0	14
HO—DMPO$^•$	CH_3CN–12.5% water (v/v), satd. by K_2CO_3	1.3	32
HO—DMPO$^•$	CH_3CN–12.5% water (v/v), [2,6-di-t-Bu-pyridine] = 60 mmol dm^{-3}	1.8	23
HO—DMPO$^•$	CH_3CN–12.5% water (v/v), [Et_4NOH] = 8.5 mmol dm^{-3}	0.22	188
HOO—DMPO$^•$	Aqueous phosphate buffer, pH 5.6	0.48	87
HOO—DMPO$^•$	Aqueous phosphate buffer, pH 7	0.83	50
HOO—DMPO$^•$	Aqueous phosphate buffer, pH 8.2	1.0	41

The upper part of Table 3.8 shows half-lives of HO—PBN$^•$, generated by photolysis of hydrogen peroxide. The half-life in unbuffered acetonitrile is short, 11 s, and is about an order of magnitude larger in an aqueous buffer between pH 6 and 7. Above pH 7 the half-life goes down steeply, presumably because the OH group ionizes and provides a rapid C—N bond cleavage pathway (equations 3.12 and 3.13).

$$\text{Ph—CH(OH)N(O}^\bullet\text{)Bu}^t + \text{HO}^- \rightleftharpoons \text{Ph—CH(O}^-\text{)N(O}^\bullet\text{)Bu}^t + \text{H}_2\text{O} \quad (3.12)$$

$$\text{Ph—CH(O}^-\text{)N(O}^\bullet\text{)Bu}^t \rightarrow \text{PhCHO} + {}^-\text{N(O}^\bullet\text{)Bu}^t \quad (3.13)$$

The middle part of Table 3.8 shows decay rates of HO—DMPO$^•$ generated by photolysis of DMPO and tetrafluorobenzoquinone in aqueous acetonitrile. Tetrafluorobenzoquinone and other strongly oxidizing quinones create acidity when dissolved in protic solvents which contributes to the short half-life in unbuffered medium via the solvolysis mechanism of equation 3.11. Addition of a base increases the half-life significantly; the reason why the mechanism of equations 3.12 and 3.13 does not come into play is presumably the lower acidity of the fully aliphatic HO—DMPO$^•$ in relation to the arylaliphatic Ph—CH(OH)N(O$^•$)Bu^t.

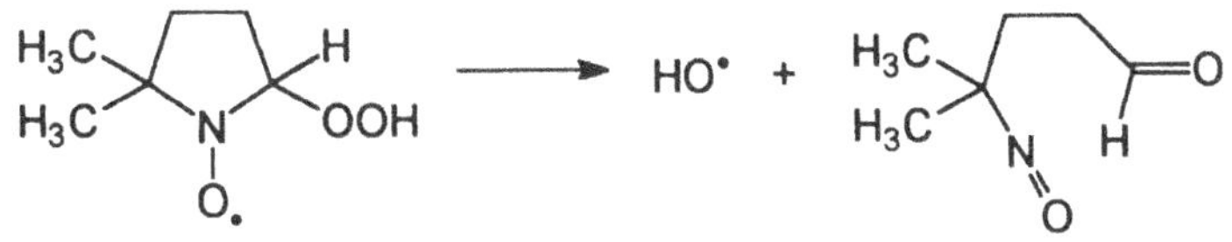

Figure 3.4 Decomposition of the DMPO hydroperoxyl spin adduct to give hydroxyl radical

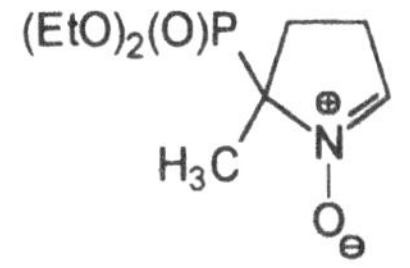

5-Diethoxyphosphoryl-5-methyl-1-pyrroline *N*-oxide, DEPMPO

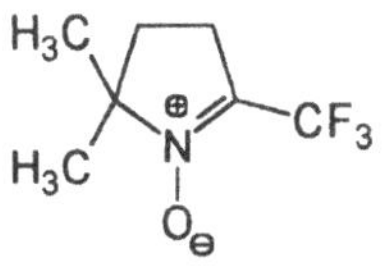

5,5-Dimethyl-2-(trifluoromethyl)-1-pyrroline *N*-oxide, TFDMPO

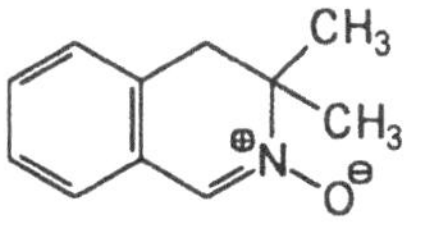

3,4-Dihydro-3,3-dimethylisoquinoline *N*-oxide, DDQO

Figure 3.5 Showing three spin traps with improved properties with regard to hydroxyl or hydroperoxyl spin adduct persistence. For HOO—DEPMPO$^•$ $\tau_{1/2}$ was 16 times that of HO—DMPO$^•$ in an aqueous buffer at pH 7 (Tuccio *et al.*, 1995); for HO—TFDMPO$^•$ $\tau_{1/2}$ was 5 times that of HO—DMPO$^•$ in water with 1% H_2O_2 (Janzen *et al.*, 1995); for HO—DDQO$^•$ $\tau_{1/2}$ was 5 times that of HO—DMPO$^•$ in an aqueous buffer at pH 7 (Thomas *et al.*, 1996)

Hydroperoxyl spin adducts HOO—ST$^•$ and/or $^-$OO—ST$^•$ are likewise shortlived species (Table 3.8) which have the additional problem of decomposing into the corresponding hydroxyl spin adducts HO—ST$^•$. The decomposition of HOO—DMPO$^•$ in aqueous buffer was shown to involve the formation of hydroxyl radical by adding ethanol which led to the trapping of the α-hydroxyethyl radical in competition with trapping of HO$^•$. The reaction of Figure 3.4 was suggested to account for the generation of HO$^•$ (Finkelstein *et al.*, 1982). A useful review of the problems associated with hydroxyl and hydroperoxyl spin-trapping has been published (Finkelstein *et al.*, 1980).

Given these short half-lives of hydroxyl and hydroperoxyl spin adducts of PBN and DMPO, much work has been committed to the synthesis of spin-traps with improved properties in this respect. Some examples are shown in Figure 3.5.

3.3.5 The acid-promoted Forrester–Hepburn mechanism

Recently (Sang *et al.*, 1996) it was noted that the thermal reaction between PBN and trichloroacetonitrile in hexane gave a mixture of three spin adducts, assigned to PBNOx (the acylaminoxyl derived from PBN), an *O*-centered spin adduct and a nitrogen-centered spin adduct, the latter being derived from the *N*-centered resonance form of the NC—$CCl_2^•$ radical (equation 3.14). This reaction type, spontaneous formation of spin adducts from a spin-trap in a system seemingly devoid of any appropriate radical reactivity, had been found in several other systems but its mechanism was not well understood. A closer examination revealed that the PBN—CCl_3CN reaction was promoted by small amounts of an acid HA, for example trichloroacetic acid which is a likely contaminant in trichloroacetonitrile (Eberson *et al.*, 1997).

$$PBN \xrightarrow[\text{dark}]{Cl_3CCN} \underset{\text{PBNOx}}{PhCON(O^\bullet)Bu^t} + XO\text{—}PBN^\bullet + Cl_2C{=}C{=}N\text{—}PBN^\bullet \qquad (3.14)$$

Thus the mechanism of equations 3.15 to 3.17 was proposed:

$$PBN + HA \rightleftharpoons PBN(H)A \qquad (3.15)$$

$$PBN(H)A + Cl_3CCN \rightarrow A\text{—}PBN^\bullet + H^+ + Cl^- + Cl_2C{=}C{=}N^\bullet \qquad (3.16)$$

$$Cl_2C{=}C{=}N^\bullet + PBN \rightarrow Cl_2C{=}C{=}N\text{—}PBN^\bullet \qquad (3.17)$$

Trichloroacetonitrile should be a weak electron transfer oxidant, capable of oxidizing the hydroxylamine intermediate to give the spin adduct corresponding to HA (which is *O*-centered if HA is a carboxylic acid) and the dichlorocyanomethyl radical which is trapped by PBN. A new acid, HCl, becomes available and continues with a new triad of reactions, giving the reactive Cl—PBN$^\bullet$ which is converted into PBNOx by solvolysis–oxidation. Similar processes presumably take place in other situations where a spin-trap and a weak electron acceptor interact thermally, for example PBN and *N*-haloimides (Kaushal and Roberts, 1989), 3-chloroperoxybenzoic acid (Janzen *et al.*, 1992), *N*-chlorosulfonamides (Evans *et al.*, 1985, 1985a), *N*-fluorodibenzenesulfonimide (Eberson and Persson, 1997) and *N*-chlorobenzotriazole. For the latter compound, autocatalysis by benzotriazole was demonstrated (Carloni *et al.*, 1996).

3.3.6 Cases of the Forrester–Hepburn mechanism in organic solvents

Table 3.9 lists typical cases of the nucleophilic addition–oxidation mechanism in organic solvents, using PBN and DMPO as spin-traps and a range of nucleophiles in combination with mild oxidants. *N*-heterocyclic bases of diazole, triazole and tetrazole type are particularly reactive in this mechanism, and allowed for a combined UV and EPR spectral kinetic study of the reaction between DMPO and benzotriazole, using $Co^{III}W_{12}O_{40}^{5-}$ as the oxidant (Alberti *et al.*, 1997). Both the disappearance of $Co^{III}W_{12}O_{40}^{5-}$ (UV) and the appearance/disappearance of the spin adduct could be monitored, and it could be shown that the reaction was first order in each of [DMPO], [benzotriazole] and [$Co^{III}W_{12}O_{40}^{5-}$] with a rate constant of about 0.3 dm^6 mol^{-2} s^{-2}. This means that the reaction is complete within 2 h at the concentration conditions prevailing in spin-trapping experiments. PBN reacted much slower under similar conditions (in the order of 100 h).

The thermal DMPO—H_2O reaction gave HO—DMPO$^\bullet$ with oxidants with redox potentials in the range of 0.53 to −0.45 V. Alcohols also reacted to give RO—DMPO$^\bullet$ with F_4Q as the oxidant, provided they were not too sterically hindered. Hindered (like *t*-BuOH) or electrophilic (like CF_3CH_2OH or $(CF_3)_2CHOH$) alcohols required photochemical activation to react, presumably via formation of DMPO$^{\bullet+}$.

As mentioned above, the HCN adducts of nitrones can be synthesized. The HCN adduct of PBN was prepared by treatment of PBN with trimethylsilyl cyanide, followed by acidic hydrolysis. This hydroxylamine was subjected to oxidation with oxidants ranging from dioxygen to Br_2 in a variety of solvents, providing the authentic

Table 3.9 Cases of the Forrester–Hepburn mechanism taking place in organic solvents (Carloni *et al.*, 1996; Alberti *et al.*, 1997; Eberson and Persson, 1997, 1997a, 1998; Eberson, 1999)

System[a]	*Solvent*	*Spin adduct(s)*
PBN, benzotriazole, Cl_4Q	PhH	Benzotriazol-1-yl-PBN$^•$ Benzotriazol-2-yl-PBN$^•$
PBN, benzotriazole, $Co^{III}W_{12}O_{40}^{5-}$	DCM	Benzotriazol-1-yl-PBN$^•$
PBN, CH_3COOH, Br_2	DCM	CH_3COO—PBN$^•$
PBN, K^+ or $Bu_4N^+(NC)_3C^-$, Br_2	AN or DCM	$(NC)_3C$—PBN$^•$
PBN, $Bu_4N^+F^-$, O_2	$CHCl_3$	F—PBN$^•$
Presynthesized NC—PBN(H), O_2, Br_2, $(PhCOO)_2$, $Co^{III}W_{12}O_{40}^{5-}$ or Tl^{III}	Various	NC—PBN$^•$
DMPO, ROH, F_4Q (R = Me, Et, isoPr)	EtOAc or AN	RO—PBN$^•$
DMPO, H_2O, F_4Q, Cl_4Q, Br_4Q, 2,3-Cl_2Q or Q	EtOAc or AN	HO—PBN$^•$
DMPO, benzotriazole, $Co^{III}W_{12}O_{40}^{5-}$	AN or DCM	Benzotriazol-1-yl-DMPO$^•$
DMPO, $(PhSO_2)_2NF$	DCM	F—DMPO$^•$
DMPO, benzotriazole, DDQ, Cl_4Q, 2,3-Cl_2Q, Q or Me_4Q	AN	Benzotriazol-1-yl-DMPO$^•$
DMPO, H_2O_2	Pyridine	HOO—DMPO$^•$
DMPO, H_2O, ferricinium ion	EtOAc	HO—DMPO$^•$
DMPO, H_2O, $FeCl_3$, 2,6-di-*t*-Bu-pyridine	EtOAc	HO—DMPO$^•$

a Q = 1,4-benzoquinone.

NC—PBN$^•$ spin adduct previously observed in a number of situations involving both inverted spin-trapping and nucleophilic addition–oxidation. A separate study of the generation and possible spin-trapping of the cyano radical indicated that this high-energy species (see Table 3.2) can only be trapped under special conditions, like photolysis of *N*-cyanosuccinimide with PBN in 1,1,1,3,3,3-hexafluoropropan-2-ol, and then gives the *N*-centered isocyano spin adduct, :C═N—PBN$^•$ (Eberson and Persson, 1998).

3.3.7 Cases of the Forrester–Hepburn mechanism in aqueous medium

If anything, this mechanism should take place with greater ease and lead to a higher concentration of spin adduct in a predominantly aqueous system than in an aqueous organic solvent. The equilibrium of equation 3.9 should be more displaced to the right, and the spin adduct should decompose at a slower rate in the aqueous medium (Table 3.8). Thus there is little doubt that many of the reactions listed in Table 3.9 can be extended to solutions containing predominantly water, solubility of the various components permitting. Relatively few studies have included the Forrester–Hepburn mechanism in the discussion of spin-trapping in aqueous solutions, but existing results point to a rather complex situation, especially when the Fenton reaction is employed as a source of HO$^•$. Table 3.10 lists some typical cases, most of them with bearing on the Fenton reaction.

The Fenton reaction typically consists of mixing hydrogen peroxide and Fe^{II} solution in water at pH 2 which for long was assumed to give Fe^{III} and HO$^•$ (equation 3.18), the latter being used to sustain various types of radical reactions, including spin-trapping.

Table 3.10 Cases of the Forrester–Hepburn mechanism taking place in aqueous media (Janzen *et al.*, 1978; Makino *et al.*, 1990, 1992; Hanna *et al.*, 1992; Monroe and Eaton, 1996)

System[a]	*Medium*	*Spin adduct*
4-PyOBN, H_2O_2	Water, pH 2	HO—POBN•
DMPO, 2-methyl-1,4-naphthoquinone	Aq. phosphate buffer, pH 7.0	HO—DMPO•
DMPO, $FeCl_3$	Water	HO—DMPO•
DMPO, $FeCl_3$	Water–10% MeOH	MeO—DMPO•
DMPO, $FeCl_3$	$H_2^{17}O$	$H^{17}O$—DMPO•
DMPO, $FeCl_3$	Water–20% EtOH	EtO—DMPO•
DMPO, $CuCl_2$	Water	HO—DMPO•
DMPO, KCN	Water, H_2O_2	NC—DMPO•

a PyO = 1-oxidopyridino.

$$Fe^{II}(OH_2)_6 + H_2O_2 \xrightarrow{k = 41\ dm^3\ mol^{-1}\ s^{-1}} (H_2O)_5Fe^{III}(OH) + HO^\bullet \tag{3.18}$$

Presently, some controversy surrounds the mechanism (Walling, 1998; MacFaul *et al.*, 1998) after the suggestion that free $HO^\bullet$ is not formed but that instead an Fe^{II}—OOH complex acts as the $HO^\bullet$ carrying intermediate (Sawyer *et al.*, 1996). Notwithstanding this problem, since Fe^{III} is an end product of equation 3.18, the question arises whether the formation of spin adducts from Fenton-type reactions is due to the Forrester–Hepburn mechanism. Indeed, the addition of $FeCl_3$ to an aqueous solution of DMPO gave HO—DMPO• (Makino *et al.*, 1990, 1992). The addition of a competing nucleophile, methanol, in 2% concentration produced in addition MeO—DMPO•, which became the predominant species at 10% MeOH. Similar results were obtained by addition of ethanol. Thus nucleophilic addition of an alcohol to DMPO takes place giving RO—DMPO• after oxidation of the hydroxylamine. The results obtained in the presence of alcohols contrast with those obtained when $HO^\bullet$ is assumed to be the source of HO—DMPO•: this radical abstracts an H atom from the α-position and gives C-centered spin adducts, $HOCH_2$—DMPO• or $CH_3(HO)CH$—DMPO•. This reaction has been assumed to be a reliable diagnostic that $HO^\bullet$ is the intermediate being trapped.

Later, the scope of the Forrester–Hepburn mechanism in the DMPO–water–Fe^{III} system was further investigated (Hanna *et al.*, 1992). The previous results were confirmed and their scope extended by using Cu^{II} as the oxidant. The use of $H_2^{17}O$ showed that $FeCl_3$ alone gave a ≈100% nucleophilic reaction, whereas with $FeCl_3$ in the presence of H_2O_2 this proportion decreased to 41%. A true Fenton reaction starting from $FeCl_2$ and H_2O_2 indicated a 57% contribution from the nucleophilic reaction. For Cu salts (Cu^I also sustains a Fenton type reaction) the corresponding numbers were ≈100%, 96% and 87%. These results present a serious problem for the interpretation of spin-trapping data in biological systems, since not only Fe^{III} but also H_2O_2 were implicated as oxidants for the intermediate hydroxylamine.

The addition of chelating agents or buffers to Fenton reactions had a significant effect on Fe^{III} induced reactions and a smaller one on Cu^{II} induced reactions. Relative EPR spectral intensities for a series of reactions with different buffers and chelators are shown in Table 3.11. It is clear that buffers or chelators have a beneficial effect

Table 3.11 Relative EPR spectral intensities of HO—DMPO$^{\bullet}$ obtained from the reaction between DMPO and water with Fe^{III} or Cu^{II} in various buffers (0.1 M) or with various chelators present (Hanna *et al.*, 1992)

Buffer or chelator	*Rel. int., reaction with Cu^{II}* (E°[Cu^{II}/Cu^{I}]/V vs. SCE)	*Rel. int., reaction with Fe^{III}* (E°[Fe^{III}/Fe^{II}]/V vs. SCE)
None	100 (−0.09)	100 (0.28)
NaCl, 0.15 M	100 (0.30)	100 (0.28)
Acetate, pH 5.0	64	60
Citrate, pH 5.4	2	0 (−0.14)
Phosphate, pH 7.4	35	0
Hepes, pH 7.5	35	2
Mops, pH 7.5	95	3
Tris-HCl, pH 8.0	42	0
Borate, pH 8.5	47	3
EDTA		0 (−0.12)
NTA	0	0 (0.09)
DTPA	0	0

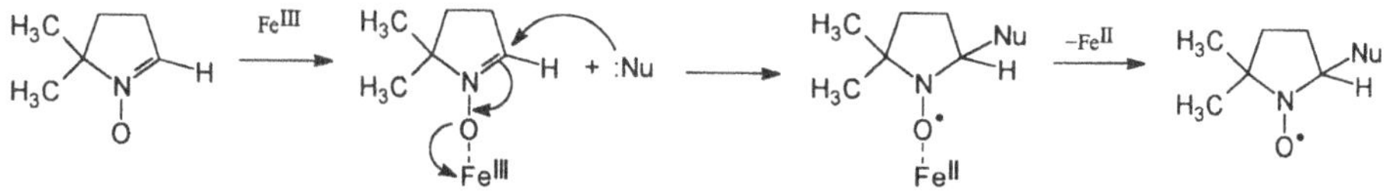

Figure 3.6 The Fe^{III} promoted nucleophilic addition–oxidation mechanism, applied to DMPO—:Nu

in removing the artefactual spin adducts, especially in the Fe^{III} reaction. This was explained by invoking an Fe^{III} promoted nucleophilic addition–oxidation to DMPO according to Figure 3.6; by complexing the iron ion, a buffer anion or a chelator will decrease the rate of this reaction. However, in other cases of the Forrester–Hepburn mechanism with water as the nucleophile there is no apparent need of promoting the addition step, and it may well be that the role of the complexing agent instead is to change the Fe^{III}/Fe^{II} redox potential. The limited data of Table 3.11 show that the complexing agents do decrease the redox potential, although more data are needed to check this hypothesis. The effect of chelators/buffers is not as pronounced for the couple Cu^{II}/Cu^{I}, but then it should be noted that even very low concentrations of halide ion increase its redox potential. It is evident that metal ions like Fe^{III} and Cu^{II} can play an important role in the development of spin adducts according to the nucleophilic addition mechanism. It seems that some control can be exerted by a buffer system or an added chelator, but it remains to establish the mechanism of this inhibition effect.

It has already been mentioned that the addition of ethanol is diagnostic for establishing the involvement of HO$^{\bullet}$ in a spin-trapping reaction. If HO$^{\bullet}$ is involved, the addition of a relatively small amount of ethanol should replace HO—ST$^{\bullet}$ by CH_3(HO)CH—ST$^{\bullet}$; if not, the nucleophilic addition–oxidation mechanism should give CH_3CH_2O—ST$^{\bullet}$. A similar use has often been made of formate, since HO$^{\bullet}$ abstracts the formyl hydrogen atom in a very fast reaction (equation 3.19) and thus can outcompete the trapping process ($k = 2.6 \times 10^9$ dm^3 mol^{-1} s^{-1}) (Castelhano *et al.*, 1983).

$$HO^{\bullet} + HCOO^{-} \xrightarrow{k = 2.9 \times 10^{9}\ dm^{3}\ mol^{-1}\ s^{-1}} H_2O + {}^{\bullet}COO^{-} \quad (3.19)$$

Thus the $^{\bullet}COO^{-}$ radical is formed and trapped by ST giving a spin adduct with quite different characteristics than HO—ST$^{\bullet}$, at least when DMPO or its congeners are used.

The last entry of Table 3.10 shows that the C-connected cyano radical adduct of DMPO can be detected in a solution of DMPO, KCN and H_2O_2 (Moreno *et al.*, 1988). The system contains only H_2O_2 as the oxidant, which is a weak one. The cyano radical adduct is a high-energy species (see Table 3.2) and already this fact militates against its formation by oxidation of cyanide ion. Most evidence points to the Forrester–Hepburn mechanism being responsible for the formation of cyano spin adducts (Eberson and Persson, 1998).

3.4 Detection of spin adducts in biochemical/biological systems

3.4.1 Is inverted spin-trapping possible?

The oxidants listed in Table 3.4 are of 'unnatural' type, apart from dioxygen and a few quinones. In biological systems the putative oxidants are redox proteins, for which redox potentials are given in Table 3.12. In general, these redox potentials are modest and therefore rule out the radical cation mechanism of spin adduct formation. One possible exception is cytochrome P450 for which the one-electron transfer mechanism has been extensively discussed, also for species which are difficult to oxidize, like sulfides and sulfoxides with E_{pa} in the range of 1.3 to 2.0 V. If such a mechanism indeed operates, the Marcus treatment of the rather scarce experimental data indicates that the redox potential of cytochrome P450 must be high, in the range of 1.3–1.8 V (Eberson, 1990).

Table 3.12 Potentials, $E°'$, of redox proteins in water at pH 7.0 and 25°C (from a compilation in: Eberson, 1985; Hayashi and Yamasaki, 1979)

System (redox couple)	$E°'$/V vs. *SCE at pH 7.0*	$E°'$/V vs. *SCE (pH)*
Horseradish peroxidase (Fe^{III}/Fe^{II})	−0.41	
Rubredoxin (Fe^{III}/Fe^{II})	−0.30	
Myoglobin (Fe^{III}/Fe^{II})	−0.19	
Haemoglobin (Fe^{III}/Fe^{II})	−0.07	
Cytochrome *c* from horse (Fe^{III}/Fe^{II})	0.01	
High-potential protein from *Chromatium vinosum* (Fe^{III}/Fe^{II})	0.11	
Plastocyanin (Cu^{II}/Cu^{I})	0.13	
Azurin (Cu^{II}/Cu^{I})	0.14	
Ceruloplasmin (Cu^{II}/Cu^{I})	0.15	
Laccase (Cu^{II}/Cu^{I})	0.18	
Cytochrome P450 (Fe^{V}/Fe^{IV} or Fe^{IV}/Fe^{III})	> 0.56	
Horseradish peroxidase Compound I (Fe^{V}/Fe^{IV})	0.67	> 0.8 (< 6)
Horseradish peroxidase Compound II (Fe^{IV}/Fe^{III})	0.70	> 0.8 (< 6)
Lignin peroxidase	~0.95	

Table 3.13 Compounds ArH which give EPR-spectroscopically detectable radical cations upon oxidation with redox proteins in aqueous buffers at pH 3.0 (Kersten *et al.*, 1990; Khindaria *et al.*, 1995)

Oxidant[a] *(redox potential/V* vs. *SCE)*	*ArH*	$E(ArH^{\bullet +}/ArH)/V$ vs. *SCE*	$\Delta G°/eV$
LP + H_2O_2 (0.95)	1,4-Dimethoxybenzene	1.41	0.46
LP + H_2O_2 (0.95)	1,2,3,4-Tetramethoxybenzene	1.25	0.30
LP + H_2O_2 (0.95)	Hexamethoxybenzene	1.24	0.29
LP + H_2O_2 (0.95)	1,2,4,5-Tetramethoxybenzene	0.81	−0.14
LP + H_2O_2 (0.95)	3,4-Dimethoxybenzyl alcohol	–	–
HRP + H_2O_2 (0.70)	1,2,4,5-Tetramethoxybenzene	0.81	0.11
Laccase + O_2 (0.18)	1,2,4,5-Tetramethoxybenzene	0.81	0.63

a LP = lignin peroxidase; HRP = horseradish peroxidase.

Some redox enzymes, in combination with hydrogen peroxide or oxygen, have been shown to generate EPR-spectroscopically detectable radical cations from organic substrates. Table 3.13 lists such experiments, showing that lignin peroxidase + H_2O_2 at pH 3 can produce EPR-detectable concentrations of the 1,4-dimethoxybenzene radical cation, with $E°(Ar^{\bullet +}/ArH) = 1.41$ V. Even more difficultly oxidizable substrates, like 1,2-dimethoxybenzene with $E°(Ar^{\bullet +}/ArH) = 1.45$ V, showed UV-spectroscopically significant changes indicating radical cation formation by lignin peroxidase + H_2O_2. The redox potential of lignin peroxidase + H_2O_2 was estimated to be *ca.* 0.25 V higher than that of horseradish peroxidase (Kersten *et al.*, 1990). Since the experiments were performed at pH 3 and the redox potentials of Table 3.11 refer to values at pH < 6, it can be estimated that the potential of lignin peroxidase + H_2O_2 should be 1.0–1.1 V. This brings a spin-trap like PBN (1.47 V) just within reach for radical cation formation in this system. However, this is a rare situation, and it appears safe to conclude that the inverted spin-trapping mechanism need not be considered in evaluating spin-trapping experiments in thermal biological systems, unless highly redox active spin-traps are employed.

3.4.2 *Is the Forrester–Hepburn mechanism possible?*

From what has been said above, it is obvious that mild oxidants with the potentials required to sustain oxidation of hydroxylamines are ubiquitous in biological systems. This mechanism thus represents a serious source of misinterpretation of spin-trapping results, and it is therefore desirable that new methods are developed to distinguish between true spin-trapping of $HO^{\bullet}$ and the Forrester–Hepburn mechanism.

3.5 Cycloaddition as a possible cause of artefacts

Nitrones undergo 1,3-dipolar cycloaddition reactions which recently have been shown to influence their properties as spin-traps. Thus nitrones like PBN add to nitriles activated by electron-withdrawing substituents (Figure 3.7), thus causing the nitrone to be rapidly depleted in some cases (Sang *et al.*, 1996; Eberson *et al.*, 1998). Cycloaddition can also take place under oxidative conditions between nitrones and olefins, in which case radical cations of isoxazolidines are formed (Eberson *et al.*,

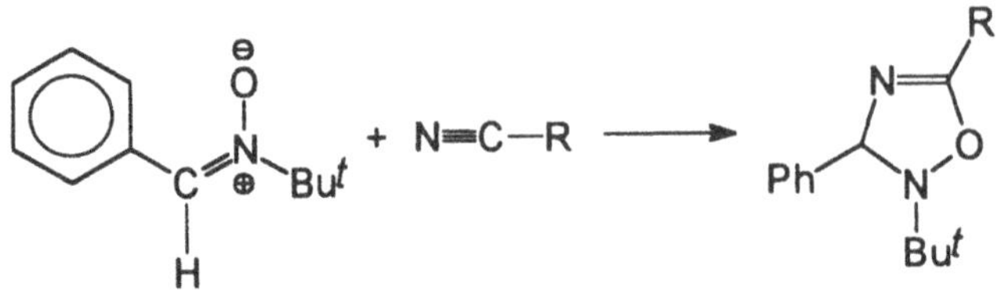

R	k_2/dm^3 mol^{-1} s^{-1}
CCl_3	4.2×10^{-6}
$CCl_2(CN)$	1.9×10^{-3}
$CCl(CN)_2$	0.30
$C(CN)_3$	2.6

Figure 3.7 The addition of some activated acetonitriles to PBN and their rate constants

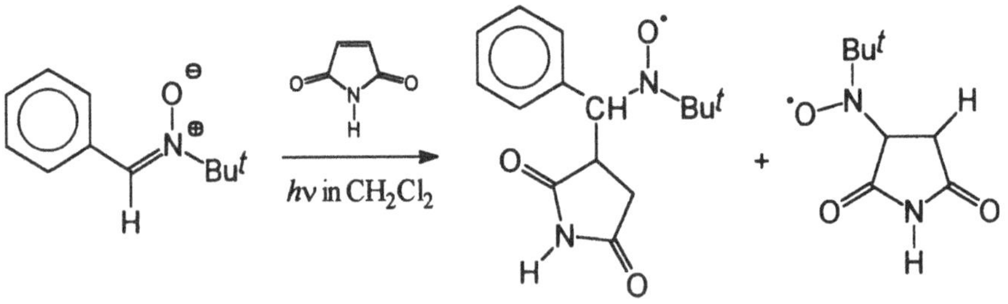

Figure 3.8 The formation of spin adducts of both PBN and *t*-BuNO from PBN and maleimide under illumination. In the dark, only the *t*-BuNO adduct is seen

1997a). The cycloaddition process can be slowed down by lowering the temperature, providing adequate conditions for spin-trapping (Eberson and McCullough, 1998).

It is sometimes observed that PBN is slowly decomposed to give *t*-BuNO, causing the appearance of both PBN and MNP spin adducts, either in parallel or in succession. Usually, this cleavage has been ascribed to hydrolytic reactions (Chamilutrat *et al.*, 1993). A new decomposition mode was recently found in the reaction between PBN and electron-deficient olefins, like maleic anhydride, maleimide or diethyl fumarate (Eberson and Persson, 1998a). As an example, the 2-succinimidyl spin adducts of both PBN and MNP were obtained from the reaction between maleic anhydride and PBN in dichloromethane (Figure 3.8), the former adduct requiring light for its formation. From the fact that PTAD, the most reactive dienophile known, reacted with PBN with almost instantaneous formation of *t*-BuNO in 74% yield and a hydrolytically very reactive tetrasubstituted olefin (Figure 3.9), it was inferred that less reactive olefins should react in a similar way with PBN (Figure 3.10) and thus give rise to a low concentration of *t*-BuNO. The latter then reacted with the olefin in what formally is a reductive coupling process, a reaction with many precedents in cathodic chemistry. This cleavage reaction took place with activated olefins with redox potentials greater than about −1.5 V. Since biologically important quinones have redox potentials in the range of −0.5 to −1.0 V, it may well be that this type of decomposition can be induced by quinones and thus produce misleading results. In particular, *t*-BuNO is prone to decompose photochemically to give *t*-Bu$^•$ and NO, a compound with many biological functions (Chamulitrat *et al.*, 1993).

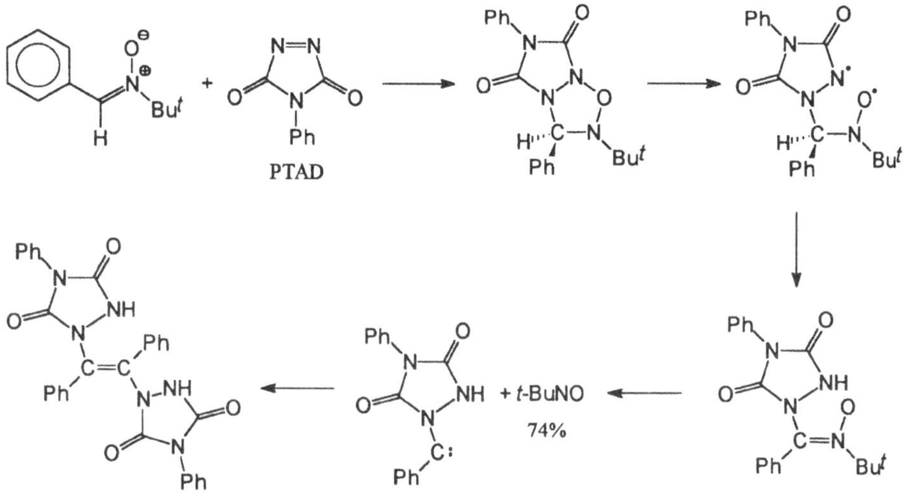

Figure 3.9 The proposed mechanism of the fast reaction between PTAD and PBN, giving a 74% yield of *t*-BuNO in 2 min

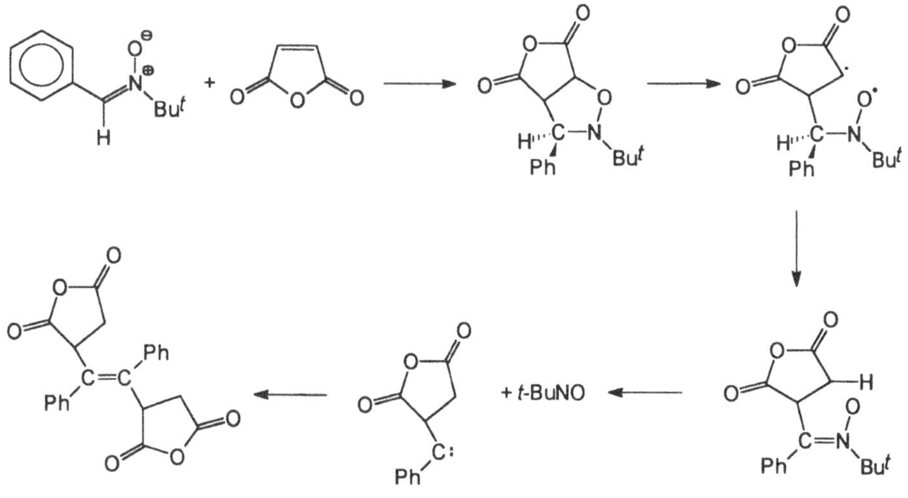

Figure 3.10 The proposed mechanism for the slow formation of *t*-BuNO from maleic anhydride and PBN

References

Alberti, A., Carloni, P., Eberson, L., Greci, L. and Stipa, P. (1997) New insights into *N-tert*-butyl-α-phenylnitrone as a spin-trap. Part 2. The reactivity of PBN and 5,5-dimethyl-4,5-dihydropyrrole *N*-oxide (DMPO) toward *N*-heteroaromatic bases, *Journal of the Chemical Society, Perkin Transactions* 2, 887–892.

Alegría, A.E., Ferrer, A. and Sepúlveda, E. (1997) Photochemistry of water-soluble quinones. Production of a water-derived spin adduct, *Photochemistry and Photobiology*, **66**, 436–442.

Astruc, D. (1995) *Electron Transfer and Radical Processes in Transition-Metal Chemistry*, Weinheim: VCH Publishers.

Bard, A.J., Gilbert, J.C. and Goodin, R.D. (1974) Application of spin-trapping to the detection of radical intermediates in electrochemical transformations, *Journal of American Chemical Society*, **96**, 620–621.

Benson, S.W. and Cohen, N. (1997) The thermochemistry of peroxides and polyoxides, and their free radicals, in Alfassi, Z. (ed.), *Peroxyl Radicals*, Chichester: Wiley, pp. 49–80.

Bhattacherjee, S., Khan, M.N., Chandra, H. and Symons, M.C.R. (1996) Radical cations from spin-traps: reaction with water to give OH adducts, *Journal of the Chemical Society, Perkin Transactions* 2, 2631–2634.

Bonnett, R., Brown, R.F.C., Clark, V.M., Sutherland, I.O. and Todd, A. (1959) Experiments toward the synthesis of corrins. Part. II. The preparation and reactions of Δ^1-pyrroline 1-oxides, *Journal of the Chemical Society*, 2094–2102.

Bordwell, F.G. and Liu, W.-Z. (1996) Equilibrium acidities and homolytic bond dissociation energies of N—H and/or O—H bonds in *N*-phenylhydroxylamine and its derivatives, *Journal of the American Chemical Society*, **118**, 8777–8781.

Breuer, E. (1989) Nitrones and nitronic acid derivatives: their structure and their roles in synthesis, in: Breuer, E., Aurich, H.G. and Nielsen, A. (eds), *Nitrones, Nitronates and Nitroxides*, Chichester: Wiley, pp. 245–312.

Buettner, G. (1993) The pecking order of free radicals and antioxidants: lipid peroxidation, α-tocopherol and ascorbate, *Archives of Biochemistry and Biophysics*, **300**, 535–543.

Carloni, P., Eberson, L., Greci, L., Sgarabotti, P. and Stipa, P. (1996) New insights on *N-tert*-butyl-α-phenylnitrone (PBN) as a spin-trap. Part 1. Reaction between PBN and *N*-chlorobenzotriazole, *Journal of the Chemical Society, Perkin Transactions* 2, 1297–1305.

Castelhano, A.L., Perkins, M.J. and Griller, D. (1983) Spin-trapping of hydroxyl in water: decay kinetics for the $^{\bullet}$OH and $CO_2^{-\bullet}$ adducts to 5,5-dimethyl-1-pyrroline-*N*-oxide, *Canadian Journal of Chemistry*, **61**, 298–299.

Chamulitrat, W., Jordan, S.J., Mason, R.P., Saito, K. and Willson, R. (1993) Nitric oxide formation during light-induced decomposition of phenyl *N-tert*-butylnitrone, *Journal of Biological Chemistry*, **268**, 11520–11527.

Davies, M.J., Gilbert, B.C., Stell, J.K. and Whitwood, A.C. (1992) Nucleophilic substitution reactions of spin adducts. Implications for the correct identification of reaction intermediates by EPR/spin-trapping, *Journal of the Chemical Society, Perkin Transactions* 2, 333–335.

Eberson, L. (1982) Electron-transfer reactions in organic chemistry, *Advances in Physical Organic Chemistry*, **18**, 82–185.

Eberson, L. (1985) The Marcus theory of electron transfer, a sorting device for toxic compounds, *Advances in Free Radical Biology & Medicine*, **1**, 19–90.

Eberson, L. (1987) *Electron Transfer Reactions in Organic Chemistry*, Heidelberg: Springer-Verlag.

Eberson, L. (1990) Calculation of the cytochrome P-450 redox potential from kinetic data by the Marcus treatment; feasible or not?, *Acta Chemica Scandinavica*, **44**, 733–740.

Eberson, L. (1992) 'Inverted spin-trapping.' Reactions between the radical cation of α-phenyl-*N-tert*-butylnitrone and ionic and neutral nucleophiles, *Journal of the Chemical Society, Perkin Transactions* 2, 1807–1813.

Eberson, L. (1994) Inverted spin-trapping. Part III. Further studies on the chemical and photochemical oxidation of spin-traps in the presence of nucleophiles, *Journal of the Chemical Society, Perkin Transactions* 2, 171–176.

Eberson, L. (1998) Spin-trapping and electron transfer, *Advances in Physical Organic Chemistry*, **31**, 91–141.

Eberson, L. (1999) Formation of hydroxyl and hydroperoxyl spin adducts *via* nucleophilic addition–oxidation to 5,5-dimethyl-1-pyrroline *N*-oxide (DMPO), *Acta Chemica Scandinavica*, **53**, 584–93.

Eberson, L. and McCullough, J.J. (1998) Generation of tricyanomethyl spin adducts of α-phenyl-*N-tert*-butylnitrone (PBN) *via* non-conventional mechanisms, *Journal of the Chemical Society, Perkin Transactions 2*, 49–58.

Eberson, L. and Persson, O. (1997) Fluoro spin adducts and their modes of formation, *Journal of the Chemical Society, Perkin Transactions 2*, 893–898.

Eberson, L. and Persson, O. (1997a) Generation of acyloxyl spin adducts from *N-tert*-butyl-α-phenylnitrone (PBN) and 4,5-dihydro-5,5-dimethylpyrrole 1-oxide (DMPO) via nonconventional mechanisms, *Journal of the Chemical Society, Perkin Transactions 2*, 1689–1696.

Eberson, L. and Persson, O. (1998) Trapping of the cyano radical: does it ever happen?, *Acta Chemica Scandinavica*, **52**, 608–621.

Eberson, L. and Persson, O. (1998a) Spin adduct from the reaction between *N*-phenyl-α-*tert*-butylnitrone (PBN) and activated olefins. A facile pathway converting PBN into 2-methyl-2-nitrosopropane (MNP), *Acta Chemica Scandinavica*, **52**, 1081–1095.

Eberson, L., Hartshorn, M.P. and Persson, O. (1996) Inverted spin-trapping. Part V. 1,1,1,3,3,3-Hexafluoropropan-2-ol as a solvent for the discrimination between proper and inverted spin-trapping, *Journal of the Chemical Society, Perkin Transactions 2*, 141–149.

Eberson, L., Hartshorn, M.P. and Persson, O. (1997a) Formation and EPR spectra of radical species derived from the oxidation of the spin-trap, α-phenyl-*N-tert*-butylnitrone (PBN), and some of its derivatives in 1,1,1,3,3,3-hexafluoropropan-2-ol. Formation of isoxazolidine radical cations, *Journal of the Chemical Society, Perkin Transactions 2*, 195–201.

Eberson, L., Lind, J. and Merenyi, G. (1994) Inverted spin-trapping. Part IV. Application to the formation of imidyl spin adducts, *Journal of the Chemical Society, Perkin Transactions 2*, 1181–1188.

Eberson, L., McCullough, J.J. and Persson, O. (1997) Spin adduct formation from the spontaneous reaction between spin-traps and weak electron acceptors, as exemplified by trichloroacetonitrile. An acid promoted version of the Forrester–Hepburn addition–oxidation mechanism, *Journal of the Chemical Society, Perkin Transactions 2*, 133–134.

Eberson, L., McCullough, J.J., Hartshorn, C.M. and Hartshorn, M.P. (1998) 1,3-Dipolar cycloaddition of α-phenyl-*N-tert*-butylnitrone (PBN) to dichloro- and dibromo-malononitrile, chlorotricyanomethane and tetracyanomethane. Structure of products and kinetics of their formation, *Journal of the Chemical Society, Perkin Transactions 2*, 41–47.

Evans, J.C., Jackson, S.K., Rowlands, C.C. and Barratt, M.D. (1985) An electron spin resonance study of radicals from chloramine-T – 1. Spin-trapping of radicals produced in acid media, *Tetrahedron*, **51**, 5191–5194.

Evans, J.C., Jackson, S.K., Rowlands, C.C. and Barratt, M.D. (1985a) An electron spin resonance study of radicals from chloramine-T – 2. Spin-trapping of photolysis products of chlormine-T at alkaline pH, *Tetrahedron*, **51**, 5195–5200.

Finkelstein, E., Rosen, G.M. and Rauckman, E.J. (1980) Spin-trapping of superoxide and hydroxyl radical: practical aspects, *Archives of Biochemistry and Biophysics*, **200**, 1–16.

Finkelstein, E., Rosen, G.M. and Rauckman, E.J. (1982) Production of hydroxyl radical by decomposition of superoxide spin adducts, *Molecular Pharmacology*, **21**, 262–265.

Forrester, A.R. and Hepburn, S.P. (1971) Spin-traps. A cautionary note, *Journal of the Chemical Society. (C)* 701–703.

Gronchi, G., Courbis, P., Tordo, P. Mousset, G. and Simonet, J. (1983) Spin-trapping in electrochemistry. Nonaqueous electrochemical behaviour of nitroso spin-traps, *Journal of Physical Chemistry*, **83**, 1343–1348.

Haire, D.L. and Janzen, E.G. (1982) Synthesis and spin-trapping kinetics of new alkyl substituted nitrones, *Canadian Journal of Chemistry*, **60**, 1514–1522.

Hanna, P.M., Chamulitrat, W. and Mason, R.P. (1992) When are metal ion-dependent hydroxyl and alkoxyl radical adducts of 5,5-dimethyl-1-pyrroline *N*-oxide artefacts?, *Archives of Biochemistry and Biophysics*, **296**, 640–644.

Hassan, A., Wazeer, M.I.M. and Ali, Sk. A. (1998) Oxidation of *N*-benzyl-*N*-methylhydroxylamines to nitrones. A mechanistic study, *Journal of the Chemical Society, Perkin Transactions 2*, 393–399.

Hayashi, Y. and Yamasaki, I. (1979) The oxidation–reduction potentials of compound I/compound II and compound II/ferric couples of horseradish peroxidases A_2 and C*, *The Journal of Biological Chemistry*, **254**, 9101–9106.

Iversen, P.E. and Lund, H. (1965) Electroorganic preparations. XVIII. Preparation and polarographic determination of some *N*-alkylhydroxylamines, *Acta Chemica Scandinavica*, **19**, 2303–2308.

Janzen, E.G. and Blackburn, B.J. (1968) Detection and identification of short-lived free radicals by an electron spin resonance trapping technique, *Journal of the American Chemical Society*, **90**, 5909–5910.

Janzen, E.G., Wang, Y.Y. and Shetty, R.V. (1978) Spin-trapping with α-pyridyl 1-oxide *N-tert*-butyl nitrones in aqueous solutions. A unique electron spin resonance spectrum for the hydroxyl adduct, *Journal of American Chemical Society*, **100**, 2923–2925.

Janzen, E.G., Hinton, R.D. and Kotake, Y. (1992) Substituent effects on the stability of the hydroxyl adduct of α-phenyl-*N-tert*-butylnitrone (PBN), *Tetrahedron Letters*, **33**, 1257–1260.

Janzen, E.G., Lin, C.-R. and Hinton, R.D. (1992) Spontaneous free-radical formation in reactions of *m*-chloroperbenzoic acid with C-phenyl-*N-tert*-butylnitrone (PBN) and 3- or 4-substituted PBN's, *Journal of Organic Chemistry*, **57**, 1633–1635.

Janzen, E.G., Zhang, Y.-K. and Arimura, M. (1995) Synthesis and spin-trapping chemistry of 5,5-dimethyl-2-(trifluoromethyl)-1-pyrroline *N*-oxide, *Journal of Organic Chemistry*, **60**, 5434–5440.

Kaushal, P. and Roberts, B.P. (1989) Cyclisation of ω-(isocyanatocarbonyl)alkyl radicals: acyclic precursors of imidyl radicals, *Journal of the Chemical Society, Perkin Transactions 2*, 1559–1568.

Kavarnos, G.J. (1993) *Fundamentals of Photoinduced Electron Transfer*, Weinheim: Verlag Chemie.

Kersten, P.J., Kalyanaraman, B., Hammel, K.E., Reinhammar, B. and Kirk, T.K. (1990) Comparison of lignin peroxidase, horseradish peroxidase and laccase in the oxidation of methoxybenzenes, *Biochemical Journal*, **268**, 475–480.

Khindaria, A., Grover, T.A. and Aust, S.D. (1995) Evidence for formation of the veratryl alcohol cation radical by lignin peroxidase, *Biochemistry*, **34**, 6020–6025.

Kotake, Y. and Janzen, E.G. (1991) Decay and fate of the hydroxyl radical adduct of α-phenyl-*N-tert*-butylnitrone in aqueous media, *Journal of American Chemical Society*, **113**, 9503–9506.

Lowry, T.H. and Richardson, K.S. (1987) *Mechanism and Theory in Organic Chemistry*, 3rd edn., New York: Harper & Row.

MacFaul, P.A., Wayner, D.D.M. and Ingold, K.U. (1998) A radical account of 'oxygenated Fenton chemistry', *Accounts of Chemical Research*, **31**, 159–162.

Makino, K., Hagiwara, T. Hagi, A., Nishi, M. and Murakami, A. (1990) Cautionary note for DMPO spin-trapping in the presence of iron ion, *Biochemical and Biophysical Research Communications*, **172**, 1073–1080.

Makino, K., Hagi, A., Ide, H., Murakami, A. and Nishi, M. (1992) Mechanistic studies on the formation of aminoxyl radicals from 5,5-dimethyl-1-pyrroline-*N*-oxide in Fenton systems. Characterization of key precursors giving rise to background ESR signals. *Canadian Journal of Chemistry*, **70**, 2818–2827.

Mann, C.K. and Barnes, K.K. (1970) *Electrochemical Reactions in Nonaqueous Systems*, New York: Dekker.

Marcus, R.A. and Sutin, N. (1985) Electron transfers in chemistry and biology, *Biochimica et Biophysica Acta*, **811**, 265–322.

McIntire, G.L., Blount, H.N., Stronks, H.J., Shetty, R.V. and Janzen, E.G. (1980) Spin-trapping in electrochemistry. 2. Aqueous and nonaqueous electrochemical characterization of spin-traps, *Journal of Physical Chemistry*, **84**, 916–921.

Monroe, S. and Eaton, S.S. (1996) Photo-enhanced production of the spin adduct 5,5,-dimethyl-1-pyrroline-*N*-oxide/$^{\bullet}$OH in aqueous menadione solutions, *Archives of Biochemistry and Biophysics*, **329**, 221–227.

Moreno, S.N.J., Stolze, K., Janzen, E.G. and Mason, R.P. (1988) Oxidation of cyanide to the cyanyl radical by peroxidase/H_2O_2 systems as determined by spin-trapping, *Archives of Biochemistry and Biophysics*, **265**, 267–271.

Murov, S.L., Carmichael, I. and Hug, G.L. (1993) *Handbook of Photochemistry*, 2nd ed., New York: Dekker.

Ononye, A.I., McIntosh, A.R. and Bolton, J.R. (1986) Mechanism of the photochemistry of *p*-benzoquinone in aqueous solution. 1. Spin-trapping and flash photolysis electron paramagnetic resonance studies, *Journal of Physical Chemistry*, **90**, 6266–6270.

Ozaki, S. and Masui, M. (1978) Oxidation of hydroxylamine derivatives. II. Anodic oxidation of phenylhydroxylamines, *Chemical Pharmaceutical Bulletin (Tokyo)*, **26**, 1364–1369.

Pearson, R.G. (1986) Ionization potentials and electron affinities in aqueous solutions, *Journal of American Chemical Society*, **108**, 6109–6114.

Sang, H., Janzen, E.G. and Poyer, J.L. (1996) Spontaneous free radical/spin adduct formation and 1,3-dipolar molecular addition in reactions of cyanohalocarbons and *C*-phenyl-*N-tert*-butyl nitrone (PBN), *Journal of the Chemical Society, Perkin Transactions 2*, 1183–1189.

Sawyer, D.T., Sobkowiak, A. and Matsushita, T. (1996) Metal [ML_x; M = Fe, Cu, Co, Mn]/hydroperoxide-induced activation of dioxygen for the oxygenation of hydrocarbons: oxygenated Fenton chemistry, *Accounts of Chemical Research*, **29**, 409–416.

Sayo, H., Ozaki, S. and Masui, M. (1973) Oxidation of *N*-alkylhydroxylamines. IV. Anodic oxidation of *N*-alkylhydroxylamines, *Chemical Pharmaceutical Bulletin (Tokyo)*, **21**, 1988–1995.

Sosonkin, I.M., Belevskii, V.N., Strogov, G.N., Domarev, A.N. and Yarkov, S.P. (1982) Selectivity of spin-traps. Oxidation–reduction reactions, *Journal of Organic Chemistry of USSR, English Translation*, **18**, 1313–1320.

Sümmermann, W. and Deffner, U. (1975) Die elektrochemische Oxidation aliphatischer Nitroxyl-Radikale, *Tetrahedron*, **31**, 593–596.

Thomas, C.E., Ohlweiler, D.F., Carr, A.A., Nieduzak, T.R., Hay, D.H., Adams, G., *et al.* (1996) Characterization of the radical trapping activity of a novel series of cyclic nitrone spin-traps, *The Journal of Biological Chemistry*, **271**, 3097–3104.

Tomasi, A. and Iannone, A. (1993) ESR spin-trapping artifacts in biological model systems, in Berliner, L.J. and Reuben, J. (eds), *Biological Magnetic Resonance. 13. EMR of Paramagnetic Molecules*, New York: Plenum Press, pp. 353–384.

Tuccio, B., Laricella, R., Fréjaville, C., Bouteiller, J.-C. and Tordo, P. (1995) Decay of the hydroperoxyl spin adduct of 5-diethoxyphosphoryl-5-methyl-1-pyrroline-*N*-oxide: an EPR kinetic study, *Journal of the Chemical Society, Perkin Transactions 2*, 295–298.

Walling, C. (1998) Intermediates in the reactions of Fenton type reagents, *Accounts of Chemical Research*, **31**, 155–157.

Zubarev, V.E. and Brede, O. (1994) Direct determination of the cation radical of the spin-trap α-phenyl-*N-tert*-butylnitrone, *Journal of the Chemical Society, Perkin Transactions 2*, 1821–1828.

Zubarev, V.E. and Brede, O. (1995) Generation of α-aminoxylcarbenium ions by electron-transfer oxidation of *N-tert*-butyl-3-phenyloxazirane and their role in nitrone spin-trapping chemistry, *Journal of the Chemical Society, Perkin Transactions 2*, 2183–2187.

4 *In vivo* spin-trapping – from chemistry to toxicology

Ronald P. Mason

For a variety of reasons, the possibility of free radical metabolism did not receive much attention for many years, although Michaelis of the Michaelis–Menten equation was interested in free radical metabolites and their importance in biochemistry in the 1930s. One reason for the late development of this area is that most biochemicals, as opposed to aromatic drugs and industrial chemicals, are not easily metabolized through free radical intermediates. Mother Nature abhors free radicals for the same reason that organic chemists did until recently (i.e. their chemistry is difficult to control). Another reason is that unless something can be demonstrated with a whole animal, there will always be some question of its actual existence in biology. To detect something as ephemeral as a free radical formed inside a whole animal is certainly not easy. Production rates of free radicals in animals relative to chemical systems are inherently slow; therefore, the highest possible sensitivity is of paramount importance. Water, with its high dielectric constant, is the worst solvent for ESR studies in that only very small samples can be studied, which decreases the molar sensitivity just when sensitivity is needed most.

Spin-trapping, in that it ideally integrates free radicals formed over time is, to my mind, the most attractive approach to the detection of free radicals *in vivo*. Spin-trapping, as developed by Janzen and Haire (1990) and others, is a technique where a diamagnetic molecule (a spin-trap) reacts with a free radical to produce a more stable radical (a radical adduct) which is readily detectable by ESR.

$$R^{\bullet} + \text{spin-trap} \rightarrow \text{radical adduct}$$

Radical adducts are substituted nitroxide free radicals, which, for free radicals, are relatively stable, although most are in fact not as stable as we would like. In rare cases some radical adducts are so stable that their structure has been determined by mass spectrometry and NMR.

There are several recent reviews which address the *in vivo* and *in vitro* applications of the spin-trapping technique (Mottley and Mason, 1989; Buettner and Mason, 1990; Knecht and Mason, 1993; DeGray and Mason, 1994). These reviews also discuss the effects of spin-traps on enzymes and more general problems of spin-trapping such as artefacts (Tomasi and Iannone, 1993) and ambiguities in the assignment of radical adduct structure.

In most biological systems, nitroso spin-traps such as *tert*-nitrosobutane (2-methyl-2-nitrosopropane) and nitrosobenzene have proved too toxic and their radical adducts too unstable to be useful. All of the *in vivo* spin-trapping work has used the nitrone spin-traps developed by Janzen.

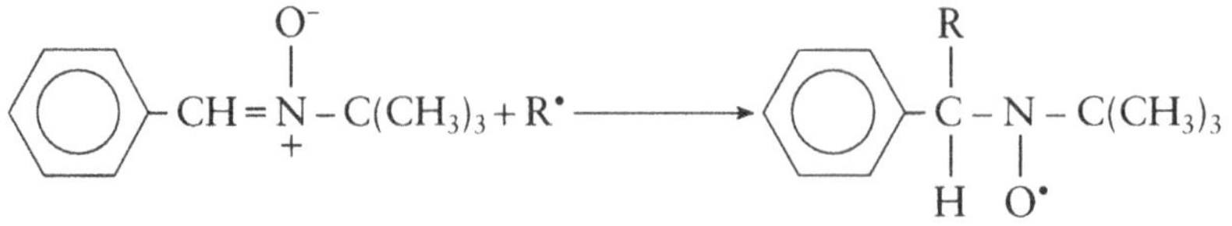

phenyl-*tert*-butylnitrone (PBN)

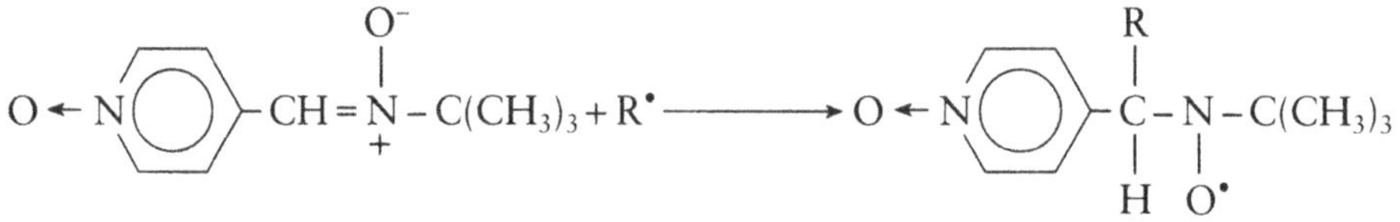

α-(4-pyridyl-1-oxide)-*N*-*tert*-butylnitrone (4-POBN)

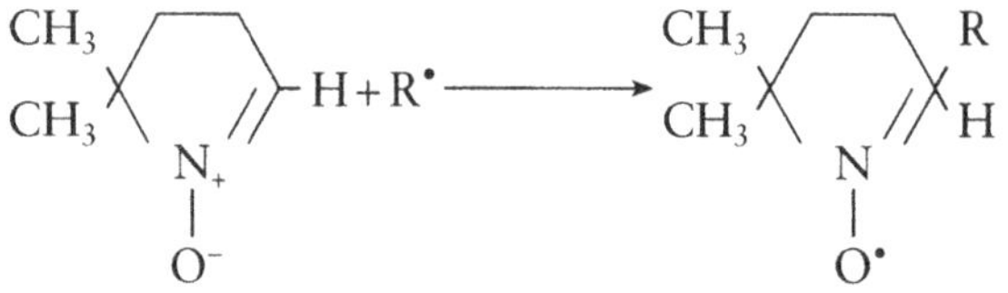

5,5-dimethyl-1-pyrroline-*N*-*oxide* (DMPO)

Since the nitrone is a foreign compound, its possible effects *in vivo*, as well as the other requirements for a successful spin-trapping experiment, have led to the following questions and guidelines.

1 Will the spin-trap participate in reactions other than those with the radical generated in the experiment? Cytochrome P450 inhibition by spin-traps? Are there any pharmacological effects unrelated to radical scavenging?
2 How readily can the spectrum be interpreted and the structure of the trapped radical be determined? A more specific aspect of this question is the availability of isotopically labeled (^{13}C, ^{2}H, ^{17}O, ^{33}S, etc.) compounds or the existence of an independent synthesis of the radical adduct for proof of structure.
3 How fast is the trapping reaction, and how stable are the radical adducts that are formed? In general, aromatic cation and anion radicals are not spin-trapped.
4 Does the appearance of a radical adduct signify a major reaction pathway, or can it be a minor side reaction? Even a minor pathway can have toxicological significance.

The first *in vivo* experiment was done by Lai *et al.* (1979). The spin-trap PBN and carbon tetrachloride were given to a rat through a stomach tube. After two hours, the liver was extracted by the method of Folch (methanol:chloroform, 2:1) and the ESR spectrum of the chloroform layer was taken. The chloroform layer was dried with anhydrous sodium sulphate and the sample evaporated under a nitrogen gas stream to a volume of 0.5 ml. The ESR spectrum of this sample was obtained using a 3 mm quartz tube.

Hyperfine coupling constants of nitrone radical adducts can be as sensitive to the solvent environment as they are to the chemical structure of the trapped radical. The chloroform layer of a Folch extract contains traces of water, methanol, and the chloroform-soluble elements of the tissue. This solvent environment is different enough from pure chloroform to noticeably affect hyperfine coupling constants. If independent synthesis of the suspected radical adduct and careful comparison of the hyperfine coupling constants is to be used in the proof of structure of a radical adduct detected *in vivo*, the solvent dependence must be addressed. One approach to this problem is the addition of the synthetic radical adduct to the control tissue and to perform the Folch extract as usual (Ghio *et al.*, 1998). The resulting chloroform layer will thus contain the synthetic radical adduct in the solvent environment, which is exactly the same as that of the radical adduct formed *in vivo*.

Kalyanaraman *et al.* (1979) challenged the assignment of the PBN/$^{\bullet}CCl_3$ radical adduct in my first publication using spin-trapping. As it turns out, a number of radical adducts are formed both *in vitro* and *in vivo*, but the PBN/$^{\bullet}CCl_3$ radical adduct is the only one extracted into chloroform. Therefore, one major limitation of organic extraction of tissue with chloroform–methanol is that only non-polar radical adducts can be detected. Radical adducts of radical ions, charged biochemicals such as glutathione, or charged macromolecules such as proteins or DNA will not be extracted and, therefore, can not be detected. The use of the Folch extraction is really a method to remove radical adducts from the aqueous environment into an organic solvent so that 3 mm sample tubes can be used, as is standard ESR practice.

The chloroform layer of a Folch extract contains sufficient water that a conventional 3 mm quartz tube cannot be tuned in the microwave cavity unless the chloroform is first dried with anhydrous sodium sulphate. Alternatively, the sample can be cooled to −70°C where the water is frozen, which lowers the dielectric constant of the sample (Albano *et al.*, 1982). In any case, the sample must be bubbled with nitrogen to deoxygenate the solution because the oxygen present in air-saturated chloroform is high enough to broaden the lines of the ESR spectrum significantly. The greatest advantage of the extraction method is that the radical adduct is transferred from a sample with a high dielectric constant, the biological tissue, to a solvent with a lower dielectric constant, the chloroform solution. This enables the use of larger sample volumes, which increases sensitivity. In addition, the sample can be easily concentrated by evaporation of the chloroform layer. In favorable cases, the sample volume can be concentrated to a few microlitres where the loop–gap resonator can greatly increase sensitivity (Kalyanaraman *et al.*, 1991).

Our approach to *in vivo* spin-trapping has been to examine biological fluids such as urine, bile, and blood (or plasma) directly for radical adducts using the TM_{110} ESR cavity and a 17 mm flat cell, which gives the largest possible aqueous sample size in the active region of the cavity, about 300 μl (Knecht and Mason, 1988). This gives us the highest available molar sensitivity for aqueous samples. Deoxygenation of samples is not usually necessary due to the low solubility of oxygen in water (approximately 250 μM O_2 in equilibrium with air), although deoxygenation will narrow the sharpest lines. No background free radical signals other than the ascorbate semidione doublet have been detected. Conceptually, the detection of radical adducts of free radical metabolites in bile, urine, or blood is little different from the detection of the stable products of drug metabolism. The primary difference is that with traditional analytical methods, the metabolite must be separated and purified before identification is possible, whereas the selectivity of ESR allows the simultaneous detection and identification of radical adducts. The sample can usually be frozen on dry ice and stored at −70°C and then, as convenient, thawed for ESR analysis, although the stability of the radical adducts needs to be verified. Often the sample is bubbled with oxygen to oxidize radical adducts that have been reduced to their respective ESR-silent hydroxylamines *in vivo* and then deoxygenated with nitrogen bubbling.

This *ex vivo* chemistry is often necessary for the success of *in vivo* spin-trapping, so we have investigated it in some detail in the case of carbon tetrachloride (Sentjurc and Mason, 1992). We unsuccessfully tried to prevent reduction by adding ascorbate oxidase or *N*-ethylmaleimide (to react with GSH) to the collection tube. This demonstrated that reduction by neither GSH nor ascorbate was occurring *ex vivo*. Janzen *et al.* (1993) found the chromatographically pure PBN/$^{\bullet}CCl_3$ radical adduct was very stable and was not reduced by ascorbate, which is very unusual for a nitroxide. Nevertheless, this radical adduct was reduced by liver microsomes and NADPH, implying the *in vivo* formation of the ESR-silent hydroxylamine metabolite of the PBN/$^{\bullet}CCl_3$ radical adduct.

In agreement with this result, when bile was collected after administration of PBN and $^{13}CCl_4$ and examined immediately, only the doublet from ascorbate semidione radical was detected in most bile samples (Figure 4.1A). Only after bubbling for 30 min with oxygen was the spectrum detected (Figure 4.1B). This EPR spectrum is the superposition of at least three spectral components. Two of them are 12-line components due to the interaction of the free electron with the N, H, and ^{13}C nuclei originating from $^{13}CCl_4$. The third, hardly visible, is a six-line component which shows no interaction of the free electron with the ^{13}C nucleus, which means that it is not an adduct of a CCl_4-derived radical. The components observed were identified previously as PBN/$^{\bullet 13}CCl_3$ (Figure 4.1C), PBN/$^{\bullet 13}CO_2^-$ (Figure 4.1D), and a carbon-centered, possibly lipid-derived radical adduct PBN/$^{\bullet}$L (Figure 4.1E). The composite spectrum obtained by computer simulation is presented in Figure 4.1B as a dotted line. No detectable spectra were obtained if rats were injected with PBN alone (Figure 4.1F). Since the signals of the adducts became significant only after bubbling with oxygen, it can be concluded that the vast majority of radical adducts were initially reduced in the liver to their corresponding, ESR-silent hydroxylamines, but were oxidized back to the paramagnetic nitroxides by molecular oxygen.

The length of oxygen bubbling has differential effects on the ESR spectra (Figure 4.2). It is evident that PBN/$^{\bullet 13}CO_2^-$ adduct was formed at a faster rate than PBN/$^{\bullet 13}CCl_3$. Its concentration reaches a maximum value after approximately 20 min of oxygen bubbling (Figure 4.2C) and decreases after 30 min. The PBN/$^{\bullet 13}CCl_3$ adduct concentration starts to increase after only 20 min but dominates at 60 min (Figure 4.2F).

After two hours, autoxidation takes place even in air (Figure 4.3). The most complete is the autoxidation of the corresponding hydroxylamine to PBN/$^{\bullet 13}CO_2^-$ (Figure 4.3A), while PBN/$^{\bullet 13}CCl_3$ formation was less pronounced. In this spectrum, an additional radical adduct, PBN/[GSH—$^{\bullet 13}CCl_3$] (Figure 4.3D), was detected which also showed a ^{13}C isotope-dependence and was not observed in experiments where oxygen bubbling was used. This species is known from *in vitro* work to be oxygen-sensitive (Connor *et al.*, 1990). Autoxidation processes are frequently found in biological systems and are usually catalyzed by trace transition metals. To examine the role of metals in this process, the bile was collected into the tubes with the iron chelators dipyridyl or Desferal. In both cases, the autoxidation is much slower and no ESR signals were detected (Figure 4.3F). Therefore, the autoxidation of reduced radical adducts was catalyzed by trace transition metals, probably iron.

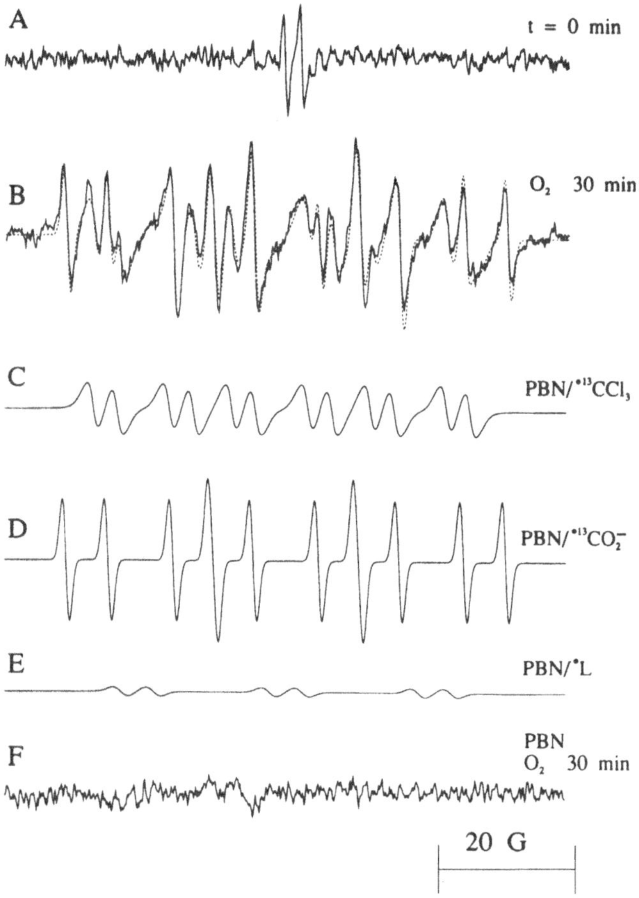

Figure 4.1 EPR spectra of bile collected from rats 105–120 min after administration of PBN and $^{13}CCl_4$. (A) Initial spectrum after thawing of bile. (B) Experimental spectrum (—) after bubbling with oxygen for 30 min. Calculated spectrum (----) obtained by addition of the spectral components presented in C, D, and E. (C) Computer simulation of PBN/$^{\bullet 13}CCl_3$ adduct. (D) Computer simulation of PBN/$^{\bullet 13}CO_2^-$ adduct. (E) Computer simulation of PBN/$^{\bullet}$L adduct. (F) Spectrum obtained if only PBN is administered i.p. to the rat. Other conditions are the same as for spectrum B (Sentjurc and Mason, 1992)

The oxidation by potassium ferricyanide is a rapid process and depends strongly on its concentration (Figure 4.4). Again, the components oxidize at different rates. The hydroxylamine of PBN/$^{\bullet 13}CO_2^-$ was fully oxidized first, followed by that of PBN/[GSH—$^{\bullet 13}CCl_3$]; the PBN/$^{\bullet 13}CCl_3$ hydroxylamine was oxidized last. At concentrations higher than 0.8 mM ferricyanide, only the PBN/$^{\bullet 13}CCl_3$ adduct was visible (Figure 4.4F). This represents a major difference between spin labels and radical adducts in that radical adducts contain a β-hydrogen and can easily be over-oxidized to ESR-silent nitrones with the loss of this β-hydrogen. Clearly, identification, not quantitation, is the major strength of spin-trapping.

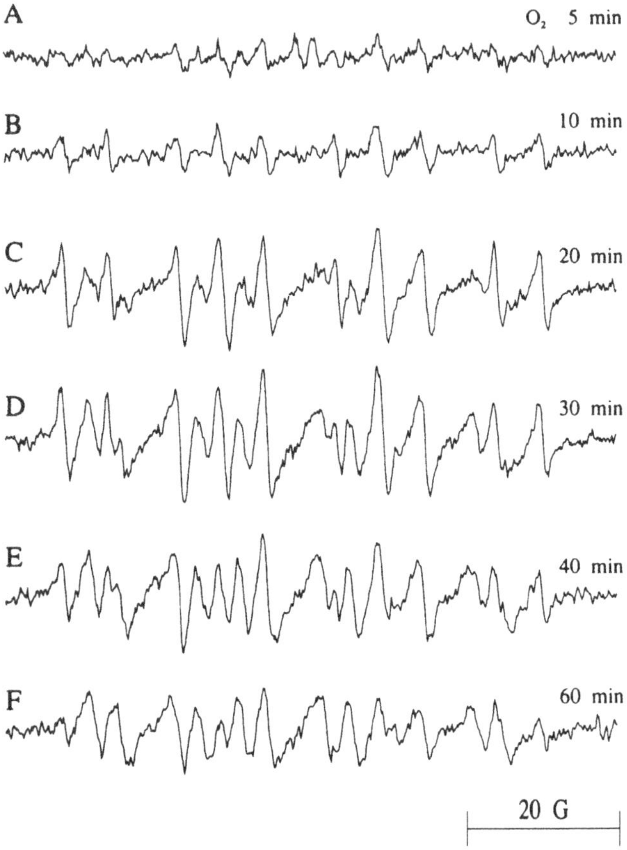

Figure 4.2 EPR spectra of bile collected from rats 105–120 min after administration of PBN and $^{13}CCl_4$. The time of oxygen bubbling through the bile samples was varied from 5 to 60 min (Sentjurc and Mason, 1992)

An additional difficulty with bile is its unique solvent environment consisting of non-polar micelles and an aqueous phase, which results in unique hyperfine coupling constants for non-polar radical adducts such as PBN/$^{\bullet}CCl_3$, but aqueous hyperfine coupling constants for PBN/$^{\bullet 13}CO_2^-$. If independent synthesis of the suspected radical adduct and careful comparison of the hyperfine coupling constants are to be used in the proof of structure, then a small volume of the synthetic radical adduct should be added to the control bile for the comparison.

To improve identification of radical adducts, the use of partially-deuterated PBN was developed by Janzen *et al.* (1987). This compound provides increased spectral resolution due to decreased line widths and allows additional coupling constants from the trapped radical to be resolved. A good example of the utility of this spin-trap is given by the volatile anesthetic halothane (2-bromo-2-chloro-1, 1,1-trifluoroethane), which has been implicated in metabolism-dependent human hepatotoxicity. As many as 20% of patients suffer a mild hepatotoxicity, which has

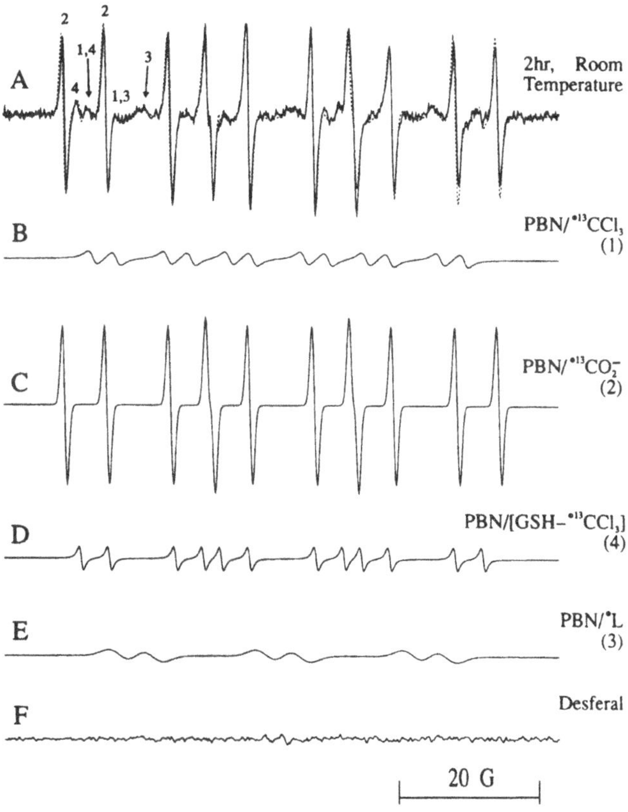

Figure 4.3 EPR spectra of bile collected from rats 105–120 min after administration of PBN and $^{13}CCl_4$. Bile samples were kept at room temperature for 2 h. (A) Experimental spectrum (—) and its calculated spectrum (----) obtained by addition of the spectral components presented in B, C, D, and E. (B) Computer simulation of PBN/$^{\bullet 13}CCl_3$ adduct (component 1). (C) Computer simulation of PBN/$^{\bullet 13}CO_2^-$ adduct (component 2). (D) Computer simulation of PBN/[GSH–$^{\bullet 13}CCl_3$] adduct (component 4); not observed previously in any *in vivo* experiment. (E) Computer simulation of PBN/$^{\bullet}$L adduct (component 3). (F) Bile collected into an Eppendorf tube containing 50 μL 1 mM Desferal (Sentjurc and Mason, 1992)

been related to reductive debromination of halothane, leading to carbon-centered free radical formation (Knecht *et al.*, 1992). A composite computer simulation of the spectrum of two radical adducts obtained in bile from halothane is shown in Figure 4.5 (upper dotted line). The solid line is the experimental spectrum. The computed component spectra (radical adducts I and II) are displayed below. For species I, the two γ-hydrogen couplings, as expected for a fatty acid alkyl β-scission fragment, were reported in an abstract (Knecht and Mason, 1990). Later we obtained ^{13}C-halothane and discovered both species I and II were halothane-derived (Knecht *et al.*, 1992). Regardless of how good the simulation, the uniqueness of ESR simulations is always

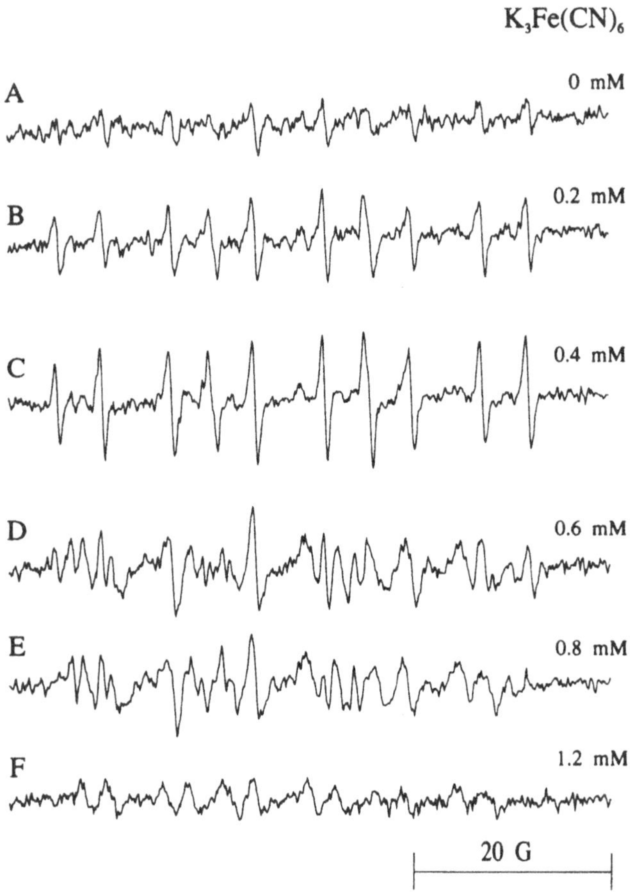

Figure 4.4 EPR spectra of bile collected from rats 105–120 min after administration of PBN and $^{\bullet 13}CCl_4$ for different concentrations of potassium ferricyanide. Potassium ferricyanide was added to the bile 4 min after thawing, and spectra initiated 5 min later (Sentjurc and Mason, 1992)

in question. Both of these radical adducts are from trapping the carbon-centered halothane-derived radical formed by reductive debromination PBN/$^{\bullet}CHClCF_3$. In the radical adduct, the 2-C carbon from halothane and the carbon α to the phenyl ring are both chiral. Thus, two distinct diasterisomers of this radical adduct are possible and will give two distinct spectra even though only one radical is trapped.

Many *in vivo* spin-trapping experiments have followed the pioneering work of Lai *et al.* (Table 4.1). Fatty acid hydroperoxides which arise in rancid fat were demonstrated to be metabolized to free radicals *in vivo* (Chamulitrat *et al.*, 1992). Another dietary source of free radical metabolites is ethanol. Chronic ethanol abuse in humans produces widespread pathological changes, including marked hepatic toxicity. The initial fatty changes in the liver were first attributed to lipid peroxidation by Di Luzio (1963). Slater (1972) suggested that the metabolism of ethanol to a free

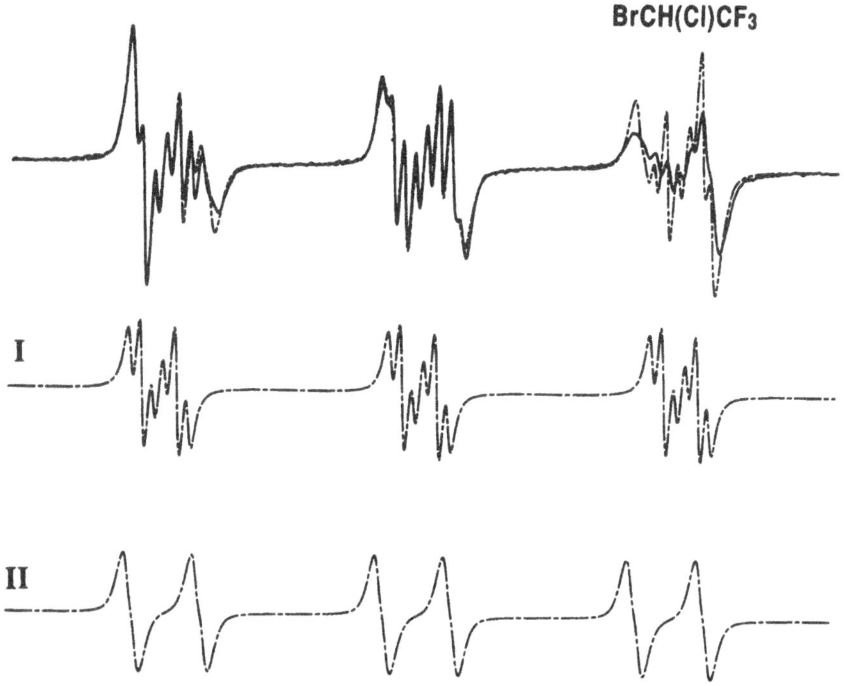

Figure 4.5 EPR spectrum of bile from rats administered partially deuterated PBN and halothane. —, EPR spectrum from bile of rat treated as described above. ----, Composite computer simulation of the spectrum composed of species I (18 lines) and II (six lines) (Knecht *et al.*, 1992)

radical could initiate lipid peroxidation, which is thought to be a factor in hepatic damage.

Figure 4.6A shows the radical adduct signal of bile from the gall bladders of six deermice which were chronically treated with an ethanol-containing, high-fat diet and given an acute dose of ethanol and of POBN (Knecht *et al.*, 1990). The central doublet with one line superimposed on the six-line species is due to the ascorbate semidione anion radical and is seen in samples from most animals. The six-line spectrum is characteristic of many free radicals trapped by POBN. The hyperfine coupling constants of the radical adduct are not definitive and could be those of the α-hydroxyethyl radical adduct or of some other radical adduct such as an ethanol-induced lipid-derived radical adduct. Coupling constants of POBN radical adducts are relatively independent of the structure of the trapped free radical, and solvent effects can greatly affect the spectra. Thus, in order to identify the radical adduct of ethanol unambiguously, ethanol labeled at the α-carbon with ^{13}C was employed. Figure 4.6B shows the spectrum of bile from six animals treated as above except that [1-^{13}C]-ethanol was used. A twelve-line spectrum is detected instead of the six-line spectrum. Thus, the radical detected is totally derived from ethanol and trapped at the α-carbon of ethanol.

Table 4.1 In vivo formation of free radical metabolites detected by *ex vivo* ESR

Year	*Radical adduct*	*Authors*
1979	PBN/$^{\bullet}CCl_3$ from CCl_4 detected in organic extract of liver	Lai *et al.*
1980	PBN/$^{\bullet 13}CCl_3$ from $^{13}CCl_4$ detected in organic extract of liver	Poyer *et al.*
1981	PBN/$^{\bullet}CHClCF_3$ from halothane detected in organic extract of liver	Poyer *et al.*
1982	PBN/$^{\bullet}CCl_3$ from CCl_4 detected in organic extract of liver	Albano *et al.*
1982	PBN/$^{\bullet}CHClCF_3$ from halothane detected in organic extract of liver	Plummer *et al.*
1984	PBN/$^{\bullet}$L from 3-methylindole detected in organic extract of lungs	Kubow *et al.*
1984	$HO(CH_3O)_2$PBN/$^{\bullet}$L and $HO(CH_3O)_2$PBN/$^{\bullet}CCl_3$ from CCl_4 detected in organic extract of liver administered $(CH_3O)_3$PBN	McCay *et al.*
1984	PBN/$^{\bullet}CHClCF_3$ from halothane detected in organic extract of liver	Fujii *et al.*
1985	PBN/$^{\bullet}$L from 3-methylindole detected in organic extract of lungs as affected by cysteine and diethylmaleate	Kubow *et al.*
1985	PBN/$^{\bullet}CHCl_2$ from chloroform detected in organic extract of liver with related radical adducts from bromoform, bromodichloromethane, and iodoform	Tomasi *et al.*
1986	PBN/$^{\bullet}CO_2^-$ from CCl_4 detected in urine	Connor *et al.*
1987	$HO(CH_3O)_2$PBN/$^{\bullet}$L from ethanol detected in organic extract of liver and heart as affected by high fat	Reinke *et al.*
1988	PBN/$^{\bullet}CCl_3$ from CCl_4 detected in organic extract of liver affected by ethanol and high fat	Reinke *et al.*
1988a	DMPO/thiyl Hb from phenylhydrazine detected in whole blood	Maples *et al.*
1988b	DMPO/thiyl Hb and PBN/thiyl Hb from phenylhydrazine and DMPO/thiyl Hb from hydrazine-based drugs in whole blood	Maples *et al.*
1988	PBN/$^{\bullet}CO_2^-$ from $CBrCl_3$ in urine	LaCagnin *et al.*
1988	PBN/$^{\bullet}CCl_3$ and PBN/$^{\bullet}CO_2^-$ from CCl_4 detected in bile	Knecht and Mason
1990	PBN-d_{14}/$^{\bullet}CHClCF_3$ from halothane detected in liver extract	Janzen *et al.*
1990a	DMPO/thiyl Hb from phenylhydroxylamine or nitrosobenzene in whole blood	Maples *et al.*
1990b	DMPO/thiyl Hb from hydroperoxides in whole blood	Maples *et al.*
1990	4-POBN/$^{\bullet}CH(OH)CH_3$ and 4-POBN/$^{\bullet}$L from ethanol and high fat detected in bile	Knecht *et al.*
1991	POBN/$^{\bullet}$? from cigarette smoke/endotoxin detected in organic extract of plasma	Murphy *et al.*
1991	PBN-d_{14}/$^{\bullet}CH(OH)CH_3$, PBN-d_{14}/$^{\bullet}CH_2R$ and PBN-d_{14}/$^{\bullet}$OL from ethanol in organic extract of liver	Reinke *et al.*
1991	PBN/$^{\bullet}CO_2^-$ from CCl_4 detected in plasma. PBN/$^{\bullet}CCl_3$ from CCl_4 detected in organic extract of plasma	Reinke and Janzen
1991	PBN/$^{\bullet}CHClCF_3$ from halothane and PBN/$^{\bullet}CCl_3$ from CCl_4 detected in bile	Hughes *et al.*
1991	PBN/$^{\bullet}CHBr_2$ from bromoform and PBN/$^{\bullet}CCl_3$ from $CBrCl_3$ in bile	Knecht and Mason
1991	PBN/$^{\bullet}CH_3$ from attack of hydroxyl radical from Fe^{2+} on DMSO detected in bile	Burkitt and Mason

Table 4.1 (cont'd)

Year	*Radical adduct*	*Authors*
1992	PBN/$^{\bullet}CH_3$ from procarbazine detected in organic extract of several organs especially blood	Goria-Gatti *et al.*
1992	PBN/$^{\bullet}CCl_2CH_3$ from 1,1,1-trichloroethane detected in organic extract of liver	Dürk *et al.*
1992	PBN/$^{\bullet}CHClCHCl_2$ from 1,1,2,2-tetrachloroethane detected in organic extract of liver	Paolini *et al.*
1992	POBN/$^{\bullet}$? From cotton smoke detected in organic extract of plasma	Yamaguchi *et al.*
1992	PBN/$^{\bullet}CO_2^-$ from CCl_4 detected in plasma affected by ethanol	Reinke *et al.*
1992	4-POBN/$^{\bullet}$? from ozone detected in organic extract of lungs	Kennedy *et al.*
1992	PBN-d_{14}/$^{\bullet 13}CHClCF_3$ from halothane detected in bile	Knecht *et al.*
1992	PBN/$^{\bullet}CCl_3$, PBN/$^{\bullet}CO_2^-$, and PBN/[GSH—$^{\bullet}CCl_3$] from CCl_4 detected in bile	Sentjurc *et al.*
1992	4-POBN/$^{\bullet}$L from oxidized fatty acids detected in bile	Chamulitrat *et al.*
1992	PBN/$^{\bullet}CH_3$ from attack of hydroxyl radical from Cu^{2+} and ascorbate on DMSO detected in bile	Kadiiska *et al.*
1993	DMPO/$^{\bullet}$OR and DMPO/$^{\bullet}$C from *E. coli*/TNF detected in extract of liver	Lloyd *et al.*
1993	PBN/$^{\bullet}CH_3$ from attack of hydroxyl radical from Fe^{2+} and ascorbate on DMSO detected in bile as affected by paraquat and desferrioxamine	Burkitt *et al.*
1993a	4-POBN/$^{\bullet}$L from Cu^{2+}-treated, vitamin E- and selenium-deficient rats detected in bile	Kadiiska *et al.*
1993b	PBN/$^{\bullet}CH_3$ from attack of hydroxyl radical from Cu^{2+} and paraquat on DMSO detected in bile	Kadiiska *et al.*
1993	PBN/$^{\bullet}CCl_3$ from CCl_4 detected in organic extract of liver unaffected by Zn^{2+}, Cr^{3+}, and metallothionein	Hanna *et al.*
1994	PBN/$^{\bullet}CCl_3$ from carbon tetrachloride detected in liver extract unaffected by age	Rikans *et al.*
1994	4-POBN/$^{\bullet}$L from Cr(VI) detected in bile	Kadiiska *et al.*
1995	4-POBN/$^{\bullet}CH(OH)CH_3$ from acute ethanol detected in bile	Moore *et al.*
1995	4-POBN/$^{\bullet}CH(OH)CH_3$ from ethanol detected in bile affected by chronic ethanol and high fat diet as mediated by Kupffer cells	Knecht *et al.*
1995	PBN/$^{\bullet}CH_3$ from attack of hydroxyl radical from dietary iron on DMSO detected in bile	Kadiiska *et al.*
1996	PBN/$^{\bullet}$OR, PBN/$^{\bullet}$C, and 4-POBN/$^{\bullet}$C from ozone detected in organic extract of lung and liver	Vincent *et al.*
1996	PBN/$^{\bullet}CH(OH)CH_3$ from ethanol detected in pancreatic secretion	Iimuro *et al.*
1997	4-POBN/$^{\bullet}CH(OH)CH_3$ and 4-POBN/$^{\bullet}$L from ethanol detected in bile as affected by dietary fats	Reinke and McCay
1997	4-POBN/$^{\bullet}CH(OH)CH_3$ and 4-POBN/$^{\bullet}$L from ethanol detected in bile as affected by chronic ethanol	Reinke *et al.*
1997a	PBN/$^{\bullet}CH_3$ from attack of hydroxyl radical from Fe^{2+} on DMSO detected in bile as affected by ascorbate and age	Kadiiska *et al.*

Table 4.1 (cont'd)

Year	*Radical adduct*	*Authors*
1997b	4-POBN/•L from air pollution particle detected in organic extract of lung	Kadiiska *et al.*
1997	4-POBN/•L from hippocampal extracellular space during kainic acid-induced seizures	Ueda *et al.*
1998	4-POBN/•L from asbestos detected in organic extract of lung	Ghio *et al.*
1998	4-POBN/•$CH(OH)CH_3$ from ethanol enhancement by endotoxin is mediated by Kupffer cells	Chamulitrat *et al.*
1998	4-POBN/•L from haloperidal detected in microdialysate of right striatum	Yokoyama *et al.*

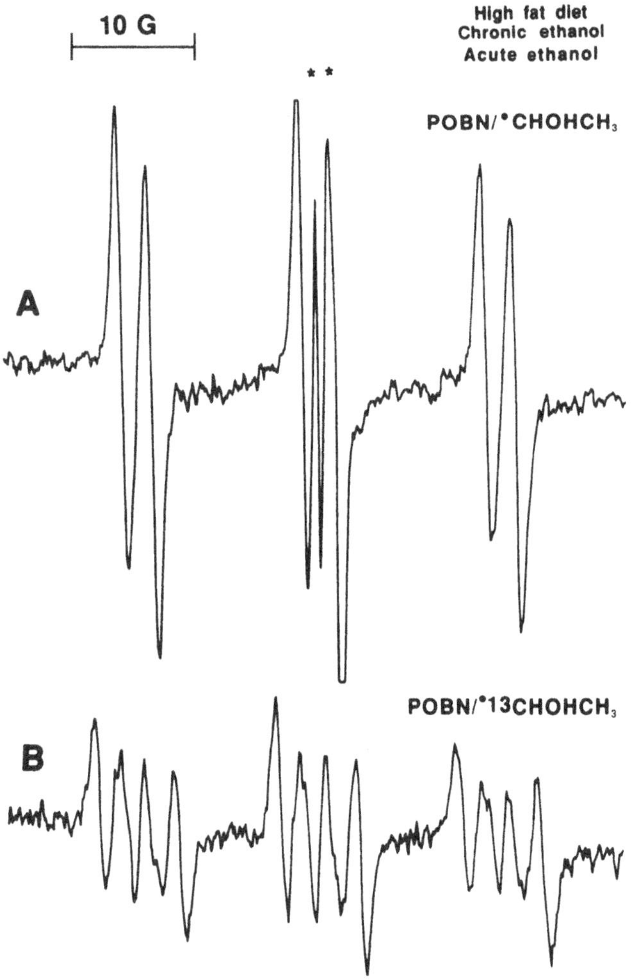

Figure 4.6 EPR spectra of dilute bile from alcohol dehydrogenase-deficient deermice. A, EPR spectrum of bile from six ethanol-induced deermice treated with acute ethanol and POBN. B, As in A, except acute [1-^{13}C] ethanol (Knecht *et al.*, 1990). The species designated with the asterisks is the ascorbate radical

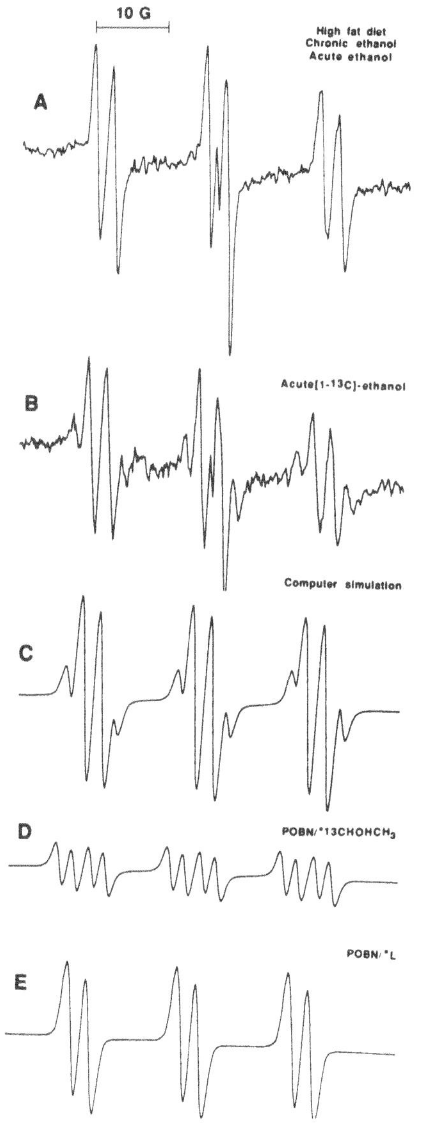

Figure 4.7 Computer simulation of EPR spectra of dilute bile from alcohol dehydrogenase-deficient deermice. A, EPR spectrum of bile from two ethanol-induced deermice treated with acute ethanol and POBN. B, As in A, except acute [1-^{13}C] ethanol. C, Computer simulation of spectrum in A. Components shown in D and E (Knecht *et al.*, 1990)

Figure 4.7A shows a spectrum from the bile of two animals treated as before, except that a 3.3-fold higher 4-POBN dose was used. This experiment was done in order to conserve alcohol dehydrogenase-deficient deermice. Figure 4.7B shows the corresponding spectrum when [1-^{13}C]-ethanol was used. Under these conditions, a new six-line spectrum overlays the 12 lines of the α-hydroxyethyl radical adduct. Figure 4.7C shows the composite computer simulation of these two species. The

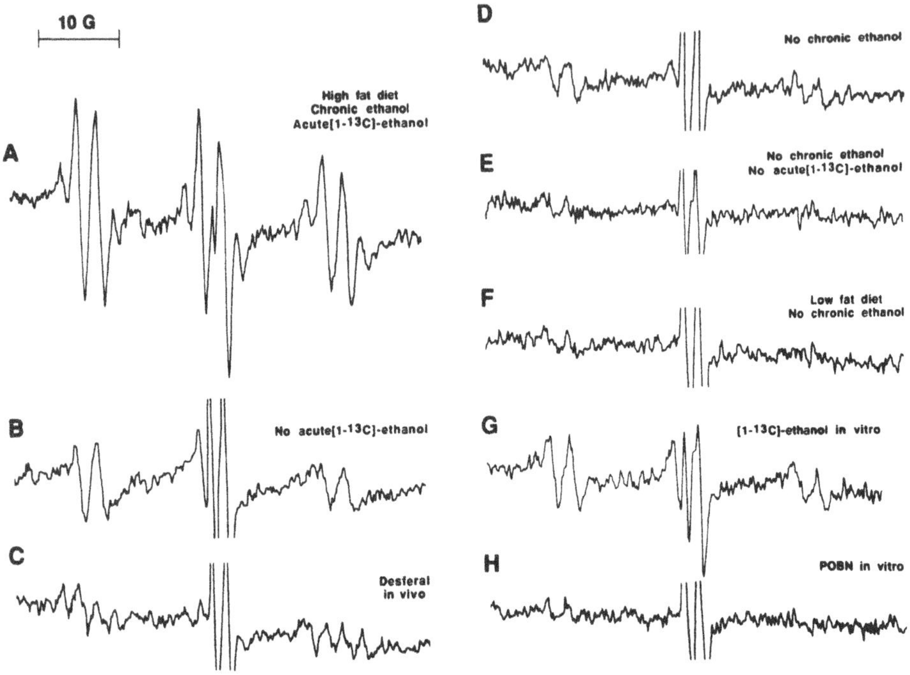

Figure 4.8 EPR spectra of dilute bile from alcohol dehydrogenase-deficient deermice. A, EPR spectrum of bile from two ethanol-induced deermice treated with acute [1-^{13}C] ethanol and POBN. B, EPR spectrum of bile from two ethanol-induced deermice pretreated with acute [1-^{13}C] ethanol, kept several hours to eliminate acute ethanol, and then treated with POBN. C, As in A, except pretreated with 15 mg/kg Desferal (intraperitoneally) 1 hr before treatment with acute [1-^{13}C] ethanol and POBN. D, As in A, except ethanol in diet replaced isocalorically with maltose/dextrin. E, As in A, except animals fed laboratory chow. G, As in A, except no acute [1-^{13}C] ethanol. [1-^{13}C] ethanol (10 mM final concentration) added to the collection vial. H, As in A, except no POBN. POBN (10 mM final concentration) added to the collection vial (Knecht *et al.*, 1990)

6-line spectrum could be a radical adduct derived from an endogenous source such as lipids. This last possibility is quite likely based on the already mentioned association between ethanol administration and lipid peroxidation. In particular, these results are similar to those of Reinke *et al.* (1987) who reported a lipid-derived radical adduct in Folch extracts of liver from rats treated chronically with ethanol. The 6-line species is detectable even in the presence of minimal levels of ethanol although the signal is stronger when a bolus dose of ethanol has been administered (compare the six-line spectra of Figure 4.8A and 4.8B). Although the coupling constants are similar to those reported in the literature for lipid-derived, carbon-centered radical adducts of POBN, the definitive identification of this radical adduct is difficult. Bile from 2 ethanol-pretreated deermice given 15 mg/kg of the iron chelator Desferal i.p. one hour before receiving a bolus dose of [1-^{13}C]-ethanol and POBN (Figure 4.8C) contained as much α-hydroxyethyl radical adduct as did corresponding bile from similarly-treated animals not receiving Desferal (Figure 4.8A). This dose of Desferal approximates the maximal recommended daily human dose for iron overload. The lack of effect of Desferal on the spectrum of α-hydroxyethyl radical adduct indicated that the formation of the radical adduct is not dependent on free iron. A concomitant disappearance of the six-line POBN/•L species indicated that sufficient quantities of the chelator were absorbed to produce a pharmacological effect. This effect of Desferal on the six-line species may reflect either the role of free iron in lipid peroxidation or that of Desferal as a chain-breaking antioxidant.

Animals administered a high-fat diet containing no ethanol still show faint traces of ethanol-dependent radical adduct in their bile after treatment with bolus doses of ethanol and spin-trap (Figure 4.8D). The bile from animals fed a low-fat diet, however, does not contain detectable quantities of this radical adduct (Figure 4.8F). Apparently, the metabolism of ethanol to the α-hydroxyethyl radical is increased in the high-fat animals. It has been reported that the feeding of lipids induces microsomal enzymes.

To test whether free radical formation occurred *ex vivo*, [1-^{13}C]-ethanol was added to the sample vial containing gall bladders from ethanol-induced, ethanol-depleted, POBN-treated deermice (Figure 4.8G). Conversely, POBN was added to the sample vial for gall bladders from ethanol-induced, [1-^{13}C]-ethanol-treated deermice (Figure 4.8H). As shown, neither system produced the twelve-line α-hydroxyethyl radical adduct, indicating that the signal detected in the bile of deermice must be produced *in vivo* prior to sample collection. The six-line spectrum seen in Figure 4.8G is the lipid radical adduct, which was dependent on chronic ethanol treatment as described in Figure 4.8B. This is due to a pathophysiological change in the deermouse that leads to free radical formation independent of acute exposure to ethanol. Although most of the work has used bile, any biological fluid, in principle, can be used. Many publications using urine, blood, and pancreatic secretions have appeared (Table 4.1). An elegant extension of this approach is the detection of the free radicals formed by brain global ischemia/reperfusion in rat striatal perfusate samples obtained by intracerebral microdialysis (Zini *et al.*, 1992).

Since the detection of radical adducts occurs *ex vivo*, the question of the possible *ex vivo* formation of these radical adducts needs to be addressed. Even in the original spin-trapping investigation of the *in vivo* formation of the trichloromethyl radical, Lai *et al.* (1979) checked for *ex vivo* radical formation by mixing Folch (chloroform–ethanol) extract from fresh liver tissue, CCl_4, and PBN and then re-extracting this mixture with chloroform–methanol. After the extraction procedure, no radical adducts were detected.

In the case of free radical formation in humans, *ex vivo* free radical formation is the only one possible because nitrone spin-traps have been administered to humans only in phase I toxicity studies. When blood taken from the coronary sinus during angioplasty was added to PBN *ex vivo*, radical adducts were detected in up to 50% of the samples taken during reperfusion (Coghlan *et al.*, 1991). Using this *ex vivo* technique, post-cardioplegia free radical production was detected in coronary sinus blood (Tortolani *et al.*, 1993). In such an experiment, the possibility that the free radical was actually formed *in vivo* but trapped *ex vivo* is precluded by the short lifetime (less than 100 ms and, in most cases, much less) characteristic of the highly reactive free radicals which can be spin-trapped. Presumably these radical adducts are formed by the trapping of free radicals formed in the blood *ex vivo*, e.g. through lipid peroxidation or other free radical chain reactions.

4-POBN radical adducts in the Folch extract from blood of transplanted livers exhibited a large six-line spectrum (Connor *et al.*, 1994). Rapid (< 1 min) extraction yielded a mixture of radicals, one with coupling constants similar to the 4-POBN/ α-hydroxymethyl adduct (4-POBN/$^{\bullet}CH_2OH$). Extraction with chloroform, however, yielded a much weaker, probably lipid-derived species. Use of ^{13}C-methanol in the Folch extracting solution yielded a 12-line EPR spectrum, indicating that a highly reactive oxidant species from blood following liver transplantation can convert the organic solvents used in tissue extractions to free radicals. Even though the POBN was given intravenously, the radical formation is clearly *ex vivo*. *Ex vivo* lipid peroxidation is clearly a concern in all *in vivo* experiments where sample handling and treatment occur and needs to be addressed no matter what technique is used to assess free radical formation. In an investigation of ozone-initiated free radical formation, α-tocopherol was added to the chloroform/methanol solution prior to extraction in an attempt to suppress *ex vivo* lipid peroxidation (Kennedy *et al.*, 1992).

Although it may not be logically possible to prove that *ex vivo* free radical formation has not occurred, it is possible to demonstrate that it is occurring and to design measures to prevent its occurrence. Trace transition metals catalyze many free radical reactions. Their presence either in the biological sample or in the collection vessel needs to taken into account. In a study of α-hydroxyethyl radical formation by ethanol-treated alcohol dehydrogenase-deficient deermice, Desferal (a very strong Fe^{3+} chelator) was used to suppress Fenton chemistry during the collection and breaking of gall bladders (Knecht *et al.*, 1990).

In an investigation of *in vivo* hydroxyl radical generation in iron overload, biliary iron caused *ex vivo* hydroxyl radical formation, which could be totally prevented when bile samples were collected into 2,2′-dipyridyl, which stabilizes Fe^{2+} (Burkitt and Mason, 1991). *Ex vivo* radical adduct formation was undetectable even when Fe^{2+} was added to the 2,2′-dipyridyl in the collection tube. This chelator also suppressed a very weak radical adduct signal obtained when animals were administered PBN alone. In a related study of acute copper toxicity (Kadiiska *et al.*, 1992), the addition of both bathocuproinedisulphonic acid (a Cu(I)-stabilizing chelator) and 2,2′-dipyridyl to the bile collection tube was found to be necessary to inhibit *ex vivo* hydroxyl radical formation.

Ideally, the radical adduct should be detected in the living animal (i.e. *in vivo* spectroscopy); sample handling would then be unnecessary. Since larger samples can be analyzed using low-frequency ESR, this method could be used to study radical adduct production directly in small animals (Table 4.2). Because of the *ex vivo* radical formation question, a low frequency L-band spectrometer was utilized to obtain an *in vivo* spectrum of the DMPO/Hb thiyl radical after administering an

Table 4.2 In vivo ESR detection of free radical metabolites

Year	*Radical adduct*	*Authors*
1995	4-POBN/$^{\bullet}CH(OH)CH_3$ from attack of hydroxyl radical from ionizing radiation on ethanol in leg of mouse	Halpern *et al.*
1995	DMPO/$^{\bullet}SO_3^-$ from dichromate oxidation of sulfite in blood of mouse	Jiang *et al.*
1996	DMPO/thiyl Hb from phenylhydrazine detected in whole neonatal rat	Jiang *et al.*

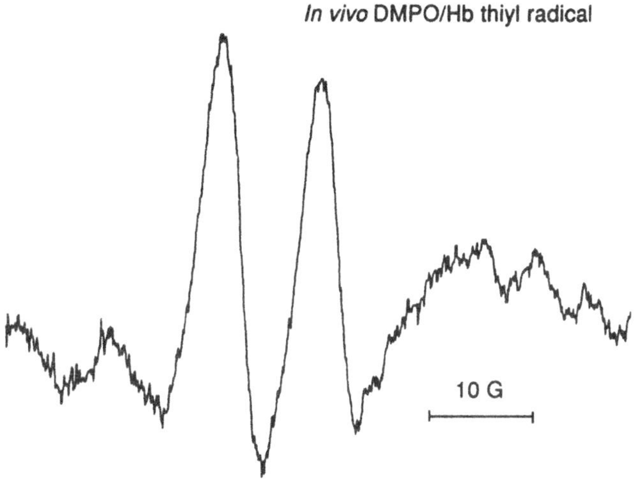

Figure 4.9 A typical 1.1 GHz EPR spectrum of the DMPO/Hb thiyl free radical detected from an intact rat. After anesthesia the rat (20 g) was treated with neat DMPO, 13 mmol/kg, *i.p.*, and an LD_{50} dose of phenylhydrazine hydrochloride, 188 mg/kg, *i.g.* The rat was then inserted into the loop-gap resonator. The spectra were collected 0.5 h after the treatment. Each spectrum was made from the average of 10 scans. Each scan took 60 s with a time constant of 0.1 s. The incident microwave power was 50 mW; the modulation amplitude was 1.6 G; scan range was 60 G (Jiang *et al.*, 1996)

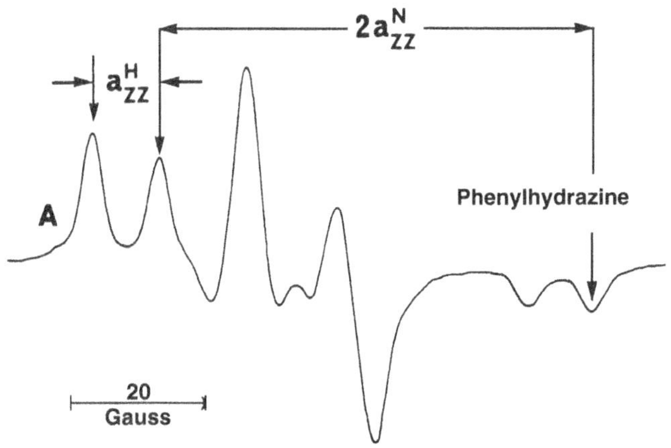

Figure 4.10 The ESR spectra obtained from the blood of rats 2 hr after being given 500 μl/kg DMPO and phenylhydrazine (188 mg/kg, p.o.). Instrumental conditions: A, 20 mW microwave power, 0.67 G modulation amplitude, 2-sec time constant, 5×10^3 receiver gain, and 12.5 G/min scan rate (Maples *et al.*, 1988b)

LD_{50} dose of phenylhydrazine. Although successful, the sensitivity of such instruments is still relatively low and the spectrum is distorted by dispersion (Figure 4.9). Nonetheless, the earlier work where the rat was bled and the DMPO/Hb thiyl radical was detected *ex vivo* was confirmed (Figure 4.10).

The observation of *ex vivo* radical adduct formation by oxidation of the corresponding ESR-silent hydroxylamines also has implications for *in vivo* spectroscopy where the rapid bioreduction of nitroxide spin-labels is well known. The facile bioreduction of radical adducts will be a major problem for *in vivo* spectroscopy even if the instrumental sensitivity problems can be overcome. In fact the DMPO/Hb thiyl radical adduct is reduced by the ascorbate present in blood (Jiang *et al.*, 1996), and much of the radical adduct may be ESR-silent *in vivo*, explaining, in part, the relatively poor signal-to-noise of the *in vivo* spectrum of DMPO/Hb thiyl radical adduct. At the present time, the *ex vivo* detection of free radicals formed and trapped *in vivo* (Table 4.1) has greater utility than *in vivo* spectroscopy (Table 4.2), although the latter approach is, in principle, the better.

References

Albano, E., Lott, K.A.K., Slater, T.F., Stier, A., Symons, M.C.R., and Tomasi, A., 1982, Spin-trapping studies on the free-radical products formed by metabolic activation of carbon tetrachloride in rat liver microsomal fractions isolated hepatocytes and *in vivo* in the rat, Biochem. J., 204, 593–603.

Buettner, G.R., and Mason, R.P., 1990, Spin-trapping methods for detecting superoxide and hydroxyl free radicals *in vitro* and *in vivo*, In Methods in Enzymology (eds L. Packer and A.N. Glazer), Academic Press, London, 186, 127–133.

Burkitt, M.J., and Mason, R.P., 1991, Direct evidence for *in vivo* hydroxyl-radical generation in experimental iron overload: An ESR spin-trapping investigation., Proc. Natl. Acad. Sci. USA, 88, 8440–8444.

Burkitt, M.J., Kadiiska, M.B., Hanna, P.M., Jordan, S.J., and Mason, R.P., 1993, Electron spin resonance spin-trapping investigation into the effects of paraquat and desferrioxamine on hydroxyl radical generation during acute iron poisoning, Mol. Pharmacol., 43, 257–263.

Chamulitrat, W., Jordan, S.J., and Mason, R.P., 1992, Fatty acid radical formation in rats administered oxidized fatty acids: *In vivo* spin trapping investigation, Arch. Biochem. Biophys., 299, 361–367.

Chamulitrat, W., Carnal, J., Reed, N.M., and Spitzer, J.J., 1998, *In vivo* endotoxin enhances biliary ethanol-dependent free radical generation, Am. J. Physiol., 274, G653–G661.

Coghlan, J.G., Flitter, W.D., Holley, A.E., Norell, M., Mitchell, A.G., Ilsley, C.D., and Slater, T.F., 1991, Detection of free radicals and cholesterol hydroperoxides in blood taken from the coronary sinus of man during percutaneous transluminal coronary angioplasty, Free Rad. Res. Comms., 14, 409–417.

Connor, H.D., Thurman, R.G., Galizi, M.D. and Mason, R.P., 1986, The formation of a novel free radical metabolite from CCl_4 in the perfused rat liver and *in vivo*, J. Biol. Chem., 261, 4542–4548.

Connor, H.D., Lacagnin, L.B., Knecht, K.T., Thurman, R.G., and Mason, R.P., 1990, Reaction of glutathione with a free radical metabolite of carbon tetrachloride, Mol. Pharmacol., 1990, 443–451.

Connor, H.D., Gao, W., Mason, R.P., and Thurman, R.G., 1994, New reactive oxidizing species causes formation of carbon-centered radical adducts in organic extracts of blood following liver transplantation, Free Rad. Biol. Med., 16, 871–875.

DeGray, J.A., and Mason, R.P., 1994, Biological spin-trapping, In Electron spin resonance (eds N.M. Atherton, M.J. Davies, and B.C. Gilbert), Royal Society of Chemistry, Cambridge, 14, 246–301.

Di Luzio, N.R., 1963, Physiologist, 6, 169.

Dürk, H., Poyer, J.L., Klessen, C., and Frank, H., 1992, Acetylene, a mammalian metabolite of 1,1,1,-trichloroethane, Biochem. J., 286, 353–356.

Fujii, K., Morio, M., Kikuchi, H., Ishihara, S., Okida, M., and Ficor, F., 1984, *In vivo* spin-trap study on anaerobic dehalogenation of halothane, Life Sci., 35, 463–468.

Ghio, A.J., Kadiiska, M.B., Xiang, Q.-H., and Mason, R.P., 1998, *In vivo* evidence of free radical formation after asbestos instillation: An ESR spin trapping investigation, Free Rad. Biol. Med., 24, 11–17.

Goria-Gatti, L., Iannone, A., Tomasi, A., Poli, G., and Albano, E., 1992, *In vitro* and *in vivo* evidence for the formation of methyl radical from procarbazine: a spin-trapping study, Carcinogenesis, 13, 799–805.

Halpern, H.J., Yu, C., Barth, E., Peric, M., and Rosen, G.M., 1995, *In situ* detection, by spin trapping, of hydroxyl radical markers produced from ionizing radiation in the tumor of a living mouse, Proc. Natl. Acad. Sci. USA, 92, 796–800.

Hanna, P.M., Kadiiska, M.B., Jordan, S.J., and Mason, R.P., 1993, Role of metallothionein in zinc(II) and chromium(III) mediated tolerance to carbon tetrachloride hepatoxicity: Evidence against a trichloromethyl radical-scavenging mechanism, Chem. Res. Toxicol., 6, 711–717.

Hughes, H.M., George, I.M., Evans, J.C., Rowlands, C.C., Powell, G.M., and Curtis, C.G., 1991, The role of the liver in the production of free radicals during halothane anaesthesia in the rat: Quantification of *N-tert*-butyl-α-(4-nitrophenyl)nitrone (PBN)-trapped adducts in bile from halothane as compared with carbon tetrachloride, Biochem. J., 277, 795–800.

Iimuro, Y., Bradford, B.U., Gao, W., Kadiiska, M., Mason, R.P., Stefanovic, B., Brenner, D.A., and Phurman, R.C., 1996, Detection of α-hydroxyethyl free radical adducts in the pancreas after chronic exposure to alcohol in the rat., Mol. Pharmacol., 50, 656–661.

Janzen, E.G., Towner, R.A., and Haire, D.L., 1987, Detection of free radicals generated from the *in vitro* metabolism of carbon tetrachloride using improved ESR spin trapping techniques, Free Rad. Res. Comms., 3, 357–364.

Janzen, E.G., Towner, R.A., Krygsman, P.H., Haire, D.L., and Poyer, J.L., 1990, Structure identification of free radicals by ESR and GC/MS of PBN spin adducts from the *in vitro* and *in vivo* rat liver metabolism of halothane, Free Rad. Res. Comms., 9, 343–351.

Janzen, E.G., and Haire, D.L., 1990, Two decades of spin-trapping, Adv. Free. Rad. Chem., 1, 253–295.

Janzen, E.G., Chen, G., Bray, T.M., Reinke, L.A., Poyer, J.L., and McCay, P.B., 1993, Study of the isolation and stability of α-trichloromethylbenzyl(*tert*-butyl)aminoxyl, the trichloromethyl radical adduct of α-phenyl-*tert*-butylnitrone (PBN), J. Chem. Soc. Perkin Trans. 2, 1983–1989.

Jiang, J., Liu, K.J., Shi, X., and Swartz, H.M., 1995, Detection of short-lived free radicals by low-frequency electron paramagnetic resonance spin trapping in whole living animals, Arch. Biochem. Biophys., 319, 570–573.

Jiang, J., Liu, K.J., Jordan, S.J., Swartz, H.M., and Mason, R.P., 1996, Detection of free radical metabolite formation using *in vivo* ESR spectroscopy: Evidence of rat hemoglobin thiyl radical formation following administration of phenylhydrazine, Arch. Biochem. Biophys., 330, 266–270.

Kadiiska, M.B., Hanna, P.M., Hernandez, L., and Mason, R.P., 1992, *In vivo* evidence of hydroxyl radical formation after acute copper and ascorbic acid intake: Electron spin resonance spin-trapping investigation, Mol. Pharmacol., 42, 723–729.

Kadiiska, M.B., Hanna, P.M., Jordan, S.J., and Mason, R.P., 1993a, Electron spin resonance evidence for free radical generation in copper-treated vitamin E- and selenium-deficient rats: *In vivo* spin-trapping investigation, Mol. Pharmacol., 44, 222–227.

Kadiiska, M.B., Hanna, P.M., and Mason, R.P., 1993b, *In vivo* ESR spin-trapping evidence for hydroxyl radical-mediated toxicity of paraquat and copper in rats, Toxicol. Appl. Pharmacol., 123, 187–192.

Kadiiska, M.B., Xiang, Q.-H., and Mason, R.P., 1994, *In vivo* free radical generation by chromium (VI): An electron spin resonance spin-trapping investigation, Chem. Res. Toxicol. 7, 800–805.

Kadiiska, M.B., Burkitt, M.J., Xiang, Q.-H., and Mason, R.P., 1995, Iron supplementation generates hydroxyl radical *in vivo*: An ESR spin-trapping investigation, J. Clin. Invest., 96, 1653–1657.

Kadiiska, M.B., Mason, R.P., Dreher, K.L., Costa, D.L., and Ghio, A.J., 1997a, *In vivo* evidence of free radical formation in the rat lung after exposure to an emission source air pollution particle, Chem. Res. Toxicol., 10, 1104–1108.

Kadiiska, M.B., Burkitt, M.J., Xiang, Q.-H., and Mason, R.P., 1997b, Effect of acute and ascorbic acid administration on free-radical generation in young and older rats: An ESR spin-trapping investigation, Envir. Nutr. Interact., 1, 143–159.

Kalyanaraman, B., Mason, R.P., Perez-Reyes, E., Chignell, C.F., Wolf, C.R., and Philpot, R.M., 1979, Characterization of the free radical formed in aerobic microsomal incubations containing carbon tetrachloride and NADPH, Biochem. Biophys. Res. Comms., 89, 1065–1072.

Kalyanaraman, B., Parthasarathy, S., Joseph, J., and Froncisz, W., 1991, EPR spectra in a loop–gap resonator for a spin-trapped radical from a low-density lipoprotein lipid, J. Magn. Reson., 92, 342–347.

Kennedy, C.H., Hatch, G.E., Slade, R., and Mason, R.P., 1992, Application of the EPR spin-trapping technique to the detection of radicals produced *in vivo* during inhalation exposure of rats to ozone, Toxicol. Appl. Pharmacol., 114, 41–46.

Knecht, K.T., and Mason, R.P., 1988, *In vivo* radical trapping and biliary secretion of radical adducts of carbon tetrachloride-derived free radical metabolites, Drug. Metab. Disposit., 16, 813–817.

Knecht, K.T., and Mason, R.P., 1990, Free radical metabolism of halothane *in vivo*: Detection of radical adducts in bile, Free Rad. Biol. Med., 9, Supp. 1, 40.

Knecht, K.T., and Mason, R.P., 1991, The detection of halocarbon-derived radical adducts in bile and liver of rats, Drug Metab. Disposit., 19, 325–331.

Knecht, K.T., and Mason, R.P., 1993, *In vivo* spin trapping of xenobiotic free radical metabolites, Arch. Biochem. Biophys., 303, 185–194.

Knecht, K.T., Bradford, B.U., Mason, R.P., and Thurman, R.G., 1990, *In vivo* formation of a free radical metabolite of ethanol, Mol. Pharmacol., 38, 26–30.

Knecht, K.T., DeGray, J.A., and Mason, R.P., 1992, Free radical metabolism of halothane *in vivo*: Radical adducts detected in bile, Mol. Pharmacol., 41, 943–949.

Knecht, K.T., Adachi, Y., Bradford, B.U., Iimuro, Y., Kadiiska, M., Xuang, Q.H., and Thurman, R.G., 1995, Free radical adducts in the bile of rats treated chronically with intragastric alcohol: Inhibition by destruction of Kupffer cells, Mol. Pharamcol., 47, 1028–1034.

Kubow, S., Janzen, E.G., and Bray, T.M., 1984, Spin-trapping of free radicals formed during *in vitro* and *in vivo* metabolism of 3-methylindole, J. Biol. Chem., 259, 4447–4451.

Kubow, S., Bray, T.M. and Janzen, E.G., 1985, Spin-trapping studies on the effects of vitamin E 2nd glutathione on free radical production induced by 3-methylindole, Biochem. Pharmacol., 34, 1117–1119.

LaCagnin, L.B., Connor, H.D., Mason, R.P., and Thurman, R.G., 1988, The carbon dioxide anion radical adduct in the perfused rat liver: Relationship to halocarbon-induced toxicity, Mol. Pharmacol., 33, 351–357.

Lai, E.K., McCay, P.B., Noguchi, T., and Fong, K.-L., 1979, *In vivo* spin-trapping of trichloromethyl radicals formed from CCl_4, Biochem. Pharmacol., 28, 2231–2235.

Lloyd, S.S., Chang, A.K., Taylor, F.B., Janzen, E.G., and McCay, P.B., 1993, Free radicals and septic shock in primates: The role of tumor necrosis factor, Free Rad. Biol. Med., 14, 233–242.

Maples, K.R., Jordan, S.J., and Mason, R.P., 1988a, *In vivo* rat hemoglobin thiyl free radical formation following phenylhydrazine administration, Mol. Pharmacol., 33, 344–350.

Maples, K.R., Jordan, S.J., and Mason, R.P., 1988b, *In vivo* rat hemoglobin thiyl free radical formation following administration of phenylhydrazine and hydrazine-based drugs, Drug Metab. Dispos., 16, 799–803.

Maples, K.R., Eyer, P., and Mason, R.P., 1990a, Aniline-, phenylhydroxylamine-, nitrosobenzene-, and nitrobenzene-induced hemoglobin thiyl free radical formation *in vivo* and *in vitro*, Mol. Pharmacol., 37, 311–318.

Maples, K.R., Kennedy, C.H., Jordan, S.J., and Mason, R.P., 1990b, *In vivo* thiyl free radical formation from hemoglobin following administration of hydroperoxides, Arch. Biochem. Biophys., 277, 402–409.

McCay, P.B., Lai, E.K., Poyer, J.L., Dubose, C.M., and Janzen, E.G., 1984, Oxygen- and carbon-centered free radical formation during carbon tetrachloride metabolism, J. Biol. Chem., 259, 2135–2143.

Moore, D.R., Reinke, L.A., and McCay, P.B., 1995, Metabolism of ethanol to 1-hydroxyethyl radicals *in vivo*: Detection with intravenous administration of α-(4-pyridyl-1-oxide)-*N-t*-butylnitrone, Mol. Pharmacol., 47, 1224–1230.

Mottley, C., and Mason, R.P., 1989, Nitroxide radical adducts in biology: Chemistry, applications and pitfalls. In Berliner, L.J. and Reuben, J. (Eds): *Biological Magnetic Resonance.* Vol. 8, New York: Plenum Publishing Corp., New York, 489–546.

Murphy, P.G., Myers, D.S., Webster, N.R., Gareth Jones, J., and Davies, M.J., 1991, Direct detection of free radical generation in an *in vivo* model of acute lung injury, Free Rad. Res. Comms., 15, 167–176.

Paolini, M., Sapigni, E., Mesirca, R., Pedulli, G.F., Corongiu, F.P., Dessi, M.A., and Cantelli-Forti, G., 1992, On the hepatoxicity of 1,1,2,2-tetrachloroethane, Toxicology, 73, 101–115.

Plummer, J.L., Beckwith, A.L.J., Bastin, F.N., Adams, J.F., Cousins, M.J., and Hall, P., 1982, Free radical formation *in vivo* and hepatoxicity due to anesthesia with halothane, Anesthesiology, 57, 160–166.

Poyer, J.L., McCay, P.B., Lai, E.K., Janzen, E.G., and Davis, E.R., 1980, Confirmation of assignment of the trichloromethyl radical spin adduct detected by spin-trapping during ^{13}C-carbon tetrachloride metabolism *in vitro* and *in vivo*, Biochem. Biophys. Res. Comms., 94, 1154–1160.

Poyer, J.L., McCay, P.B., Weddle, C.C., and Downs, P.E., 1981, *In vivo* spin-trapping of radicals formed during halothane metabolism, Biochem. Pharmacol., 30, 1517–1519.

Reinke, L.A., and Janzen, E.G., 1991, Detection of spin adducts in blood after administration of carbon tetrachloride to rats, Chem.-Biol. Interact., 78, 155–165.

Reinke, L.A., and McCay, P.B., 1997, Spin-trapping studies of alcohol-initiated radicals in rat liver: Influence of dietary fat, J. Nutr., 127, 899S–902S.

Reinke, L.A., Lai, E.K., DuBose, C.M., and McCay, P.B., 1987, Reactive free radical generation *in vivo* in heart and liver of ethanol-fed rats: Correlation with radical formation *in vitro*, Proc. Natl. Acad. Sci. USA, 84, 9223–9227.

Reinke, L.A., Lai, E.K., and McCay, P.B., 1988, Ethanol feeding stimulates trichloromethyl radical formation from carbon tetrachloride in liver, Xenobiotica, 18, 1311–1318.

Reinke, L.A., Kotake, Y., McCay, P.B., and Janzen, E.G., 1991, Spin-trapping studies of hepatic free radicals formed following the acute administration of ethanol to rats: *In vivo* detection of 1-hydroxyethyl radicals with PBN, Free Rad. Biol. Med., 11, 31–39.

Reinke, L.A., Towner, R.A., and Janzen, E.G., 1992, Spin-trapping of free radical metabolites of carbon tetrachloride *in vitro* and *in vivo*: Effect of acute ethanol administration, Toxicol. Appl. Pharmacol., 112, 17–23.

Reinke, L.A., Moore, D.R., and McCay, P.B., 1997, Free radical formation in livers of rats treated acutely and chronically with alcohol, Alcohol. Clin. Exp. Res., 21, 642–646.

Rikans, L.E., Hornbrook, K.R., and Cai, Y., 1994, Carbon tetrachloride hepatoxicity as a function of age in female Fischer 344 rats, Mech. Ageing and Dev., 76, 89–99.

Sentjurc, M., and Mason, R.P., 1992, Inhibition of radical adduct reduction and reoxidation of the corresponding hydroxylamines in *in vivo* spin-trapping of carbon tetrachloride-derived radicals, Free Rad. Biol. Med., 13, 151–160.

Slater, T.F., 1972, Free radical mechanisms in tissue injury, Pion Limited, London.

Tomasi, A., and Iannone, A., 1993, ESR spin-trapping artefacts in biological model systems, In: Biological Magnetic Resonance, 'EMR of Paramagnetic Molecules,' Vol. 13, New York: Plenum Publishing Corp., pp. 353–384.

Tomasi, A., Albano, E., Biasi, F., Slater, T.F., Vannini, V., and Dianzani, M.U., 1985, Activation of chloroform and related trihalomethanes to free radical intermediates in isolated hepatocytes and in the rat *in vivo* as detected by the ESR-spin trapping technique, Chem.-Biol. Interact., 55, 303–316.

Tortolani, A.J., Powell, S.R., Misik, V., Weglicki, W.B., Pogo, G.J., and Kramer, J.H., 1993, Detection of alkoxyl and carbon-centered free radicals in coronary sinus blood from patients undergoing elective cardioplegia, Free Rad. Biol. Med., 14, 421–426.

Ueda, Y., Yokoyama, H., Niwa, R., Konaka, R., Ohya-Nishiguchi, H., and Kamada, H., 1997, Generation of lipid radicals in the hippocampal extracellular space during kainic acid-induced seizure in rats, Epilepsy Res., 26, 329–333.

Vincent, R., Janzen, E.G., Chen, G., Kumarathasan, P., Haire, D.L., Guenette, J., Chen, J.Z., and Bray, T.M., 1996, Spin-trapping study in the lungs and liver of F344 rats after exposure to ozone, Free Rad. Res., 25, 475–488.

Yamaguchi, K.T., Stewart, R.J., Wang, H.M., Hudson, S.E., Vierra, M., Akhtar, A., Hoffman, C., and George, D., 1992, Measurement of free radicals from smoke inhalation and oxygen exposure by spin-trapping and ESR spectroscopy, Free Rad. Res. Comms., 16, 167–174.

Yokoyama, H., Kasai, N., Ueda, Y., Niwa, R., Konaka, R., Mori, N., Tsuchihashi, N., Matsue, T., Ohya-Nishiguchi, H., and Kamada, H., 1998, *In vivo* analysis of hydrogen peroxide and lipid radicals in the striatum of rats under long-term administration of a neuroleptic, Free Rad. Biol. Med., 24, 1056–1060.

Zini, I., Tomasi, A., Grimaldi, R., Vannini, V., and Agnati, L.F., 1992, Detection of free radicals during brain ischemia and reperfusion by spin-trapping and microdialysis, Neuorosci. Lett., 138, 279–282.

5 *Ex vivo* detection of free radicals

Aldo Tomasi, Anna Iannone, Marco Ferrari and Valentina Quaresima

5.1 Background

The beginning of what is now universally known as free radical pathology dates back to the paper authored by the late T.F. Slater, 'Necrogenic action of carbon tetrachloride in the rat: a speculative mechanism based on activation', which appeared in 1966 in *Nature* (Slater, 1966). The toxicological significance of highly reactive free radical intermediates became quickly and widely recognised. In a few years, the progress in the field was so wide and thorough that Slater was able to assemble, in a milestone book (Slater, 1978) scientific contributions from more than 30 laboratories studying the 'Biochemical Mechanism of Liver Injury', as far as free radical intermediates were involved. The number of laboratories involved in this research field increased since exponentially. About 100 papers were published in 1976 on the subject of free radical in biomedicine, the number increased to around 1,000 in 1986 and 4,000 appeared in 1996 (Medline, 1997).

Aim of this contribution is to review and discuss the relevant literature following the 1994 article of Mason and Knecht on a similar subject (Mason and Knecht, 1994). The focus will be on *ex vivo* detection; since the *in vivo* and *ex vivo* approaches are often associated in the same paper, also some *in vivo* results will be discussed.

5.1.1 In vivo *vs.* ex vivo *free radical detection*

The term *in vivo*, in this paper, will be referred to studies where the detection of free radicals was actually carried out in the living experimental animal. Therefore, the only way to do so, is direct electron spin resonance spectroscopy (EPR) of the whole animal or part of it. For this purpose, spectrometers tuned at a frequency lower than the usual X-band (around 9 GHz) are used. L- or S-band and recent EPR spectrometers working at NMR frequencies have been developed. Spectra or images may be obtained in this way. Overviews on these new and recent developments may be found in (Gallez *et al.*, 1996; Kuppusamy *et al.*, 1996a; Lurie, 1996; Quaresima *et al.*, 1996; Quine *et al.*, 1996; Utsumi *et al.*, 1996).

In *ex vivo* studies the free radical formation occurs, or better should occur, *in vivo*, while the detection is performed on a sample obtained from the experimental animal. The sample is usually blood, plasma, urine, or tissue.

5.1.2 Electron spin resonance spectroscopy and spin-trapping

The free radical research crux lies in the overt difficulties in demonstrating the presence, identifying, and assigning a role to very short lived free radicals. Electron spin resonance spectroscopy can give the right answer to the difficult question. It detects and identifies free radicals with high sensitivity. A major problem, however, arises when EPR is applied to biological model systems: water present in the sample, absorbing the irradiating microwave, reduces enormously the sensitivity of the spectrometer (Symons, 1978).

The spin-trapping (ST) technique has been developed to overcome this drawback. With this technique, a higher steady state concentration of free radicals is achieved, which can overcome the sensitivity problem. It has been widely applied to a multitude of different biological problems where free radicals are thought to be involved. For review see (Janzen, 1980; Mottley and Mason, 1989). Various spin-trapping agents, both in *in vitro* and *in vivo* models, easily trap carbon-, sulphur-, and oxygen-centred radicals. Nitrone- or nitroso- compounds have been used as spin-trapping agents. Recently a new class of phosphorylated nitrones, which gives long-lived oxygen adducts, has been introduced (Frejaville *et al.*, 1995; Tuccio *et al.*, 1996). Briefly, short-lived free radicals react with a nitrone- or nitroso-based diamagnetic compound to form a relatively stable free radical adduct. The adduct is detected with high sensitivity by EPR spectroscopy. The technique, initially developed for studies *in vitro*, has been applied to *ex vivo* animal and human studies (Buettner and Mason, 1990; Mason and Knecht, 1994; Reszka *et al.*, 1996). Pitfalls and artefacts of the technique have been also described (Mottley and Mason, 1989; Tomasi and Iannone, 1993).

5.2 *Ex vivo* spin-trapping of short-lived free radicals

Most of the reported spin-trapping investigations have used nitrone spin-traps phenyl-*tert*-butylnitrone (PBN), 4-pyridine-*N*-oxide *N*-*tert*-butyl nitrone (POBN), α-2,4,6-trimethoxy-PBN $(CH_3O)_3$, 5,5,-dimethyl-1-pyrroline *N*-oxide (DMPO), and *N*-*tert*-butyl-alpha-(2-sulfophenyl)-nitrone. Rarely, nitroso-based spin-trapping agents such as 3,5-dibromo-4-nitrosobenzene sulfonate (DBNDS) have been used in *ex vivo/in vivo* studies, mostly because of their chemical instability and toxicity.

The most common experimental design requires the *in vivo* administration of the spin-trap. This introduces a major interference problem in the experimental model, since the spin-trap may affect the system by inhibiting or stimulating metabolic pathways, could produce toxicity, and exhibit pharmacological effects. To overcome this problem in some experimental protocols, based on trapping free radicals in the circulating blood, the spin-trap is not administered *in vivo*, and the effluent blood is immediately infused with an isotonic saline solution containing the spin-trap (Coghlan *et al.*, 1993; Coghlan *et al.*, 1991; Mergner *et al.*, 1991).

Because the detection of radical adducts occurs *ex vivo*, the question of possible *ex vivo* formation of the adducts needs to be addressed, and the experimental protocol should contemplate appropriate tests. For instance, it is possible that trace transition metals could catalyse radical reactions. Their presence either in the biological sample or in the collection vessel needs to be taken into account.

Particular care should also be taken when, having administered *in vivo* the spin-trap, tissues or blood undergo Folch extraction, to increase sensitivity. In this case, possible water-soluble adducts are lost, and the water/methanol layer should be examined utilising a flat cell.

Sample handling and treatment could also induce *ex vivo* lipid peroxidation. An appropriate protocol should include iron chelating agent, antioxidant and peroxidases in order to test for the *ex vivo* occurrence of lipid peroxidation. It is however true that in most cases the *ex vivo* free radical-forming reactions are closely related to those presumably occurring *in vivo*.

An important problem in trapping experiments is the rapid bioreduction of the radical adduct to an EPR silent hydroxylamine. The radical adduct can be detected only after oxidation of the hydroxylamine by different means, such as air, oxygen, or ferricyanide (Sentjurc and Mason, 1992; Stoyanovsky *et al.*, 1996).

5.2.1 Pharmacological effect of spin-trapping agent

Spin-trapping agents not only form stable adducts with free radicals, but also exert a pharmacological and toxic activity. The spin-trap PBN prolongs the life span of mice (Edamatsu *et al.*, 1995), reverses age-related maze learning performance and motor activity deficits in c57 BL/6 mice (Fredriksson and Archer, 1996), protects the brain from oxidative damage (Carney and Floyd, 1991), inhibits the breakdown of nitric oxide (Inanami and Kuwabara, 1995), and inhibits the induction of nitric oxide synthase in endotoxin-induced shock in mice (Miyajima and Kotake, 1995). PBN also protects against stroked-induced damage, so much that new spin-trapping agents, analogues of PBN, are now being synthesised and tested for their *in vitro* and *in vivo* neuroprotective properties (Thomas *et al.*, 1994). These nitrones minimised seizures and delayed the time to death in mice following central injection of ferrous iron, suggesting that they may be useful in the treatment of acute and chronic neurodegeneration. Furthermore, the stability of the spin-trap adducts of these nitrones provides a new tool for the study of oxidative tissue injury (Thomas *et al.*, 1997). The mechanism of action by which PBN attenuates glutamate excitotoxicity in rat striatum has been recently studied. The results indicate that PBN neuroprotection depends on its free radical scavenging activity (Lancelot *et al.*, 1997).

A newly synthesised conformationally constrained cyclic analogue of the PBN, 3,4-dihydro-3,3-dimethyl-isoquinoline-2-oxide, is currently being evaluated in stroke models as a means to ameliorate the oxidative insult associated with reperfusion injury (Dage *et al.*, 1997).

DMPO derivatives are also studied for their pharmacological properties. For instance, 2-phenyl-DMPO resulted active in adriamycin-induced cardiotoxicity (Piccinini *et al.*, 1995).

On the other hand, the nitrone spin-trap 4-POBN can enhance the peroxidase-mediated oxidation of desferrioxamine (Desferal), possibly via the formation of a transient 4-POBN radical species. This suggests the importance of assessing the potential for nitrone spin-traps to both inhibit and enhance biological oxidation prior to their use as potential pharmacological agents (McCormick *et al.*, 1995).

5.2.2 Toxicological studies

Ex vivo spin-trapping has been particularly useful in toxicological studies, where the role of free radical intermediates has to be assessed.

The *in vivo* administration of the neurotoxic compound 1-methyl-4-phenyl-1,2,3,6-tetrahydropyridine (MPTP) produces ATP depletion, excitotoxicity and free radical generation. Agents which improve mitochondrial function or free radical scavengers should attenuate MPTP neurotoxicity. The administration of the spin-trap *N-tert*-butyl-α-(2-sulfophenyl)-nitrone was effective in protecting from MPTP toxicity which produced mild dopamine depletion (Schulz *et al.*, 1995).

Chromium (VI) mediated free radical formation was studied using the spin-trap 4-POBN collected in the bile of animals given an intragastric dose of potassium dichromate. Upon the administration of Cr(VI) and 4-POBN, the EPR spectrum of the radical adduct was detected in the bile and exhibited hyperfine coupling constants typical of a carbon-centred radical, mostly derived from endogenous lipids (Kadiiska *et al.*, 1994).

The mechanism of ethanol toxicity and the formation of ethanol-derived free radical have been studied from any possible point of view (for review see Albano *et al.*, 1993). Free radical products have been detected in rodents after chronic feeding with an ethanol-containing, high-fat diet, where serious hepatic damage occurs in rats after continuous intragastric administration of ethanol (Knecht *et al.*, 1990). Rats treated with ethanol for at least 2 weeks were administered the spin-trap 4-POBN. A six-line radical adduct spectrum was detected in the bile of the treated rats, where little or no radical adduct signal was detected with chow-fed animals. Destruction of Kupffer cells by chronic treatment with gadolinium chloride decreased by about 50% the radical adduct formation in rats fed the ethanol-containing, high-fat diet. The radical species was largely ethanol derived, because addition of [^{13}C]ethanol produced a 12-line spectrum, indicating the formation of α-hydroxyethyl radical. The signal intensity correlates with hepatotoxicity and the decrease in free radical formation produced by $GdCl_3$ implicates Kupffer cells in the development of alcoholic liver injury (Knecht *et al.*, 1995).

In similar studies, the *in vivo* pre-treatment of rats with a flavonoid (silybin)–phosphatidylcholine complex is able to decrease the trapping of the hydroxyethyl radical in microsomes from chronic alcohol-fed rats; the effect is related to the capacity of the silybin molecule to scavenge the hydroxyethyl radical. The flavonoid administration also decreases the hydroxyethyl radical adduct detected in the bile of rats acutely treated with ethanol. The study suggests that drugs modulating the free radical activation of ethanol are potentially useful in counteracting ethanol-mediated injury and liver damage (Comoglio *et al.*, 1995).

An important study on the pathophysiological role played by free radical activation of ethanol was recently published. Rats chronically fed ethanol were injected intravenously with the spin-trap 4-POBN, and pancreatic secretions were collected. A six-line free radical adduct spectrum, indicative of a carbon-centred free radical, was detected in pancreatic secretions and in Folch extracts of pancreatic tissue. Control experiments ruled out *ex vivo* radical formation. When [^{13}C]ethanol was administered, a 12-line spectrum was detected in pancreatic secretions, demonstrating that the α-hydroxyethyl radical adduct was formed in the pancreas from [^{13}C]ethanol: a direct demonstration that free radicals are formed in the pancreas during chronic alcohol intake in rats (Iimuro *et al.*, 1996).

A particular aspect of iron-enhanced ethanol toxicity was investigated in rats with chronic dietary iron loading. A secondary radical spin-trapping technique was used where hydroxyl radical forms methyl radical upon reaction with DMSO. The methyl radical was then detected with the spin-trap PBN. This adduct was detected in the bile of rats 10 week after being fed an iron-loading diet and 40 min after the i.p. injection of the spin-trap PBN dissolved in DMSO. Identification of the EPR spectrum of the major radical adduct as that of PBN/(CH_3)—C— provides evidence for the generation of the hydroxyl radical during iron supplementation (Kadiiska *et al.*, 1995).

5.2.3 Free radical intermediates in circulation

5.2.3.1 Blood flow and ischaemia–reperfusion

Many recent studies on *ex vivo* spin-trapping address the problem of free radical mediated damages in the ischaemia–reperfusion syndrome both in animal model systems and in human studies. Tissue damage observed after the readmission of blood in a previously ischaemic organ is thought to be free radical related, with relevant implications on the therapeutic approach.

An interesting study on the mechanisms underlying production of vascular free radical was undertaken by Laurindo *et al.* Aortas isolated from rabbits were perfused with the spin-trap PBN. Sequential perfusion at 2, 7.5, and 12 ml/min changed cumulative vascular PBN radical adduct yields. Superoxide dismutase (SOD) completely blocked flow-induced free radical production, whereas inactivated SOD, indomethacin, and the nitric oxide synthetase (NOS) antagonist nitro-L-arginine methyl ester (L-NAME) had no effect. It was demonstrated that increases in blood flow trigger vascular free radical generation; such a response seems to involve endothelium-derived superoxide radicals unrelated to cyclooxygenase or nitric oxide synthetase activity (Laurindo *et al.*, 1994).

In an excellent study, Tang *et al.* administered an intravenous infusion of the spin-trap PBN to dogs undergoing a 15-min coronary artery occlusion and 3 h of reperfusion. The production of PBN adducts in plasma exhibited an initial burst immediately after the onset of reflow and remained elevated until 10 min after reperfusion. Dogs treated with the hydrophilic vitamin E analogue, MDL 74,405 demonstrated a marked decrease in PBN adduct production. An elegant way to demonstrate that the observed salutary effects of this vitamin E analogue on myocardial stunning are due to attenuation of oxidative stress (Tang *et al.*, 1995).

Recent studies suggest that oxygen-derived free radicals are involved in mediating renal reperfusion injury. EPR spectroscopy and spin-trapping, with the spin-traps DMPO and PBN, were used to detect and quantitate the formation of hydroxyl radicals in rat kidney after ischaemia–reperfusion *in vivo* and *in vitro* in the absence of leukocytes. EPR analysis of homogenised kidneys and venous samples did not detect radical adducts with either spin-trap. When DMPO was used as the spin-trap in kidneys perfused without albumin in the perfusate, EPR signals characteristic of hydroxyl and carbon-centred radical adducts were detected during early reperfusion following ischaemia. Addition of 15 mM dimethylthiourea abolished formation of these adducts following ischaemia–reperfusion but did not prevent a reduction in glomerular filtration rate (Kadkhodaee *et al.*, 1996).

Hypovolaemic shock, usually caused by a major haemorrhage, is an extremely serious condition, and death ensues in the absence of immediate treatment. The effect of ACTH-(1–24) on the blood levels of highly reactive free radicals was studied in a volume-controlled haemorrhagic shock produced in adult rats under general anaesthesia by stepwise bleeding until mean arterial pressure stabilised at 20–23 mmHg. Rats were intravenously (i.v.) treated with either ACTH-(1–24), or equivolume saline. Free radicals were measured in arterial blood by EPR spectrometry using an *ex vivo* method where the spin-trapping agent PBN was not injected into the animal, in order to avoid the known anti-shock properties of PBN (Miyajima and Kotake, 1995; Novelli *et al.*, 1985). In this study, the spin-trapping agent was preloaded in the syringe used for blood withdrawal, allowing the immediate mixing of PBN and blood outside of the organism. A major increase in trapped radicals was observed after bleeding termination. All rats treated with saline died within 30 min, having a still higher blood level of free radicals. Treatment with ACTH-(1–24) produced a rapid and sustained restoration of arterial pressure and respiratory function, with 100% survival at the end of the observation period (2 h); this was associated with the disappearance in the blood of trapped free radicals. These results provide a direct demonstration that in haemorrhagic shock there is a rapid and massive production of highly reactive free radicals, and that the sustained restoration of cardiovascular and respiratory functions induced by the i.v. injection of ACTH-(1–24) is associated with a substantial reduction of free radical blood levels. It is suggested that ACTH-(1–24) prevents the burst of free radical generation during blood mobilisation and subsequent tissue reperfusion, an important component of its mechanism of action in effectively preventing death for haemorrhagic shock (Guarini *et al.*, 1996).

The role of free radical intermediates in inflammation is a burning issue and the following study has used an ingenious approach to address the problem of demonstrating free radical in human synovial tissue. Specimens of fresh synovial tissue obtained from patients undergoing primary total knee joint replacement and graded according to the degree of inflammation present were subjected to an *ex vivo* hypoxia–reoxygenation cycle in the presence of the nitroso based spin-trap DBNBS. The degree of oxidation of the spin-trap to a stable free radical was determined and followed with time. Control samples were subjected to hypoxia only. Only the more inflamed specimens, from both rheumatoid arthritis and osteoarthritis patients, demonstrated increased production of reactive oxygen species when subjected to a hypoxia–reoxygenation cycle. This change was reduced by both competitive and non-competitive inhibitors of the endothelial based enzyme xanthine oxidase (Singh *et al.*, 1995).

5.2.3.2 Free radicals in muscle pathology

Muscle gives a superb model system for our understanding of free radical pathogenesis of damage. The hypothesis that free radicals are formed during ischaemia–reperfusion of a skeletal muscle was tested in an *ex vivo* EPR–ST study. The experimental model was a rat pedicled rectus femoris muscle flap. The femoral artery and vein were cannulated to inject the spin-trap and collect the effluent flow. Three injections and collections were made: i) before ischaemia; ii) after ischaemia of 15, 30, 60, 120, and 180 minutes, but before blood flow had been restored; and iii) after blood flow had been restored. No EPR signal was detected either before the ischaemic period or within 15 minutes ischaemia. PBN radical adducts were detected at 30, 60, 120, and 180 minutes of ischaemia. This study demonstrated the unambiguous presence of free radicals in an *in vivo* intact skeletal muscle ischaemia–reperfusion model (De Santis and Pinelli, 1994).

A similar study was performed on a post-ischaemic muscular reperfusion in a halothane anaesthetised horse. The authors used PBN as a spin-trap agent to observe *ex vivo* free radical generation (Serteyn *et al.*, 1994).

Studies were performed on an experimental model of muscle fatigue where free radicals can be released in the extracellular moiety. Because muscle fatigue may be an important factor in respiratory failure, the authors tested the hypothesis that increased concentrations of free radicals could be detected in the blood of animals undergoing a severe inspiratory resistive loading. Free radical levels in the form of PBN-adducts were found to rise significantly over the control group after 2.5–3 h of inspiratory resistive loading with 70% supplemental inspired oxygen. This study presents direct evidence that free radicals are produced *ex vivo* and that they can be detected in the systemic circulation due to excessive resistive loading of the respiratory muscles (Hartell *et al.*, 1994).

5.2.4 *New spin-trapping agents*

To ameliorate scavenging properties, specificity and stability of the adducts, new spin-trapping agents are continuously devised for *in vitro* and *ex vivo* studies. *N-tert*-butyl-α-(4-[F-18]fluorophenyl)nitrone ([F-18]FPBN) has been synthesised for *in vivo* detection of free radicals (Bormans and Kilbourn, 1995). A new phosphorylated nitrone, 5-diethoxyphosphoryl-5-methyl-1-pyrroline N-oxide (DEPMPO), has been demonstrated to efficiently trap oxygen centred radicals *in vivo* (Frejaville *et al.*, 1994, 1995).

5.2.5 *Low frequency EPR – spin-trapping*

Low frequency EPR ameliorates the sensitivity of the spectroscopy in the presence of water and has allowed the recent development of truly *in vivo* EPR studies. The considerable developments over the last 10 years (Quine *et al.*, 1996) with the combined use of spin-trapping made it possible to measure and detect a variety of oxygen- or carbon-centred free radicals. In most of the cases the instability of the *in vivo* formed spin adducts and the limited sensitivity of *in vivo* EPR spectroscopy restrict the applicability of the spin-trapping technique only when the radical adduct is relatively stable. The absence of the EPR signal does not necessarily indicate that the spin adducts are not formed.

Low frequency EPR can also detect directly in the skin of hairless mice free radicals after topical application of anthralin (Mader *et al.*, 1995). Usually however, a spin-trapping agent has to be present in order to stabilise short-lived free radicals. This technique was used to detect the hemoglobin thiyl free radical in living rats. The hemoglobin thiyl free radical was formed following the intragastric administration of phenylhydrazine; the hemoglobin thiyl free radical was then trapped by preinjected DMPO, which formed the DMPO/hemoglobin thiyl-free radical adduct in the blood. The DMPO/hemoglobin thiyl free radical was detected in blood samples using 9.5 GHz (X-band) and 1.1 GHz (L-band) EPR at room temperature and 77 K. This adduct is susceptible to the reduction induced by ascorbate and the thiol-blocking agent diethylmaleate. DMPO has also been used to trap sulfur trioxide anion free radical in whole mice (Jiang *et al.*, 1995).

The stability and the distribution of the spin-trap and the spin adducts of DMPO were investigated in plasma, whole blood, peritoneal fluid and homogenised heart tissue of the rat (Liu *et al.*, 1996). Halpern *et al.* (1995) first measured the free radical production in the tissue of a living animal. Hydroxyl radicals, produced during radiation in the leg tumour of a living mouse, were detected by a low frequency 260 MHz spectrometer and deuterated DMPO as spin-trap, which, being deuterated DMPO, was useful in that it also increased the EPR signal intensity.

5.3 Nitrogen monoxide (nitric oxide, $NO^{\bullet}$, NO)

Recently, the unique role of nitric oxide as a signal molecule and a mediator of either physiological or pathological functions has increasingly been recognised (Henrich, 1991; Koshland, 1993). NO is a neurotransmitter and regulator of the vascular tone, it exhibits protective effects in oxidative stress as antioxidant, but also modulates lymphocytes, monocytes and neutrophils adhesion to endothelium, as well as platelet aggregation. It is synthesised on demand, after enzyme activation, by the constitutive endothelial and neuronal NO synthases (NOS) for short periods of time, while the killer molecule is synthesised by an inducible NOS (iNOS) that produces NO for a long time (Curtis and Pabla, 1997; Kröncke *et al.*, 1997). New sources of NO have been also recently identified and the possibility of NO formation arising from nitrate–nitrite reduction has been envisaged (Zweier *et al.*, 1995). NO is involved in damaging effects acting as a cytotoxic molecule in cardiac ischaemia–reperfusion (Muijsers *et al.*, 1997; Nonami, 1997), haemorrhagic (Guarini *et al.*, 1997) and septic shock (Cobb and Danner, 1996; Thiemermann, 1997), splanchnic artery occlusion shock (Squadrito *et al.*, 1994), brain ischaemia (Iadecola, 1997; Zini *et al.*, 1994), and cardiovascular diseases (Darley Usmar *et al.*, 1997; White *et al.*, 1997). In short, the evidence collected suggests that NO is protective or destructive depending on the stage of evolution of the pathological process and on the cellular source of NO.

5.3.1 NO detection using EPR spectroscopy

The role of NO in *ex vivo* and *in vivo* experimental models is complex. NO is a relatively long-lived molecule in biological systems (Archer, 1993) and reacts at high rate constant with molecular oxygen, superoxide anion, thiol- and iron-proteins (Kerwin and Heller, 1994). These characteristics render difficult the determination of NO in a biological environment. Evaluations of NO stable by-products, like nitrate and nitrite, or NO_x-induced secondary messengers, like cGMP and citrulline, have been used as indirect methods to demonstrate changes in NO synthesis in plasma (Kumura *et al.*, 1994) and brain (Kader *et al.*, 1993). Therefore development of new methods for a direct assessment of NO formation has been attempted.

5.3.1.1 Iron chelates

NO has been trapped using iron-centred spin-trapping agents, such as the iron chelate (Fe^{2+})–sodium N,N-diethyldithiocarbamate trihydrate (DETC–Fe) or (Fe^{2+})–*N*-methylglucamine dithiocarbamate (MGD–Fe). The first resulting in the formation of lipophilic, the second in water soluble EPR-active nitrosyl complexes, both stable at room-temperature (Komarov and Lai, 1995; Lai and Komarov, 1994; Quaresima *et al.*, 1996; Reinke *et al.*, 1996; Zweier *et al.*, 1995).

The problem of major interference introduced in the model system arises in the case of the lipophilic DETC–Fe. Here, iron citrate and DETC have to be injected separately to the animal in order to obtain the DETC–Fe complex *in situ* in the living animal. Toxicity related to the injection of iron citrate is well known. In addition, the toxicological relevance of the DETC–Fe and MDG–Fe complex has not yet been addressed and the side effects related to the administration *in vivo* of these compounds remains unknown. New water-soluble iron complexes, such as *N*-(dithiocarboxy)sarcosine–Fe (DTCS–Fe), are under study in order to improve NO trapping capability and to reduce side-effects (Yoshimura *et al.*, 1996).

5.3.1.2 Hemoglobin and myoglobin

Physiologically occurring compounds, such as hemoglobin or myoglobin efficiently trap NO (Kozlov *et al.*, 1996; Murphy and Noack, 1994) giving, at low oxygen tension, a stable nitroso–hemoglobin compound (NO–Hb), at higher oxygen tension, methemoglobin (MET–Hb).

NO–Hb gives a distinctive EPR signal at liquid nitrogen temperature (Oda *et al.*, 1975), while MET–Hb can be easily measured spectrophotometrically. There is a major advantage in using this approach, since the physiologic trapping agent does not interfere with the experimental model.

5.3.1.3 Nitronyl nitroxide

Nitrone and nitroso spin-traps including PBN, 4-POBN, DMPO, 2-methyl-2-nitrosopropane (MNP) and DBNBS form an adduct with NO. Nitroso spin-traps (MNP, DBNBS) are better suited for the identification of NO-related signals, than the nitrones. However this technique is prone to many artefactual results and this approach has been rapidly dismissed (Arroyo and Kohno, 1991).

Nitronyl nitroxides, a group of organic compounds with nitronyl and nitroxide functional groups, produce imino nitroxides when reacting with NO. The EPR spectra of nitronyl nitroxides and imino nitroxides are characteristic and distinctly different (Joseph *et al.*, 1993). Nitrite ion and nitrate are also produced in this reaction (Hogg *et al.*, 1995).

A few applications of this trapping method have been reported in the literature. Nitronyl nitroxides have been used as probes to study the mechanism of vasodilatory action of nitrovasodilators, nitrone spin-traps, nitroxides, and nitrosothiol compounds (Konorev *et al.*, 1995a, 1995b).

The reason of the scarce utilisation of the technique rests in the low rate constants (about $10^4\ M^{-1}s^{-1}$) of the reaction between NO and nitronyl nitroxides to give iminonitroxides. Iminonitroxides, at their turn, are very rapidly reduced in biological environment (Woldman *et al.*, 1994). Recently it has also been shown that superoxide-mediated reduction of the nitroxide group can prevent detection of nitric oxide by nitronyl nitroxides (Haseloff *et al.*, 1997).

5.3.2 Ex vivo *NO detection: applications*

NO formation was measured on the effluent of isolated rat hearts, subjected either to normal perfusion or to reperfusion after 30 min of ischaemia, in the presence of the NO trapping agent MGD–Fe. Trace signals were present before ischaemia, and a prominent NO adduct was seen during the first 2 min of reflow (Wang and Zweier, 1996). NO was also measured in isolated rat hearts subjected to global ischaemia, using the same trapping agent. The NO production was about 10 times higher in the heart after 30 min of ischaemia than that observed in normally perfused hearts. With increased duration of ischaemia, NO formation and trapping was also increased (Zweier *et al.*, 1995). The spatial distribution of the NO generated in the ischaemic myocardium was mapped using L-band EPR imaging. The three-dimensional images of the trapped NO with the complex MGD–Fe clearly showed that NO is formed throughout the myocardium. Kinetic experiments showed that maximum NO formation and trapping occurred at the mid myocardium and spread out to endocardium and epicardium of the left ventricle (Kuppusamy *et al.*, 1996b).

Intracerebral NO production was demonstrated in basal conditions and after intrastriatal injection of the vasoconstrictor peptide endothelin-1 using locally injected hemoglobin as a NO trapping agent. In the absence of local Hb injection, no signal related to endogenous NO was detected in the neostriatum. After Hb injection, nitrosyl–Hb signal was detected in neostriatal homogenates. The signal increased significantly after endothelin 1 infusion, used to cause neuronal loss in the neostriatum (Kozlov *et al.*, 1995, 1996). NO increase was also demonstrated, in the brain cortex of rat in a similar model of brain ischemia (Sato *et al.*, 1994; Tominaga *et al.*, 1994).

The NO scavenger MGD–Fe was used to characterise spontaneous and agonist-stimulated NO activity arising from rat aortic endothelium. The addition of MGD–Fe to aortic ring segments suspended contracted with phenylephrine elicited a rapid additional increment in tension which was not altered by indomethacin. This increase in tension was absent in rings pre-treated with L-nitroarginine or in rings without endothelium. Acetylcholine-produced relaxation was completely eliminated in the presence of L-nitroarginine or upon removal of the endothelium. Higher concentrations of MGD–Fe caused a further decrease in the maximum relaxation to acetylcholine. This effect was mimicked by using another NO trapping agent, carboxy–PTIO. These data suggest that MGD–Fe scavenges agonist-stimulated NO, but also reveals a NO synthase-dependent component, which is unavailable to interact with MGD–Fe (Pieper and Lai, 1996).

Production of NO in the renal cortex and medulla was studied with an *in vivo* microdialysis technique. Oxyhemoglobin (OxyHb) was perfused through the dialysis system to trap tissue NO. MetHb was formed by NO oxidation of OxyHb in the dialysate and was assayed spectrophotometrically. NO concentration was significantly higher in the medulla than in the cortex. Intravenous infusion of L-arginine produced a two- to three-fold increase in cortical and medullary NO; L-NAME, an inhibitor of the iNOS decreased NO in the renal cortex and medulla. The degradation products of NO, nitrite, and nitrate, were also measured in this study using *in vivo* microdialysis. The results indicate that the OxyHb–NO microdialysis trapping technique is a highly sensitive *in situ* method for detecting regional tissue NO concentration and changes in the NOS activity in the kidney (Zou and Cowley, 1997).

Lipopolysaccharide (LPS), the purified *E. coli* endotoxin, causes endotoxic shock. Shock was induced in rats, after cannulation of the bile duct, MGD–Fe was administered by intravenous injection, and samples of bile were collected for EPR analyses. The EPR spectra of bile from LPS-pre-treated rats contained characteristic three-line signals of NO trapped by the MGD–Fe complex, while bile from control rats (not treated with LPS) did not contain similar EPR signals. Only weak signals from NO could be detected in plasma or urine under these conditions. The administration of PBN (which also inhibits iNOS), or L-NAME to LPS-treated rats resulted in a significant decrease in NO levels (Reinke *et al.*, 1996).

Studies were carried out in the blood of rats subjected to severe volume-controlled hemorrhagic shock and to a less severe pressure-controlled hemorrhagic shock (Guarini *et al.*, 1997; Westenberger *et al.*, 1990). NO production was measured as nitrosyl-Hb formation. NO increased significantly during shock onset and continued to increase during the duration of the experiment. The administration of resuscitating drugs caused an immediate drop in NO–Hb levels suggesting a causative relationship between NO concentration and gravity of the shock.

5.3.3 In vivo *NO studies*

As mentioned above, many studies use both *in vivo* and *ex vivo* approaches in measuring NO. What follows is the report on some recent developments. The first *in vivo* detection of nitric oxide free radical goes back to 1993 (Komarov *et al.*, 1993). NO was trapped in the blood circulation using MGD–Fe and detected in the tail of a conscious mouse by an S-band spectrometer operating at 3.5 GHz. The same method was used to monitor trapped NO in septic-shock mice (Lai and Komarov, 1994). Using the same experimental animal model and MGD–Fe as a trap, a signal was measured by L-band over the head and the abdominal region of the mouse, 30 min after the injection of the MGD–Fe–NO complex (Fujii *et al.*, 1997; Guiberteau, 1997). The signal in the liver region was higher than that in the brain. The estimated concentration in the brain (3–5 μM) was due only to the blood MGD–Fe–NO complex that is not able to cross the blood brain barrier (Fujii *et al.*, 1997). The spin adduct signal was observed in the liver region 60 min after injection of the trapping agent and increased with time reaching a plateau at 150–180 min.

The intracellular production of NO was investigated in the tissues of mice in septic shock using DETC–Fe (Quaresima *et al.*, 1996). Because of its hydrophobicity, the NO-trap complex was mainly localised in the lipophilic compartments. The complex was detected in the upper abdomen of the mouse by an L-band (1–2 GHz) spectrometer. Tissues were homogenised and examined at 20°C; the highest signal intensity of the trapped NO was found in the liver homogenates. Kidneys retained 40% of the signal found in the liver; brain and lung, *ca.* 10%. The signal was stable for about one hour.

NO in the brain, during ischaemia–hypoxia, was trapped following systemic administration of DETC–Fe complex to the rat, and imaged in the right hemisphere of the frozen brain. An L-band EPR imaging instrumentation was used in this experiment. The results demonstrated that NO radical was produced and trapped in the areas known to have high NOS, such as cortex, hippocampus, hypothalamus, amygdala, and substantia nigra. NO trapped in the in the cerebellum was approximately 30% of that in the cerebrum (Kuppusamy *et al.*, 1995).

The water-soluble iron complex DTCS–Fe has been successfully employed to trap NO *in vivo* in the mouse treated with LPS. In *ex vivo* experiments performed on the isolated organs, higher NO levels were detected in the liver and kidney. This finding supports the assumption that NO detected in LPS-treated mice is mainly produced in the liver, and it does not reflect NO-adduct complex accumulated in the liver via blood circulation (Yoshimura *et al.*, 1996).

5.4 Conclusion

Free radicals formed *in vivo* can be trapped and detected by EPR *ex vivo*. The many different, sometimes ingenious and sometimes cumbersome, techniques developed mirror the inadequacy of the *in vivo* EPR spectroscopy. Both reactive free radicals and, more recently, nitric oxide can be demonstrated using various trapping agents. Problems encountered and cautions which have to be taken into account in the application of these techniques have been delineated and discussed.

The *in vivo* EPR spectroscopy is now being continuously improved in both sensitivity and quality, and it can be foreseen that many *ex vivo* experiments will be soon performed *in vivo*.

References

Albano, E., Tomasi, A., Parola, M., Comoglio, A., Ingelman-Sundberg, M., and Dianzani, M.U., 1993, Mechanisms responsible for free radical formation during ethanol metabolism and their role in causing oxidative injury by alcohol. In Free Radicals and Antioxidants in Nutrition, F.P. Corongiu, S. Banni, A. Dessì and C. Rice-Evans, eds. (London: Richelieu), 77–96.

Archer, S., 1993, Measurement of nitric oxide in biological models. FASEB J 7: 349–360.

Arroyo, C.M., and Kohno, M., 1991, Difficulties encountered in the detection of nitric oxide (NO) by spin-trapping techniques. A cautionary note. Free Rad Res Comms 14: 145–55.

Bormans, G., and Kilbourn, M.R., 1995, Synthesis of n-tert-butyl-alpha-(4-[f-18]fluorophenyl) nitrone ([f-18]FPBN) for in vivo detection of free radicals. J Label Comp Radiopharmac 36: 103–10.

Buettner, G.R., and Mason, R.P., 1990, Spin-trapping methods for detecting superoxide and hydroxyl free radicals in vitro and in vivo. In Methods in Enzymology Oxygen Radical in Biological Systems Part B., L. Packer and A.N. Glazer, eds. (New York: Academic Press), 127–33.

Carney, J.M., and Floyd, R.A., 1991, Protection against oxidative damage to CNS by alpha-phenyl-tert-butyl nitrone (PBN) and other spin-trapping agents – a novel series of nonlipid free radical scavengers. J Mol Neurosci 3: 47–57.

Cobb, J.P., and Danner, R.L., 1996, Nitric oxide and septic shock. JAMA 275: 1192–6.

Coghlan, J.G., Flitter, W.D., Holley, A.E., Norell, M., Mitchell, A.G., Ilsley, C.D., and Slater, T.F., 1991, Detection of free radicals and cholesterol hydroperoxides in blood taken from the coronary sinus of man during percutaneous transluminal coronary angioplasty. Free Rad Res Comms 14: 409–17.

Coghlan, J.G., Flitter, W.D., Clutton, S.M., Ilsley, C.D.J., Rees, A., and Slater, T.F., 1993, Lipid peroxidation and changes in vitamin-e levels during coronary artery bypass grafting. J Thor Cardiovasc Surg 106: 268–74.

Comoglio, A., Tomasi, A., Malandrino, S., Poli, G., and Albano, E., 1995, Scavenging effect of silipide, a new silybin-phospholipid complex, on ethanol-derived free radicals. Biochem Pharmacol 50: 1313–16.

Curtis, M.J., and Pabla, R., 1997, Nitric oxide supplementation or synthesis block: which is the better approach to treatment of heart disease. Trends Pharmacol Sci 18: 239–44.

Dage, J.L., Ackermann, B.L., Barbuch, R.J., Bernotas, R.C., Ohlweiler, D.F., Haegele, K.D., and Thomas, C.E., 1997, Evidence for a novel pentyl radical adduct of the cyclic nitrone spin-trap MDL 101,002. Free Radic Biol Med 22: 807–12.

Darley Usmar, V.M., McAndrew, J., Patel, R., Moellering, D., Lincoln, T.M., Jo, H., Cornwell, T., Digerness, S., and White, C.R., 1997, Nitric oxide, free radicals and cell signalling in cardiovascular disease. Biochem Soc Trans 25: 925–9.

De Santis, G., and Pinelli, M., 1994, Microsurgical model of ischemia reperfusion in rat muscle: evidence of free radical formation by spin-trapping. Microsurgery 15: 655–9.

Edamatsu, R., Mori, A., and Packer, L., 1995, The spin-trap N-tert-alpha-phenyl-butylnitrone prolongs the life span of the senescence accelerated mouse. Biochem Biophys Res Commun 211: 847–9.

Fredriksson, A., and Archer, T., 1996, Alpha-phenyl-tert-butyl-nitrone (PEN) reverses age-related maze learning performance and motor activity deficits in c57 BL/6 mice. Behav Pharmacol 7: 245–53.

Frejaville, C., Karoui, H., Tuccio, B., leMoigne, F., Culcasi, M., Pietri, S., Lauricella, R., and Tordo, P., 1994, 5-diethoxyphosphoryl-5-methyl-1-pyrroline n-oxide (DEPMPO): a new phosphorylated nitrone for efficient in vitro and in vivo spin-trapping of oxygen-centred radicals. J. ChemSoc, Chem Commun 1793–4.

Frejaville, C., Karoui, H., Tuccio, B., Le Moigne, F., Culcasi, M., Pietri, S., Lauricella, R. and Tordo, P., 1995, 5-(Diethoxyphosphoryl)-5-methyl-1-pyrroline N-oxide: a new efficient phosphorylated nitrone for the in vitro and in vivo spin-trapping of oxygen-centered radicals. J Med Chem 38: 258–65.

Fujii, H., Koscielniak, J., and Berliner, L.J., 1997, Determination and characterisation of nitric oxide generation in mice by in-vivo EPR spectroscopy. Magn Reson Med 38: 565–84.

Gallez, B., Mader, K., and Swartz, H.M., 1996, Noninvasive measurement of the pH inside the gut by using pH-sensitive nitroxides. An in vivo EPR study. Magn Reson Med 36: 694–7.

Guarini, S., Bazzani, C., Ricigliano, G.M., Bini, A., Tomasi, A., and Bertolini, A., 1996, Influence of ACTH-(1–24) on free radical levels in the blood of haemorrhage-shocked rats: direct ex vivo detection by electron spin resonance spectrometry. Br J Pharmacol 119: 29–34.

Guarini, S., Bini, A., Bazzani, C., Ricigliano, G.M., Cainazzo, M.M., Tomasi, A., and Bertolini, A., 1997, Adrenocorticotropin normalises the blood levels of nitric oxide in hemorrhage-shocked rats. Eur J Pharmacol 336: 15–21.

Guiberteau, T. 1997, In vivo detection of nitric oxide by L-band EPR spectroscopy. In Proc. International Society Magnetic Resonance in Medicine 5th scientific meeting, Vancouver, Canada, April 12–18, p. 2126.

Halpern H.J., Yu, C., Barth, E., Peric, M., and Rosen, G.M., 1995, In situ detection, by spin-trapping, of hydroxyl radical markers produced from ionizing radiation in the tumor of a living mouse. Proc Natl Acad Sci USA 92: 796–800.

Hartell, M.G., Borzone, G., Clanton, T.L., and Berliner, L.J., 1994, Detection of free radicals in blood by electron spin resonance in a model of respiratory failure in the rat. Free Rad Biol Med 17: 467–72.

Haseloff, R.F., Zollner, S., Kirilyuk, I.A., Grigor'ev, I.A., Reszka, R., Bernhardt, R., Mertsch, K., Roloff, B., and Blasig, I.E., 1997, Superoxide-mediated reduction of the nitroxide group can prevent detection of nitric oxide by nitronyl nitroxides. Free Rad Res 26: 7–17.

Henrich, W.L., 1991, The Endothelium a Key Regulator of Vascular Tone. Am J Med Sci 302: 319–28.

Hogg, N., Singh, R.J., Joseph, J., Neese, F., and Kalyanaraman, B., 1995, Reactions of nitric oxide with nitronyl nitroxides and oxygen: prediction of nitrite and nitrate formation by kinetic simulation. Free Rad Res 22: 47–56.

Iadecola, C., 1997, Bright and dark sides of nitric oxide in ischemic brain injury. Trends Neurosci 20: 132–9.

Iimuro, Y., Bradford, B.U., Gao, W., Kadiiska, M., Mason, R.P., Stefanovic, B., Brenner, D.A., and Thurman, R.G., 1996, Detection of alpha-hydroxyethyl free radical adducts in the pancreas after chronic exposure to alcohol in the rat. Mol Pharmacol 50: 656–61.

Inanami, O., and Kuwabara, M., 1995, alpha-phenyl n-tert-butyl nitrone (PBN) increases the cortical cerebral blood flow by inhibiting the breakdown of nitric oxide in anesthetised rats. Free Rad Res 23: 33–9.

Janzen, E.G., 1980, A critical review of spin-trapping in biological systems. In Free Radicals in Biology, V.A. Pryor, ed. (New York: Elsevier), 115–53.

Jiang, J., Liu, K.J., Shi, X., and Swartz, H.M., 1995, Detection of short-lived free radicals by low-frequency electron paramagnetic resonance spin-trapping in whole living animals. Arch Biochem Biophys 319: 570–3.

Joseph, J., Kalyanaraman, B., and Hyde, J.S., 1993, Trapping of nitric oxide by nitronyl nitroxides: an electron spin resonance investigation. Biochem Biophys Res Commun 192: 926–34.

Kader, A., Frazzini, V.I., Solomon, R.A., and Triffiletti, R.R., 1993, Nitric oxide production during focal cerebral ischemia in rats. Stroke 24: 1709–16.

Kadiiska, M.B., Xiang, Q.H., and Mason, R.P., 1994, In vivo free radical generation by chromium(VI): an electron spin resonance spin-trapping investigation. Chem Res Toxicol 7: 800–5.

Kadiiska, M.B., Burkitt, M.J., Xiang, Q.H., and Mason, R.P., 1995, Iron supplementation generates hydroxyl radical in vivo – an ESR spin-trapping investigation. J Clin Invest 96: 1653–7.

Kadkhodaee, M., Hanson, G.R., Towner, R.A., and Endre, Z.H., 1996, Detection of hydroxyl and carbon-centred radicals by EPR spectroscopy after ischaemia and reperfusion of the rat kidney. Free Rad Res 25: 31–42.

Kerwin, J.F.J., and Heller, M., 1994, The arginine-nitric oxide pathway: A target for new drugs,. Med Res Rev 14: 23–74.

Knecht, K.T., Bradford, B.U., Mason, R.P., and Thurman, R.G., 1990, In vivo formation of a free radical metabolite of ethanol. Mol Pharmacol 38: 26–30.

Knecht, K.T., Adachi, Y., Bradford, B.U., Iimuro, Y., Kadiiska, M., Xuang, Q.H., and Thurman, R.G., 1995, Free radical adducts in the bile of rats treated chronically with intragastric alcohol: inhibition by destruction of Kupffer cells. Mol Pharmacol 47: 1028–34.

Komarov, A., Mattson, D., Jones, M.M., Singh, P.K., and Lai, C.S., 1993, In vivo spin-trapping of nitric oxide in mice. Biochem Biophys Res Commun 195: 1191–8.

Komarov, A.M., and Lai, C.S., 1995, Detection of nitric oxide production in mice by spin-trapping electron paramagnetic resonance spectroscopy. Biochim Biophys Acta 1272: 29–36.

Konorev, E.A., Tarpey, M.M., Joseph, J., Baker, J.E., and Kalyanaraman, B., 1995a, Nitronyl nitroxides as probes to study the mechanism of vasodilatory action of nitrovasodilators, nitrone spin-traps, and nitroxides: role of nitric oxide. Free Rad Biol Med 18: 169–77.

Konorev, E.A., Tarpey, M.M., Joseph, J., Baker, J.E., and Kalyanaraman, B., 1995b, S-nitrosoglutathione improves functional recovery in the isolated rat heart after cardioplegic ischemic arrest-evidence for a cardioprotective effect of nitric oxide. J Pharmacol Exp Ther 274: 200–6.

Koshland, D.E., 1993, Molecule of the year. Science 262: 1953.

Kozlov, A.V., Biagini, G., Tomasi, A., and Zini, I., 1995, Ex vivo demonstration of nitric oxide in the rat brain: effects of intrastriatal endothelin-1 injection. Neurosci Lett 196: 140–4.

Kozlov, A., Bini, A., Iannone, A., Zini, I., and Tomasi, A., 1996, Electron paramagnetic resonance characterisation of rat neuronal NO production *ex vivo*. In Nitric Oxide (Part A) Sources and Detection of NO; NO Synthase., L. Packer, ed. (San Diego: Academic Press), 229–36.

Kröncke, K.-D., Fehsel, K., and V.K.-B., 1997, Nitric oxide: cytotoxicity versus cytoprotection – how, why, when, and where? Nitric Oxide 1: 107–20.

Kumura, E., Kosaka, H., Shiga, T., Yoshimine, T., and Hayakawa, T., 1994, Elevation of plasma nitric oxide end products during focal cerebral ischemia and reperfusion in the rat. J Cereb Blood Flow Metab 14: 487–91.

Kuppusamy, P., Ohnishi, S.T., Numagami, Y., Ohnishi, T., and Zweier, J.L., 1995, Three-dimensional imaging of nitric oxide production in the rat brain subjected to ischemia-hypoxia. J Cereb Blood Flow Metab 15: 899–903.

Kuppusamy, P., Wang, P., Samouilov, A., and Zweier, J.L., 1996, Spatial mapping of nitric oxide generation in the ischemic heart using electron paramagnetic resonance imaging. Magn Reson Med 36: 212–8.

Kuppusamy, P., Wang, P.H., Zweier, J.L., Krishna, M.C., Mitchell, J.B., Ma, L., Trimble, C.E., and Hsia, C.J.C., 1996, Electron paramagnetic resonance imaging of rat heart with nitroxide and polynitroxyl-albumin. Biochemistry 35: 7051–7.

Lai, C.S. and Komarov, A.M., 1994, Spin-trapping of nitric oxide produced in vivo in septic-shock mice. FEBS Lett 345: 120–4.

Lancelot, E., Revaud, M.I., Boulu, R.G., Plotkine, M., and Callebert, J., 1997, Alpha-phenyl-N-tert-butylnitrone attenuates excitotoxicity in rat striatum by preventing hydroxyl radical accumulation. Free Rad Biol Med 23: 1031–4.

Laurindo, F.R., Pedro, M.dA., Barbeiro, H.V., Pileggi, F., Carvalho, M.H., Augusto, O., and da Luz P.L., 1994, Vascular free radical release. Ex vivo and in vivo evidence for a flow-dependent endothelial mechanism. Circ Res 74: 700–9.

Liu, K.J., Jiang, J.J., Ji, L.L., Shi, X., and Swartz, H.M., 1996, An HPLC and EPR investigation on the stability of DMPO and DMPO spin adducts in vivo. Res Chem Intermed 22: 499–509.

Lurie, D.J., 1996, Commentary: electron spin resonance imaging studies of biological systems. Br J Radiol 69: 983–4.

Mader, K., Bacic, G., and Swartz, H.M., 1995, In vivo detection of anthralin-derived free radicals in the skin of hairless mice by low-frequency electron paramagnetic resonance spectroscopy. J Invest Dermatol 104: 514–7.

Mason, R.P., and Knecht, K.T., 1994, In vivo detection of radical adducts by electron spin resonance. In Oxygen Radicals in Biological Systems, Pt C, L. Packer, ed. (San Diego: Academic Press), 112–17.

McCormick, M.L., Buettner, G.R., and Britigan, B.E., 1995, The spin-trap alpha-(4-pyridyl-1-oxide)-n-tert-butylnitrone stimulates peroxidase-mediated oxidation of deferoxamine – implications for pharmacological use of spin-trapping agents. J Biol Chem 270: 29265–9.

Medline, 1997, Medline search results.

Mergner, G.W., Weglicki, W.B., and Kramer, J.H., 1991, Postischemic free radical production in the venous blood of the regionally ischemic swine heart – effect of deferoxamine. Circulation 84: 2079–90.

Miyajima, T., and Kotake, Y., 1995, Spin-trapping agent, phenyl N-tert-butyl nitrone, inhibits induction of nitric oxide synthase in endotoxin-induced shock in mice. Biochem Biophys Res Commun 215: 114–21.

Mottley, C., and Mason, R.P., 1989, Nitroxide radical adducts in biology: chemistry, applications, and pitfalls. In Biological Magnetic Resonance, L.J. Berliner, ed. (New York: Plenum Press), 489–546.

Muijsers, R.B.M., Folkerts, G., Henricks, P.A., Sadeghihashjin, G., and Nijkamp, F.P., 1997, Peroxynitrite: a two faced metabolite of nitric oxide. Life Sciences 60: 1833–45.

Murphy, M.E., and Noack, E., 1994, Nitric oxide assay using hemoglobin method. In Oxygen Radicals in Biological Systems, Pt C, L. Packer, ed. (San Diego: Academic Press), 240–50.

Nonami, Y., 1997, The role of nitric oxide in cardiac ischemia reperfusion injury. Japa Circul J (English Edition) 61: 119–32.

Novelli, G.P., Angiolini, R., Tani, G., Consales, L., and Bordi, 1985, Phenyl-t-butyl-nitrone is active against traumatic shock in rats. Free Rad Res Comms 1: 321–7.

Oda, H., Kusumoto, S., and Nakajima, T., 1975, Nitrosyl-hemoglobin formation in the blood of animals exposed to nitric oxide. Arch Environ Health 30: 453–6.

Piccinini, F., Bradamante, S., Monti, E., Zhang, Y.K., and Janzen, E.G., 1995, Pharmacological action of a new spin-trapping compound, 2-phenyl DMPO, in the adriamycin-induced cardiotoxicity. Free Rad Res 23: 81–7.

Pieper, G.M., and Lai, C.S., 1996, Evaluation of vascular actions of the nitric oxide-trapping agent, N-methyl-D-glucamine dithiocarbamate-Fe^{2+}, on basal and agonist-stimulated nitric oxide activity. Biochem Biophys Res Commun 219: 584–90.

Quaresima, V., Takehara, H., Tsushima, K., Ferrari, M., and Utsumi, H., 1996, *In vivo* detection of mouse liver nitric oxide generation by spin-trapping electron paramagnetic resonance spectroscopy. Biochem Biophys Res Commun 221: 729–34.

Quine, R.W., Rinard, G.A., Ghim, B.T., Eaton, S.A., and Eaton, G.R., 1996, 1–2 GHz pulsed and continuous wave electron paramagnetic resonance spectrometer. Rev Sci Instrum 67: 2514–27.

Reinke, L.A., Moore, D.R., and Kotake, Y., 1996, Hepatic nitric oxide formation: spin-trapping detection in biliary efflux. Anal Biochem 243: 8–14.

Reszka, K.J., Bilski, P., Chignell, C.F., and Dillon, J., 1996, Free radical reactions photosensitised by the human lens component, kynurenine: an EPR and spin-trapping investigation. Free Rad Biol Med 20: 23–34.

Sato, S., Tominaga, T., Ohnishi, T., and Ohnishi, S.T., 1994, Electron paramagnetic resonance study on nitric oxide production during brain focal ischemia and reperfusion in the rat. Brain Res 647: 91–6.

Schulz, J.B., Henshaw, D.R., Matthews, R.T., and Beal, M.F., 1995, Coenzyme Q10 and nicotinamide and a free radical spin-trap protect against MPTP neurotoxicity. Exp Neurol 132: 279–83.

Sentjurc, M., and Mason, R.P., 1992, Inhibition of radical adduct reduction and reoxidation of the corresponding hydroxylamines in in vivo spin-trapping of carbon tetrachloride-derived radicals. Free Rad Biol Med 13: 151–60.

Serteyn, D., Pincemail, J., Mottart, E., Caudron, I., Deby, C., Deby Dupont, G., Philippart, C., and Lamy, M., 1994, Direct approach for demonstrating free radical phenomena during equine postanesthetic myopathy: preliminary study. Can J Vet Res 58: 309–12.

Singh, D., Nazhat, N.B., Fairburn, K., Sahinoglu, T., Blake, D.R., and Jones, P., 1995, Electron spin resonance spectroscopic demonstration of the generation of reactive oxygen species by diseased human synovial tissue following ex vivo hypoxia-reoxygenation. Ann Rheum Dis 54: 94–9.

Slater, T.F., 1966, Necrogenic action of carbon tetrachloride in the rat: a speculative mechanism based on activation. Nature 209: 36–40.

Slater, T.F., 1978, Biochemical Mechanism of Liver Injury. (New York: Academic Press).

Squadrito, F., Altavilla, D., Canale, P., Ioculano, M., Campo, G.M., Ammendolia, L., Ferlito, M., Zingarelli, B., Squadrito, G., Saitta, A., *et al.*, 1994, Participation of tumour necrosis factor and nitric oxide in the mediation of vascular dysfunction in splanchnic artery occlusion shock. Br J Pharmacol 113: 1153–8.

Stoyanovsky, D.A., Goldman, R., Jonnalagadda, S.S., Day, B.W., Claycamp, H.G., and Kagan, V.E., 1996, Detection and characterisation of the electron paramagnetic resonance-silent glutathionyl-5,5-dimethyl-1-pyrroline n-oxide adduct derived from redox cycling of phenoxyl radicals in model systems and HL-60 cells. Arch BIochem Piophys 330: 3–11.

Symons, M.C.R., 1978, Chemical and Biochemical Aspects of Electron Spin Resonance Spectroscopy (New York: Nostrand Reinhold Company).

Tang, X.L., McCay, P.B., Sun, J.Z., Hartley, C.J., Schleman, M., and Bolli, R., 1995, Inhibitory effect of a hydrophilic alpha-tocopherol analogue, MDL 74,405, on generation of free radicals in stunned myocardium in dogs. Free Rad Res 22: 293–302.

Thiemermann, C., 1997, Nitric oxide and septic shock. Gen Pharmacol 29: 159–66.

Thomas, C.E., Carney, J.M., Bernotas, R.C., Hay, D.A., and Carr, A.A., 1994, *In vitro* and *in vivo* activity of a novel series of radical trapping agents in model systems of CNS oxidative damage. Ann NY Acad Sci 738: 243–9.

Thomas, C.E., Huber, E.W., and Ohlweiler, D.F., 1997, Hydroxyl and peroxyl radical trapping by the monoamine oxidase-B inhibitors deprenyl and MDL 72,974A: implications for protection of biological substrates. Free Radic Biol Med 22: 733–7.

Tomasi, A., and Iannone, A., 1993, ESR spin-trapping artefacts in biological model systems. In Biological Magnetic Resonance. EMR of Paramagnetic Molecules, L. J. B. a. J. Reuben, ed. (New York: Plenum Press), 353–84.

Tominaga, T., Sato, S., Ohnishi, T., and Ohnishi, S.T., 1994, Electron paramagnetic resonance (EPR) detection of nitric oxide produced during forebrain ischemia of the rat. J Cereb Blood Flow Metab 14: 715–22.

Tuccio, B., Zeghdaoui, A., Finet, J.P., Cerri, V., and Tordo, P., 1996, Use of new b-phosphorylated nitrones for the spin-trapping of free radicals. Res Chem Intermed 22: 393–404.

Utsumi, H., Takeshita, K., Ichikawa, K., Matsumoto, K., Chung, Y.S., Han, J.Y., Yamada, K., and Kawai, S., 1996, *In vivo* ESR measurements of free radical reactions in living mice. J Toxicol Sci 21: 293–5.

Wang, P., and Zweier, J.L., 1996, Measurement of nitric oxide and peroxynitrite generation in the postischemic heart. Evidence for peroxynitrite-mediated reperfusion injury. J Biol Chem 271: 29223–30.

Westenberger, U., Thanner, S., Ruf, H.H., Gersonde, K., Sutter, G., and Trentz, O., 1990, Formation of free radicals and nitric oxide derivative of hemoglobin in rats during shock syndrome. Free Rad Res Comms 11: 167–78.

White, C.R., Darleyusmar, V. and Oparil, S., 1997, Gender and cardiovascular disease: recent insights. Trends Cardiovasc Med 7: 94–100.

Woldman, Y.Y., Khramtsov, V.V., Grigorev, I.A., Kiriljuk, I.A., and Utepbergenov, D.I., 1994, Spin-trapping of nitric oxide by nitronylnitroxides: measurement of the activity of NO synthase from rat cerebellum. Biochem Biophys Res Commun 202: 195–203.

Yoshimura, T., Yokoyama, H., Fujii, S., Takayama, F., Oikawa, K., and Kamada, H., 1996, In vivo EPR detection and imaging of endogenous nitric oxide in lipopolysaccharide treated mice. Nature Biotechnology 14: 992–4.

Zini, I., Kozlov, A.V., Biagini, G., Tomasi, A., Fuxe, K., and Agnati, L.F., 1994, Et-1-induced brain ischemia and nitric oxide levels: possible implications for post-ischemia depressive syndrome. Wenner Gren Symposium 17–20 April 1994.

Zou, A.P., and Cowley, A.W., Jr, 1997, Nitric oxide in renal cortex and medulla. An in vivo microdialysis study. Hypertension 29: 194–8.

Zweier, J.L., Wang, P., and Kuppusamy, P. 1995, Direct measurement of nitric oxide generation in the ischemic heart using electron paramagnetic resonance spectroscopy. J Biol Chem 270: 304–7.

Zweier, J.L., Wang, P., Samouilov, A., and Kuppusamy, P., 1995, Enzyme independent formation of nitrix oxide in biological tissues. Nature Medicine 1: 804–8.

Part II

Nitroxides

6 The metabolism of nitroxides in cells and tissues

Harold M. Swartz and Graham S. Timmins

6.1 Introduction and overview

The aim of this chapter is to provide an overview of the interactions of nitroxides with cells and tissues, with a focus on how the study of these reactions may be used as research tools and also as potential diagnostic and therapeutic agents using some relevant examples. Nitroxides were developed initially to allow the use of electron paramagnetic resonance spectroscopy (EPR, also termed electron spin resonance, ESR) in a wide range of systems, by providing stable free radicals whose EPR spectra reflect their immediate environment, a technique known as spin-labeling. The use of sterically hindered nitroxides proved extremely successful for both this original purpose and also for a wide range of other uses. Initially, the fact that nitroxides were not completely stable in biological systems was seen as a limitation, but interesting and potentially very productive uses of nitroxides developed from studying their metabolism in functioning biological systems. A recent book reviews the biological uses of nitroxides in detail, to which the reader is referred for additional details and a comprehensive bibliography (Kocherginsky and Swartz, 1995).

The study of reactions of nitroxides has proven valuable because they respond to variables that are of great biological interest but can be difficult to measure by other means (e.g. redox metabolism, concentration of oxygen, and presence of oxidizing agents) and because of the sensitivity and specificity of EPR spectroscopy for free radicals such as nitroxides (assuming that other factors which could affect the EPR signal could be controlled, as is usually the case). Another advantage of EPR spectroscopy is the relative transparency of tissues and dense cell suspensions to the microwave frequencies used *cf.* optical techniques. The recent development of low-frequency *in vivo* EPR techniques has yet further expanded the potential applications of this approach, including the possibility of direct clinical use (Swartz and Walczak, 1996).

The occurrence of reactions of nitroxides in cell systems was noted soon after the successful use of nitroxides as spin labels to measure motion in model membranes led to the adoption of this approach in the membranes of viable cellular systems. Many investigators (Calvin *et al.*, 1969; Raison *et al.*, 1971; Schara *et al.*, 1977) noted that in cellular systems there was a gradual diminution of the EPR signal intensity, and presumed this was due to chemical reactions of the nitroxides, and it was indeed shown that this loss of signal was due to reversible reduction of nitroxide to the hydroxylamine (Giotta and Wang, 1972). This was demonstrated by the use of mild oxidizing agents such as potassium ferricyanide and perdeuterated Tempone (PDT), which can oxidize hydroxylamines to nitroxides but are not capable of oxidizing more reduced nitroxide derivatives such as amines. When applied to such cell systems, these oxidants rapidly restored most or all of the initial intensity of the EPR signal (Eriksson *et al.*, 1987; Kaplan *et al.*, 1973; Melhorn and Packer, 1982; Rosantsev, 1970). Even in the most complex systems studied, the total amount of nitroxide plus hydroxylamine (as defined by the increase in EPR signal of the nitroxide after the use of mild oxidizing agents) has been found to be greater than 90% of the total amount of nitroxide administered. A key question is whether the remaining few percent represent experimental difficulties in the assay system, or the production of small quantities of other metabolites. This has been investigated extensively in cell cultures (Chen and Swartz, 1989) and to a lesser extent *in vivo* (Couet *et al.*, 1984; Eriksson *et al.*, 1987a, 1987b) and no evidence of any metabolites other than hydroxylamines was found, although there are reports that water soluble nitroxides were only partially reoxidized after their reduction by cells indicating that other reactions are also possible (Belkin *et al.*, 1987).

In studies *in vivo* and perhaps also in some cellular studies, an additional consideration is whether some of the non-nitroxide functionalities of such compounds are metabolically transformed by drug detoxification mechanisms. If such transformations occur without affecting the nitroxide or hydroxylamine groups, they might go undetected by assays which depend simply on total intensity of the EPR spectra, although such transformations might be detected by their effect on hyperfine coupling(s). To our knowledge there are only a few reports of metabolic conversions of nitroxides to derivatives that still have a paramagnetic nitroxide group (Dodd *et al.*, 1976; Gutierez *et al.*, 1985; Schimmack and Summer, 1978) although wide-ranging chemical conversion of nitroxides has been used to synthesize a range of derivatives (Aurich, 1989). Such metabolic conversion could be of use in effecting specific compartmentalization, e.g. intracellular delivery by the use esters hydrolyzed by intracellular esterases (Hu *et al.*, 1989).

In addition to being able to reduce nitroxides to hydroxylamines, cell systems can also oxidize hydroxylamines to nitroxides (Chen and Swartz, 1988; Nettleton *et al.*, 1989). In some cases the rate of hydroxylamine oxidation may be sufficiently rapid to interfere with measurements of reduction of nitroxides. The situation becomes especially complex *in vivo*, where hydroxylamines and/or nitroxides may be moving between different tissue compartments, and eliminated via the renal and biliary systems and, in the case of volatile nitroxides, via the lungs. It has been reported that over a period of time, with continuous administration (Ishida *et al.*, 1989) or in some cases with repeated administrations of the nitroxides (Lukiewicz *et al.*, 1985), a pseudo steady state can be established in which the net rate of reduction of nitroxides and hydroxylamine oxidation, plus the rate of transport into and out of the circulatory system can result in a steady state blood level, or even a steady state whole body level of nitroxide.

In summary, the possible reactions of nitroxides in biological systems are summarized in Figure 6.1, with the available evidence indicating that under most circumstances, in most cellular systems, the predominant reaction of nitroxides is reduction to the hydroxylamine. There is no definitive evidence that oxidation of the nitroxides or nitroxide-radical addition occurs as a result of 'normal' cellular metabolism, although this may be an important reaction in systems where nitroxides protect against oxidative stress (see Mitchell *et al.*, Chapter 7). There is no substantiated

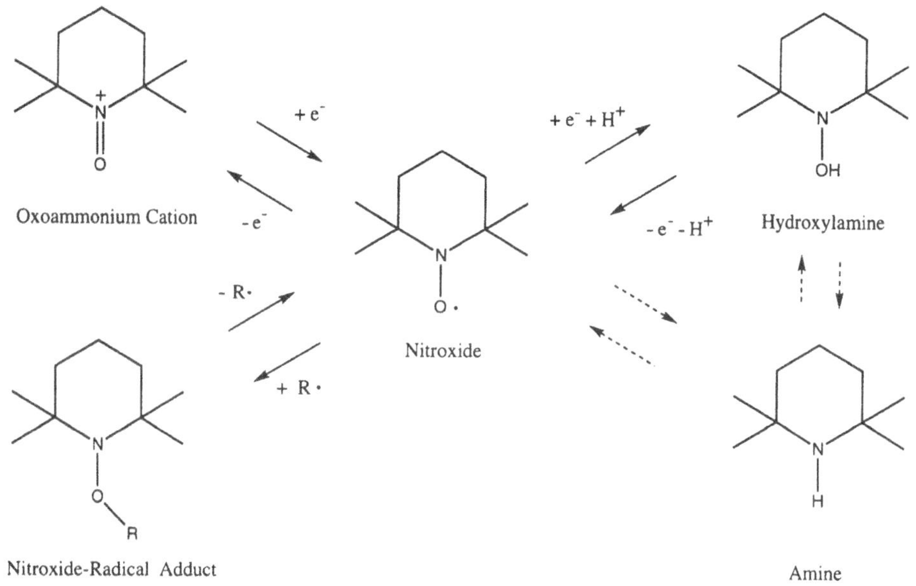

Figure 6.1 Possible reactions of nitroxides in biological systems; dotted arrows signify reactions that are not thought to usually occur to any extent in these systems

evidence that cells can reduce nitroxides beyond the hydroxylamine to the amine, and unless *explicitly* noted otherwise in the balance of this chapter, loss of nitroxide is considered to be due to reduction to the hydroxylamine.

6.2 Reduction of nitroxides in mammalian cells

The reduction of nitroxides has been demonstrated in all cellular systems in which it has been investigated, to a greater or lesser extent unless specific precautions have been taken (see later). Several factors, that have been shown to affect the rate of reduction of nitroxides, are considered in this section.

The experimental assessment of the extent of reduction that a nitroxide has undergone in cells and tissues usually starts with a determination of how much nitroxide remains and how much hydroxylamine has been produced using an assay based on EPR measurement of the amount of nitroxide present before and after treatment with a mild oxidizing agent. The usual precautions such as use of 1st derivative signal intensities and double integration must be observed (Poole, 1981). A commonly used oxidizing agent is potassium ferricyanide, $K_3Fe(CN)_6$ which rapidly and completely oxidizes (essentially all relevant) hydroxylamines to their corresponding nitroxides. The use of ferricyanide has the additional advantage that it usually does not cross cell membranes (Kaplan *et al.*, 1973; Keith and Snipes, 1974) and it can therefore be used to probe the location of hydroxylamines by using both intact and freeze-thawed cells (Swartz *et al.*, 1986). If the total amount of nitroxide plus ferricyanide-oxidized hydroxylamine in the system is known, it then is possible to estimate the residual amount of hydroxylamine that is not accessible to ferricyanide. Such studies may be complemented by the use of a membrane-permeable oxidizing agent such as

^{15}N-PDT, which will oxidize hydroxylamines regardless of membrane compartmentalization effects (Chen *et al.*, 1989).

In experiments in systems with more than one compartment it can be helpful to use high concentrations (e.g. 60 mM) of charged paramagnetic ions which do not penetrate cellular membranes easily (such as ferricyanide, $NiCl_2$ or potassium trioxalatochromiate) in order to broaden the spectra of nitroxides present in the same compartment as the ion. This approach is based on the fact that the paramagnetic ions broaden the EPR spectral absorption lines due to dipolar and exchange interactions between the nitroxide and the ions (Keith *et al.*, 1977). Some practical details should be considered in the use of broadening agents: if a distinct separation of redox and paramagnetic line-broadening effects is required $NiCl_2$ is a suitably redox-inactive agent; the maintenance of medium osmolality should be considered if high concentrations of paramagnetic species are used; and the potential for biological effects of the broadening agent considered.

6.2.1 *How the chemical and physical properties of nitroxides affect their rate of reduction by cells*

In view of the large number of both nitroxides and cellular systems that have been used, it is not surprising that there is significant variation in reports of the absolute and even relative rates of reduction of nitroxides of different structures. Some generalizations on the effects of structure on rates of reduction of nitroxides in cells, however, seem to be clear. There are three principal factors that affect relative rates of reduction of nitroxides: the nature of the ring in which the nitroxide group is located, the lipid solubility of the nitroxide, and the charge of the nitroxide.

The type of ring is the most important chemical factor for the relative resistance to bioreduction: pyrrolidines $\cong$ pyrrolines $\gg$ piperidines $\cong$ oxazolidines (Belkin *et al.*, 1987; Couet *et al.*, 1985a; Keana *et al.*, 1987; Morris *et al.*, 1991; Swartz *et al.*, 1986). The rates of reduction of nitroxides by cells generally parallel those observed in model systems, e.g. reduction by ascorbate in solutions (Belkin *et al.*, 1987; Couet *et al.*, 1985b; Morris *et al.*, 1991) or liposome suspensions (Schara *et al.*, 1990; Sentjurc *et al.*, 1990) unless modified by factors which affect either their permeation into cells or cellular metabolic activity.

The rates of reduction of lipophilic and hydrophilic nitroxides differ primarily because they localize in different parts of the cell; lipophilic nitroxides being reduced primarily in membrane compartments, and hydrophilic nitroxides reduced in aqueous compartments. The most extensively used lipophilic nitroxides are analogs of stearic acid with a doxyl group at different positions along the chain of the fatty acid (Marsh, 1981; Seelig, 1976) and their rates of reduction can vary. In anoxic cell systems the rate of reduction of doxyl stearates with the doxyl ring near the surface of the membrane is greater than for those with the ring located deeper in the membrane (Baldassare *et al.*, 1974; Chen *et al.*, 1988; Sentjurc *et al.*, 1986). In aerated samples, the net rate of reduction is much slower and is independent of the position of the doxyl group on acyl chain. The reaction order of this reduction also changes with the location of the nitroxide; nitroxides near the surface of the membrane have first order kinetics while those deeper in the membrane exhibit zero order kinetics with respect to nitroxide concentration. Nitroxides such as 16-doxyl stearate, which appears to reside both near the surface of membranes and in the hydrocarbon core, can exhibit a combination of zero and first order kinetics.

In suspensions of intact cells, highly charged hydrophilic nitroxides do not appear to be rapidly reduced, but once cellular membranes are disrupted, their rate of reduction seems to be similar to other nitroxides of the same ring structure (Sentjurc *et al.*, 1986; Swartz *et al.*, 1986), indicating that the reduction of nitroxides by cells is primarily intracellular. Nitroxides that are partially-ionized at physiological pH may exhibit kinetics of reduction in which the rate of entrance into the cells is the rate limiting step, and therefore the rate of reduction may be used to study this process (Swartz *et al.*, 1986).

The other major effect of charge is determined by its sign: in systems where ascorbate may play a significant role in reduction (erythrocytes, hepatocytes, kidney), nitroxides with negative charges are more resistant to reduction, probably due to electrostatic repulsion effects with this anionic reductant (Belkin *et al.*, 1987; Craescu *et al.*, 1982; Morris *et al.*, 1987; Sentjurc, 1990).

6.2.2 *Effects of oxygen on the rate of reduction of nitroxides by cells*

The discovery that the rate of reduction of some nitroxides in cells is dependent on the oxygen concentration has raised a number of interesting possibilities for both controlling undesired losses of the nitroxides and for exploiting this phenomenon as a way of measuring oxygen concentrations and related redox metabolism (Chen *et al.*, 1989; Lai *et al.*, 1986; Pals and Swartz, 1987; Swartz, 1986, 1987; Swartz *et al.*, 1986). For some nitroxides the presence of oxygen can decrease the rate of reduction greater than 10-fold, as shown with 5-doxyl stearate, with only low concentrations of oxygen needed to slow down the rate of reduction in susceptible nitroxides; the calculated K_m for this reaction is less than 1 μM (Chen *et al.*, 1989). This decrease in rate of reduction of nitroxides by oxygen has been observed not only for lipophilic nitroxides as mentioned above, but also for hydrophilic nitroxides such as Tempone (Pals and Swartz, 1987; Swartz *et al.*, 1986). The type of ring does not seem important for the effects of oxygen, as similar effects were found with doxyl, pyrrolidine, and piperidine rings. The effect of oxygen on the rate of reduction is larger for 5-doxyl stearate compared to 10-doxyl stearate, and this suggests that the site of reduction for lipophilic nitroxides may be close to the surface of the membrane and is consistent with the different type of kinetics observed for the reduction of the various doxyl stearates (Chen *et al.*, 1988).

The effect of oxygen could occur either from direct interactions with nitroxides and/or hydroxylamines, or via effects on the redox state of the electron transport system or other redox systems in the cell. This was investigated by varying the redox state of the cell while keeping oxygen concentration constant (118 μM) and, in a complimentary experiment, varying the concentration of oxygen while keeping the redox state constant (Chen *et al.*, 1989). The results indicated that the important variable is the redox state of the electron transport chain (also discussed later in section 6.2.4 of this chapter). It has also been reported that oxidation of the hydroxylamines subsequent to the reduction of the nitroxides was partially inhibited by superoxide dismutase, indicating that superoxide radicals may be involved in this process (Belkin *et al.*, 1987).

6.2.3 *Effect of cell metabolism and physiology on the rate of reduction of nitroxides*

Although the rate of reduction of nitroxides has been measured in many different cellular systems, there are few studies that directly compare the rates of reduction in different types of cells under strictly comparable conditions (Kaplan *et al.*, 1973). Judging from studies with tissue homogenates, cells with greater numbers of mitochondria such as liver, heart, and kidney appear to be more active in reducing nitroxides, while the rates of reduction of doxyl stearates in erythrocytes and human platelets are quite slow (Lai *et al.*, 1986b; Sentjurc *et al.*, 1994). The principal sites of reduction may be different in some cell lines: in most cell lines the main site of reduction is the mitochondria (in accord with the importance of the electron transport chain detailed above in 6.2.2). In hepatocytes, however, it has been suggested that microsomal reducing systems also are important (Chen *et al.*, 1988; Iannone *et al.*, 1990; Keana *et al.*, 1987). Reduction by ascorbate should also should be taken into account for cells such as hepatocytes and kidney cells which are rich in this compound (Apte *et al.*, 1988; Keana *et al.*, 1987).

There have been several reports of physiological variables that can affect the rate of reduction of nitroxides, with substances such as hormones, drugs and viruses (Bobell *et al.*, 1993; Schara *et al.*, 1990; Sentjurc *et al.*, 1983) having been noted to influence the rate of reduction of nitroxides by cells. There appear to be no systematic rules about the effect of such changes, but it is evident that the rate of reduction can depend on the physiological state of cells, and that different chemicals influence it. It is not completely clear what the causes of these changes might be, and it would be desirable to investigate this further. One possibility is that these substances directly influence the activity of enzymes which are responsible for reduction, for example the redox enzymes system of mitochondria, or can indirectly influence parameters such as the oxygen consumption of cells and thereby affect important variables such as the concentration of oxygen in cell suspensions.

The effect of the metabolic state of cells on reduction has been less extensively investigated, although it seems intuitive that a cellular reduction process should be linked to and affected by the metabolic state of the cell, as has been shown for different stages of sporulation and germination of bacterial cells (Stewart *et al.*, 1980). As more details are learned about the mechanism of reduction of nitroxides by cells, it should become more apparent as to what kinds of metabolic changes are likely to affect the rate of reduction. In preliminary experiments we have noted moderate changes in the rate of reduction of nitroxides due to pH and the nature of the energy source of the cells. Inhibitors of the tricarboxylic cycle and electron transfer chain also decrease the rate of reduction of nitroxides (Sentjurc *et al.*, 1986). The effects of active cell division are not clear, however. It was noted (Gutierrez *et al.*, 1985) that L-1210 leukemia cells in log phase growth reduced nitroxides more rapidly than those cells in stationary phase. This contrasts with the report that stimulation of division of lymphocytes by phytohemagglutinin decreased the rate of reduction of nitroxides (Hedrick *et al.*, 1982).

Cell permeability is another important factor in the rate of reduction; nitroxides which do not readily enter cells are not reduced very rapidly. While the failure of a nitroxide to enter a cell usually can be explained on the basis of the nitroxide being charged, or located on a very large molecule or other structure such as a liposome, these factors may not account for all circumstances. Nitroxides that otherwise would not be expected to permeate the lipid bilayer of a cell membrane may do so if they undergo carrier-mediated transport, if the membrane loses its ability to exclude some types of molecules, or if they are located in or bound to structures (such as some liposomes) which can fuse with cell membranes or enter into the cell by endocytosis (Chan *et al.*, 1988; Sentjurc *et al.*, 1991). In order to use nitroxides as drugs or probes in actively metabolizing systems it is often desirable to modify their rates of reduction and/or their distribution, and lessons learned from the study of nitroxide metabolism can be used. Within simple cellular systems physical modifications of nitroxides can change the rates of metabolism by several mechanisms, altering the uptake of the nitroxides into the cells (Chan *et al.*, 1988) or the availability of the nitroxides to the reducing mechanisms. The latter can occur by changing the intracellular distribution of the nitroxide either by affecting the location in the membrane (such as by attaching a doxyl stearate to a phospholipid), or by providing a barrier to the free access of the nitroxide to the reducing site (e.g. by physically blocking access due to location within a liposome or on a macromolecule such as albumin) (Chan *et al.*, 1988, 1990; Sentjurc *et al.*, 1991).

6.2.4 Mechanisms of enzymatic and non-enzymatic reduction of nitroxides in cells

Studies of the kinetics of reduction of nitroxides in cells have indicated that these can be consistent with enzymatic mechanism (Goldberg *et al.*, 1977; Sentjurc *et al.*, 1983) and can be described by the usual expression for enzymatic reactions:

$$dC/dt = -V_{max}C/(K_m - C)$$

where C is the concentration of nitroxide, V_{max} is the maximum velocity of the reaction and K_m is the Michaelis–Menten constant. When K_m is much greater than C the reaction can be approximated as a first order reaction and when K_m is much less than C a zero order reaction is a good approximation for the reduction of nitroxides by cells. Reduction of many nitroxides follows first order kinetics in regard to nitroxide concentrations.

An interesting exception to the usual first order kinetics is the cellular reduction of some doxyl stearates, where in some cells this depends upon the position of the doxyl group (Chen *et al.*, 1988). When the doxyl group is on the fifth carbon of the stearate alkyl chain and presumably more accessible to the surface of the membrane, its reduction follows first order kinetics but nitroxides with the doxyl group located at the tenth or twelfth carbon atom in the chain have zero order kinetics. Similar zero order kinetics have been observed with nitroxide analogs of phorbol esters which also are located deep in membranes (Schara *et al.*, 1983, 1985). These data are interpreted as indicating that the reducing equivalents are available in the aqueous phase at the surface of the membrane, and because the rate of reduction of 10- and 12-doxyl stearates appears to be independent of their concentration, the rate limiting step is the transmembrane diffusion of the reducing equivalents to the nitroxide center, or vice versa.

A number of different enzyme systems are capable of reducing nitroxides, such as microsomal enzymes, mitochondrial enzymes, glutathione reductase (in the presence of NADPH and oxidized glutathione), dihydrolipoamide dehydrogenase (requires NADH and lipoamide), DT-diaphorase (requires NAD(P)H), and thioredoxin reductase (Chen *et al.*, 1988; Gascoyne *et al.*, 1987; Rauckman *et al.*, 1984; Schallreuter and Wood, 1988, 1989; Sentjurc *et al.*, 1986). However, the critical question is which of these are important in the observed rates of reduction of a given nitroxide in a particular cell type.

Some insights as to the nature of the important enzymes have been obtained by the use of various inhibitors (Chen *et al.*, 1988, 1989) with these data being interpreted as indicating that the reduction of doxyl stearates is enzymatic and occurs primarily at the level of ubiquinone in mitochondria in intact cells. Agents which affected the activity of key microsomal enzymes did not have large effects on the rates of reduction in intact cells. The effects of oxygen on the rate of reduction of nitroxides in intact cells and on different enzyme systems are also compatible with the principal site of reduction being in the mitochondrial electron transport chain and not consistent with reduction in microsomes being a dominant process (Chen *et al.*, 1988). In hepatocytes however, which contain a larger proportion of such systems, microsomal enzymes may have a more significant role in the reduction of nitroxides (Iannone *et al.*, 1990; Keana *et al.*, 1987). Inhibitors of sulfhydryl groups have been shown to decrease the rate of reduction of nitroxides by cells but it is not clear whether this is a direct effect on thiol-containing enzymes or a more indirect effect due to the multiple changes that occur when a cell is treated with a non-specific thiol-group inhibitor (Baldassare *et al.*, 1974; Chen and McLaughlin, 1985; Giotta and Wang, 1972; Sentjurc *et al.*, 1986). Freeze-thawing of cells, which presumably decreased barriers for the inhibitors to reach the reduction site(s) did not change the general relationships observed in intact cells, and nor was there any general trend for the effects of inhibitors on the effects of oxygen on reduction.

Recent studies indicate that there also are non-enzymatic mechanisms of reduction of nitroxides in mammalian cells. This is not surprising in view of the presence of potential reducing substances within the cells. The most likely nitroxide-reducing substances within the cell are ascorbic acid, free thiols, NAD(P)H and vitamin E. Ascorbic acid appears to be the principal non-enzymatic reducing agent in many cells (Keana *et al.*, 1987; Rauckman *et al.*, 1984) but the levels of ascorbate in some cells are relatively low and probably insufficient to account for all of the non-enzymatic reduction that can be observed in most cells. In systems which are rich in ascorbate such as muscle extracts (Perkins *et al.*, 1980), and kidney and liver homogenates (Eriksson *et al.*, 1987b; Keana *et al.*, 1987), the rates of non-enzymatic reduction did correlate with ascorbate levels. Glutathione, the most abundant low molecular weight thiol, is not a good reductant of nitroxides by itself (Craescu *et al.*, 1982) but, in the presence of metal ions such as ferric iron, it can be quite effective (Rauckman *et al.*, 1984). NADPH (Keana *et al.*, 1987) and tocopherol (Craescu *et al.*, 1982) also have been reported to be ineffective in reducing nitroxides: however, there are recent reports that lipoprotein bound tocopherol can be an effective nitroxide reducing agent in a complicated reaction involving loss of tocopherol (Witting *et al.*, 1997).

Overall, while the data on the site(s) of reduction in cells sometimes appear to be disparate and even contradictory, the bulk of the evidence indicates that for most cells the principal site of reduction is the mitochondrion. Under some circumstances reduction by microsomes and ascorbate also can be significant. The role of thiols is unclear and perhaps most or all of the effects that have been observed can best be accounted for on the basis of indirect effects on the sites where the reduction usually takes place. It is important to recognize that a number of other factors, such as modifiers of the electron transport chain and the presence of highly reactive species such as metabolic free radicals, also can affect the rates and where these can occur they need to be taken into account. The effects of oxygen on rates of reduction and on oxidation of the hydroxylamines are further important complicating factors.

6.3 Cellular oxidation of hydroxylamines

The oxidation of hydroxylamines to nitroxides has been demonstrated in a number of different types of cells and presumably can occur in many actively functioning mammalian cells (Chen and Swartz, 1988; Kveder *et al.*, 1991; Rosen *et al.*, 1982; Sentjurc *et al.*, 1991; Swartz *et al.*, 1986) and also has been observed in some bacteria (Biniukov *et al.*, 1989). This oxidation appears to be an enzymatic process inasmuch as it can be inhibited by heating cells to 60°C for five minutes. The rates of oxidation of some lipophilic hydroxylamines are comparable to the rates of reduction of the corresponding nitroxides, while hydroxylamines formed by reduction of aqueous soluble nitroxides are not oxidized by cells, except for slight oxidation of some pyrrolidine derivatives due to autoxidation. Since oxidation occurs much more rapidly for lipophilic nitroxides than for aqueous soluble nitroxides, this suggests that the site of oxidation may be within cellular membranes. The kinetics of the oxidation of the hydroxylamines of different doxyl stearates located at various depths within the membrane are all first-order, in contrast to the kinetics of their reduction, which also is consistent with the oxidation site being located within the membranes. It is inhibited by cyanide but not with antimycin A or SKF-525A, indicating that it might occur in mitochondria at cytochrome c oxidase (Chen and Swartz, 1988; Rosen *et al.*, 1982). The nature of the relationship between oxygen concentration and the rate of oxidation of hydroxylamines indicates that oxygen is an active participant in the reaction (Chen *et al.*, 1989); this may make this reaction a very useful tool for measuring the concentration of oxygen in complex biological systems. The relationship differs completely from that for reduction, indicating that the mechanism for oxidation of hydroxylamines probably is not simply the reverse of the reaction for reduction of nitroxides.

6.4 The toxicity of nitroxides

This recently has been reviewed in detail (Kocherginsky and Swartz, 1995) and hence is not covered extensively here. It would generally appear that nitroxides are minimally or even non-toxic under most plausible circumstances, especially with regard to shorter term effects. The data on long term effects are limited but generally do not indicate that this is likely to be a significant problem. In cellular systems in which relatively large amounts of lipophilic nitroxides are present, there can be significant effects on the membranes due to locally high concentrations of the nitroxides which cause physical perturbations. Otherwise there appear to be few detectable effects of concentrations of nitroxides up to several millimolar. The evidence for mutagenic effects of nitroxides is limited and also suggests that this unlikely to be a very significant effect. Although nitroxides are free radical species, it is important to recognize the very different reactivities of these *stable* free radicals compared to reactive species such as the hydroxyl radical, and not assume that their toxicities will be similar.

6.5 Applications of the metabolism of nitroxides

The properties of nitroxides described in the previous sections suggest that there may be some potentially very productive applications for the study of physiology and pathophysiology. This has become especially attractive with the development of techniques that use EPR in functional biological systems, including *in vivo*, because of the specificity and sensitivity of EPR for unpaired electron species. The key to the productive use of the interactions of nitroxides is the ability to obtain 'better' data than can be obtained by other approaches. In this context 'better' can mean unique, or more accurate, or more readily obtained. The potentially useful areas that may be investigated by use of nitroxides include their physical interactions to reflect the environment (e.g. motion, pH, amount of oxygen) and their chemical interactions which usually reflect redox metabolism. The use of the physical interactions of nitroxides is a large and still growing area which is beyond the scope of this chapter, but it should be noted that successful use of those applications often depends on a good understanding of the chemical interactions which the nitroxides may undergo in functioning biological systems. In the balance of this chapter we summarize some uses of nitroxides which particularly reflect the chemical interactions. We will not cover uses of the nitroxides as tracers and pharmacological probes because these usually are not based on chemical interactions but it should be noted that these can be very powerful tools for the study of the theme of this book (Gallez *et al.*, 1996; Mader *et al.*, 1996a, 1996b, 1996c). These techniques can be useful because of the sensitivity of EPR spectroscopy, its specificity for paramagnetic species (with nitroxides an essentially clear background is typical of nearly all biological systems), and its ability to probe deep within heterogeneous biological systems.

There are several inter-related applications of the chemical interactions of nitroxides. These generally involve observing changes in the amount of the nitroxide at a site and relating these to factors which affect their rate of metabolism; especially redox metabolism and the concentration of oxygen. Both of these parameters are potentially important in a number of processes and also difficult to measure in viable systems. Some uses also are based on taking advantage of knowledge of the sites of metabolism of the nitroxides and therefore being able to follow permeation (e.g. the entrance of nitroxides into cells where they become reduced; for readily reduced nitroxides such as those based on the pyrimidine ring the rate of entry into the cell is the rate limiting step for the observable changes in concentration of the nitroxides).

6.5.1 Transmembrane transport in erythrocytes studied by metabolism of nitroxides

Based on the assumption that the transport step is the rate-limiting step (i.e. that reduction occurs quickly once the nitroxide enters the cell), the disappearance of nitroxides has been used to measure transport across the erythrocyte membrane (Bartosz and Gwodzinski, 1983; Gwodzinski *et al.*, 1981a, 1982; Ross and McConnell, 1975). Such studies indicate that anionic nitroxides are transported actively across erythrocyte membranes by at least two different mechanisms while cations are transported much less effectively. By measuring the rate of reduction in nitroxides after various treatments of erythrocytes, effects on the permeability of the membrane of treatments such as ionizing radiation have been studied (Bartosz and Gwodzinski, 1983; Bogdanova *et al.*, 1985a, 1985b; Eriksson *et al.*, 1986; Gwodzinski, 1985, 1986; Gwodzinski *et al.*, 1981a, 1981b, 1982). It is worth noting that the permeability through the membrane itself is not a rate-limiting step for most neutral nitroxides and those charged nitroxides which can penetrate through the membrane as uncharged, neutral species due to their pK being close to 7, such as the protonated form of nitroxides with a carboxylic group (pK ~5.0) or the unprotonated form of nitroxides with an NH_2 group (pK 9.4). For nitroxides with essentially entirely charged phosphate or trimethylammonium groups, transport is the rate-limiting step (Eriksson *et al.*, 1986). Nitroxides with negative charges can penetrate into the erythrocytes by an active transport, probably via band 3 protein, and this can be stopped by the corresponding inhibitors: for the cationic nitroxide Cat_1 no transport into erythrocytes was observed (Bartosz and Gwodzinski, 1983; Ross and McConnell, 1975).

6.5.2 *Cell physiology studied by changes in nitroxide metabolism*

The rate of reduction of nitroxides has been used as a probe for changes in cellular metabolism, especially to search for differences associated with malignant transformation. Initially studies of malignant transformation looked for changes of the line shapes of EPR spectra, which reflect the motion of nitroxides, anticipating that there would be characteristic changes in the fluidity of membranes of normal and transformed cell lines: the results, however, were disappointing. In some cell lines no detectable changes of membrane fluidity were observed, for some others there was a decrease in the membrane fluidity after transformation of the cells, but in the majority of cases there was an increase in membrane fluidity (Bales *et al.*, 1977; Barnett *et al.*, 1974; Chen and McLaughlin, 1985; Gaffney, 1975; Schara *et al.*, 1977; Sentjurc *et al.*, 1983, 1990).

It was observed, however, that cells reduced the nitroxides fairly readily, and consequently, changes in the rate of reduction of nitroxides have been investigated to probe changes in metabolism associated with malignant transformation. Chen and McLaughlin (1985) found increased rates of reduction of 5-doxyl stearate and Tempol in undifferentiated malignant mouse neuroblastoma cells compared to differentiated nontumorigenic cells, and concluded that this reflected a difference in the content and permeation of thiol-compounds from cytosol to membrane. An increased rate of reduction in malignant cells also has been observed and it was suggested that enhanced oxidoreductase activity may be a general property of malignant cells and that increased reduction of nitroxides by malignant cells was a reflection of this phenomenon (Gascoyne *et al.*, 1987). Cells transformed by adenoviruses have given somewhat different results however: studies of both the rate of reduction and the effects of oxygen on nitroxide metabolism in AD5 virus transformed hamster kidney cells demonstrated that transformed cell reduced Tempol slower than untreated cells (Bobell *et al.*, 1993). On the other hand, in Kirsten murine sarcoma virus transformed rat kidney cells, a dramatic increase in the rate of reduction at 37°C was found for doxyl palmitic acids while no changes were seen at 18°C. A similar effect was not seen with the methyl esters of the same nitroxides; perhaps because the methyl esters remained deeper in the membranes. Later the same laboratory reported that membranes of malignant tumors are more fluid than those of nonmalignant tumors and described a decreased rate of reduction of the methyl ester of 5-doxyl palmitate in malignant tumors of brain tissue slices (Sentjurc *et al.*, 1990). It therefore appears that a consistent effect of malignant transformation on nitroxide metabolism has not been demonstrated thus far.

The effects of both drugs and chemicals on cellular reduction of nitroxides have been investigated, especially in regard to their use in treatment of tumor cells. Zenser and coworkers noted that various hormones decreased the rate of reduction of nitroxides by thymocytes (Zenser *et al.*, 1976). Also examined have been the effects of antimetastatic drugs on transformed cells, by comparing the rate of reduction of nitroxides in L-1210 leukemic cells from both treated mice and control animals (Sentjurc *et al.*, 1983). They found that drug treatment decreased both the rate of reduction of the methyl ester of 5-doxyl palmitate and the fluidity of the plasma membrane, which then approached the fluidity of normal leukocytes. They also found both an increase in the reduction of Tempo after treatment and its increased partitioning into the membrane leading to the conclusion that the alteration in the rate of reduction is related to the increased concentration of Tempo in the membrane, where the reduction occurs.

The rate of reduction of Tempone, Tempo, and PCA in lymphocytes from normal and tumor bearing animals, using phytohemagglutinin (PHA) stimulation, and following the rates of reduction of nitroxides has been studied (Hedrick *et al.*, 1982). Lymphocytes from tumor bearing animals reduced Tempone and Tempol at significantly slower rates compared to those from control animals. These differences were seen only after stimulation by PHA, and all PHA stimulated lymphocytes had lower rates of reduction compared to unstimulated lymphocytes. It had previously been demonstrated that the reduction of PCA was approximately 30 times slower than for Tempone, and that it occurred inside the normal cells without PHA with a decay constant of 0.25 min^{-1}. Treatment with *N*-ethylmaleimide reduced the rate by a factor of four, while a 10°C decrease in temperature decreased the rate of reduction by a factor of 2.7 for Tempone, demonstrating that the reduction is an enzymatic process. The implication was that the intracellular metabolism of lymphocytes from a tumor animal is selectively altered by PHA (Hedrick *et al.*, 1979). Note that PHA increased the rate of reduction of nitroxides with ascorbate in the case of lymphocytes with spin-labeled cerebroside and decreased the rate for spin-labeled phosphatidylcholine (Curtain *et al.*, 1980), consistent with the postulated clustering of sphingolipids after the treatment with the ligand.

A correlation between the ability of cells to reduce nitroxides and their metabolic activity might lead to practical applications. For example, the rate of reduction of Tempo by intact human spermatozoa is sensitive to the quantity and quality of sperm (Bahl *et al.*, 1988). Cold shock of the sperm had no effect on the rate of reduction but inhibitors of the electron transport chain such as rotenone, antimycin A, KCN, and sodium azide and bubbling N_2 and O_2 changed the rate of reduction. The ability of spermatozoa to reduce nitroxides has been used as a parameter to follow and understand maturation, capacitation, and calcium uptake of sperm obtained from Holstman strain rats (Fernandes *et al.*, 1992).

The results cited indicate that various biologically active substances can change the rate of reduction significantly although their mechanisms are not completely clear.

These substances may directly influence the activity of enzymes which are responsible for reduction, for example the redox system of mitochondria, or may act more indirectly, e.g. by affecting cellular oxygen consumption cells and hence the oxygen concentration. Many other modes of action are possible, however, because the rate of reduction depends on the concentration of substrates, the activity of the sites of reduction as well as on the concentration of spin probe at the site of reduction, which is determined by the diffusion rate or distribution of the spin probe in membrane and solution, and by the permeability across the membranes.

While it still seems that the reduction of nitroxides could be a sensitive and potentially useful tool for investigating some of the biophysical phenomena associated with redox processes and antioxidant status in various processes, including transformation of cells, the varied results found so far indicate that it is too early to make any general conclusions, and that each system is best studied individually.

6.5.3 Cellular production of oxidizing species studied by their reaction with nitroxides

Although the study of reduction of nitroxides has received most attention in biological systems, the study of oxidation of nitroxides can also be useful. As shown in Figure 6.1 there are two possible initial reactions of the nitroxide moiety with oxidizing species typically found in biological systems: one-electron oxidation to the oxoammonium radical cation, or radical–radical reaction to form alkoxyamine adducts. The loss of nitroxide by such processes is readily differentiated from the more common reduction to hydroxylamine, as it cannot simply be reversed by addition of mild oxidant such as ferricyanide.

The initial oxoammonium cation product of oxidation of the stable di-*t*-alkyl nitroxides commonly used can either be relatively stable or undergo further rearrangement (Aurich, 1989). An interesting consequence of this reactivity is the ability of nitroxides to act as superoxide dismutases (discussed by Mitchell *et al.*, Chapter 7), which may be important in their functions as antioxidants. The addition of radicals to nitroxides is rapid in the case of less thermodynamically stable carbon centered species, although the reaction of nitroxides with peroxyl radicals is much slower. A related reaction that also has been used in studies of oxidants in biological systems is the oxidation of diamagnetic di-*t*-alkylhydroxylamines to nitroxides readily detected by EPR.

The oxidation of nitroxides has been used to investigate oxidative intermediates in a wide range of biological systems. Examples of systems studied include the species formed by macrophages as part of their mechanism of cell killing (see review by Melhorn and Packer, 1984, which also details earlier studies) and the production of hydroperoxide-derived radicals in murine skin (Timmins and Davies, 1993). It must be stated however, that unambiguous identification of the oxidant or radical involved in such reactions is not generally possible by EPR alone because of the lack of characteristic hyperfine couplings of the resulting adduct (*cf.* spin-trapping). These techniques are perhaps best used as a sensitive assay for general 'oxidative stress' in biological systems (analogous to such uses of chemiluminescence).

A number of new approaches to the detection of radicals and radical production have been developed based upon the use of derivatized nitroxides and trapping of the initial radical by its addition to the nitroxide to form a diamagnetic adduct.

While this means EPR cannot be used to characterize or detect such adducts directly, the general approach in these studies has been the use of bifunctional nitroxides also containing a fluorescent label in chemical and biological systems (either as the bifunctional nitroxide itself or using subsequent nitroxide derivatization). The paramagnetic nitroxide functional group in these compounds results in a marked intramolecular quenching of fluorescence, so that radical addition to the nitroxide center causes a great increase in observed fluorescence intensity (Batz *et al.*, 1997; Gerlock *et al.*, 1990; Kieber and Blough, 1990; Pou *et al.*, 1993, 1995). Another technique is the use of liquid chromatographic/mass spectrophotometric analysis of derivatized nitroxide adducts, which can also allow identification of the initial radical (Johnson *et al.*, 1996; Kieber *et al.*, 1992).

Such studies would appear to show great potential in complementing spin-trapping EPR studies, for instance by allowing fluorescence visualization of radical generation by confocal microscopy. Also important could be the elimination of some of the problems associated with spin-trapping on biological systems (e.g. see Davies and Timmins, 1996; Silvester *et al.*, 1997; Timmins *et al.*, 1997) because of the very different chemistries involved in both adduct formation and adduct decomposition/metabolism.

6.5.4 Measurements of the concentration of oxygen and/or state of redox metabolism

Although nitroxides have been very successfully used to study the concentration of oxygen in biological systems *in vitro* by studying the broadening of their EPR lines by spin exchange, these techniques cannot be applied to all the biological systems one might wish to study, as a calibration curve of line width *vs.* oxygen concentration must first be made (and also the nitroxide concentration at the time of measurement determined), and this may not be possible in the system of interest.

As stated previously, however, the rate of reduction of nitroxides can be very dependent upon the oxygen concentration in many biological systems, and this reduction can be readily measured. In addition, because the rate of reduction is measured and compared as a function of oxygen concentration, the initial concentration of nitroxide need not to be accurately defined (*cf.* spin-exchange measurements) making this approach more easily applied to complex *in vitro* and *in vivo* cases. Because the oxygen-dependent changes in the rate of reduction of nitroxides really occur only when the oxygen concentration is very low (the apparent K_m for oxygen is approx. 1 μM) this technique is most sensitively applied to study systems with low oxygen concentrations. This has been applied to cellular studies through observation of the more rapid rate of reduction of nitroxides such as 5-doxyl stearate in such areas of low oxygen concentration (Chen *et al.*, 1989).

The determination of low oxygen concentrations using this technique has considerable potential value because a range of important pathologies are associated with such low concentrations of oxygen, including tissues undergoing an ischemic episode, and in tumors. In tumors, it is well established that there can be considerable heterogeneity in pO_2, and that very hypoxic regions may be the determining factor for the response to therapy, because cells in hypoxic environments are resistant to radiation and some chemotherapeutics (Hockel *et al.*, 1996a, 1996b; Vaupel, 1996). While this has been known for a long time in regard to the effects of oxygen on the acute effects of radiation, it recently has been shown that the hypoxia itself also can induce changes in properties of cells (increasing the tumor's malignancy) which makes it even more important to have methods to detect low concentrations of oxygen (Hockel *et al.*, 1996a, 1996b; Vaupel, 1996). The effects of oxygen on the rate of reduction of 5-doxyl stearate have been used to obtain oxygen dependent EPR images, using the differential rate of metabolism of 5-doxylstearate by TB cells (Bacic *et al.*, 1988). The resulting changes in the concentration of the nitroxide gave image intensities that reflected the amount of reduction that had occurred. The development of this and related EPR approaches for the *in vivo* determination of tumor oxygenation shows great promise in tailored treatment of tumors (Swartz and Walczak, 1996).

6.5.5 Use of nitroxides to study drug delivery into skin, especially liposomal systems

The encapsulation of charged, readily bioreducible nitroxides within liposomes has provided an effective system for following liposomal transport and metabolism in biological systems. The technique is based upon the fact that these liposome-enclosed nitroxides will be effectively protected from common biological reductants, such as ascorbate, due to the inability of nitroxide and reductant to diffuse through the liposomal membrane (Chan *et al.*, 1989) until the integrity of the liposome is damaged, when the nitroxide is 'released' and can react with reductants. The events causing release of the nitroxide can be the damage of integrity of the liposomes induced by the cell or tissue causing extracellular nitroxide release; fusion of liposomes with the cell plasma membrane causing intracellular nitroxide release; or endocytosis of the liposome resulting in its release into the intracellular lysosomal system (Chan *et al.*, 1988).

An application of interest using this technique has been to simultaneously study the spatial and temporal distribution of nitroxide containing liposomes by EPR when they are applied to skin. This system is therefore an excellent model for studying the cutaneous and percutaneous delivery of both drugs and other substances, such as cosmetic preparations to the skin (Gabrijelcic *et al.*, 1990, 1991, 1994, 1995). Since the concentration of reductants in the stratum corneum is low and that in the keratinocytes beneath it is high (Fuchs *et al.*, 1997), the rate of reduction of nitroxides provides an excellent measure of the liposomal delivery of nitroxides through the stratum corneum.

Other uses of nitroxides in studies of skin permeation include a related series of studies by Fuchs and co-workers, concentrating upon the development and use of X-band EPR microscopy of nitroxides in skin, with a resolution better than 10 μM being reported (Fuchs *et al.*, 1992; Herrling *et al.*, 1996). This technique has been used to study the dermal penetration of spin-labeled dihydrolipoate and also its effects upon skin membrane polarity (Freisleben *et al.*, 1994). Studies of the oxygen permeability of the stratum corneum have also been conducted using nitroxide-labeled samples (Hatcher and Plachy, 1993).

Acknowledgment

This work was supported by NIH Grant P41 RR11602, a NIH supported resource center, and NIH Grant PO1 GM51630.

References

Apte, D.V., Sentjurc, M., MacAllister, L. and Swartz, H.M., Reduction of nitroxides in whole blood, erythrocytes and Plasma, *Biophys. J.*, **53**, 80a, 1988.

Aurich, H.G., Nitroxides, in *Nitroxides, Nitronates and Nitrones*, Patai, S. and Rappoport, Z., eds., John Wiley and Sons, Chichester, 1989, 313.

Bacic, G., Demsar, F., Zolnai, Z. and Swartz, H.M., Contrast enhancement in ESR imaging: role of oxygen, *Mag. Reson. Med. Biol.*, **1**, 55, 1988.

Bahl, S., Bawa, S.R., Srivastava, S., Phadke, R.S. and Govil, G., Magnetic resonance studies of intact human spermatozoa, *Physiol. Chem. Phys. Med.*, **20**, 183, 1988.

Baldassare, J.J., Robertson, D.E., McAfee A.G. and Ho, C., A spin-label study of energy coupled active transport in Escherichia coli membrane vesicles, *Biochemistry*, **13**, 5210, 1974.

Bales, B., Lesin, E. and Oppenheiner, S., On cell membrane lipid fluidity and plant lectin agglutinability – a spin label study of mouse ascites tumor cells, *Biochim. Biophys. Acta.*, **465**, 400, 1977.

Barnett, R.E., Furcht, L. and Scott, R., Differences in membrane fluidity and structure in contact-inhibited and transformed cells, *Proc. Natl. Acad. Sci. USA*, **71**, 1992, 1974.

Bartosz, B. and Gwodzinskin, K., Aging of the erythrocyte XXIII. Changes in the permeation of spin labeled electrolyte, *Am. J. Hematol.*, **14**, 377, 1983.

Batz, M., Korth, H.G. and Sustmann, R., A novel method for detecting nitric oxide (NO) by formation of fluorescent products based on cheletropic spin-traps, *Ange. Chemie.*, **36**, 1501, 1997.

Belkin, S., Melhorn, R.J., Hideg, K., Hanskovsky, O. and Packer, L., Reduction and destruction rates of nitroxide spin probes, *Arch. Biochem. Biophys.*, **256**, 232, 1987.

Biniukov, V.I., Kharatian, E.F., Ostrovsky, D.N. and Shaskov, A.S., Derivatives of glutamic acid and trehalose which can be transformed into long-living radicals by the loss of an electron are identified in some bacteria, *Biofactors*, **2**, 95, 1989.

Bobell, J.R., Luchter, L. and Morse, II, P.D., Viral transformation of baby hamster kidney cells affects the oxygen sensitivity of oxygen reduction, *Appl. Radiat. Isotopes*, **44**, 465, 1993.

Bogdanova, N.P., Cherednichenko, A.D., Avrutskaya, I.A. and Fioshin, M.Y., Electrochemical behaviour of the radical fragment of the piperidine nitroxide radicals, *Electrokhimya*, **21**, 1070, 1985.

Bogdanova, N.P., Petrova, L.G., Avrutskaya, I.A. and Fioshin, M.Y., Electrochemical reduction of 1-oxyl-2,2,6,6-tetramethyl-4-oximinopiperidine, *Electrokhimya*, **21**, 1369, 1985.

Calvin, M., Wang, H.H., Entine, G., Gill, D., Ferruti, P., Harpold, M.A. and Klein, M.P., Biradical spin labelling for nerve membranes, *Proc. Natl. Acad. Sci. USA*, **63**, 1, 1969.

Chan, H.C., Magin, R.L. and Swartz, H.M., Delivery of nitroxide spin labels to cells in culture by liposomes, *Mag. Reson. Med.*, **8**, 160, 1988.

Chan, H.C., Magin, R.L. and Swartz, H.M., Rapid assessment of liposomal stability in blood by an aqueous nitroxide spin label, *Biochem. Biophys. Methods*, **18**, 271, 1989.

Chan, H.C., Sun, K., Magin, R.L. and Swartz, H.M., Potential of albumin labelled with nitroxides as contrast agents for MRI and MRS, *Bioconjugate Chem.*, **1**, 32, 1990.

Chen, K., Glockner, J.F., Morse II, P.D. and Swartz, H.M., Effects of oxygen on the metabolism of nitroxide spin labels in cells, *Biochemistry*, **28**, 2496, 1989.

Chen, K. and McLaughlin, M.G., Differences in reduction kinetics of incorporated spin labels in undifferentiated and differentiated mouse neuroblastoma cells, *Biochem. Biophys. Acta.*, **845**, 189, 1985.

Chen, K., Morse II, P.D. and Swartz, H.M., Kinetics of enzyme-mediated reduction of lipid soluble nitroxide spin labels by living cells, *Biochim, Biophys. Acta.*, **943**, 477, 1988.

Chen, K. and Swartz, H.M., The products of reduction of doxyl stearates in cells are hydroxylamines as shown by oxidation by 15N-perdeuterated Tempone, *Biochim, Biophys. Acta.*, **992**, 131, 1989.

Chen, K. and Swartz, H.M., Oxidation of hydroxylamines to nitroxide spin labels in living cells, *Biochim, Biophys. Acta.*, **970**, 270, 1988.

Couet, W.R., Brasch, R.C., Sosnovsky, G., Lukszo, J., Prakash, I., Gnewuch, C.T and Tozer, T.N., Influence of chemical structure of nitroxyl spin labels on their reduction by ascorbic acid, *Tetrahedron*, **41**, 1165, 1985.

Couet, W.R., Brasch, R.C., Sosnovsky, G. and Tozer, T.N., Factors affecting nitroxide reduction in ascorbate solution and tissue homogenates, *Mag. Reson. Imaging*, **3**, 83, 1985.

Couet, W.R., Eriksson, U., Tozer, T.N., Tuck, L.D., Wesbey, G.E., Nitecki, D. and Brasch, R.C., Pharmacokinetics and metabolic fate of two nitroxides potentially useful as contrast agents of magnetic resonance imaging, *Pharm. Res.*, 203, 1984.

Craescu, C.T., Baracu, I., Grecu, N., Busca, L. and Niculscu-Duvaz, I., On the reduction of free radicals by ascorbic acid in solution and erythrocte suspension, *Rev. Roum. Biochim.*, **19**, 15, 1982.

Curtain, C.C., Looney, F.D. and Smelsorius, J.A., Lipid domain formation and ligand induced lymphocyte membrane changes, *Biochim. Biophys. Acta.*, **596**, 42, 1980.

Davies, M.J. and Timmins, G.S., EPR Spectroscopy of biologically-relevant free radicals, in Clark, R.J.H. and Hester, R.E., eds., *Biomedical Applications of Spectroscopy*, John Wiley and Sons, Chichester, 1996, 217.

Dodd, N.J.F., Harcus, R.G. and Preston, P., Synthesis and electron spin resonance study of spin-labelled compounds related to tumour growth inhibitory nitroarylaziridines, *Z. Naturforsch.*, **31c**, 328, 1976.

Eriksson, U.G., Tozer, T.N., Sosnovsky, G., Lukszo, J. and Brasch, R.C., Human erythrocyte membrane permeability and nitroxyl spin-label reduction, *J. Pharm. Sci.*, **75**, 334, 1986.

Eriksson, U.G., Ogan, M.D., Peng, C.T., Brasch, R.C. and Tozer, T.N., Metabolic fate in the dog of the nitroxyl moiety in a spin label with potential utility as a contrast agent in MRI, *Mag. Reson. Med.*, **5**, 73, 1987(a).

Eriksson, U.G., Brasch, R.C. and Tozer, T.N., Non-enzymatic bioreduction in rat liver and kidney of nitroxyl spin labels, potential contrast agents in MRI, *Drug Metab. Dispos.*, **15**, 155, 1987(b).

Fernandes, J., Hegde, U.C., Srivasta, S., Phadke, R.S. and Govil, G., Maturation of rat spermatozoa: ESR spin labelling studies, *Physiol., Chem. Phys. Med., NMR*, **24**, 63, 1992.

Freisleben, H.J., Groth, N., Fuchs, J., Rudolph, P., Zimmer, G. and Herrling, T., Penetration of spin-labelled dihydrolipoate into the skin of hairless mice-modification of epidermal and dermal polarity, *Arzneimittel Forschung*, **44**, 1047, 1994.

Fuchs, J., Groth, N., Herrling, T., Milbradt, R., Zimmer, G. and Packer, L., Electron paramagnetic resonance (EPR) imaging in skin-Biophysical and biochemical microscopy, *J. Inv. Dermatol.*, **98**, 713, 1992.

Fuchs, J., Groth, N., Herrling, T. and Zimmer, G., Electron paramagnetic resonance studies on nitroxide radical 2,2,5,5-tetramethyl-4-piperidin-1-oxyl (TEMPO) redox reactions in human skin, *Free Rad. Biol. Med.*, **22**, 967, 1997.

Gabrijelcic, V. and Sentjurc, M., Influence of hydrogels on liposome stability and on the transport of liposome-entrapped substances into the skin, *Int. J. Pharm.*, **118**, 207, 1995.

Gabrijelcic, V., Sentjurc, M. and Kristl, J., Evaluation of liposomes as drug carriers into the skin by one-dimensional EPR imaging, *Int. J. Pharm.*, **62**, 75, 1990.

Gabrijelcic, V., Sentjurc, M. and Schara, M., Liposome entrapped molecule penetration into the skin measured by nitroxide reduction kinetic imaging, *Period. Biol.*, **93**, 245, 1991.

Gabrijelcic, V., Sentjurc, M. and Schara, M., The measurement of liposome entrapped molecules penetration into the skin – an 1D-EPR and kinetic imaging study, *Int. J. Pharm.*, **102**, 151, 1994.

Gaffney, B.J., Fatty acid side chain flexibility in the membranes of normal and transformed fibroblasts, *Proc. Natl. Acad. Sci. USA*, **72**, 664, 1975.

Gallez, B., Mader, K. and Swartz, H.M., Noninvasive measurement of the pH inside the gut using pH-sensitive nitroxides. An In Vivo EPR study, *Magn. Reson. Med.*, **36**, 694, 1996.

Gascoyne, P., Pethig, R. and Szent-Gyorgi, A., Electron spin resonance studies of oxidoreductases with 2,6-dimethoxy-p-quinone and semiquinone, *Biochim. Biophys. Acta.*, **923**, 257, 1987.

Gerlock, J.L., Zacmanidis, P.J., Bauer, D.R., Simpson, D.J., Blough, N.V. and Salmeen, I.T., Fluorescence detection of free radicals by nitroxide scavenging, *Free Rad. Res. Commun.*, **10**, 119, 1990.

Giotta, G.J. and Wang, H.H., Reduction of nitroxide free radicals by biological materials, *Biochem Biophys. Res. Commun.*, **46**, 1576, 1972.

Goldberg, J.S., Rauckman, E.J. and Rosen, G.M., Bioreduction of nitroxides by *Staphylococcus aureus*, *Biochem. Biophys. Res. Commun.*, **79**, 198, 1977.

Gutierrez, P.L., Konienczy, M. and Sosnovsky, G., In the search for new anticancer drugs II antiumor activity, toxicity and electron spin resonance of spin-labeled thio-TEPA derivatives, *Z. Naturforsch*, **36b**, 1612, 1981.

Gutierrez, P.L., Cohen, B.E., Sosnovsky, G., Davis, T.A. and Egorin, M.J., On the search for new anticancer drugs XIV The plasma pharmacokinetics and tissue distribution of spin-labeled thio-TEPA (SL-O-TT), *Cancer Chemother. Pharmacol.*, **15**, 185, 1985.

Gwodzinski, K., Effect of gamma radiation on the transport of electrolyte spin labels across the fish erythrocyte membrane, *Stud. Biophys.*, **106**, 43, 1985.

Gwodzinski, K., Effect thiol reactive reagents and ionizing radiation on the permeability of erythrocyte membrane for non-electrolyte spin labels, *Radiat. Environ. Biophys.*, **25**, 107, 1986.

Gwodzinski, K., Bartosz, B. and Leyko, W., Effect of γ-irradiation on the transport of non-electrolyte spin-labels across the human erythrocyte membrane, *Stud. Biophys.*, **86**, 187, 1981.

Gwodzinski, K., Bartosz, B. and Leyko, W., Effect of γ-irradiation on the transport of spin-labeled compounds across the erythrocyte, *Radiat. Environ. Biophys.*, **19**, 275, 1981.

Gwodzinski, K., Bartosz, B. and Leyko, W., Effect of γ-irradiation on the transport of electrolyte spin-labels across the human erythrocyte membrane, *Stud. Biophys.*, **89**, 141, 1982.

Hatcher, M.E. and Plachy, W.Z., Dioxygen diffusion in the strtum corneum – an EPR spin-label study, *Biochim. Biophys. Acta.*, **1149**, 73, 1993.

Hedrick, W.R., Mathew, A., Zimbric, J.D. and Whaley, T.W., Intracelular viscosity of lymphocytes determined by an N-15 spin label probe, *J. Magn. Reson.*, **36**, 207, 1979.

Hedrick, W.R., Zimbrick, J.D. and Mathew, A., Phytohaemagglutinin-induced changes in the spin label reduction in lymphocytes from tumor-bearing rats, *Biochem. Biophys. Res. Commun.*, **109**, 180, 1982.

Herrling, T.E., Groth, N.K. and Fuchs, J., Biochemical EPR imaging of skin, *Appl. Mag. Reson.*, **11**, 471, 1996.

Hockel, M., Schlenger, K., Aral, B., Mitze, M., Schaffer, U. and Vaupel, P. Association between tumor hypoxia and malignant progression in advanced cancer of the uterine cervix, *Cancer Res.*, **6**, 4509, 1996(a).

Hockel, M., Schlenger, K., Mitze, M., Schaffer, U. and Vaupel, P., Hypoxia and radiation response in human tumors, *Sem. Radiat. Oncol.*, **6**, 3, 1996(b).

Hu, H., Sosnovsky, G., Li, S.W., Rao, N.U.M., Morse II, P.D. and Swartz, H.M., Development of nitroxides for selective localization inside cells, *Biochim. Biophys. Acta.*, **1014**, 211, 1989.

Iannone, A, Tomasi, A., Vannini, V. and Swartz, H.M., Metabolism of nitroxide spin lables in subcellular fractions of rat liver I. Reduction by microsomes, *Biochim. Biophys. Acta.*, **1034**, 285, 1990.

Ishida, S., Kumashiro, H., Tsuchihashi, N., Ogata, T., Ono, M., Kamada, H. and Yoshida, E., *In vivo* analysis of nitroxide radicals injected into small animals by L-band ESR technique, *Phys. Med. Biol.*, **349**, 1317, 1989.

Johnson, C.G., Caron, S. and Blough, N.V., Combined liquid chromatography mass spectrometry of the radical adducts of a fluorescamine-derivatised nitroxide, *Anal. Biochem.*, **68**, 867, 1996.

Kaplan, J., Canonico, P.G. and Caspary, W.J., Electron spin resonance studies of spin-labeled mammalian cells by detection of surface-membrane signals, *Proc. Natl. Acad. Sci. USA*, **70**, 66, 1973.

Keana, J.F.W., Pou, S. and Rosen, G.M., Nitroxides as potential contrast agents for MRI application: influence of structure on the rate of reduction by rat hepatocytes, whole liver homogenate, subcellular fractions and ascorbate, *Magn. Reson. Med.*, **5**, 525, 1987.

Keith, A.D. and Snipes, W., Viscosity of cellular protoplasm, *Science*, **183**, 666, 1974.

Keith, A.D., Snipes, W., Melhorn, R.J. and Gunter, T., Factors affecting the diffusion of water soluble spin labels, *Biophys. J.*, **19**, 205, 1977.

Kieber, D.J. and Blough, N.V., Detection of carbon-centered radicals in aqueous solution, *Free Rad. Res. Commun.*, **10**, 109, 1990.

Kieber, D.J., Johnson, C.G. and Blough, N.V., Mass-spectrometric identification of the radical adducts of a fluorescamine-derivatized nitroxide, *Free Rad. Res. Commun.*, **16**, 35, 1992.

Kocherginsky, N. and Swartz, H.M., *Nitroxide Spin Labels*, CRC Press, Boca Raton, 1995.

Kveder, M., Pecar, S., Nemec, M. and Schara, M., Nitroxide radicals as a model system for biotransformations of radicals in tissues, *Acta. Pharm. Jugosl*, **41**, 391, 1991.

Lai, C.S., Crossley, C., Sridhar, R., Misra, H.P., Janzen, E.G. and McCay, P.B., *In vivo* spin-trapping of free radicals generated in brain, spleen and liver during γ-iradiation, *Arch. Biochem. Biophys.*, **244**, 156, 1986(a).

Lai, C.S., Wirt, M.D., Yin, J.J., Frocisz, W., Feix, J.B., Kunicki, T.J. amd Hyde, J.S., Lateral diffusion of lipid probes in the surface membrane of human platelets – an electron-electron double resonance (ELDOR) study, *Biophys. J.*, **50**, 503, 1986(b).

Lukiewicz, S.J., Sochanik, A.S. and Lukiewicz, S.G., Monitoring of accumulation of NMR contrast agents in murine tumors using *in vivo* ESR spectroscopy, in *Proc. 4th Meet. Society of Magnetic Resonance in Medicine*, Barbican, London, 1985, 1240.

Mader, K., Gallez, B., Liu, K.J. and Swartz, H.M., Noninvasive in vivo characterization of release processes in biodegradable polymers by low frequency electron paramagnetic resonance spectroscopy, *Biomaterials*, **17**, 457, 1996(a).

Mader, K., Domb, A. and Swartz, H.M., Gamma sterilization induced radicals in biodegradable drug delivery systems, *Appl. Radiat. Isotopes*, **47**, 1669, 1996(b).

Mader, K., Gallez, B. and Swartz, H.M., In vivo EPR: An effective new tool for studying pathophysiology, physiology, and pharmacology, *Appl. Radiat. Isotop.*, **47**, 1663, 1996(c).

Marsh, D., Electron spin resonance: spin labels, in *Membrane Spectroscopy*, Grell, E., ed., Springer-Verlag, Berlin, 1981, 1.

Melhorn, R.J. and Packer, L., Nitroxide destruction on flavin-photosensitized damage in inner mitochondrial membranes, *Can. J. Chem.*, **56**, 1452, 1982.

Melhorn, R.J. and Packer, L., Electron spin resonance destruction methods for radical detection, *Methods. Enzymol.*, **105**, 215, 1984.

Morris, G., Sosnovsky, G., Hui, B., Huber, C.O., Rao, N.U.M. and Sawrtz, H.M., Chemical and electrochemical reduction rates of cyclic nitroxides (nitroxyls), *J. Pharm. Sci.*, **80**, 149, 1991.

Nettleton, D.O., Morse, II, P.D. and Swartz, H.M., Exchange and shuttling of electrons by nitroxide spin labels, *Arch. Biochem. Biophys.*, **271**, 414, 1989.

Pals, M.A. and Swartz, H.M., Oxygen dependant metabolism of potential magnetic resonance contrats agents, *Invest. Radiol.*, **22**, 497, 1987.

Perkins, R.C., Beth, A.H., Wilkerson, L.S., Serafin, W., Dalton, L.R., Park, C.R. and Park, J.H., Enhancement of free radical production by elevated concentrations of ascorbic acid in avian dystrophic muscle, *Proc. Natl. Acad. Sci. USA*, **77**, 190, 1980.

Poole, C.P., *Electron Spin Resonance*, 2nd. ed., John Wiley and Sons, New York, 1983.

Pou, S., Huang, Y.I., Bhan, A., Bhadti, V.S., Hosmane, R.S., Wu S.Y., Cao, G.L. and Rosen, G.M., A fluorophore-containing nitroxide as a probe to detect superoxide and hydroxyl radical generated by stimulated neutrophils, *Anal. Biochem.*, **212**, 85, 1993.

Pou, S., Bhan, A., Bhadti, V.S., Hosmane, R.S., Wu, S.Y. and Rosen, G.M., The use of fluorophore-containing spin-traps as potential probes to localize free-radicals in cells with fluorescence imaging methods, *FASEB J.*, **9**, 1085, 1995.

Raison, J.K., Lyons, J.M., Melhorn, R.J. and Keith, A.D., Temperature-induced phase changes in mitochondrial membranes detected by spin labelling, *J. Biol. Chem.*, **246**, 4036, 1971.

Rauckman, E.J., Rosen, G.M. and Griffeth, L.K., Enzymatic reactions of spin labels, in *Spin Labelling in Pharmacology*, Holtzmann, J.L., ed., Academic Press, New York, 1984, 175.

Rosantsev, E.G., *Free Nitroxyl Radicals*, Plenum Press, New York, 1970.

Rosen, G.M., Finkelstein, E. and Rauckman, E.J., A method for detection of superoxide in biological systems, *Arch. Biochem. Biophys.*, **215**, 367, 1982.

Ross, A.H. and McConnell, H.M., Permeation of a spin-label phosphate into the human erythrocyte, *Biochemistry*, **14**, 2973, 1975.

Schallreuter, K.U. and Wood, J.M., The activity and purification of membrane associated thioredoxin reductase from human metastatic melanotic melanoma, *Biochim. Biophys. Acta.*, **967**, 103, 1988.

Schallreuter, K.U. and Wood, J.M., Free radical reduction in the human epidermis, *Free Rad. Biol. Med.*, **6**, 519, 1989.

Schara, M., Sentjurc, M., Cotic, L., Pecar, S., Palcic, B. and Mont-Bragadin, C., Spin label study of normal and Ki-MSV transformed rat kidney cell membranes, *Stud. Biophys.*, **62**, 141, 1977.

Schara, M., Pecar, S., Sentjurc, M., Nemec, M., Hecker, E. and Sorg, B., Role of the phorbol moiety in binding of spin-labelled phorbol esters to the erythrocyte membrane, *Period. Biol.*, **85**, 116, 1983.

Schara, M., Sentjurc, M., Nemec, M. and Pecar, S., Binding of phobol ester to the membranes of bovine leukocytes, *Acta. Pharm. Jugosl.*, **35**, 81, 1985.

Schara, M., Pecar, S. and Svetek, J., Reactivity of hydrophobic nitroxides in lipid bilayers, *Colloids Surf.*, **45**, 303, 1990.

Schimmack, W. and Summer, K.H., Reaction of the nitroxyl radical TAN with glutathione, *Int. J. Radiat. Biol.*, **34**, 293, 1978.

Seelig, J., Anisotropic motion in liquid crystalline structures, in *Spin Labelling. Theory and Applications*, Berliner, L.J., ed., Academic Pree, New York, 1976, 373.

Sentjurc, M., Mechanism of nitroxide reduction to hydroxylamines, *Pharm. J. Jugosl. (Farm. Vestn.)*, **41**, 309, 1990.

Sentjurc, M., Schara, M., Nemec, M., Sava, G. and Giraldi, T., Influence of antimetastatic drugs on L-1210 cell membranes, *Pharm. J. Jugosl. (Farm. Vestn.)*, **34**, 229, 1983.

Sentjurc, M., Morse II, P.D. and Swartz, H.M., Influence of metabolic inhibitors on the nitroxide reduction in cells, *Period. Biol.*, **88**, 202, 1986.

Sentjurc, M., Bacic, G. and Swartz, H.M., Reduction of doxyl stearates by ascorbate in unilamellar liposomes, *Arch. Biochem. Biophys.*, **282**, 207, 1990.

Sentjurc, M., Gabrijelcic, V. and Likovic, M., Interaction of liposomes with leukocytes as studied by EPR, *Period. Biol.*, **93**, 327, 1991a.

Sentjurc, M., Pecar, S., Chen, K., Wu, M. and Swartz, H.M., Cellular metabolism of proxyl nitroxides and hydroxylamines, *Biochim. Biophys. Acta.*, **1073**, 329, 1991b.

Sentjurc, M., Apte, D.V., MacAllister, L. and Swartz, H.M., Reduction of nitroxides in whole blood, erythrocytes and plasma, *Curr. Top. Biophys.*, **18**, 81, 1994.

Silvester, J.A., Wei, X.D., Davies, M.J. and Timmins, G.S., A study of photochemically generated protein radical spin adducts on bovine sevum albumin: the detection of genuine spin-trapping and artefactual, non-radical addition in the same molecule, *Redox Rep.*, **3**, 225, 1997.

Stewart, G.S.A.B., Eaton, M.W., Johnstone, K., Barrett, M.D. and Ellar, D.J., An investigation of membrane fluidity changes during sporulation and germination of *Bacillus megaterium* K.M. measured by electron spin and nuclear magnetic resonance spectroscopy, *Biochim. Biophys. Acta.*, **600**, 270, 1980.

Swartz, H.M., Interactions between cells and nitroxides and their implications for their uses as biophysical probes and as metabolically responsive contrast agents for *in vivo* NMR, *Bull. Magn. Reson.*, 8, 172, 1986.

Swartz, H.M., Use of nitroxides to measure redox metabolism in cells and tissues, *J. Chem. Soc., Faraday Trans. 1*, **83**, 191, 1987.

Swartz, H.M. and Walczak, T., An overview of considerations and approaches for developing In Vivo EPR for clinical applications, *Res. Chem. Intermed.*, **22**, 511, 1996.

Swartz, H.M., Chen, K., Pals, M., Sentjurc, M. and Morse, II, P.D., Hypoxia sensitive NMR contrats agents, *Magn. Reson. Med.*, **3**, 169, 1986.

Swartz, H.M., Sentjurc, M. and Morse, II, P.D., Cellular metabolism of water soluble nitroxides: effect on rate of reduction of cell/nitroxide ratio, oxygen concentrations, and permeability of nitroxides, *Biochim. Biophys. Acta.*, **888**, 92, 1986.

Timmins, G.S. and Davies, M.J., Free radical formation in murine skin treated with tumour promoting organic peroxides, *Carcinogenesis*, **14**, 1499, 1993.

Timmins, G.S., Barlow, G.K., Silvester, J.A., Wei, X. and Whitwood, A.C., Use of isotopically labelled spin-traps to determine definitively the presence or absence of non-radical addition artefacts in EPR spin-trapping systems, *Redox Rep.*, **3**, 125, 1997.

Vaupel, P., Is there a critical tissue oxygen tension for bioenergetic status and cellular pH regulation in solid tumors? *Experientia*, **52**, 464, 1996.

Witting, P.K. and Stocker, R., Reduction of a nitroxide spin label by native and partially oxidized human low-density lipoprotein, *Mag. Reson. Chem.*, **35**, 100, 1997.

Zenser, T.V., Petrella, V.J. and Hughes, F., Spin labelled stearates as probes for microenvironments of murine thymocyte adenylate cylase-cyclic adenosine 3′,5′-monophosphate system, *J. Biol. Chem.*, **251**, 7341, 1976.

7 Clinical uses of nitroxides as superoxide-dismutase mimics

James B. Mitchell, Murali C. Krishna, Amram Samuni, Periannan Kuppusamy, Stephen M. Hahn, and Angelo Russo

7.1 Introduction

Cellular metabolism can generate free radicals and other oxygen derived oxidants. If left unchecked, 'oxidative stress' in biological systems may ultimately lead to cellular damage. During normal aerobic metabolism, oxidative stress can occur as a result of spillage of reducing equivalents from the mitochondrial respiratory chain which can result in incomplete reduction of oxygen (Chance *et al.*, 1979). These incomplete reductive products of oxygen are typically superoxide, hydrogen peroxide (H_2O_2), and hydroxyl radical as well as secondary carbon-centered radicals. In addition, redox active transition metals, by reacting with hydrogen peroxide can induce oxidative damage in biological systems. Therefore, for cells and ultimately organisms to avoid or limit damage in an aerobic environment, elaborate antioxidant systems are necessary. Normally, adequate antioxidant defense is provided by enzymatic as well as non-enzymatic systems. For example, accumulation of superoxide concentration is limited by the enzyme superoxide dismutase (SOD); whereas, hydrogen peroxide can be scavenged by catalase (CAT) or glutathione peroxidase. Organic peroxides are detoxified by the glutathione peroxidase/glutathione reductase system. In addition, free radicals such as hydroxyl radicals, alkyl, alkoxyl, and alkyl peroxyl radicals are scavenged by non-enzymic antioxidants in a stoichiometric manner. Therefore, enzymatic and non-enzymatic antioxidants provide the required defense. In spite of these layers of protective systems, some of the oxidants escape the detoxification barriers and have the potential to create lesions in vital cellular targets. If damage does occur, there is a need to repair or eliminate damaged molecules or macromolecular structures. Systems for repair of lesions on DNA and membrane and molecular structures effectively function to repair or eliminate damaged and non-functional biomolecules. Appropriate functioning of all protective systems is required for the survival of aerobic life.

Adequate antioxidant levels function to limit the cellular damage caused by oxidants and maintain homeostatic balance. However, conditions such as aging may compromise this balance, which can lead to an inadequate antioxidant status and ultimate tissue/cellular deterioration. Under acute conditions such as exposure to ionizing radiation or environmental pollutants that are capable of generating free radicals, supplementation by exogenous antioxidants is necessary, to limit oxidative damage to normal tissue. Therefore, antioxidant supplementation represents a valuable and inexpensive means of preventive therapeutic strategies that can be effectively utilized in numerous chronic and acute conditions. Since multiple chemical species are capable of mediating oxidative damage in many pathophysiological conditions, a desirable property of an antioxidant would be to participate in as many detoxification reactions as possible. Lipid peroxidation chain reactions are commonly involved in many of the pathophysiological conditions. Agents which can either directly scavenge the initiating species or scavenge the chain propagating species would represent effective antioxidants, given that the agent can partition into the lipid bilayer of a membrane.

While there are many classes of antioxidants being developed, tested, and utilized, the present study concentrates on a new class of agents called the nitroxide radicals, that can function both as *catalytic* and *stoichiometric* antioxidants. Stable nitroxide free radicals are commonly employed as tools to probe various biophysical and biochemical processes including paramagnetic contrast agents in MRI (Bennett *et al.*, 1987a, 1987b), probes of membrane structure (Berliner, 1979), and as sensors of oxygen in biological systems (Strzalka *et al.*, 1990). However, nitroxides also react with several biologically relevant free radicals by one-electron transfer reactions (Mehlhorn and Packer, 1984; Chateauneuf *et al.*, 1988; Nilsson *et al.*, 1989; Belkin *et al.*, 1987). The metabolism of stable nitroxides and the corresponding hydroxylamines has been studied in various cellular and subcellular fractions (Sentjurc *et al.*, 1989; Iannone *et al.*, 1989, 1990a, 1990b).

By undergoing one-electron transfer reactions, nitroxides can be either reduced to the corresponding hydroxylamines or oxidized to oxoammonium cations (Krishna *et al.*, 1992). The hydroxylamines on the other hand can be oxidized to the nitroxide, whereas the oxoammonium cation can be reduced to the nitroxide or the hydroxylamine. Consequently, all three forms can be present simultaneously. The cycling between the three states is schematically shown in Figure 7.1, top. However, the nitroxide and the hydroxylamine forms predominate in most cases. In this chapter, we will review studies that have evaluated the protective effects of stable nitroxides in mammalian cells, isolated organs, and whole animals subjected to various types of oxidative damage. Nitroxides have been shown to protect biological targets both *in vitro* and *in vivo*. The chemical mechanism(s) underlying these observations will be discussed.

7.2 Structural features and chemical properties of nitroxides

Nitroxides described in this chapter belong to several ring types such as (A) the five-membered oxazolidine ring, (B) five-membered saturated pyrrolidine ring, (C) five-membered unsaturated pyrroline ring, and (D) the six-membered piperidine ring (Figure 7.1, bottom). At least four oxidation states exist for a given ring structure namely, amine, hydroxylamine, nitroxide, and oxoammonium. While, in general, the amines and the hydroxylamines are chemically stable, nitroxide radicals within the ring structure are stabilized by the bulky substituents at their α-position. The observed antioxidant effects of the nitroxides can be explained in terms of the shuttling between three oxidation states as shown in Figure 7.1, top. The nitroxide ($RR'NO^{\bullet}$) acts as a reducing agent to provide detoxification by becoming oxidized by one electron to the corresponding oxoammonium cation ($RR'NO^{+}$).

$$RR'NO \text{ (oxidation)} \rightarrow RR'NO^{+} \qquad (7.1)$$

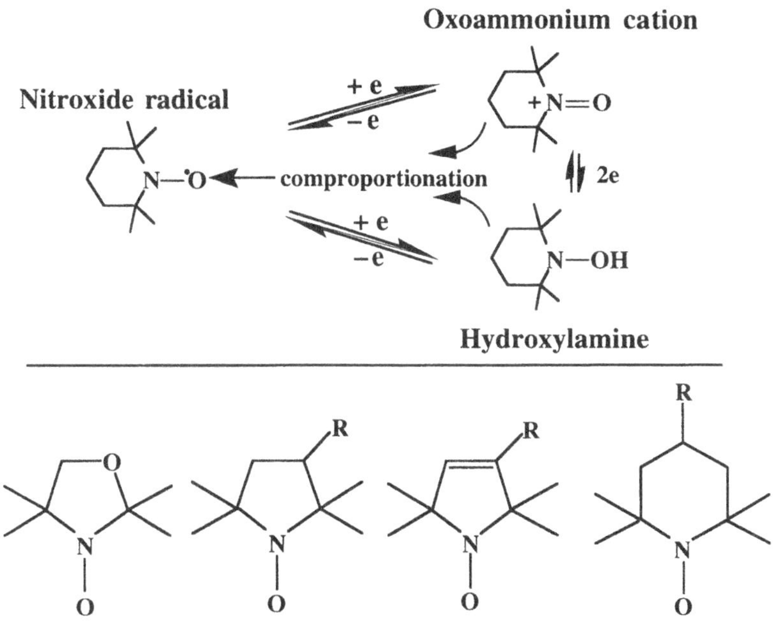

A B C D

Figure 7.1 Redox transitions of piperidine ring between the nitroxide, hydroxylamine, and oxoammonium cation (top). Four different classes of nitroxide derivatives: (A) the five-membered oxazolidine ring, (B) five-membered saturated pyrrolidine ring, (C) five-membered unsaturated pyrroline ring, and (D) the six-membered piperidine ring (bottom)

The oxoammonium cation can undergo reduction to either the nitroxide by a one electron reducing agent,

$$RR'NO^+ \text{ (one-electron reduction)} \rightarrow RR'NO^\bullet \tag{7.2}$$

or to the hydroxylamine by two electron reducing agents,

$$RR'NO^+ \text{ (two-electron reduction)} \rightarrow RR'NOH \tag{7.3}$$

or undergo comproportionation to give the nitroxides,

$$RR'NOH + RR'NO^+ \text{ (comproportionation)} \rightarrow 2RR'NO^\bullet \tag{7.4}$$

These redox reactions effectively provide detoxification, and more importantly, allow a dynamic equilibrium to be established for the corresponding nitroxide. The hydroxylamine is a weak reducing agent that can donate an H-atom and in the process is converted to the nitroxide.

$$RR'NOH \text{ (one-electron oxidation)} \rightarrow RR'NO^\bullet \tag{7.5}$$

Cyclic voltammetric techniques have shown that unlike the nitroxide/hydroxylamine couple, the nitroxide/oxoammonium redox couple is a reversible redox couple that mediates the catalytic dismutation of superoxide in a pH dependent manner (Krishna *et al.*, 1992; Krishna *et al.*, 1996a).

$$RR'NO + HO_2^{\bullet} + H^+ \rightarrow RR'NO^+ + H_2O_2 \quad (7.6)$$

$$RR'NO^+ + {}^{\bullet}O_2^- \rightarrow RR'NO + O_2 \quad (7.7)$$

Summing equations 6 and 7 yields:

$$2{}^{\bullet}O_2^- + 2H^+ \rightarrow H_2O_2 + O_2 \quad (7.8)$$

which is the dismutation reaction for superoxide catalyzed by the enzyme SOD. Since nitroxides also catalyze reaction 7.8 without themselves being consumed, nitroxides represent the first class of metal independent SOD mimics (Samuni *et al.*, 1988). In the SOD mimetic reactions, the univalent oxidation of the nitroxide, primarily by the protonated form of superoxide (equation 7.6) is a pH dependent reaction. The reduction of the oxoammonium cation by superoxide is diffusion-controlled, and the reaction rate constant has been estimated to be 1.2×10^{10} $M^{-1}s^{-1}$. For several piperidine nitroxides examined at pH values in the range of 5.5–8.0, the catalytic rate constants were found to be in the range of 10^5–1.2×10^8 $M^{-1}s^{-1}$; consistent with equations 7.6 and 7.7, the reaction is more efficient at lower pH (Krishna *et al.*, 1996a).

Another important antioxidant effect exerted by nitroxides is the detoxification of hydrogen peroxide associated toxicity. Organic peroxides and hydrogen peroxide can react with heme proteins such as hemoglobin, myoglobin, cytochrome C, etc., and convert the heme iron into the ferryl state (+4 oxidation state). In the ferryl state, the heme protein possesses oxidizing capabilities similar to peroxidase enzymes such as horseradish peroxidase. The higher valence oxidation states of heme iron can inflict significant biologic damage by oxidizing critical targets, particularly at the cell membrane (Dalaris *et al.*, 1989). Nitroxides function as antioxidants in a catalytic manner by directly detoxifying the hypervalent heme iron through reduction of the ferryl heme (Krishna *et al.*, 1996b).

In addition to the catalytic pathways of detoxification, nitroxides and hydroxylamines have been shown to inhibit lipid peroxidation stoichiometrically. A single reactive oxidizing species can greatly amplify cellular damage by initiating and propagating the lipid peroxidation process. Inhibiting these free radical processes is a desirable feature of an effective antioxidant.

The lipid peroxidation process can be classified into three stages.

a) *Initiation*:

$$LH + X^{\bullet} \rightarrow L^{\bullet} + XH \quad (7.9)$$

where LH is a lipid molecule and $X^{\bullet}$ is a reactive free radical species.

b) *Propagation*:

$$L^\bullet + O_2 \rightarrow LOO^\bullet \quad (7.10)$$

$$LOO^\bullet + LH \rightarrow L^\bullet + LOOH \quad (7.11)$$

$$LOOH \rightarrow LO^\bullet + {}^-OH \quad (7.12)$$

$$LO^\bullet + LH \rightarrow L^\bullet + LOH \quad (7.13)$$

Together, reactions 7.9–7.13 start and propagate a chain process whereby the initial lesion on a lipid molecule is significantly amplified in the presence of oxygen and redox-active transition metals. Reactions in the propagation chain (equations 7.10–7.13) are dependent on oxygen and substantial damage can be inflicted with consumption of oxygen.

c) *Termination*:
Chain termination reactions predominate at low oxygen levels as represented below.

$$L^\bullet + L^\bullet \rightarrow LL \quad (7.14)$$

$$LOO^\bullet + LOO^\bullet \rightarrow LOOL + O_2\ ({}^1O_2\ ?) \quad (7.15)$$

$$LOO^\bullet + L^\bullet \rightarrow LOOL \quad (7.16)$$

However, the initiation and propagation reactions can be effectively inhibited by antioxidants to limit biologic damage. Most of the antioxidants are stoichiometric in inhibiting lipid peroxidation. Nitroxides and hydroxylamines on the other hand can inhibit lipid peroxidation.

$$X^\bullet + RR'NO^\bullet \rightarrow RR'NOX \quad (7.17)$$

$$X^\bullet + RR'NOH \rightarrow RR'NO^\bullet + XH \quad (7.18)$$

$$L^\bullet + RR'NO^\bullet \rightarrow RR'NOL \quad (7.19)$$

$$L^\bullet + RR'NOH \rightarrow RR'NO^\bullet + LH \quad (7.20)$$

$$LO^\bullet + RR'NOH \rightarrow RR'NO^\bullet + LOH \quad (7.21)$$

Thus, these agents can inhibit at the initiation stage (equations 7.17–7.18) as well as the propagation stage (equations 7.19–7.21). Since nitroxides and hydroxylamines are inter convertible (e.g. reactions 7.7, 7.9 and 7.10), effective protection against lipid peroxidative damage is observed even at low concentrations of either the hydroxylamine or the nitroxide. Miura *et al.* (1993) suggest that hydrophilic nitroxides and hydroxylamines act as preventive antioxidants by inhibiting initiation reactions in lipid peroxidation such as reaction 7.9, where $X^\bullet$ is the initiating species. The lipophilic nitroxides and hydroxylamines interrupt chain propagating lipid peroxidation reactions as shown in reactions 7.19–7.21.

7.2.1 Oxidation of transition metal ions

Low molecular weight complexes of transition metals such as copper and iron are thought to induce peroxide associated free radicals that subsequently result in biologic damage via Fenton reactions (equation 7.23) or metal catalyzed Haber–Weiss reactions (equations 7.22, 7.23).

$$M^n + O_2^{-\bullet} \rightarrow M^{n-1} + O_2 \quad (7.22)$$

$$M^{n-1} + H_2O_2 \rightarrow OH^- + {}^\bullet OH + M^n \quad (7.23)$$

When transition metal ions are associated with critical targets such as DNA, site specific free radical generation can exert significant toxicity. Metal chelators such as desferrioxamine are used to inhibit reaction 7.22 and thereby limit the generation of hydroxyl radical or similar oxidants. Nitroxides can also limit metal catalyzed free radical generation by oxidizing the reduced metals to preempt reaction 7.23.

$$M^{n-1} + RRNO^\bullet \rightarrow M^n + RRNOH \quad (7.24)$$

Nitroxides have been effective in inhibiting reaction 7.23 (Mitchell *et al.*, 1990, Samuni *et al.*, 1991a), when the metal is associated with either DNA or protein, thereby preventing the Fenton reaction (equation 7.23) and its consequences.

7.3 *In vitro* and *in vivo* toxicity

The major concern related to the use of nitroxides as antioxidants stems from the fact that they are free radicals and free radicals are known to be toxic. However, in the case of nitroxides, the toxicity, at the levels necessary to exert antioxidant effects, has been found not to be an issue for several reasons. The main reason for free radical mediated toxicity is the high reactivity of the free radicals. In the case of nitroxide, the bulky substituents around the radical site confer high chemical stability and hence diminished reactivity. However, discriminate electron transfer reactions are still facilitated, which presumably are governed by redox potentials as well as the ring size. The *in vitro* toxicity has been evaluated by several research groups in various cell types. At 1–2 hour exposure times, nitroxides at concentrations up to 10 mM were well tolerated, without any adverse effects. Longer exposure times of up to 24 h were also found to be tolerated by cells at nitroxide concentrations of 100 μM. The toxicity of nitroxides (or the lack of) in various cell types has been summarized by (Kocherginsky and Swartz, 1995).

As in *in vitro* studies, the toxicity of nitroxides in several *in vivo* models has been studied and found to be acceptable for use as antioxidants. A summary of the nitroxide toxicity in *in vivo* models is summarized (Kocherginsky and Swartz, 1995).

7.3.1 Partitioning of nitroxides

Using EPR spectrometry, it was found that nitroxides partition into cells instantaneously (Mitchell *et al.*, 1990; Samuni *et al.*, 1991a). The ease by which most nitroxides enter into cells allows these agents to localize at critical intracellular sites at effective concentrations to provide protection. Therefore, nitroxides would be expected to better protect against intracellular oxidative damage than exogenously added SOD, which may be restricted from intracellular entry because of its large size. The partition of various nitroxides between aqueous and lipid compartments indicates that based on the ring substituents, a wide range of partition coefficients can be expected. Therefore, these agents offer the potential to be directed to specific sub-cellular locations based on the ring substituents.

7.3.2 Mutagenicity of nitroxides

Free radical species generated by ionizing radiation are hypothesized to cause the mutagenic effects of ionizing radiation (Hsie *et al.*, 1986). Though nitroxides have been shown to possess beneficial chemical modes to confer antioxidant effects, a concern remains regarding to their inherent free radical nature and the potential to be mutagenic in mammalian cells. The mutagenic effects of selected nitroxides have been evaluated in mammalian cells. First, Tempol treatment to cells for times of up to 1 h was not mutagenic nor toxic at 5 or 10 mM (DeGraff *et al.*, 1992a). Secondly, Tempol was found to act as an anti-mutagenic agent to cells exposed to hypoxanthine/xanthine oxidase (HX/XO) and hydrogen peroxide. Tempol treatment provided near complete protection against mutations at the XPRT locus in AS52 cells induced by either HX/XO or hydrogen peroxide treatment. In addition, Tempol was shown to protect against X-ray-induced mutations in the same cell system which suggests that the protection of nitroxides results from protection against DNA damage (DeGraff *et al.*, 1992b). Such observations, when viewed objectively indicate that although nitroxides are free radicals, they do not have any detectable mutagenic properties in mammalian cells.

7.3.3 Protection against superoxide mediated cytotoxicity

Since nitroxide radicals by themselves were not mutagenic in mammalian cells and reversed the mutagenic effects of ionizing radiation in mammalian cells, their effects in providing cellular cytoprotection was evaluated using the clonogenic cell survival assay. When V79 cells were exposed to $O_2^{-\bullet}$ generated by the HX/XO reaction, it was found that the six-membered nitroxide Tempol (5 mM) conferred cytoprotection when applied immediately before the induction of oxidative stress (Figure 7.2A) (Mitchell *et al.*, 1990). Catalase which detoxifies H_2O_2 and Desferal (DF) which is a ferric ion chelator also provided protection; whereas, the enzyme SOD had no effect. Pretreating the cells with SOD for 6 hours did not result in cytoprotection. Based on these experiments, the following conclusions were drawn. First, extracellular $O_2^{-\bullet}$ generated by the HX/XO reaction is in and of itself not toxic to cells since SOD which, in V79 cells can only localize in the extracellular space, had no protective effects. Second, even though $O_2^{-\bullet}$ is generated by the HX/XO reaction, H_2O_2 initiates the damage since $O_2^{-\bullet}$ dismutates to produce H_2O_2 which can freely diffuse through the cell membrane to localize at critical sites. Catalase completely prevented the damage induced by the HX/XO reaction by intercepting H_2O_2. Third, the damage mediated by H_2O_2 is transition metal-dependent, since DF completely prevented the cytotoxicity induced by the HX/XO reaction. There was no change in the metabolism of H_2O_2 by V79 cells in the presence and absence of nitroxides (Mitchell

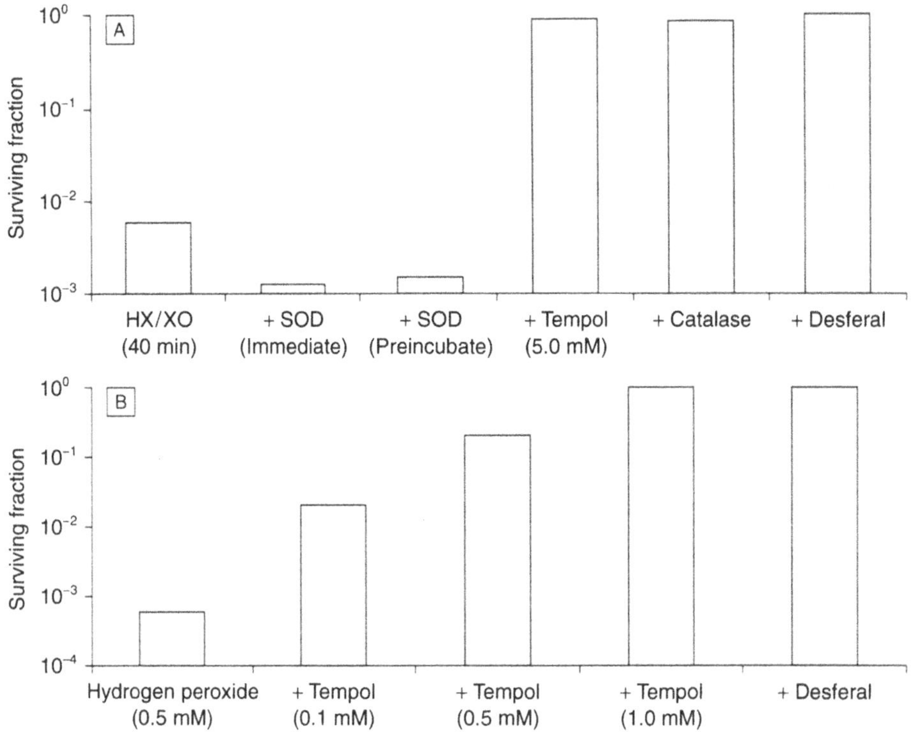

Figure 7.2 **A** Survival of Chinese hamster V79 cells exposed to hypoxanthine/xanthine oxidase (HX/XO). Chinese hamster V79 cells in full medium at 37°C were exposed to 0.05 unit/ml XO + 0.5 mM HX for 40 min in the presence of various additives: 100 units/ml SOD added simultaneously with HX/XO; 100 units/ml SOD, preincubated for 6 hr with cells prior to addition of HX/XO; 5 mM Tempol; 100 units/ml catalase added simultaneously with HX/XO, or 500 μM Desferal. **B** Survival of Chinese hamster V79 cells following a 1 hr exposure to 500 μM H_2O_2 in the absence or presence of varying concentrations of Tempol or Desferal (500 μM). (Data adapted from Mitchell *et al.*, 1990, with permission)

et al., 1990). Lastly, the mode of nitroxide protection against the cytotoxicity induced by HX/XO is by oxidizing reduced transition metals or terminating intracellular free radical chain reactions (Mitchell *et al.*, 1990). The dismutation of $O_2^{-\bullet}$ by nitroxides, though a potential reaction pathway, may not be the major detoxification pathway in cells exposed to extracellularly generated $O_2^{-\bullet}$ radicals. However, the reactions with redox active transition metal complexes might be at least in part responsible for the observed protective effects of the nitroxides.

7.3.4 Protection against hydrogen peroxide mediated cytotoxicity

In studies designed to evaluate the protective effects of stable nitroxides in V79 cells exposed to exogenously supplied H_2O_2, Tempol was found to inhibit completely the damage in a concentration dependent manner (Figure 7.2B). In the study a range of Tempol concentrations provided protection against H_2O_2 mediated cytotoxicity, with a concentration of 1 mM providing near complete protection. The ferric ion chelator DF was also found to be protective. Therefore, stable nitroxides can limit transition metal mediated H_2O_2-induced damage, presumably by inhibiting the generation of site specific hydroxyl radicals (Mitchell *et al.*, 1990). Similar protective effects against H_2O_2 induced damage have been observed in beating cardiomyocytes (Samuni *et al.*, 1991c) and mutant bacterial cells which are repair deficient and also hypersensitive to H_2O_2 (Samuni *et al.*, 1991a). Reddan *et al.* evaluated the ability of Tempol to block hydrogen peroxide toxicity of cultured rabbit lens epithelial cells and found that Tempol blocked or decreased H_2O_2-induced inhibition of cell growth and decreased the induction of single strand DNA breaks (Reddan *et al.*, 1992, 1993).

In another study, nitroxides have been shown to protect cells against peroxide-mediated apoptosis (Slater *et al.*, 1995). Exposure of rat thymocytes to the glucocorticoid, methylprednisolone (MPS), resulted in a progressive increase in intracellular peroxides and a decrease in glutathione (GSH). These changes accompanied the onset of apoptosis of rat thymocytes exposed to MPS. The increase in peroxide content was found to be restricted to apoptotic cells, while a significant depletion of GSH and reduced protein thiol was detected in both pre-apoptotic and fully apoptotic cells. To investigate the biological significance of these redox changes, the nitroxide-radical antioxidant 2,2,6,6-tetramethyl-1-piperidinyl-1-oxyl (Tempo) was tested as an inhibitor of thymocyte apoptosis. The cell shrinkage and DNA fragmentation induced by four different initiators of apoptosis were reduced by Tempo. Tempo inhibition of both etoposide- and MPS-induced thymocyte DNA fragmentation was also found to correlate with an increase in intracellular GSH, providing support for the proposal that its antioxidant properties were responsible for the observed protective activity (Slater *et al.*, 1995).

7.3.5 Protection against organic hydroperoxide mediated cytotoxicity

Nitroxides have also been shown to protect monolayered cells exposed to organic hydroperoxides such as *t*-butyl hydroperoxide (TBH) (Samuni *et al.*, 1991b). Damage to cells by TBH is independent of both $O_2^{-\bullet}$ and H_2O_2. Protection has been proposed to result from the reaction of nitroxides with the alkyl, alkoxyl, and alkyl peroxyl radicals formed by the decomposition of TBH (Samuni *et al.*, 1991b). The reaction of these radicals with nitroxides has been documented by others (Belkin *et al.*, 1987; Chateauneuf *et al.*, 1988; Mehlhorn and Packer, 1984; Nilsson *et al.*, 1989).

7.3.6 Hyperbaric oxygen induced oxidative damage

Several model studies examining the protective effects of nitroxides at isolated organ and whole body level also show that nitroxides offer protection against diverse types of oxidative damage. Hyperbaric oxygen exposure has long been associated with free radical toxicity (Gerschman *et al.*, 1954). Rats exposed to hyperbaric oxygen (0.5 mPa O_2) can be assessed for neurologic damage such as seizures by using electroencephalography (EEG). The duration of the latent time for electrical discharges from the brain to appear after the administration of hyperbaric oxygen has been used to quantitate neurotoxicity. Neurological toxicity associated with exposure to hyperbaric oxygen in rats was reversed by the intraperitoneal administration of the nitroxides, Tempol and Tempo, as measured by the changes in the EEG at doses as low as 2.5 mg/kg body weight (Bitterman and Samuni, 1995).

7.3.7 *Reperfusion injury*

Reoxygenation after periods of ischemia has been shown to generate free radicals such as superoxide, which ostensibly can cause cardiac tissue injury (McCord, 1987). Low molecular weight antioxidants, as well as antioxidant enzymes such as SOD and catalase, have been applied experimentally during reperfusion to minimize injury associated with the reoxygenation of ischemic tissue (Simpson *et al.*, 1987). Tempo, a lipophilic stable nitroxide, was used to test the protective role of stable nitroxides in regional post-ischemic reperfusion injury in isolated rat hearts. Gelvan *et al.* demonstrated that 0.4–1 mM Tempo diminished the post-ischemic release of LDH and reduced the duration of reperfusion damage induced arrhythmias (Gelvan *et al.*, 1991).

7.3.8 *Mechanical trauma*

Mechanical trauma to the brain has been shown to alter the antioxidant status of the central nervous system and cause accumulation of reactive oxygen species (Siesjo *et al.*, 1989). Free radical scavengers are being screened for their ability to limit neural damage. Nitroxides, when administered 4 hours after mechanical trauma, at a dose of 50 mg/kg body weight, were shown to provide protection to rat brains, as measured by neurologic severity scores and the integrity of the blood brain barrier. Such results suggest that the nitroxides may be useful clinically to lessen the morbidity resulting from head trauma (Yannai *et al.*, 1996).

7.3.9 *Ulcerative colitis and gastric mucosal injury*

Evidence supporting the association between free radicals and the pathogenesis of colitis and gastric mucosal injury is growing. The effect of intravenous or intraperitoneal administration of Tempol on the gastric mucosal damage induced by ethanol, indomethacin, and aspirin was studied by Rachmilewitz *et al.*, by measuring the size of mucosal lesions and the levels of mediators of inflammatory response such as myeloperoxidase and leukotrienes (Rachmilewitz *et al.*, 1994). Tempol given at a dose of 100 mg/kg body weight in rats 5 min prior to the administration of ethanol or indomethacin completely prevented the mucosal lesions and decreased the levels of myeloperoxidase and leukotrienes (Rachmilewitz *et al.*, 1994).

In another study, Karmeli *et al.* examined the effect of antioxidants in a rat model of experimental colitis induced by acetic acid or trinitrobenzene sulfonic acid (Karmeli *et al.*, 1995). Nitroxides, administered intragastrically at a dose of 0.5 g/kg body weight, were studied for efficacy in this model. Nitroxide treatment led to a decrease in the size of the ulcerative lesion and a decrease in the levels of leukotrienes (Karmeli *et al.*, 1995).

7.4 *In vitro* radioprotection by nitroxides

Radiation-induced cellular damage and cytotoxicity are critically influenced by oxygen free radicals. Hence, cells exposed to ionizing radiation under aerobic conditions are

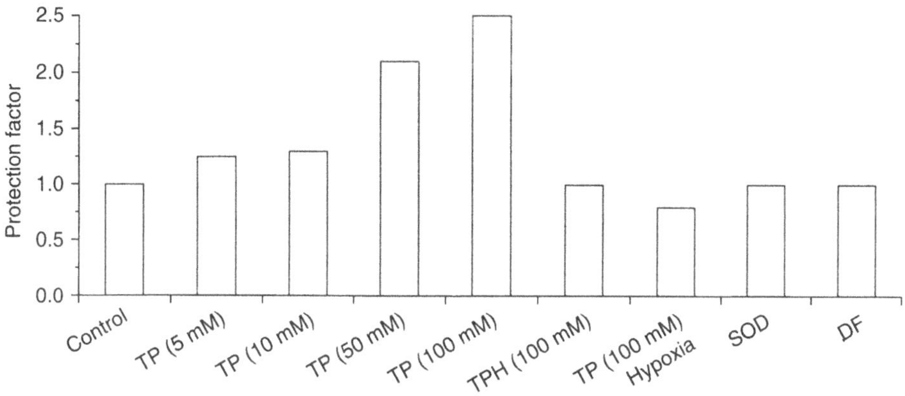

Figure 7.3 Protection of Chinese hamster V79 cells exposed to X-rays treated with Tempol (TP) at concentrations of 5, 10, 50 and 100 mM (aerobic and hypoxic); Tempol-H (TPH) at 100 mM, SOD (100 μg/ml), Desferal (500 μM). Protection factors were calculated at the 10% survival by dividing the radiation dose of control cells by the radiation dose of treated cells. (Data adapted from Mitchell *et al.*, 1991, with permission)

more sensitive to radiation-induced cytotoxicity when compared to cells exposed to radiation under hypoxic conditions (Hall, 1994). Efforts over the years have been directed toward modulating the effects of radiation by enhancing or suppressing cytotoxicity mediated by free radicals (Alexander and Charlesby, 1954; Bacq, 1965). Sulfhydryl compounds are excellent biological scavengers of free radicals (Bacq *et al.*, 1953) and by extension, the classic group of radiation protectors are the aminothiol compounds including cysteine, cysteamine, and WR-2721 (Amifostine) (Patt *et al.*, 1949; Yuhas and Storer, 1969).

Chinese hamster V79 cells in the presence of the nitroxide, Tempol, were protected against the aerobic lethal damage induced by ionizing radiation in a concentration dependent manner (Mitchell *et al.*, 1991). From the data presented in Figure 7.3 and subsequent studies, several points regarding the effects of Tempol on the radiation response can be made (Mitchell *et al.*, 1991). Firstly, Tempol-mediated radioprotection of V79 cells was concentration dependent and required mM concentrations to exert significant radioprotection. It should be noted that the requirement for mM concentrations of Tempol to provide radioprotection is similar to that required for other radioprotectors such as aminothiols; however, aminothiols in general are more efficient radioprotectors on a concentration basis compared to Tempol. Secondly, the hydroxylamine form of Tempol, did not provide radioprotection. Thirdly, Tempol modestly radiosensitized cells irradiated under hypoxic conditions, in agreement with previous studies (Millar *et al.*, 1977, 1978, 1983). Lastly, neither exogenous SOD nor DF altered the radiation response. This last point suggests that it is unlikely (although it can not be ruled out) that the SOD mimetic action of nitroxides is responsible for aerobic radioprotection. Previous reports suggested that SOD protected bone marrow progenitor cells against radiation cytotoxicity *in vitro* and *in vivo* (Petkau *et al.*, 1975; Petkau and Chelack, 1984; Petkau, 1987); however, not all authors have been able to reproduce these findings (Abe *et al.*, 1981). We have repeatedly demonstrated *in vitro* that exogenously applied SOD does not afford aerobic radioprotection. Likewise, it is unlikely that Tempol radioprotection results from its effect upon metal reduction since the metal chelator, Desferal, does not provide *in vitro* radioprotection.

Nitroxides could provide biologic radioprotection by inhibiting the damage mediated by radiation-induced reactive species ($X^{\bullet}$) to biologically important molecules (BIM) such as DNA by at least two modes.

Damage

$$X^{\bullet} + BIM\text{—}H \rightarrow XH + BIM^{\bullet} \qquad (7.25)$$

Protection by radical scavenging

$$H^{+} + X^{\bullet} + RR'NO^{\bullet} \rightarrow HX + RR'NO^{+} \qquad (7.26)$$

$$X^{\bullet} + RR'NO^{\bullet} \rightarrow RR'NO\text{—}X \qquad (7.27)$$

Protection by repair

$$X^{\bullet} + BIM\text{—}H \rightarrow XH + BIM^{\bullet} \qquad (7.28)$$

$$H^{+} + BIM^{\bullet} + RR'NO^{\bullet} \rightarrow BIM\text{—}H + RR'NO^{+} \qquad (7.29)$$

Radiation exposure can result in the formation of carbon-centered radicals on BIM (in the case of radiation, the most likely critical intracellular target is the DNA) as shown in equation 7.25. As proposed in equations 7.26 and 7.27, nitroxides could scavenge radiation-induced reactive species. Likewise, equations 7.28 and 7.29 demonstrate that nitroxides have the capability to restore a carbon-centered radical on BIM to normal by donating an electron (BIM—H). It is possible that a combination of equations 7.26, 7.27 and 7.28, 7.29 operate to provide radioprotection.

Radiation-induced, unrepaired, DNA double strand breaks and subsequent production of chromosome aberrations have been closely linked with radiation-induced cell killing (Puck, 1958; Carrano, 1973; Bedford *et al.*, 1978). DeGraff *et al.* (1992b) showed a direct correlation between Tempol-mediated *in vitro* radio-protection (cell survival) and reduced DNA double strand breaks. Johnstone *et al.* (1995) showed that Tempol significantly reduced the frequency of radiation-induced chromosome aberrations in human peripheral blood lymphocytes. Both of these studies strongly suggest that Tempol-mediated radioprotection is accompanied by a reduction in DNA damage.

7.5 *In vivo* radioprotection by nitroxides

The protective effects of nitroxides in *in vitro* experiments, prompted the study of nitroxides using *in vivo* models. In order to screen stable nitroxides as *in vivo* radioprotectors, the toxicity, pharmacology, and *in vivo* radioprotective effects of

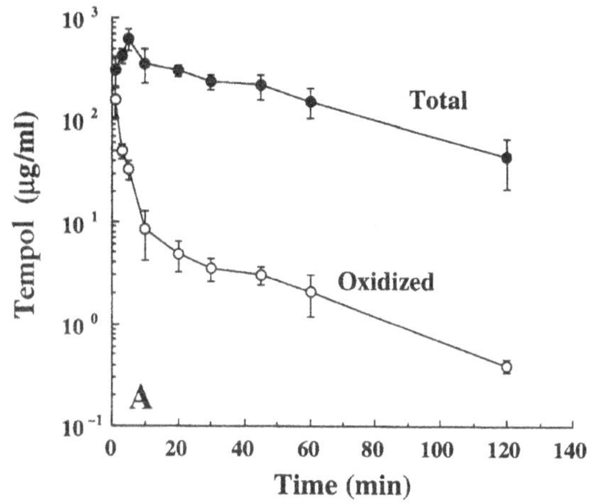

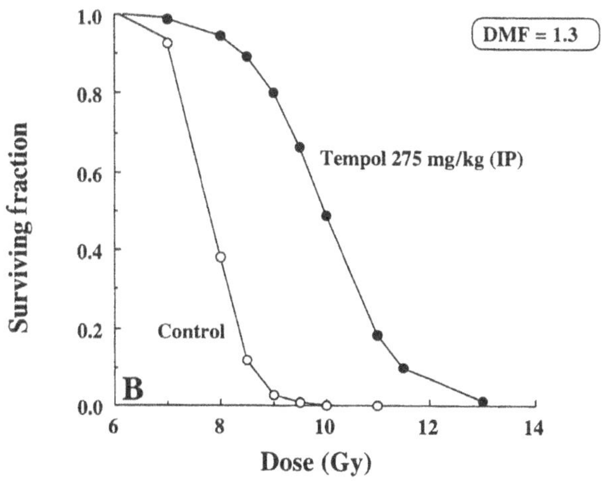

Figure 7.4 *In vivo* studies of Tempol in C3H mice. **A** The time dependence of Tempol in the whole blood of mice. Mice were injected with Tempol, 275 mg/kg, IP and at various times after injection, whole blood levels were measured with EPR spectroscopy. The closed symbols represent the total Tempol concentration (oxidized plus reduced). The open symbols represent only the oxidized form. **B** Survival of mice 30 days after whole body radiation. Control mice received saline IP and the treated mice received Tempol 275 mg/kg IP 10 minutes prior to radiation. The dose modification factor (DMF) is 1.3. (Data from Hahn *et al.*, 1992, with permission)

Tempol were studied in C3H mice (Hahn *et al.*, 1992). Tempol was injected intraperitoneally into mice at the maximally tolerated dose (275 mg/kg), and whole blood levels of nitroxide were measured by EPR spectroscopy (Figure 7.4A). The peak Tempol concentration was observed 10 min after the intraperitoneal injection. Nitroxide was not detected in whole blood 24 hr after injection. Since the peak Tempol concentration occurred 10 min after injection, mice were injected intraperitoneally with Tempol (275 mg/kg) 10 min prior to whole body radiation

(7.0–13.0 Gy). Tempol provided a dose modification factor of 1.3 at the $LD_{50/30}$ level as shown in Figure 7.4B. *In vivo*, Tempol is rapidly reduced to the hydroxylamine, a form shown not to be radioprotective *in vitro* (see above). Despite the rapid reduction of Tempol *in vivo*, significant radioprotection was observed. It is expected that *in vivo* radioprotection would be increased if higher concentrations of the oxidized form of the nitroxide were available at the time of irradiation. Currently our laboratory has initiated a search for other nitroxides that are bio-reduced at a slower rate and hence would yield a higher concentration of the oxidized form at the time of irradiation.

The ability to selectively protect normal tissues in cancer patients receiving radiation treatment would be most advantageous. If selective protection of normal tissues were possible, higher radiation doses could be delivered to the tumor accompanied with higher local control rates. The key, however, is *selective* normal tissue protection because if a systemic radioprotector also protects the tumor, no advantage would be realized. To test whether systemic administration of Tempol would protect against local irradiation delivered to a tumor, Hahn *et al.* (1997) used a RIF-1 transplantable rodent tumor to evaluate the effect of Tempol on local tumor control by radiation. As shown in Figure 7.5A, the administration of Tempol to tumor bearing mice at the same concentration and timing as shown in Figure 7.4B resulted in no protection of tumor. To identify a mechanism for the apparent differential protection of the hematopoietic system (see above) and tumor tissue, pharmacologic studies were carried out which revealed that the percentage of oxidized compound was approximately 2 fold greater in the bone marrow compartment, compared to RIF-1 tumor (see Figure 7.5B) at the time of irradiation. Greater bioreduction of Tempol occurs in RIF-1 tumor. Such a result may provide at least a partial explanation for the absence of tumor radioprotection, since it previously had been demonstrated that the oxidized form of Tempol is the active radioprotector (see above). These preliminary data imply that a potential difference exists between normal and tumor tissues with respect to bio-reduction. Future studies will focus on other normal tissues and different tumor systems to determine the generality of the findings.

Another application of nitroxides as radioprotectors which may have benefit in a clinical setting would be to protect against radiation-induced alopecia, a common radiotherapeutic problem. Hair loss as a result of irradiation especially from whole brain irradiation often leads to cosmetic, social, and psychological problems for the radiotherapy patient. Clinically, no successful interventions are available. Topical application of Tempol was evaluated for possible protective effects against radiation-induced alopecia using guinea pig skin as a model. For single acute doses up to 30 Gy, Tempol, when topically applied 15 min prior to irradiation, provided a marked increase in the rate and extent of new hair recovery when compared to untreated irradiated skin (Goffman *et al.*, 1992). By EPR spectroscopy Tempol was detected in treated skin specimens. EPR measurements of blood samples and brain tissue failed to show any systemic nitroxide signal resulting from topical application to skin or scalp. Recently these studies have been extended to evaluate fractionated radiation delivery with multiple applications of Tempol or Tempo (Cuscela *et al.*, 1996). Topical administration of Tempol or Tempo 15 min before each radiation treatment (daily fractions of 7 Gy, for a total of 8 treatments over 10 days) resulted in statistically significant radioprotection with respect to hair loss and regrowth of hair in the

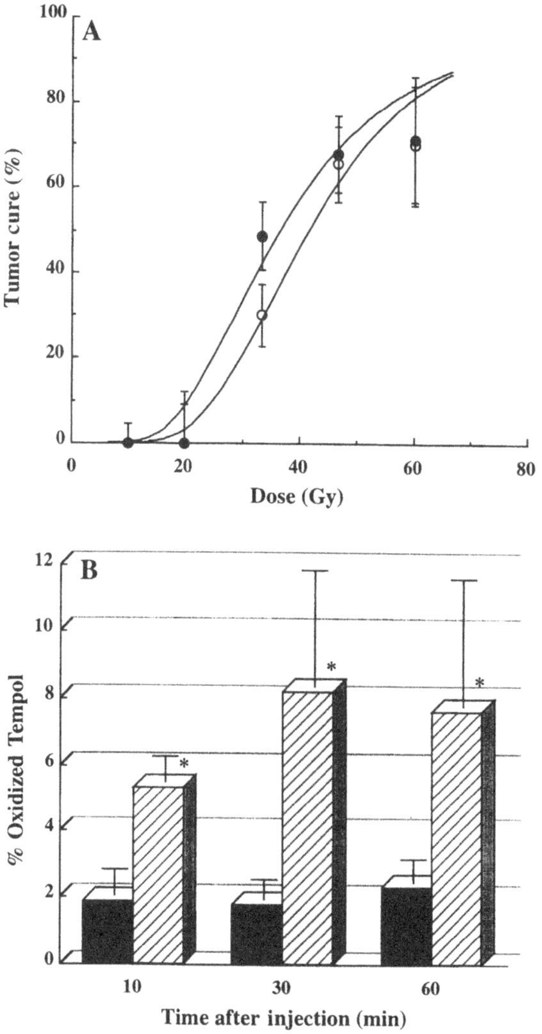

Figure 7.5 **A** Radiation tumor control curves for animals treated in the absence or presence of Tempol. Closed circles: Tempol-treated mice; open circles: PBS-treated mice. There was no statistical difference between the two curves ($p = 0.54$). The calculated $TCD_{50/30}$ values for Tempol-treated and PBS-treated mice were 36.7 and 41.8 Gy, respectively. There was no statistical difference between these values ($p = 0.32$). The error bars represent one standard error above and below the value shown. **B** The percent oxidized Tempol with time in RIF-1 tumor versus bone marrow cells harvested simultaneously from C3H mice. Hatched bars: bone marrow; solid bars: RIF-1 tumor. Error bars represent the standard error of the mean. A greater percentage of oxidized Tempol was present in bone marrow compared to RIF-1 tumor cells (*$p \leq 0.004$, as analyzed by Student's t test). (Data adapted from Hahn *et al.*, 1997, with permission)

treatment field (Cuscela *et al.*, 1996). Histological evaluation showed that radiation treatment without nitroxide resulted in a marked decrease in the number of hair follicles and poor development of remaining follicles; however, nitroxide pretreatment resulted in no appreciable decrease in hair follicles and hair follicles appeared mature, similar to unirradiated controls (Cuscela *et al.*, 1996). These studies suggest that the topical application of nitroxides may be useful clinically to reduce the undesirable toxicity of radiation-induced alopecia. Likewise, topical application of nitroxides may have utility in other sites such as rectum and bladder, two normal tissues at risk from radiation treatment of prostate and/or cervix cancer. The advantage of topical application is that high concentrations of the nitroxide could be used directly on the tissue at risk. The resulting systemic levels of the drug after topical administration would hopefully be inadequate to protect the tumor or cause systemic toxicity.

7.6 Protection against redox-cycling chemotherapy drugs

7.6.1 Streptonigrin

Some chemotherapeutic agents undergo bioreductive activation by enzymatic processes to produce intermediates which can damage DNA, thereby exerting tumoricidal effects. Quinone based drugs (Q) such as adriamycin, mitomycin C, and the benzotriazine based drug SR 4233 are used clinically. In addition to these agents, another quinone based antibiotic, streptonigrin is known to exert cytotoxic effects by free radical generation via bioreductive activation to the semiquinone ($SQ^{-\bullet}$) as shown below.

$$\text{Q (bioreductive activation)} \rightarrow SQ^{-\bullet} \tag{7.30}$$

$$SQ^{-\bullet} + O_2 \rightarrow Q + {}^{\bullet}O_2^- \tag{7.31}$$

$$2{}^{\bullet}O_2^- + 2\,H^+ \rightarrow H_2O_2 + O_2$$

$$H_2O_2 + \text{transition metals or heme proteins} \rightarrow \text{Oxidants} \tag{7.32}$$

For many neoplastic drugs, a free radical mechanism for the anti-proliferative effects as well as the side effects is gaining acceptance. In the case of adriamycin, however, the mechanism of tumoricidal action is less clear. While inhibition of topoisomerase II has been suggested as its mechanism of cancer cell cytotoxicity, oxyradicals generated by the redox cycling of the semiquinone may be responsible for the cardiac toxicity of adriamycin. Since nitroxides are cell-permeable and can participate in radical–radical recombination reactions, and have been shown to react with semiquinone radicals (DeGraff *et al.*, 1994) as well as autooxidation products such as superoxide, nitroxides may be useful as research tools to evaluate free radical pathways of chemotherapy drug cytotoxicity. The effect of Tempol on the cytotoxicity of adriamycin and streptonigrin in mammalian cells was studied in an effort to determine the importance of free radical mechanisms on the cytotoxicity of these drugs (DeGraff *et al.*, 1994). While Tempol was effective in completely inhibiting the cytotoxicity and the DNA double strand breaks induced by streptonigrin

(Figure 7.6A), it had no effect on either the cytotoxicity or DNA double strand breaks induced by adriamycin (Figure 7.6B). While Tempol is cell permeable and reacts with chemically generated semiquinone radicals of streptonigrin and adriamycin, the lack of protection against adriamycin induced cytotoxicity and DNA double strand breaks suggests that free radical pathways do not have a major role in the anti-proliferative effects of adriamycin.

7.6.2 *Mitomycin C*

Mitomycin C (MMC) is a quinone containing antineoplastic agent where the semiquinone radical intermediate has been implicated in DNA inter strand cross links that lead to cytotoxicity. Nitroxides have been shown to be effective in reversing the hypoxic cytotoxicity of MMC *in vitro* (Krishna *et al.*, 1991) (Figure 7.6C). The cytoprotective effects provided by the nitroxides presumably result from reoxidizing the semiquinone of MMC, which has been implicated to cause DNA interstrand cross links (equation 7.33).

$$SQ^{-\bullet} + RR'NO \rightarrow Q + RR'NOH \tag{7.33}$$

An example of the use of nitroxides to prevent damage mediated by MMC was recently evaluated in a swine model of chemotherapy extravasation (Hahn *et al.*, 1996). Extravasation tissue injury from chemotherapeutic drugs such as MMC is a serious clinical problem with no effective antidotes available. Several nitroxides were screened as potential protectors of MMC induced skin necrosis. The nitroxide 3-carbamoyl-PROXYL (2,2,5,5-tetramethylpyrrolidine-*N*-oxyl or 3-CP) protected when given within 5 minutes after MMC extravasation. Histologic sections of the 3-CP- and MMC-treated pig skin showed marked reduction in acute inflammation and the absence of deep dermal scarring when compared to MMC alone.

7.6.3 *Adriamycin-induced cardiotoxicity*

The cumulative dose limiting effect of adriamycin, clinically, is irreversible myocardial damage. Oxygen radicals, generated by the redox cycling of the drug through the semiquinone intermediate, have been proposed as the cause of cardiotoxicity and agents that will inhibit such redox cycling are being investigated (Monti *et al.*, 1991). In a recent study, Tempol was evaluated in isolated rat heart model for protective effects against adriamycin-induced acute cardiotoxicity (Monti *et al.*, 1996). When rat hearts were perfused with 100 μg/ml of adriamycin for 60 minutes, significant impairment of the contractile function as well as elevated lipid peroxidation was observed. Co-perfusion of adriamycin with Tempol (2.5 mM) significantly inhibited the contractile impairment of the cardiac tissue and decreased lipid peroxidation. The chemical basis for the observed protective effects was proposed to involve free radical scavenging by nitroxides.

7.6.4 *6-Hydroxydopamine*

An analogue of the neurotransmitter dopamine, 6-hydroxydopamine, has been proposed for clinical use in the treatment of neuroblastoma. This neurotransmitter

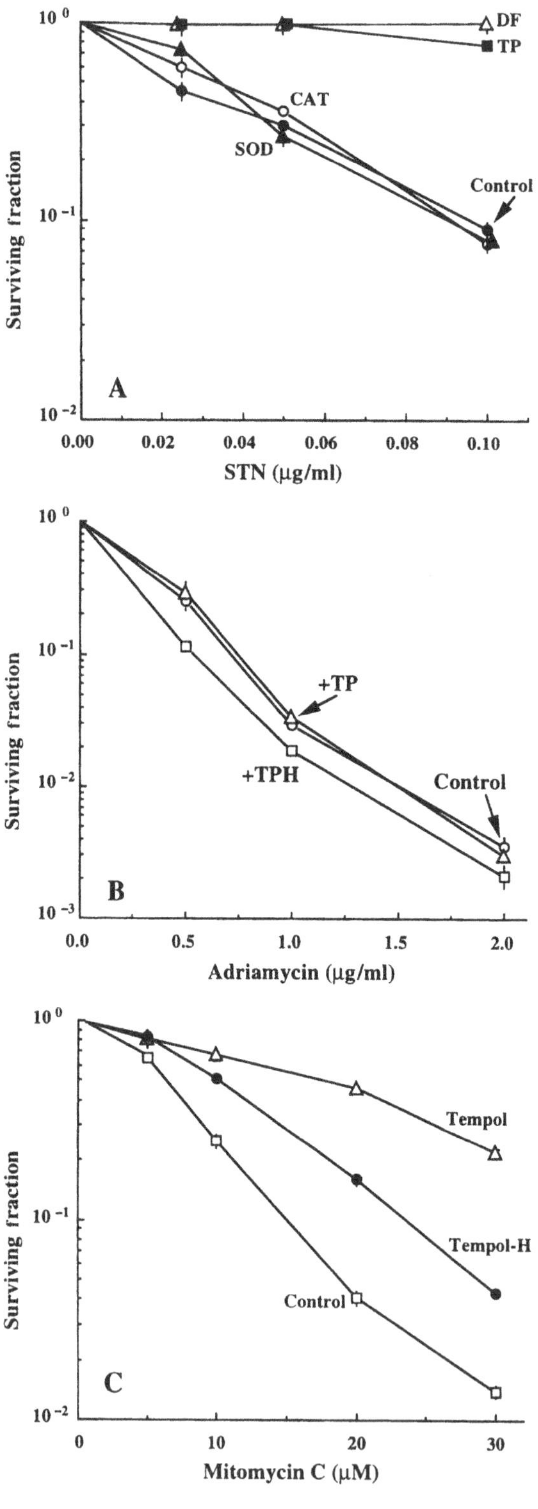

DF
TP
CAT
SOD
Control
Surviving fraction
A
STN (μg/ml)
+TP
+TPH
Control
Surviving fraction
B
Adriamycin (μg/ml)
Tempol
Tempol-H
Control
Surviving fraction
C
Mitomycin C (μM)

analogue localizes in tumor cells by the dopamine uptake system in the cell membrane. However, systemic toxicity, particularly of the sympathetic nervous system, limits its therapeutic usefulness. Reactive oxygen species generated by the autooxidation of the drug have been shown to mediate the systemic toxicity though not the tumoricidal effects (Sachs and Johnson, 1975). Adjunctive use of antioxidants such as the radioprotector WR 2721, to limit the systemic toxicity have not been successful. A recent study (Purpura *et al.*, 1996) using a murine model showed that the mortality observed upon administration of toxic doses of 6-hydroxydopamine is decreased by prior intraperitoneal administration of Tempol. In mice bearing experimental neuroblastoma tumors, Tempol alone had no effect on tumor growth; however, treatment of these animals with the combination of 6-hydroxydopamine and Tempol resulted in enhanced tumor response when compared to treatment with 6-hydroxydopamine alone. The protective effects of the nitroxide were attributed to its ability to scavenge radicals generated by the autooxidation of 6-hydroxydopamine.

7.7 Nitroxides as functional imaging probes

Nitroxides, as discussed earlier, are electron paramagnetic resonance (EPR) detectable and the one-electron reduced product, the hydroxylamine, is diamagnetic and hence EPR-silent. Nitroxides, when administered *in vivo*, are converted by cellular redox processes to the hydroxylamine, while the hydroxylamine is converted back to the nitroxide. The ratio of the nitroxide to the hydroxylamine depends on several parameters such as pO_2 and the redox status of the tissue being examined. Swartz and co-workers have used this principle to obtain metabolic information in tissue using *in vivo* EPR spectroscopy on live animals (Bacic *et al.*, 1989). Such information can be spatially encoded using magnetic field gradients and a spatial image of differences in tissue metabolism can be obtained based on the differences in metabolism of nitroxide.

Since solid tumors are known to differ in redox status, as well as oxygen status compared to normal tissue (Thomlinson and Gray, 1955; Hockel *et al.*, 1996), EPR imaging studies by using nitroxides as redox sensitive spin probes might provide valuable metabolic information non-invasively. Issues that have been addressed recently were: 1) whether nitroxides could be detected in tumor tissue, and 2) whether differential reduction rates of nitroxide probes could be discerned in tumor versus normal tissues (Hahn *et al.*, 1997; Kuppusamy *et al.*, 1998). EPR imaging (EPRI)

Figure 7.6 (*opposite*) Survival of Chinese hamster V79 cells exposed under aerobic conditions to three different redox-active chemotherapy drugs. **A** Survival of V79 cells exposed to varying concentrations of mitomycin C for 1 hr in the absence or presence of Tempol (10 mM) or Tempol-H (10 mM). **B** Survival of V79 cells exposed to varying concentrations of streptonigrin (STN) for 1 hr in the absence or presence of Tempol (TP, 10 mM), catalase (CAT, 100 U/ml), SOD (100 μg/ml) or Desferal (DF, 500 μM). **C** Survival of V79 cells exposed to varying concentrations of Adriamycin for 1 hr in the absence or presence of Tempol (10 mM) or Tempol-H (10 mM). (Data adapted from Krishna *et al.*, 1991; DeGraff *et al.*, 1994, with permission)

experiments have been performed using an EPR spectrometer operating at 1.2 GHz corresponding to a magnetic field of 40 mT (Kuppusamy *et al.*, 1994). A specially built bridged-loop surface resonator was used (Walczak and Swartz, 1989). The open structure of the resonator is ideal for localized study of metabolic activity in large objects. With this imaging system, a cylindrical volume of 10 mm diameter and 5 mm depth could be probed. Mice bearing ~1 cm diameter tumors were anesthetized and the tail vein cannulated with a heparin-filled 30 gauge catheter for nitroxide 3-CP infusion (160 mg/kg). Either the right leg with tumor or the left leg with normal tissue (muscle, skin) was utilized for the imaging studies. The presence of nitroxide in normal and tumor tissue was readily detected by using EPRI. A two dimensional spatial image of the distribution of this nitroxide in normal muscle and RIF-1 tumor as a function of time is shown in Figure 7.7. The panels in the top row show the clearance of the nitroxide in normal muscle as a function of time and the corresponding images in the bottom row from tumors. The data from the images indicate that the rate of clearance of nitroxide in tumors is faster than in normal tissue. This is consistent with previous observations that tumors provide a strong reducing environment when compared to normal tissue and result in faster reduction of the nitroxide (Hahn *et al.*, 1997; Kuppusamy *et al.*, 1998). These observations agree with earlier studies that suggest hypoxic cells within tumors reduce nitroxides more efficiently than well oxygenated normal tissue (Swartz, 1990). Estimates of oxygen concentration in the tissues shown in Figure 7.7 using EPRI/oximetry indeed confirmed that the tumor tissue was much lower in oxygen concentration than the normal tissue (Kuppusamy *et al.*, 1998). A three dimensional image of the normal tissue and tumor after nitroxide administration provides information on the physical architecture of the tissue based on the nitroxide distribution. Figure 7.8 (top) shows the nitroxide distribution in 0.3 mm adjacent slices in normal tissue (leg muscle) and the bottom panels show the corresponding slices from RIF-1 tumors. It is clear from these spatial images that the nitroxide distribution differs between the normal and tumor tissue which may reflect differences in the vasculature of the microenvironment associated with tumors.

Collectively, the studies shown in Figures 7.7 and 7.8 establish the feasibility of detecting nitroxides in tumors and clearly show major differences in nitroxide distribution and metabolism between normal and tumor tissue. As the technique evolves and becomes more sensitive, EPRI may play a useful role in advancing functional imaging in clinical medicine.

7.8 Summary

The use of nitroxides has been greatly expanded over the past few years. The ability of nitroxides to function as mimics of antioxidant enzymes such as SOD and catalase and to act as efficient radical scavengers confers unique protective capabilities which are shown schematically in Figure 7.9. The observed protective effects of nitroxides at the cellular and animal level against diverse oxidative insults have made nitroxides potentially useful as agents to assess free radical pathways of cytotoxicity and as a new class of antioxidants. Lastly, the use of nitroxides as *in vivo* functional imaging probes to non-invasively report on the oxygen status and redox properties of tissue may have utility in clinical biomedical research.

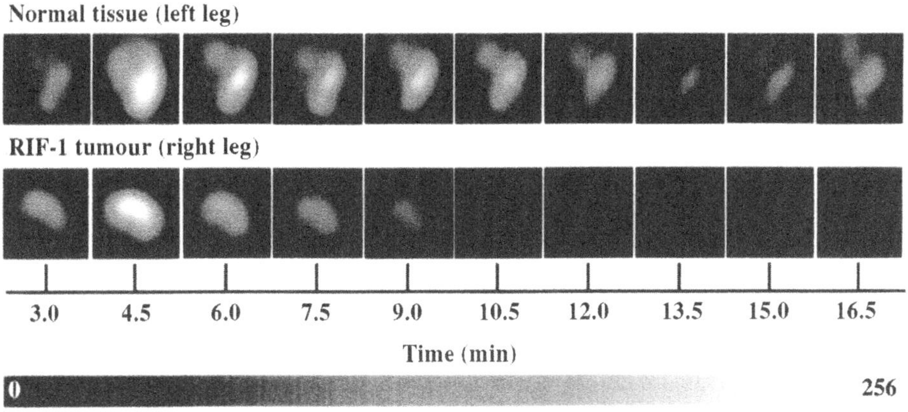

Figure 7.7 Spatially-resolved clearance of nitroxide in normal and tumor tissue. Following a tail vein infusion of 160 mg/kg of 3-CP, a series of two-dimensional images of the nitroxide from normal muscle (top) and tumor (bottom) were measured using L-band EPR imaging instrumentation. The nitroxide in normal tissue persisted for longer than 16 min, while in tumor it was cleared within 10 min of infusion. Image acquisition parameters: projections, 16; gradient, 15 G/cm; acquisition time, 1.5 min. (Data from Kuppusamy *et al.*, 1998, with permission)

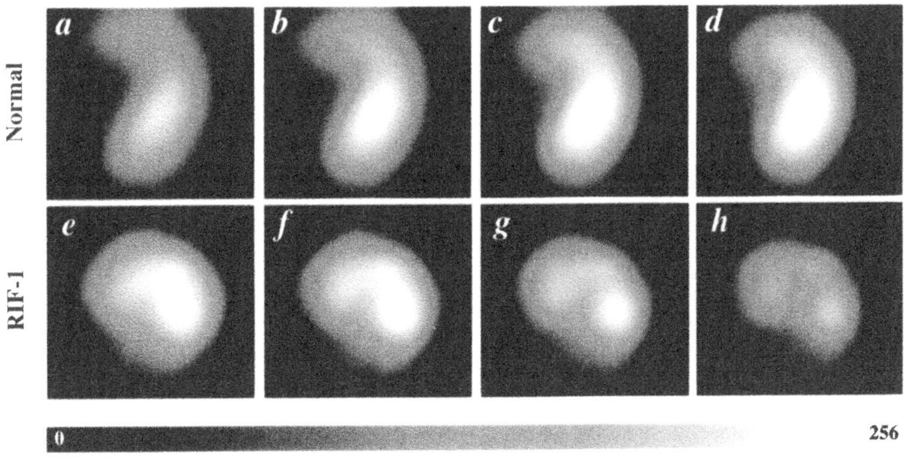

Figure 7.8 Visualization of tumor heterogeneity monitored by nitroxide uptake. Following a tail vein infusion of 160 mg/kg of 3-CP, three-dimensional EPR images of the nitroxide in normal muscle and tumor were measured. A few selected adjacent slices of 0.3 mm thickness obtained from the 3D image of normal (a–d) and tumor (e–f) tissue are shown. The images show some differences between the two tissues in terms of anatomy and physical architecture. The tumor tissue shows significant heterogeneity of nitroxide uptake compared to the normal tissue. Image acquisition parameters: projections, 100; gradient, 15 G/cm; acquisition time, 10 min. (Data from Kuppusamy *et al.*, 1998, with permission)

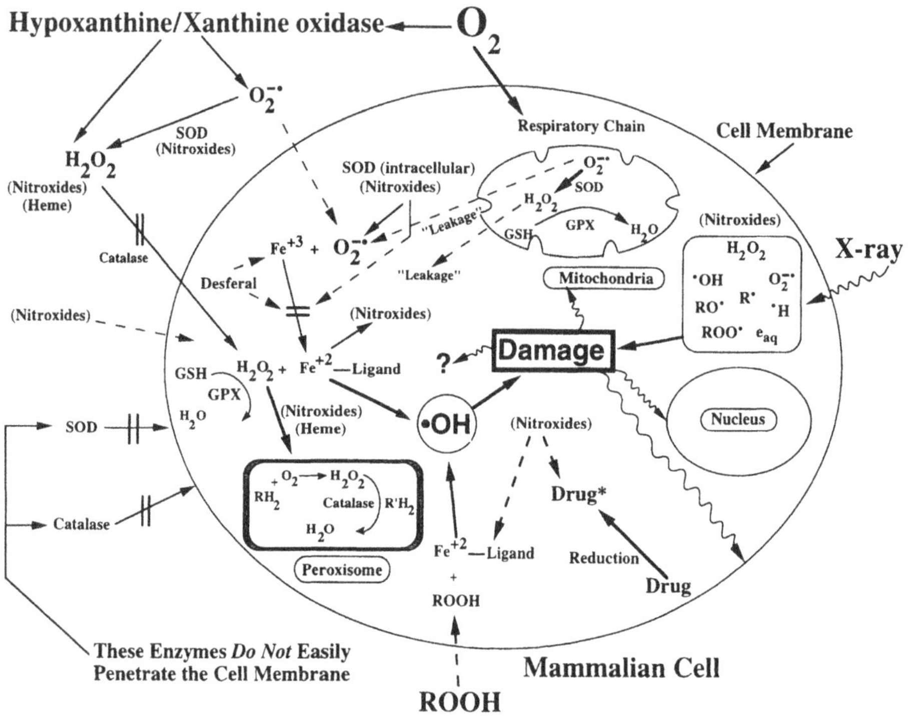

Figure 7.9 Proposed scheme for the cellular protection of nitroxides. The damage by various modes is indicated. The sites of interception of cytotoxic oxidants both intracellularly and extracellularly by nitroxides (shown in parenthesis) are indicated. Also included are the endogenous antioxidant enzymes and Desferal to illustrate the locations of action of these agents. Nitroxides may be protective to cellular and tissue systems by: 1) reacting with superoxide; 2) inducing catalase-like activity in heme proteins; 3) oxidizing reduced transition metals and preventing the reduction of hydrogen peroxide and organoperoxides to hydroxyl radical; 4) scavenging of alkyl, alkyl, and alkyl peroxyl radicals (produced by X-rays and lipid peroxidation) and 5) reacting with bioactivated xenobiotics whose cytotoxicity is mediated by radical intermediates

References

Abe, M., Nishidai, T., Yukawa, Y., *et al.* (1981) Studies on the radioprotective effects of superoxide dismutase in mice. Int J Radiat Oncol Biol Phys, 7, 205–209.

Alexander, P., Charlesby, A. (1954) Physico-chemical methods of protection aganist ionizing radiations. Radiobiology Symposium, 49.

Bacic, G., Nilges, M.J., Magin, R.L., Walczak, T., Swartz, H.M. (1989) In vivo localized spectroscopy reflecting metabolism. Magn Reson Med, 10, 266–272.

Bacq, Z.M., Dechamps, G., Fischer, P., *et al.* (1953) Protection against X-rays and therapy of radiation sickness with β-mercaptoethylamine. Science, 117, 633–636.

Bacq, Z.M. (1965) Chemical Protection Against Ionizing Radiation, C.C. Thomas Publ., Springfield, IL.

Bedford, J.S., Mitchell, J.B., Griggs, H.G., Bender, M.A. (1978) Radiation-induced cellular reproductive death and chromosome aberrations. Radiat Res, 76, 573–586.

Beit-Yannai, E., Zhang, R., Trembovler, V., Samuni, A., Shohami, E. (1996) Cerebroprotective effect of stable nitroxide radicals in closed head injury in the rat. Brain Res, 717, 22–28.

Belkin, S., Mehlhorn, R.J., Hideg, K., Hankovsky, O., Packer, L. (1987) Reduction and destruction rates of nitroxide spin probes. Arch Biochem Biophys, 256, 232–243.

Bennett, H.F., Brown III, R.D., Koenig, S.H., Swartz, H.M. (1987a) Effects of nitroxides on the magnetic field and temperature dependence of 1/T1 of solvent water protons. Magnetic Resonance in Medicine, 4, 93–111.

Bennett, H.F., Swartz, H.M., Brown III, R.D., Koenig, S.H. (1987b) Modification of relaxation of lipid protons by molecular oxygen and nitroxides. Invest Radiol, 22, 502–507.

Berliner, L.J. (1979) Spin Labelling II: Theory and applications, Academic Press, New York.

Bitterman, N., Samuni, A. (1995) Nitroxide stable radicals protect against hyperoxic-induced seizures in rats. Undersea Hyperb Med, 22, 47–48.

Carrano, A.V. (1973) Chromosome aberrations and radiation-induced cell death: II. Predicted and observed cell survival. Mutat Res, 17, 355–366.

Chance, B., Sies, H., Boveris, A. (1979) Hydroperoxide metabolism in mammalian organs. Physiol Rev, 59, 527–605.

Chateauneuf, J., Lusztyk, J., Ingold, K.U. (1988) Absolute rate constants for the reactions of some carbon-centered radicals with 2,2,6,6-tetramethylpiperidine-N-oxyl. J Org Chem, 53, 1629–1632.

Cuscela, D., Coffin, D., Lupton, G., *et al.* (1996) Protection from radiation-induced alopecia with topical application of nitroxides: fractionated studies. Cancer J Sci Am, 2, 273–278.

Dalaris, D., Eddy, L., Arduini, A., Cadenas, E., Hochstein, P. (1989) Mechanisms of reoxygenation injury in myocardial infarction: Implications of a myoglobin redox cycle. Biochem Biophys Res Commun, 160, 1162–1168.

DeGraff, W.G., Krishna, C.M., Russo, A., Mitchell, J.B. (1992a) Antimutagenicity of a low molecular weight superoxide dismutase mimic against oxidative mutagens. Environ Molec Mutagen, 19, 21–26.

DeGraff, W.G., Krishna, M.C., Kaufman, D., Mitchell, J.B. (1992b) Nitroxide-mediated protection against x-ray- and neocarzinostatin-induced DNA damage. Free Radic Biol Med, 13, 479–487.

DeGraff, W., Hahn, S.M., Mitchell, J.B., Krishna, M.C. (1994) Free radical modes of cytotoxicity of adriamycin and streptonigrin. Biochem Pharmacol, 48, 1427–1435.

Gelvan, D., Saltman, P., Powell, S. (1991) Cardiac reperfusion damage prevented by a stable nitroxide free radical. Proc Natl Acad Sci USA, 88, 4680–4684.

Gerschman, R., Gilbert, D.L., Nye, S.W., Dwyer, P., Fenn, W.O. (1954) Oxygen poisoning and X-irradiation: A mechanism in common. Science, 119, 623–626.

Goffman, T., Cuscela, D., Glass, J., *et al.* (1992) Topical application of nitroxide protects radiation induced alopecia in guinea pigs. Int J Radiat Oncol Biol Phys, 22, 803–806.

Hahn, S.M., Tochner, Z., Krishna, C.M., *et al.* (1992) Tempol, a stable free radical, is a novel murine radiation protector. Cancer Res, 52, 1750–1753.

Hahn, S.M., Sullivan, F.J., DeLuca, A.M., *et al.* (1997) Protection of mitomycin C induced skin extravasation with the nitroxide, 3-carbamoyl-PROXYL (3-CP). Int J Oncol, 10, 119–123.

Hahn, S.M., Sullivan, F.J., DeLuca, A.M., *et al.* (1997) Evaluation of tempol radioprotection in a murine tumor model. Free Radic Biol Med, 22, 1211–1216.

Hall, E.J. (1994) The oxygen effect and reoxygenation. In: Radiobiology for the Radiologist, J. B. Lippincott Co., Philadelphia, PA, 133–152.

Hockel, M., Schlenger, K., Aral, B., Mitze, M., Schaffer, U., Vaupel, P. (1996) Association between tumor hypoxia and malignant progression in advanced cancer of the uterine cervix. Cancer Res, 56, 4509–4515.

Hsie, A.W., Recio, L., Katz, D.S., Lee, C.Q., Wagner, M., Schenley, R.L. (1986) Evidence for reactive oxygen species inducing mutations in mammalian cells. Proc Natl Acad Sci USA, 83, 9616–9620.

Iannone, A., Bini, A., Swartz, H.M., Tomasi, A., Vannini, V. (1989) Metabolism in rat liver microsomes of the nitroxide spin probe Tempol. Biochem Pharm, 38, 2581–2586.

Iannone, A., Tomasi, A., Vannini, V., Swartz, H.M. (1990a) Metabolism of nitroxide spin labels in subcellular fractions of rat liver. II. Reduction by microsomes. Biochimica Biophysica Acta, 1034, 285–289.

Iannone, A., Tomasi, A., Vannini, V., Swartz, H.M. (1990b) Metabolism of nitroxide spin labels in subcellular fractions of rat liver. I. Reduction in the cytosol. Biochimica Biophysica Acta, 1034, 290–293.

Johnstone, P.A.S., DeGraff, W.G., Mitchell, J.B. (1995) Protection of radiation-induced chromosomal aberrations by the nitroxide Tempol. Cancer, 75, 2323–2327.

Karmeli, F., Eliakim, R., Okon, E., Samuni, A., Rachmilewitz, D. (1995) A stable nitroxide radical effectively decreases mucosal damage in experimental colitis. Gut, 37, 386–393.

Kocherginsky, N., Swartz, H. (1995) Toxicity and the use of nitroxides as drugs. Nitroxide Spin Labels Reactions in Biology and Chemistry, CRC Press, Boca Raton, FL.

Krishna, C.M., DeGraff, W., Tamura, S., *et al.* (1991) Mechanisms of hypoxic and aerobic cytotoxicity of Mitomycin C in Chinese hamster V79 cells. Cancer Res, 51, 6622–6628.

Krishna, M.C., Grahame, D.A., Samuni, A., Mitchell, J.B., Russo, A. (1992) Oxoammonium cation intermediate in the nitroxide-catalyzed dismutation of superoxide. Proc Natl Acad Sci, 89, 5537–5541.

Krishna, M.C., Russo, A., Mitchell, J.B., Goldstein, S., Dafni, H., Samuni, A. (1996a) Do nitroxides antioxidants act as scavengers of superoxide or as SOD mimics? J Biol Chem, 271, 26026–26031.

Krishna, M.C., Samuni, A., Taira, J., Goldstein, S., Mitchell, J.B., Russo, A. (1996b) Stimulation by nitroxides of catalase-like activity of hemeproteins. J Biol Chem, 271, 26018–26025.

Kuppusamy, P., Chzhan, M., Vij, K., *et al.* (1994) Three-dimensional spectral-spatial EPR imaging of free radicals in the heart: a technique for imaging tissue metabolism and oxygenation. Proc Natl Acad Sci USA, 91, 3388–3392.

Kuppusamy, P., Afeworki, M., Shankar, R.A., *et al.* (1998) In vivo electron paramagnetic resonance imaging of tumor heterogeneity and oxygenation in a murine model. Cancer Res., 58, 1562–1568.

McCord, J.M. (1987) Oxygen-derived radicals: a link between reperfusion injury and inflammation. Fed Proceed, 46, 2402–2406.

Mehlhorn, R.J., Packer, L. (1984) Electron paramagnetic resonance spin destruction methods for radical detection. Methods in Enzymology, 105, 215–220.

Millar, B.C., Fielden, E.M., Smithen, C.E. (1977) Polyfunctional radiosensitzers III, effect of the biradical (Ro-03-6061) in combination with other radiosensitizers on the survival of hypoxic V-79 cells. Radiat Res, 69, 489–499.

Millar, B.C., Fielden, E.M., Smithen, C.E. (1978) Polyfunctional radiosensitizers IV. The effect of contact time and temperature on sensitization of hypoxic Chinese hamster cells in vitro by bifunctional nitroxyl compounds. Br J Cancer, 37, 73–79.

Millar, B.C., Jenkins, T.C., Fielden, E.M., Jinks, S. (1983) Polyfunctional radiosensitizers. VI. Dexamethasone inhibits shoulder modification by uncharged nitroxyl biradicals in mammalian cells irradiated in vitro. Radiat Res, 96, 160–172.

Mitchell, J.B., Samuni, A., Krishna, M.C., *et al.* (1990) Biologically active metal-independent superoxide dismutase mimics. Biochemistry, 29, 2802–2807.

Mitchell, J.B., DeGraff, W., Kaufman, D., *et al.* (1991) Inhibition of oxygen-dependent radiation-induced damage by the nitroxide superoxide dismutase mimic, Tempol. Arch Biochem Biophys, 289, 62–70.

Miura, Y., Utsumi, H., Hamada, A. (1993) Antioxidant activity of nitroxide radicals in lipid peroxidation of rat liver microsomes. Arch Biochem Biophys, 300, 148–156.

Monti, E., Paracchini, L., Perletti, G., Piccinni, F. (1991) Protective effects of spin-trapping agents on adriamycin-induced cardiotoxicity in isolated rat atria. Free Radic Res Commun, 14, 41–45.

Monti, E., Cova, D., Guido, E., Morelli, R., Oliva C (1996) Protective effects of the nitroxide Tempol against the cardiotoxicity of adriamycin. Free Radic Biol Med, 21, 463–470.

Nilsson, U.A., Olsson, L.I., Carlin, G., Bylund-Fellenius, A.C. (1989) Inhibition of lipid peroxidation by spin labels. Relationships between structure and function. J Biol Chem, 264, 11131–11135.

Patt, H.M., Tyree, E.B., Staube, R.L., Smith, D.E. (1949) Cysteine protection against X-irradiation. Science, 110, 213–214.

Petkau, A., Chelack, W.S., Pleskach, S.D., Meeker, B.E., Brady, C.M. (1975) Radioprotection of mice by superoxide dismutase. Biochem Biophys Res Commun, 65, 886–893.

Petkau, A., Chelack, W.S. (1984) Radioprotection by superoxide dismutase of macrophage progenitor cells from mouse bone marrow. Biochem Biophys Res Commun, 119, 1089–1095.

Petkau, A. (1987) Role of superoxide dismutase in modification of radiation injury. Br J Cancer Suppl, 8, 87–95.

Puck, T.T. (1958) Action of radiation on mammalian cells: III. Relationships between reproductive death and induction of chromosome anomalies by X-irradiation of euploid human cells in vitro. Proc Natl Acad Sci USA, 44, 772–280.

Purpura, P., Westman, L., Will, P., *et al.* (1996) Adjunctive treatment of murine neuroblastom with 6-hydroxydopamine and Tempol. Cancer Res, 56, 2336–2342.

Rachmilewitz, D., Karmeli, F., Okon, E., Samuni, A. (1994) A novel antiulcerogenic stable radical prevents gastric mucosal lesions in rats. Gut, 35, 1181–1188.

Reddan, J., Sevilla, M., Giblin, F., Padgaonkar, V., Dziedzic, D., Leverenz, V. (1992) Tempol and deferoxamine protect cultured rabbit lens epithelial cells from H2O2 insult: insight into the mechanism of H2O2-induced injury. Lens Eye Toxic Res, 9, 385.

Reddan, J.R., Sevilla, M.D., Giblin, F.J., *et al.* (1993) The superoxide dismutase mimic TEMPOL protects cultured rabbit lens epithelial cells from hydrogen peroxide insult. Exp Eye Res, 56, 543.

Sachs, A., Johnson, G. (1975) Mechanisms of action of 6-hydroxydopamine. Biochem Pharmacol, 24, 1–8.

Samuni, A., Krishna, C.M., Riesz, P., Finkelstein, E., Russo, A. (1988) A novel metal-free low molecular weight superoxide dismutase mimic. J Biol Chem, 263, 17921–17924.

Samuni, A., Godinger, D., Aronovitch, J., Russo, A., Mitchell, J.B. (1991a) Nitroxides block DNA scission and protect cells from oxidative damage. Biochemistry, 30, 555–561.

Samuni, A., Mitchell, J.B., DeGraff, W., Krishna, C.M., Samuni, U., Russo, A. (1991b) Nitroxide SOD-mimics: Modes of action. Free Rad Res Comms, 12–13, 187–194.

Samuni, A., Winkelsberg, D., Pinson, A., Hahn, S.M., Mitchell, J.B., Russo, A. (1991c) Nitroxide stable radicals protect beating cardiomyocytes against oxidative damage. J Clin Invest, 87, 1526–1530.

Sentjurc, M., Pecar, S., Chen, K., Wu, M., Swartz, H.M. (1989) Cellular metabolism of proxyl nitroxides and hydroxylamines. Biochimica Biophysica Acta, 1073, 329–335.

Siesjo, B.K., Agardh, C.D., Bengtsson, F. (1989) Free radicals and brain damage. Cereb Brain Metab Rev, 1, 165–211.

Simpson, P.J., Mickelson, J.K., Luchesi, B.R. (1987) Free radical scavengers in myocardial ischemia. Federation Proceedings, 46, 2413–2421.

Slater, A.F., Nobel, C.S., Maellaro, E., Bustamante, J., Kimland, M., Orrenius, S. (1995) Nitrone spin-traps and a nitroxide antioxidant inhibit a common pathway of thymocyte apoptosis. Biochem J, 306, 771–778.

Strzalka, K., Walczak, T., Sarna, T., Swartz, H.M. (1990) Measurement of time-resolved oxygen concentration changes in photosynthetic systems by nitroxide-based EPR oximetry. Arch Biochem Biophys, 281, 312–318.

Swartz, H.M. (1990) Principles of the metabolism of nitroxides and their implications for spin-trapping. Free Radic Res Commun, 9, 399–405.
Thomlinson, R.H., Gray, L.H. (1955) The histological structure of some human lung cancers and the possible implications for radiotherapy. Br J Cancer, 9, 539–549.
Walczak, T., Swartz, H.M. (1989) A 1GHz in vivo ESR spectrometer with a surface probe. Physica Medica, 5, 195–202.
Yuhas, J.M., Storer, V.B. (1969) Differential chemoprotection of normal and malignant tissues. J Natl Cancer Inst, 42, 331–335.

8 Nitroxide skin toxicity

Jürgen Fuchs

Abstract

No data are available on the irritant and sensitizing effects of nitroxide free radicals and nitrone spin-traps in skin. Nitroxides and nitrones are important biomedical skin probes used in electron paramagnetic resonance spectroscopy/imaging and potential therapeutic applications are emerging. The attempts to use nitroxide free radicals and nitrone spin-traps topically in skin require analysis of their potential cutaneous adverse effects. Our purpose was to study the skin irritation potential and the sensitizing potency of different nitroxide free radical structures and nitrone spin-traps in the skin of guinea pigs and to assess the cutaneous tolerance to nitroxides in humans. We investigated the following unsubstituted nitroxides: 2,2,6,6-tetramethyl-1-piperidinoxyl (Tempo), 2,2,5,5-tetramethyl-3-oxazolidinoxyl (Doxo), 2,2,5,5-tetramethyl-1-dihydro-pyrrolinoxyl (Proxo), 2,2,3,4,5,5-hexamethyl imidazoline-1-yloxyl (Imidazo), and the nitrones: 5,5-dimethyl-1-pyrroline-*N*-oxide (DMPO) and *N-tert*-butyl-phenylnitrone (PBN). In the guinea pig cutaneous irritation was determined following the modified Draize protocol. The response was evaluated visually as well as by a biophysical method analyzing transepidermal water loss (TEWL). The sensitizing effect was evaluated in the animals according to the Magnusson and Kligman test. In human skin the cutaneous irritation was determined after a single application and after repetitive applications of the test substances in comparison to the standardized irritant sodium lauryl sulfate (SLS). The response was evaluated clinically as well as by analyzing transepidermal water loss (TEWL). In the guinea pigs the nitroxides and nitrones were classified clinically from non-irritant (Proxo, Imidazo, DMPO; 100 mM) to slightly irritant (Tempo, Doxo, PBN; 100 mM) according to the Draize protocol. In good agreement with the clinical scoring, the TEWL values were significantly increased by Tempo, Doxo and PBN. TOLH (2,2,6,6-tetramethyl-1-hydroxypiperidine), the hydroxylamine of Tempo and one of its major skin metabolites, did not cause skin irritation. No cutaneous hypersensitivity to all nitroxides and nitrones were found, indicating a weak sensitizing potential. In human skin the nitroxides were classified clinically from non-irritant (Imidazo, Proxo; 100 mM), to slightly irritant (Doxo; 100 mM), or moderately irritant (Tempo; 100 mM) after a single application. The TEWL values were significantly increased by Doxo and Tempo. In the cumulative irritation test Tempo (10 mM) was scored as a slight irritant. TOLH did not cause skin irritation after a single (100 mM) or repetitive applications (10 mM). The order of nitroxide irritation potency (Tempo > Doxo ≫ Imidazo = Proxo) is inverse to the order of nitroxide biostability in murine and human skin (Imidazo = Proxo ≫ Doxo > Tempo). Nitroxide free radicals are classified as non-irritant to moderately irritant in human skin. Particularily the pyrroline and imidazoline type nitroxides have a low potential to cause acute or subacute skin toxicity.

8.1 Introduction

Nitroxide free radicals have been shown to function as superoxide dismutase mimics (Samuni *et al.*, 1988) and to protect mammalian cells against oxidative stress (Samuni *et al.*, 1991; Monti *et al.*, 1996; Zhang *et al.*, 1997). In view of their potential clinical (Hahn *et al.*, 1994; Cuscela *et al.*, 1996) and experimental diagnostic applications in skin (Rehfeld *et al.*, 1990; Nishi *et al.*, 1991; Taira *et al.*, 1992; Fuchs and Packer, 1991; Jurkiewicz and Buttner, 1994, 1996), the low systemic toxicity of nitroxides (Ehmann *et al.*, 1986; Afzal *et al.*, 1984), and the development of the new and emerging technique of EPR skin imaging (Herrling, 1991; Fuchs *et al.*, 1991, 1992, 1994; Herrling *et al.*, 1996) it is essential to know the cutaneous tolerance to nitroxides. Small molecular nitroxides (Fuchs *et al.*, 1992) and nitrones (Jurkiewicz and Buettner, 1994) readily cross through the epidermal barrier and penetrate well into the skin. Although nitroxides are classified as stable free radicals, they are chemically not inert and react with skin constituents. The predominant reaction of nitroxides in skin is reduction to the hydroxylamine (Fuchs *et al.*, 1989, 1993); formation of covalent bonds with nucleophiles was not observed in biological material (Swartz *et al.*, 1995). Recently the secondary amine of Tempo was identified as one of the major metabolites besides the hydroxylamine inside human keratinocytes (Kroll and Borchert, 1997). Spin-traps are usually one electronic oxidation step above the nitroxides, which upon reacting with free radicals, become converted to nitroxides. They have been shown to protect cells from oxidative stress (Hearse and Tosaki, 1987) and are presumably fairly stable in cells.

Free radicals have been suggested to be mediators of cutaneous inflammation (Viluksela, 1991; Trenam *et al.*, 1991, 1992), and the radical mechanism has gained interest in the discussion of the mechanism of hapten-protein binding (Schmidt *et al.*, 1990). Studies indicating that radical reactions are important for haptens containing hydroperoxide groups have been published (Gäfvert *et al.*, 1994; Lepoittevin and Karlberg, 1994). The low chemical reactivity of most nitroxide compounds with diamagnetic reactants would imply a low allergenic potential of nitroxides. Through the reduction of the nitroxides to hydroxylamines and amines, nitroxides use up cellular reducing equivalents. Piperidine nitroxides diminish reduced glutathione and ascorbate in human keratinocytes and fibroblasts, and thus may change the redox status in these cells (Fuchs *et al.*, 1997). The nitrone spin-trap PBN was also reported to decrease cellular glutathione (Albano *et al.*, 1986). The cellular redox status modulates the expression of proinflammatory cytokines. Particularly the upregulation of interleukin-8 expression through an oxidant stress indicates that reactive oxidants participate in proinflammatory cytokine activation in human cells (Deforge *et al.*, 1993). This would imply some irritant potential of nitroxide free radicals and nitrone spin-traps. In the literature no data are available on the skin sensitizing and irritant potential of nitroxide free radicals and nitrone spin-traps. The most important compounds containing the nitroxide group are those of imidazolines, pyrrolines, piperidines, and oxazolidines, which slightly differ in their physicochemical properties and biostability. The objective of this study was to investigate the potential for cutaneous irritation and sensitization of unsubstituted imidazoline (Imidazo), pyrroline (Proxo), piperidine (Tempo), and oxazolidine nitroxides (Doxo) and two frequently used nitrone spin-traps (DMPO, PBN) in guinea pigs according to well established and standardized techniques and to study the potential for cutaneous irritation of the nitroxide compounds in humans according to well established and standardized techniques.

8.2 Materials and methods

8.2.1 Chemicals

2,4-dinitro-1-chlorobenzene (DNCB), Tempo (2,2,6,6-tetramethyl-1-piperidinoxyl), sodium lauryl sulfate (SLS), Freunds complete adjuvant, DMPO (5,5-dimethyl-1-pyrroline-*N*-oxide) and PBN (*N-tert*-butyl-phenylnitrone) were purchased from Sigma Chemical Company, Munich, Germany. Doxo (2,2,5,5-tetramethyl-3-oxazolidinoxyl) and the hydroxylamine of Tempo, TOLH (2,2,6,6-tetramethyl-1-hydroxypiperidine) were synthesized as described (Fuchs *et al.* 1990; Mehlhorn, 1991). Proxo (2,2,5,5-tetramethyl-1-dihydropyrrolinoxyl) was kindly donated by Dr K. Hideg, Pecs, Hungary. Imidazo (2,2,3,4,5,5-hexamethyl-imidazoline-1-yloxyl) was a gift of Dr Volodarsky, Novosibirsk, Russia. The structural formulas of the nitroxides and spin-traps are shown in Figure 8.1. The substances were applied on the skin using Finn chambers (Hermal Chemie, Hamburg).

8.2.2 Animals

Guinea pigs, weighing 300–400 g, approximately 7–8 weeks old (Charles River Wiga, Sulzfeld, Germany) were used for the maximized Magnusson and Kligman test and for the Draize protocol. The animals were housed separately in stainless-steel cages and identified by tags attached to one ear. Water and food pellets were available *ad libitum*. The environment was controlled with a 12h light/dark cycle, a room temperature of 22°C, and a relative humidity of 60%.

8.2.3 Cutaneous irritation in the guinea pig

In the modified Draize protocol (Draize *et al.*, 1944; Zisu, 1995) six animals were used in each test group, a total of six groups (four nitroxide groups, two spin-trap groups) were investigated. The test substance (0.5 ml/10 mM, or 100 mM final concentration) in petrolatum (petroleum jelly) was applied by an occlusive patch for 24 hours on the flank, carefully shaved the day before. The substances were also applied on the contralateral scarified flank. Petrolatum was used as a negative control, and an aqueous solution of 5% SLS as a positive control in each animal. Animals were observed for erythema and oedema, the response was evaluated according to the classification outlined below. The scores 24 hours and 72 hours after application of the substances were added (non-scarified and scarified zones). The primary irritation values are calculated by averaging values for erythema from all sites (abraded and non-abraded), averaging the values for edema from all sites, and adding the average values. The total was divided by 4, corresponding to the sum of two scores (erythema and oedema) obtained after 24 and 72 hours. The mean is termed

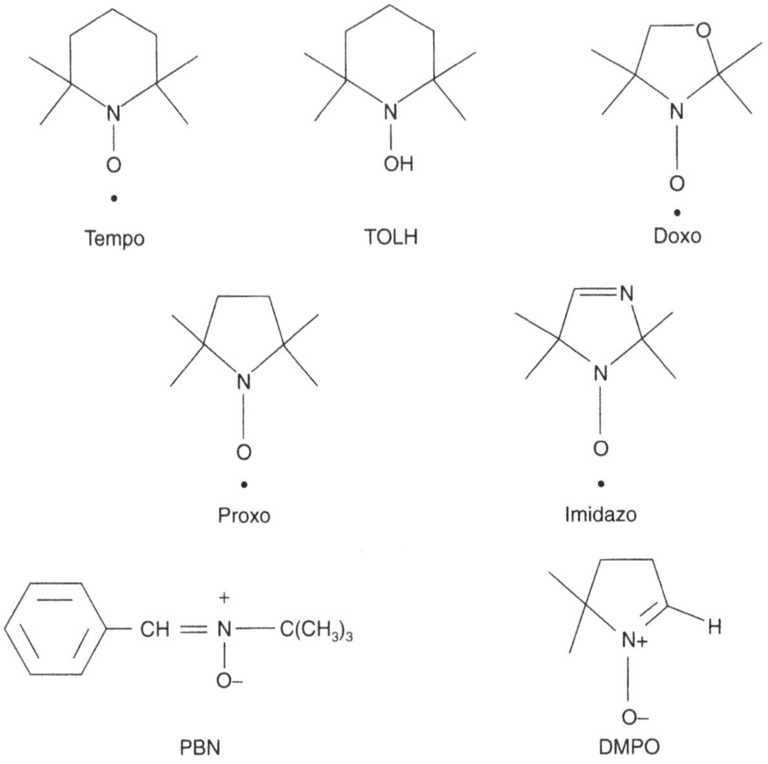

Figure 8.1 Structural formulas of the nitroxide free radicals and nitrone and spin-traps investigated in this study

Table 8.1 Clinical scoring and classification of cutaneous irritation

Score	*Erythematous lesion*	*Oedematous lesion*
0	No erythema	No oedema
1	Very slight erythema	Very slight oedema
2	Well defined erythema	Well defined oedema
3	Erythema very marked	Moderate oedema (raised < 1 mm)
4	Severe erythema	Severe oedema (raised > 1 mm)

Index of primary irritation	*Classification*
< 0.5	Non irritant
0.5–1.9	Slightly irritant
2.0–4.9	Moderately irritant
5.0–8.0	Severely irritant

the primary cutaneous irritation index (Table 8.1) and may be used to classify the test substance (Zisu, 1995). The degree of irritation was also evaluated by measurement of transepidermal water loss (TEWL) after 24 hours at the non-scarified skin area. TEWL was measured with an Evaporimeter EP1 (Servo Med, Stockholm, Sweden) after a 30 minute acclimatization period at 20°C room temperature and 40–50% humidity. Statistical evaluation of the biophysical data was performed using the Student t-test to compare TEWL values recorded from irritated skin with the control site.

8.2.4 Cutaneous sensitization in the guinea pig

In the standardized performance according to Magnusson and Kligman (Magnusson and Kligman, 1969; Schlede and Eppler, 1995) 10 animals were used in each experimental group, a total of eight groups were investigated (four nitroxide groups, two spin-trap groups, one positive control group (DNCB) and one negative control group (petrolatum)).

1 Intradermal induction: In the intradermal induction treatment, the animals are given the following injections at the beginning of the first week in the shoulder region on both the right and left sides. 1) Cranial injections: 0.1 ml Freund's complete adjuvant (1:1 with physiological saline). 2) Medial injections: 0.1 ml of the test substance at a locally and systemically well tolerated concentration (10–100 mM in physiological saline, see results Draize test). 3) Caudal injections: 0.1 ml Freund's complete adjuvant + test substance (1:1), final concentration of the test substance as for the medial injection.
2 Dermal induction: In the dermal induction experiment on day eight, the test substance was applied at a concentration causing slight to moderate irritation in petrolatum to a filter paper and fixed to the site of application for 48 hours. If the test substance caused no irritation (see results Draize test), the shaved area to be exposed to the test substance was pretreated with 5% sodium lauryl sulfate for 24 hours to cause skin irritation. The animals were then left to rest until the challenge test. The negative controls are treated in the same way as the test animals, except that during the induction experiments, application of the test material is omitted. The positive controls were exposed to the skin sensitizer DNCB (0.01%).
3 Challenge: In the challenge experiment on day 24, the test substance in the vehicle (petrolatum) was applied at a non-irritant concentration (see results Draize protocol) to the sheared flank on a filter paper (2 × 2 cm), fixed to the skin for 24 hours under an occlusive covering.
4 Evaluation of the test results: The treated area of the skin was cleaned and inspected 24, 48 and 72 hours after removal of the occlusive patch. The skin reaction after challenge was assessed according to the Draize scale. Evaluation of the sensitizing potential of the substances is based on the number of guinea pigs showing a positive degree of sensitization (score > 1) in comparison to the total numbers of animals tested (Table 8.2).

8.2.5 Cutaneous tolerance to nitroxides in human skin

The irritant properties of nitroxides were tested in human skin by a single application (a), and by repetitive applications (b).

(a): The nitroxide compounds to be tested were applied for 24 hours in different concentrations (10 mM, 100 mM) to the flexor site of the distal arm of human

Table 8.2 Skin irritation and sensitization in the guinea pig to nitroxides and nitrones

Compounds	*Cutaneous irritation index*	*Transepidermal water loss (g/m²h)*	*Significance*	*Sensitization rate*
negative control	0 non irritant	13 ± 4		0/10
positive control SLS 5%	2.5 moderately irritant	32 ± 10	$p < 0.001$	
positive control DNCB 0.01%				10/10
Tempo 100 mM	1.7 slightly irritant	30 ± 8	$p < 0.001$	
Tempo 10 mM	0 non irritant	16 ± 4	n.s.	0/10
TOLH 100 mM	0 non irritant	15 ± 5	n.s.	0/10
Doxo 100 mM	0.9 slightly irritant	24 ± 5	$p < 0.05$	
Doxo 10 mM	0 non irritant	14 ± 4	n.s.	0/10
Proxo 100 mM	0 non irritant	16 ± 5	n.s.	0/10
Imidazo 100 mM	0 non irritant	15 ± 6	n.s.	0/10
PBN 100 mM	1.4 slightly irritant	25 ± 3	$p < 0.001$	
PBN 10 mM	0 non irritant	13 ± 3	n.s.	0/10
DMPO 100 mM	0 non irritant	15 ± 4	n.s.	0/10

volunteers. Nitroxide solution in water, 50 μl, was pipetted onto a filter disk, which was then placed in a Finn chamber on the skin. Water was used as a negative control and SLS 5% as a positive control in the experiment. Twenty healthy volunteers (21–57 years old) participated in the study after informed consent was obtained. In each individual, four different nitroxides were tested at 10 mM and 100 mM concentration, plus a positive and a negative control. Patients were free of eczema and had no history of atopic dermatitis or respiratory atopy. The degree of irritation was evaluated clinically (Table 8.1) and by measurement of transepidermal water loss before the application and after 24 hours. The nitroxides were also studied in scarified skin (Frosch and Kligman, 1976). The skin on the volar aspect of the contralateral forearm was scarified in each test areal with a fine needle eight times in a vertical chessboard-like pattern. Subsequently the nitroxide solutions were applied as described above. For macroscopical evaluation the exposed skin was observed for erythema and oedema, the response was evaluated according to a standardized classification (Zisu, 1995). Biophysical measurements were carried out before the application and after 72 hours at the non-scarified site.

(b): The nitroxide solutions were exposed to the inner site of the proximal arm in different concentrations (1 mM, 10 mM) in Finn chambers on 20 subjects over 21 days according to the protocol described (Lanman *et al.*, 1968). The intensity of the cutaneous reaction is scored clinically on a five point scale (0, no visible reaction; 1+, faint, patchy erythema; 2+, erythema and oedema; 3+, erythema and oedema with papules; 4+, erythema and oedema with vesicles; 5+, erythema with bulla). The scores were recorded daily as the test material was renewed. SLS 0.5% and water were used as a positive and negative control in each individual. The scores were added and divided by 21 corresponding to the sum of 21 scores obtained during the three week test period. Irritants were classified on mean values: score 0–0.4 = very low irritant, 0.5–1.4 = slight irritant, 1.5–2.4 = moderate irritant, 2.5–3.4 strong irritant, 3.5–4.5 very strong irritant. Biophysical measurements were carried out before the application and after 21 days.

TEWL was measured with an Evaporimeter EP1 (Servo Med, Stockholm, Sweden) after a 30 minute acclimatization period at 20°C room temperature and 40–50% humidity. Statistical evaluation of the data was performed using the Student t-test to compare TEWL values recorded from irritated skin with the control site.

8.3 Results

8.3.1 Preclinical studies

For most analytical purposes nitroxide and nitrone concentrations are used in the μM or low mM region. At concentrations of 100 mM and above self quenching of nitroxide radicals results in electron paramagnetic resonance (EPR) signal distortion, which makes EPR measurements not feasible. In practice, for studies in which EPR spectroscopy is a major goal of the investigation, the effective concentration of the nitroxide is likely to be within an order of magnitude of 0.1 mM. In dermatologic EPR applications (e.g. skin imaging) we use nitroxide concentrations up to 10 mM. We therefore tested skin irritation at concentrations between 1 and 100 mM. At 1 mM and 10 mM no signs of skin irritation were detected by clinical and biophysical measurement for all nitroxides and nitrones (data not shown for 1 mM concentration). At 100 mM concentration Tempo (score 1.7), Doxo (score 0.9) and PBN (score 1.4) were scored clinically as slightly irritant in the skin irritation test according to the Draize protocol when compared to the moderate irritant SLS 5% (score 2.5) (Table 8.2). Imidazo, Proxo and DMPO were evaluated as non-irritant at 100 mM (score 0). The biophysical measurement showed significantly increased values of TEWL for the positive control SLS 5% (32 ± 10 g/m^2 h), for Tempo 100 mM (30 ± 8 g/m^2 h), for Doxo 100 mM (24 ± 5 g/m^2 h) and for PBN (25 ± 3) in comparison to the negative control water (13 ± 4 g/m^2 h) (Table 8.2). In the skin sensitization test according to Magnusson and Kligman no cases of positive skin reactions were observed for the nitroxides and nitrones tested, in comparison to 10 cases of reactions in 10 animals exposed to the positive control DNCB 0.01% (Table 8.2). TOLH (100 mM), the hydroxylamine of Tempo, did neither cause skin irritation nor sensitization (Table 8.2).

8.3.2 Clinical studies

In the single application test Tempo 100 mM was scored moderately irritant (score 2.3) and Doxo 100 mM slightly irritant (score 0.9), in comparison to the moderate irritant SLS 5% (score 3.1) (Table 8.3). Imidazo 100 mM and Proxo 100 mM were evaluated as non-irritant (score 0) in human skin after a single application. In good agreement with other reports the TEWL increased from 5 ± 2 g/m^2 h in non-treated skin (data not shown) to 15 ± 4 g/m^2 h immediately after removal of the water patch (negative control) (Agner, 1993). This is interpreted as a transient damage to the skin

Table 8.3 Human skin irritation after a single application of nitroxides

Nitroxide	*Irritation index score*	*TEWL (g/m²h)*	*Significance*
control	0 non irritant	15 ± 4	
SLS 5%	3.1 moderately irritant	40 ± 18	p < 0.001
Tempo 100 mM	2.3 moderately irritant	33 ± 10	p < 0.001
Tempo 10 mM	0 non irritant	17 ± 5	n.s.
TOLH 100 mM	0 non irritant	13 ± 5	n.s.
Doxo 100 mM	0.9 slightly irritant	24 ± 5	p < 0.05
Doxo 10 mM	0 non irritant	17 ± 3	n.s.
Proxo 100 mM	0 non irritant	17 ± 4	n.s.
Proxo 10 mM	0 non irritant	14 ± 3	n.s.
Imidazo 100 mM	0 non irritant	16 ± 4	n.s.
Imidazo 10 mM	0 non irritant	15 ± 3	n.s.

Table 8.4 Human skin irritation after cumulative application of nitroxides

Nitroxide	*Irritation index score*	*TEWL (g/m²h)*	*Significance*
control	0 non irritant	12 ± 4	
SLS 0.5%	2.2 moderately irritant	35 ± 8	p < 0.001
Tempo 10 mM	1.8 slightly irritant	29 ± 6	p < 0.01
Tempo 1 mM	0 non irritant	15 ± 3	n.s.
TOLH 10 mM	0 non irritant	11 ± 5	n.s.
Doxo 10 mM	0 non irritant	15 ± 4	n.s.
Doxo 1 mM	0 non irritant	10 ± 6	n.s.
Proxo 10 mM	0 non irritant	14 ± 4	n.s.
Proxo 1 mM	0 non irritant	11 ± 3	n.s.
Imidazo 10 mM	0 non irritant	13 ± 5	n.s.
Imidazo 1 mM	0 non irritant	15 ± 3	n.s.

barrier by the occlusive water patch. The bioengineering measurement showed significantly increased values of TEWL for the positive control SLS 5% (40 ± 18 g/m^2 h), for Tempo 100 mM (33 ± 10 g/m^2 h) and for Doxo 100 mM (24 ± 5 g/m^2 h) in comparison to the negative control water (15 ± 4 g/m^2 h) (Table 8.3).

In the cumulative irritation test Tempo 10 mM was scored as a slight irritant (score 1.8) in comparison to the moderate irritant SLS 0.5% (score 2.2) after repetitive applications over 21 days. All the other nitroxides (10 mM) were scored non irritant (score 0) after repetitive applications (Table 8.4). The bioengineering measurement showed significantly increased TEWL values for Tempo 10 mM (29 ± 6 g/m^2 h) and SLS 0.5% (35 ± 8 g/m^2 h) in comparison to the negative control water (12 ± 4 g/m^2 h) (Table 8.4). TOLH (100 mM) did not cause skin irritation after a single (Table 8.3) or repetitive applications (Table 8.4) as assessed macroscopically or biophysically by TEWL measurements.

8.4 Discussion

8.4.1 Preclinical studies

To evaluate the potential of nitroxide free radicals and nitrone spin-traps to produce skin irritation from a single application in the guinea pig, a skin irritation test based on the Draize method (Draize *et al.*, 1944) was used, which is based on visual scoring. The Draize test is very valuable in the identification of strong irritants, but it may not be sensitive enough to separate mild from moderate irritants (Patil *et al.*, 1996). Skin irritation results in a disturbed barrier function which can be analyzed by biophysical techniques. Measurement of transepidermal water loss (TEWL) was suggested to be a suitable non-invasive technique for determination of skin barrier function (Pinnagoda and Tupker, 1995), and the sensitivity of TEWL measurements as a screening technique for early and weak signs of irritancy was found to be superior as compared with visual scoring (Agner and Serup, 1990). The nitroxides were compared with the positive control sodium lauryl sulfate, which causes skin irritation through its action as surfactant and membrane perturbator. Sodium lauryl sulfate is suitable for the purpose of irritant patch testing, because of its ability to influence the barrier function of the skin as well as to cause skin inflammation. Measurement of TEWL is well suited for quantification of irritant patch testing and shows a linear dose response relationship between dose of SLS and skin response evaluated by clinical assessment of skin inflammation (Agner and Serup, 1990). The TEWL is profoundly influenced by a variety of physiological and environmental factors and the recording technique needs to be standardized (Van Sam *et al.*, 1994). We therefore standardized the TEWL measurements for equilibration time, humidity, room temperature and anatomic site.

The visual scoring data as well as the results obtained by the biophysical method for determination of skin irritation both indicate that the nitroxides and spin-traps selected here are either non irritants (Imidazo, Proxo, DMPO) or slight irritants (Tempo, Doxo, PBN) (Table 8.2). The order of skin biostability of the nitroxides (Proxo = Imidazo ≫ Doxo > Tempo) determined in the hairless mouse (Fuchs *et al.*, 1993), is inverse to the oder of the skin irritation potential in the guinea pig (Tempo > Doxo ≫ Imidazo = Proxo).

The results of the standardized guinea pig maximation test according to Magnusson and Kligman show that none of the nitroxide free radicals and nitrone spin-traps induced an allergic type reaction (Table 8.2), indicating a weak sensitizing potential. The well known skin sensitizer DNCB (positive control) caused skin sensitization in 10 out of 10 animals (Table 8.2), indicating adequate sensitivity of the test animals. The Magnusson and Kligman test is suitable for allergenicity assessment of neat chemicals as well as end-use products, and is sensitive enough to identify weak contact allergens. However this test can only predict the sensitizing potential of a compound, but it cannot predict whether a positive compound will act as a weak or strong sensitizer in human skin (Maurer *et al.*, 1994).

TOLH, one of the major skin metabolites of TEMPO, did not cause skin irritation or sensitization. Metabolic processes resulting in reduction of the nitroxide into its corresponding hydroxylamine (Fuchs *et al.*, 1997) or to the secondary amine (Kroll and Borchert, 1997), may be involved in the mechanism of cutaneous irritation caused by Tempo. PBN, which also caused skin irritation in the guinea pig, has been shown to decrease cellular glutathione and affect cell integrity as indicated by leakage of marker enzymes (Albano *et al.*, 1986). DMPO, which caused no skin irritation in this study, is described in the literature as having a low toxicity in cells (Sentjurc *et al.*, 1995).

8.4.2 *Clincal trial*

A single application of nitroxides on human skin caused no clinical sign of irritation (Imidazo, Proxo; 100 mM), slight irritation (Doxo; 100 mM), or moderate irritation (Tempo; 100 mM), as assessed by visual scorring. Biophysical measurements of TEWL confirmed the clinical results, Doxo and Tempo caused a significant increase of TEWL at 100 mM concentration.

Low level irritants can only be differentiated after repetitive exposure at a relatively low concentration. To assess the irritant potency of nitroxides after repetitive applications, the cumulative irritation test according to Lanman was applied. In the cumulative irritation assay nitroxides were tested at a maximum concentration of 10 mM. Macroscopic investigations of cutaneous irritation after repetitive applications over 21 days showed that only the piperidine nitroxide Tempo caused slight skin irritation. The bioengineering technique showed a significant increase in TEWL after repetitive applications of Tempo compared to the negative control.

In human skin the order of nitroxide irritation potency in the acute exposure test (Tempo > Doxo ≫ Proxo = Imidazo) was inverse to the order of nitroxide biostability in human keratinocytes and skin biopsies (Imidazo = Proxo ≫ Doxo > Tempo), (Fuchs *et al.*, 1993, 1997). TOLH did not cause cutaneous irritation after acute or repetitive administration.

8.5 Conclusion

The available data from studies in both *in vitro* and *in vivo* systems, show that at the concentrations that are usually used in functional biological systems (1 μM–1 mM), nitroxides usually do not cause readily detectable toxicity and upon closer analysis they usually appear to be quite nontoxic. To the extent that toxic effects have been noted, the pyrrolidine nitroxides seem to be less cytotoxic than piperidine ring nitroxides (Swartz *et al.*, 1995). Our results are in good agreement with these reports. As is the case for all drugs, nitroxide concentrations have been found which were lethal to the test animals. Systemic toxicity of piperidine and pyrroline type nitroxides have been investigated in the rat (Afzal *et al.*, 1984) and mouse (Morgan *et al.*, 1985; Ehman, 1985). In the rat LD_{50} values of 15 mmol/kg body weight (pyrroline type nitroxide intravenously) or 2.3–4.3 mmol/kg body weight (piperidine type nitroxide intravenously), and in the mouse LD_{50} values of 2–6 mmol/kg body weight (piperidine type nitroxide intraperitoneally) were published. To the best of our knowledge there are no reports in the literature about percutaneous nitroxide toxicity. In our experiments no systemic toxicity of nitroxides or nitrones following cutaneous application was observed. The 'therapeutic margin', which indicates how close the dose used is to a dose that has a high probability of a specified toxic effect of, for example, the pyrroline type nitroxide Proxo and the spin-trap DMPO in practical EPR spectroscopy is large (1 mM versus > 100 mM). This is in good agreement with the findings of Lovin *et al.*, who noted that the LD_{50} for some nitroxides is 100 times higher than the dose needed to obtain renal enhancement when these are used as contrast agents for magnetic resonance imaging (Lovin *et al.*, 1985).

While there are some reports of potential long-term effects of nitroxides as reflected by studies of mutagenicity, most such studies show that nitroxides are not very effective mutagens (Sosnovsky *et al.*, 1992; Swartz *et al.*, 1995). There may be an inverse correlation between mutagenicity and resistance of the nitroxide to reduction (Gallez *et al.*, 1992), indicating that piperidine type nitroxides possess some mutagenic activity. However, under some circumstances nitroxides may decrease mutagenic activity of some carcinogens (De Graff *et al.*, 1992), and can be effective antioxidants, which protect cells from oxidative stress (Mitchell *et al.*, 1990). The risk profile of piperidine type nitroxides, which are versatile spin probes because of their redox behaviour, should be analyzed in human skin in comparison to the toxicity and mutagenicity potential of topical drugs used routinely in dermatologic therapy. Anthralin, fluorouracil, tar preparations, photochemotherapy employing 8-methoxypsoralene and UVA light, or photodynamic therapy employing 5-aminolevulinic acid and red light are such regimens, which have been shown to possess some mutagenic activity in animals and human cells. However, all these drugs and procedures are used safely in humans when some precautions are followed.

Considering all the reports in the literature about nitroxide toxicity (Sosnovsky *et al.*, 1992; Swartz *et al.*, 1995), and the results of this study, we conclude that pyrroline and imidazoline type nitroxides can be used safely as biomedical skin probes in humans. Although piperidine type nitroxides caused moderate skin irritation after a single application (100 mM), they may be used at lower concentration (10 mM) for topical applications in human skin without the risk for serious cutaneous side effects. In human subjects no clinical signs of allergies have developed so far after extended cutaneous testing of unsubstituted piperidine and pyrroline type nitroxides (Herrling *et al.*, 1997). Substituted nitroxide compounds, e.g. nitroxides carrying maleimido, isothiocyano, iodoacetamido and other groups or lipophilic side chains, may however have different cutaneous toxicity profiles and must be evaluated individually.

References

Afzal, F., Brasch, R.C., Nitecki, D.E., Wolff, S. Nitroxyl spin label contrast enhancers for magnetic resonance imaging. Invest Radiol 19: 549–552, 1984.

Agner, T., Serup, J. Sodium lauryl sulphate for irritant patch testing – a dose-response study using bioengineering methods for determination of skin irritation. J Invest Dermatol 95: 543–549, 1993.

Albano, E., Cheeseman, K.H., Tomasi, A., Carini, R., Dianzini, M.U., Slater, T.F. Effect of spin-traps in isolated rat hepatocytes and liver microsomes. Biochem Pharmacol 35: 3955–3962, 1986.

Cuscela, D., Coffin, D., Lupton, G.P., Cook, J.A., Krishna, M.C., Bonner, R.F., Mitchell, J.B. Protection from radiation-induced alopecia with topical application of nitroxides: fractionated studies. Cancer J Sci Am 2: 273–278, 1996.

Deforge, L.E., Preston, A.M., Takeuchi, E., Kenny, J., Boxter, L.A., Remick, D.G. Regulation of interleukin 8 gene expression by oxidant stress. J Biol Chem 268: 25568–25576, 1993.

De Graff, W.G., Krishna, M.C., Russo, A., Mitchell, J.B. Antimutagenicity of a low molecular weight superoxide dismutase mimic against oxidative mutagens. Environ Mol Mutagenesis 19: 21–26, 1992.

Draize, J.H., Woodard, G., Calvery, H.O. Methods for the study of irritation and toxicity of substances applied topically to the skin and mucous membranes. J Pharmacol Exp Ther 821: 377–390, 1944.

Ehman, R.L., Brasch, R.C., McNamara, M.T., Erikkson, U., Sosnovsky, G., Lukszo, J., Li, S.W. Diradical nitroxyl spin label contrast agents for magnetic resonance imaging. Invest Radiol 21: 125–131, 1986.

Ehman, R.L., Wesbey, G., Moon, K., Williams, R., McNamara, M., Couet, W., Tozer, T., Brasch, R. Enhanced MRI of tumors utilizing a new nitroxyl-spin label contrast agent. Magn Reson Imaging 3: 89–94, 1995.

Frosch, P.J., Kligman, A.M. The chamber scarification test for irritancy. Contact Dermatitis 2: 314–318, 1976.

Fuchs, J., Mehlhorn, R.J., Packer, L. Free radical reduction mechanisms in mouse epidermis and skin homogenates. J Invest Dermatol 93: 633–640, 1989.

Fuchs, J., Nitschmann, W.H., Packer, L., Hankovszky, O., Hideg, K., pK values and partition coefficients of nitroxide spin probes for membrane bioenergetic measurements. Free Rad Res Commun 10: 315–323, 1990.

Fuchs, J., Milbradt, R., Groth, N., Herrling, N., Zimmer, G., Packer, L. One- and two dimensional EPR (Electron Paramagnetic Resonance) imaging in skin. Free Rad Res Commun 15: 245–253, 1991.

Fuchs, J., Packer, L. Electron paramagnetic resonance (EPR) in dermatologic research with particular reference to photodermatology. Photodermatol Photoimmunol Photomed 7: 229–232, 1991.

Fuchs, J., Milbradt, R., Groth, N., Herrling, T., Zimmer, G., Packer, L. Electron Paramagnetic Resonance (EPR) imaging in skin: biophysical and biochemical microscopy. J Invest Dermatol 98: 713–719, 1992.

Fuchs, J., Freisleben, H.J., Podda, M., Zimmer, G., Milbradt, R., Packer, L. Nitroxide radical biostability in skin. Free Rad Biol Med 15: 415–423, 1993.

Fuchs, J., Groth, N., Herrling, T., Packer, L. In-vivo EPR skin imaging. Methods Enzymology 203: 140–149, 1994.

Fuchs, J., Groth, N., Herrling, T., Zimmer, G. Electron paramagnetic resonance studies on nitroxide radical 2,2,5,5-tetramethyl-4-piperdine-1-oxyl (Tempo) redox reactions in human skin. Free Rad Biol Med 22: 967–976, 1997.

Gallez, B., De Meester, C., Debuyst, R., Dejehet, F., Dumont, P. Mutagenicity of nitroxyl compounds: structure activity relationship. Toxicol Lett 63: 35–45, 1992.

Gäfvert, E., Shao, L.P., Karlberg, A.T., Nilson, U., Nilson, J.L.G. Contact allergy to resin acid hydroperoxides. Hapten binding via free radicals and epoxides. Chem Res Toxicol 7: 260–266, 1994.

Hahn, S.M., Krishna, C.M., Samuni, A., Degraff, W., Cuscela, D.O., Johnstone, P., Mitchell, J.B. Potential use of nitroxides in radiation oncology. Cancer Res 54 (Suppl): 2006S–2010S, 1994.

Hearse, D.J., Tosaki, A. Free radicals and reperfusion-induced arrythmias: Protection by spin-trap agent PBN in the rat heart. Circ Res 60: 375–383, 1987.

Herrling, T. Modulated field gradient: instrumentation. In: EPR imaging and in-vivo EPR, Eaton, G.R., Eaton, E.E., Ohno, K., eds, CRC Press, Boca Raton, 35–47, 1991.

Herrling, T.E., Groth, N., Fuchs, J. Biochemical EPR imaging of skin. Applied Magnetic Resonance 11: 471–486, 1996.

Herrling, T., Fuchs, J., Zastrow, L., Malenke, B., Groth, N. In-vivo application of electron paramagnetic resonance (EPR) in human skin. 1st Asia-Pacific EPR/ESR Symposium, City University of Hong Kong, Hong Kong, 20–24 January, 1997.

Jurkiewicz, B.A., Buettner, G.R. EPR detection of free radicals in UV irradiated skin: mouse versus human. Photochem Photobiol 64: 918–935, 1996.

Jurkiewicz, B.A., Buettner, G.R. Ultraviolet light induced free radical formation in skin: an electron paramagnetic resonance study. Photochem Photobiol 59: 1–4, 1994.

Kroll, C., Borchert, H.H. Metabolism of nitroxide spin probes with piperidine structure in a human keratinocyte cell line. International Workshop on ESR (EPR) Imaging and in-vivo ESR spectroscopy. Yamagata, Japan, October 12–16, 1997.

Lanman, B.M., Elvers, W.B., Howard, C.J. The role of human patch testing in a product development program. Proc Joint Conf Cosmetic Science, The Toillet Goods Association, Washington DC, 135, 1968.

Lepoittevin, J.P., Karlberg, A.T. Interaction of allergic hydroperoxides with proteins. A radical mechanism. Chem Res Toxicol 7: 130–133, 1994.

Lovin, J.D., Wesbey, G.E., Engelstad, B.L., Sosnovsky, G., Moseley, M., Tuck, D.L., Brasch, R.C. Magnetic field dependence of spin lattice relaxation enhancement using piperidinyl nitroxyl spin labels. Magn Reson Imaging 3: 73–78, 1985.

Magnusson, B., Kligman, A.M. The identification of contact allergens by animal assay. The guinea pig maximation test. J Invest Dermatol 52: 268–276, 1969.

Maurer, T., Arthur, A., Bentley, P. Guinea pig contact sensitization assays. Toxicology 93: 47–54, 1994.

Mehlhorn, R.J. Ascorbate- and dehydroascorbic acid mediated reduction of free radicals in the human erythrocyte. J Biol Chem 266: 2724–2731, 1991.

Mitchell, J.B., Samuni, A., Krishna, M.C., De Graff, W.O., Samuni, U., Russo, A. Biologically active metal-independent superoxide dismutase mimics. Biochemistry 29: 2802–2807, 1990.

Monti, E., Cova, D., Guido, E., Morelli, R., Oliva, C. Protective effect of the nitroxide tempol against cardiotoxicity of adriamycin. Free Rad Biol Med 21: 463–470, 1996.

Morgan, D.D., Mendenhall, C.L., Bobst, A.M., Rouster, S.D. Incorporation of the spin-trap DMPO into cultured fetal mouse liver cells. Photochem Photobiol 42: 93–97, 1985.

Nishi, J., Ogura, R., Sugiyama, M., Hidaka, T., Kohno, M. Involvement of active oxygen in lipid peroxide radical reaction of epidermis following ultraviolet light exposure. J Invest Dermatol 97: 115–119, 1991.

Patil, S., Patrick, E., Maibach, H.I. Animal, human and in vivo test methods for predicting skin irritation. In: Dermatotoxicology, 5th edition, Marzulli, F.N., Maibach, H.I., eds, Taylor & Francis, San Francisco, 411–436, 1996.

Pinnagoda, J., Tupker, R.A. Measurement of the transepidermal water loss. In: Handbook of non-invasive methods and the skin, Serup, J., Jemec, G.B.E., eds, CRC Press, Boca Raton, 173–178, 1995.

Rehfeld, S.J., Plachy, W.Z., Hou, S.Y.E., Elias, P.M. Localization of lipid microdomains and thermal phenomena in murine stratum corneum and isolated membrane complexes: an electron spin resonance study. J Invest Dermatol 95: 217–223, 1990.

Samuni, A., Krishna, C.M., Riesz, P., Finkelstein, E., Russo, A. A novel metal free low molecular weight superoxide dismutase mimic. J Biol Chem 263: 17921–17926, 1988.

Samuni, A., Winkelsberg, D., Pinson, A., Hahn, S.M., Mitchell Jbm, Russo, A. Nitroxide stable radicals protect beating cardiomyocytes against oxidative damage. J Clin Invest 87: 1526, 1991.

Schlede, E., Eppler, R. Testing for skin sensitization according to the notification procedure for new chemicals: the Magnusson and Kligman test. Contact Dermatitis 32: 1–4, 1995.

Schmidt, R.J., Khan, L., Chung, L.Y. Are free radicals and not quinones the haptenic species derived from urushiols and other contact allergenic mono- and dihydric alkylbenzenes? The significance of NADH, glutathione, and redox cycling in the skin. Arch Dermatol Res 282: 56–64, 1990.

Sentjurc, M., Swartz, H.M., Kocherginsky, N. Metabolism, toxicity and distribution of spin-traps. In: Nitroxide spin labels. Reactions in biology and chemistry, Kocherginsky, N., Swartz, H.M., eds, CRC Press, Boca Raton, 199–206, 1995.

Sosnovsky, G. A critical evaluation of the present status of toxicity of aminoxyl radicals. J Pharm Sci 81: 496–499, 1992.

Swartz, H.M., Sentjurc, M., Kocherginsky, N. Toxicity and the use of nitroxides as drugs. In: Nitroxide spin labels. Reactions in biology and chemistry, Kocherginsky, N., Swartz, H.M., eds, CRC Press, Boca Raton, 175–198, 1995.

Taira, J., Mimura, K., Yoneya, T., Hagi, A., Murakami, A., Makino, K. Hydroxyl radical formation by UV irradiated epidermal cells. J Biochem (Tokyo) 111: 693–695, 1992.

Trenam, C.W., Dabbagh, A.J., Morris, C.J., Blake, D.R. Skin inflammation induced by reactive oxygen species (ROS): an in-vivo model. Br J Dermatol 125: 325–329, 1991.

Trenam, C.W., Dabbagh, A.J.D., Morris, C.J. The role of iron in an acute model of skin inflammation induced by reactive oxygen species (ROS). Br J Dermatol 126: 250–256, 1992.

Van Sam, V., Passet, J., Maillols, H., Guillot, B., Guilhou, J.J. TEWL measurement standardization: Kinetic and topographic aspects. Acta Derm Venereol (Stockh) 74: 168–170, 1994.

Viluksela, M. Characteristics and modulation of dithranol (anthralin) induced skin irritation in the mouse ear model. Arch Dermatol Res 283: 262–268, 1991.

Zhang, R., Pinson, A., Samuni, A. Both hydroxylamine and nitroxide protect cardiomyocytes from oxidative damage. J Mol Cell Cardiol 29: A101, 1997.

Zisu, D. Experimental study of cutaneous tolerance to glycol ethers. Contact Dermatitis 32: 74–77, 1995.

Part III

Toxic role of specific agents

9 Biological oxidations catalyzed by iron released from ferritin

Christopher A. Reilly and Steven D. Aust

9.1 Introduction

Iron is the most abundant transition metal found in biological systems. The total amount of iron present in a normal adult human has been estimated to be 3–4 grams (Emery, 1991). The unique ability of iron to vary oxidation state, spin state, and redox properties in response to varying ligand environments has permitted it to participate as a multi-functional constitutent of a plethora of biochemical processes (Crichton, 1991). Iron constitutes the active centers of a variety of enzymes, including the cytochromes and other heme proteins, iron–sulfur proteins, ribonucleotide reductase, etc. These proteins are essential for a variety of functions, including respiration and energy production, oxygen transport, antimicrobial defense, xenobiotic metabolism, and cellular reproduction (Crichton, 1991). Enzymes that use iron as a cofactor exploit the redox properties of iron while limiting undesirable side reactions with a variety of nonsubstrate molecules, but iron that is not effectively sequestered is potentially toxic.

Oxidation reactions are involved in many fundamental biological processes. Transition metals, such as iron, facilitate the enzymatic and non-enzymatic oxidation of chemicals and biomolecules by oxygen; the metabolically uncoupled oxidation of biomolecules may be detrimental to cells (Miller *et al.*, 1990). Toxicities and pathologies associated with the oxidation of biomolecules by iron may be included in the subject collectively termed 'oxidative stress' (Ryan and Aust, 1992). In this chapter, we will describe the mechanisms by which iron serves as a catalyst for undesirable macromolecule oxidation, discuss the potential physiological sources of iron for these oxidations, and describe mechanisms potentially involved in protection from iron-mediated oxidation of molecules.

9.2 Iron-catalyzed oxidations

9.2.1 *Iron chemistry*

To understand the basic mechanisms by which iron serves as a functional component in biological systems as well as promotes oxidative stress, a general understanding of its chemical properties must be established. Iron is a third row group VIII transition metal having six valence electrons in its 3d subshell. Iron can exist in several oxidation states ranging from −2 to +6 depending upon the chemical environment. Under physiological conditions, the most common oxidation states for iron are the ferrous, Fe^{II}, and ferric, Fe^{III}, forms, though under some conditions, iron may also exist in the ferryl, Fe^{IV}, state (Crichton, 1991). Only the ferric and ferrous states of iron are stable in an aqueous environment, therefore we will focus our discussion on these. In solution, iron forms hexacoordinate complexes with a variety of ligands. With water as the only available ligand, ferrous iron exists as the hydrated ion, $[Fe(H_2O)_6]^{2+}$, at pH 7.0. The hydrated ferric ion, $[Fe(H_2O)_6]^{3+}$, is only stable in an acidic environment and will readily precipitate as hydrated iron oxide, or rust, at pH 7.0 (Lippard, 1986). Thus, the solubilities of the ferrous and ferric forms of iron are 10^{-1} M and 10^{-17} M at pH 7.0, respectively (Spiro and Saltman, 1974). It is important to note that ferrous iron is unstable in aqueous media in the presence of oxygen and will readily reduce oxygen to yield ferric iron. Additionally, whenever water is displaced from coordination sites on iron by another ligand type, the properties of the iron are drastically affected. Thus, the inclusion of any chemical, including buffers, in an aqueous solution of iron may affect the properties and reactivity of the iron.

For its effective use as a cofactor within an enzyme, coordinating ligands must chelate iron to shield it from molecular oxygen and the surrounding aqueous environment (Crichton, 1991; Miller *et al.*, 1990). The relative stability of an iron chelate is dependent upon the ability of the complexing agent to compete with and displace water from coordination sites (Smith and Martell, 1975). Displacement of water from coordination sites on iron by another ligand in solution occurs by a substitution reaction (Langford and Gray, 1965; Swaddle, 1974). The mechanism of a substitution reaction is represented in equation 9.1, where M represents iron and X and Y are potential ligands.

$$\mathrm{M{-}X + Y \rightleftharpoons M{-}Y + X} \qquad (9.1)$$

The equilibria for these reactions are represented by stability constants which depict the relative binding affinity of a ligand over the solvent molecules (Smith and Martell, 1975). The type of atom through which the ligand binds to iron can enhance, or diminish, the stability of the ferrous or ferric ions. Negatively charged, weak field ligands that bind through oxygen atoms (or halides) promote oxidation of ferrous iron, and are generally good chelators of ferric iron (Crichton, 1991; Smith and Martell, 1975). Strong field ligands that are capable of electron donation that bind through either nitrogen or sulfur, will tend to chelate and stabilize the ferrous form of iron. For example, in the presence of chelators such as desferrioxamine and EDTA, which predominately bind through oxygen atoms, the solubility of ferric iron is significantly increased. Likewise, stable ferrous iron chelators such as phenanthroline and ferrozine, which bind through nitrogen atoms, will slow or prevent the oxidation of ferrous iron thus maintaining its stability and solubility. The redox properties of iron can also be altered and 'fine-tuned' to permit specific functions depending upon the nature of the coordinating ligands and the stability of the iron complex.

In biological systems, enzymes that catalyze electron transfer contain cofactors or prosthetic groups, such as transition metals and flavins. The ability to modify the redox potential of iron permits its use in a wide variety of redox processes. The redox potential is a measure of the relative thermodynamic stability of a particular oxidation state under a given set of conditions. The standard reduction potential (E'_0) is measured at pH 7.0 at 298 K in reference to the hydrogen electrode. Using this convention, a more positive E'_0 value indicates that the reduced form is more stable than the oxidized form. The standard redox couple for Fe^{II}/Fe^{III} is 0.77 V, at pH 1.0

indicating that the ferrous form is more stable (Miller *et al.*, 1990). This value can change significantly upon changes in pH and liganding, decreasing in value with increasing pH ($E_0' = 0.11$ V at pH 7.0) (Reed, 1985). The reduction potential for ferrioxamine-chelated ferric iron is –0.45 V while the reduction potential is 1.1 V for phenanthroline-chelated ferrous iron at pH 7.0 (Reed, 1985). Although the oxidation of chemicals *in vivo* is generally catalyzed by an enzyme and the mechanisms are to a large extent dependent upon protein structure, the fundamental concepts of liganding and its effects on the properties of iron are analogous to simple iron complexes (Sheldon and Kochi, 1981). To best demonstrate how the ligand environment within a protein can 'fine-tune' the redox properties of iron, we will compare the redox potential and liganding of several iron containing proteins.

The Fe^{II}/Fe^{III} redox couple can vary hundreds of millivolts between iron containing proteins (Moore *et al.*, 1986; Reed, 1985). In heme proteins, such as the peroxidases and P450 mixed function oxidases, more electron donating ligands such as nitrogen containing and cysteinyl ligands tend to stabilize the reduced form of iron and thus drive the reduction potential to more positive values and decrease the ability of the enzyme to catalyze reduction reactions (Moore *et al.*, 1986; Reed, 1985). Likewise, addition of electron donating substituents on the heme porphyrin ring will drive the E_0' to more negative values decreasing the oxidizing nature of the heme (Moore *et al.*, 1986; Reed, 1985). For example, the redox potential of heme in cytochrome c is 0.26 V, heme peroxidases is about 1.0 V, and the redox potential of the iron–sulfur cluster in ferredoxin is –0.20 V (Reed, 1985). As a result of the variable ligand environment provided by these proteins, the redox potential as well as the function of these iron-containing proteins are extemely diverse (Moore *et al.*, 1986). Similar ligand effects on the redox properties of organic prosthetic groups such as flavins are not observed because they undergo electron transfer via a radical intermediate.

9.2.2 Metal-catalyzed activation of dioxygen

Dioxygen is thermodynamically a powerful oxidant. However, direct oxidation of biomolecules by dioxygen is kinetically unfavorable, and occurs slowly ($< 10^{-5}$ M^{-1} s^{-1}) (Miller *et al.*, 1990). The slow rate of reaction is due to the relative strength of the O—O bond, the electronic character of the ground state of dioxygen, or both (Atkins, 1983; Miller *et al.*, 1990; Woodward and Hoffman, 1970). Dioxygen, in its ground state, is a biradical having a spin multiplicity of three, thus termed a 'triplet' state, ($2S + 1 = 3$, where $S = 1/2$ for each unparied electron), with its two unpaired electrons in different orbitals having parallel spin. The vast majority of biomolecules are non-radical species and exist in the 'singlet' state. If the reaction of dioxygen, in its ground state, is to occur with biomolecules, there must be a change of spin of one of the unpaired electrons of dioxygen at some stage of the reaction (Atkins, 1983; Gutteridge and Halliwell, 1990; Miller *et al.*, 1990; Woodward and Hoffman, 1970). The reaction of dioxygen with a singlet molecule requires that dioxygen exist in the singlet, or doublet (one unpaired electron or radical) state. However, the energy required to change the spin of an electron is too formidable to be reached by thermal activation (Atkins, 1983; Miller *et al.*, 1990; Woodward and Hoffman, 1970). Direct reaction between triplet and singlet molecules is therefore spin restricted. The limited, direct reaction of triplet and singlet reactants is also a consequence of the non-crossing rule of Woodward and Hoffman (1970).

Activation of dioxygen to the singlet or doublet state may occur via photoactivation, enzymatic generation, or by sequential one-electron reductions to reduced oxygen species, including the superoxide anion radical ($O_2^{-\bullet}$), hydrogen peroxide (H_2O_2), and the hydroxyl radical ($^{\bullet}OH$) (Miller *et al.*, 1990). The photoactivation of triplet dioxygen to its singlet state permits direct reaction with biomolecules, such as lipids, to produce endoperoxides. The energetic requirement for this reaction is provided by light. Enzymatic generation of partially reduced species of oxygen requires a cofactor or prosthetic group capable of catalyzing electron transfer reactions, a function most often provided by transition metals (Wrigglesworth and Baum, 1980). In solution, the sequential reduction of dioxygen is catalyzed by transition metals, particularly iron (Gutteridge and Halliwell, 1990; Miller *et al.*, 1990). Transition metals, such as iron, are able to interact with dioxygen because they exist in variable spin states which is dependent upon liganding. Transition metal complexes relieve the spin restriction for reaction of dioxygen with biomolecules by forming superoxo–metal complexes with dioxygen which causes a delocalization of the unpaired electron density and decreases the triplet character of dioxygen (Sheldon and Kochi, 1981). This is represented by reaction 9.2.

$$Fe^{II} + O_2 \leftrightharpoons Fe^{III}{-}O_2^{\bullet} \leftrightharpoons Fe^{III} + O_2^{-\bullet} \qquad (9.2)$$

The delocalization of electron density into the transition metal orbitals can also result in spin-orbit coupling which can further facilitate reaction (Atkins, 1983; Wrigglesworth and Baum, 1980).

The sequential reduction of dioxygen by iron is best exemplified by the iron-catalyzed Haber–Weiss reaction, represented by the following reactions.

$$Fe^{II} + O_2 \leftrightharpoons Fe^{III}{-}O_2^{\bullet} \leftrightharpoons Fe^{III} + O_2^{-\bullet} \qquad (9.2)$$

$$2O_2^{-\bullet} + 2H^+ \rightarrow O_2 + H_2O_2 \qquad (9.3)$$

$$Fe^{II} + H_2O_2 \rightarrow Fe^{III} + {}^{-}OH + {}^{\bullet}OH \qquad (9.4)$$

The iron catalyzed decomposition of H_2O_2 (equation 9.4) produces $^{\bullet}OH$ ($E'_0 = 2.31$ V) (Koppenol, 1987), which reacts nonspecifically with almost any molecule at diffusion limited rates (10^8–10^9 M^{-1} s^{-1}) (Sheldon and Kochi, 1981).

In addition to the generation of reduced oxygen species, iron and oxygen may also combine to generate complexes of high oxidation potential such as oxo-ferryl intermediates, or the perferryl ion (Fe O_2^{+}). The perferryl ion has been shown to have an oxidation potential equal to, or greater than, that of the hydroxyl radical (Koppenol and Leibman, 1984). The reactions of dioxygen with iron are represented in the following equations.

$$Fe^{II} + O_2 \leftrightharpoons Fe^{III}{-}O_2^{\bullet} \leftrightharpoons Fe^{III} + O_2^{-\bullet} \qquad (9.2)$$

$$Fe^{III}{-}O_2^{\bullet} + Fe^{II} \rightarrow Fe^{III}{-}O{-}O{-}Fe^{III} \qquad (9.5)$$

$$Fe^{III}{-}O{-}O{-}Fe^{III} \rightarrow 2Fe^{IV}{=}O \qquad (9.6)$$

$$Fe^{IV}{=}O + Fe^{II} \rightarrow Fe^{III}{-}O{-}Fe^{III} \qquad (9.7)$$

It has also been proposed that iron and oxygen can combine to form an Fe^{II}—O_2—Fe^{III} complex capable of oxidizing fatty acids (Aust *et al.*, 1984; Bucher *et al.*, 1983b). However, the involvement of molecular oxygen in the complex cannot be demonstrated since the oxidation of lipid only occurs in the presence of oxygen. The reactions outlined above provide evidence that iron, in the presence of oxygen, will catalyze the generation of potent oxidants capable of oxidizing a variety of molecules found in biological systems.

9.2.3 Metal-catalyzed oxidation of biomolecules

Some chemicals are said to 'autoxidize' or react with dioxygen by a direct one- or two- electron transfer to produce $O_2^{-\bullet}$ or H_2O_2, respectively (Uri, 1961). However, the one-electron reduction of dioxygen is thermodynamically unfavorable ($E'_0 = -0.33$ V) (Koppenol and Butler, 1985). The large energy requirements for the reduction of triplet dioxygen are attributed to the energy barriers associated with the addition of an electron to the partially filled π^* orbitals. When coupled to an energetically favorable process (i.e., ≥ 0.33 V), the one-electron reduction of dioxygen is thermodynamically possible. The only biological molecules capable of catalyzing the reduction of dioxygen are the reduced flavins (Miller *et al.*, 1990). Ascorbic acid has been said to 'autoxidize', and directly reduce dioxygen to superoxide, which is thermodynamically unfavorable (Boyer and McClearly, 1987; Scarpa *et al.*, 1983). However, the two-electron reduction of dioxygen to H_2O_2 is favorable (Miller *et al.*, 1990). But direct two-electron reduction of dioxygen by ascorbate does not occur due to the kinetic barrier imposed by spin restrictions. The 'autoxidation' of ascorbic acid, as well as other chemicals including thiols, catechols, and biogenic amines, occurs only in the presence of transition metals such as iron (Aust *et al.*, 1985; Buettner, 1986, 1988; Miller *et al.*, 1990; Miller and Aust, 1988; Ryan *et al.*, 1993). Ferric iron is readily reduced by chemicals, such as ascorbate, to ferrous iron which oxidizes in the presence of dioxygen and promotes the Haber–Weiss reactions. The reductant of iron then exists as the one-electron oxidized form, or radical. Radicals can readily react with the unpaired electrons of dioxygen without the kinetic barriers imposed by spin restrictions. The result, in many cases, is the reduction of dioxygen by the radical or the addition of oxygen to the radical. The cycle established between the reductant and iron continues until all of the reductant is oxidized and iron can no longer be reduced.

A prototypical oxidative reaction involving transition metals which is of toxicological significance is the oxidation of polyunsaturated fatty acids (PUFA) constituting lipid membranes. The toxicological and pathological ramifications of lipid peroxidation, or addition of oxygen to PUFA, are complex. Lipid peroxidation in membranes may alter various processes associated with cellular homeostasis including specific transport mechanisms and ion gradients (Ackworth and Bailey, 1995). Oxidation of arachidonic acid can interfere with the production of prostaglandins, leukotrienes, and thromboxanes which are local mediators involved in the regulation of blood flow, smooth muscle tone, and neuronal transmission (Ackworth and Bailey, 1995). The aldehyde byproducts of lipid peroxidation have also been shown to be mutagenic and act as potent inhibitors of various enzymes (Ackworth and Bailey, 1995; Ueda *et al.*, 1985). In addition, damage to red cell membranes will affect oxygen transport, promote premature cell turnover, and ultimately manifest itself as anemia (Ackworth and Bailey, 1995).

Lipid peroxidation is a free-radical mediated process composed of three steps: initiation, propagation, and termination (Ryan and Aust, 1992). The initiation of lipid peroxidation occurs with formation of an alkyl radical ($L^{\bullet}$). There are two hypotheses on how the initiation of lipid peroxidation occurs. The most commonly proposed mechanism involves hydrogen abstraction from PUFA by $^{\bullet}OH$ generated from the iron-dependent decomposition of H_2O_2 (i.e. the Fenton reaction) (Halliwell and Gutteridge, 1990). This reaction is illustrated by equations 9.4 and 9.8.

$$Fe^{II} + H_2O_2 \rightarrow Fe^{III} + {}^{-}OH + {}^{\bullet}OH \tag{9.4}$$

$$^{\bullet}OH + LH \rightarrow L^{\bullet} + H_2O \tag{9.8}$$

Alternatively, the other mechanism proposed for initiation of lipid peroxidation has a more direct involvement of iron, independent of the Fenton reaction (Minotti and Aust, 1987). This hypothesis includes a requirement for mixed valences of iron in the presence of oxygen (Aust *et al.*, 1989; Bucher *et al.*, 1983b; Minotti and Aust, 1987). It was proposed that the initiator of lipid peroxidation was an $Fe^{II}—O_2—Fe^{III}$ complex, since it was found that maximal rates of lipid peroxidation occurred when a 1:1 ratio of $Fe^{II}:Fe^{III}$ was generated in the presence of molecular oxygen *in vitro* (Aust *et al.*, 1989; Bucher *et al.*, 1983b; Minotti and Aust, 1986, 1987). Proof of the existance of the $Fe^{II}—O_2—Fe^{III}$ complex is, however, debated (Aruoma *et al.*, 1989). The propogation of lipid peroxidation, a radical-mediated chain reaction, involves the incorporation of molecular oxygen by the lipid alkyl radical ($L^{\bullet}$) to generate a lipid alkyl peroxyl radical ($LOO^{\bullet}$) capable of abstracting an additional hydrogen from neighboring fatty acids. In addition, ferrous iron may reduce lipid hydroperoxides (LOOH) to generate $LO^{\bullet}$, which can also propagate the reaction. Propagation reactions of lipid peroxidation are outlined by reactions 9.9–9.12.

$$L^{\bullet} + O_2 \rightarrow LOO^{\bullet} \tag{9.9}$$

$$LOO^{\bullet} + LH \rightarrow LOOH + L^{\bullet} \tag{9.10}$$

$$LOOH + Fe^{II} \rightarrow Fe^{III} + LO^{\bullet} + {}^{-}OH \tag{9.11}$$

$$LO^{\bullet} + LH \rightarrow L^{\bullet} + LOH \tag{9.12}$$

The termination of lipid peroxidation occurs when lipid radicals react and generate nonradical products. There are several mechanisms of termination including radical quenching by lipophilic chain-breaking antioxidants such as α-tocopherol (Niki *et al.*, 1984; Witting, 1980) and β-carotene (Burton and Ingold, 1984), radical dismutation, or cyclization. Vitamin E (α-tocopherol), is preferentially oxidized by $L^{\bullet}$, $LO^{\bullet}$, and $LOO^{\bullet}$ during the propagation stage of lipid peroxidation and therefore serves as an antioxidant (Witting, 1980). Oxidation of α-tocopherol generates the α-tocopherol radical which is thermodynamically relatively stable and cannot readily oxidize additional lipids, thus inhibiting or terminating propagation. It would be reasonable to assume that termination of lipid peroxidation during the propagation stage serves as a secondary mechanism to protect lipid membranes from oxidation; the primary mechanism being inhibition of initiation.

However, as mentioned in the previous paragraph, there are two proposed mechanisms for the initiation of lipid peroxidation. The first mechanism requires the production of the •OH by the Fenton reaction (Gutteridge and Halliwell, 1990; Halliwell and Gutteridge, 1990). The •OH is a strong, non-specific oxidant that will react with any molecule in close proximity to the site of its generation. Initiation of lipid peroxidation by this mechanism would have to be a site-specific reaction requiring the abstraction of a hydrogen atom from a carbon atom on the lipid side chain. The ability of •OH to diffuse into the hydrophobic lipid bilayer and abstract a hydrogen atom such that it may initiate lipid peroxidation is debated. The most likely site of •OH generation would be at the polar head group located on the surface of the phospholipid bilayer rather than in the hydrophobic interior of the membrane. This is because the phosphate group is likely to be the site of iron chelation, an essential component for •OH generation. Hydroxyl radical generation at the membrane surface would probably not be favorable for the initiation of lipid peroxidation. This may explain why the •OH is not capable of initiating lipid peroxidation in liposomes; •OH can initiate lipid peroxidation in detergent dispersed lipids (Bucher *et al.*, 1983a; Fukuzawa *et al.*, 1988). In addition, catalase has been shown to stimulate lipid peroxidation under certain conditions (Ryan and Aust, 1992). If iron-dependent decomposition of H_2O_2 to generate •OH was, in fact, required to initiate lipid peroxidation, catalase would actually inhibit this process by eliminating H_2O_2. How can catalase stimulate lipid peroxidation?

A second possible mechanism for the initiation of lipid peroxidation requires the presence of both Fe^{II} and Fe^{III}, and oxygen (Aust *et al.*, 1984, 1989; Bucher *et al.*, 1983b; Minotti and Aust, 1986). Using various methods of oxidation or reduction of iron, it was shown that maximal rates of lipid peroxidation occurred at a 1:1 ratio of Fe^{II}:Fe^{III}. No lipid peroxidation occurred if either Fe^{II} or Fe^{III} was absent. For example, when ADP-chelated ferric iron was used to initiate lipid peroxidation, no reaction occurred. When ascorbate was added to reduce the iron, lipid peroxidation ensued with a maximum rate observed when the concentration of ascorbate was enough to keep approximately half of the ferric iron reduced. Addition of excess ascorbate to further increase the concentration of Fe^{II} decreased the rate of lipid peroxidation; no lipid peroxidation was observed when enough ascorbate was added to reduce all of the iron (Aust *et al.*, 1989; Miller and Aust, 1988). Conversely, when H_2O_2 was used as the oxidant of Fe^{II}, the maximum rate of lipid peroxidation occurred when sufficient H_2O_2 was available to maintain a 1:1 ratio of Fe^{II}:Fe^{III} (Minotti and Aust, 1986). The data from these experiments are presented in Figures 9.1 and 9.2, respectively.

Obtaining and maintaining the optimal 1:1 ratio of Fe^{II}:Fe^{III} is dependent on the type and concentration of iron chelator present, with some chelators being much more efficient promoters of lipid peroxidation (Bucher *et al.*, 1983a). It has been shown that ADP-chelated ferric iron is significantly better at promoting lipid peroxidation than EDTA- or ferrioxamine-chelated iron (Aust *et al.*, 1989; Miller, 1990; Ryan *et al.*, 1990; Tien *et al.*, 1982). It was observed that ADP-chelated ferrous iron oxidized slowly, relative to both EDTA- and 1,2-diaminopropane-N,N,N′,N′-tetraacetic acid (PDTA)-chelated iron (Aust *et al.*, 1989; Ryan *et al.*, 1990). Addition of small amounts of PDTA in proportion to the concentration of ADP stimulated lipid peroxidation, however, increasing the ratio of PDTA to ADP inhibited lipid peroxidation (Ryan and Aust, 1992; Ryan *et al.*, 1990). This occurred

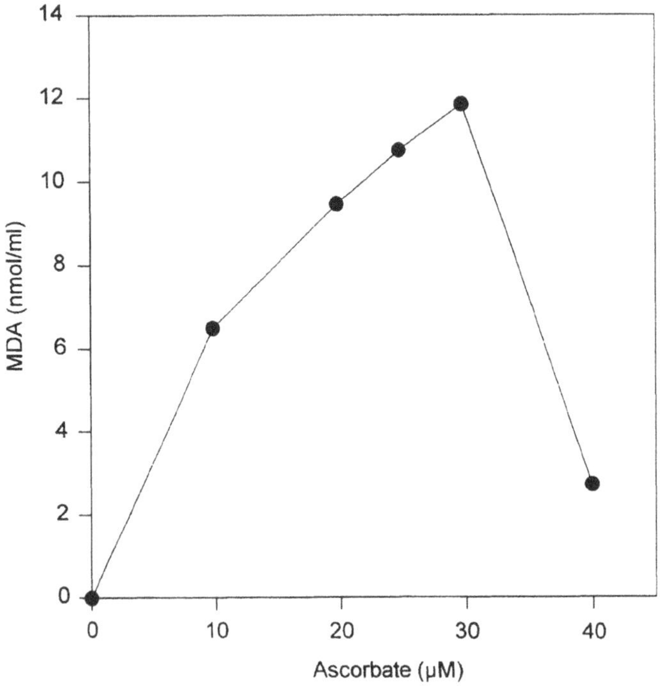

Figure 9.1 The effect of ascorbate concentration on ADP:Fe^{III}-dependent lipid peroxidation. Reactions contained liposomes (1 mM lipid phosphate), and ADP:Fe^{III} (50 μM Fe^{III}) in Chelex treated 50 mM NaCl, pH 7.0. Samples were incubated at 37°C for 2 minutes prior to the addition of ascorbate. Data points represent the amount of MDA formed in 10 min, determined by the thiobarbituric acid test. Data from Miller and Aust (1988), with permission

because PDTA promoted rapid oxidation of ferrous iron and the Fenton reaction (Ryan and Aust, 1992; Ryan *et al.*, 1990). As observed in the ascorbate dependent lipid peroxidation system, catalase stimulated lipid peroxidation in systems having both PDTA and ADP by inhibiting additional iron oxidation by H_2O_2, generated by $O_2^{-\bullet}$ dismutation. These data are presented in Table 9.1. It can also be concluded from the data presented in Table 9.1 that lipid peroxidation was independent of $^\bullet OH$ generation since the rates of lipid peroxidation did not correlate with the rates of $^\bullet OH$ generation. The initiation of lipid peroxidation was contingent upon generating and maintaining a ratio of Fe^{II} to Fe^{III}.

To further address the issue of whether iron was in fact the initiator of lipid peroxidation or if the formation of oxygen radicals was necessary, Aust *et al.* (1989) used ceruloplasmin, a ferroxidase, to oxidize ferrous iron without the simultaneous generation of oxygen radicals. Ceruloplasmin catalyzes the one-electron oxidation of ferrous iron with concomitant complete reduction of dioxygen to water (Rydén, 1981). The experimental data obtained for ceruloplasmin and ferrous iron-dependent lipid peroxidation are presented in Figure 9.3. It was observed that there was an optimum ceruloplasmin concentration for obtaining the maximum rate of lipid peroxidation. This indicated that an optimal ratio of Fe^{II}:Fe^{III} was needed for

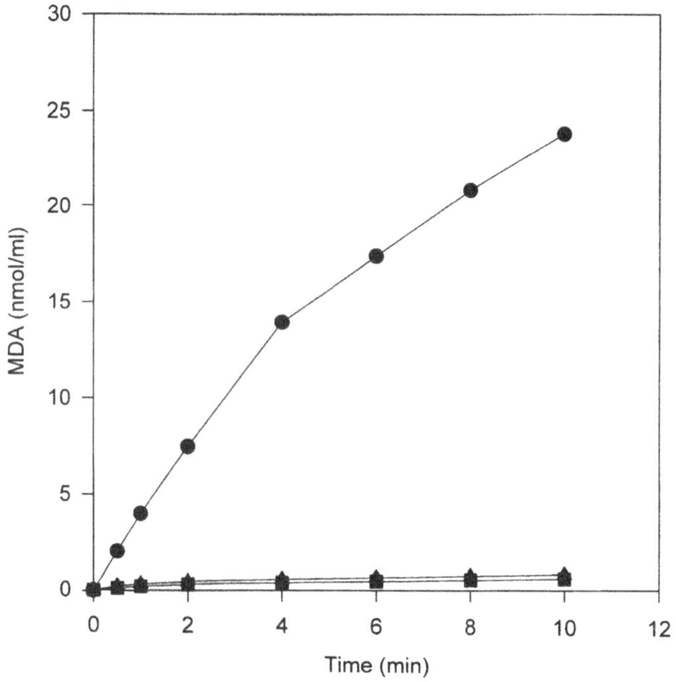

Figure 9.2 Ferrous iron and hydrogen peroxide-dependent lipid peroxidation. Reaction mixtures contained liposomes (1 mM lipid phosphate) in Chelex treated 50 mM NaCl, pH 7.0, at 37°C. Reactions were initiated by the addition of 0.1 mM H_2O_2 (▲); 0.2 mM $FeCl_2$ (■); 0.1 mM H_2O_2 and 0.1 mM FeCl (●). At specific time points, aliquots were removed and assayed for MDA using the thiobarbituric acid test. Data from Minotti and Aust (1986), with permission

Table 9.1 The effects of chelation on $O_2^{-\bullet}$ and iron-dependent lipid peroxidation, and production of $^{\bullet}OH$[a]

Iron chelate	*MDA (nmol/ml)*[b]	*$^{\bullet}$OH radical production*[c] *(relative ESR intensity)*
ADP:Fe(III) (5:1)	9.0	5.4
PDTA:Fe(III) (1.1:1)	1.7	10.5
ADP:Fe(III) (5:1) + 25 μM PDTA	18.6	4.5
ADP:Fe(III) (5:1) + 200 μM PDTA	1.0	4.4
ADP:Fe(III) (5:1) + 25 μM PDTA + catalase	23.4	0

a Data from Ryan and Aust (1992), with permission.

b Reaction mixtures contained liposomes (1 μM lipid phosphate), ADP:Fe^{III} (250 μM ADP:50 μM $FeCl_3$), xanthine (0.33 M), xanthine oxidase (0.05 U/ml), the indicated amount of PDTA, and where included catalase (1000 U/ml) in 50 mM NaCl, pH 7.0. Incubation time was 30 minutes at 37°C.

c Incubations for detection of $^{\bullet}OH$ contained ADP:Fe^{III} (250 μM ADP:50 μM $FeCl_3$), xanthine (0.33 M), xanthine oxidase (0.05 U/ml), DMPO (100 mM), catalase (1000 U/ml, where indicated), and the indicated amount of PDTA in 50 mM NaCl, pH 7.0, at room temperature.

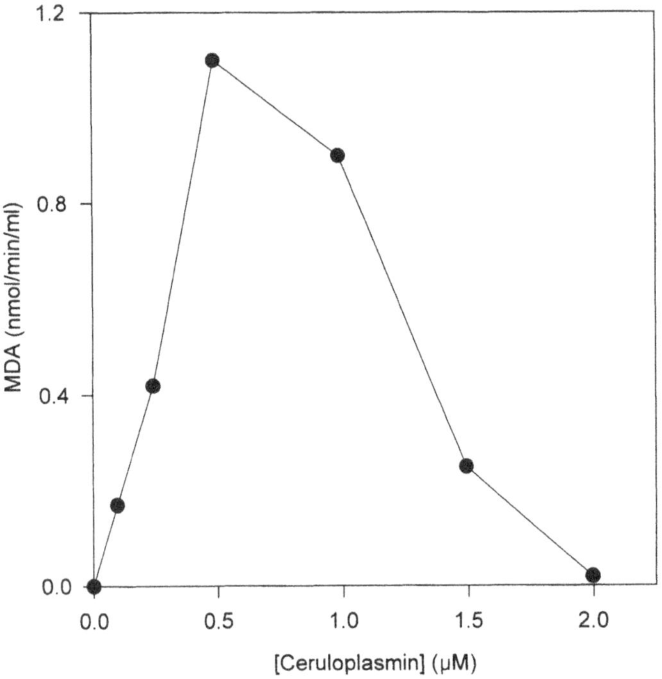

Figure 9.3 The effect of ceruloplasmin on ferrous iron-dependent lipid peroxidation. Samples contained liposomes (1 mM lipid phosphate) and increasing concentrations of ceruloplasmin in Chelex treated 50 mM NaCl, pH 7.0. Samples were preincubated at 37°C for 2 min prior to addition of ferrous chloride (200 μM). Data points represent the amount of MDA produced in 25 min, as determined by the thiobarbituric acid test. Data from Aust *et al.* (1989), with permission

maximum rates of lipid peroxidation. Using ceruloplasmin to generate the ratio of Fe^{II}:Fe^{III} required to initiate lipid peroxidation also provided evidence that oxygen radicals were not involved. It will be shown later that ceruloplasmin may also serve as a potent inhibitor of lipid peroxidation.

9.3 Biological sources of catalytic iron

9.3.1 Control of iron in vivo

The catalysis of oxidative reactions by 'free' iron, iron not contained within proteins, has been implicated in a host of diseases and disorders including rheumatoid arthritis (Halliwell and Gutteridge, 1984; Rowley *et al.*, 1984), Parkinson's disease (Dexter *et al.*, 1989; Halliwell and Gutteridge, 1984), atherosclerosis (Halliwell and Gutteridge, 1984; Heinecke *et al.*, 1984), cancer (Halliwell and Gutteridge, 1984; Stevens *et al.*, 1988), Alzheimer's disease, amyotrophic lateral sclerosis, multiple sclerosis (Gerlach *et al.*, 1994; Halliwell and Gutteridge, 1984), and post ischemic reperfusion injury (Flaherty and Weisfeldt, 1988; Halliwell and Gutteridge, 1984). Additionally, the toxicity of various chemicals has been attributed, at least in part, to reactions catalyzed by iron. Due to the inherent reactivity of iron as well as the limited solubility of Fe^{II} and Fe^{III}, 'free' iron must be chelated under physiological conditions. This type of iron is termed the 'low molecular weight chelate pool' of iron, one of which may be responsible for the catalysis of redox reactions contributing to the toxicity of certain chemicals (Thomas and Aust, 1986a). Attempts to isolate and identify the molecules responsible for chelation of iron *in vivo* have been futile thus far, however nucleotides and other small molecules such as citrate have been proposed (Bartlett, 1976a, 1976b; Jacobs, 1977; Jones *et al.*, 1980; Morley and Bezkorovainy, 1983; Weaver and Pollack, 1989).

Several hypotheses concerning the nature of the physiological low molecular weight chelators of iron suggest that several criteria must be met (Miller *et al.*, 1990). Chelation must be strong, so that the chelator can effectively compete for liganding of the iron, and sufficiently stable such that redox cycling with dioxygen and other biomolecules is prevented. The latter requirement arises since the low molecular weight chelates of iron probably exists in the ferrous form due to the inherent reducing environment within cells. Another criterion is that iron must remain bound by a specific chelator until delivered to its ultimate destination where it would be released in a 'safe' manner. In general, however, the intracellular concentrations of non-heme, non-protein bound, low molecular weight chelated iron would be expected to be extremely low and transient. High concentrations of non-physiological low molecular weight chelated iron would be disadvantageous because of the deleterious reactions that may be catalyzed by this type of iron. Only under specific pathological conditions, oxidative stress, or disease states such as hemochromatosis, are large deposits of 'free' iron found in cells.

Iron can elicit toxicity as a result of both an acute or chronic exposure. Acute iron toxicity generally occurs as a result of ingestion of a large dose of iron-containing medications, primarily by children who mistake them for candy. It has been estimated that nearly 2000 children are poisoned each year from ingestion of iron-containing supplements (Goyer, 1991). Symptoms of iron poisoning may include bloody vomit, ulceration, dark stool, metabolic acidosis, liver damage, renal dysfunction, and death (Goyer, 1991). The adverse symptoms and pathologies observed due to acute iron poisoning are proposed to be the result of saturation of the normal mechanisms for the control of iron in the body and subsequent oxidation of macromolecules by 'free' iron. The adverse effects of acute iron poisoning are mitigated by desferrioxamine, a stable and water soluble ferric iron chelator. In general, however, the acute dose of iron is relatively low since the primary source of iron for mammals is the diet. Under normal conditions, a healthy individual can maintain the proper balance in absorption and elimination of iron even when the dietary intake is ten-fold the needed amount (Ganong, 1991). However, prolonged exposure to excess iron, as well as certain genetic factors, can result in the accumulation of iron in tissues. This condition is termed hemochromatosis. Hemochromatosis is characterized by abnormal skin pigmentation, increased incidence of diabetes and hepatic carcinoma, and general tissue damage. Genetic, or idiopathic hemochromatosis, results from a defect in the control of absorption of iron from the small intestine. The increase in susceptibility to certain diseases in individuals with hemochromatosis is attributed to the increase in total iron stored in the body.

As a result of the potential for toxicity associated with excess iron *in vivo*, as well as the disorders associated with altered iron metabolism, nature has devised

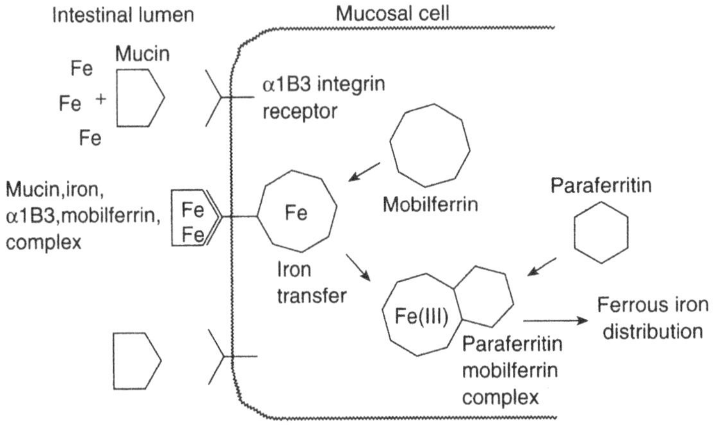

Figure 9.4 Schematic representation of iron absorption by intestinal mucosal cells. Adopted from Conrad *et al.* (1993) and Umbreit *et al.* (1996)

precise mechanisms for the control of iron. The concentration of iron in the body is regulated by a balance in absorption and elimination, and is responsive to total iron stores; therefore, subjects with iron deficiency generally absorb more iron (McCance and Widdowson, 1937; Turnbull, 1974). Iron, particularly in the ferrous form (Wollenberg and Rummel, 1987), is absorbed from the lumen of the upper small intestine into mucosal cells. Approximately 10% of the total iron ingested is absorbed to replace iron that is lost through sloughing off of cells, bleeding, and excretion in bile (Crichton, 1991; Emery, 1991). The mechanism of iron absorption is extremely complex and not clearly defined. In 1943, Hahn *et al.* proposed that a receptor was responsible for the uptake of iron by mucosal cells of the intestinal lumen. Recently, however, Conrad *et al.* (1993) isolated a 56 kDa protein with high affinity for iron from the cytosolic fraction of intestinal mucosal cells. This protein was named mobilferrin. In addition, the extracellular adhesive glycoprotein mucin, and the membrane bound α1β3 integrin receptor have been found to be associated with mobilferrin, and are postulated to play an integral role in the absorption of iron (Conrad *et al.*, 1993). Mucin is located in the intestinal lumen and exhibits high affinity for iron and the α1β3 integrin receptor. Transfer of mucin-bound iron to mobilferrin was proposed to occur through the α1β3 receptor. It was demonstrated that there was a requirement for complex formation between mobilferrin and the α1β3 receptor in order for mobilferrin to obtain iron, and that mobilferrin later became complexed with a flavin monoxygenase called paraferritin (Umbreit *et al.*, 1996). Paraferritin was found to possess ferrireductase activity and was proposed to be instrumental in the reduction of iron and to promote further distribution of iron to other proteins (Umbreit *et al.*, 1996). This was proposed since most iron-requiring biochemical pathways appear to utilize ferrous iron. The model proposed by Conrad *et al.* (1993) is presented schematically by Figure 9.4.

Distribution of iron to cells occurs via transferrin. The transferrins are a family of proteins which control the concentration of 'free' iron in body fluids by binding, sequestering, and transporting ferric iron (Baker, 1994; Brock, 1985; Chasteen and Woodworth, 1990). There are four types of transferrin: serum transferrin, ovotransferrin, lactoferrin, and melanotransferrin. Serum transferrin is responsible for binding iron in the plasma, and transporting it from its site of absorption to cells for immediate use or storage. Transferrin has also been proposed to be involved in the absorption of iron from the gut. However, transferrin receptors are only found on the serosal side of mucosal cells indicating that it serves solely as a mechanism of iron delivery to these cells (Huebers *et al.*, 1983). Transferrin synergistically binds two atoms of ferric iron per transferrin molecule forming a ternary complex with carbonate or bicarbonate (Bailey *et al.*, 1988). The coordination geometry of transferrin-bound ferric iron is octahedral, consisiting of four amino acid ligands and two ligands from the bidentate (bi)carbonate anion (Bailey *et al.*, 1988). The binding site for iron exhibits tight, but reversible, binding for ferric iron (Baker, 1994).

Cells deficient in iron will express the transferrin receptor such that the uptake of the transferrin–iron complex may occur. The concentration of low molecular weight complexes of iron control the translation of mRNA coding for the transferrin receptor as well as the iron storage protein ferritin (Bailey *et al.*, 1988; Munro and Linder, 1978; Rao *et al.*, 1985, 1986; Zahringer *et al.*, 1976). The synthesis of the transferrin receptor and ferritin are reciprocally regulated (Alberts *et al.*, 1994; Cavill *et al.*, 1975). When the concentration of the low molecular weight chelate 'pool' of iron is low, the iron-response element binding protein (IRE-BP), or cytosolic aconitase, blocks translation of ferritin mRNA by binding to the untranslated 5′ iron-response element (IRE) (Alberts *et al.*, 1994; Aziz and Munro, 1987; Rouault *et al.*, 1988; Zahringer *et al.*, 1976). Meanwhile, IRE-BP binds and stabilizes the transferrin receptor mRNA by binding the 3′ IRE, thus increasing translation (Alberts *et al.*, 1994; Koeller *et al.*, 1989; Müllner *et al.*, 1989). When the low molecular weight chelate 'pool' of iron is in excess, iron is bound to the IRE-BP. Iron binding to IRE-BP induces a conformational change in the IRE-BP and causes it to dissociate from the IRE of the mRNA for both the transferrin receptor and ferritin. Dissociation of IRE-BP from the IREs promotes degradation of the transferrin receptor mRNA and limits iron uptake. Conversely, dissociation of IRE-BP from the ferritin mRNA permits ferritin synthesis in order to increase the iron storage capacity within the cell.

Following receptor-mediated endocytosis of the transferrin–iron complex, iron becomes dissociated from transferrin in endocytic vesicles. The exact mechanism of iron release from transferrin is not clear. However, several factors have been shown to contribute to the release of iron: competitive chelation, reduction, acidification, and binding of transferrin to the transferrin receptor (Bali and Aisen, 1991). In general, it is thought that acidification of the endocytic vesicle causes iron to be released from transferrin, and both the transferrin receptor and apotransferrin are recycled (Alberts *et al.*, 1994). It is at this stage where a low molecular weight chelator for iron must be involved in order to prevent redox reactions and precipitation of iron during its transfer to iron-requiring biosynthetic pathways or storage in ferritin. Iron which is not used immediately for the biosynthesis of iron-containing proteins is stored in ferritin. The precise mechanism by which iron is transported across the vesicle membrane into the cytosol is not known. Some researchers have proposed that reduction of the iron must occur prior to translocation to cytosolic pathways and proteins (Nunez *et al.*, 1992). Although a reductive process has not been shown for the transferrin-dependent pathway of iron absorption, an NADH-dependent ferrireductase has been postulated to be instrumental in the absorption of non-transferrin bound iron and storage of this iron in ferritin (Inman *et al.*, 1994; Jordan and Kaplan, 1994;

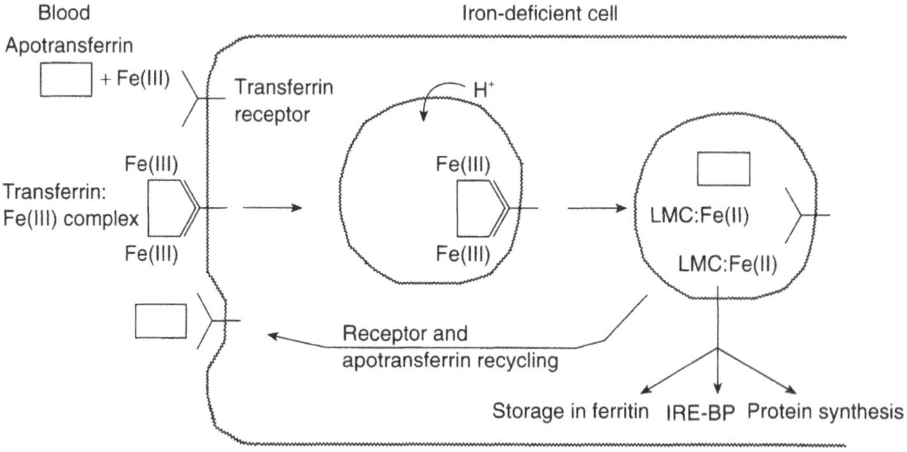

Figure 9.5 Schematic representation of iron absorption by cells expressing the transferrin receptor

Riedel *et al.*, 1995; Scheiber and Goldenberg, 1993, 1996; Sun *et al.*, 1987). Since absorption of non-transferrin bound iron occurs at the plasma membrane and endocytic vesicles are derived from this membrane a similar protein may participate in both pathways. The transferrin-dependent iron absorption pathway is represented schematically by Figure 9.5.

Recent studies on the absorption of iron by yeast cells (*S. cerevisiae*) may provide insight into the process of iron absorption by mammalian cells. Under iron-deplete growth conditions, yeast cells express a high-affinity iron uptake system composed of several proteins including a reductase (Fre1p or 2p), a copper-containing oxidase (Fet3p), and a transport protein (Ftr1p) (reviewed in DeSilva *et al.*, 1996; Kaplan and O'Halloran, 1996; Winzerling and Law, 1997). First, extracellular iron is reduced by Fre1p or 2p followed by oxidation of the ferrous iron by Fet3p and concomitant translocation into cell by Ftr1p. The presence of a cell surface ferrireductase on mammalian cell membranes has prompted some investigators to propose that the yeast high-affinity uptake system may be an accurate model for iron absorption by mammalian cells (DeSilva *et al.*, 1996; Jordan and Kaplan, 1994; Winzerling and Law, 1997). Additional support for this hypothesis was presented by Reilly and Aust, who demonstrated the presence of a membrane-bound ferroxidase capable of loading iron into ferritin (Reilly and Aust, 1998). This protein was shown to have enzymatic properties nearly identical to both Fet3p and serum ceruloplasmin (DeSilva *et al.*, 1997; Reilly and Aust, 1998).

Another proposed mechanism for the absorption of iron and other divalent metal ions involves the membrane-bound divalent cation transporter (DCT 1). DCT 1 exhibits broad substrate specificity, facilitating transport of Fe^{2+}, Zn^{2+}, Mn^{2+}, Co^{2+}, Cd^{2+}, Cu^{2+}, Ni^{2+}, and Pb^{2+} (Gunshin *et al.*, 1997). The energetic requirement for translocation of metal ions across the membrane was supplied by the cell membrane potential, since active transport of the metal ions was found to be coupled to proton pumping (Gunshin *et al.*, 1997). Although expression of this protein was greatest in mucosal cells of the proximal duodenum, its presence was ubiquitous (Gunshin *et al.*,

1997). This observation prompt these researchers to propose that DCT 1 may be instrumental in the absorption of divalent metal ions from the intestinal lumen as well as by other cell types.

Regardless of how cells absorb iron, the majority of iron found in the body is complexed with proteins that require iron for function, or sequestered in ferritin. It has been estimated that approximately 65% of the total body iron is found in hemoglobin, an additional 15% resides in various other iron-containing proteins, while the remaining 20% is stored within ferritin (Ganong, 1991). The percentage of total body iron in ferritin is variable depending upon age, diet, genetic factors, and sex. Ferritin is a ubiquitous, intracellular iron storage protein that can store up to 2500 atoms of iron in its 80 Å hollow internal cavity as a polymer of ferric oxyhydroxide complexed with phosphate (Harrison and Arosio, 1996; Theil, 1990). The ferritin molecule is composed of variable proportions of 24 H and L subunits. The proportion of H and L subunits in a ferritin molecule is dependent upon age, tissue of origin, and diet (Bomford *et al.*, 1981). The 24 subunits assemble with 4:3:2 symmetry to form a 450 KDa globular protein molecule with several pores (Harrison and Arosio, 1996; Theil, 1990). The H and L subunits of ferritin are immunologically distinct (Otsuka *et al.*, 1981) and differ significantly in their function. Ferritin isolated from tissues with high iron turnover, such as the liver or spleen, consists primarily of the L subunit, and generally contains more iron than ferritin isolated from tissues with low iron burden such as the heart and brain. Ferritin isolated from heart and brain generally contains less iron and is constituted of mostly the H chain. Due to this observation, as well as the specific induction of L chain ferritin upon iron supplementation, the L chain of ferritin has been postulated to be instrumental in the storage of iron (Bomford *et al.*, 1981). Recent studies however, contradict this hypothesis and suggest that iron storage capacity is independent of subunit composition (Juan and Aust, 1998), and that the induction of specific subunits of ferritin by iron is primarily dependent upon the type or chelate of iron supplied (Bomford *et al.*, 1981; Lin and Girotti, 1997, 1998). The role of the H subunit of ferritin appears to be involved in iron loading (Boyd *et al.*, 1985; Guo *et al.*, 1996) and will be discussed in detail later. Independent of the source of ferritin and its relative subunit composition, the iron contained within ferritin represents the single most concentrated 'pool' of iron *in vivo* (Samokyszyn *et al.*, 1987). For this reason, we will focus our discussion on ferritin as the source of iron for deleterious redox reactions catalyzed by iron.

9.3.2 Mobilization of iron from ferritin

Although iron stored within ferritin has been proposed to be transient, storing and providing iron for biochemical pathways, the mechanism by which iron is removed from ferritin *in vivo* is not known. Studies on the mobilization of iron *in vitro* indicate that efficient mobilization of iron from ferritin requires both reduction of the ferric iron and chelation of the resultant ferrous iron (Theil, 1983). The physiological reductant of the iron in ferritin and the chelator of the ferrous iron is not known since physiological reductants (e.g. ascorbate and GSH) are not capable of mobilizing iron *in vitro* (Boyer and McClearly, 1987; Dognin and Crichton, 1975). Several non-physiological molecules including simple reductants such as thioglycolate and dithionite as well as the radicals of several chemicals have been shown to reduce iron in ferritin *in vitro* (Ahmad *et al.*, 1995; Oteiza *et al.*, 1995; Reif *et al.*, 1989; Thomas and Aust,

1986b, 1986c; Thomas *et al.*, 1985; Watt *et al.*, 1985). Iron has been mobilized from ferritin by reduced flavins (Dognin and Crichton, 1975; Jones *et al.*, 1978; Sirivech *et al.*, 1974), superoxide anion radical (Thomas *et al.*, 1985), alloxan semiquinone radical (Reif *et al.*, 1989), and redox cycling xenobiotics such as paraquat (Thomas and Aust, 1986b), diquat (Thomas and Aust, 1986b), and adriamycin and daunomycin) (Thomas and Aust, 1986c). The iron stored in ferritin has a reduction potential of −0.19 V to −0.23 V at pH 7.0, the variation depending upon the extent of iron loading (Watt *et al.*, 1985) and possibly the amount of phosphate associated with the iron core (Hoy *et al.*, 1974). In general, the ability of a radical to reduce iron in ferritin will be dictated by its reduction potential, though the size of the reductant appears to also limit the reduction of iron in ferritin (Thomas and Aust, 1986c; Watt *et al.*, 1988).

There is considerable evidence which suggests that the toxicity of certain redox-cycling xenobiotics is contingent upon iron-dependent oxidative processes (Horton and Fairhurst, 1987). Specific examples include the bipyridyl herbicides (paraquat and diquat) (Kohen and Chevion, 1985) and the anthracycline antibiotics (adriamycin and daunomycin) (Herman and Ferrans, 1983). Paraquat is a bipyridyl whose herbicidal properties are attributed to reduction within the chloroplasts of plants and subsequent redox cycling with dioxygen. Diquat, another bipyridyl herbicide, is thought to act in a similar manner. Both paraquat and diquat are toxic to mammals, probably also due to their ability to redox cycle (Bus *et al.*, 1974). Another class of chemicals thought to exert their toxic effects through redox cycling are the anthracycline antibiotics (Bachur *et al.*, 1979). These chemicals are isolated from *Streptomyces* and are exploited medicinally for their antitumorogenic activity, though the therapeutic use of anthracyclines is hindered due to their cardiotoxic effects (Lenaz and Page, 1976).

Paraquat primarily manifests its toxic effects in the lung while diquat affects the liver and adriamycin the heart (Clark *et al.*, 1966; Herman and Ferrans, 1983; Kohen and Chevion, 1985; Lenaz and Page, 1976). The tissue-specific toxicity of these chemicals stems in part from the accumulation of these compounds in the tissue as well as tissue-specific deficiencies in antioxidant defenses. However, the mechanism of bioactivation is quite similar. The proposed mechanism by which these chemicals are activated such that they may promote toxicity in mammals involves the facile one-electron reduction to a radical by the microsomal flavoprotein NADPH-cytochrome P450 reductase (Bachur *et al.*, 1979; Bus *et al.*, 1974). Under 'normal' conditions, P450 reductase will transfer two-electrons in two distinct one-electron transfer steps from the flavin to the heme prosthetic group of the cytochrome P450 mixed-function oxidase system. Redox-cycling chemicals serve as alternate electron acceptors and become reduced by one electron to their respective radical forms. The radical generated by one-electron reduction will readily reduce dioxygen to generate superoxide, provided that the reaction is thermodynamically favorable, with the xenobiotic radical being reoxidized back to the parent compound. This process has been termed redox cycling (Kappus and Sies, 1981), and is schematically represented by Figure 9.6.

Redox cycling of diquat and paraquat has been shown to promote reductive release of iron from ferritin *in vitro* (Thomas and Aust, 1986b). The mechanism involves reduction of iron in ferritin by $O_2^{-\bullet}$ (E'_o = −0.33 V) or directly by the respective bipyridyl cation radical species. Experimental data showed that the reduction of iron

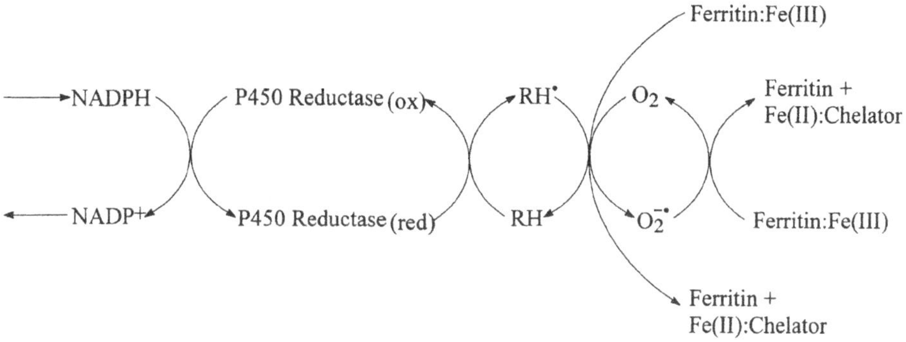

Figure 9.6 Schematic representation of redox cycling and mobilization of iron from ferritin by reducing radicals and superoxide anion radical

Table 9.2 Rates of iron release from ferritin by redox-active compounds, and the effects of superoxide dismutase[a]

System	*Iron release (nmol/min/ml)*		
	Aerobic	*Aerobic + SOD*	*Anaerobic (glucose + glucose oxidase)*
NADPH-cytochrome P450 reductase[b]	0.09	0.04	0.23
+ Paraquat[b]	0.45	0.18	22.2
+ Diquat[b]	0.63	0.23	27.8
+ Adriamycin[b]	0.41	0.29	7.3
+ Daunomycin[b]	0.44	0.17	11.4
GSH[c]	0.00	0.00	0.00
+ Alloxan[c]	0.90	0.86	0.87

a Data from Reif *et al.* (1989), Thomas and Aust (1986b, 1986c), with permission.
b Reaction mixtures contained NADPH-cytochrome P450 reductase (0.1 U), catalase (500 U), horse spleen ferritin (500 µM Fe^{III}), NADPH (0.5 mM), chemicals (all at 0.25 mM), bathophenanthroline disulfonic acid (1 mM) and where indicated SOD (500 U), glucose (5 mM), and glucose oxidase (10 U). Release of iron was monitored spectrophotometrically at 530 nm for formation of the ferrous phenanthroline complex. For adriamycin and daunomycin, ortho-phenanthroline (1 mM) was used to monitor formation of the ferrous:phenanthroline complex at 510 nm.
c Reaction mixtures consisted of 50 mM NaCl, pH 7.2, GSH (1 mM), alloxan (1 mM), rat liver ferritin (200 µM Fe^{III}), and bathophenanthroline sulfonate (25 µM) with addition of SOD (10 U/ml), or glucose (5 mM), and glucose oxidase (15 U/ml) where indicated. Iron release was monitored by spectrophotometric detection of the ferrous phenanthroline complex at 530 nm.

in ferritin by redox-cycling bypyridyls was only partially inhibited by superoxide dismutase (SOD), indicating that the xenobiotic radical itself can indeed reduce iron stored in ferritin (Table 9.2). This may also explain why conflicting data exist concerning the ability of SOD to protect against paraquat toxicity (Autor, 1974; Patterson and Rhodes, 1982). Diquat (E'_0 = −0.35 V), a potent hepatotoxin which has been shown to reduce iron stored in ferritin *in vitro* by a similar mechanism to paraquat, has also been shown to significantly alter the distribution of iron in rat liver (Reif *et al.*, 1988). It was shown that administration of diquat to rats resulted in increased concentrations of low molecular weight chelatable iron while the amount of iron associated with ferritin decreased. The decrease in iron stored in ferritin was not a result of ferritin synthesis, since a concomitant increase in ferritin content in the liver was not observed. These results indicate that the ability of diquat to reduce and mobilize iron stored in ferritin *in vitro* may also occur *in vivo*. It was also demonstrated during *in vitro* studies on the mobilization of iron from ferritin by reduced bipyridyl herbicides, that the radical form of the compound was a better, more efficient reductant of iron in ferritin than $O_2^{-\bullet}$ (Table 9.2). It was also shown that rapid and complete release of iron from ferritin could occur under anaerobic conditions. Approximately 1.37 and 1.10 moles of NADPH were oxidized per mole of iron reduced and subsequently released by paraquat and diquat, respectively (Thomas and Aust, 1986b). These values indicate that the reduction of iron in ferritin by the bipyridyl herbicides was tightly coupled. During redox cycling of these chemicals in cells, iron may be released from ferritin thus saturating the physiological or 'normal' low molecular weight chelate pool of iron. Similar to cases of acute iron intoxication, saturation of this pool would overwhelm the endogenous protective mechanisms of cells and promote various redox reactions of iron, ultimately causing deleterious oxidations of biological macromolecules.

In similar studies, the anthracycline antibiotics adriamycin and daunomycin were used as the electron acceptors from P450 reductase and the reductant of iron in ferritin (Thomas and Aust, 1986c). The cardiotoxicity of adriamycin has been proposed to be dependent upon the production of oxygen radicals, particularly $O_2^{-\bullet}$. This is due in part to the observation that the heart is low in SOD activity. Iron is also implicated in the toxicity of adriamycin since iron chelators such as desferrioxamine and ICRF-187 have been shown to mitigate the cardiotoxic effects (Herman and Ferrans, 1983). Reduction of adriamycin and daunomycin by P450 reductase produced the respective semiquinone radicals which readily reduced dioxygen to superoxide. In the presence of NADPH, P450 reductase, adriamycin or daunomycin, and ferritin, iron is reduced and mobilized from ferritin (Table 9.2). Similar to the paraquat cation radical under anaerobic conditions, the adriamycin and daunomycin semiquinone radicals promoted rapid and complete release of the iron stored in ferritin (Thomas and Aust, 1986c). The efficiency of iron release from ferritin by adriamycin and daunomycin, determined as the moles of NADPH oxidized per mole of iron released, was found to be 3.79 and 3.96, respectively (Thomas and Aust, 1986c). These values were considerably higher than the values observed for the reduced bipyridyl compounds. The redox potential for the diquat cation radical and the adriamycin semiquinone anion radical are approximately the same ($E_0' = -0.33$ V and $E_0' = -0.35$ V, respectively), indicating that differences in the reduction potential do not adequately account for the differences in the ability of these compounds to reduce the iron stored in ferritin. It was concluded that the size of the reductant able to access the iron stored within ferritin was limited by the diameter of the channel that traverses the protein shell (Thomas and Aust, 1986c).

To determine if iron released from ferritin was suitable to promote oxidation of biological macromolecules, the effects of ferritin on the peroxidation of lipids by various redox-cycling xenobiotics was determined. The model system consisted of phospholipid liposomes, P450 reductase, NADPH, paraquat, and ferritin. It was observed that iron mobilized from ferritin by the redox cycling of paraquat was capable of initiating lipid peroxidation (Figure 9.7) (Saito *et al.*, 1985). It can be concluded from the data presented in Figure 9.7 that the rate of lipid peroxidation

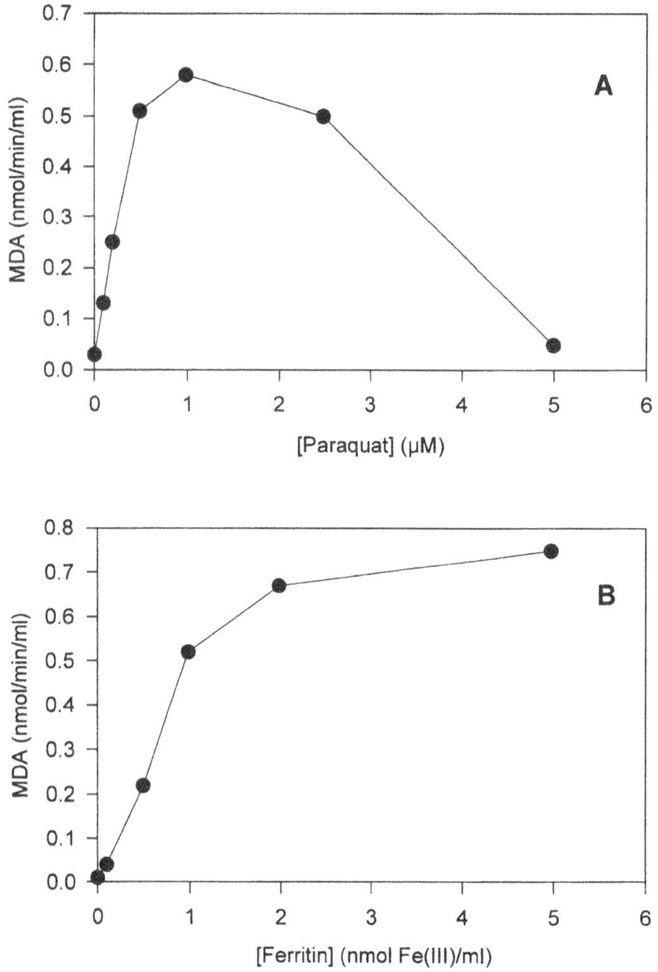

Figure 9.7 The effects of paraquat (A) and ferritin (B) concentrations on rates of lipid peroxidation. Reactions contained liposomes (1 mM lipid phosphate), ADP (2.5 mM), NADPH (0.2 mM), and NADPH-cytochrome P450 reductase (0.1 U/ml) in 0.25 mM NaCl, pH 7.0, at 37°C. In (A) ferritin was at a concentration of 1 mM Fe^{III}, and in (B) paraquat was at 0.5 mM. MDA content was measured at 0, 10, 20, and 30 minutes as determined by the thiobarbituric acid test. Data from Saito *et al.* (1985), with permission

was affected by the rate of iron release which is ultimately dependent upon the concentrations of paraquat, P450 reductase, and ferritin. The rate of iron release was an important factor for initiation of lipid peroxidation because the ratio of $Fe^{II}:Fe^{III}$ was dictated by the rate and amount of Fe^{II} released from ferritin. The rate of lipid peroxidation was also stimulated by catalase. This indicated that the oxidation of ferrous iron by H_2O_2, produced from dismutation of superoxide, limited the generation and maintainance of the optimal ratio of $Fe^{II}:Fe^{III}$ required to initiate lipid peroxidation. Removal of H_2O_2 by catalase also prevented formation of the $^{\bullet}OH$ confirming that paraquat and ferritin dependent lipid peroxidation was initiated by iron complexes. Superoxide dismutase also inhibited lipid peroxidation, presumably by preventing $O_2^{-\bullet}$ dependent mobilization of iron from ferritin as well as disrupting the ratio of $Fe^{II}:Fe^{III}$ by increasing the rate of H_2O_2 production. Like paraquat, reductive release of iron from ferritin by adriamycin also initiated lipid peroxidation (Li, 1991).

Alloxan is a chemical frequently used to induce diabetes in experimental animals. The diabetogenic action of alloxan is thought to involve redox cycling and the generation of oxygen radicals capable of damaging the cells producing insulin in the pancreas (Heikkila *et al.*, 1976). Although the precise mechanism by which alloxan elicits toxicity is not known, some have proposed that lipid peroxidation may be involved (Fee *et al.*, 1975; Reif *et al.*, 1989). In the presence of a reductant, alloxan is reduced to the semiquinone radical ($E_0' = -0.44$ V), which is capable of reducing molecular oxygen and iron in ferritin (Reif *et al.*, 1989). It was found that the reduction of iron stored in ferritin by the alloxan semiquinone radical initiated lipid peroxidation, and was not inhibited by either SOD or catalase. This indicates that the mechanism of iron release from ferritin was independent of $O_2^{-\bullet}$, and that lipid peroxidation was initiated by iron complexes. Lipid peroxidation was, however, inhibited by ceruloplasmin, a ferrous iron oxidase, which indicated that low molecular weight chelates of iron were essential for alloxan- and ferritin-dependent lipid peroxidation.

It can be concluded from the previous discussions that deleterious oxidative reactions catalyzed by iron may be a direct result of its mobilization from ferritin to provide a deleterious low molecular weight chelate of iron. Conditions necessary for the release of iron from ferritin by various chemicals were found to be ideal for the initiation of lipid peroxidation as well as the oxidation of other biomolecules not discussed (Kukielka and Cederbaum, 1994). The *in vitro* studies discussed previously are significant since subsequent *in vivo* studies provide support for ferritin serving as the source of redox active iron involved in the toxicity of redox-cycling xenobiotics that produce oxidative stress. Therefore, it is reasonable to suggest that ferritin may be a source of the low molecular weight chelates of iron *in vivo* needed for catalysis of oxidative damage, and that processes promoting the release of iron from ferritin may cause oxidative damage.

9.4 Antioxidant defenses for iron-mediated toxicity

9.4.1 Enzymatic removal of reduced oxygen species

Several enzymes including NADH oxidase, urate oxidase, glycolate oxidase, glucose oxidase, D-amino acid oxidase, SOD, and xanthine oxidase constituatively produce $O_2^{-\bullet}$ and/or H_2O_2 (Halliwell and Gutteridge, 1984). Superoxide has been labelled an endogenous toxin (Fridovich, 1983) and has previously been shown to promote iron release form ferritin (Saito *et al.*, 1985; Thomas *et al.*, 1985). Iron released from ferritin was capable of initiating lipid peroxidation (Thomas *et al.*, 1985) and catalyzing the Haber–Weiss reactions leading to the damage of DNA (Kukielka and Cederbaum, 1994). Fortunately, nature has equipped cells with a battery of enzymes capable of removing specific reduced species of oxygen to protect themselves from continual damage.

Superoxide dismutase catalyzes the dismutation of $O_2^{-\bullet}$ to H_2O_2 and H_2O. The rate constant for dismutation of $O_2^{-\bullet}$ in the absence of a catalyst at physiological pH, is approximately $5 \times 10^5\ M^{-1}\ s^{-1}$ (Halliwell and Gutteridge, 1984). The rate constant for dismutation of $O_2^{-\bullet}$ in the presence of SOD is approximately 1.6×10^9 M^{-1} s^{-1} at physiological pH (Halliwell and Gutteridge, 1984), approximately 3200 times faster than the non-enzymatic reaction. In addition, a manganese-dependent SOD resides in the mitochondria, and maintains the concentration of $O_2^{-\bullet}$ below < 10^{-11} M (Halliwell and Gutteridge, 1984). Superoxide dismutase has been shown to inhibit superoxide-dependent iron release from ferritin and subsequent lipid peroxidation. Unfortunately, SOD is inactivated by the dismutation product, H_2O_2, over time; hence, H_2O_2 must be removed as well.

Catalase, a heme protein located in the peroxisomes of most cell types, catalyzes the rapid decomposition of H_2O_2 to H_2O and O_2, exhibiting a second order rate constant greater than 10^7 M^{-1} s^{-1} (Halliwell and Gutteridge, 1984). Catalase prevents the production of $^\bullet OH$ by iron-dependent decomposition of H_2O_2. Additionally, removal of H_2O_2 inhibits the oxidation of the low molecular weight chelates of ferrous iron, and prevents the generation of a mixture of Fe^{II} and Fe^{III} shown previously to initiate lipid peroxidation. Although seemingly essential for protection against oxidative damage, genetic defects resulting in decreased catalase activity are not lethal (Halliwell and Gutteridge, 1984). A second cooperative mechanism for the removal of H_2O_2 is achieved by reduction of H_2O_2 by glutathione peroxidase (Halliwell and Gutteridge, 1984; Sunde, 1994). Glutathione peroxidase is a selenium-containing enzyme found in the cytosol and mitochondria of cells, particularly in tissues where catalase activity is low (Halliwell and Gutteridge, 1984). Glutathione peroxidase reduces H_2O_2 by two electrons to water with concomitant oxidation of two molecules of reduced glutathione. The rate constant for decomposition of H_2O_2 by glutathione peroxidase has been reported as 5×10^7 M^{-1} s^{-1} (Halliwell and Gutteridge, 1984). The concentration of reduced glutathione is maintained at a relatively high concentration in the cytosol (~10^{-3} M) by glutathione reductase, a flavoprotein which utilizes NADPH for reducing equivalents (Halliwell and Gutteridge, 1984; Hill, 1994). Defects in the gene coding for glucose-6-phosphate dehydrogenase cause decreased production of NADPH by the pentose phosphate pathway. A lower concentration of reducing equivalents will limit reduction of glutathione by glutathione reductase, ultimately limiting glutathione peroxidase activity. Individuals with such a defect are extremely susceptible to hemolytic anemia induced by redox cycling of chemicals such as primaquine, since red blood cells depend highly upon this system to prevent oxidative damage catalyzed by partially reduced species of oxygen (Halliwell and Gutteridge, 1984; Smith, 1991). Similar symptoms also result from selenium deficiency. However, SOD, catalase, and glutathione peroxidase do not inhibit all oxidative processes, such as iron-dependent lipid peroxidation. Accumulation of iron in a low molecular weight fraction can continually promote redox reactions unless effectively chelated, or removed and stored in a stable non-reactive form in ferritin.

9.4.2 Regulation of 'free' iron

During the course of investigations on the mechanism of iron-dependent lipid peroxidation, it was found that ceruloplasmin, a copper-containing ferroxidase, possessed both pro- and anti-oxidant properties (Aust *et al.*, 1989; Reif *et al.*, 1989; Samokyszyn *et al.*, 1989). It was shown that oxidation of ferrous iron by ceruloplasmin in the absence of ferritin generated the proper ratio of Fe^{II}:Fe^{III} required to initiate lipid peroxidation (Aust *et al.*, 1989). The uncoupled oxidation of iron by

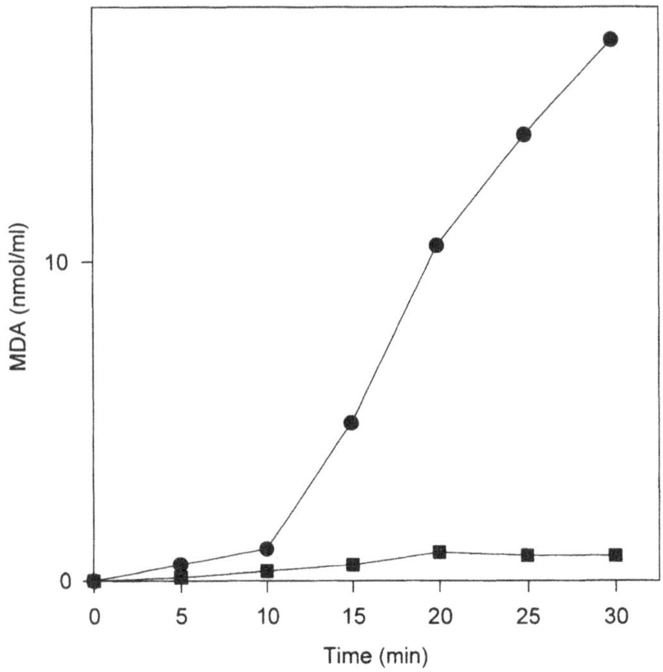

Figure 9.8 The effects of apoferritin on ceruloplasmin-dependent lipid peroxidation. Reactions contained liposomes (1 mM lipid phosphate) and ceruloplasmin (0.5 μM) in the presence (■) or absence (●) of apoferritin (0.1 μM) in 50 mM NaCl, pH 7.0, at 37°C. Reactions were initiated by addition of $FeCl_2$ (200 μM) after a 2 min incubation, and MDA determined by the thiobarbituric acid test. Data from Samokyszyn *et al.* (1989), with permission

ceruloplasmin would not be a favorable process *in vivo*, therefore is likely not to have evolved. When apoferritin was included in the ceruloplasmin-dependent lipid peroxidation system, the pro-oxidant effect of ceruloplasmin was eliminated (Figure 9.8) (Samokyszyn *et al.*, 1989). Likewise, when ceruloplasmin was included in the ferritin- and superoxide- or alloxan-dependent model lipid peroxidation systems, it was found to be a potent inhibitor of lipid peroxidation (Reif *et al.*, 1989; Samokyszyn *et al.*, 1989). The mechanism of inhibition appeared to involve the inhibition of iron release from ferritin due to the sequestration of the reduced iron in ferritin catalyzed by ceruloplasmin (Samokyszyn *et al.*, 1989). Together, ferritin and ceruloplasmin constituted a highly effective means of preventing oxidative damage catalyzed by 'free' iron.

Incorporation of iron into ferritin requires ferrous iron, and through an oxidative process ferric iron is deposited into ferritin (Theil, 1987). There exist two main hypotheses on the mechanism by which the oxidation of ferrous iron with subsequent deposition of the ferric iron into ferritin occurs. The first proposed mechanism relies on the 'ferroxidase' activity of ferritin, while the second requires an exogenous ferroxidase. Probably the most frequently utilized system to load iron into ferritin *in vitro* relies on the proposed 'ferroxidase' activity of ferritin (Bakker and Boyer,

Table 9.3 Stoichiometry of iron incorporation into rat liver apoferritin[a]

System	Fe^{II}/O_2^d
HEPES + ferrous ammonium sulfate[b]	2.3 ± 0.2
HEPES + ferrous ammonium sulfate + catalase[b]	3.8 ± 0.2
NaCl + ceruloplasmin + histidine:Fe^{II}[c]	3.8 ± 0.3
NaCl + ceruloplasmin + histidine:Fe^{II} + catalase[c]	3.9 ± 0.3

a Data from DeSilva and Aust (1992), with permission.
b Reaction mixtures contained 0.1 mM apoferritin in 50 mM HEPES buffer, pH 7.0, and 50 μM ferrous ammonium sulfate in the presence or absence of catalase (500 U/ml).
c Reaction mixtures contained 0.1 mM apoferritin, 0.1 μM ceruloplasmin, 50 μM histidine-chelated ferrous iron (5:1), and 50 mM NaCl, pH 7.0. Where indicated catalase (500 U/ml) was added.
d Oxygen consumption was monitored using a Gilson Oxy 5/6 oxygraph equipped with a Clark-type electrode. Oxidation of ferrous iron was monitored by quenching the reaction at 1 min intervals with 1 mM ferrozine in 50 mM NaCl, pH 7.0, and the ferrous iron quantitated by detection of the ferrous: ferrozine complex (ε_{564} = 27,900 M^{-1} cm^{-1}). Reactions were carried out at 37°C and Fe^{II}/O_2 ratios determined at the 5 min time point.

1986; Chasteen *et al.*, 1985; Lawson *et al.*, 1989; Theil, 1987). In this system, ferrous iron is oxidized and deposited into ferritin using HEPES or another appropriate Good buffer. It has been found that the ferritin loaded using its own 'ferroxidase' activity was damaged by oxidation of basic amino acids on ferritin (DeSilva *et al.*, 1992). Additionally, the ferritin exhibited iron storage properties that were significantly different from ferritin isolated from tissue (DeSilva *et al.*, 1992). In this process, HEPES effects the redox properties of iron resulting in the oxidation of ferrous iron with subsequent partial reduction of dioxygen. The stoichiometry observed for this reaction was one mole of dioxygen reduced to two moles of ferrous iron oxidized (Table 9.3) (DeSilva and Aust, 1992; Xu and Chasteen, 1991). The two-electron reduction of molecular oxygen generates H_2O_2, which, in the presence of ferrous iron, generated $^{\bullet}OH$. When glutamine synthetase, an enzyme sensitive to inactivation by $^{\bullet}OH$ (Rivett and Levine, 1990; Stadtman, 1990), was included in this reaction mixture, it was rapidly inactivated (Abedi, 1997). The oxidation of amino acids on ferritin during the oxidation of iron in the presence of HEPES was prevented by catalase. Replacement of HEPES buffer with saline eliminated the ferroxidase activity of ferritin (DeSilva *et al.*, 1992).

It has been shown that ferritin may also be loaded with iron enzymatically using the ferroxidase activity of ceruloplasmin (Figure 9.9) (Boyer and Schori, 1983; DeSilva and Aust, 1992; DeSilva *et al.*, 1992). This reaction was first described in 1983 by Boyer and Schori who found that in the presence of ferritin, the ferroxidase activity of ceruloplasmin was stimulated and ferritin was loaded. The stoichiometry observed for this reaction is one mole of dioxygen reduced for four moles of ferrous iron oxidized (Table 9.3). The ability of ceruloplasmin to load iron into ferritin is unique, other ferroxidases such as tyrosinase are unable to mimic ceruloplasmin in this process. Removal of copper from ceruloplasmin to generate the apo-enzyme negated loading of iron into ferritin by ceruloplasmin, indicating that this process is dependent upon the ferroxidase activity of ceruloplasmin (Samokyszyn *et al.*, 1989). Ferritin loaded using ceruloplasmin as the ferroxidase is not damaged by oxidation of amino acids by $^{\bullet}OH$ and exhibited iron storage properties that were nearly identical to those observed for ferritin isolated from tissue (DeSilva *et al.*, 1992).

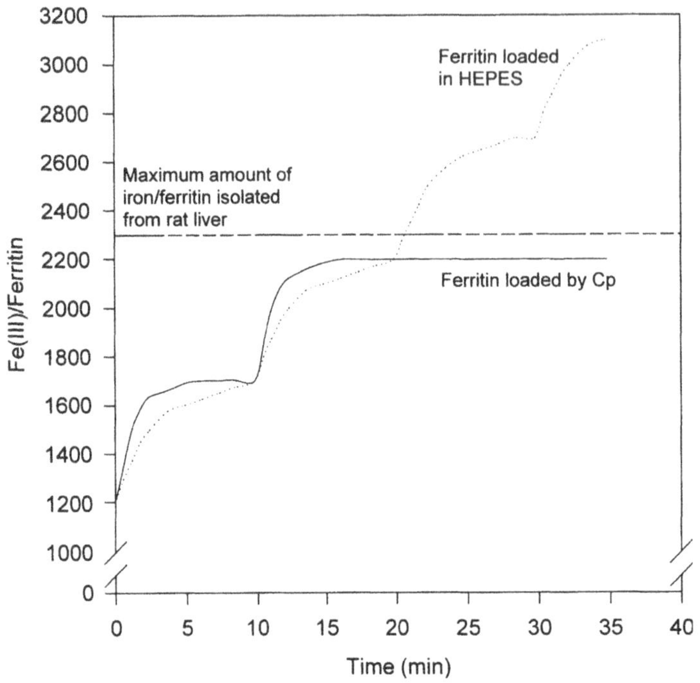

Figure 9.9 Sequential loading of rat liver ferritin with iron by ceruloplasmin or in HEPES buffer. Samples contained rat liver ferritin containing ~ 1000 Fe^{III}/ferritin (0.1 μM) in either 50 mM HEPES buffer, pH 7.0 (··········) or Chelex treated 50 mM NaCl, pH 7.0, containing rat ceruloplasmin (Cp, 0.15 μM) (—) at 37°C. Reactions were initiated by addition of 500 atoms Fe^{II}/ferritin as ferrous ammonium sulfate for the HEPES system, or histidine:Fe^{II} (5:1) for the ceruloplasmin system. Loading was monitored spectrophotometrically as an increase in absorbance at 380 nm. Data from DeSilva and Aust (1992), with permission

Ceruloplasmin-dependent iron loading into ferritin has been proposed to occur via a specific protein–protein complex such that iron oxidized by ceruloplasmin is directly conveyed into ferritin and does not precipitate out of solution (Guo *et al.*, 1996; Juan *et al.*, 1997; Reilly and Aust, 1997; Reilly *et al.*, 1998). This reaction has been shown to be dependent upon the presence of the H subunit of ferritin and appears to be highly specific (Guo *et al.*, 1996; Juan *et al.*, 1997). Research utilizing recombinant homopolymers of ferritin has provided evidence that the H chain of ferritin possessed both the recognition site for ceruloplasmin association as well as the channel necessary for passage of iron into the core of ferritin (Guo *et al.*, 1996). The H chain of ferritin is unique from L chain in that it has a conserved 1 Å intrasubunit four α-helix bundle channel which permits transfer of ferric iron into ferritin through the protein shell. Engineering an opened α-helix bundle channel into L chain permitted iron loading into ferritin, but the ferroxidase activity of ceruloplasmin was not stimulated by L chain or the L chain mutant (Guo *et al.*, 1996). This indicated that stimulation of the ferroxidase activity was independent of iron loading, and was dependent upon binding of ceruloplasmin to a recognition site

Table 9.4 The effect of the molar ratio of ceruloplasmin to H chain ferritin on the efficiency of iron loading into ferritin[a]

Molar ratio	*1 H chain/ferritin*		*2 H chain/ferritin*	
	iron/ferritin[b]	*ferrous iron*[b] *remaining*	*iron/ferritin*[b]	*ferrous iron*[b] *remaining*
0.5:1	330 ± 10	17.4 ± 0.4	–	–
1:1	450 ± 20	5.2 ± 0.5	465 ± 5	3.9 ± 0.4
2:1	415 ± 6	0	490 ± 6	0
3:1	392 ± 4	0	435 ± 8	0
4:1	–	–	410 ± 10	0

a Data from Juan *et al.* (1997), with permission.

b Reactions (1 ml) contained ferritin variants (0.1 μM) and various concentrations of rat ceruloplasmin to obtain the specified molar ratios of ferritin to ceruloplasmin in 50 mM NaCl, pH 7.0, at 37°C. Reactions were initiated by addition of 500 atoms Fe^{II}/ferritin as histidine:Fe^{II} (5:1). Iron per ferritin was determined by the total iron assay and ferrous iron remaining in solution quantitated using ferrozine (ε_{564} = 27,900 M^{-1} cm^{-1}).

Table 9.5 Affinity chromatography of ceruloplasmin using immobilized ferritin[a]

Eluent	*Activity (units)*[b,c]
50 mM NaCl, pH 7.0	–
50 mM NaCl, pH 7.0, + 1% w/v N-octyl glucoside	0.036 ± 0.006
75 mm NaCl, pH 7.0	0.159 ± 0.008

a Data from Reilly *et al.* (1998), with permission.

b 0.18 ± 0.02 units of ceruloplasmin were loaded onto the column.

c Activity was determined by oxidation of p-phenylenediamine (10 mM) in 100 mM sodium acetate buffer, pH 5.0. One unit is defined as the amount of oxidase required to cause a change in A_{570} = 1.0/hr at 37°C.

on the H chain. Optimal iron loading efficiency requires that a 1:1 stoichiometric ratio of H chain of ferritin to ceruloplasmin exists, with excess ceruloplasmin resulting in accumulation of ferric iron in solution (Table 9.4) (Juan *et al.*, 1997). Additional evidence for the interaction between ferritin and ceruloplasmin was obtained by retention of ceruloplasmin on an affinity column using ferritin as an immobilized ligand (Table 9.5) (Reilly *et al.*, 1998). It was also found that ferrous iron significantly increased the affinity of ceruloplasmin for ferritin (Reilly *et al.*, 1998). Although ceruloplasmin is commonly used to load transferrin with iron, evidence for an association between these two proteins could not be shown. Ferritin and other proteins (Walker and Fay, 1990) that interact with ceruloplasmin stimulate the oxidase activity of ceruloplasmin, however, transferrin does not (Reilly and Aust, 1997). These studies suggest that formation of a protein–protein complex is essential for loading of iron into ferritin by ceruloplasmin.

Although there exists much evidence for ceruloplasmin acting as the ferroxidase responsible for loading ferritin with iron, one major conflict exists: the physiological function of ceruloplasmin is not yet well defined. Various investigators have proposed a variety of functions for ceruloplasmin, including loading transferrin with iron (Osaki *et al.*, 1966), mobilization of iron from cells (Osaki *et al.*, 1971; Osaki and Johnson, 1969; Young *et al.*, 1997), delivery of copper to cells (Harris and

Percival, 1989), scavenging superoxide anion radical (Goldstein *et al.*, 1982; Halliwell and Gutteridge, 1984) and serving as a nonspecific oxidase catalyzing the oxidation of various phenols, biogenic amines, and catechols (Rydén, 1981). The latter is apparently not the case since the oxidation products of the proposed substrates have not been found in high concentrations *in vivo*. The vast majority of the reactions involving these chemicals with ceruloplasmin have been shown to be due entirely to redox cycling of contaminating iron, and were inhibited by iron chelators which prevent the reduction of iron by the chemicals (Curzon and O'Reilly, 1960; McDermott *et al.*, 1968). However, ceruloplasmin has been shown to exhibit high specificity for ferrous iron (Rydén, 1981; Curzon and O'Reilly, 1960; McDermott *et al.*, 1968) with reported K_m values around 5–10 μM for histidine-chelated ferrous iron (Ryan *et al.*, 1992). As a result of this specificity, the role of ceruloplasmin as a specific iron oxidase has gained much support in recent literature. Also, the coupling of the oxidation of iron by ceruloplasmin to its loading into ferritin effectively controls redox active iron and is therefore physiologically advantageous. However, ceruloplasmin is often regarded as a serum protein and storage of iron in ferritin occurs within cells. Serum ferritin has been isolated and described as a glycosylated form of ferritin having low iron content (Worwood *et al.*, 1976). If ceruloplasmin does, in fact, load ferritin with iron *in vivo*, serum ferritin should contain iron. However, serum ferritin has been shown to consist primarily of a peptide that closely resembles the L chain of liver and spleen ferritin (Worwood *et al.*, 1976; Arosio *et al.*, 1977) and L-chain homopolymers of ferritin have been shown to be incapable of loading iron using ceruloplasmin as the ferroxidase (Guo *et al.*, 1996). In addition, it is unlikely that excess amounts of ferrous iron exists in the serum such that it must be loaded into ferritin.

Ceruloplasmin is synthesized in the liver and secreted into the circulatory system. However, it has been shown using immunohistological staining techniques that a cross-reactive protein exists in many tissues (Puchkova *et al.*, 1990), and that a protein with ceruloplasmin-like activity may be isolated (Linder and Moor, 1977; Reilly and Aust, 1998). It was proposed by Linder and Moor (1977) that ceruloplasmin was internalized by receptor-mediated endocytosis for delivery of copper to cells, since ceruloplasmin contains up to 95% of the total serum copper. However, this cannot be true because individuals with aceruloplasminemia exhibit normal intracellular copper distribution despite the absence of circulating ceruloplasmin (Harris and Gitlin, 1996; Harris *et al.*, 1998). In addition, studies on the delivery of copper to cells by ceruloplasmin demonstrated that the protein was not internalized (Harris and Percival, 1989). Recently, an intracellular form of ceruloplasmin has been proposed after detection of ceruloplasmin mRNA in extrahepatic tissues including the retina (Levin and Geszvain, 1998) heart, brain, kidney (Gaitskhoki *et al.*, 1990), and lung (Fleming and Gitlin, 1990; Fleming *et al.*, 1991). A ceruloplasmin analogue has also been isolated from brain astrocytes (Patel and David, 1997) and horse heart (Reilly and Aust, 1998), indicating tissue-specific synthesis of ceruloplasmin or a ceruloplasmin-like enzyme. The ceruloplasmin isolated from astrocytes was proposed to be involved in iron metabolism and protection from iron-mediated free-radical reactions and the subsequent development of neurodegenerative disorders such as Alzheimer's disease; ceruloplasmin concentrations decreased in the brains of individuals with Alzheimer's disease (Connor *et al.*, 1993). It was also shown that ceruloplasmin mRNA in the heart was inducible by adriamycin (Ryan, 1991) which has been shown to promote oxidative stress through reductive release of iron from ferritin *in vivo* (Reif *et al.*, 1988). Ceruloplasmin is also induced by exposure to hyperoxic conditions, inflammation, and cellular injury where potentially catalytic iron may be released during the degradation of proteins in dying cells (Fleming *et al.*, 1991; Gitlin, 1988; Levin and Geszvain, 1998; Moak and Greenwald, 1984; Okumura *et al.*, 1991). This suggests that the concentration of ceruloplasmin may be regulated, in part, by iron, and is increased when control of 'free' iron is needed.

Disorders in copper metabolism as well as dietary copper deficiency also suggest a specific role for ceruloplasmin in iron metabolism. Copper deficiency is often associated with hepatic iron overload and anemic conditions (Owen, 1973). In a recent study, iron and copper deficient animals that developed symptoms consistent with anemia were not responsive to treatment with iron, however copper or ceruloplasmin supplementation completely reversed the symptoms (Lee *et al.*, 1968). In addition, a deficiency in active ceruloplasmin (aceruloplasminemia) has been shown to have profound effects on iron metabolism (Harris and Gitlin, 1996; Harris *et al.*, 1995, 1998; Yoshida *et al.*, 1995). Aceruloplasminemia is characterized by hemosiderosis, diabetes mellitus, neurologic and ophthalmic degeneration, and altered pigmentation. Excessive iron deposition resembling hemosiderosis in liver, brain, and pancreas has been observed upon histopathological examination of patients diagnosed with aceruloplasminemia (Harris and Gitlin, 1996; Harris *et al.*, 1995, 1998; Yoshida *et al.*, 1995). Furthermore, individuals with aceruloplasminemia exhibited a substantial increase in thiobarbituric acid reactive substances (TBARS) in plasma. This increase was due to increased lipid peroxidation, presumably due to altered systemic iron homeostasis, and was ameliorated by treatment with exogenous ceruloplasmin (Miyajima *et al.*, 1996). Collectively, these observations indicate that ceruloplasmin or a protein functioning in a similar manner to ceruloplasmin is essential for normal iron metabolism and storage within cells.

9.5 Conclusions

In summary, transition metals such as iron are highly reactive and have the potential to catalyze a series of deleterious reactions both *in vitro* and *in vivo*. Since iron is capable of catalyzing the oxidation of biomolecules and is an essential component of various biochemical processes, precise mechanisms for uptake, distribution, and storage of iron have evolved. Iron is stored in ferritin which can be released in the presence of redox-cycling xenobiotics, $O_2^{-\bullet}$, and other small chemical reductants, and a ferrous iron chelator. Iron released from ferritin can serve as the iron required to promote a series of deleterious reactions ultimately causing the oxidation of biomolecules. Although cells generally possess the necessary protective enzymes to prevent oxidative damage, certain conditions, including redox cycling of xenobiotics, can overwhelm the normal cellular defenses responsible for preventing these reactions. Aside from the protection of cellular components from damage by partially reduced oxygen species, cells may possess a system to efficiently scavenge and store iron in a stable non-redox active form within ferritin. This system consists of ceruloplasmin, or a ceruloplasmin-like protein, and ferritin which together have been shown to constitute a potent antioxidant system. Although the presence of ceruloplasmin in tissue is debated, the probability that it or a functional analogue exists and is essential in iron metabolism and the loading ferritin *in vivo* is increasing.

Acknowledgments

The authors would like to acknowledge NIH grants ES05056 and DK52823 for providing financial support for this work. We would also like to thank Dr Will McCarthy, Terri Maughan, and Marc Van Eden for their assisance in preparation of this manuscript.

References

Abedi, M. (1997) *Recombinant Expression and Characterization of Rat Liver Light Chain Ferritin Homopolymers*, M.S. thesis, pp. 1–45, Logan: Utah State University.

Ackworth, I.N., and Bailey, B. (1995) *The Handbook of Oxidative Metabolism*, Chelmsford: ESA, Inc.

Ahmad, S., Singh, V., and Rao, G.S. (1995) 'Release of iron from ferritin by 1,2,4-benzenetriol', *Chemico-Biol. Int.* **96**, 103–111.

Alberts, B., Bray, D., Lewis, J., Raff, M., Roberts, K., and Watson, J.D. (1994) *The Molecular Biology of the Cell*, 3rd edn, New York: Garland Publishers.

Arosio, P., Yokota, M., and Drysdale, J.W. (1977) 'Characterization of serum ferritin in iron overload: possible identity to natural apoferritin', *Br. J. Haematol.* **36**, 199–207.

Aruoma, O.I., Halliwell, B., Laughton, M.J., Quinlan, G.J., and Gutteridge, J.M.C. (1989) 'The mechanism of initiation of lipid peroxidation. Evidence against a requirement for an iron(II)-iron(III) complex', *Biochem. J.* **258**, 617–620.

Atkins, P.W. (1983) *Molecular Quantum Mechanics*, 2nd edn, New York: Oxford University Press.

Aust, S.D., Bucher, J.R., and Tien, M. (1984) 'Evidence for the initiation of lipid peroxidation by a ferrous-dioxygen-ferric chelate complex', in Bors, W., Saran, M., and Tait, S.D. (eds.) *Proceedings of the Third International Conference on Oxygen Radicals in Chemistry and Biology*, Berlin: Walter de Gruyter & Co., pp. 147–154.

Aust, S.D., Morehouse, L.A., and Thomas, C.E. (1985) 'Role of metals in oxygen radical reactions', *J. Free Rad. Biol. Med.* **1**, 3–25.

Aust, S.D., Miller, D.M., and Samokyszyn, V.M. (1989) 'Role of iron in model lipid peroxidation systems', in Beaumont, P.C., Deeble, D.J., Parsons, B.J., and Rice-Evans, C. (eds.) *Free Radicals, Metal Ions, and Biopolymers*, London: Richelieu Press, pp. 251–282.

Autor, A.P. (1974) 'Reduction of paraquat toxicity by superoxide dismutase', *Life Sci.* **14**, 1309–1319.

Aziz, N., and Munro, H.N. (1987) 'Iron regulates ferritin mRNA translation through a segment of its 5′ untranslated region', *Proc. Natl. Acad. Sci. USA* **84**, 8478–8482.

Bachur, N.R., Gordon, S.L., Gee, M.V., and Kon, H. (1979) 'NADPH cytochrome P-450 reductase activation of quinone anticancer agents to free radicals', *Proc. Natl. Acad. Sci. USA* **76**, 954–957.

Bailey, S., Evans, R.W., Garratt, R.C., Gorinsky, B., Hasnain, S., Horsburgh, C., Jhoti, H., Lindley, P.F., Mydin, A., Serra, R., and Watson, J.L. (1988) 'Molecular structure of serum transferrin at 3.3-A resolution', *Biochemistry* **27**, 5804–5812.

Baker, E.N. (1994) 'Structure and reactivity of transferrins', in Sykes, A.G. (ed.) *Advances in Inorganic Chemistry, vol. 41*, San Diego: Academic Press, pp. 389–463.

Bakker, G.R., and Boyer, R.F. (1986) 'Iron incorporation into apoferritin. The role of apoferritin as a ferroxidase', *J. Biol. Chem.* **261**, 13182–13185.

Bali, P.K., and Aisen, P. (1991) 'Receptor-modulated iron release from transferrin: differential effects on N- and C-terminal sites', *Biochemistry* **30**, 9947–9952.

Bartlett, G.R. (1976a) 'Phosphate compounds in rat erythrocyte and reticulocytes', *Biochem. Biophys. Res. Comm.* **70**, 1055–1062.

Bartlett, G.R. (1976b) 'Iron nucleotides in human and rat red cells', *Biochem. Biophys. Res. Comm.* **70**, 1063–1070.

Bomford, A., Conlon-Hollingshead, C., and Munro, H.N. (1981) 'Adaptive responses of rat tissue isoferritins to iron administration. Changes in subunit synthesis, isoferritin abundance, and capacity for iron storage', *J. Biol. Chem.* **256**, 948–955.

Boyd, D., Vecoli, C., Belcher, D.M., Jain, S.K., and Draysdale, J.W. (1985) 'Structural and functional relationships of human ferritin H and L chains deduced from cDNA clones', *J. Biol. Chem.* **260**, 11755–11761.

Boyer, R.F., and McClearly, C.J. (1987) 'Superoxide ion as a primary reductant in ascorbate-mediated ferritin iron release', *Free Rad. Biol. Med.* **3**, 389–395.

Boyer, R.F., and Schori, B.E. (1983) 'The incorporation of iron into apoferritin as mediated by ceruloplasmin', *Biochem. Biophys. Res. Comm.* **116**, 244–250.

Brock, J.H. (1985) 'Transferrins', in Harrison, P.M. ed. *Metalloproteins: Part 2 Metal proteins with non-redox roles*, Basingstoke: MacMillan Press, pp. 183–262.

Bucher, J.R., Tien, M., Morehouse, L.A., and Aust, S.D. (1983a) 'Redox cycling and lipid peroxidation: The central role of iron chelates', *Fund. Appl. Toxicol.* **3**, 222–226.

Bucher, J.R., Tien, M., and Aust, S.D. (1983b) 'The requirement for ferric in the initiation of lipid peroxidation by chelated ferrous iron', *Biochem. Biophys. Res. Comm.* **111**, 777–784.

Buettner, G.R. (1986) 'Ascorbate autoxidation in the presence of iron and copper chelates', *Free Rad. Res. Comm.* **1**, 349–353.

Buettner, G.R. (1988) 'In the absence of catalytic metals ascorbate does not autoxidize at pH 7: ascorbate as a test for catalytic metals', *J. Biochem. Biophys. Meth.* **16**, 27–40.

Burton, G.W., and Ingold, K.U. (1984) 'beta-Carotene: an unusual type of lipid antioxidant', *Science* **224**, 569–573.

Bus, J.S., Aust, S.D., and Gibson, J.E. (1974) 'Superoxide- and singlet oxygen-catalyzed lipid peroxidation as a possible mechanism for paraquat (methyl viologen) toxicity', *Biochem. Biophys. Res. Comm.* **58**, 749–755.

Cavill, I., Worwood, M., and Jacobs, A. (1975) 'Internal regulation of iron absorption', *Nature* **256**, 328–329.

Chasteen, N.D., and Woodworth, R.C. (1990) 'Transferrin and lactoferrin', in Ponka, P., Schulman, H.M., and Woodworth, R.C., (eds.) *Iron Transport and Storage*, Boca Raton: CRC Press, Inc, pp. 67–79.

Chasteen, N.D., Antanaitis, B.C., and Aisen, P. (1985) 'Iron deposition in apoferritin. Evidence for the formation of a mixed valence binuclear iron complex', *J. Biol. Chem.* **260**, 2926–2929.

Clark, D.G., McElligott, T.F., and Hurst, E.W. (1966) 'The toxicity of paraquat', *Br. J. Ind. Med.* **25**, 126–132.

Connor, J.R., Tucker, P., Johnson, M., and Snyder, B. (1993) 'Ceruloplasmin levels in the human superior temporal gyrus in aging and Alzheimer's disease', *Neurosci. Lett.* **159**, 88–90.

Conrad, M.E., Umbreit, J.N., Peterson, R.D.A., Moore, E.G., and Harper, K.P. (1993) 'Function of integrin in duodenal mucosal uptake of iron', *Blood* **81**, 517–521.

Crichton, R.R. (1991) *Inorganic Biochemistry of Iron Metabolism*, England: Ellis Horwood Press.

Curzon, G., and O'Reilly, S. (1960) 'A coupled iron-ceruloplasmin oxidation system', *Biochem. Biophys. Res. Comm.* **2**, 284–286.

DeSilva, D.M., and Aust, S.D. (1992) 'Stoichiometry of Fe(II) oxidation during ceruloplasmin-catalyzed loading of ferritin', *Arch. Biochem. Biophys.* **298**, 259–264.

DeSilva, D.M., Reif, D.W., Miller, D.M., and Aust, S.D. (1992) '*In vitro* loading of apoferritin', *Arch. Biochem. Biophys.* **293**, 409–415.

DeSilva, D.M., Askwith, C.C., and Kaplan, J. (1996) 'Molecular mechanisms of iron uptake in eukaryotes', *Physiol. Rev.* **76**, 31–47.

DeSilva, D.M., Davis-Kaplan, S., Fergestead, J., and Kaplan, J. (1997) 'Purification and characterization of Fet3 protein, a yeast homologue of ceruloplasmin', *J. Biol. Chem.* **272**, 14208–14213.

Dexter, D.T., Carter, C.J., Wells, F.R., Javoy-Agid, F., Agid, Y., Lees, A., Jenner, P., and Marsden, C.D. (1989) 'Basal lipid peroxidation in substantia nigra is increased in Parkinson's disease', *J. Neurochem.* **52**, 381–389.

Dognin, J., and Crichton, R.R. (1975) 'Mobilization of iron from ferritin fractions of defined iron content by biological reductants', *FEBS Lett.* **54**, 234–236.

Emery, T. (1991) *Iron and Your Health: Facts and Falacies*, Boca Raton: CRC Press.

Fee, J.A., Bergamini, R., and Briggs, R.G. (1975) 'Observations on the mechanism of the oxygen/dialuric acid-induced hemolysis of vitamin E deficient rat red blood cells and the protective roles of catalase and superoxide dismutase', *Arch. Biochem. Biophys.* **169**, 160–167.

Flaherty, J.T., and Weisfeldt, M.L. (1988) 'Reperfusion injury', *J. Free Rad. Biol. Med.* **5**, 409–419.

Fleming, R.E., and Gitlin, J.D. (1990) 'Primary structure of rat ceruloplasmin and analysis of tissue-specific gene expression during development', *J. Biol. Chem.* **265**, 7701–7707.

Fleming, R.E., Whitman, I.P., and Gitlin, J.D. (1991) 'Induction of ceruloplasmin gene expression in rat lung during inflammation and hyperoxia', *Am. J. Physiol.* **260**, L68–L74.

Fridovich, I. (1983) 'Superoxide radical: An endogenous toxicant' in George, R., Okun, R., and Cho, A.K. (eds.) *Annual Review of Pharmacology and Toxicology, Vol. 23*, Palo Alto: Annual Reviews, pp. 239–257.

Fukuzawa, K., Tadokoro, T., Kishikawa, K., Mukai, K., and Gebicki, J.M. (1988) 'Site-specific induction of lipid peroxidation by iron in charged micelles', *Arch. Biochem. Biophys.* **260**, 146–152.

Gaitskhoki, V.S., Voronina, O.V., Denezhkina, V.V., Pliss, M.G., Puchkova, L.V., Shvartsman, A.L., and Neifakh, S.A. (1990) 'Expression of ceruloplasmin gene in various organs of the rat', *Biokhimiia* **55**, 927–937.

Ganong, W.F. (1991) *Review of Medical Physiology*, 15th edn, Norwalk: Appleton and Lange.

Gerlach, M., Ben-Shachar, D., Riederer, P., and Youdim, M.B. (1994) 'Altered brain metabolism of iron as a cause of neurodegenerative diseases?', *J. Neurochem.* **63**, 793–807.

Gitlin, J.D. (1988) 'Transcriptional regulation of ceruloplasmin gene expression during inflammation', *J. Biol. Chem.* **263**, 6281–6287.

Goldstein, I.M., Kaplan, H.B., Edelson, H.S., and Weissmann, G. (1982) 'Ceruloplasmin: An acute phase reactant that scavenges oxygen-derived free radicals', *Ann. N. Y. Acad. Sci.* **389**, 368–379.

Goyer, R.A. (1991) 'Toxic effects of metals' in Amdur, M.O., Doull, J.D., and Klaassen, C.D. (eds.) *Toxicology: The Basic Science of Poisons*, 4th edn, New York: McGraw-Hill, Inc., pp. 623–680.

Gunshin, H., Mackenzie, B., Berger, U.V., Gunshin, Y., Romero, M.F., Boron, W.F., Nussberger, S., Gollan, J.L., and Hediger, M.A. (1997) 'Cloning and characterization of a mammalian proton-coupled metal-ion transporter', *Nature* **338**, 482–488.

Guo, J.-H., Abedi, M., and Aust, S.D. (1996) 'Expression and loading of recombinant heavy and light chain homopolymers of rat liver ferritin', *Arch. Biochem. Biophys.* **335**, 197–204.

Gutteridge, J.M.C., and Halliwell, B. (1990) 'Iron and oxygen: A dangerous mixture' in Ponka, P., Schulman, H.M., and Woodworth, R.C. (eds.) *Iron Transport and Storage*, Boca Raton: Academic Press, Inc., pp. 55–65.

Halliwell, B., and Gutteridge, J.M.C. (1984) *Free Radicals in Biology and Medicine*, 2nd edn, New York: Oxford University Press, Inc.

Halliwell, B., and Gutteridge, J.M.C. (1990) 'Role of free radicals and catalytic metal ions in human disease: An overview', *Meth. Enzymol.* **186**, 1–85.

Harris, Z.L., and Gitlin, J.D. (1996) 'Genetic and molecular basis for copper toxicity', *Am. J. Clin. Nutr.* **63**, 836S–841S.

Harris, E.D., and Percival, S.S. (1989) 'Copper transport: Insights into a ceruloplasmin-based delivery system', *Adv. Exp. Med. Biol.* **258**, 95–102.

Harris, Z.L., Takahashi, Y., Miyajima, H., Serizawa, M., MacGillivray, R.T., and Gitlin, J.D. (1995) 'Aceruloplasminemia: Molecular characterization of this disorder of iron metabolism', *Proc. Nat. Acad. Sci. USA* **92**, 2539–2543.

Harris, Z.L., Klomp, L.W., and Gitlin, J.D. (1998) 'Aceruloplasminemia: An inherited neurodegenerative disease with impairment of iron homeostasis', *Am. J. Clin. Nutr.* **67**, 972S–977S.

Harrison, P.M., and Arosio, P. (1996) 'The ferritins: Molecular properties, iron storage function and cellular regulation', *Biochim. Biophys. Acta* **1275**, 161–203.

Heikkila, R.E., Winston, B., Cohen, G. (1976) 'Alloxan-induced diabetes-evidence for hydroxyl radical as a cytotoxic intermediate', *Biochem. Pharmacol.* **25**, 1085–1092.

Heinecke, J.W., Rosen, H., and Chait, A. (1984) 'Iron and copper promote modification of low density lipoprotein by human arterial smooth muscle cells in culture', *J. Clin. Invest.* **74**, 1890–1894.

Herman, E.H., and Ferrans, V.J. (1983) 'ICRF-187 reduction of chronic daunomycin and doxorubicin cardiac toxicity in rabbits, beagle dogs and miniature pigs', *Drugs Exp. Clin. Res.* **9**, 483–490.

Hill, K.E. (1994) 'Selenium status and glutathione metabolism' in Burk, R.F. (ed.) *Selenium in Biology and Human Health*, New York: Springer-Verlag, pp. 151–168.

Horton, A.A., and Fairhurst, S. (1987) 'Lipid peroxidation and mechanisms of toxicity' in Goldberg, L., (ed.) *Critical Reviews in Toxicology, Vol. 18*, Boca Raton: CRC Press, pp. 27–79.

Hoy, T.G., Harrison, P.M., Shabbir, M., and Macara, I.G. (1974) 'The release of iron from horse spleen ferritin to 1,10-phenanthroline', *Biochemistry* **137**, 67–70.

Huebers, H.A., Huebers, E., Csiba, E., Rummel, W., and Finch, C.A. (1983) 'The significance of transferrin for intestinal iron absorption', *Blood* **61**, 283–290.

Inman, R.S., Coughlan, M.M., and Wessling-Resnick, M. (1994) 'Extracellular ferrireductase activity of K562 cells is coupled to transferrin-independent iron storage', *Biochemistry* **33**, 11850–11857.

Jacobs, A. (1977) 'Low molecular weight intracellular iron transport compounds', *Blood* **50**, 433–439.

Jones, R.L., Peterson, C.M., Grady, R.W., and Cerami, A. (1980) 'Low molecular weight iron-binding factor from mammalian tissue that potentiates bacterial growth', *J. Exp. Med.* **151**, 418–428.

Jones, T., Spencer, R., and Walsh, C. (1978) 'Mechanism and kinetics of iron release from ferritin by dihydroflavins and dihydroflavin analogues', *Biochemistry* **17**, 4011–4017.

Jordan, I., and Kaplan, J. (1994) 'The mammalian transferrin-independent iron transport system may involve a surface ferrireductase activity', *Biochem. J.* **302**, 875–879.

Juan, S.-H., and Aust, S.D. (1998) 'Iron and phosphate content of rat ferritin heteropolymers', *Arch. Biochem. Biophys.* **357**, 293–298.

Juan, S.-H., Guo, J.-H., and Aust, S.D. (1997) 'Loading of iron into recombinant rat liver ferritin heteropolymers by ceruloplasmin', *Arch. Biochem. Biophys.* **341**, 280–286.

Kaplan, J., and O' Halloran, T.V. (1996) 'Iron metabolism in eukaryotes: Mars and Venus at it again', *Science* **271**, 1510–1512.

Kappus, H., and Sies, H. (1981) 'Toxic drug effects associated with oxygen metabolism: Redox cycling and lipid peroxidation', *Experientia* **37**, 1233–1241.

Koeller, D.M., Casey, J.L., Hentze, M.W., Gerhardt, E.M., Chan, L.-N., Klausner, R.D., and Harford, J.B. (1989) 'A cytosolic protein binds to structural elements within the iron regulatory region of the transferrin receptor mRNA', *Proc. Natl. Acad. Sci. USA* **86**, 3574–3578.

Kohen, R., and Chevion, M. (1985) 'Paraquat toxicity is enhanced by iron and reduced by desferrioxamine in laboratory mice', *Biochem. Pharmacol.* **34**, 1841–1843.

Koppenol, W.H. (1987) 'Thermodynamics of reactions involving oxyradicals and hydrogen peroxide', *Bioelectrochem. Bioenerg.* **18**, 3–11.

Koppenol, W.H., and Butler, J. (1985) 'Energetics of interconversion reactions of oxy radicals', *Adv. Free Rad. Biol. Med.* **1**, 91–131.

Koppenol, W.H., and Leibman, J.F. (1984) 'The oxidizing nature of the hydroxyl radical. A comparison with the ferryl ion (FeO^{2+})', *J. Phys. Chem.* **88**, 99–101.

Kukielka, E., and Cederbaum, A.I. (1994) 'Ferritin stimulation of hydroxyl radical production by rat liver nuclei', *Arch. Biochem. Biophys.* **308**, 70–77.

Langford, C.H., and Gray, H.B. (1965) *Ligand Substitution Process*, New York: Benjamin.

Lawson, D.M., Treffry, A., Artymiuk, P.J., Harrison, P.M., Yewdall, S.J., Luzzago, A., Cesareni, G., Levi, S., and Arosio, P. (1989) 'Identification of the ferroxidase centre in ferritin', *FEBS Lett.* **254**, 207–210.

Lee, G.R., Nacht, S., Lukens, J.N., and Cartwright, G.E. (1968) 'Iron metabolism in copper-deficient swine', *J. Clin. Invest.* **47**, 2058–2069.

Lenaz, L. and Page, J.A. (1976) 'Cardiotoxicity of adriamycin and related anthracyclines', *Cancer Treat. Rev.* **3**, 111–120.

Levin, L.A., and Geszvain, K.M. (1998) 'Expression of ceruloplasmin in the retina: Induction after optic nerve crush', *Invest. Opthalmol. Vis. Sci.* **39**, 157–163.

Li, J. (1991) *Iron Release from Rat Liver Ferritin and Lipid Peroxidation by Adriamycin*, M.S. thesis, Logan: Utah State University, pp. 1–51.

Lin, F., and Girotti, A.W. (1997) 'Elevated ferritin production, iron containment, and oxidant resistance in hemin-treated leukemia cells', *Arch. Biochem. Biophys.* **346**, 131–141.

Lin, F., and Girotti, A.W. (1998) 'Hemin-enhanced resistance of human leukemia cells to oxidative killing: Antisense determination of ferritin involvement', *Arch. Biochem. Biophys.* **352**, 51–58.

Linder, M.C., and Moor, J.R. (1977) 'Plasma ceruloplasmin. Evidence for its presence in and uptake by heart and other organs of the rat', *Biochim. Biophys. Acta* **499**, 329–336.

Lippard, S.J. (1986) 'The bioinorganic chemistry of rust', *Chem. Br.* **22**, 222–229.

McCance, R.A., and Widdowson, E.M. (1937) 'Absorption and excretion of iron', *Lancet* **2**, 680–684.

McDermott, J.A., Huber, C.T., Osaki, S., and Freiden, E. (1968) 'Role of iron in the oxidase activity of ceruloplasmin', *Biochim. Biophys. Acta* **151**, 541–557.

Miller, D.M. (1990) *Redox Reactions of Iron Chelates and Lipid Peroxidation*, Ph.D. dissertation, Logan: Utah State University, pp. 131–163.

Miller, D.M., and Aust, S.D. (1988) 'Studies of ascorbate-dependent, iron-catalyzed lipid peroxidation', *Arch. Biochem. Biophys.* **271**, 113–119.

Miller, D.M., Buettner, G.R., and Aust, S.D. (1990) 'Transition metals as catalysts of "autoxidation" reactions', *Free Rad. Biol. Med.* **8**, 95–108.

Minotti, G., and Aust, S.D. (1986) 'The requirement for iron (III) in the initiation of lipid peroxidation by iron (II) and hydrogen peroxide', *J. Biol. Chem.* **262**, 1098–1104.

Minotti, G., and Aust, S.D. (1987) 'The role of iron in the initiation of lipid peroxidation', *Chem. Phys. Lipids* **44**, 191–208.

Miyajima, H., Takahashi, Y., Serizawa, M., Kaneko, E., and Gitlin, J.D. (1996) 'Increased plasma lipid peroxidation in patients with aceruloplasminemia', *Free Rad. Biol. Med.* **20**, 757–760.

Moak, S.A., and Greenwald, R.A. (1984) 'Enhancement of rat serum ceruloplasmin levels by exposure to hyperoxida', *Proc. Soc. Exp. Biol. Med.* **177**, 97–103.

Moore, G.R., Pettigrew, G.W., and Rogers, N.K. (1986) 'Factors influencing redox potentials of electron transfer proteins', *Proc. Natl. Acad. Sci. USA* **83**, 4998–4999.

Morley, C.G., and Bezkorovainy, A. (1983) 'Identification of the iron chelate in hepatocyte cytosol', *IRCS Med. Sci. Biochem.* **11**, 1106–1107.

Müllner, E.W., Neupert, B., and Kühn, L.C. (1989) 'A specific mRNA binding factor regulates the iron-dependent stability of cytoplasmic transferrin receptor mRNA', *Cell* **58**, 373–382.

Munro, H.N., and Linder, M.C. (1978) 'Ferritin: Structure, biosynthesis, and role in iron metabolism', *Physiol. Rev.* **58**, 317–396.

Niki, E., Saito, T., Kawakami, A., and Kamiya, Y. (1984) 'Inhibition of oxidation of methyl linoleate in solution by vitamin E and vitamin C', *J. Biol. Chem.* **259**, 4177–4182.

Nunez, M.T., Escobar, A., Ahumada, A., and Gonzalez-Supelveda, M. (1992) 'Sealed reticulocyte ghosts. An experimental model for the study of Fe^{2+} transport', *J. Biol. Chem.* **267**, 11490–11494.

Okumura, M., Fujinaga, T., Yamashita, K., Tsunoda, N., and Mizuno, S. (1991) 'Isolation, characterization, and quantitative analysis of ceruloplasmin from horses', *Am. J. Vet. Res.* **52**, 1979–1985.

Osaki, S., and Johnson, D.A. (1969) 'Mobilization of liver iron by ferroxidase (ceruloplasmin)', *J. Biol. Chem.* **244**, 5757–5758.

Osaki, S., Johnson, D.A., and Frieden, E. (1966) 'The possible significance of the ferrous oxidase activity of ceruloplasmin in normal human serum', *J. Biol. Chem.* **241**, 2746–2751.

Osaki, S., Johnson, D.A., and Frieden, E. (1971) 'The mobilization of iron from the perfused mammalian liver by a serum copper enzyme, ferroxidase I', *J. Biol. Chem.* **246**, 3018–3023.

Oteiza, P.I., Kleinman, C.G., Demasi, M., and Bechara, E.J. (1995) '5-Aminolevulinic acid induces iron release from ferritin', *Arch. Biochem. Biophys.* **316**, 607–611.

Otsuka, S., Maruyama, H., and Listowsky, I. (1981) 'Structure, assembly, conformation, and immunological properties of the two subunit classes of ferritin', *Biochemistry* **20**, 5226–5232.

Owen, C.A., Jr. (1973) 'Effects of iron on copper metabolism and copper on iron metabolism in rats', *Am. J. Physiol.* **224**, 514–518.

Patel, B.N., and David, S. (1997) 'A novel glycosylphosphatidylinositol-anchored form of ceruloplasmin is expressed by mammalian astrocytes', *J. Biol. Chem.* **272**, 20185–20190.

Patterson, C.E., and Rhodes, M.L. (1982) 'The effect of superoxide dismutase on paraquat mortality in mice and rats', *Toxicol. Appl. Pharmacol.* **62**, 65–72.

Puchkova, L.V., Denezhkina, V.V., Zakharova, E.T., Gaitskhoki, V.S., and Neifakh, S.A. (1990) 'Biosynthesis of ceruloplasmin in various rat organs', *Biokhimiia* **55**, 2095–2102.

Rao, K.K., Shapiro, D., Mattia, E., Bridges, K., and Klausner, R.D. (1985) 'Effects of alterations in cellular iron on biosynthesis of the transferrin receptor in K562 cells', *Mol. Cell Biol.* **5**, 595–600.

Rao, K.K., Harford, J.B., Rouault, T., McClelland, A., Ruddle, F., and Klausner, R.D. (1986) 'Transcriptional regulation by iron of the gene for the transferrin receptor', *Mol. Cell Biol.* **6**, 236–240.

Reed, C.A. (1985) 'Oxidation states, redox potentials and spin rates', in Dunford, H.B., Dolphin, D., Raymond, K.N., Seiker, L. (eds.) *The Biological Chemistry of Iron*, Boston: D. Reidel, pp. 25–42.

Reif, D.W., Beales, I.L.P., Thomas, C.E., and Aust, S.D. (1988) 'Effect of diquat on the distribution of iron in rat liver', *Toxicol. Appl. Pharmacol.* **93**, 506–510.

Reif, D.W., Samokyszyn, V.M., Miller, D.M., and Aust, S.D. (1989) 'Alloxan and glutathione dependent ferritin iron release and lipid peroxidation', *Arch. Biochem. Biophys.* **269**, 407–414.

Reilly, C.A., and Aust, S.D. (1997) 'Stimulation of the ferroxidase activity of ceruloplasmin during iron loading into ferritin', *Arch. Biochem. Biophys* **347**, 242–248.

Reilly, C.A., and Aust, S.D. (1998) 'Iron loading into ferritin by an intracellular ferroxidase', *Arch. Biochem. Biophys.* **359**, 69–76.

Reilly, C.A., Sorlie, M., and Aust, S.D. (1998) 'Evidence for a protein-protein complex during iron loading into ferritin by ceruloplasmin', *Arch. Biochem. Biophys* **354**, 165–171.

Riedel, H.D., Remus, A.J., Fitscher, B.A., and Stremmel, W. (1995) 'Characterization and partial purification of a ferrireductase from human duodenal microvillus membranes', *Biochem. J.* **309**, 745–748.

Rivett, A.J., and Levine, R.L. (1990) 'Metal-catalyzed oxidation of *Escherichia coli* glutamine synthetase: Correlation of structural and functional changes', *Arch. Biochem. Biophys.* **278**, 26–34.

Rouault, T.A., Hentze, M.W., Caughman, S.W., Harford, J.B., and Klausner, R.D. (1988) 'Binding of a cytosolic protein to the iron-responsive element of human ferritin messenger RNA', *Science* **241**, 1207–1210.

Rowley, D., Gutteridge, J.M.C., Blake, D., Farr, M., and Halliwell, B. (1984) 'Lipid peroxidation in rheumatoid arthritis: Thiobarbituric acid-reactive material and catalytic iron salts in synovial fluid from rheumatoid patients', *Clin. Sci.* **66**, 691–695.

Ryan, T.P. (1991) *Rat Ceruloplasmin: Characterization, Gene Expression, and Mechanism of Action in Iron Catalyzed Oxidations*, Ph.D. thesis, Logan: Utah State University, pp. 164–196.

Ryan, T.P., and Aust, S.D. (1992) 'The role of iron in oxygen-mediated toxicities', *Crit. Rev. Toxicol.* **22**, 119–141.

Ryan, T.P., Grover, T.A., and Aust, S.D. (1992) 'Rat ceruloplasmin: Resistance to proteolysis and kinetic comparison with human ceruloplasmin', *Arch. Biochem. Biophys.* **293**, 1–8.

Ryan, T.P., Miller, D.M., and Aust, S.D. (1993) 'The role of metals in the enzymatic and nonenzymatic oxidation of epinephrine', *J. Biochem. Toxicol.* **8**, 33–39.

Ryan, T.P., Samokyszyn, V.M., Dellis, S., and Aust, S.D. (1990) 'The effects of (+)-1,2-bis(3,5-dioxopiperzin-1-yl)propane (ADR-529) on iron catalyzed lipid peroxidation', *Chem. Res. Tox.* **3**, 384–390.

Rydén, L. (1981) 'Ceruloplasmin', in Contie, R. (ed.) *Copper Proteins and Copper Enzymes*, Boca Raton: CRC Press, pp. 37–100.

Saito, M., Thomas, C.E., and Aust, S.D. (1985) 'Paraquat and ferritin-dependent lipid peroxidation', *J. Free Rad. Biol. Med.* **1**, 179–185.

Samokyszyn, V.M., Thomas, C.E., Reif, D.W., Saito, M., and Aust, S.D. (1987) 'Release of iron from ferritin and its role in oxygen radical toxicities', *Drug Metabol. Rev.* **19**, 283–303.

Samokyszyn, V.M., Miller, D.M., Reif, D.W., and Aust, S.D. (1989) 'Inhibition of superoxide and ferritin-dependent lipid peroxidation by ceruloplasmin', *J. Biol. Chem.* **264**, 21–26.

Scarpa, M., Stevanato, R., Viglino, P., and Rigo, A. (1983) 'Superoxide ion as active intermediate in the autoxidation of ascorbate by molecular oxygen. Effect of superoxide dismutase', *J. Biol. Chem.* **258**, 6695–6697.

Scheiber, B., and Goldenberg, H. (1993) 'NAD(P)H:ferric iron reductase in endosomal membranes from rat liver', *Arch Biochem. Biophys.* **305**, 225–230.

Scheiber, B., and Goldenberg, H. (1996) 'Uptake of iron by isolated rat hepatocytes from a hydrophilic impermeant ferric chelate, Fe(III)-DTPA', *Arch Biochem. Biophys.* **326**, 185–192.

Sheldon, R.A., and Kochi, J.K. (1981) *Metal-Catalyzed Oxidations of Organic Compounds*, New York: Academic Press.

Sirivech, S., Freiden, E., and Osaki, S. (1974) 'The release of iron from horse spleen ferritin by reduced flavins', *Biochem. J.* **143**, 311–315.

Smith, R.M., and Martell, A.E. (1975) *Critical Stability Constants*, New York: Plenum.

Smith, R.P. (1991) 'Toxic responses of the blood', in Amdur, M.O., Doull, J., and Klaasen, C.D. (eds.) *Casarett and Doull's Toxicology: The Basic Science of Poisons*, 4th edn, New York: McGraw Hill, Inc., pp. 257–281.

Spiro, T.G., and Saltman, P. (1974) 'Inorganic chemistry', in Jacobs, A., and Worwood, M. (eds.) *Iron in Biochemistry and Medicine*, London: Academic Press, pp. 1–26.

Stadtman, E.R. (1990) 'Metal ion-catalyzed oxidation of proteins: Biochemical mechanism and biological consequences', *Free Rad. Biol. Med.* **9**, 315–325.

Stevens, R.G., Jones, D.Y., Micozzi, M.S., and Taylor, P.R. (1988) 'Body iron stores and the risk of cancer', *N. Eng. J. Med.* **319**, 1047–1052.

Sun, I.L., Navas, P., Crane, F.L., Morré, D.J., and Löw, H. (1987) 'NADH diferric transferrin reductase in liver plasma membrane', *J. Biol. Chem.* **262**, 15915–15921.

Sunde, R.A. (1994) 'Intracellular glutathione peroxidases – structures, regulation, and function', in Burk, R.F. (ed.) *Selenium in Biology and Human Health*, New York: Springer-Verlag, pp. 45–78.

Swaddle, T.W. (1974) 'Activation parameters and reaction mechanisms in octahedral substitution', *Coor. Chem. Rev.* **14**, 218–268.

Theil, E.C. (1983) 'Ferritin: Structure, function, and regulation', in Theil, E.C., Eichhorn, G.L., and Marzilli, L.G. (eds.) *Advances in Inorganic Biochemistry,* Vol. 5, New York: Elsevier, pp. 1–38.

Theil, E.C. (1987) 'Ferritin: Structure, gene regulation, and cellular function in animals, plants, and microorganisms', *Ann. Rev. Biochem.* **56**, 289–315.

Theil, E.C. (1990) 'The ferritin family of iron storage proteins', in Meister, A. (ed.) *Advances in Enzymology and Related Areas of Molecular Biology,* Vol. 63, New York: John Wiley & Sons, Inc., pp. 421–449.

Thomas, C.E., and Aust, S.D. (1986a) 'Free radicals and environmental toxins', *Ann. Emerg. Med.* **15**, 1075–1083.

Thomas, C.E., and Aust, S.D. (1986b) 'Reductive release of iron from ferritin by paraquat and related bipyridyls', *J. Biol. Chem.* **261**, 13064–13070.

Thomas, C.E., and Aust, S.D. (1986c) 'Release of iron from ferritin by cardiotoxic anthracycline antibiotics', *Arch. Biochem. Biophys.* **248**, 684–689.

Thomas, C.E., Morehouse, L.A., and Aust, S.D. (1985) 'Ferritin and superoxide-dependent lipid peroxidation', *J. Biol. Chem.* **260**, 3275–3280.

Tien, M., Morehouse, L.A., Bucher, J.R., and Aust, S.D. (1982) 'The multiple effects of EDTA in several model lipid peroxidation systems', *Arch. Biochem. Biophys.* **218**, 450–458.

Turnbull, A. (1974) 'Iron absorption', in Jacobs, A., and Worwood, M. (eds.) *Iron in Biochemistry and Medicine*, New York: Academic Press, pp. 370–403.

Ueda, K., Kobayashi, S., Morita, J., and Komano, T. (1985) 'Site-specific DNA damage caused by lipid peroxidation products', *Biochem. Biophys. Acta* **824**, 341–348.

Umbreit, J.N., Conrad, M.E., Moore, E.G., Desai, M.P., and Turrens, J. (1996) 'Paraferritin: A protein complex with ferrireductase activity is associated with iron absorption in rats', *Biochemistry* **35**, 6460–6469.

Uri, N. (1961) 'Physico-chemical aspects of autoxidation', in Lundberg, W.O. (ed.) *Autoxidation and Antioxidants,* Vol. 1, New York: Interscience, pp. 55–106.

Walker, F.J., and Fay, P.J. (1990) 'Characterization of an interaction between protein C and ceruloplasmin', *J. Biol. Chem.* **265**, 1834–1836.

Watt, G.D., Frankel, R.B., and Papaefthymiou, G.C. (1985) 'Reduction of mammalian ferritin', *Proc. Natl Acad. Sci. USA* **82**, 3640–3643.

Watt, G.D., Jacobs, D., and Frankel, R.B. (1988) 'Redox activity of bacterial and mammalian ferritin: Is reductant entry into the ferritin interior a necessary step for iron release?', *Proc. Natl Acad. Sci.* **85**, 7457–7461.

Weaver, J., and Pollack, S. (1989) 'Low-Mr iron isolated from guinea pig reticulocytes as AMP-Fe and ATP-Fe complexes', *Biochem. J.* **261**, 787–792.

Winzerling, J.J., and Law, J.H. (1997) 'Comparative nutrition of iron and copper', *Ann. Rev. Nutr.* **17**, 501–526.

Witting, L.A. (1980) 'Vitamin E and lipid antioxidants in free-radical initiated reactions', in Pryor, W. (ed.) *Free Radicals in Biology,* Vol. 4, New York: Academic Press, pp. 295–319.

Wollenberg, P., and Rummel, W. (1987) 'Dependence of intestinal iron absorption on the valency state of iron', *Naunyn Schmiedebergs Arch. Pharmacol.* **336**, 578–582.

Woodward, R.B., and Hoffman, R. (1970) *The Conservation of Orbital Symmetry*, Germany: Verlag Chemie, GmbH.

Worwood, M., Dawkins, S., Wagstaff, M., and Jacobs, A. (1976) 'The purification and properties of ferritin from human serum', *Biochem. J.* **157**, 97–103.

Wrigglesworth, J.M., and Baum, H. (1980) 'The biochemical functions of iron', in Jacobs, A., and Worwood, M. (eds.) *Iron in Biochemistry and Medicine*, Vol. 2, Boca Raton: Academic Press, Inc., pp. 29–86.

Xu., B., and Chasteen, N.D. (1991) 'Iron oxidation chemistry in ferritin. Increasing Fe/O_2 stoichiometry during core formation', *J. Biol. Chem.* **266**, 19965–19970.

Young, S.P., Fahmy, M., and Golding, S. (1997) 'Ceruloplasmin, transferrin and apotransferrin facilitate iron release from human liver cells', *FEBS Lett.* **411**, 93–96.

Yoshida, K., Furihata, K., Takeda, S., Nakamura, A., Yamamoto, K., Morita, H., Hiyamuta, S., Ikeda, S.-I., Shimizu, N., and Yanagisawa, N. (1995) 'A mutation in the ceruloplasmin gene is associated with systemic hemosiderosis in humans', *Nature Genet.* **9**, 267–272.

Zahringer, J., Baliga, B.S., and Munro, H.N. (1976) 'Novel mechanism for translational control in regulation of ferritin synthesis by iron', *Proc. Natl Acad. Sci. USA* 73, 857–861.

10 The toxicology of iron

Rodney Bilton

10.1 Introduction

10.1.1 Iron and the origins of life

Iron presents a contradiction in its role in the biosphere. It is as fully essential to life as it is exquisitely toxic, in its reduced form, to all life forms when free oxygen is present.

It has been postulated in a controversial hypothesis (Maden, 1995) to have provided the energy to initiate the life process via the oxidation of H_2S. This drove the fixation of CO_2 in prebiotic reactions, leading to the synthesis of more complex polymers. This process was thought to be associated with the FeS crystal surface acting as a positively charged template for polymer binding and extension (Huber and Wächtershäuser, 1998).

Iron plays a central role in the energy and redox reactions of both the anaerobic and aerobic life-forms, largely via iron–sulphur clusters and porphyrin complexes.

Following the appearance of oxygenic organisms some 3×10^9 years ago, a rapid oxidation of oceanic Fe^{2+} occurred, associated with the deposition of the banded iron ore deposits (Licari and Cloud, 1968). It is tempting to speculate that the explosive increase in species diversity that occurred during the period was, in part, a consequence of the greatly increased rates of Fenton-driven mutations caused by the ingress of oxygen and reactive oxygen species (ROS) into the lower Fe^{2+}-rich layers of the oceans. The ubiquity and antiquity of the protective enzymes superoxide dismutase (SOD) and catalase and the iron sequestering proteins support this contention (Joshi and Dennis, 1993) and also underpin the discussion of contemporary O_2/Fe-mediated cellular damage here and in Chapter 9.

A study of the toxicology of iron is essentially a study of redox interfaces and gradients and a comprehensive review is beyond the scope of this chapter, which will present selected examples to illustrate how iron uptake and transfer is controlled, the toxic effects in iron overload, and neurological disorders, inflammation, cancer and the consequences of ischaemia/reperfusion. Indeed, the general statement can be made that if a disease or pathological condition involves the participation of reactive oxygen species (ROS) particularly $O_2^{\bullet-}$, then iron will be involved in the initiation and/or progression of that process.

10.1.2 Biomarkers for assessment of ROS-FR/Fe mediated tissue damage

Table 10.1 gives a selection of the more common conditions in which ROS-FR and Fe are thought to be aetiological agents. The classification is arbitrary where several of the diseases could appear under more than one heading.

10.1.2.1 At the level of the membrane

It has long been known that lipid peroxidation in membranes requires iron in the initiation process (Hochstein *et al.*, 1964), but the involvement of $^{\bullet}OH$ or $Fe^{2+}{\cdot}Fe^{3+}O_2$ complexes is still not resolved (Minolti and Aust, 1987) (Chapter 9).

However, the measurement of lipid peroxidation products remains a useful tool, although the assay of thiobarbituric reactive substances (TBARS) is now largely discarded, as a measure of whole body or cellular lipid peroxidation.

The cytotoxic aldehydes, malondialdehyde (MDA) and 4-hydroxynonenal (HNE) are products of lipid peroxidation and can react with amino groups in proteins to form Schiff's bases. The lipoprotein B in LDL presents an important target for these aldehydes (Jürgens *et al.*, 1986).

MDA-LDL and HNE-LDL in oxidised LDL can be visualised in rabbit atherosclerotic plaques using immunocytochemistry (Esterbauer *et al.*, 1990).

Methodological problems associated with the measurement of lipid peroxidation products are reviewed in Halliwell and Chirico (1993).

10.1.2.2 F_2-isoprostanes

These prostaglandin-like compounds are formed during the peroxidation of arachidonic acid, and are structurally similar to prostaglandin $F_{2\alpha}$. Elevated serum F_2-isoprostanes are thought to be involved in the pathology of hepatorenal syndrome, and urinary levels are elevated in scleroderma patients and smokers (Morrow and Roberts, 1996; Bachi *et al.*, 1996). These isoprostanes and their metabolites can be assayed in human urine. If the compounding effects of dietary derived sources can be resolved, then these compounds may give a very useful estimation of whole-lipid peroxidation (Aruoma, 1998).

Trans,trans-2,4-decadienal (DDE) is an important breakdown product following lipid peroxidation, and its 2′-deoxyadenosine adduct has been proposed as a biomarker, possibly linking membrane and DNA damage (Carvalho *et al.*, 1998).

Loss of membrane integrity as a function of ROS-FR/Fe mediated damage can be conveniently measured by ethidium bromide uptake of loss of LDH. In some cases the kinetics of membrane damage may reflect the nature of the DNA damaging properties of ROS-FR/Fe (Lowe *et al.*, 1999).

10.1.2.3 Cytosolic markers

In view of the multiplicity of potential end products of radical damage to proteins it is not surprising that this area of research has been relatively neglected until quite recently when compared to studies on lipid peroxidation and DNA oxidation (Fu *et al.*, 1998).

The best studied ROS induced protein cross link is that of bi-tyrosine (Prütz *et al.*, 1983), however the usual method of detection by fluorescence measurements can

Table 10.1 A partial list of disease and degenerative states where ROS-FR and Fe are implicated in the initiation and/or potentiation of the conditions

Ageing	Accumulated tissue damage Alzheimer's disease Parkinson's disease Macular degeneration
Alcoholism	
Amylotrophic lateral sclerosis	
Brain	Vitamin E deficiency Hyperbaric oxygen damage in premature infants Hypertensive cerebrovascular injury Post-traumatic injury Accident/surgery sequelae Stroke sequelae
Cancer	GI cancers Hepatocarcinoma Pulmonary carcinomas
Cardiovasculature	Atherosclerosis Cardiovascular Heart disease Cardiomyopathy Keshan disease Adriamycin Cardiomyopathy
Gastrointestinal tract	Crohn's disease Ulcerative colitis Iron overdose Endotoxin insult Combination ascorbate/Fe therapy Pancreatitis Diabetes
Eye	Photic retinopathy Ocular haemorrhage Cataractogenesis Retinal degeneration Premature retinopathy
Inflammatory immune/ autoimmune conditions	Rheumatoid arthritis Glomerular nephritis HIV/AIDS Vasculitis Autoimmune nephrotic syndromes
Iron overload disorders	See Table 10.3 for details
Ischaemia/Reperfusion	Myocardial infarction Organ transplantation Circulatory obstruction in crash victims Thrombophlebitis
Kidney	Fe-mediated nephrotoxicity Aminoglycoside toxicity
Lung	Bronchial dysplasia Mineral dust pneumoconiosis Hypoxia Cigarette smoke Emphysema ARDS (adult respiratory distress syndrome)

yield erroneous data, and HPLC separation following complete hydrolysis is now the method of choice (Francis *et al.*, 1993).

Attempts to predict site specificity of protein damage are beset with confounding variables involving sequence and conformational data, reactivity of neighbouring groups (e.g. ability of amino acids to bind redox active metals) and not least the rate at which the oxidised proteins are degraded *in vivo*. Methods for evaluating these problems are now greatly improved, particularly with respect to the chemical characterisation of oxidised moieties, but at this time it is still not possible to propose a specific panel of markers as unequivocal indicators of ROS-FR/Fe damage in proteins (Dean *et al.*, 1994).

Recent studies on radical damage to proteins mediated by Fenton-like reactions with iron, haematin and haemoglobin have revealed the generation of different ROS (Kim *et al.*, 1998) and protein-modified products. ESR and UV/VIS spectroscopy detected the DMPO spin adduct of the ${}^{\bullet}OH$ radical from Fe/EDTA and the ferryl radical generated from met Hb. The ferryl species derived from haematin was found to be more effective for protein modification.

Corroboration of this work will have an important bearing on understanding the role of free Fe and haem Fe in free radical damage in biological systems.

10.1.3 ROS-FR/Fe mediated damage in the nucleus

Despite the strong affinity of ferritin and the transferrins which serve to sequester and compartmentalise iron there must be a small pool of redox active iron available for participation in Fenton-type reactions (Griffiths and Lunec, 1998). Single strand breaks (ssb) in DNA elicited by redox cycling quinones and xanthine/XO challenge of single cells yield linear dose responses in the Comet Assay, indicating that DNA-bound redox active iron is available for generation of ${}^{\bullet}OH$ and is not rate limiting (Lowe *et al.*, 1999; Woods *et al.*, 1997; Booth *et al.*, 1997). Moreover, preincubation of the cells with O-phenanthroline reduces ssb formation in cells subsequently challenged with ROS. The Comet assay can be further modified to detect oxidative damage by post-ROS treatment with Endonuclease III and FaPy glycosylase which cleave DNA preferentially at oxidised bases (Collins *et al.*, 1993).

10.1.4 Direct measurement of oxidative DNA damage

Of the DNA bases modified by ROS attack, 8-OHdG has been most extensively studied (Wiseman and Halliwell, 1996). It can be separated by either HPLC and quantitated by electrochemical detection, or GC/MS (Ames, 1989). 8-OHG has sometimes been analysed by its level, which in urine can be affected by sources from cooked foods, or from oxidative RNA damage.

Although urinary 8-OHdG levels are not supplemented by uptake from the gut, an additional source may be from the oxidation of the dGTP pool (Sakumi *et al.*, 1993). Sample work-up in air leads to oxidative DNA damage particularly when phenol is used to remove protein (Claycamp, 1992).

Elimination of background oxidative DNA damage is virtually impossible but perhaps the use of techniques common to anaerobic bacteriology (i.e. extraction in H_2/CO_2 filled anaerobic cabinets) could reduce damage levels further, and allow the production of cellular fractions sufficiently low in endogenous oxidised species to refine the study of peroxidation mechanisms.

10.1.5 Iron uptake and homeostasis

The metabolism of iron in mammals is essentially conservative with no known mechanism for excretion other than through sloughing of skin and intestinal epithelium, loss of red blood cells and from complexes in bile and urine. A human male of 75 kg, loses approximately 1 mg/day and absorbs this amount from the diet.

The richest source of dietary iron is red meat. Vegetables contain less iron, which exists predominantly as Fe(III) complexes and is less well absorbed, particularly in the absence of ascorbate which serves to reduce acid solubilised Fe(III) in the stomach. However, the iron content of most vegetarian diets is adequate, i.e. the incidence of iron-deficiency anaemia is no greater than in populations consuming mixed or high meat diets. Some vegan diets contain a high concentration of phytate and fibre which can result in a greater incidence of iron deficiency. However, in the context of protection against CHD and colorectal cancer phytate may play a crucial role in the chelation of iron and prevention of excessive uptake and/or reduction of radical induced damage to the vasculature and intestinal mucosa (Graf and Eaton, 1985). Moreover, stool bulk is increased by fibre in vegetarian diets and this will tend to increase intestinal motility, dilute out the iron concentration and therefore reduce uptake. This may be one of the more important aspects of the vegetarian diet in protecting against CHD and colorectal cancers.

The average omnivorous diet contains 10–20 mg iron per day, mainly as Fe(III)–protein and haem–protein complexes, but with some in the inorganic form.

If body iron stores are normal then an adult will absorb approximately 5–10% of the total uptake, i.e. 0.5–2 mg/day, in the upper half of the small intestine, following dissolution in the acid pH of the stomach and chelation by mucins in the duodenum. The uptake mechanism, which is still not completely understood, requires energy and is a saturable, electrogenic process. Mucin bound iron is presented to the mucosal cell where it binds to a β3 integrin membrane receptor. Upon internalisation the iron complex associates with a 56 kDa protein mobilferrin resulting in the transfer of the Fe(III) across the apical membrane into a larger 520 kDa complex referred to as paraferritin which contains Fe(III), integrin, mobilferrin and a flavin monooxygenase. This complex has ferrireductase activity and serves to provide Fe^{2+} for incorporation into haem proteins. The paraferritin complex also contains β2-microglobulin whose function is, as yet, unknown (Umbreit *et al.*, 1998). The mobilferrin–integrin pathway appears to be shared by other biologically important transition metals (Cu, Zn, Mn, Co, Pb) which competitively bind mobilferrin. However, this does not lead to metal deficiencies in iron-replete cells because certain of these metals use an integrin pathway different from that used by iron. Haem is the preferred source of iron and is not subject to the mucosal block which stringently controls the uptake of free iron. Once inside the cell it is cleaved by haem oxygenase to yield free iron.

Absorption of iron does not remain constant but is inversely related to the size of the iron stores and directly related to the rate of erythropoiesis.

10.1.6 Role of iron responsive elements (IRE) and iron response elements binary proteins (IRP-1 and IRP-2) in iron homeostasis

As the iron requirements of the cell change iron is shuttled between the major store in ferritin and the mobile transferrin (see Chapter 9). When iron depletion occurs mechanisms are brought into play which inhibit ferritin synthesis and at the same time stimulate transferrin synthesis; iron homeostasis is maintained through the post-transcriptional regulation of genes coding for ferritin and the transferrin receptor (TFR) – a protein that delivers iron to the cytoplasm by receptor mediated endocytosis (Addess *et al.*, 1997). The iron response element (IRE) is a 30 nucleotide RNA hairpin that is located in the 5′ untranslated region of all ferritin mRNAs and in the 3′ untranslated region of all transferrin receptors mRNAs. These IREs are bound by two related IRE-binding proteins (IRPs). IRP-1 is a bifunctional cytosolic protein which requires a bound [4Fe-4S] cluster to confer aconitase activity (Pantopoulos *et al.*, 1996). When the cell is depleted of iron, IRP-1 loses its iron cluster and binds with high affinity to the IREs. This serves to: (i) repress the expression of ferritin by preventing binding of the translation initiation factors to the 5′ cap site of the ferritin RNA (Gray and Hentze, 1994); and (ii) simultaneously bind it to the 3′ untranslated region and increases the expression of the TfR gene by protecting the TfR mRNA against degradation by cellular ribonucleases (Figure 10.1) (Binder *et al.*, 1994). In contrast to IRP-1, IRP-2 lacks aconitase activity even though 16 of the 18 aconitase active site residues are conserved (Guo *et al.*, 1994). The 3 cys residues that coordinate the Fe–S cluster are conserved but IRP-2 regulation does not appear to require the formation of the IRP-1 type cubane Fe–S cluster (Iwai *et al.*, 1995). IRP-2 degradation in iron-replete cells requires the translation of another protein and its accumulation in iron-deficient cells also requires *de novo* protein synthesis (Henderson and Kühn, 1995). A cys rich element in the 73 amino acid insertion which mediates iron-dependent degradation of IRP-2 may possibly involve a proteosome pathway (Guo *et al.*, 1995).

10.1.7 Modulation of IRP-1 and IRP-2 by ROS and RNS

The effect of NO on IRP-1 and IRP-2 responses is remarkably similar to those elicited by iron depletion (Pantopoulos *et al.*, 1996). IRP-1 induction is slow and post-translational whilst IRP-2 induction is also slow but requires *de novo* protein synthesis. These responses require the continuous presence of inducing conditions throughout the entire phase. By contrast H_2O_2 induces rapid and sustained post-translational activation of IRP-1. A minimum period of induction with H_2O_2 of 15 minutes is required for complete activation of IRP-1 within 1 hour. The activation is then sustained for several hours after H_2O_2 removal. IRP-2 appears not to be responsive to H_2O_2 modulation. The differential susceptibilities of IRP-1 and IRP-2 to H_2O_2 may be involved in the tissue-specific regulation of IRE-containing mRNAs. Moreover, IRP-1 and mitochondrially generated ROS may interact to give a novel regulatory feedback mechanism for the control of mitochondrial energy metabolism (Gray *et al.*, 1996).

The response of IRP-1 to H_2O_2 by raising intracellular iron concentrations is highly toxic via the stimulation of Fenton-type reactions and may explain the I/R injury following heart attacks and strokes. Presumably this mechanism has evolved to allow a rapid response to the presence of pathogenic agents. The process would be autocatalytic, cumulative and ultimately fatal were it not for a compensatory mechanism in which surrounding cells may gain hyperresistance to ROS-FR and iron mediated damage by overproduction of the H-chain of ferritin. HL-60 cells become hyperresistant to ROS challenge 24 hours after a one hour incubation with haemin.

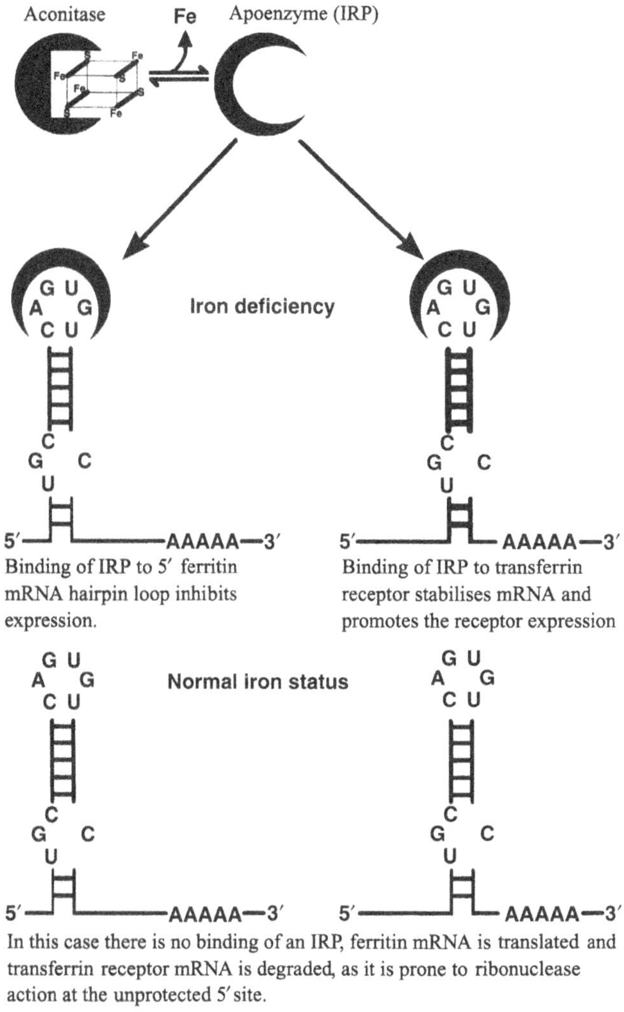

Figure 10.1 The role of IRP/aconitase in iron homeostasis

When the cells were challenged with ROS immediately after the one hour exposure to haemin they were less resistant than the controls. In addition to the delayed hyperresistance there was also a strong induction of haem oxygenase, indicating a role for free iron in the induction of hyperresistance (Lin and Girotti, 1998).

10.2 Human conditions and diseases involving iron toxicity

10.2.1 Hereditary haemochromatosis – a model for the role of iron in disease progression

For a comprehensive review of iron overload conditions see *Seminars in Haematology*, Vol. 35, No. 1 (January 1998).

Table 10.2 Presenting signs of HH

Hepatomegaly
Hyperpigmented, fine wrinkled skin
Diabetes mellitus (Bronze diabetes)
Arthritis
Cardiac failure or arrhythmia
Testicular atrophy

Hereditary haemochromatosis (HH) is one of the commonest inherited diseases in Caucasians of European descent. Genetic studies indicate that 10–13% of this population are heterozygous for the haemochromatosis mutation and that the homozygous condition occurs with a frequency of approximately 5 per 1000. The high frequency of this mutation suggests that, in the past, it may have offered a selective advantage rather than a liability, particularly in areas deficient in iron, e.g. Jamtland County, Central Sweden, where 12.8% of the population are gene carriers (Olsson *et al.*, 1983). Homozygous and perhaps heterozygous females with this gene are better prepared to meet the increased iron demands of pregnancy. Moreover, aggressive males might well have been able to recover more effectively from repeated blood loss following conflicts than their 'normal' adversaries. Indeed, the discovery of the HFE mutation and its ubiquity gives a whole new significance to the once derided practice of bleeding patients for a variety of ailments. Perhaps the northern European physicians of yore noted that a significant proportion of affluent, fee paying patients actually derived benefit from the leeches. Has HH been exascerbated by affluence?

Features of this disease include: arthropathy, cirrhosis of the liver, diabetes, hypermelanotic pigmentation of the skin, heart failure and often hepatocarcinoma in the later stages. In the absence of regular phlebotomy, hepatocellular carcinoma, complicating cirrhosis, is responsible for 30% of deaths in affected homozygotes, and *post mortem* examinations invariably reveal massive iron deposition in the liver.

10.2.1.1 Diagnosis

Hepatic iron is the most sensitive index of the preclinical disease, whereas non-invasive tests involving serum ferritin are unreliable. However, elevated serum transferrin saturation correlates well with hepatic iron content (Edwards *et al.*, 1977), and a transferrin saturation of more than 62% is taken as a positive predictor for the disease state (Edwards *et al.*, 1988).

Past studies of the aetiology of HH have been beset with a mass of conflicting variables involving the genetic bases of iron overload and the disease sequelae. There are several presenting signs for HH (Table 10.2).

Several epidemiological studies have attempted to relate the genotypic characteristics with the disease sequelae and invariably there is a substantial proportion of HH gene carrying patients who are asymptomatic, indicating that the particular gene defect in HH requires to be potentiated by other as yet unidentified genetic variants, interacting with factors such as: bioavailable iron in the diet, age and sex of the subject (losses via menstruation, pregnancy and regular blood donation) (Bothwell and MacPhail, 1998).

10.2.1.2 The genetics of HH

The initial haemochromatosis mutation is thought to have occurred on chromosome 6 bearing the HLA-A3 allele, and HLA typing in HH homozygotes has shown that the A3 allele occurs with a much higher frequency than can be accounted for by chance. The presence of a conserved haplotype on haemochromatosis chromosomes has enabled several groups to use positional cloning to target the haemochromatic gene (Ajioka *et al.*, 1997; Edwards *et al.*, 1998).

The gene, originally designated HLA-H was identified in a 250 kilobase region of DNA located approximately 4.5–5 megabases telomeric from HLA-A (Feder *et al.*, 1996).

This gene has now been designated HFE and encodes an HLA class-1-like protein. Most HH homozygotes contain a cys 282 tyr mutation (C282Y) in the HFE gene. Studies of the frequencies of this gene in different HH populations have found it to be Utah 89%, France (Brittany) 92%, USA 82%, Australia 100% (Beutler *et al.*, 1996; Jazwinska *et al.*, 1996; de Sousa *et al.*, 1994).

The highly conserved region of HLA Class I genes where the C282Y mutation occurs is concerned with β2 microglobulin (β2M) binding. Crucial experiments with β2 microglobulin knockout mice have revealed that the pattern of iron overload that these mice develop with ageing is identical to the progression of human HH (Rothenberg and Voland, 1996).

Whether HFE has a role in signalling or transport is still unclear but recently mobilferrin has been reported to bind human class-1-β2 microglobulin heterodimer (Sadasivan *et al.*, 1996) and HFE may have a role in iron delivery to this system.

The failure of C282Y mutant HFE protein to associate with endogenous β2M in human embryonic kidney cells stably transfected with the mutant cDNA confirms the prediction that the C288Y mutation disrupts a critical disulphide bond in the α-3 loop of the HFE protein. This prevents the binding of this protein to β2M and its subsequent transport and presentation to the cell surface (Feder *et al.*, 1996). Further sophisticated studies with a bone marrow transplant model in β2M knockout mouse indicate that genetically normal hepatic macrophages retain iron when body stores are elevated. In the absence of β2M they release iron into the plasma. This iron release also occurs in β2M knockout mice and HH humans leading to iron overload in the hepatocyte rather than the hepatic macrophage. This is presumably due to disruption of β2M binding by the C282Y mutation (Santos *et al.*, 1996).

It is possible that the enterocyte regulates iron release into the plasma depending on total body iron loading, via a similar HFE-mediated mechanism.

10.2.2 The pathology of iron overload disorders

Those anaemias not associated with the HH allele show clinical features associated with ineffective erythropoiesis (I/E) and are listed in Table 10.3. They are characterised by progressive iron loading to the reticuloendothelial system and the parenchymatous tissues. This category of iron overload (I/O) is referred to as erythropoietic haemachromatosis (Bottomley, 1998). Long term transfusional I/O is inevitable in patients who are dependent on regular transfusions and concurrent therapy with desferrioxamine is still the only available treatment.

Table 10.3 Iron overload disorders

Condition	*Overload mechanism(s)*
Heriditary haemochromatosis (HH)	Enhancement of intestinal iron uptake by HFE gene with recessive mutation
Porphyria cutanea tarda	Associated with HH alleles
Haemolytic anaemias (some)	Associated with HH alleles
Defects of iron transport and metabolism	
Hereditary atransferrinaemia	Increase absorption, deposition in non-erythroid tissues
Heriditary caeruloplasmin deficiency	Lack of ferrooxidase activity leading to increased intestinal absorption
Iron-loading anaemias	Defective erythropoiesis exacerbated by red cell (rbc) transfusion
Thalassaemia syndromes	
Sideroblastic anaemias	
Congenital dyserythropoietic anaemias	
Excessive iron intake	
Red cell transfusion	Induction of haem oxygenase. Release of haemoglobin iron
Elemental iron ($FeSO_4$)	Prolonged ingestion of medicinal iron. Saturation of mucosal iron stores. Enhancement by ascorbate?
Iron dextran	Prolonged parenteral iron therapy
African (Bantu) iron overload	Excessive uptake as soluble complexes from local beers. Role of genetic factors distinct from HH alleles?
Focal haemosiderosis	Release of alveolar rbc. Chronic intravascular haemolysis. Lack of dietary antioxidants contribute to pathology
Idiopathic pulmonary haemosiderosis	
Renal haemosiderosis	
Kwashiorkor	
Liver disease	
Alcoholic liver disease	Enhanced gut permeability
Non-alcoholic steatohepatitis	
Viral hepatitis	
Portacaval shunt	
Neonatal	
Haemochromatosis	Enhanced uptake of iron from foetal haemoglobin degradation?

I/O resulting from hereditary caeruloplasmin deficiency highlights the crucial role of Cu in Fe metabolism. Increased lipid peroxidation is a characteristic of this syndrome and desferrioxamine has been used successfully to reduce the clinical symptoms of I/O (Miyajima *et al.*, 1996, 1997).

10.3 From oxygen via iron to the hydroxyl radical

10.3.1 Underlying mechanisms linking tissue damage, ischaemia/reperfusion and inflammation

10.3.1.1 Xanthine dehydrogenase/oxidase (XD/XO)

In the healthy animal, XD plays a crucial role in the elimination of purines via uric acid. Reducing equivalents are conserved in the formation of NADH/H^+. However, when tissues are damaged or there is severe restriction of blood flow leading to hypoxia or ischaemia, the enzyme is modified via SH oxidation and/or proteolysis and loses its NAD requirement. Instead following reperfusion O_2 replaces NAD as the cofactor and superoxide is produced. Furthermore, the substrate hypoxanthine is provided following ATP degradation during the period of anoxia. The reason for this conversion is not clear, but the author suggests that it probably evolved as a first line of defence in intestinal cells as a rapid response to generate $O_2^{-\bullet}$ as a chemoattractant, causing neutrophils to swarm from the lamina propria to the site of damage and prevent intestinal bacteria colonising the wound. A luminal concentration of 10^{10}–10^{11} bacteria per g plus the effect of peristalsis must require superlative protective strategies in the intestinal mucosa.

This XD/XO conversion is not restricted to the intestine but its kinetics can vary considerably between different tissues and animal species, making the design of relevant animal model studies more difficult (Halliwell and Gutteridge, 1998). Recruitment and priming of neutrophils leads to increased production of $O_2^{-\bullet}$ which can be further supplemented from ischaemia damaged mitochondria. This radical flux together with the secretion of interleukins and proteases by the activated neutrophils leads to the initiation and enhancement of the local inflammatory response. Incremental vascular permeabilisation by $O_2^{-\bullet}$ allows rbc extravasation and haemolysis. Finally the conversion of $O_2^{-\bullet}$ to the H_2O_2 by SOD will allow the release of redox active iron from the haemoglobin to enable localised Fenton driven tissue damage to occur. In whole or part, the above processes can initiate autoimmune reactions, carcinogenesis and other ROS-FR induced conditions listed in Table 10.1.

10.3.1.2 Colorectal cancer (CRC)

The role of iron and redox regulation in the induction of carcinogenesis has been reviewed by Toyokuni (1996).

The colonic environment is extremely complex with respect to the variability of the resident microflora and nutrient input. The potential interactions between carcinogens, radicals, mitogens and nutrients place considerable limitations on study design and attempts to correlate CRC and individual dietary or other lifestyle components tend to produce conflicting results (Helzlsouer *et al.*, 1994). A study of colonic morphology suggests that a unique aetiology exists where the initial damaging events may occur in the mature colonocytes and that the products must be transported via the microcirculation to initiate the process of carcinogenesis in the stem cells which are protected from luminal genotoxins at the base of the crypts by an outflow of mucus from the goblet cells (Booth and Bilton, 1999). Mutagenicity studies using the Ames and SOS Chromotests have revealed DNA damaging synergies between vitamin K, ferrous complexes and the secondary bile acid lithocholate, all of which can accumulate in the colon as a result of high fat/red meat, low fibre diets (Blakeborough *et al.*, 1989).

In vitro studies have demonstrated that iron supplemented human faecal microflora can generate large concentrations of ROS when incubated under aerobic conditions. This led to a hypothesis that iron-mediated redox cycling reactions at the highly aerobic mucosal surface may lead to the initiation of CRC (Babbs, 1990). Moreover, a similar rationale could explain the aetiology of ulcerative colitis – a condition which increases the risk of CRC at least threefold (Babbs, 1992).

Recently, the effect of high fat/low fibre diets on the radical generating capacity of human gut microflora has been studied. The mean hydroxyl radical production was thirteen-fold greater in faeces of patients consuming a high fat/low fibre diet than in controls and this also correlated with an increased iron concentration (Erhardt *et al.*, 1997).

Early epidemiological studies on the aetiology of CRC centred on the role of dietary fat, bile acids and lack of fibre. Originally red meat was considered a risk factor by virtue of its fat rather than iron content (Reddy, 1981). Although currently available evidence implicates red meat and saturated fat, or fats in general, in the aetiology of CRC, the picture is by no means clear. Two fairly recent case control studies have failed to identify any correlation between diet and CRC (Giovannucci *et al.*, 1994; Fredrickson *et al.*, 1995).

10.3.1.3 Animal studies

Many studies have been carried out using the rat colon where tumours have been chemically induced. However, the physiology and microbiology is so different from the human colon that it is difficult to extrapolate from results in a rat model. Such studies indicate that the genotoxic effect for iron and its complexes may be in promotion rather than initiation of CRC. Additional facets to the role of iron in cancer have been demonstrated in rat models where mammary development is greatly accelerated by s.c. iron administration (Diwan *et al.*, 1997).

Iron stimulation of lipid peroxidation in the rat colon was observed following dietary supplementation with iron and polyunsaturated fat (Chin and Carpenter, 1997). A dietary combination of high lipid and haem-iron promotes tumour growth in N-Nitroso-N-methyl urea initiated F344 female rats (Sawa *et al.*, 1998).

From the above brief account it can be seen that iron alone can potentially act at a variety of levels in the aetiology of CRC. It may have an additional and unique role in the development of CRC via the influence of I/O on development of type I diabetes mellitus (IDDM), and to a lesser extent on insulin resistance (NIDDM) as a result of hepatic I/O (Bothwell and MacPhail, 1998). Evidence is accumulating to support the notion that insulin receptors may play a crucial role in the development of CRC via their role in controlling obesity and the growth rate and increased glucose uptake of tumour cells (Kim, 1998).

The conflicting claims in this complex field require a unifying hypothesis to account for the interacting factors which may contribute to increased CRC risks, such as: obesity, physical inactivity, alcohol consumption, predisposing diseases (IBD, acromegaly, diabetes), and high-risk Western diets – high in red meat, saturated fat and refined carbohydrates and low in fibre, fruit and vegetables. This approach may serve to reveal crucial factors that have confounded previous single issue studies and may bring into prominence the importance of sugar and other refined carbohydrates (McKeown-Eyssen, 1994; Giovannucci, 1995), which may act in synergy with iron to induce and/or promote carcinogenesis.

10.3.2 Iron and brain damage

The brain is particularly prone to oxidant damage due to high concentrations of iron and PUFAs, along with low ascorbate levels and low activity for catalase SOD and glutathione peroxidase. Moreover, cerebrospinal fluid has a poor iron binding capacity. Conditions appear to be optimised for lipid peroxidation, following the ischaemia/reperfusion events in stroke victims and bleeding after head injury (Halliwell and Gutteridge, 1998). The age related diseases of the brain are now considered to invoke iron in their pathology.

10.3.2.1 Alzheimer's disease

Alzheimer's disease (AD) is the most common form of adult onset dementia with approximately four million sufferers in the USA. Its prevalence increases with age, e.g. 65–74 years old – 3% incidence, 75–84 years old – 18.7% and over 85 years the incidence is 47.2% (Markesbery, 1997). The original finding of high Al concentrations in the grey matter of AD patients following post mortem analysis implicated this metal in the pathogenesis of senile dementia. Currently the role of Al is not clearly understood, but recent work has implicated iron and free radical damage in the disease progression (Smith and Perry, 1995). Amyloid-β, the main component of senile plaques, has been found to mediate its neurotoxicity via free radicals (Behl *et al.*, 1994).

Iron accumulates in the neurofibrillary tangles, perhaps as a result of increased haem oxygenase activity (Smith *et al.*, 1995), and elevated Al concentration in the same neurons has a stimulatory effect on lipid peroxidation (Gutteridge *et al.*, 1985).

Recently in a rat cortical study Al was taken up by a transferrin-independent iron uptake mechanism (Tf-IU) which could interfere with iron metabolism (Oshiro *et al.*, 1998). Al markedly upregulated Tf-IU activity. Moreover, Al decreased the transferrin receptor TfR m-RNA, as does iron. Finally, Al has been shown to replace iron lost from the 4Fe–4S cluster in aconitase, regenerating the enzyme activity in a dose dependent manner, and also reduces the binding of IRP to the IRE. These data suggest that in AD patients Al competes with iron, both in the uptake and regulation processes.

10.3.2.2 Parkinson's disease (PD)

Although $^{\bullet}NO$ can be neuroprotective in certain circumstances, strong evidence exists supporting its role in the pathology of a variety of neurological disorders. The role of oxidative stress and $^{\bullet}NO$ derived DNA damage in PD has recently been reviewed (Jenner and Olanow, 1996; Simonian and Coyle, 1996). The effect of $^{\bullet}NO$ on iron homeostasis (see section 10.1.6) may contribute to oxidative DNA damage, and 8-OHG residues have been identified in the brain tissue of PD victims (Alam *et al.*, 1997).

This brief review can only give an indication of the truly universal role that iron occupies in the aetiologies of many human diseases and conditions. Its toxicity is almost totally unknown to the general public, many of whom believe that more iron must be good for you – rather like the ancient Greeks, who dissolved iron filings in their vinegar (McCord, 1998). This ignorance of iron and lack of government regulation of its distribution poses health threats to the community, particularly in the supplement market where Fe/ascorbate tablets are readily available.

The press and many clinicians encourage megadoses of vitamin C as an antiviral prophylactic. This should be viewed with caution as several recent publications highlight potential damage to DNA following excessive uptake of ascorbate (Singh, 1997; Rehman *et al.*, 1998; Podmore *et al.*, 1998). This will be exacerbated in HH and thalassaemic patients with clinical I/O, and there are calculated to be about 50,000 homozygous HH carriers in the UK alone.

The redox cycling anthraquinone laxatives (emodin, rhein) have recently been routinely prescribed to pregnant women following constipation caused by excessive $FeSO_4$ ingestion. Such compounds cause extensive DNA damage in HT29 colon carcinoma cell lines in the presence of Fe^{2+} (Roberts, J. and Bilton, R.F., unpublished data). The widespread supplementation of children's foods (particularly cereals) with iron may not be appropriate for a population that is largely iron-replete. In a recent literature search the authors found 846 papers on I/O. Inserting the free radical descriptor yielded none. Clearly more research is required to investigate the role of ROS in I/O disorders.

Finally, if we take McCord's recent analysis (McCord, 1998) to its logical conclusion, then voluntary blood donation by healthy, mature males (at the proscribed rate of 2 units per annum) could reduce risk of all the iron-related disorders particularly CHD, diabetes, Parkinson's disease and small cell carcinoma of the lung, to the lower rates associated with pre-menopausal women. Moreover, it would at a stroke, relieve the supply problems of the Blood Transfusion Service, provide an abundant supply for hospitals and the poorer sections of society where anaemia is more common. This could be the ultimate health incentive – improve the health of others by first improving your own. A message for the millennium.

References

Addess, K.J., Basolion, J.P., Klausner, R.D. Rouault, T.A. and Pardi, A. (1997). Structure and dynamics of the iron responsive element RNA: implications for binding of the RNA by iron regulatory binding proteins, *J. Mol. Biol.*, **274**, 72–83.

Ajioka, R.S., Yu, P., Gruen, J.R., Edwards, C.Q., Griffen, L.M. and Kushner, J.P. (1997). Recombinations defining centrometric and telomeric borders for the hereditary haemochromatosis locus, *J. Med. Genet.*, **34**, 28–33.

Alam, Z.I., Jenner, A., Daniel, S.E., Lees, A.J., Cairns, N., Marsden, C.D. *et al.* (1997). Oxidative DNA damage in the parkinsonian brain: an apparent selective increase in 8-hydroxyguanine levels in substantia nigra, *J. Neurochem.*, **69**, 1196–1203.

Ames, B.N. (1989). Endogenous oxidative DNA damage, aging, and cancer, *Free Rad. Res. Comm.*, 7, 121–128.

Aruoma, O.I. (1998). Free radicals, oxidative stress, and antioxidants in human health and disease, *JAOCS*, 75, 199–212.

Babbs, C.F. (1990). Free radicals and the etiology of colon cancer, *Free Rad. Biol. Med.*, **8**, 191–200.

Babbs, C.F. (1992). Oxygen radicals in ulcerative colitis, *Free Rad. Biol. Med.*, **13**, 169–181.

Bachi, A., Zuccato, E., Baraldi, M., Fanelli, R. and Chiabrando, C. (1996). Measurement of urinary 8-epi-prostaglandin F2α, a novel index of lipid peroxidation *in vivo*, by immunoaffinity extraction/gas chromatography–mass spectrometry. Basal levels in smokers and nonsmokers, *Free Rad. Biol. Med.*, **206**, 619–624.

Behl, C., Davis, J.B., Lesley, R. and Schubert, D. (1994). Hydrogen peroxide mediates amyloid β protein toxicity, *Cell*, **77**, 817–827.

Beutler, E., Gelbart, T., West, C., Lee, P., Adams, M., Blackstone, R. *et al.* (1996). Mutation analysis in hereditary hemochromatosis, *Blood Cells, Molecules & Diseases*, **22**, 187–194.

Binder, R., Horowitz, J.A., Basilion, J.P., Koeller, D.M., Klausner, R.D. and Harford, J.B. (1994). Evidence that the pathway of transferrin receptor mRNA degradation involves an endonucleolytic cleavage within the 3′ UTR and does not involve poly(A) tail shortening, *EMBO J.*, **13**, 1969–1980.

Blakeborough, M.H., Owen, R.W. and Bilton, R.F. (1989). Free radical generating mechanisms in the colon: their role in the induction and promotion of colorectal cancer?, *Free Rad. Res. Comm.*, **6**, 359–367.

Booth, L.A. and Bilton, R.F. (1999). Gentoxic potential of secondary bile acids: A role for reactive oxygen species, in Aruoma, O.I. and Halliwell, B. (eds), *DNA and Free Radicals: Techniques, Mechanisms and Applications.*

Booth, L.A., Gilmore, I.T. and Bilton, R.F. (1997). Secondary bile acid induced DNA damage in HT29 cells: are free radicals involved?, *Free Rad. Res.*, **26**, 135–144.

Bothwell, T.H. and MacPhail, A.P. (1998). Hereditary hemochromatosis: etiologic, pathologic, and clinical aspects, *Semin. Hematol.*, **35**, 55–71.

Bottomley, S.S. (1998). Secondary iron overload disorders, *Seminars in Haematology*, **35**, 77–86.

Carvalho, V.M., Di Mascio, P., de Arruda Campos, I.P., Douki, T., Cadet, J. *et al.* (1998). Formation of 1,*N*6-etheno-2′-deoxyadenosine adducts by *trans,trans*-2,4-decadienal, *Chemical Research in Toxicology*, **11**, 1042–1047.

Chin, J. and Carpenter, C.E. (1997). Effects of dietary fat type and iron level on *in vivo* lipid peroxidation in rat colon, *Nutrition Research*, **17**, 1381–1389.

Claycamp, H.G. (1992). Phenol sensitization of DNA to subsequent oxidative damage in 8-hydroxyquanine assays, *Carcinogenesis*, **13**, 1289–1292.

Collins, A.R., Duthie, S.J. and Dobson, V.L. (1993). Direct enzymic detection of endogenous oxidative base damage in human lymphocyte DNA, *Carcinogenesis*, **14**, 1733–1735.

Dean, R.T., Armstrong, S.G., Fu, S. and Jessop, W. (1994). Oxidised proteins and their enzymic proteolysis in eukaryotic cells: a critical appraisal, in Nohl, H., Esterbaur, H. and Rice-Evans, C. (eds), *Free Radicals in the Environment, Medicine and Toxicology*, London: Richelieu Press.

de Sousa, M., Reimao, R., Lacerda, R., Hugo, P., Kaufmann, S.H. and Porto, G. (1994). Iron overload in β2-microglobulin-deficient mice, *Immunology Letters*, **39**, 1005–1011.

Diwan, B.A., Anderson, L.M. and Ward, J.M. (1997). Proliferative lesions of oviduct and uterus in CD-1 mice exposed prenatally to tamoxifen, *Carcinogenesis*, **18**, 2009–2014.

Edwards, C.Q., Carroll, M., Bray, P. and Cartwright, G.E. (1977). Hereditary hemochromatosis. Diagnosis in siblings and children, *N. Engl. J. Med.*, **297**, 7–13.

Edwards, C.Q., Griffen, L.M., Goldgar, D., Drummond, C., Skolnick, M.H. and Kushner, J.P. (1988). Prevalence of hemochromatosis among 11,065 presumably healthy blood donors, *N. Engl. J. Med.*, **318**, 1355–1362.

Edwards, C.Q., Griffen, L.M., Ajioka, R.S. and Kushner, J.P. (1998). Screening for hemochromatosis: phenotype versus genotype, *Seminars in Hematology*, **35**, 72–76.

Erhardt, J.G., Lim, S.S., Bode, J.C. and Bode, C. (1997). A diet rich in fat and poor in dietary fiber increases the *in vitro* formation of reactive oxygen species in human feces, *J. Nutr.*, **127**, 706–709.

Esterbauer, H., Dieber-Rotheneder, M., Waeg, G., Puhl, H. and Tatzber, F. (1990). Endogenous antioxidants and lipoprotein oxidation, *Biochem. Soc. Trans.*, **18**, 1059–1061.

Feder, J.N., Gnirke, A., Thomas, W. *et al.* (1996). A novel MHC class I-like gene is mutated in patients with hereditary haemochromatosis, *Nature Genetics*, **13**, 399–408.

Francis, G.A., Mendez, A.J., Bierman, E.L. and Heinecke, J.W. (1993). Oxidative tyrosylation of high density lipoprotein by peroxidase enhances cholesterol removal from cultured fibroblasts and macrophage foam cells, *Proc. Natl. Acad. Sci. USA*, **90**, 6631–6635.

Fredrickson, M., Hardell, L., Bengtsson, N.O. and Axelson, O. (1995). Colon cancer and dietary habits. A case controlled study, *Int. J. Oncology*, **7**, 113–141.

Fu, S., Davies, M.J. and Dean, R.J. (1998). Molecular aspects of free radical damage to proteins, in Aruoma, O.I. and Halliwell, B. (eds), *Molecular Biology of Free Radicals in Human Diseases*, Saint Lucia: OICA International.

Giovannucci, E. (1995). Insulin and colon cancer, *Cancer Causes Control*, **6**, 164–179.

Giovannucci, E., Rimm, E.B., Stampfer, M.J., Colditz, G.A., Ascherio, A. and Willett, W.C. (1994). Intake of fat, meat, and fiber in relation to risk of colon cancer in men, *Cancer Res.*, **54**, 2390–2397.

Graf, E. and Eaton, J.W. (1985). Dietary suppression of colonic cancer. Fiber or phytate?, *Cancer*, **56**, 717–718.

Gray, N.K. and Hentze, M.W. (1994). Iron regulatory protein prevents binding of the 43S translation pre-initiation complex to ferritin and eALAS mRNAs, *EMBO J.*, **13**, 3882–3891.

Gray, N.K., Pantopoulos, K., Dandekar, T., Ackrell, B.A. and Hentze, M.W. (1996). Translational regulation of mammalian and Drosophila citric acid cycle enzymes via iron-responsive elements, *Proc. Natl Acad. Sci. USA*, **93**, 4925–4930.

Griffiths, H.R. and Lunec, J. (1998). Molecular aspects of free radical damage in inflammatory autoimmune pathology, in Aruoma, O.I. and Halliwell, B. (eds), *Molecular Biology of Free Radicals in Human Diseases*, Saint Lucia: OICA International.

Guo, B., Yu, Y. and Leibold, E.A. (1994). Iron regulates cytoplasmic levels of a novel iron-responsive element-binding protein without aconitase activity, *J. Biol. Chem.*, **269**, 24252–24260.

Guo, B., Philips, J.D., Yu, Y. and Leibold, E.A. (1995). Iron regulates the intracellular degradation of iron regulatory protein 2 by the proteasome, *J. Biol. Chem.*, **270**, 21645–21651.

Gutteridge, J.M., Quinlan, G.J., Clark, I. and Halliwell, B. (1985). Aluminium salts accelerate peroxidation of membrane lipids stimulated by iron salts, *Biochim. Biophys. Acta*, **835**, 441–447.

Halliwell, B. and Chirico, S. (1993). Lipid peroxidation: its mechanism, measurement, and significance, *Am. J. Clin. Nutr.*, **57**, Suppl. 715S–724S; discussion 724S–725S.

Halliwell, B. and Gutteridge, J.M.C. (1998). *Free Radicals in Biology and Medicine*, 3rd edition, Oxford: Oxford University Press.

Helzlsouer, K.J., Block, G., Blumberg, J., Diplock, A.T., Levine, M., Marnett, L.J. *et al.* (1994). Summary of the round table discussion on strategies for cancer prevention: diet, food, additives, supplements and drugs, *Cancer Res.*, **54**, Suppl. 2044S–2051S.

Henderson, B.R. and Kühn, L.C. (1995). Differential modulation of the RNA-binding proteins IRP-1 and IRP-2 in response to iron. IRP-2 inactivation requires translation of another protein, *J. Biol. Chem.*, **270**, 20509–20515.

Hochstein, P., Nordenbrand, K. and Ernster, L. (1964). Evidence for the involvement of iron in the ADP-activated peroxidation of lipids in the microsomes and mitochondria, *Biochem. Biophys. Res. Commun.*, **14**, 323–328.

Huber, C. and Wächtershäuser, G. (1998). Peptides by activation of amino acids with CO on (Ni,Fe)S surfaces: implications for the origin of life, *Science*, **281**, 670–672.

Iwai, K., Klausner, R.D. and Rouault, T.A. (1995). Requirements for iron-regulated degradation of the RNA binding protein, iron regulatory protein 2, *EMBO J.*, **14**, 5350–5357.

Jazwinska, E.C., Cullen, L.M., Busfield, F., Pyper, W.R., Webb, S.I., Powell, L.W. *et al.* (1996). Haemochromatosis and HLA-H, *Nature Genetics*, **14**, 249–251.

Jenner, P. and Olanow, C.W. (1996). Oxidative stress and the pathogenesis of Parkinson's disease, *Neurology*, **47**, Suppl. 3, S161–S170.

Joshi, P. and Dennis, P.P. (1993). Structure, function, and evolution of the family of superoxide dismutase proteins from halophilic archaebacteria, *J. Bacteriol.*, **175**, 1572–1579.

Jürgens, G., Lang, J. and Esterbauser, H. (1986). Modification of human low-density lipoprotein by the lipid peroxidation product 4-hydroxynonenal, *Biochim. Biophys. Acta*, **875**, 103–114.

Kim, Y.I. (1998). Diet, lifestyle, and colorectal cancer: is hyperinsulinemia the missing link?, *Nutr. Rev.*, **56**, 275–279.

Kim, Y.M., Kim, S.S., Kang, G., Yoo, Y.M., Kim, K.M., Lee, M.E. *et al.* (1998). Comparative studies of protein modification mediated by Fenton-like reactions of iron, hematin, and hemoglobin: Generation of different reactive oxidizing species, *J. of Biochemistry and Molecular Biology*, **31**, 161–169.

Licari, G.R. and Cloud, P.E., Jr (1968). Reproductive structures and taxonomic affinities of some nanofossils from gunflint iron formations, *Proc. Natl Acad. Sci. USA*, **59**, 1053–1060.

Lin, F. and Girotti, A.W. (1998). Hemin-enhanced resistance of human leukemia cells to oxidative killing: antisense determination of ferritin involvement, *Arch. Biochem. Biophys.*, **352**, 51–58.

Lowe, G.M., Booth, L.A., Young, A.J. and Bilton, R.F. (1999). Lycopene and β-carotene protect against oxidative damage in HT29 cells at low concentrations but rapidly lose this capacity at higher doses, *Free Rad. Res.*, **30**, 141–151.

Maden, B.E. (1995). No soup for starters? Autotrophy and the origins of metabolism, *Trends Biochem. Sci.*, **20**, 337–341.

Markesbery, W.R. (1997). Oxidative stress hypothesis in Alzheimer's disease, *Free Rad. Biol. Med.*, **23**, 134–147.

McCord, J.M. (1998). Iron, free radical, and oxidative injury, *Seminars in Haematology*, **35**, 5–12.

McKeown-Eyssen, G. (1994). Epidemiology of colorectal cancer revisited: are serum triglycerides and/or plasma glucose associated with risk?, *Cancer Epidemiol. Biomarkers Prev.*, **3**, 687–695.

Minolti, G. and Aust, S.D. (1987). The role of iron in the initiation of lipid peroxidation, *Chemistry and Physics of Lipids*, **44**, 191–208.

Miyajima, H., Takahashi, Y., Serizawa, M., Kaneko, E. and Gitlin, J.D. (1996). Increased plasma lipid peroxidation in patients with aceruloplasminemia, *Free Rad. Biol. Med.*, **20**, 757–760.

Miyajima, H., Takahashi, Y., Kamata, T., Shimizu, H., Sakai, N. and Gitlin, J.D. (1997). Use of desferrioxamine in the treatment of aceruloplasminemia, *Am. Neurol.*, **41**, 404–407.

Morrow, J.D. and Roberts, L.J. (1996). The isoprostanes. Current knowledge and directions for future research, *Biochem. Pharmacol.*, **51**, 1–9.

Olsson, K.S., Ritter, B., Rosén, U., Heedman, P.A. and Staugård, F. (1983). Prevalence of iron overload in central Sweden, *Acta Med. Scand.*, **213**, 145–150.

Oshiro, S., Kawahara, M., Mika, S. *et al.* (1998). Aluminium taken up by transferrin-independent iron uptake affects the iron metabolism in rat cortical cells, *J. Biochem. (Tokyo)*, **123**, 42–46.

Pantopoulos, K., Weiss, G. and Hentze, M.W. (1996). Nitric oxide and oxidative stress (H_2O_2) control mammalian iron metabolism by different pathways, *Mol. Cell Biol.*, **16**, 3781–3788.

Podmore, I.D., Griffiths, H.R., Herbert, K.E., Mistry, N., Mistry, P. and Lunec, J. (1998). Vitamin C exhibits pro-oxidant properties, *Nature*, **392**, 695.

Prütz, W.A., Butler, J. and Land, E.J. (1983). Phenol coupling initiated by one-electron oxidation of tyrosine units in peptides and histone, *Int. J. Radiat. Biol. Relat. Stud. Phys. Chem. Med.*, **44**, 183–196.

Reddy, B.S. (1981). Diet and excretion of bile acids, *Cancer Res.*, **41**, 3766–3768.

Rehman, A., Collis, C.S., Yang, M., Kelly, M., Diplock, A.T., Halliwell, B. *et al.* (1998). The effects of iron and vitamin C co-supplementation on oxidative damage to DNA in healthy volunteers, *Biochem. Biophys. Res. Comm.*, **246**, 293–298.

Rothenberg, B.E. and Voland, J.R. (1996). β2 knockout mice develop parenchymal iron overload: A putative role for class I genes of the major histocompatability complex in iron metabolism, *Proc. Natl Acad. Sci. USA*, **93**, 1529–1534.

Sadasivan, B., Lehner, P.J., Ortmann, B., Spies, T. and Cresswell, P. (1996). Roles for calreticulin and a novel glycoprotein, tapasin, in the interaction of MHC class I molecules with TAP, *Immunity*, **5**, 103–114.

Sakumi, K., Furuichi, M., Tsuzuki, T., Kakuma, T., Kawabata, S., Maki, H. *et al.* (1993). Cloning and expression of cDNA for a human enzyme that hydrolyzes 8-oxo-dGTP, a mutagenic substrate for DNA synthesis, *J. Biol. Chem.*, **268**, 23524–23530.

Santos, M., Schilham, M.W., Rademakers, L.H., Marx, J.J., de Sousa, M. and Clevers, H. (1996). Defective iron homeostasis in β2-microglobulin knockout mice recapitulates hereditary hemochromatosis in man, *Journal of Experimental Medicine*, **184**, 1975–1985.

Sawa, T., Akaike, T., Kida, K., Fukushima, Y., Takagi, K. and Maeda, H. (1998). Lipid peroxyl radicals from oxidized oils and heme-iron: implication of a high-fat diet in colon carcinogenesis, *Cancer Epidemiol. Biomarkers Prev.*, **7**, 1007–1012.

Simonian, N.A. and Coyle, J.T. (1996). Oxidative stress in neurodegenerative diseases, *Annu. Rev. Pharmacol. Toxicol.*, **36**, 83–106.

Singh, N.P. (1997). Sodium ascorbate induces DNA single strand breaks in human cells *in vitro*, *Mutation Res.*, **375**, 195–203.

Smith, M.A. and Perry, G. (1995). Free radical damage, iron, and Alzheimer's disease, *J. Neurol. Sci.*, **134**, Suppl., 92–94.

Smith, M.A., Sayre, L.R., Monnier, V.M. and Perry, G. (1995). Radical Ageing in Alzheimer's disease, *Trends Neurosci.*, **18**, 172–176.

Toyokuni, S. (1996). Iron-induced carcinogenesis: the role of redox regulation, *Free Rad. Biol. Med.*, **20**, 553–566.

Umbreit, J.N., Conrad, M.E., Moore, E.G. and Latour, L.F. (1998). Iron absorption and cellular transport: the mobilferrin/paraferrin paradigm, *Semin. Hematol.*, **35**, 13–26.

Wiseman, H. and Halliwell, B. (1996). Damage to DNA by reactive oxygen and nitrogen species: role in inflammatory disease and progression to cancer, *Biochem. J.*, **313**, 17–29.

Woods, J.A., Young, A.J., Gilmore, I.T., Morris, A. and Bilton, R.F. (1997). Measurement of menadione-mediated DNA damage in human lymphocytes using the comet assay, *Free Rad. Res.*, **26**, 113–124.

11 The role of free radicals in the toxicology of airborne particulate pollutants

Peter Evans

But what is dust, save time's most lethal weapon . . . ?

Sir Osbert Sitwell

11.1 Introduction

The suffocating dense city smog episodes of 'the fifties', and the occupational dust hazards of the many thousands of employed coalminers, are now largely fading memories of British social and industrial life. The closure, several years ago, of the Medical Research Council Air Pollution and Pneumoconiosis Research Units, testifies to the demise of the then perceived prevalence and importance of the hazards of airborne pollutants in the communal and workers' environment.

Respiratory diseases induced by the inhalation of dusts have been a long-time occupational hazard for large numbers of workers employed in a variety of jobs. The pneumoconioses are characterised by a range of inflammatory, fibrotic, emphysematous and cancerous pathology related to the extent and nature of the inhaled particles. Anthracosis, silicosis and asbestosis, indicate the specific lung fibrotic pathology produced by coal, silica and asbestos dusts. Asbestos has also been shown to cause tumours of the lung and mesotheliomas of the pleura.

However, as is the cyclical or spiralling nature of life, and environmental and medical issues, public and governmental concerns related to problems of air pollution have re-emerged in recent years to be a matter of paramount public health and political concern, and indeed controversy. The issue has been driven by a number of scientific and social forces: the active voices of the environmentalist 'green' movement, the public's direct experience of traffic pollution, and electoral concern about the environment and health, together with the consequent legislative policy promulgated by the UK and the EU parliaments.

The burgeoning awareness and anxiety of the role of particulates, especially the microparticulate fraction of ambient urban air, in the aetiology of respiratory and related cardiovascular disease, has stimulated renewed scientific research effort into the epidemiology, toxicology, clinical and public health aspects of the topic. Parameters of particle size, shape, composition, solubility, and surface charge and reactivity, are all crucial, complex and interacting properties which determine the biological, cellular and biochemical toxic reactions. An understanding of the diverse mechanisms of particulate toxicity is a prerequisite to the identification and implementation of effective preventive public health action and remedial medical management and treatment.

In this chapter, the specific biochemical and cellular mechanisms implicating the pathogenic role of reactive oxidant species (ROS): namely superoxide anion ($O_2^{-\bullet}$) and hydroxyl ($OH^\bullet$) free radicals and related metabolites, e.g. hydrogen peroxide (H_2O_2), in mediating various particulate-induced tissue injury, will be reviewed.

11.2 Silica toxicity

Silicosis, caused by the inhalation of stone silica dust, is identified histologically by characteristic fibrotic collagenous nodules in the lung. The crystalline nature of the silica particles is considered to play a critical role in its toxic action.

11.2.1 Surface activity

The radical reactive surface properties of silica are directly demonstrable by electron spin resonance (ESR) studies. Using a DMPO spin-trap reagent, it has been shown that freshly fractured silica compared to 'air aged' silica exhibits an enhanced $OH^\bullet$ activity in aqueous media. This reactivity is similarly directly related to increased cell toxicity to alveolar macrophages (AM) and to erthyrocytes (Valyathan *et al.*, 1988). Such findings shed evidential light on the early ideas concerning the 'freshly fractured surface' pathogenic theory of silicosis (Wright, 1950), and the associated oxidative processes (Marasas and Harington, 1960). Further studies with the antioxidant enzymes superoxide dismutase (SOD) and catalase have demonstrated an additional role of $O_2^\bullet$ in silica generated free radicals and related lipid peroxidation reactions (Shi *et al.*, 1994). Similarly, electron paramagnetic resonance (EPR) studies comparing different silica polymorphs, namely DQ-12, Min-U-Sil 5 and cristabolite, revealed significant radical activity related to quartz crystal purity (Fubini *et al.*, 1990). Analogous detailed studies using varieties of alpha-quartz, hydrofluoric acid etched samples, and crystalline quartz cristobalite and tridymite samples, were able to discriminate between the capacity to generate $OH^\bullet$ radicals as detected by ESR, and the ability to induce oxidation of DNA and thymine glycol production. The zeta potential of the samples were similar, and such differences in activity were still apparent when the samples were normalised for particle surface area (Daniel *et al.*, 1996). The inhibition of silica and H_2O_2 mediated DNA strand breakage by the iron chelator desferrioxamine, indicates that Fenton reaction oxidation products of trace iron contaminants may also play a role in silica toxicity (Daniel *et al.*, 1993).

11.2.2 Phagocyte interactions

The defensive biological bactericidal function of phagocytes in the production of ROS may in certain specific circumstances, as with the chronic activation by indigestible mineral particulates, be turned against the organism itself in an act of autodestruction (Babior, 1984).

In vitro cellular interactions of silica with AM induced the production of lipid peroxides (Gabor *et al.*, 1975) and ROM as monitored by luminol-enhanced chemiluminescence (Vilim *et al.*, 1984). The exogenic production of ROS by silica-stimulated phagocytes as measured by nitroblue-tetrazolium (NBT), was shown to be greater for polymorphonuclear leukocytes (PMN) than for AM cells (Gusev *et al.*, 1993). By comparison, in a similar *in vitro* study, a reduction in intracellular AM ferritin accompanied a paradoxical decrease in AM protein carbonyl formation (Connor *et al.*, 1996). Silica surface-mediated free radical generation has also been associated with the stimulated production of the pro-inflammatory eicosanoid mediators: prostaglandin E2, thomboxane E2, and leukotriene B4 from exposed AM *in vitro* (Kuhn *et al.*, 1992).

11.2.3 In vivo *studies*

Experimental *in vivo* studies in rats exposed to silica by intratracheal injection, demonstrated an increase in tissue lipid peroxidation and activities of SOD and catalase (Zsoldos *et al.*, 1983). Pulmonary exposure by inhalation in rats to silica and other dusts, showed that increases in lung tissue MnSOD mRNA levels were predictive of inflammatory cellular changes measured by bronchoalveolar lavage (BAL) (Janssen *et al.*, 1994). Increased lipid peroxidation, upregulation of SOD and glutathione peroxidase, and kinetic rate changes in clearance of the stable nitroxide radical TEMPO in rats exposed to quartz, illustrate the enhanced oxidative stress in silicotic tissue (Vallyathan *et al.*, 1997). Studies into the temporal changes occuring in rats exposed to silica, indicated that the generation of free radical oxidants is a primary event influencing the production of tumour necrosis factor (TNF)-alpha production by AM, and can be prevented by pretreatment with the free radical scavenger *N-tert*-butyl-α-phenylnitrone (BAPN) (Gossart *et al.*, 1996).

11.3 Coal

Pneumoconiosis in coal workers is characterised primarily by fibrotic and emphysematous reactions to the inhaled dust. The composition of coal dust is highly heterogeneous, comprising a complex mixture of organic chemicals and inorganic mineral components. Miners are generally exposed to a mixture of both coal and stone dust and the resultant respiratory disease may also consequently exhibit aspects of the more pronounced silicotic reactions to the inhaled toxic silica particles. As the prevalence of the occupational hazards to coal workers has diminished in recent decades due to economic decline and closure of deep pit mining, the resurgence of open cast mining and associated exposure of neighbouring communities to environmental dust is currently presenting increasing concern.

11.3.1 Free radical activity

Using an ESR spin-trap technique, both bituminous and anthracite coal dust have been shown to possess free radical activity, being greater in the latter and especially so when freshly ground, and exhibiting a direct correlation with the dust toxicity. Autopsy lung tissue from coal miners also exhibited similar free radical activity (Dalal *et al.*, 1989). Subsequent related studies have shown that various samples of bituminous coal dust taken from different mining areas, generate $OH^{\bullet}$ when incubated with H_2O_2. Free radical activity was inhibited by catalase and desferrioxamine, and correlated with the surface iron content, indicating the primary involvement of the Fenton reaction (Dalal *et al.*, 1995). Oxidant generation by coal dust iron is stimulated by the presence of humic coal components and correlates with coal miners' lung tissue hydroxyproline, a collagen marker (Ghio and Quigley, 1994). Interaction of coal dust with human macrophages *in vitro* resulted in the activation and resultant generation of oxyradicals, as monitored using lucigenin-enhanced chemiluminescence and NBT assays, together with evidence of associated lysosomal protease release (Maly, 1988). Investigations undertaken on blood samples obtained from coalworkers, have found evidence of decreased serum sulphydryl levels (Thomas and Evans, 1975) and also perturbed erythrocyte antioxidant catalase and SOD enzyme activities (Engelen *et al.*, 1990; Nadif *et al.*, 1998), indicative of elevated systemic oxidative stress present in the more severe inflammatory fibrotic categories of coalworkers' pneumoconiosis.

11.4 Asbestos

Asbestos has been used extensively for insulation and a wide variety of other industrial and construction purposes. Asbestos is a fibriform silicate which occurs in a number of different mineral types, namely chrysotile (white asbestos), and the amphiboles: crocidolite (blue), amosite (brown) and anthophyllite. Inhalation of the fine asbestos fibres in the mining and manufacturing industries particularly, has resulted in an epidemic of asbestosis, lung cancers and mesotheliomas. Despite the strict hygiene control measures now in place, it is expected that due to the long latency period of tumour development, the incidence of asbestos-related cancers will continue to increase in the future (Peto *et al.*, 1995).

Experimental studies into the pathogenic mechanisms of asbestos toxicity have indicated that particle fibre shape, size (length and diameter), and trace metal composition, are all factors which play a role in the fibrotic and carcinogenic action.

11.4.1 Iron-catalysed Fenton reactions

The property of the transitional element iron to participate in ferrous/ferric redox reactions and catalyse the formation of $OH^{\bullet}$ in the Fenton reaction, enables it to play a key role in free radical chemistry. Iron is normally sequestered in the body, bound tightly to transfer and storage proteins, namely transferrin and ferritin. The physiological availability of 'free' or 'catalytic' iron to participate in adverse Fenton reactions is thus under strict biological control.

Iron is a significant component in several types of asbestos, especially the amphibole asbestos types: crocidolite, amosite, and anthophyllite. The crystal lattice location, amount, availability and redox nature of the iron varies with each asbestos type. The hypothetical involvement of a decontrolled-iron pathogen in the carcinogenic activity of asbestos has been proposed (Kon, 1978). As such, the part this particular reaction plays in asbestos toxicity has been a topic of detailed research. Studies utilising several types of iron-containing particles: asbestos, non-toxic iron oxides, and model particles with surface exposed iron, have demonstrated that only a small fraction of the iron, in specific coordination and redox states, is apparently involved in the toxic effect (Fubini, *et al.*, 1995). The extraction of ferric iron from the fibre surface is not affected by the presence of the physiological iron-binding proteins transferrin and lactoferrin, but is released by $O_2^{-\bullet}$ generated by interaction of the particles with activated phagocytic cells (Ghio *et al.*, 1994). The ability of the nutrient and physiological reductant ascorbic acid to mobilise iron from asbestos implies that it may have, paradoxically, a potentially deleterious effect on iron-mediated toxicity *in vivo* (Lund and Aust, 1990). The associated production of H_2O_2 may also induce the generation of $OH^{\bullet}$ in the presence of active iron-rich amosite and crocidolite asbestos particles (Weitzman and Graceffa, 1984). Evidence of the significance of role of iron present in amosite and crocidolite asbestos is presented by findings demonstrating the inhibition of fibre-induced DNA damage by the the iron chelator desferrioxamine and the $OH^{\bullet}$ scavenger mannitol (Donaldson, *et al.*, 1995). Comparable studies have indicated that amosite-induced DNA strand breaks, an $OH^{\bullet}$ mediated reaction, are also prevented by the iron chelator phytic acid (Kamp *et al.*, 1995). Investigation of the direct interaction of asbestos fibres with cells *in vitro* using high resolution X-ray photelectron spectroscopy (XPS) have revealed alterations in the disposition of iron at the fibre/cell interface (Seal *et al.*, 1996).

Autopsied lung samples from asbestos workers have been shown, using Mössbauer spectroscopy, to contain considerable quantities of iron (Stroink *et al.*, 1987). Asbestos 'ferruginous' bodies are formed in the lung following exposure to asbestos, and are characterised by accumulated deposits of the iron storage protein ferritin on the fibres (Pooley, 1972). Mobilisation of ferritin bound iron and hence further exacerbation of Fenton $OH^{\bullet}$ oxidant stress, is facilitated by $O_2^{-\bullet}$ generated by activation of the respiratory burst of phagocytic cells (Biemond *et al.*, 1984).

11.4.2 Phagocyte oxidative interactions

The capacity of asbestos particles to directly catalyse oxyradical interactions via the Fenton iron-mediated reaction, is complemented by the ability to stimulate $O_2^{-\bullet}$ generation and associated luminol-enhanced chemiluminescence by the activation of the respiratory burst NADPH oxidase of phagocytic cell membranes, namely AM (Hatch *et al.*, 1980) and PMN (Doll *et al.*, 1982). Both phagocyte cell types infiltrate into the lung consequent to the inhalation of dusts (Rola-Pleszcynski *et al.*, 1984). Experimental *in vivo* exposure of rats to crocidolite and titanium dioxide – a nonfibrous dust, caused a marked difference in the lung inflammatory response, with crocidolite producing pulmonary fibrosis, increased levels of the lung collagen hydroxyproline, and elevated levels of malondialdehyde (MDA), marker of lipid peroxidation, in the bronchoalveolar lavage (BAL) cells and fluid, but not lung tissue (Petruska *et al.*, 1991). While the relationship, as determined *in vitro*, between the phagocytic chemiluminescent response and the pathogenicity of mineral dusts of varying toxicity is apparently not a simple one (Gormley *et al.*, 1985), the evidence concerning the pathogenic role of phagocyte-derived oxyradicals in pneumoconiotic and dust-induced carcinogenic diseases is substantial. *In vitro* ESR DMPO spin-trap studies on the interaction of PMN with crocidolite using various free radical scavengers, namely SOD, catalase and dimethyl sulphoxide, indicated that $O_2^{-\bullet}$, H_2O_2 and $OH^{\bullet}$ are involved in the stimulatory mechanism (Ishizaki *et al.*, 1994).

Activation of AM nitric oxide (NO) synthase and the generation of NO following esposure to crocidolite and chrysotile asbestos particles has been also demonstrated, indicating the potential involvement of other potent toxic free radicals, including peroxynitrite ($ONOO^-$) formed by the the interaction of NO with $O_2^{-\bullet}$, in the inflammatory response (Thomas *et al.*, 1994).

11.4.3 Tissue and cell reactions in vitro

Asbestos exposure has been linked to carcinoma of the lung and characteristically to the occurrence of mesotheliomas, tumours of the mesothelial cell lining the pleura. Investigation into the biochemical and cellular mechanisms of asbestos pathogenicity have utilised a variety of *in vitro* cell and tissue model systems. Exposure of human mesothelial explants to crocidolite in organ culture produced a proliferative cellular response (Rajan *et al.*, 1972). However, despite evidence that rat mesothelial cells are capable of phagocytosing chrysotile asbestos fibres (Jaurand *et al.*, 1979), experiments using amosite asbestos exposed to human mesothelial cells (Gabrielson *et al.*, 1986) and to mesothelial cell lines (Kinnula *et al.*, 1994), failed to detect evidence of free radical generation or oxidant damage. The cytotoxicity of crocidolite and chrysotile asbestos to tracheal epithelial cells was reduced by SOD but not by catalase, the long fibre chrysotile samples being especially toxic (Mossman and Landesman, 1983). Transfection experiments of tracheal epithelial cells by MnSOD cDNA, have demonstrated an increased resistance to asbestos toxicity, providing further evidence of the crucial role which oxyradicals play in the pathological process (Mossman *et al.*, 1996).

11.4.4 Fibre dimensions and shape

Proposals for the pathogenic mechanism related to the specific toxicity of fibrous particles, longer than about 8 microns – the 'Stanton hypothesis' (Stanton *et al.*, 1981) – have been varied. The concept of 'frustrated phagocytosis' postulating the partial endocytosis of long fibres by phagocytic AM and PMN cells and the associated extracellular release of ROS, is of special interest (Archer and Dixon, 1979). Electron microscopical studies have revealed that several macrophages can be associated with what is apparently a vain attempt to ingest a long asbestos fibre (Johnson and Davies, 1981). *In vitro* AM-generated $O_2^{-\bullet}$ is increased by exposure to asbestiform when compared to non-fibrous particulates (Hansen and Mossman, 1987), and, similarly, comminution of both crocidolite asbestos and glass fibre particle size by milling, caused a reduction in the oxyradical generation from PMN *in vitro* as measured by luminol-enhanced chemiluminescence (Yano *et al.*, 1991). The increased pathogenicity to Chinese hamster ovary cells *in vitro* of long as compared to short fibre amosite asbestos, is exacerbated by pretreatment with buthionine sulphoximine and resultant depletion of levels of glutathione, the primary cellular thiol antioxidant; and is therefore indicative of the role which oxidant stress may play in promoting the toxic effects exhibited by fibriform particulates (Donaldson and Golyasnya, 1995).

11.4.5 Cigarette smoke cocarcinogenesis

Epidemiological studies have indicated that asbestos workers who also smoke have an increased lung cancer mortality rate (Selikoff *et al.*, 1968). A basic interaction is the potential direct adsorption of carcinogenic polycyclic hydrocarbons from cigarette smoke onto asbestos fibres and thus their colocalised carriage and deposition in the lung (Harvey *et al.*, 1984). A number of mechanisms have been proposed which also directly implicate asbestos interaction with ROS (Kamp *et al.*, 1992). Metabolic oxidative activation of benzo(a)pyrene via 6-hydroxybenzo(a)pyrene to form the reactive intermediate 6-oxobenzo(a)pyrene, is catalysed by asbestos (Graceffa and Weitzman, 1987), a process mediated by the microsomal system and involving $O_2^{-\bullet}$ reactions (Byczkowski and Gessner, 1987). Cellular interactions using AM have demonstrated that particles can enhance the benzo(a)pyrene-induced mutation rate of Chinese hamster cells (Romert and Jenssen, 1983), a finding consistent with the proposed capacity of stimulated phagocytes to cause oxidative bioactivation of procarcinogens via the oxyradicals generated during the respiratory burst (Roman-Franco, 1982).

Cell-free studies have shown that asbestos-bound iron catalyses the production of $OH^{\bullet}$ by interaction with cigarette smoke-borne free radicals, which can also directly contribute to DNA damage (Jackson *et al.*, 1987). Similar findings, using spin-traps and electron paramagnetic resonance techniques, have demonstrated enhanced $OH^{\bullet}$ generation by the interactions of asbestos with smoke tar components, hence indicative of potential cocarcinogenic mechanisms (Valavanidis *et al.*, 1995). Direct reaction of chrysotile with physiological redox metabolites, namely adrenaline and ascorbate, and monitored using a lucigenin-enhanced chemiluminescence system, indicated that both $OH^{\bullet}$ and $O_2^{-\bullet}$ radicals are generated (Suslova *et al.*, 1994).

In addition to the relatively large amount of iron present in asbestos, particular in the brown amosite and blue crocidolite types, asbestos also contains vaying quantities of other trace elements including chromium, cobalt, manganese, nickel, titanium, and zinc (Timbrell, 1969). The ability of heavy metal contaminants to inhibit lung tissue hydroxylative detoxification of benzo(a)pyrene and related polycyclic aromatic hydrocarbon carcinogens present in cigarette smoke, indicates a potential cocarcinogenic mechanism (Dixon *et al.*, 1970). Intraperitoneal and intratracheal instillation of mineral and metallic particulates into experimental rats demonstrated the carcinogenic activity of nickel and nickel subsulphide (Pott *et al.*, 1987), a process which may be related to the capacity of such materials to stimulate the production of PMN-derived ROS (Evans *et al.*, 1992a).

11.5 Man-made mineral fibres

The production and use of man-made mineral fibre (MMMF) materials, primarily rockwool, glass wool and ceramic fibres, has considerably expanded over recent decades for a range of industrial purposes, frequently as a substitute for the now diminishing use of asbestos. The toxicity of asbestos has implicated a variety of pathogenic mechanisms related to the size, shape, composition, solubility and surface chemical and electrical properties. While differing in such properties, the potential for MMMF to likewise induce fibrotic and carcinogenic pulmonary disease has been a matter of considerable concern.

A general review of the findings from several epidemiological studies involving workers in the manufacture and use of MMMF concluded that the evidence linking MMMF exposure to chronic lung disease and cancer was uncertain. This contrasted with experimental data indicating the fibrogenic action of intratracheal administration, the carcinogenic effect of intrapleural injection, and the absence of effects by inhalation delivery (Saracci, 1985). Cohort worker studies have indicated that MMMF workers exhibited a limited excess mortality from lung cancer, being higher for rock or slag wool than for glass wool, with exposure to small-diameter glass fibres having an increased relative risk (Doll, 1987). However, the importance of determining adequate data on MMMF exposure levels, smoking habits and other confounding factors in accounting for increased lung cancer risk, has recently been emphasised (Wong and Musselman, 1995). The solubility of types of MMMF *in vivo* is variable, with long glass fibres dissolving more rapidly than short fibres, and thus the biopersistance of fibres is a significant factor in their toxicity (Morgan, 1994). *In vitro* cytotoxicity assays using a human mesothelial cell line have demonstrated increased toxicity of thin compared to coarse fibres, with glass wool being more toxic than rock wool (Pelin *et al.*, 1992).

11.5.1 Oxidant reactions

The role of oxidative processes in MMMF toxicity has been studied using a range of MMMF samples in relation to iron surface chemical reactivity, and a correlation with epidemiological findings of excess lung cancer has been identified (Pezerat *et al.*, 1992). Estimation of free radical activity by various MMMF samples, using plasmid DNA assay, indicates a consistent but low level of reactivity of ceramic and glass wool fibres compared to amosite and crocidolite asbestos. This is despite some particular insulation wool and refractory ceramic fibre samples releasing comparatively large amounts of ferric iron into aqueous solution at physiologically neutral pH. However, addition of desferrioxamine did not reduce MMMF $OH^{\bullet}$ mediated DNA damage (Gilmour *et al.*, 1995). Subsequent studies utilising AM cells *in vitro* to examine the mechanism of fibre induced oxyradical interactions, indicated that while a reduction in intracellular reduced glutathione occurred with MMMF ceramic and glass fibres, and amosite asbestos, only amosite was able to upregulate nuclear transcription factor. Enhancement of cellular glutathione levels with cysteine methyl ester eliminated the upregulation, while reduction of glutathione with the inhibitor buthionine sulphoximine increased the transcription factor activation by amosite (Gilmour *et al.*, 1997b).

11.5.2 Modulatory effects

However, extrapolation of *in vitro* findings to the *in vivo* situation is complicated, as illustrated by the differential effects which MMMF rat immunoglobulin opsonisation exhibits on the *in vitro* stimulated activation of the respiratory burst in treated AM (Donaldson *et al.*, 1995). The complex interaction of the many protein and lipid components of the lung lining surfactant fluid with the airborne 'naked' fibres is problematic to adequately reproduce *in vitro*. Nevertheless, it has been shown that pretreatment of MMMF with rat and sheep lung lining fluid causes an increase in the release of iron, and an enhanced ability to stimulate the AM oxidative burst (Donaldson *et al.*, 1997). It has also been demonstrated that the although the direct adsorption of the carcinogen benzo(a)pyrene to rock, slag and glass wool is limited, the binding is enhanced in the presence of phosphotidylcholine, a principal component of the lung lipoprotein surfactant (Gerde and Scholander, 1988).

There is also evidence of a synergistic interaction of MMMF with smoking, as indicated by the potentiating effect of rock wool on the Fenton generation of $OH^{\bullet}$ and the formation of 8-hydroxyguanosine in exposed isolated calf thymus DNA (Leanderson and Tagesson, 1989).

11.6 Zeolites

Zeolites are aluminosilicates characterised by an open lattice microporous structure, which, because of their unique ion-exchange and catalytic properties, are extensively used in industry for a wide variety of purposes (Newsam, 1986). The epidemiological discovery in Cappodocia, central Turkey, that environmental community exposure to tuff rock dust containing zeolite minerals is associated with a very high prevalence of mesothelioma (Baris *et al.*, 1981), and that sample particulates were shown to be extremely potent carcinogens in experimental animals (Wagner *et al.*, 1985), prompted subsequent cellular studies into the possible causal mechanism. Detailed mineralogical analysis of the Cappodocian tuff rock dust samples by electron microscopy and X-ray diffraction analysis confirmed the presence of the zeolite fibrous mineral erionite (Spurney, 1983).

11.6.1 Oxidant interactions

The activity of various mineral dusts, including the natural zeolite clinoptilolite, to stimulate the production of AM-derived luminol-enhanced chemiluminescence *in vitro* demonstrates the possible role of ROM in zeolite toxicity, and the accompanying finding that zeolite-induced chromosomal aberrations in exposed human cells are inhibited by catalase, is an indication of the genotoxic activity being mediated by oxidant damage (Durnev *et al.*, 1990). Similar *in vitro* investigations using a chemiluminescence technique to monitor oxyradical generation, revealed that erionite was highly active in stimulating the PMN respiratory burst, acting via both surface-active and phagocytosis-dependent mechanisms (Urano *et al.*, 1991). Using the same technique with a number of natural and synthetic isometric and fibriform zeolite particulate samples of modified ionic sodium and calcium composition, revealed the complex interaction of the parameters of shape and composition in the activation process (Evans *et al.*, 1989). Comparable phagocytic oxyradical interactions of model aluminosilicates particulates with brain macrophage-type microglial cells *in vitro* has also been demonstrated (Evans *et al.*, 1992b). The analogous oxyradical generation from chronically reactive inflammatory microglia, stimulated by aluminosilicate and colocalised fibrous beta-amyloid deposits as identified in the pathognomic cerebral plaques present in the brains from Alzheimer subjects, have been proposed to play a neurotoxic role in the pathogenesis of neurodegenerative senile dementia, so-called 'cephaloconiosis' (Evans *et al.*, 1995). The capacity of a range of particulates to induce a variety of oxyradical-mediated pathogenic responses in the lung and in other various body organs and tissues – the 'conioses' – has been reviewed (Evans, 1995).

11.7 Smoke

The combustion of materials is accompanied not only by the production of heat and light, but also, dependent on the composition of the burning substance, the temperature and thermal conditions, the production of a diverse and large number of phasic components and chemical species. The investigation of these complex mixtures of substances in the flames, smoke and gases generated during burning has led to the identification of many types of reactive free radicals.

11.7.1 Cigarette smoke

The combustion of cigarette tobacco produces many smoke free-radical species. Both polyphenol related $O_2^{-\bullet}$ oxyradical and H_2O_2 have been identified in the particulate smoke phase using the spin-trapping agent 5,5-dimethyl-1-pyrroline-*N*-oxide (DMPNO), with nicotine producing a synergistic effect. In the vapour phase, carbonyl sulphide has been shown to participate in the generation of $OH^•$ from H_2O_2 (Kodama *et al.*, 1997). Comparable ESR spin-trapping techniques identified principally alkoxy free radicals, with about equal concentrations present in both main and side stream cigarette smoke. Although reactive, the radicals are long lived, and represent the slow oxidation of nitric oxide to nitrogen dioxide and subsequent reactions with organic molecules. Tar, derived from phenolic leaf pigments, present in main and sidestream smoke contains persistent free radicals extractable in *tert*-butyl benzene which exhibits a broad single-line ESR spectrum (Pryor *et al.*, 1983).

The role of cigarette smoke free-radicals in carcinogenesis can involve participation at both the initiation and promotion stage of the multistage tumour formation process. Oxidative processes are involved in the intiation stage during both the activation of the procarcinogen benzo(a)pyrene to the dihydrodiol epoxy ultimate carcinogen form, and also in the binding of the carcinogen to DNA itself. Formation of benzo(a)pyrene-7,8-dihydrodiol-9,10-epoxide, the ultimate carcinogenic species, is mediated by the cytochrome P450 and peroxyl radical catalysing systems, the relative active contributions of which varies with the organ, individual and species (Sindhu *et al.*, 1995). Aqueous extracts of tar containing the catechol quinone–hydroquinone–semiquinone redox system react with oxygen to generate tissue injurious $O_2^{-•}$, $OH^•$ and H_2O_2 oxidant species (Pryor, 1997).

Oxyradicals may exert their effect on tumour promotion by exacerbating the chronic inflammatory cellular changes (Trush and Kensler, 1991). In addition to the direct toxic effect of the free radicals in cigarettes contributing to carcinogenic transformations, smokers have been shown to exhibit increased blood PMN activation and decreased blood plasma vitamin C (Theron *et al.*, 1994). Oxidative inactivation of the methionine residue of the α1-proteinase inhibitor by smoke and phagocyte-derived ROM is related to enhanced elastase activity and development of emphysema in smokers (Carp *et al.*, 1982). Surprisingly, however, chronic exposure of experimental animals to cigarette smoke produced relatively little alteration in the levels of lung lipid peroxidation markers. Adaptive metabolic responses involving accumulation of lung tissue vitamin E levels and increased AM antioxidant SOD activities, evidently contribute to redox homeostasis (Chow *et al.*, 1993). *In vitro* studies using explant rat tracheal tissue have demonstrated that exposure to cigarette smoke caused membrane blebbing and focal cell necrosis, with histochemical evidence using the redox reagent NBT of $O_2^{•}$ and H_2O_2 production at the apical cell membrane (Hobson *et al.*, 1991).

11.7.2 Fire smoke

It is a testament to the apparent paradoxes and puzzles of human behaviour that individuals who would naturally strive energetically to escape from the smoke of burning forest and building fires, may nevertheless voluntarily choose to directly expose their lungs to the known hazardous and carcinogenic effects of cigarette and cigar smoke!

Within the UK, following the implementation of the Clean Air Act in 1956 and subsequent changes in fuel use from coal to gas and electricity, the worst excesses of air pollution caused by domestic and industrial coal burning have been largely eliminated. However, in some other countries in Europe and the developing world, with less effective legislation or with other customary energy resources, the practise of burning coal, peat, wood and dried dung for fuel creates not only a general environmental atmospheric hazard, but also a particular major domestic respiratory hazard to people exposed within their own homes.

While differing in precise chemical composition and free radical reactivity, there are nevertheless common features in the smoke generated from all organic materials. The free radical moieties generated arise not only immediately in the combustion process *per se*, but also in the later chemical reactions that occur in the smoke phase itself (Pryor, 1992). The dangers of smoke inhalation, particularly to occupational fire-fighters, is not confined solely to the visible smoke. While longer lived (half-life of days) organic free radicals, collected on filter samples and detected by ESR, correlated with the atmospheric smoke concentration, short lived reactive free radicals measured by a chemiluminescence methodology, are detectable when no smoke can be seen, thus representing a particular danger to the unsuspecting in the locality (Jankovic *et al.*, 1993).

The recent environmental catastrophe in south east Asia, particularly Indonesia and Malaysia, involving the massive areas of forest fires burning out of control and worsened by the drought conditions caused by the El Niño system oscillations of the Pacific Ocean warm currents, is a matter of great human and, indeed, global concern (World Wide Web, 1997). The acute effects on health of 'The Haze', the dense smoke consisting of fine and ultrafine organic particles which covered thousands of square miles and persisted for several months during the autumn of 1997, have been major and extensive, with pollution levels exceeding those of the notorious London 1952 smog. However, the delayed and chronic effects of such a disastrous conflagration and smoke episode on human lungs and health, together with the Earth's ecosystem and atmosphere, have yet to be realised. The spectre of a public health epidemic of particle-induced and oxyradical-related respiratory diseases is a real one, demanding urgent research and remedial protective medical action.

11.7.3 Geogenic ash

In addition to anthropogenic sources, the Earth itself, by volcanic activity as exemplified by the Mount St Helens eruption, expells massive amounts of gas, smoke and ash into the atmosphere. Particular eruptions, such as Mount Pinatubo, can even project vast amounts of particulates high into the stratosphere enabling it to be rapidly dispersed over the whole world. The continual volcanic activity of Mount Sakurajima in Kyushu island, Japan, is a specific hazard to the neighbouring citizens, although epidemiological and toxicity investigations have indicated that the particular dust generated is fortunately not especially toxic (Yano *et al.*, 1985). Airborne particulates generated from both natural and anthropogenic sources, namely power plants, incinerators, refining industry, biomass, vehicles, etc., all contribute to the total global and regional load of toxic trace-element airborne particles (Schroeder *et al.*, 1987).

11.8 Traffic air pollution

Particulate air pollution, once primarily a particular risk of occupationally dust-exposed workers, has now exploded into the mainstream as an ubiquitous and principal ambient environmental hazard to society as a whole. An increasing number of epidemiological investigations, in the USA (Schwartz and Dockery, 1992; Dockery *et al.*, 1993), in Sao Paulo, Brazil (Saldiva *et al.*, 1995), and in Birmingham, UK, (Wordley *et al.*, 1996), have produced evidence implicating the airborne micro-particulate PM_{10} fraction (particles smaller than 10 microns) as causing increased lung cancer mortality, cardiopulmonary disease, elderly mortality and hospital admissions in exposed urban populations. The dose response was almost linear with no evidence of a 'safe' threshold limit. The capacity of smaller particles (about 1 micron) to penetrate deep into the lung renders them directly available to the alveoli, and provides the basis for the selective monitoring of the smaller $PM_{2.5}$ fraction. The capacity for the ultrafine diesel particles to form chain-like aggregates and larger agglomerates adds additional complexity to the measurement and health assessment of urban particulates (BéruBé *et al.*, 1997). Intratracheal instillation of well-characterised fine and ultrafine particles have shown that the contribution which particle size and surface chemistry makes to epithelial inflammatory changes is a complex one (Murphy *et al.*, 1998). The increasing generation of microparticles by vehicular emissions, particularly by diesels, is now considered to be a major pollution issue (Pope *et al.*, 1995) demanding radical counteractive measures.

A series of reports covering a range of air pollutants by various expert scientific and medical panels have been published which contain scientific assessments and advice for action on air quality standards. Of these, there have been several produced by the Advisory Groups and Committees of the Department of Health (DH, 1992a; 1995a) and the Expert Panels of the Department of the Environment (DE, 1993b; 1995; 1996a), which relate specifically to the role of particulates in causing repiratory disorders. Within the context of the present chapter concerning the effect of airborne free-radical hazards, the risks associated with the direct and indirect synergistic interactions of gaseous oxidants, namely oxides of nitrogen and sulphur, and photochemical ozone oxidants, have also been the topic of UK government reports (DH, 1992b; DE, 1993a). Strategies for implementing air quality standards have been published by the World Health Organisation (WHO, 1987) and the UK government (DE, 1996b).

11.8.1 Oxidant reactions

The generation of free-radical oxidant diesel exhaust products (DEP) by car engines, and their reaction with important biomolecules, namely ascorbic acid, cysteine, glutathione and serum albumin, is illustrative of their potential toxic effect. However, it is of interest to note that these reactions are decreased when using a three-way catalytic converter (Blaurock *et al.*, 1992). The ability of PM_{10} particles to promote free-radical oxidant mediated scission of supercoiled DNA, together with its partial inhibition by mannitol and the iron chelator desferrioxamine, indicates that a Fenton reaction production of $OH^{\bullet}$ is invoved in its toxic action. The enhanced release of particulate iron at pH 4.6, the equivalent acidity of phagolysosomes, further implicates the macrophage in the pathogenic oxidative response (Gilmour *et al.*, 1996). Subsequent investigation, involving centrifugation of the PM_{10} sample, has shown that most of the free-radical activity is associated with the ultrafine (sub 0.01 micron) fraction (Gilmour *et al.*, 1997a). Although a small proportionate fraction on a weight basis, such ultrafine particles contribute massively to the total particulate number and surface area. Methanol extracts of diesel exhaust particles, probably quinoid or nitroaromatic compounds, have been shown to generate $O_2^{-\bullet}$ and $OH^\bullet$ in the presence of P450 reductase (Kumagai *et al.*, 1997).

However, exposure of airborne urban particulates to purified human blood monocytes stimulated *in vitro* with soluble phorbol 12-myristate 13-acetate (PMA) or Zymosan particles, as measured by SOD inhibitable reduction of cytochrome C, caused a decrease in the respiratory burst, due to a proposed toxic effect on the cell membrane (Fabiani *et al.*, 1997). By comparison, *in vitro* exposure of AM to urban air particles produced a small increase in chemiluminescence but with significant increases in TNF and IL-6 cytokine secretion (Becker *et al.*, 1996). It has been hypothesised that the fine and ultrafine particles exert their specific toxic effect by the induction of an inflammatory cellular response, involving the generation of cytokines and haemostatic mediators, thus leading to cardiovascular and cerebrovascular disorders (Seaton *et al.*, 1995).

11.8.2 In vivo *investigations*

In vivo experiments, in which PM_{10} particles have been instilled via the trachea into rats, have shown increased oxidative stress as demonstrated in altered BAL glutathione redox status and increased production of BAL leukocyte NO generation. Support for the hypothesis that the effect was related to the ultrafine particle size PM_{10} fraction, was indicated by the analogous exacerbated inflammatory changes induced by a sample of ultrafine as compared to a fine carbon-black particle suspension (Li *et al.*, 1997). The capacity of DEP to induce pulmonary tumours in experimental rats has been related to the increase in oxidative DNA damage and 8-hydroxydeoxyguanosine formation in mouse lung tissue, a process exacerbated by high intake of dietary fat (Nagashima *et al.*, 1995). Intratracheal instillation of DEP in rats increases NO in exhaled air with a concomitant decrease in CuZn-SOD and Mn-SOD in airway epithelium (Lim *et al.*, 1998).

11.8.3 Asthma incidence

The evident increase in the incidence of childhood asthma over recent years has raised questions about the role of atmospheric pollution. Epidemiological studies into the prevalence of childhood asthma have suggested links with traffic flow, although whether urban air causes the disease remains uncertain (Edwards *et al.*, 1994). Particular concerns about the adverse effect of traffic air pollution in contributing to the evident rise in asthma have prompted detailed evaluation (MRC, 1994; DH, 1995b). Presently available evidence indicates that, while urban air pollution can exacerbate asthmatic symptoms, indoor allergic pollutants such as house mite dust may be more directly implicated in the aetiology of the disease.

The findings that asthmatic subjects display an increase in the oxidative respiratory burst of their circulating eosinophils (Sedgewick *et al.*, 1990) and BAL cells of asthmatic patients have reduced SOD activity levels (Smith *et al.*, 1997), demonstrate a potential role for ROS in the pathogenesis of the disease. The additional findings that histamine enhances $O_2^{-\bullet}$ generation by particle stimulated macrophages (Diaz *et al.*, 1979) and that H_2O_2 diminishes the β-adrenergic response of guinea pig trachea (Cornelis *et al.*, 1990), provides additional evidence of oxyradical involvement.

Studies in groups of humans variably exposed to air pollutants in Mexico City showed that blood serum oxidant markers, SOD and lipid peroxidation were related to duration of human exposure (Hicks *et al.*, 1996). However, blood plasma antioxidant levels, as measured by the Trolux equivalent antioxidant capacity (TEAC), are decreased in both chronic and acute asthmatic subjects (Rahman *et al.*, 1996).

11.9 Nutritional and physiological antioxidants

There is a growing body of evidence, epidemiological, clinical and experimental, that the involvement of dietary antioxidants in respiratory function and the prevention of pulmonary disorders, pneumoconiotic, fibrotic, emphysematous and cancerous diseases, is a significant one. The lung as an organ is uniquely vulnerable to the environmental effect of oxidant mediated damage, being both directly, intimately and extensively exposed to gaseous and particulate toxins. It is fortuitous that the lung has accordingly evolved and developed a wide range of physiological antioxidant mechanisms to combat the injurious effect of pollutant particulates and oxyradicals, which are dependent directly or indirectly on the ready availability of antioxidant or pro-antioxidant nutrients (White and Repine, 1985; Evans *et al.*, 1987).

Physical protection is provided initially by the entrapment of inhaled large particles by the nasal hairs and the elimination of small particles, which are deposited within the distal respiratory tract, by AM phagocytosis and subsequent transportation to the pharynx, via the tracheal ciliary mechanism, where they are either expectorated or ingested. The presence of the antioxidant uric acid in nasal fluid has been proposed to provide initial protection against environmental oxidative stress (Housley *et al.*, 1996). Within the tracheobronchial mucous, the extracellular glycoproteins (Cross *et al.*, 1984) and the lung surfactant alveolar fluid catalase (Cantin *et al.*, 1990), provide a secondary defence mechanism to oxidative injury. The antioxidant enzymes catalase and SOD reduce asbestos-induced mutagenicity to human–hamster hybrid cells in culture (Khei *et al.*, 1995). The sequestration of tissue 'catalytic' iron by ferritin protects vascular epithelia against the toxic effects of Fenton oxyradical generation (Balla *et al.*, 1992) and is one of several mechanisms by which tissues are protected against oxidative stress (Halliwell and Gutteridge, 1990).

Epidemiological studies of the general population have revealed that dietary vitamin C is directly correlated with lung function as measured by the forced expiratory volume (FEV1) (Schwartz and Weiss, 1994), whereas there was no observed additional independent effect of the lipid soluble vitamin E intake (Britton *et al.*, 1995). *In vitro* studies have demonstrated that ascorbate protects against asbestos-induced lipid peroxidation and reduced glutathione-S-transferase in lung microsomes (Khan *et al.*, 1990). In experimental investigations, in addition to its free radical scavenging role, vitamin E inhibits the phagocytic respiratory burst by its action on protein kinase C (Sakamoto *et al.*, 1990). It is of interest to note that hyperoxia-induced lung damage was more effectively ameliorated by zinc than vitamin E in zinc-deficient and energy-restricted rats, an effect apparently unrelated to CuZn–SOD or metallothionein activity, and possibly dependent on thiol group protection (Taylor *et al.*, 1997). Additionally, other micronutrients may exhibit antioxidant functions. Selenium, via its cofactor role in glutathione peroxidase, and thiol amino acids protect against oxygen toxicity (Forman *et al.*, 1983), and non-free radical scavenging vitamins, nicotinamide (Nadeau *et al.*, 1989) and retinol (Mossman *et al.*, 1980), inhibit mineral dust cytotoxicity to pulmonary macrophages and tracheal epithelia.

Blood plasma exposed to cigarette smoke-borne free radicals, which include NO, is depleted in a number of antioxidants including ascorbate, α-tocopherol, β-carotene and several other carotenoids (Eiserich *et al.*, 1995). The adequacy of the actual dietary intake, and indeed the formal reference dietary values, of vitamin C, to enable the lung to effectively cope with the additional burden of oxidant toxins generated by cigarette smoke, and by the general populace exposed to community atmospheric pollutants, has been questioned (Menzel *et al.*, 1992). The decreases in blood plasma Trolox equivalent antioxidant capacity (TEAC) and protein sulphydryl levels found in chronic obstructive pulmonary disease, and the related role of antioxidant therapy in the treatment of smoking-induced pulmonary diseases, are findings and matters of current concern (Rahman and McNee, 1996).

The issue is especially relevant to developing children, and indeed particularly pertinent to Scotland where the high national prevalence of smoking coincides with a low dietary intake of fruit and vegetables. A doubling of fruit and vegetable consumption has been recommended as the single most important dietary target to improve Scottish public health (Scottish Office, 1996). At the other end of the life span spectrum, the capacity of maintaining an effective redox homeostatic balance with advancing age may be an important factor in modulating the injurious oxidant stress of cigarette smokers in the elderly (Kondo *et al.*, 1994).

The dietary and experimental evidence of the potential beneficial effect of antioxidants provided the rationale for two major intervention trials using vitamin supplements: the Alpha-Tocopherol Beta-Carotene Cancer Prevention Study Group (ATBCCPSG, 1994) and the Beta-Carotene and Retinol Efficacy Trial (CARET) (Omenn *et al.*, 1996). However, the unexpected finding of an increase in the risk of lung cancer in smokers has raised critical questions and cautionary attitudes. The precise antioxidant composition and the relatively high dosage used may not have been optimal; the possibility of adverse direct reactions with smoke components and the potential for actually enhancing the promotion stage of precancerous lung lesions have been posited as plausible explanations of the apparently paradoxical results.

11.10 Pharmacological antioxidants

The evidence linking the value of physiological antioxidants in alleviating the effects of exacerbated oxidant stress induced by various environmental pollutants, has prompted complementary studies into the potential efficacy of analogous pharmacological agents. It is of interest to note that a proposed mechanism by which the agent polyvinylpyridine-*N*-oxide (PVPNO) exerts its long established anti-silicotic effect is as a free radical scavenger (Gulumian and Van Wyk, 1987).

Thiol drugs have a long pedigree in the prevention and treatment of free-radical related tissue injury, particularly in relation to their role in radioprotection. The broncholytic agent *N*-acetyl cysteine (NAC), a precursor of the major physiological tripeptide thiol antioxidant glutathione, prevents oxidant-mediated toxicity in AM when exposed to oxidant gases, tobacco smoke and silica in culture (Voisin *et al.*, 1987), and similaly for PMN-induced cytotoxicity to co-cultured lung epithelial cells (Simon and Suttorp, 1983). Clinical use of NAC in silicotic patients has been reported to be of benefit (Margolis and Margolis, 1974), despite findings indicating that NAC treatment in patients with chronic obstructive pulmonary disease do not produce a sustained increase in glutathione levels in BAL fluid, although blood plasma cysteine concentrations were raised (Bridgeman *et al.*, 1994). However, alternative thiol agents, namely, oxathiazolidine-4-carboxylate (Taylor *et al.*, 1992), erdosteine (Ciaccia *et al.*, 1992), mercaptoproprionylglycine (Ayenne *et al.*, 1993), and *N*-acetylcysteinyl-lysine (NAL) (Gillissen *et al.*, 1997), may offer more efficacious antioxidant activity. Delivery by aerosol directly to the lung, rather than by the oral route, can further enhance the thiol agent's effectiveness (Buhl *et al.*, 1990). Whether thiol drugs act via a direct scavenging mechanism, are dependent on conversion to glutathione, or directly inhibit PMN oxyradical generation as demonstrated by the effect of *N*-5-thioxo-L-propyl-1-cysteine (Meloni *et al.*, 1994), remains to be clarified.

Non-thiol antioxidant agents may also prove of value. Pentoxifylline has been shown to reduce PMN-induced tissue injury in perfused lungs (McDonald, 1991). The Lazeroid 21-aminosteroid drug inhibits diesel smoke-induced AM oxyradical generation (Wang *et al.*, 1997). Iron chelators, namely desferrioxamine, which inhibits crocidolite-induced cell toxicity (Mossman and Marsh, 1989), 2,3-dihydroxybenzoate, which inhibits PMN oxyradical production (Boxer *et al.*, 1978), and phytic acid, which reduced amosite-induced PMN lung influx and fibrosis in exposed experimental animals (Kamp *et al.*, 1995), also offer some therapeutic promise. The development of new clinically safe and effective iron chelating drugs is an avenue of special interest in the potential treatment of a large number of oxyradical-mediated diseases and disorders (Hoffbrand and Wonke, 1997).

11.11 Conclusion

Recent decades have been marked by a heightened scientific, medical and public awareness of the real and potential problems associated with air pollution and health. While more stringent legislative policies, enforcement controls, and technical facilities and expertise, may have ensured that the worst of the smog episodes have evidently passed, the nature and extent of air pollution has changed rather than been eliminated. In the so-called developed world, the major air pollution problems arise largely from traffic emissions, rather than from the smoke stacks of the earlier industrial era. In the developing world, the accelerating increase in largely uncontrolled vehicular emissions, the rapid industrial development and associated burning of fossil fuel, and the major but largely neglected problem of indoor household respiratory disease linked to the biomass burning of wood, peat and sun-dried dung, pose massive health problems and intractable economic issues.

An examination of the causal mechanisms of particulate air pollution requires the adoption of a 'twin track' strategy approach. Namely, the prevention of particulate production and human exposure, complemented by the adequate supply of dietary antioxidants and essential micronutrient vitamins and trace elements, primarly present in fresh fruits and vegetables. Public health concerns for the dietary availability and cost of fresh fruit and vegetables, especially for deprived and vulnerable society groups, have been voiced. In this context, the potential preventative role of supplementary nutritional antioxidants is highlighted by concerns of the adequacy of the reference dietary values to combat community exposure to elevated environmental oxidative stress. Awareness of the value of pharmacological antioxidant interventions in the treatment of pulmonary disease is growing, and provides an important medical adjunct to the management of diseases related to air pollution.

The public health challenge to prevent human exposure to airborne pollution, particularly the injurious microparticulate toxins, is matched by the enormous political awareness, wisdom and determination that will be required to resolve the present difficulties and predicted future disasters. The efficient and healthy management of traffic and associated vehicular exhaust is currently a contentious topic on the political agenda throughout the world. In addition, as recently illustrated on a catastrophic scale in Australasia, the wilful and accidental forest burning not only massively generates hazardous airborne particulates and gaseous free radicals, but also simultaneously contributes – directly by production of the greenhouse gas carbon dioxide and indirectly by loss of restorative carbon dioxide 'sink' greenery – to incipient adverse climatic changes. The global consequence for public health related to current and unresticted PM_{10} production in the future has been predicted to exceed half a million avoidable deaths annually (Working Group on Public Health and Fossil-Fuel Combustion, 1997).

The massive endeavours to reach local and national resolutions of the problem are reflected and magnified by the international conflicts which emerged at the United Nations Framework Convention on Climate Change held in Kyoto, Japan. The partial success in reaching agreement on targets to reduce the world's atmospheric pollution over the coming decades, and to establish reliable monitoring and enforcement procedures, exemplifies the major public health issues and political problems that await resolution in the new millennium.

References

Alpha-Tocopherol Beta-Carotene Cancer Prevention Study Group. The effect of vitamin E and beta carotene on the incidence of lung cancer and other cancers in male smokers. N Engl J Med 1994;330:1029–1035.

Archer, V.E., Dixon, W.C. Carcinogenicity of fibers and films: a theory. Med Hypoth 1979;5:1257–1262.

Ayene, I.S., Al-Mehdi, A.B., Fisher, A.B. Inhibition of lung tissue oxidation during ischemia/reperfusion by 2-mercaptopropionyl glycine. Arch Biochem Biophys 1993;303:307–312.

Babior, B.M. Oxidants from phagocytes: agents of defence and destruction. Blood 1984;64:959–966.

Balla, G., Jacob, H.S., Balla, J., Rosenberg, M., Nath, K., Apple, F., Eaton, J.W., Vercellotti G.M. Ferritin: a cytoprotective antioxidant strategem of endothelium. J Biol Chem 1992;267:18148–18153.

Baris, Y.I., Simonato, L., Saracci, R., Skidmore, J.W., Artvinli, M. Malignant mesothelioma and radiological chest abnormalities in two villages in central Turkey. Lancet 1981;i:984–987.

Becker, S., Soukop, J.M., Gilmour, M.I., Devlin, R.B. Stimulation of human and rat alveolar macrophages by urban air particulates: effects on oxidant radical generation and cytokine production. Toxicol Appl Pharmacol 1996;141:637–648.

BéruBé, K.A., Jones, T.P., Williamson, B.J. Electron microscopy of urban airborne particulate matter. Micros Analysis 1997;61:11–13.

Biemond, P., Van Eijk, H.G., Swaak, A.J.G., Koster, J.F. Iron mobilization from ferritin by superoxide derived from stimulated polymorphonuclear leukocytes. Possible mechanism in inflammation diseases. J Clin Invest 1984;73:1576–1579.

Blaurock, B., Hippeli, S., Metz, N., Elstner, E.F. Oxidative destruction of biomolecules by gasoline engine exhaust products and detoxifying effects of three-way catalytic converters. Arch Toxicol 1992;66:681–687.

Boxer, L.A., Allen, J.M., Baehner, R.L. Potentiation of polymorphonuclear leukocyte motile functions by 2,3-dihydroxybenzoic acid. J Lab Clin Med 1978;92:730–736.

Bridgeman, M.M.E., Marsden, M., Selby, C., Morrison, D., MacNee, W. Effect of N-acetyl cysteine on the concentration of thiols in plasma, brochoalveolar lavage fluid, and lung tissue. Thorax 1994;49:670–675.

Britton, J.R., Pavord, I.D., Richards, K.A., Knox, A.J., Wisniewski, J.F., Lewis, S.A., Tattersfied, A.E., Weiss, S.T. Dietary antioxidant vitamin intake and lung function in the general population. Amer J Resp Crit Care Med 1995;15:1383–1387.

Buhl, R., Vogelmeier, C., Critenden, M., Hubbard, R.C., Hoyt, R.F., Wilson, E.M., Cantin, A.M., Crystal, R.G. Augmentation of glutathione in the fluid lining the epithelium of the lower respiratory tract by directly administering glutathione aerosol. Proc Natl Acad Sci 1990;87:4063–4067.

Byczkowski, J.Z., Gessner, T. Asbestos-catalysed oxidation of benzo(a)pyrene by superoxide-peroxidised microsomes. Bull Environ Contam Toxicol 1987;39:312–317.

Cantin, A.M., Fells, G.A., Hubbard, R.C., Crystal, R.G. Antioxidant macromolecules in the epithelial lining fluid of the normal human lower respiratory tract. J Clin Invest 1990;86:962–971.

Carp, H., Miller, F., Hoidal, J.R., Janoff, A. Potential mechanism of emphysema: alpha1-proteinase inhibitor recovered from lungs of cigarette smokers contains oxidised methionine and has decreased elastase inhibitory capacity. Proc Natl Acad Sci USA 1982;79:2041–2045.

Chow, C.K., Rylander, R., Pryor, W. Cigarette smoking and oxidative damage in the lung. Ann NY Acad Sci 1993;686:289–298.

Ciaccia, A., Papi, A., Tschirky, B., Fregnan, B. Protection of erdosteine on smoke-induced peripheral neutrophil dysfunction both in healthy and in bronchitic smokers. Fundam Clin Pharmacol 1992;6:375–382.

Connor, J.R., Snyder, B., Stauffer, J.L., Gaydos, L.J., Demers, L.M., Kuhn, D.C. Mineral dust promotes alterations in ferritin isoforms and oxidative damage to proteins in rat and human alveolar macrophages. Appl Occup Environ Hyg 1996;11:969–972.

Cornelis, J.A., Doelman, R.L., Oosterom, W.C., Bast, A. Mineral dust exposure and free radical-mediated lung damage. Exp Lung Res 1990;16:41–55.

Cross, C.E., Halliwell, B., Allen, A. Antioxidant protection: a function of tracheobronchial and gastrointestinal mucus. Lancet 1984;i:1328–1330.

Dalal, N.S., Suryan, M.M., Vallyathan, V., Green, F.H.Y., Jafari, B., Wheeler, R. Detection of free radicals in fresh coal mine dust and their implication for pulmonary injury. Ann Occup Hyg 1989;33:79–84.

Dalal, N.S., Newman, J., Pack, D., Leonard, S., Vallyathan, V. Hydroxyl generation by coal mine dust: possible implication to coal workers' pneumoconioisis (CWP). Free Rad Biol Med 1995;18:11–20.

Daniel, L.N., Mao, Y., Saffiotti, U. Oxidative DNA damage by crystalline silica. Free Rad Biol Med 1993;14:463–472.

Daniel, L.N., Mao, Y., Wang, T.-C.L., Markey, C.J., Markey, S.P., Shi, X., Saffiotti, U. DNA strand breakage, thymine glycol production, and hydroxyl radical generation induced by different samples of crystalline silica in vitro. Environ Res 1996;71:60–73.

Department of the Environment. Photochemical Oxidants Review Group. Ozone in the United Kingdom. Third report. London. HMSO. 1993a.

Department of the Environment. Expert Panel on Air Quality Standards. Diesel vehicle emissions and urban air quality. Second report. London. HMSO. 1993b.

Department of the Environment. Expert Panel on Air Quality Standards. Particles. London. HMSO. 1995.

Department of the Environment. Quality of Urban Air Review Group. Airborne particulate matter in the United Kingdom London. HMSO. 1996a.

Department of the Environment. The United Kingdom national air quality strategy. London. HMSO. 1996b.

Department of Health. Advisory Group on the Medical Aspects of Air Pollution Episodes. Sulphur dioxide, acid aerosols and particulates. Second Report. London. HMSO. 1992a.

Department of Health. Advisory Group on the Medical Aspects of Air Pollution Episodes. Oxides of nitrogen. Third report. London. HMSO. 1992b.

Department of Health. Advisory Group on the Medical Aspects of Air Pollution. Non-biological particles and health. London. HMSO. 1995a.

Department of Health. Advisory Group on the Medical Aspects of Air Pollution. Asthma and outdoor air pollution. London. HMSO. 1995b.

Diaz, P., Jones, D.G., Kay, A.B. Histamine-coated particles generate superoxide and chemiluminescence. Nature 1979;278:454–456.

Dixon, J.R., Lowe, D.B., Richards, D.E., Cralley, L.J., Stokinger, H.E. The role of trace metals in chemical carcinogenesis: asbestos cancers. Cancer Res 1970;30:1068–1074.

Dockery, D.W., Pope, C.A., Xu, X., Spengler, J.D., Ware, J.H., Fay, M.E., Ferris, B.G., Speizer, F.E. An association betwen air pollution and mortality in six U.S. cities. N Eng J Med 1993;329:1753–1759.

Doll, N.J., Stankus, R.P., Goldbach, S., Salvaggio, J.E. In vitro effect of asbestos fibers on polymorphonuclear leukocyte function. Int Arch Allergy Appl Immunol 1982;68:17–21.

Doll, R. Symposium on MMMF. Overview and conclusions. Ann Occup Hyg 1987;31:805–819.

Donaldson, J., Golyasnya, N. Cytogenic and pathogenic effects of long and short amosite asbestos. J Pathol 1995;177:303–307.

Donaldson, K., Hill, I.M., Beswick, P.H. Superoxide anion release by alveolar macrophages exposed to respirable industrial fibres: modifying effect of fibre opsonisation. Exp Toxicol Pathol 1995;47:229–231.

Donaldson, K., Gilmour, P., Beswick, P.H., Fisher, C., Brown, D.M. Effect of the pH and biochemical milieu of the lung on release of iron from the surface of respirable industrial fibres and ability to cause a respiratory burst in alveolar macrophages. Ann Occup Hyg 1997;41(S1):334–339.

Durnev, A.D., Suslova, T.B., Cheremisina, Z.P., Dubovskaia, O., Nigarova, E.A., Korkina, L.G., Seredenin, S.B., Velichovskii, B.T. The mutagenic action of of natural zeolites and chrysotile asbestos. Eksp Onkol 1990;12:21–24.

Edwards, J., Walters, S., Griffiths, R.K. Hospital admissions for asthma in preschool children: relationship to major roads in Birmingham, United Kingdom. Arch Env Health 1994;49:223–227.

Eiserich, J.P., Van der Vliet, A., Handelman, D.J., Halliwell, B., Cross, C.E. Dietary antioxidants and cigarette smoke-induced biomolecular damage: a complex interaction. Amer J Clin Nutr 1995;62:1490–1500.

Engelen, J.J., Borm, P.J., Van Sprudel, M., Leenaerts, L. Blood anti-oxidant parameters at different stages of pneumoconiosis in coal workers. Env Health Perspect 1990;84:165–172.

Evans, P.H. The conioses: organotypic inflammatory oxidative responses to environmental particulate pollutants. In 'Immunopharmacology of free radical species' Eds Blake, D., Winyard, P.G. Academic Press, London (1995) pp 247–262.

Evans, P.H., Campbell, A.K., Yano, E., Goodman, B. Phagocytic oxidant stress and antioxidant interactions in the pneumoconioses and dust-induced tumourigenic lung disease. In 'Free radicals, oxidant stress and drug action' Ed. Rice-Evans, C. Richelieu, London (1987) pp 213–235.

Evans, P.H., Klinowski, J., Yano, E., Urano, N. Alzheimer's disease: a pathogenic role for aluminosilicate-induced phagocytic free radicals. Free Rad Res Comm 1989;6:317–321.

Evans, P.H., Morgan, L.G., Yano, E., Urano, N. Chemiluminescent detection of free radical generation by stimulated polymorphonuclear leukocytes: in vitro effect of nickel compounds. In 'Nickel and human health: current perspectives' Eds Nieboer, E., Nriagu, J.O. Wiley, New York (1992a) pp 363–373.

Evans, P.H., Peterhans, E., Bürge, T., Klinowski, J. Aluminosilicate-induced free radical generation by murine brain glial cells in vitro: potential significance in the etiopathogenesis of Alzheimer's dementia. Dementia 1992b;3:1–6.

Evans, P.H., Peterhans, E., Bürge, T., Klinowski, J., Yano, E. Senile neurodegeneration: pathogenic role of microglia-derived free radicals. In 'Oxidative stress and aging' Eds Cutler, R.G., Packer, L., Bertram, J., Mori, A. Birkhäuser, Basel (1995) pp 319–330.

Fabiani, R., Pampanella, L., Minelli, A., Mezzasoma, I., Vecchiarelli, A., Morozzi, G. Effect of airborne particulate extracts on monocyte oxidative metabolism. J Environ Pathol Toxicol Oncol 1997;16:195–199.

Forman, H.J., Rotman, E., Fisher, A.B. Roles of selenium and sulfur-containing amino acids in protection against oxygen toxicity. Lab Invest 1983;49:148–153.

Fubini, B., Giamello, E., Volante, M., Bolis, V. Chemical functionalities at the silica surface determining its reactivity when inhaled. Formation and reactivity of surface radicals. Toxicol Indust Health 1990;6:571–598.

Fubini, B., Mollo, L. Role of iron in the reactivity of mineral fibres. Toxicol Lett 1995;82/83:951–960.

Gabor, S., Anca, Z. Zugravu, E. In vitro action of quartz on alveolar macrophage lipid peroxides. Arch Environ Health 1975;30:499–501.

Gabrielson, E.W., Rosen, G.M., Grafstrom, R.C., Strauss, K.E., Harris, C.C. Studies on the role of oxygen radicals in asbestos-induced cytopathology of cultured human lung mesothelial cells. Carcinogenesis 1986;7:1161–1164.

Gerde, P., Scholander, P. Adsorption of benzo(a)pyrene on to asbestos and manmade mineral fibres in an aqueous solution and in a biological model solution. Br J Indust Med 1988;45:682–688.

Ghio, A.J., Quigley, D.R. Complexation of iron by humic-like substances in lung tissue: role in coalworkers' pneumoconioisis. Amer J Physiol 1994;267:173–179.

Ghio, A.J., Stonehuerner, J., Steele, M.P., Crumblis, A.L. Phagocyte-generated superoxide reduces Fe^{3+} to displace it from the surface of asbestos. Arch Biochem Biophys 1994;315:219–225.

Gillissen, A., Jaworska, M., Orth, M., Coffiner, M., Maes, P., App, E.M., Cantin, A.M., Schultze-Werninghaus, G. Nacystelyn, a novel lysine salt of N-acetylcysteine, to augment cellular antioxidant defence in vitro. Resp Med 1997;91:159–168.

Gilmour, P.S., Beswick, P.H., Brown, D.M., Donaldson, K. Detection of surface free radical activity of respirable industrial fibres using supercoiled Ø X174 RF1 plasmid DNA. Carcinogenesis 1995;16:2973–2979.

Gilmour, P.S., Brown, D.M., Beswick, P.H., Benton, E., MacNee, W., Donaldson, K. Surface free radical activity of PM_{10} and ultrafine titanium dioxide: a unifying factor in their toxicity? Ann Occup Hyg 1997a;41(S1):32–38.

Gilmour, P.S., Brown, D.M., Beswick, P.H., MacNee, W., Rahman, I., Donaldson, K. Free radical activity of industrial fibers: role of iron in oxidative stress and activation of transcription factors. Environ Health Perspect 1997b;105(S5):1313–1317.

Gilmour, P.S., Brown, D.M., Lindsay, T.G., Beswick, P.H., MacNee, W., Donaldson, K. Adverse health effects of PM_{10} particles: involvement of iron in generation of hydroxyl radical. Occup Environ Med 1996;53:817–822.

Gormley, I.P., Kowolik, M.J., Cullen, R.T. The chemiluminescent response of human phagocytic cells to mineral dusts. Br J Exp Path 1985;66:409–416.

Gossart, S., Cambon, C., Orfila, C., Seguelas, M.-H., Lepert, J.-C., Rami, J., Carre, P., Pipy, B. Reactive oxygen intermediates as regulators of TNF-alpha production in rat lung inflammation induced by silica. J Immunol 1996;156:1540–1548.

Graceffa, P., Weitzman, S.A. Asbestos catalyses the formation of 6-oxobenzo(a)pyrene radical from 6-hydroxybenzo(a)pyrene. Arch Biochem Biophys 1987;257:481–484.

Gulumian, M., Van Wyk, A. Free radical scavenging properties of polyvinylpyridine-N-oxide: a possible mechanism for its action in pneumoconiosis. Med del Lavoro 1987;78:124–128.

Gusev, V.A., Danislovskaja, Y.V., Vatolkina, O.Y., Lomonosova, O.S., Velichkovsky, B.T. Effect of quartz and alumina dust on generation of superoxide radicals and hydrogen peroxide by alveolar macrophages, granulocytes, and monocytes. Br J Indust Med 1993;50:732–735.

Halliwell, B., Gutteridge, J.M.C. The antioxidants of human extracellular fluids. Arch Biochem Biophys 1990;280:1–8.

Hansen, K., Mossman, B.T. Generation of superoxide ($O_2^{-\bullet}$) from alveolar macrophages exposed to asbestiform and nonfibrous particles. Cancer Res 1987;47:1681–1686.

Harvey, G., Pagé, M., Dumas, L. Binding of environmental carcinogens to asbestos and mineral fibres. Br J Indust Med 1984;41:396–400.

Hatch, G.E., Gardner, D.E., Menzel, D.B. Stimulation of oxidant production in alveolar macrophages by pollutant and latex particles. Environ Res 1980;23:121–136.

Hicks, J.J., Medina-Navarro, R., Guzman-Grenfell, A., Wacher, N., Lifshitz, A. Possible effects of air pollutants (Mexico City) on superoxide dismutase activity and serum lipid peroxides in the human adults. Arch Med Res 1996;27:145–149.

Hobson, J., Wright, J., Churg, A. Histochemical evidence for generation of active oxygen species on the apical surface of cigarette-smoke exposed tracheal explants. Amer J Pathol 1991;139:573–580.

Hoffbrand, A.V., Wonke, B. Iron chelation therapy. J Intern Med 1997;242:37–41.

Housley, D.G., Eccles, R., Richards, R.J. Gender difference in the concentration of the antioxidant uric acid in human nasal lavage. Acta Oto-Laryng 1996;116:751–754.

Ishizaki, T., Yano, E., Urano, N., Evans, P.H. Crocidolite-induced reactive oxygen metabolite generation from human poymorphonuclear leukocytes. Environ Res 1994;66:208–216.

Jackson, J.H., Schraufstatter, I.U., Hyslop, P.A., Vosbeck, K., Sauerheber, R., Weitzman, S.A., Cochrane, C.G. Role of oxidants in DNA damage. J Clin Invest 1987;80:1090–1095.

Jankovic, J., Jones, W., Castranova, V., Dalal, N. Measurement of short-lived reactive species and long-lived free radicals in air samples from stuctural fires. Appl Occup Environ Hyg 1993;8:650–654.

Janssen, Y.M.W., Marsh, J.P., Driscoll, K.E., Borm, P.J.A, Obersdorster, G., Mossman, B.T. Increased expression of manganese-containing superoxide dismutase in rat lungs after inhalation of inflammatory and fibrogenic minerals. Free Rad Biol Med 1994;16:315–322.

Jaurand, M.C., Kaplan, H., Thiollet, J., Pinchon, M.C., Bernaudin, J.F., Bignon, J. Phagocytosis of chrysotile fibers by pleural mesothelial cells in culture. Am J Pathol 1979;84:529–538.

Johnson, N.F., Davies, R. An ultrastructural study of the effect of asbestos fibres on cultured peritoneal macrophages. Br J Exp Path 1981;62:559–570.

Kamp, D.W., Graceffa, P., Pryor, W.A., Weitzman, S.A. The role of free radicals in asbestos-induced diseases. Free Rad Biol Med 1992;12:293–315.

Kamp, D.W., Israbian, V.A., Yeldandi, A.V., Panos, R.J., Graceffa, P., Weitzman, S.A. Phytic acid, an iron chelator, attenuates pulmonary inflammation and fibrosis in rats after intratracheal instillation of asbestos. Toxicol Pathol 1995;23:689–695.

Khan, S.G., Ali, S., Rahman, Q. Protective role of ascorbic acid against asbestos induced toxicity in rat lung: in vitro study. Drug Chem Toxicol 1990;13:249–256.

Khei, T.K., He, Z.Y., Suzuki, K. Effects of antioxidants on fiber mutagenesis. Carcinogenesis 1995;16:1573–1578.

Kinnula, V.L., Aalto, K., Raivio, K.O., Walles, S., Linnainmaa, K. Cytotoxicity of oxidants and asbestos fibers in cultured human mesothelial cells. Free Rad Biol Med 1994;16:169–176.

Kodama, M., Kaneko, M., Aida, M., Inoue, F., Nakayama, T., Akimoto, H. Free radical chemistry of cigarette smoke and its implications in human cancer. Anticancer Res 1997;17:433–437.

Kon, S.H. Biological autoxidation. 1. Decontrolled iron: an ultimate carcinogen and toxicant: An hypothesis. Med Hypoth 1978;4:445–471.
Kondo, T., Tagami, S., Yoshioka, A., Nishimura, M., Kawakami, Y. Current smoking in elderly men reduces antioxidants in alveolar macrophages. Amer J Resp Crit Care Med 1994;149:178–182.
Kuhn, D.C., Demers, L.M. Influence of mineral dust surface chemistry on eicosanoid production by alveolar macrophage. J Toxicol Environ Health 1992;35:39–50.
Kumagai, Y., Arimoto, T., Shinyashiki, M., Shimojo, N., Nakai, Y., Yoshikawa, T., Sagai, M. Generation of reactive oxygen species during interaction of diesel exhaust particle components with NADPH-cytochrome P450 reductase and involvement of the bioactivation in DNA damage. Free Rad Biol Med 1997;22:479–487.
Leanderson, P., Tagesson, C. Cigarette smoke potentiates the DNA-damaging effect of manmade mineral fibers. Amer J Indust Med 1989;16:697–706.
Li, X.Y., Gilmour, P.S., Donaldson, K., MacNee, W. *In vivo* and *in vitro* proinflammatory effects of particulate air pollution (PM_{10}). Environ Health Perspect 1997;105:1279–1283.
Lim, H.-B., Ichinose, T., Miyabara, Y., Takano, H., Kumagai, Y., Shimojyo, N., Devalia, J.L., Sagai, M. Involvement of superoxide and nitric oxide on airway inflammation and hyperresponsiveness induced by diesel exhaust particles in mice. Free Rad Biol Med 1998;25:635–644.
Lund, L.G., Aust, A.E. Iron mobilization from asbestos by chelators and ascorbic acid. Arch Biochem Biophys 1990;278:60–64.
Maly, E.R. Generation of free oxygen radicals from human mononuclear cells, treated with quartz DQ12 or coal dust TF1 – new aspects in pathogenesis of pneumoconiosis. Zentralbl Bakteriol Mikcrobiol Hyg 1988;187:142–165.
Marasas, L.W., Harington, J.S. Some oxidative and hydroxylative actions of quartz: their possible relationship to the development of silicosis. Nature 1960;188:1173–1174.
Margolis, J., Margolis, L.S. Silicosis – can the course of the disease be prevented? Chest 1974;66:107.
McDonald, R.J. Pentoxifylline reduces injury to isolated lungs perfused with human neutrophils. Am Rev Resp Dis 1991;144:1347–1350.
Medical Research Council. Institute for Environment and Health. Report on air pollution and health: understanding the uncertainties. (Report R1) 1994.
Meloni, F., Ballabio, P., Leo, G., Gorrini, M., Manzardo, S., Coppi, G., Luisetti, M. Interactions of P1507, a new antioxidant agent, with phagocyte functions. Agents Actions 1994; 43:24–28.
Menzel, D.B., Colby, F., Machlin, J. Antioxidant vitamins and prevention of lung disease. Annals NY Acad Sci 1992;669:141–155.
Morgan, A. In vivo evaluation of chemical biopersistence of man-made mineral fibres. Environ Health Perspect 1994;102(S5):127–131.
Mossman, B.T., Craighead, J.E., MacPherson, B.V. Asbestos-induced epithelial changes in organ cultures of hamster trachea: inhibition by retinyl methyl ether. Science 1980;207:311–313.
Mossman, B.T., Landesman, J.M. Importance of oxygen free radicals in asbestos-induced injury to airway epithelial cells. Chest 1983;83(S5):50–51.
Mossman, B.T., Marsh, J.P. Evidence supporting a role of active oxygen species in asbestos-induced toxicity and lung disease. Environ Health Perspect 1989;81:91–94.
Mossman, B.T., Surinrut, P., Brinton, B.T., Marsh, J.P., Heintz, N.H., Lindau-Shepard, B., Shaffer, J.B. Transfection of manganese-containing superoxide dismutase gene into hamster tracheal epithelial cells ameliorates asbestos-mediated cytotoxicity. Free Rad Biol Med 1996;21:125–131.
Murphy, S.A., BéruBé, K.A., Pooley, F.D., Richards, R.J. The response of lung epithelium to well characterised fine particles. Life Sciences 1998;62:1789–1799.
Nadeau, D., Lane, D.A., Paradis, D., Fouquette, L. Effects of nicotinamide on the cytotoxicity of mineral dusts towards pulmonary alveolar macrophages. In 'Effects of mineral dusts on cells' Eds Mossman, B.T., Begin, R.O. Springer-Verlag, Berlin (1989) pp 115–122.

Nadif, R., Bourgkard, E., Dusch, M., Bernadac, P., Bertrand, J.-P., Mur, J.-M., Pham, Q.-T. Relations between occupational exposure to coal mine dusts, erythrocyte catalase and Cu/Zn superoxide dismutase activities and the severity of coal workers' pneumoconiosis. Occup Environ Med. 1998;55:533–540.

Nagashima, M., Kasai, H., Yokota, J., Nagamachi, Y., Ichinose, T., Sagai, M. Formation of an oxidative DNA damage, 8-hydroxydeoxyguanosine, in mouse lung DNA after intratracheal instillation of diesel exhaust particles and effects of high dietary fat and beta-catotene on this process. Carcinogenesis 1995;16:1441–1445.

Newsam, J.M. The zeolite cage structure. Science 1986;231:1093–1099.

Omenn, G.S., Goodman, G.E., Thornquist, M.D., Balmes, J., Cullen, M.R., Glass, A., Keogh, J.P., Meyskens, F.L., Valanis, B., Williams, J.H., Barnhart, S., Hammar, S. The effects of a combination of alpha-tocopherol and beta-carotene and vitamin A on lung cancer and cardiovascular disease. N Engl J Med 1996;334:1150–1155.

Pelin, K., Husgafvel-Pursiainen, K., Vallas, M., Vanhala, E., Linnainmaa, K. Cytotoxicity and anaphase aberrations induced by mineral fibres in cultured human mesothelial cells. Toxicol in vitro 1992;6:445–450.

Peto, J., Hodgson, J.T., Mathews, F.E., Jones, J.R. Continuing increase in mesothelioma mortality in Britain. Lancet 1995;345:535–539.

Petruska, J.M., Leslie, K.O, Mossman, B.T. Enhanced lipid peroxidation in lung lavage of rats after inhalation of asbestos. Free Rad Biol Med 1991;11:425–432.

Pezerat, H., Guignard, J., Cherrie, J.W. Man-made mineral fibres and lung cancer: an hypothesis. Toxicol Indust Health 1992;8:77–87.

Pooley, F.D. Asbestos bodies, their formation, composition and character. Environ Res 1972;5:363–379.

Pope, C.A., Bates, D.V., Raizenne, M.E. Health effects of particulate pollution: time for reassessment? Environ Health Perspect 1995;103:472–480.

Pott, F., Ziem, U., Reiffer, F.J., Huth, F., Ernst, H., Mohr, U. Carcinogenicity studies on fibres, metal compounds, and some other dusts in rats. Exp Pathol 1987;32:129–152.

Pryor, W.A. Cigarette smoke and the involvement of free radical reactions in chemical carcinogenesis. Br J Cancer 1987;55:19–23.

Pryor, W.A. Biological effects of cigarette smoke, wood smoke, and the smoke from plastics: the use of electron spin resonance. Free Rad Biol Med 1992;13:659–676.

Pryor, W.A. Cigarette smoke radicals and the role of free radicals in chemical carcinogenicity. Environ Health Perspect 1997;105(S4):875–882.

Pryor, W.A., Prier, D.G., Church, D.F. Electron-spin resonance studies of mainstream and sidestream cigarette smoke: nature of the free radicals in gas-phase smoke and in cigarette tar. Environ Health Perspect 1983;47:345–355.

Rahman, I., MacNee, W. Role of oxidants/antioxidants in smoking-induced lung disease. Free Rad Biol Med 1996;21:669–681.

Rahman, I., Morrison, D., Donaldson, K., MacNee, W. Systemic oxidative stress in asthma, COPD, and smokers. Am J Resp Crit Care Med 1996;154:1055–1060.

Rajan, K.T., Wagner, J.C., Evans, P.H. The response of human pleura in organ culture to asbestos. Nature 1972;238:346–347.

Rola-Pleszczynski, M., Gouin, S., Bégin, R. Asbestos-induced lung inflammation. Inflammation 1984;8:53–62.

Roman-Franco, A.A. Non-enzymic extramicrosomal activation of chemical carcinogens by phagocytes: a proposed new pathway. J Theor Biol 1982;97:543–555.

Romert, L., Jenssen, D. Rabbit alveolar macrophage-mediated mutagenesis of polycyclic aromatic hydrocarbons in V79 Chinese hamster cells. Mutat Res 1983;111:245–252.

Sakamoto, W., Fujie, K., Handa, H., Ogihara, T., Mino, M. In vivo inhibition of superoxide production and protein kinase C activity in macrophages from vitamin E-treated rats. Int J Vit Nutr Res 1990;60:338–342.

Saldiva, P.H.N., Pope, C.A., Schwartz, J., Dockery, D.W., Lichtenfels, A.J., Salge, J.M., Barone, I., Bohm, G.M. Air pollution and mortality in elderly people: a time-series study in Sao Paulo, Brazil. Arch Environ Health 1995;50:159–163.

Saracci, R. Man-made mineral fibres and health. Scand J Work Environ Health 1985;11:215–222.

Schroeder, W.H., Dobson, M., Kane, D.M., Johnson, N.D. Toxic trace elements associated with airborne particulate matter: a review. J Air Pollution Control Assoc 1987;37:1267–1285.

Schwartz, J., Dockery, D.W. Particulate air pollution and daily mortality in Steubenville. Am J Epidemiol 1992;135:12–19.

Schwartz, J., Weiss, S.T. Relationship between dietary vitamin C intake and pulmonary function in the first National Health and Nutrition Examination Survey (NHANES I) Amer J Clin Nutr 1994;59:110–114.

Scottish Office. Scotland's health: a challenge to us all. Eating for health: a diet action plan for Scotland. HMSO. 1996.

Seal, S., Krezoski, S., Barr, T.L., Petering, D.H., Klinowski, J., Evans, P.H. Surface chemistry and biological pathogenicity of silicates: an X-ray photoelectron spectroscopic study. Proc Roy Soc 1996;B263:943–951.

Seaton, A., MacNee, W., Donaldson, K., Godden, D. Particulate air pollution and acute health effects. Lancet 1995;345:176–178.

Sedgewick, J.B., Geiger, K.M., Busse, W.W. Superoxide generation by hypodense eosinophils from patients with asthma. Am Rev Resp Dis 1990;142:120–125.

Selikoff, I.J., Hammond, E.G., Churg, J. Asbestos exposure, smoking and neoplasia. J Am Med Assoc 1968;204:106–112.

Shi, X., Mao, Y., Daniel, L.N., Saffiotti, U., Dalal, N.S., Vallyathan, V. Silica radical-induced DNA damage and lipid peroxidation. Env Health Perspect 1994;102:149–154.

Simon, L.M., Suttorp, N. Lung cell oxidant injury: decrease in polymorphonuclear leucocyte mediated cytotoxicity by N-acetyl cysteine. Am Rev Resp Dis 1983;127:286.

Sindhu, R.K., Rasmussen, R.E., Kikkawa, Y. Effect of environmental tobacco smoke on the metabolism of (-)-trans-benzo(a)pyrene-7,8-dihydrodiol in juvenile lung and liver. J Toxicol Environ Health 1995;45:453–464.

Smith, L.J., Shamsuddin, M., Sporn, P.H.S., Denenberg, M., Anderson, J. Reduced superoxide dismutase in lung cells of patients with asthma. Free Rad Biol Med 1997;22:1301–1307.

Spurney, K.R. Natural fibrous zeolites and their carcinogenicity – a review. Sci Total Environ 1983;30:147–166.

Stanton, M.F., Layard, M., Tegeris, A., Miller, E., May, M., Morgan, E., Smith, A. Relation of particle dimension to carcinogenicity in amphibole asbestoses and other fibrous minerals. J Nat Cancer Inst 1981;67:965–975.

Stroink, G., Lim, D., Dunlap, R.A. A Mössbauer-effect study of autopsied lung tissue of asbestos workers. Phys Med Biol 1987;32:203–211.

Suslova, T.B., Cheresmisina, Z.P., Korkina, J.G. Free radical generation during interaction of chrysotile asbestos with natural compounds. Environ Res 1994;66:222–234.

Taylor, C.G., Bauman, P.F., Sikorski, B., Bray, T.M. Elevation of lung glutathione by oral supplementation of L-2-oxathiazolidine-4-carboxylate protects against oxygen toxicity in protein-energy malnourished rats. FASEB J 1992;6:3101–3107.

Taylor, C.G., McCutcheon, T.L., Boermans, H.J., DiSilvestro, R.A., Bray, T.M. Comparison of Zn and vitamin E for protection against hyperoxia-induced lung damage. Free Rad Biol Med 1997;22:543–550.

Theron, A.J., Richards, G.A., Myers, M.S., Van Antwerpen, V.L., Sluis-Cremer, G.K., Wolmarans, L., Van der Mewe, C.A., Anderson, R. Investigations of the relative contributions of cigarette smoking and mineral dust to activation of circulating phagocytes, alterations

in plasma concentrations of vitamin C, vitamin E, and beta carotene, and pulmonary dysfunction in South African gold mines. Occup Env Med 1994;51:564–567.
Thomas, J., Evans, P.H. Serum protein changes in coalworker's pneumoconiosis. Clin Chim Acta 1975;60:237–247.
Thomas, G., Ando, T., Verma, K., Kagan, E. Asbestos-induced nitric oxide production: synergistic effect with interferon-gamma. Ann NY Acad Sci 1994;725:207–212.
Timbrell, V. Characteristics of the International Union Against Cancer standard reference samples of asbestos. In 'Pneumoconiosis' Proceedings International Conference, Johannesburg (1969) pp 28–36.
Trush, M.A., Kensler, T.W. An overview of the relationship between oxidative stress and chemical carcinogenesis. Free Rad Biol Med 1991;10:201–209.
Urano, N., Yano, E., Evans, P.H. Reactive oxygen metabolites produced by the carcinogenic fibrous mineral erionite. Environ Res 1991;54:74–81.
Valavanidis, A., Zarodimos, J., Makropoulou, E., Balomenou, E. Asbestos, cigarette smoke and free radicals. Studies upon their cocarcinogenic effect. Rev Clin Pharmacol Pharmokin 1995;12:83–94.
Vallyathan, V., Leonard, S., Kuppusamy, P., Pack, D., Chzhan, M., Sanders, S.P., Zweir, J.L. Oxidative stress in silicosis: evidence for the enhanced clearance of free radicals from whole lungs. Mol Cell Biochem 1997;168:125–132.
Vallyathan, V., Shi, X., Dalal, N.S., Irr, W., Castranova, V. Generation of free radicals from freshly fractured silica dust. Potential role in acute silica-induced lung injury. Amer Rev Resp Dis 1988;138:1213–1219.
Vilim, V., Wilhelm, J., Brzak, P., Hurych, J. The chemiluminescence of rabbit alveolar macrophages induced by quartz dust particles. Immunol Lett 1984;8:69–73.
Voisin, C., Aerts, C., Wallaert, B. Prevention of in vitro oxidant-mediated alveolar macrophage injury by cellular glutathione and precursors. Clin Resp Physiol 1987;23:309–313.
Wagner, J.C., Skidmore, J.W., Hill, R.J., Griffiths, D.M. Erionite exposure and mesotheliomas in rats. Br J Cancer 1985;51:727–730.
Wang, S., Lantz, R.C., Rider, E., Chen, G.J., Breceda, V., Hays, A.M., Robledo, R.F., Tollinger, B.J., Dinesh, S.V.R., Witten, M.L. A free radical scavenger (Lazeroid U74512E) attenuates tumour necrosis factor-alpha generation in a rabbit model of smoke-induced lung injury. Respiration 1997;64:358–363.
Weitzman, S.A., Graceffa, P. Asbestos catalyses hydroxyl and superoxide radical generation from hydrogen peroxide. Arch Biochem Biophys 1984;228:373–376.
White, C.W., Repine, J.E. Pulmonary antioxidant defense mechanisms. Exp Lung Res 1985;8:81–96.
Wong, O., Musselman, R.P. An epidemiological and toxicological evaluation of the carcinogenicity of man-made vitreous fiber, with a consideration of coexposures. J Environ Pathol Toxicol Oncol 1995;13:169–180.
Wordley, J., Walters, S., Ayres, J.G. Short term variations in hospital admissions and mortality and particulate air pollution. Occup Environ Med 1996;54:108–116.
Working Group on Public Health and Fossil-Fuel Combustion. Short-term improvements in public health from global-climate policies on fossil-fuel combustion: an interim report. Lancet 1997;350:1341–1349.
World Health Organisation. Air Quality Guidelines for Europe. WHO Regional Publications. European Series No. 21,1987.
Wright, B.M. 'Freshly fractured surface' theory of silicosis. Nature 1950;166:538–540.
World Wide Web site. 'The Haze' air pollution episode S E Asia. http://WWW.Vensara.com/haze/
Yano, E., Takeuchi, A., Nishii, S., Koizumi, A., Poole, A., Brown, R.C., Johnson, N.F., Evans, P.H., Yukiyama, Y. In vitro biological effects of volcanic ash from Mount Sakurajima. J Toxicol Environ Health 1985;16:127–135.

Yano, E., Urano, N., Evans, P.H. Reactive oxygen metabolite production induced by mineral fibres. In 'Mechanisms in fibre carcinogenesis' Eds Brown, R.C., Hoskins, J.A., Johnson, N.F. Plenum, New York (1991) pp 433–438.

Zsoldos, T., Tigyi, A., Montsko, T., Puppi, A. Lipid peroxidation in the membrane damaging effect of silica-containing dust on rat lungs. Expt Pathol 1983;23:73–77.

12 Free radical mechanisms of ethanol toxicity

Emanuele Albano

12.1 Introduction

The association between excess consumption of alcoholic beverages and the development of human diseases has been recognised for about three thousand years, being mentioned in the Ayur Veda, a medical textbook of ancient India. Although alcoholic liver disease is the most common medical consequence of chronic alcohol abuse, several organs are injured by ethyl alcohol. The most commonly affected organs include the gastrointestinal tract, the pancreas, the heart, the skeletal muscles, the haemopoietic bone marrow, the testes and the nervous system.

Epidemiological studies have demonstrated that alcohol-related diseases are an important cause of morbidity and mortality in most of the well developed countries. For instance, in 1986–87, alcohol-related mortality in the USA and in Western Europe has been estimated to range from 4.9% to 6.1% (Kato *et al.*, 1996). Alcoholic cirrhosis has a prevalence of 3.6/1,000 in the USA (Dufour *et al.*, 1993) and is estimated as the ninth most frequent cause of death in the general population. Similar data hold for Europe, where the prevalence of liver cirrhosis in different countries correlates quite well to the national per capita alcohol consumption (Lelbach, 1975). This relationship is also reflected in how changes in the drinking pattern of the population correspond to later changes in the incidence of liver cirrhosis (Savolainen *et al.*, 1992).

For a long time the chronic diseases associated with alcohol abuse have been attributed to malnutrition and, indeed, some alcoholics suffer from nutritional deficiencies, such as thiamine and folic acid deficiencies. However, most alcohol abusers have adequate diets and the great majority of alcohol-related diseases should be attributed to the biochemical changes induced by ethanol or its metabolites in different tissues. Ethanol, in fact, is readily absorbed from the gastrointestinal tract and is largely metabolized (90–98%) in the body, mainly in the liver. With the exception of the stomach, extrahepatic metabolism of ethanol is small. This relative organ specificity explains why ethanol toxicity mostly involves the liver. The major pathway for ethanol metabolism involves its oxidation to acetaldehyde by the activity of alcohol dehydrogenase (ADH), an NAD-dependent zinc metalloenzyme for which five different classes have been distinguished in human tissues (Lieber, 1994). These classes of the enzyme arise from the association of eight different types of subunits with active dimeric molecules. Ethanol metabolism in the liver largely relies on Class I isoenzyme activity, while a Class IV isoenzyme is mainly responsible for alcohol oxidation in the gastric mucosa (Lieber, 1994). Ethanol can also be oxidized by a cytochrome

P450-dependent pathway, known as microsomal ethanol metabolizing system (MEOS), that relies on the activity of cytochrome P450 2E1 isoenzyme (CYP2E1) (Lieber, 1994). CYP2E1 is mostly present in the centrilobular areas of the liver, but low levels of the enzyme are also detectable in the gastrointestinal tract, in the kidney, in the lung and in the brain (Ronis *et al.*, 1996). Although the K_m for ethanol of hepatic ADH (0.2–2 mmol/l) is lower than that of CYP2E1 (8–10 mmol/l), the latter is inducible by alcohol. Chronic alcohol exposure causes, in fact, a 5 to 20 fold stimulation of CYP2E1 activity that involves both enzyme stabilization and increased gene expression (Ronis *et al.*, 1996). The toxic action of acetaldehyde along with the metabolic disorders consequent to the excess production of NADH have been proposed to be responsible for causing the adverse effects of alcohol (Lieber, 1994). However, in spite of several decades of research, no definitive conclusion has been reached on the mechanisms of alcohol toxicity. In recent years a growing interest has concerned the possibility that the free radical mediated oxidative damage might play a role in the pathogenesis of alcohol-related injury to the liver as well as to other organs (Nordmann *et al.*, 1992).

12.2 Mechanisms involved in causing oxidative damage by alcohol

Several free radical species, i.e. oxygen-derived radicals, hydroxyethyl radicals, nitric oxide, and lipid-derived radicals, have been proposed to play a role in causing ethanol-mediated oxidative tissue damage (Table 12.1). These radical species can be produced by parenchymal cells as well as by tissue macrophages, endothelial cells and by infiltrating phagocytes. The impairment of cellular antioxidant defences is also a common feature in tissues exposed to alcohol. Thus the combination of increased free radical production and decreased cellular antioxidants is probably responsible for the development of oxidative injury associated with alcohol abuse.

Table 12.1 Free radical species possibly responsible for alcohol-induced oxidative stress

Species	*Chemical structure*	*Possible sources*
Superoxide anion	$O_2^{\cdot-}$	Cytochrome P450 2E1 Mitochondrial respiratory chain Xanthine oxidase Aldehyde oxidase Oxidative burst of phagocytes
Hydroxyl radical	$OH^{\bullet}$	Iron-mediated degradation of H_2O_2
Nitric oxide	NO	Nitroxide synthetase of parenchymal cells, macrophages and endothelial cells
Peroxynitrite	$ONOO^{\bullet}$	Reaction of NO with $O_2^{\cdot-}$
α-Hydroxyethyl radical	$CH_3C^{\bullet}\ H_2OH$	Ethanol oxidation by CYP2E1 Reaction with $O_2^{\cdot-}$ or $OH^{\bullet}$
Methylcarbonyl radical	$CH_3C^{\circ}{=}O$	Acetaldehyde oxidation by xanthine oxidase or aldehyde oxidase

12.2.1 Formation of reactive oxygen species

The formation of reactive oxygen species such as superoxide anion (O_2^-) and hydrogen peroxide (H_2O_2) represents an important cause of oxidative injury in many diseases associated with free radical formation. Several enzymatic systems, including microsomal monoxygenase system, mitochondrial respiratory chain, cytosolic xanthine and aldehyde oxidase have been proposed to produce O_2^- and H_2O_2 in cells exposed to ethanol (Table 12.1). The mechanisms responsible for the generation of reactive oxygen species are discussed in detail later. However, it should be considered that most of the evidence so far available has been obtained *in vitro*, while comparatively little is known about the capacity of ethanol to induce the formation of oxygen free radicals *in vivo*. Nonetheless, preliminary data obtained by measuring hydroxyl radical-mediated conversion of acetylsalicylic acid to 2,3-dihydrobenzoic acid are consistent with an increased formation of reactive oxygen species in alcoholic patients (Thome *et al.*, 1997).

12.2.1.1 Role of cytochrome P4502E1

The induction of CYP2E1-dependent monoxygenase activity by chronic alcohol exposure can represent an important source of oxygen radicals since, even in the absence of substrates, CYP2E1 has an especially high NADPH oxidase activity, leading to an extensive production of O_2^- and H_2O_2 (Ronis *et al.*, 1996). Indeed, liver microsomes obtained from rats chronically exposed to alcohol are more active than microsomes from untreated animals in producing O_2^- and H_2O_2 and $OH^\bullet$ and also show an enhanced susceptibility to lipid peroxidation (Cederbaum, 1989; Persson *et al.*, 1990). Oxygen radical production and lipid peroxidation can be selectively reduced by antibodies directed against CYP2E1 (Ekström and Ingelman-Sundberg, 1989). Despite the fact that human liver microsomes produce O_2^- and H_2O_2 at rates 20–30% lower than those observed in rat liver microsomes (Rabshabe-Step and Cederbaum, 1994), a positive correlation can be observed between the CYP2E1 content and the NADPH-oxidase activity (Ekström *et al.*, 1989). Interestingly, in both human and rat liver microsomes NADH is equally effective as NADPH in promoting the production of reactive oxygen species (Dicker and Cederbaum, 1992; Rabshabe-Step and Cederbaum, 1994). Such a peculiarity can be important during ethanol intoxication because alcohol metabolism leads to an excess formation of NADH (Lieber, 1994). Experiments performed using rats chronically treated with alcohol by the Tsukamoto–French model of intragastric feeding support the role of CYP2E1 in promoting ethanol-mediated oxidative stress. In this experimental model CYP2E1 induction by ethanol shows a positive correlation with the stimulation of lipid peroxidation, whereas compounds that interfere with CYP2E1 induction significantly reduce peroxidative damage (French *et al.*, 1993; Morimoto *et al.*, 1995; Albano *et al.*, 1996).

12.2.1.2 Role of mitochondria

The mitochondrial respiratory chain represents one of the main sources of superoxide anion in cells (Forman and Boveris, 1988). Acute alcohol exposure promotes O_2^- production by liver sub-mitochondrial particles (Sinaceur *et al.*, 1985), while similar effects can be observed following chronic alcohol intake in brain sub-mitochondrial particles (Ribiére *et al.*, 1994). Kukielka and co-workers (1994) have reported that chronic alcohol consumption increases the production of reactive oxygen species in intact liver mitochondria incubated with NADH or NADPH by stimulating the activity of a rotenone-insensitive NADH-cytochrome c reductase on the outer mitochondrial membranes. The importance of this enzyme in causing ethanol-induced oxidative injury to the mitochondria could be even greater than that of the respiratory chain, since it does not require the transfer of NADH through the mitochondrial membranes. Accumulation of lipid peroxidation products and oxidative modifications of mitochondrial proteins and DNA (mtDNA) (Kamimura *et al.*, 1992; Weiland and Lautemburg, 1995; Cahill *et al.*, 1997) can be observed following both acute and chronic exposure of rats to ethanol, confirming that ethanol-mediated oxidative injury of mitochondria can take place *in vivo*.

12.2.1.3 Role of cytosolic enzymes

The oxidation by xanthine oxidase of acetaldehyde present in the tissues during ethanol metabolism has been suggested as an alternative pathway for the generation of O_2^- and hydroxyl radicals during ethanol metabolism (Shaw, 1989). Indeed, alcohol treatment stimulates the conversion of xanthine dehydrogenase to the O_2^--producing oxidase form (Sultatos, 1988; Abbondanza *et al.*, 1989), while the inhibition of xanthine oxidase by allopurinol prevents ethanol-induced lipid peroxidation (Kato *et al.*, 1990). Although the K_m of xanthine oxidase for acetaldehyde is very high (30 mmol/l), Puntarulo and Cederbaum (1989) have reported the formation of reactive oxygen species at concentrations of acetaldehyde close to those present in the liver following alcohol intake (about 0.1 mmol/l) (Stowel *et al.*, 1980). Acetaldehyde oxidation by aldehyde oxidase has been also implicated in the alcohol-mediated production of O_2^- (Shaw and Jayatilleke, 1990a). However, Mira and co-workers (1995) have recently shown that NADH is a better substrate (K_m 28 μmol/l) than acetaldehyde (K_m 1 mmol/l) for O_2^- generation by aldehyde oxidase. The possible contribution of aldehyde oxidase in promoting oxidative injury by alcohol is supported by the observation that the enzyme stimulates lipid peroxidation when added to rat liver microsomes incubated with ethanol in the presence of NAD^+ and alcohol dehydrogenase (Mira *et al.*, 1995), while menadione, an inhibitor of aldehyde oxidase, significantly decreased lipid peroxidation in isolated rat hepatocytes incubated with ethanol or acetaldehyde (Shaw and Jayatilleke, 1990b).

12.2.1.4 Role of phagocytes

The activation of phagocytic cells (resident macrophages, infiltrating monocytes and polymorphonuclear granulocytes) might contribute to the generation of reactive oxygen species in tissues damaged by alcohol. Short term *in vitro* exposure of alveolar macrophages to low doses of ethanol is, in fact, capable of directly stimulating O_2^- production (Dorio *et al.*, 1988). Moreover, Bautista and Spitzer (1992) have observed that both acute and chronic ethanol administration stimulates O_2^- production by Kupffer cells. Interestingly, an increased O_2^- formation has been demonstrated in perfused livers during the recovery period after 12 hour of continuous ethanol infusion (Bautista and Spitzer, 1996). Such a 'post-binge' O_2^- generation involves

Kupffer and endothelial cells and can be significantly attenuated by depleting the liver of Kupffer cells by rat treatment with gadolinium chloride (Bautista and Spitzer, 1996). Nonetheless, it is possible that liver sinusoidal endothelial cells might also contribute the formation of reactive oxygen species associated with alcohol intake, since in another study the inactivation of Kupffer cells by gadolinium chloride does not appear to affect O_2^- generation during ethanol infusion in *in situ* perfused livers (Nakano *et al.*, 1995).

The infiltration by granulocytes and monocytes is common in many tissues (liver, pancreas, stomach) acutely damaged by alcohol and might substantially contribute to oxygen radical production. Powerful chemiotactic agents, such as interleukin-8 (IL-8) (Shiratori *et al.*, 1993), leucotriene B_4 (Shirley *et al.*, 1992) and macrophage inflammatory protein-2 (MIP_2) (Bautista, 1997) are produced following the exposure to ethanol. The action of these mediators can explain the sequestration of neutrophis observed in the liver of ethanol-treated animals as well as their priming to O_2^- production (Bautista *et al.*, 1992).

12.2.2 Free radical species derived from ethanol or acetaldehyde

The metabolic conversion of ethanol to carbon-centred 1-hydroxyethyl free radical was first demonstrated in 1987 by two independent studies applying electron spin resonance (ESR) spectroscopy and spin-trapping techniques to the analysis of the NADPH-dependent pathway of ethanol metabolism in rat liver microsomes (Albano *et al.*, 1987; Reinche *et al.*, 1987). These observations have been confirmed by several other reports (Albano *et al.*, 1988; Reinche *et al.*, 1990; Rao *et al.*, 1996) and by the demonstration that hydroxyethyl radical can be generated *in vivo* in the liver of ADH-deficient deer-mice (Knecht *et al.*, 1993) or of alcohol-fed rats (Knecht *et al.*, 1990; Moore *et al.*, 1995). Free radical intermediates are similarly produced during microsomal oxidation of propanol, butanol and pentanol, indicating a common metabolic pathway for radical production from aliphatic alcohols (Albano *et al.*, 1994a). CYP2E1 is mostly responsible for the formation of hydroxyethyl radicals in rat and human liver microsomes and anti-CYP2E1 antibodies or CYP2E1 inhibitors greatly reduce the spin-trapping of these radicals (Albano *et al.*, 1991, 1994a). So far, the mechanisms responsible for hydroxyethyl radical production by CYP2E1 have not yet been completely elucidated. The results of *in vitro* experiments indicate that the interaction of reactive oxygen species (O_2^-, $OH^•$) originating as a result of NADPH-oxidase activity of CYP2E1 with iron might be responsible for hydroxyethyl radical formation (Figure 12.1) (Albano *et al.*, 1988; Knecht *et al.*, 1993; Rao *et al.*, 1996; Reinke *et al.*, 1997). Nonetheless it is also possible that these radicals could originate at the active site of the enzyme, possibly by the interaction of ethanol with ferric cytochrome P450–oxygen complex (CYP2E1–$Fe^{3+}O_2^-$) (Albano *et al.*, 1991).

Hydroxyethyl free radicals are quite reactive species and can interact with GSH, ascorbic acid, α-tocopherol, proteins and DNA (Schöneich *et al.*, 1989; Schuessler *et al.*, 1992; Stoyanovshy *et al.*, 1998). Upon incubation of rat liver microsomes with NADPH and radioactive ethanol, hydroxyethyl radical residues can be recovered covalently bound to microsomal proteins (Albano *et al.*, 1993; Moncada *et al.*, 1994). In human liver microsomes or in isolated rat hepatocytes incubated *in vitro* with ethanol at least four microsomal proteins (apparent molecular weights 78 kD, 60 kD, 52 kD and 40 kD) are targets for hydroxyethyl radical attack (Clot *et al.*, 1996). The

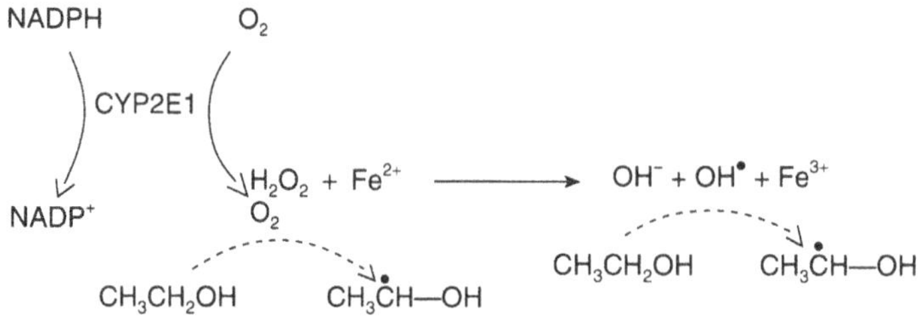

Figure 12.1 Proposed mechanisms for the formation of ethanol-derived hydroxyethyl free radicals in hepatocytes exposed to ethanol. The NADPH-oxidase activity of CYP2E1 is responsible for the formation of O_2^- and H_2O_2. The interaction of iron with H_2O_2 produces hydroxyl radicals ($OH^•$) that, by reacting with ethanol, would give rise to hydroxyethyl free radicals. Alternatively, a direct one-electron oxidation of ethanol by O_2^- might also account for the formation of ethanol-derived radical species

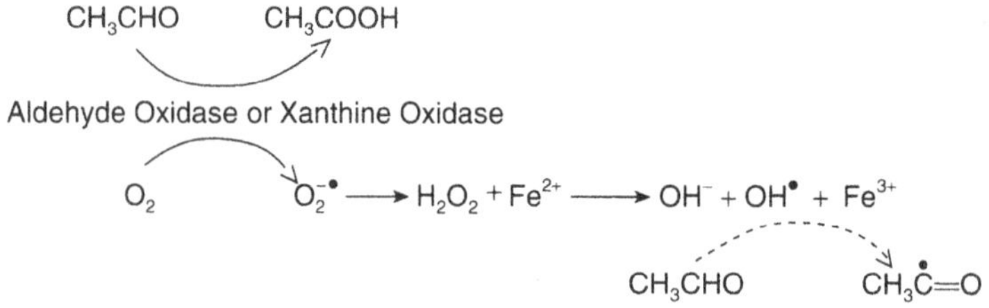

Figure 12.2 Postulated mechanism for the formation of free radical species from acetaldehyde. Several studies have demonstrated that that O_2^- and H_2O_2 are produced during the oxidation of acetaldehyde by xanthine oxidase or aldehyde oxidase. In the presence of iron, O_2^- and H_2O_2 can lead to the formation of $OH^•$ radicals that are then responsible for attacking another molecule of acetaldehyde giving rise to methyl carbonyl radical (Albano *et al.*, 1994b). Thus, acetaldehyde might act at the same time as source of reactive oxygen species, being a substrate for xanthine oxidase, as well as a target for $OH^•$ radicals

52 kD protein has been identified by combined immunoblotting and immunoprecipitation techniques as CYP2E1 (Clot *et al.*, 1996), while the identity of 78 kD, 60 kD, and 40 kD proteins is presently unknown. It is possible that these proteins are intimately associated with CYP2E1, although the 78 kD protein is not NADPH-cytochrome P450 reductase. The formation of hydroxyethyl-CYP2E1 adducts has been shown to occur also *in vivo* and can be detected in immunoblots of microsomal proteins obtained from rats acutely treated with a large dose of ethanol as well as in microsomes from rats receiving ethanol by intragastric feeding (Clot *et al.*, 1996).

As mentioned above, xanthine oxidase and aldehyde oxidase metabolize acetaldehyde with the formation of reactive oxygen species. During the oxidation of acetaldehyde by xanthine oxidase a carbon-centred free radical, identified as methyl carbonyl species ($CH_3C^{\circ}{=}O$) is also produced by the abstraction of a hydrogen atom from the acetaldehyde molecule (Albano *et al.*, 1994b; Santiard *et al.*, 1995) (Figure 12.2). Interestingly, the formation of methyl carbonyl radicals by xanthine oxidase is evident at concentrations of the aldehyde as low as 0.1 mM (Albano *et al.*, 1994b). This suggests that radical species originating from acetaldehyde might contribute to the covalent binding of acetaldehyde to proteins (Nicholls *et al.*, 1992) which is implicated in causing cell toxicity and immunological reactions associated with alcohol abuse (Lieber, 1994; Paronetto, 1993).

12.2.3 Nitrogen oxide and alcohol-induced oxidative injury

The possibility that ethanol intake might stimulate NO production has been suggested by Wang and co-workers (1995) who have observed that chronic alcohol treatment alone or in combination with an acute endotoxin administration increases NO metabolites in the plasma and in the liver perfusate. They have also demonstrated that Kupffer cells are responsible for the increased NO formation in the liver of alcohol fed animals, while hepatocytes mostly account for the post-endotoxins increase (Wang *et al.*, 1995). However, the actual capacity of alcohol intake to stimulate NO production, has been questioned by more recent studies demonstrating that ethanol can interfere at the transcriptional and post-transcriptional level with inducible NO-synthetase (iNOS) (Zhao *et al.*, 1997). The reaction of NO with O_2^- leads to the formation of highly reactive peroxynitrite radical ($ONOO^•$) that is capable to inactivate a number of enzymes, impairing mitochondrial functions and causing cytotoxicity (Beckman and Koppenol, 1996). The actual role of NO and $ONOO^•$ in the development of tissue injury by ethanol is still controversial. In a recent study by Chamulitrat and Spitzer (1996) rats chronically exposed to ethanol generate 2–3 times more NO than naive rats when receiving 12 hours endotoxin infusion. Endotoxin treatment of alcohol-fed rats also promotes liver transaminase release that can be attenuated by the inhibition of iNOS with aminoguanidine (Chamulitrat and Spitzer, 1996). On the other hand, other evidence indicates that NO formation might counteract the development of oxidative injury. For instance, the inhibition of NO production *in vivo* enhances O_2^- release by Kupffer cells (Bautista and Spitzer, 1994), while the stimulation of hepatocyte NO formation is associated with an increased resistance to oxygen radical-mediated injury (Harbrecht and Billiar, 1995). The capacity of NO to decrease cytosolic low molecular weight iron complexes has been proposed to account for the resistance to ethanol-induced oxidative stress observed in isolated hepatocytes in which NO production is triggered by endotoxin (Sergent *et al.*, 1997). Inhibition of CYP2E1 activity by NO might also be important in this respect (Gergel *et al.*, 1997), considering the ability of CYP2E1 to generate reactive oxygen species. The observation that the treatment with NO-inhibitor, *N*-nitro-L-arginine methyl ester (L-NAME), potentiates liver damage in rats receiving intragastric alcohol feeding (Nanji *et al.*, 1995a) supports the possible role of NO in preventing alcohol liver injury. However, further studies are needed to clarify the actual role of NO in the development of alcoholic damage.

12.2.4 Role of iron in ethanol-mediated oxidative damage

In the presence of trace amounts of transition metals, most frequently iron, O_2^- and H_2O_2 generated from either enzymatic or non-enzymatic sources undergo the so-called metal-catalyzed Haber–Weiss reaction, producing highly reactive hydroxyl radicals ($OH^•$) (Aust *et al.*, 1985). Furthermore, the propagation of lipid peroxidation is also favoured by the degradation of lipid hydroperoxides in the presence of iron (Aust *et al.*, 1985). Alcohol abuse in humans is often associated with an impaired utilization and with an increased deposition of iron in the liver (Chapman *et al.*, 1983; Irving *et al.*, 1988). Experiments performed in rats have shown that dietary supplementation with iron increases alcohol toxicity (Tsukamoto *et al.*, 1995; Stål *et al.*, 1996) and enhances ethanol-induced lipid peroxidation (Tsukamoto *et al.*, 1995). Conversely, the administration of an oral iron chelator (1,2-dimethyl-3-hydroxypyrid-4-one) has been reported to reduce non-heme iron levels, lipid peroxidation and fat accumulation during intragastric ethanol feeding of rats (Sadrzadeh *et al.*, 1994a). The form of iron associated with free radical reactions involves a small pool of low molecular weight non-protein iron complexes (Minotti *et al.*, 1991). Acute ethanol intake increases low molecular weight iron content in liver and in the cerebellum by increasing the uptake of transferrin (Rouach *et al.*, 1994). Moreover, experiments *in vitro* have shown that the rise in the cytosolic levels of NADH (Tophan *et al.*, 1989) or the generation of O_2^- can release catalytically active iron from ferritin (Shaw and Jayatilleke, 1990a). Consistently, the addition of ferritin to liver microsomes from ethanol-fed rats greatly stimulates both NADPH and NADH dependent lipid peroxidation. This effect is prevented by the addition of superoxide dismutase, iron chelators and anti-CYP2E1 antibodies (Kulielka and Cederbaum, 1996), indicating that CYP2E1-generated O_2^- might also contribute in mobilizing iron from ferritin. However, a role of iron in causing ethanol-induced oxidative stress of intact tissues has not been unequivocally proven. For instance, in the liver, the cytosolic levels of low molecular weight iron appear to increase following acute alcohol administration (Rouach *et al.*, 1990), but they are decreased (Rouach *et al.*, 1997) or not modified (Kamimura *et al.*, 1992) in rats chronically treated with ethanol by intragastric feeding. These latter findings might be explained considering that hepatocyte iron homeostasis is finely tuned by cytosolic iron regulatory protein (IRP). By interacting with the metal, IRP changes its affinity for the iron responsive elements (IRE) present in the mRNAs of, respectively, ferritin and transferrin receptor. Thus, in the presence of low molecular weight iron IRP does not bind to IRE and this leads to the stimulation of ferritin synthesis and to iron sequestration, while iron uptake by the transferrin receptor is depressed (Mascotti *et al.*, 1995). Recently, Cairo and co-workers (1996) have reported that O_2^- and H_2O_2 are able to reversibly inhibit IRP binding to IRE, suggesting that an increased formation of reactive oxygen species can stimulate iron chelation by ferritin in order to prevent the spreading of iron-dependent oxidative events.

12.2.5 Lowering of antioxidant defences in ethanol-induced oxidative injury

The lowering of liver antioxidant defences might significantly contribute to the development of ethanol-induced oxidative damage. A decrease in glutathione (GSH) levels can be observed in the liver, heart, kidney, pancreas and brain of experimental animals subjected to acute alcohol intoxication (Guerri and Grisolia, 1980; Videla and Valenzuela, 1982; Israel *et al.*, 1992; Altomare *et al.*, 1996). The loss of liver GSH is not associated with a significant elevation of oxidized glutathione (GSSG) levels (Lauterburg *et al.*, 1984), but is rather caused by the combination of an increased GSH efflux from hepatocytes and an impaired GSH re-synthesis (Lauterburg *et al.*, 1984; Speisky *et al.*, 1985). In contrast to acute intoxication, chronic alcohol intake does not appreciably affect total hepatocyte GSH content in rats (Israel *et al.*, 1992). However, a decrease in liver GSH is a common feature in ethanol-fed baboons (Shaw *et al.*, 1981) as well as in alcoholic patients, in which GSH loss is independent from the nutritional status or the degree of liver disease (Shaw *et al.*, 1983; Jewell *et al.*, 1986; Situnayake *et al.*, 1990). Recent studies have demonstrated that rats receiving alcohol chronically either by traditional pair feeding (Fernandez-Checa *et al.*, 1987) or by intragastric nutrition (Takeshi *et al.*, 1992) undergo a progressive decrease in the GSH pool of liver mitochondria. Such a selective depletion of mitochondrial GSH appears to depend upon a defect in the transfer of the tripeptide from cytosol to the mitochondrial matrix (Fernandez-Checa *et al.*, 1991) due to a decreased efficiency of an ATP dependent GSH transporter in the inner mitochondrial membrane (Colell *et al.*, 1997). Mitochondrial GSH depletion is more evident in centrilobular hepatocytes (Garcia-Ruitz *et al.*, 1994) and precedes the development of lipid peroxidation and of functional alterations in ATP production (Takeshi *et al.*, 1992). This suggests that the effects of ethanol on mitochondrial GSH homeostasis might significantly contribute to the development of oxidative damage in these organelles (Fernandez-Checa *et al.*, 1997). The importance of GSH homeostasis in preventing alcohol toxicity and oxidative injury is further supported by the observation that the liver GSH depletion favours lipid peroxidation and acute alcohol toxicity (Kera *et al.*, 1989; Strubelt *et al.*, 1987), while stimulation of GSH re-synthesis by treatment with *S*-adenosyl-L-methionine (SAME) reduces alcohol hepatoxicity (Vendemiale *et al.*, 1989; Lieber *et al.*, 1990). A relationship between GSH depletion and cellular injury has been also documented in gastric mucosa using either canine chambered stomach preparations (Victor *et al.*, 1991) or mucosa biopsies obtained from the stomach of healthy human volunteers after the exposure to ethanol (Loguercio *et al.*, 1991). The lowering of liver and plasma levels of the liposoluble antioxidant vitamin E is often detectable during chronic alcohol administration to rats (Bjørneboe *et al.*, 1987; Kawase *et al.*, 1989; Sadrzadeh *et al.*, 1994b; Rouach *et al.*, 1997) and involves both the α-tocopherol (20–50% decrease) and γ-tocopherol (65–75% decrease) isoforms of the vitamin (Sadrzadeh *et al.*, 1994b). A similar reduction in the plasmatic and hepatic levels of α-tocopherol has been also documented in patients with alcohol abuse with or without overt signs of liver disease (Tanner *et al.*, 1986; Bell *et al.*, 1992; Lecomte *et al.*, 1994; Clot *et al.*, 1994). The mechanisms responsible for vitamin E decrease during alcohol intake have not jet been completely elucidated. However, Kawase and co-workers (1989) have proposed that an increased oxidation of α-tocopherol to α-tocopherol quinone might account for the loss of this antioxidant during alcohol exposure. Such an interpretation is consistent with the presence of an inverse correlation between the levels of α-tocopherol and lipid peroxidation products in either the liver of intragastric ethanol-fed rats (Sadrzadeh *et al.*, 1994a; Rouach *et al.*, 1997) or the plasma of patients with alcoholic cirrhosis (Clot *et al.*, 1994). Nonetheless, the actual importance of vitamin E loss in the development of alcohol toxicity is still uncertain. Vitamin E deficient rats appear more susceptible to alcohol toxicity (Sadrzadeh *et al.*, 1994b), but the supplementation with high doses of α-tocopherol acetate does not prevent the development of liver damage (Sadrzadeh *et al.*, 1995). Interestingly, administration of α-tocopherol contributes in reducing the severity of hepatic injury upon discontinuation of alcohol feeding (Nanji *et al.*, 1996).

A number of studies have also investigated the effect of ethanol on the enzymes devoted to the detoxification of reactive oxygen species. The results of these studies are rather inconclusive. Acute alcohol intoxication lowers catalase, superoxide dismutase and glutathione S-transferase activities in several tissues, but these effects are not constantly observed following chronic alcohol treatment (Nordmann *et al.*, 1992; Nordmann, 1994). More recent investigations using the intragastric alcohol nutrition model have shown a significant decline in either the enzymatic activity and the immunoreactive protein concentrations of liver (Cu–Zn)-superoxide dismutase, catalase and glutatione peroxidase (Rouach *et al.*, 1997; Polavarapu *et al.*, 1998). These changes appear to be inversely correlated with the extent of, respectively, lipid peroxidation and hepatic injury (Polavarapu *et al.*, 1998). Nonetheless, using the same experimental model Nanji and co-workers (1995b) have reported an induction in mRNA expression of liver glutathione peroxidase and catalase. This suggests the possibility that ethanol might interfere at the post-transcriptional level with the synthesis of antioxidant enzymes or might stimulate the intracellular degradation of antioxidant enzymes.

12.3 Free radical-mediated processes in organ damage by alcohol

12.3.1 Alcohol and oxidative damage to the liver

The contribution of oxidative injury to ethanol hepatotoxicity was first proposed by Di Luzio in the early 1960s following the observation that the pre-treatment of rats with antioxidants alleviated ethanol-induced liver fat accumulation (Di Luzio, 1963). In subsequent years a number of experimental results have confirmed the presence of ethanol-induced oxidative damage in the liver of rats (see Dianzani, 1985; Nordmann *et al.*, 1992 for review), mini-pigs and baboons chronically fed with alcohol (Niemelä *et al.*, 1995; Lieber *et al.*, 1997). The possible implication of oxidative damage in human alcoholic liver disease is supported by several clinical studies. In particular, it has been observed that indices of oxidative stress, namely lipid peroxidation products and protein carbonyls, are higher in the liver biopsies or in the serum obtained from alcoholic patients as compared to specimens from non-drinker subjects or patients with non-alcoholic liver diseases (Suematzu *et al.*, 1981; Shaw *et al.*, 1983; Situnayake *et al.*, 1990; Baldi *et al.*, 1993; Lecomte *et al.*, 1994; Grattagliano *et al.*, 1996; Aleynik *et al.*, 1998). Patients with alcoholic cirrhosis also exhale more pentane, a volatile end product of lipid peroxidation (Letteron *et al.*, 1993). We have observed (Clot *et al.*, 1994) that among patients with alcoholic liver disease, markers of lipid peroxidation such as blood levels of lipid hydroperoxides and malondialdehyde (MDA), are about three fold higher in subjects drinking more than 100 g ethanol/day than in those drinking below 100 g ethanol/day, irrespective of the extent of liver injury (Figure 12.3). Nonetheless the most relevant contribution in establishing a connection between free radical-mediated oxidative damage and alcoholic liver disease has been obtained in recent years by the use of a new experimental model of alcohol toxicity based on the continuous intragastric administration of high amounts of alcohol along with a liquid diet rich in fat and poor in carbohydrates (Tsukamoto *et al.*, 1985, 1986). Using this experimental model it is possible to reproduce in rats several pathological features of human alcoholic liver disease, including steatosis, inflammatory infiltrates, focal necrosis and, after 16 weeks of treatment, liver fibrosis (Tsukamoto *et al.*, 1985, 1986). The studies performed using rats receiving ethanol by intragastric feeding have shown that the development

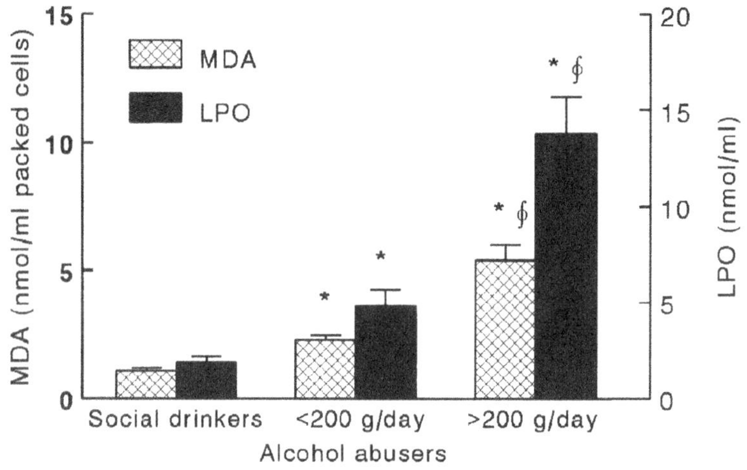

Figure 12.3 Blood markers of oxidative stress in patients with alcoholic liver disease and different alcohol consumption. The erythrocyte content of malondialdehyde (MDA) and plasma levels of lipid hydroperoxides (LPO) have been evaluated in 20 healthy social drinkers (ethanol intake below 60 g/day) and in 30 patients with alcoholic liver disease. These latter were sub-grouped according to the estimated daily alcohol consumption of less (10 subjects) or more (20 subjects) than 200 g ethanol/day. Statistical significance: * $p < 0.001$ as compared to control; ∮ < 0.05 as compared to patients drinking less than 200 g ethanol per day

of histological signs of liver damage is associated with an increase in lipid peroxidation and protein carbonyls (Kamimura *et al.*, 1992; Nanji *et al.*, 1994a; Rouach *et al.*, 1997). Furthermore, the replacement of corn oil with ω-3 unsaturated fatty acid-rich fish oil stimulates lipid peroxidation and worsens liver pathology in intragastric ethanol-fed rats (Nanji *et al.*, 1994b). Immunohistochemical analysis performed in the livers of ethanol-fed rats have shown that aldehydes derived from lipid peroxidation are present in the areas of fatty infiltration, focal necrosis and fibrosis (Tsukamoto *et al.*, 1995). Similar findings have been also obtained in liver biopsies from patients with alcoholic liver disease (Niemelä *et al.*, 1994), further supporting the concept of a causal relationship between oxidative events and the development of alcoholic liver injury.

12.3.2 Induction of oxidative stress by alcohol in extrahepatic tissues

12.3.2.1 Gastrointestinal tract

Beside alcoholic liver disease, injury to the gastrointestinal tract and to the pancreas represents the most common consequence of alcohol abuse (Kato, 1996). The involvement of oxidative injury in causing gastric mucosal damage associated with alcohol abuse comes from the observation that oral administration of absolute ethanol to rats results in a decrease of non-protein sulphydrils and in an increase of lipid peroxidation in the stomach mucosa (Mizui and Doteuchi, 1986; Pihan *et al.*, 1987).

Conversely, the pre-treatment of rats with antioxidants prevents the formation of gastric lesions caused by orally administered ethanol (Mizui and Doteuchi, 1986; Szelenyi and Brune, 1988). Experiments performed with cultured gastric mucosal cells have confirmed that superoxide anion is actually produced following the addition of ethanol and that O_2^- generation increases with the amount of ethanol used (Mutoh *et al.*, 1990). Ethanol is actively metabolized by gastric mucosa (Lin *et al.*, 1994), and xanthine oxidase or aldehyde oxidase-mediated acetaldehyde oxidation might be responsible for O_2^- production. Consistently, rat pre-treatment with allopurinol or oxypurinol, to block xanthine oxidase, protects against the appearance of haemorrhagic lesions due to alcohol administration (Smith *et al.*, 1987). Similar mechanisms can also be involved in causing ethanol damage to the gut. For instance, the catabolism by xanthine oxidase located in intestinal epithelial cells (Grossrau *et al.*, 1990) of acetaldehyde originating from bacteria-mediated ethanol oxidation (Salaspuro, 1997) might lead to O_2^- production in the gut. CYP2E1 might also be a source of reactive oxygen species in the gut, since chronic ethanol feeding induces CYP2E1 expression in duodenal and jejunal villous cells and in the surface epithelium of proximal colon (Shimizu *et al.*, 1990).

12.3.2.2 Pancreas

The possible involvement of oxidative stress in the pathogenesis of alcoholic pancreatitis is suggested by several studies. Altomare and co-workers (1996) have reported that acute ethanol administration to rats increases MDA, protein carbonyls and GSSG content in the pancreas, while pancreatic GSH levels are decreased. Moreover, patients with alcohol-related chronic pancreatitis have plasma levels of vitamin E lower than healthy controls (Van Gossum *et al.*, 1996). Spin-trapping experiments performed in rats exposed for 4 weeks to ethanol by intragastric feeding have revealed the presence of hydroxyethyl free radicals in the pancreatic secretions (Iimuro *et al.*, 1996). However, the mechanisms responsible for free radical formation in the pancreas have not been characterized in detail. In perfused canine pancreas the infusion of acetaldehyde associated with the conversion of xanthine dehydrogenase to xanthine oxidase by short term ischemia is capable of causing damage to tissues that are protected by superoxide dismutase, catalase and dithiothreitol (Nordback *et al.*, 1995). This suggests the possibility that the oxidation of acetaldehyde by xanthine oxidase might contribute to the formation of reactive oxygen species. On the other hand, it can not be excluded that CYP2E1 present in pancreatic acinar cells (Foster *et al.*, 1993) might contribute to the formation of reactive oxygen species and hydroxyethyl radical. Recent studies have shown, in fact, that chronic alcohol administration to rats increases by about five-fold the CYP2E1 content in the pancreas (Norton *et al.*, 1998; Kessova *et al.*, 1998).

12.3.2.3 Heart

Alcoholic cardiomyopathy is not uncommonly seen among alcohol abusers and is responsible for congestive heart failure (Preedy *et al.*, 1996). Several mechanisms might account for the progressive degeneration of myocardiocytes exposed to alcohol and a role for free radicals can not be excluded (Preedy *et al.*, 1996). Lipid peroxidation products, such as lipofuscins, are, in fact, increased in the myocardium of chronic alcohol-fed rats and of alcoholic patients (Jaatinen *et al.*, 1993, 1994). Moreover, using the spin-trapping technique, Reinche and co-workers (1987) have observed the presence of lipid-derived free radicals in the hearts of rats chronically exposed to ethanol after receiving a further acute alcohol load. It has been postulated that a decrease in myocardiocyte GSH content (Guerri and Grisolia, 1980) along with the conversion of xanthine dehydrogenase to xanthine oxidase and the activation of acyl coA-oxidase (Panchenko *et al.*, 1987) might contribute to free radical damage of the heart.

12.3.2.4 Nervous system

In recent years increasing evidence has demonstrated a role for oxidative mechanisms in the development of various neurological diseases (Jenner and Olanow, 1996; Multhaup *et al.*, 1997). The observation that the addition of ethanol to brain synaptosomes obtained from chronic alcohol-fed rats stimulates the formation of reactive oxygen species and of lipid peroxidation products (Montoliu *et al.*, 1994) suggests a possible contribution of oxidative stress to alcohol-related brain damage. Indeed, lowering of antioxidant levels and stimulation of lipid peroxidation are appreciable in the brain and in the peripheral nerves of rats chronically exposed to ethanol (Nordmann *et al.*, 1992; Boveris *et al.*, 1997; Bosch-Morell *et al.*, 1998). MDA accumulation is detectable in several brain areas and particularly in the cortex, in the striatum, in the hippocampus and in the cerebellum (Renis *et al.*, 1996). The stimulation in O_2^- production by mitochondria (Ribière *et al.*, 1994) and the increase in low-molecular weight iron content (Ruoach *et al.*, 1990) have been proposed as mechanisms for the pro-oxidant effect of ethanol in CNS. More recently, Montoliu and colleagues (1995) have reported that the induction of CYP2E1 by ethanol stimulates oxidative stress in primary cultures of astrocytes. Since CYP2E1 can be detected in several types of brain cells, including the Bergmann glial cells of the cerebellum, the pyramidal cells of the hippocampus and of the cortex and the dopaminergic neurons of the striatum (Hansson *et al.*, 1990), it is conceivable that the role of CYP2E1 in promoting oxidative injury might not be limited to astrocytes.

12.3.2.5 Testes

Testicular atrophy is often observed in male alcoholics and is associated with a decrease in testosterone secretion and with an impaired spermatogenesis. Rosenblum and co-workers (1989) have reported that the appearance of testicular atrophy in rats chronically exposed to alcohol is accompanied by an increase in lipid peroxidation and a decrease in GSH content of the testes. They have also observed that testicular atrophy in alcohol-fed animals is correlated with the levels of either MDA and GSH and can be prevented by dietary supplementation with vitamin A that also lowers oxidative damage (Rosenblum *et al.*, 1989). On these bases they have proposed the contribution of oxidative stress alcohol toxicity in the testes. The above findings have been recently confirmed by Grattagliano and colleagues (1997) who have reported an increase in MDA and protein carbonyls and a decrease of GSH, α-tocopherol and ascorbic acid content in the testes of rats exposed for eight weeks to a low dose of ethanol (3%) in the drinking water. Little is known about the mechanisms responsible for alcohol-mediated oxidative damage to the testis, but the detection of CYP2E1 in this organ (Ronis *et al.*, 1996) might suggest it possible involvement in causing ethanol toxicity.

12.4 Role of free radical-mediated mechanisms in alcohol toxicity

12.4.1 Oxidative mechanisms in ethanol-induced cellular injury

Hepatic steatosis is one of the most constant consequences of alcohol intoxication (Ishak *et al.*, 1991). The impairment of mitochondrial fatty acid oxidation is regarded as one of key factors in causing hepatocyte fat accumulation (Dianzani, 1991; Fromenty and Pessayre, 1995). Early works by Di Luzio have shown that rat pre-treatment with antioxidants partially prevents fatty liver caused by acute alcohol poisoning (Di Luzio, 1963). This effect might be related to the prevention of oxidative mitochondrial damage, since oxidative stress impairs the activity of the enzymes involved in mitochondrial fatty acid β-oxidation (Fromenty and Pessayre, 1995). Moreover, by causing mutations in mtDNA, reactive oxygen species might affect NADH oxidation by the respiratory chain enzymes, thus contributing to the metabolic imbalances responsible for hepatocyte steatosis (Fromenty and Pessayre, 1995). Consistently, either single or multiple deletions of liver mtDNA are eight times more frequent in alcoholics than in age-matched controls (Mansouri *et al.*, 1997). Interestingly, mtDNA deletions have a very high prevalence (about 85% of the cases) in alcoholics with microvesicular steatosis (Fromenty *et al.*, 1995; Mansouri *et al.*, 1997), a liver lesion that is ascribed to an impairment of mitochondrial β-oxidation of fatty acids (Fromenty and Pessayre, 1995). However, it can not be excluded that oxidative damage can also contribute to alcoholic steatosis by affecting lipoprotein secretion. Evidence indicates, in fact that aldehyde end-products of lipid peroxidation can be responsible for the impairment of lipoprotein glycosylation in the Golgi apparatus (Marinari *et al.*, 1987; Cottalasso *et al.*, 1996).

The role played by oxidative stress in causing ethanol cytotoxicity is receiving increasing attention. A recent study using a human hepatoblastoma cell line (HepG2) transfected with the human CYP2E1 gene has shown the development of ethanol cytotoxicity only in the cells expressing CYP2E1 gene, but not in those infected with the retrovirus lacking CYP2E1 cDNA, and the possibility to prevent cell killing by the addition of CYP2E1 inhibitors, antioxidants and superoxide dismutase or catalase (Wu and Cederbaum, 1996). In a similar manner, the formation of the superoxide anion appears to contribute to the damage of cultured gastric mucosal cells (Mutoh *et al.*, 1990). As discussed above, oxidative modifications of protein and mitochondrial DNA might be responsible for the depression in the levels of mitochondrially-encoded sub-units of the electron transport chain and for the decreased efficiency in ATP production observed in animals exposed to ethanol (Coleman *et al.*, 1994). In agreement with these observations, experiments *in vivo* have shown that a decrease of mitochondrial GSH pool precedes the development of functional alterations of mitochondria during chronic ethanol feeding, while restoring mitochondrial GSH by rat supplementation with SAME partially prevents the loss of mitochondrial functions (Fernandez-Checa *et al.*, 1997). A possible link between the impairment of mitochondrial functions and the development of ethanol cytotoxicity has been recently observed in cultured rat hepatocytes (Kurose *et al.*, 1997a). In these experiments, a 50% decrease of mitochondrial membrane potential does not result in cytotoxicity.

However, hepatocyte death was appreciable when extensive mitochondrial damage was promoted by the combined treatment with ethanol and the GSH-depleting agent diethlymaleate (Kurose *et al.*, 1997a). In a similar way, oxidative stress contributes to the collapse mitochondrial membrane potential and promotes ethanol cytotoxicity in cultured gastric mucosal cells (Hirokawa *et al.*, 1998).

In recent years an increasing interest has been devoted to capacity of low level of oxidants to induce apoptosis rather than necrosis (Buttke and Sanderstrom, 1994; Slater *et al.*, 1995). The possible involvement of ethanol-induced oxidative stress in triggering cell apoptosis comes from the observation that lipid peroxidation and hepatocyte apoptosis are significantly increased in the liver of rats fed with ethanol plus a corn oil- or fish oil-rich diet, but not in those receiving ethanol in combination with saturated fat (Yacoub *et al.*, 1995). More recently Kurose and co-workers (1997b) have reported that the formation of reactive oxygen species by CYP2E1 promotes apoptotic changes in cultured hepatocytes exposed *in vitro* to ethanol. In this study, lowering hepatocyte GSH increases the number of cells showing apoptotic modifications, while the addition of oxygen radical scavengers prevents these effects (Kurose *et al.*, 1997b). Although these observations suggest a link between ethanol-induced oxidative stress and liver cell apoptosis, the mechanisms involved have not jet been elucidated. Increasing evidence obtained in different experimental systems suggests that the stimulation of oxygen radical formation within the mitochondria triggers the release of cytochrome c and of apoptosis inducing factor (AIF), probably by inducing mitochondria permeability transition (Green and Kroemer, 1998). The release in the cytosol of cytochrome c and AIF can then activate caspases 9 and 3 that are responsible for the progression of the apoptotic programme (Green and Kroemer, 1998). The release of cytochrome c from the mitochondrial matrix is prevented by the overexpression of anti-apoptotic protein Bcl-2 that is also able to inhibit the interaction of cytochrome c with caspases (Kroemer, 1997). An increase in Bcl-2 has been reported in the liver of intragastric ethanol-fed rats with a good correlation with lipid peroxidation, but Bcl-2 expression was mainly localized in bile duct epithelial cells and in infiltrating inflammatory cells (Yacoub *et al.*, 1995), since hepatocytes do not express Bcl-2 (Patel and Gores, 1995).

12.4.2 Hydroxyethyl free radicals as a trigger for immune reactions induced by alcohol

Alcoholic liver disease is associated with the development of immune reactions towards the liver (Paronetto, 1993). This immunological response has been ascribed to the binding of acetaldehyde to proteins (Nicholls *et al.*, 1992), since either experimental animals exposed to alcohol (Israel *et al.*, 1986) or alcoholic patients have high titres of immunoglobulins reacting with acetaldehyde-protein adducts (Koskinas *et al.*, 1992; Teare *et al.*, 1993; Viitala *et al.*, 1997). However, the antibody response toward acetaldehyde adducts can not completely explain the immuno-allergic reactions associated with alcoholic liver disease, since anti-acetaldehyde antibodies can also be found in patients with liver diseases unrelated to alcohol (Worrall *et al.*, 1990).

Proteins alkylated by hydroxyethyl radicals are also immunogenic and lead to the formation of antibodies that specifically recognise hydroxyethyl radical epitopes. These antibodies have been detected in rats chronically fed with ethanol (Albano *et al.*, 1996) and in the sera of patients with alcoholic cirrhosis, but not in patients with non-alcoholic liver diseases (Clot *et al.*, 1995). In alcoholic patients the formation of anti-hydroxyethyl radical antibodies shows a good correlation with CYP2E1 activity, as measured by chloroxazone hydroxylation (Dupont *et al.*, 1998). Experiments using immunofluorescence and laser confocal microscopy have demonstrated that the anti-hydroxyethyl radical IgG from alcoholic patients reacts with epitopes present in the outer side of the plasma membrane of intact hepatocytes incubated *in vitro* with ethanol (Clot *et al.*, 1997). Western blot analysis of plasma membrane proteins from these cells has allowed the recognition of three main plasma membrane protein bands, one of which corresponds to CYP2E1-hydroxyethyl radical adducts (Clot *et al.*, 1997). Hydroxyethyl radical-CYP2E1 adducts on hepatocyte plasma membranes can also be detected by the co-localization of the immunofluorescence following combined cell immunostaining with anti-hydroxyethyl radical and anti-CYP2E1 antibodies (Clot *et al.*, 1997). Such a plasma membrane localization of CYP2E1-hydroxyethyl radical adducts is consistent with that of trifluoroacetyl-CYP2E1 and tienilic acid-CYP2C11 adducts on hepatocyte surface (Elliasson and Kenna, 1996; Robin *et al.*, 1996). The presence of these plasma membrane adducts appears to be critical for the development of immuno-mediated cytotoxicity, since the sera of patients with halothane-, diclofenac- or isaxonine-induced hepatitis stimulate human peripheral blood mononuclear cells to kill hepatocytes treated with the same drugs (van Pelt *et al.*, 1995; Manns and Obermayer-Straub, 1997). In a similar way, isolated rat hepatocytes exposed *in vitro* to ethanol can be killed by antibody-dependent cell-mediated cytotoxic (ADCC) reactions upon the addition of sera from alcoholic patients and normal human blood mononuclear cells (Clot *et al.*, 1997). This suggests that, during alcohol abuse, the development of immuno-toxic reactions towards hydroxyethyl radical-derived antigens might contribute to liver damage (Figure 12.4). In this respect, a clinical survey among alcoholic patients has associated the presence of antibodies reacting with alcohol-modified hepatocytes with an increased risk of developing liver cirrhosis (Takase *et al.*, 1993).

12.4.3 Oxidative mechanisms in the onset of alcohol-induced fibrosis

Liver fibrosis and cirrhosis represent the terminal stage of alcoholic liver disease and one of the main causes of death among patients with alcohol abuse. Moreover, pancreatic fibrosis is also a common outcome of chronic alcoholic pancreatitis. Increasing evidence indicates that oxidative stress is involved in the evolution of fibrotic processes associated with chronic inflammatory lung diseases, atherosclerosis as well as in liver fibrosis induced by iron or copper overload and chronic cholestasis (Poli and Parola, 1996; Olaso and Friedman, 1998). In all these conditions, the modulation by intracellular redox changes of nuclear factor kB (NFkB) and of activator protein-1 (AP-1) in tissue macrophages has been shown to induce the transcriptional up-regulation of genes encoding for pro-inflammatory or fibrogenetic cytokines. The secretion of granulocyte-macrophage colony stimulating factor (GM-CSF), tumor necrosis factor-β (TNF-β), interleukin-6 (IL-6) and transforming growth factor β1 (TGF-β1) by activated macrophages would then lead to the stimulation of matrix-producing cells (Poli and Parola, 1996; Olaso and Friedman, 1998). In the liver, the main connective tissue-producing cells are represented by hepatic stellate cells (perisinusoidal fat-storing cells or Ito cells), that are also activated under the influence of TGF-β1 and platelet-derived growth factor (PDGF) transforming to

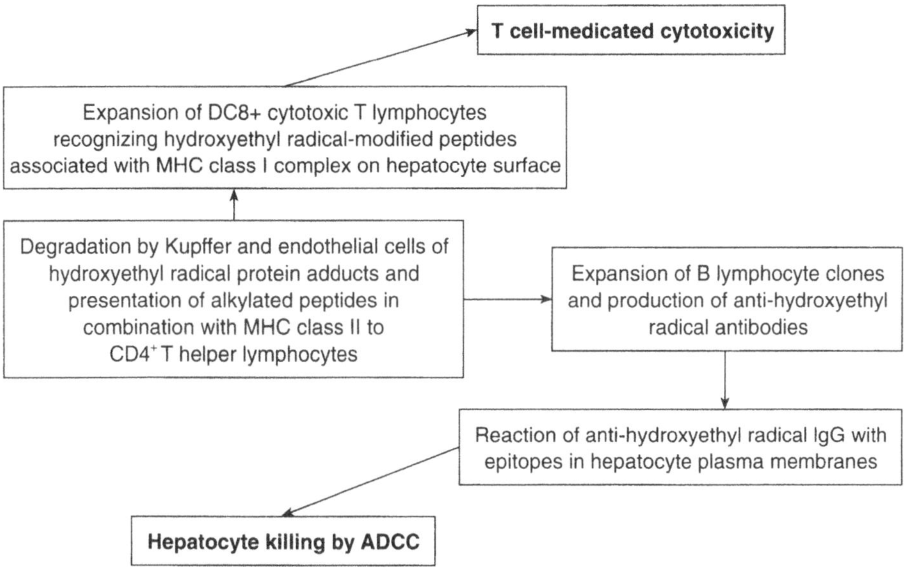

Figure 12.4 Possible role of hydroxyethyl radical protein of adducts in the development of immunotoxic reactions towards liver cells. Hydroxyethyl radicals bind covalently with CYP2E1 and other proteins in hepatocyte endoplasmic reticulum. Following hepatocyte death, the degradation of alkylated proteins by endothelial cells and liver phagocytes can lead to the presentation of the modified peptides in association with the major histocompatibility complex (MHC) class II molecules to $CD4^+$ helper T lymphocytes. The activation of $CD4^+$ lymphocyte may then provide help for further events involving either $CD8^+$ T lymphocytes or immature B lymphocyte clones. B lymphocytes can be stimulated to differentiate in plasmocytes secreting immunoglobulins of IgA and IgG class directed toward hydroxyethyl radical-derived antigens. The reaction of IgG with hydroxyethyl radical-modified proteins on the plasma membranes of hepatocytes can then trigger antibody-mediated cell-dependent cytotoxic (ADCC) reactions involving hepatocyte killing by macrophages and NK cells. Alternatively, the intracellular degradation of hydroxyethyl radical modified proteins may lead to the presentation on the hepatocyte surface of alkylated peptides in association with MHC class I proteins. The recognition as non-self of these peptides by cytotoxic $CD8^+$ T lymphocytes in association with the help of stimulated $CD4^+$ cells leads to the clonal expansion of T lymphocytes that will be then capable of killing all hepatocytes expressing hydroxyethyl modified peptides

myofibroblast-like cells (Olaso and Friedman, 1998). Recent studies using cultured human and rat hepatic stellate cells have shown that MDA and 4-HNE derived from lipid peroxidation are able to stimulate gene transcription and protein synthesis of collagene type 1 (Poli and Parola, 1996). The binding of 4-HNE with 46 and 54 k KD isoforms of c-Jun terminal kinase and their subsequent translocation to the nucleus of human hepatic stellate cells has been proposed to lead to AP-1 activation and the stimulation of procollagen gene expression (Parola *et al.*, 1998). The relevance of these mechanisms to the *in vivo* conditions comes from the observation that the biochemical and immunohistochemical detection of hepatic MDA and 4-HNE precedes the appearance of collagen deposition in experimental models of alcoholic liver fibrosis (Kamimura *et al.*, 1992; Kaminura and Tsukamoto, 1995; Niemelä *et al.*, 1995). Moreover, using intragastric alcohol-fed rats, Kaminura and Tsukamoto (1995) have observed that the stimulation of lipid peroxidation is associated with a marked induction of TNF-α, IL-6 and TGF-β1 production by Kupffer cells. The same group has also reported that dietary supplementation with carbonyl iron greatly stimulates alcoholic fibrosis. This effect is closely associated with the promotion of MDA and 4-HNE formation and with an increase in the levels of TGF-β1 and procollagen-α-1 mRNA in the whole liver and in freshly isolated hepatic stellate cells (Tsukamoto *et al.*, 1995). Although these data are far from conclusive, the possibility that ethanol-induced oxidative stress might contribute to the induction of liver fibrogenesis through mechanisms involving free radical damage is very attractive.

12.4.4 Oxidative stress and foetal alcohol syndrome

The *in utero* exposure of human foetus to ethanol due to alcohol consumption by the mother is often accompanied by growth retardation, craniofacial malformations, heart defects and mental deficit (Clarren *et al.*, 1978). Despite foetal alcohol syndrome having been recognized for about 25 years, the mechanisms of foetus damage by alcohol are still largely unknown. Studies *in vitro* using foetal rat hepatocytes and mouse cranial neural crest cells have shown that ethanol blocks the replicatory response to epidermal growth factor and exerts direct cytotoxicity (Devi *et al.*, 1993; Chen and Sulik, 1996). These effects are associated with the stimulation of O_2^- production and lipid peroxidation and can be prevented by antioxidants (Devi *et al.*, 1993; Chen and Sulik, 1996). The cytotoxic effect of oxidative damage in the cranial neural crest is of special interest, since craniofacial malformations associated with the foetal alcohol syndrome have been related to alterations of the foetal structures derived from the neural crest. Mitochondria appear to be the main target of oxidative damage in foetal cells, suggesting that the impairment of mitochondrial functions might be responsible for cell growth inhibition and cytotoxicity in the developing tissues. These observations are supported by *in vivo* experiments demonstrating that two days' exposure of pregnant rats to ethanol results in the stimulation of lipid peroxidation in either the foetal brain or liver (Henderson *et al.*, 1995). Interestingly, comparative analysis of maternal tissues has not revealed appreciable signs of oxidative stress. The enhanced susceptibility of foetal tissues to free radical attacks can been related to a lower content of antioxidants as well as to a relative inefficiency of the enzymes involved in the detoxification of lipid peroxidation products (Henderson *et al.*, 1995; Chen *et al.*, 1997).

12.5 Conclusions

A growing body of experimental and clinical studies has linked the exposure to alcohol with the presence of free radical-mediated oxidative damage in several tissues. In many instances, the lowering of cellular antioxidant defences along with the production of reactive oxygen species by different enzymatic sources, including mitochondrial electron transport chain, alcohol-inducible cytochrome P4502E1 (CYP2E1), xanthine oxidase, aldehyde oxidase and activated phagocytes have been proposed as mechanisms for ethanol-induced oxidative stress. Ethanol oxidation by

CYP2E1 is instead responsible for the conversion of ethanol itself to hydroxyethyl free radical intermediates. Although no definitive conclusions have yet been reached, much evidence suggests that oxidative modifications of cellular components and the stimulation of lipid peroxidation are likely to represent the mechanisms by which free radical-mediated reactions can contribute to the development of alcohol cytotoxicity, as well as to promote tissue fibrosis. In addition, immunological reactions triggered by hydroxyethyl radical adducts with liver proteins are able to induce immune reactions towards hepatocytes that might contribute to alcohol toxicity.

References

Abbondanza, A., Battelli, M.G., Soffritti, M. and Cessi, C. (1989) Xanthine oxidase status in ethanol-intoxicated rat liver. *Alcohol. Clin. Exp. Res.*, **13**, 841–844.

Albano, E., Tomasi, A., Goria-Gatti, L., Poli, G., Vannini, V. and Dianzani, M.U. (1987) Free radical metabolism of alcohols in rat liver microsomes. *Free Rad. Res. Communs.*, **3**, 243–249.

Albano, E., Tomasi, A., Goria-Gatti, L. and Dianzani, M.U. (1988) Spin-trapping of free radical species produced during the microsomal metabolism of ethanol. *Chem. Biol. Inter.*, **65**, 223–234.

Albano, E., Tomasi, A., Goria-Gatti, L., Persson, J.O., Terelius, Y. and Ingelman-Sundberg, M. and Dianzani, M.U. (1991) Role of ethanol-inducible cytochrome P-450 (P450IIE1) in catalyzing the free radical activation of aliphatic alcohols. *Biochem. Pharmacol.*, **41**, 1895–1902.

Albano, E., Parola, M., Comoglio, A. and Dianzani, M.U. (1993) Evidence for the covalent binding of hydroxyethyl radicals to rat liver microsomal proteins. *Alcohol Alcohol.*, **28**, 453–459.

Albano, E., Tomasi, A. and Ingelman-Sundberg, M. (1994a) Spin-trapping of alcohol-derived radicals in microsomes and recostituted systems by electron spin resonance. *Meth. Enzymol.*, **233**, 117–127.

Albano, E., Clot, P., Comoglio, A., Dianzani, M.U. and Tomasi, A. (1994b) Free radical activation of acetaldehyde and its role in protein alkylation, *FEBS Lett.*, **384**, 65–70.

Albano, E., Clot, P., Morimoto, M., Tomasi, A., Ingelman-Sundberg, M. and French, S.W. (1996) Role of cytochrome P4502E1-dependent formation of hydroxyethyl free radicals in the development of liver damage in rats intragastrically fed with ethanol. *Hepatology*, **23**, 155–163.

Aleynik, S.I., Leo, M.A., Aleynik, M.K. and Lieber, C.S. (1998) Increased circulating products of lipid peroxidation in patients with alcoholic liver disease. *Alcohol. Clin. Exp. Res.*, **22**, 192–196.

Altomare, E., Grattagliano, I., Vendemiale, G., Palmieri, V. and Palasciano, G. (1996) Acute ethanol administration induces oxidative changes in rat pancreatic tissue. *Gut*, 38, 742–746.

Aust, S.D., Morehouse, L.A. and Thomas, C.E. (1985) Role of metals in oxygen radical reactions, *J. Free Rad. Biol. Med.*, **1**, 3–25.

Baldi, E., Burra, P., Plebani, M. and Salvagnini, M. (1993) Serum malondialdehyde and mitochondrial aspartate amino transferase activity as markers of chronic alcohol intake and alcoholic liver disease. *Ital. J. Gastroenterol.*, **25**, 429–432.

Bautista, A. (1997) Chronic alcohol intoxication induces hepatic injury through enhanced macrophage inflammatory protein-2 production and intracellular adesion molecule-1 expression in the liver. *Hepatology*, **25**, 335–342.

Bautista, A.P. and Spitzer, J.J. (1992) Ethanol intoxication stimulates superoxide anion production by in situ perfused rat liver, *Hepatology*, **15**, 892–898.

Bautista, A.P. and Spitzer, J.J. (1994) Inhibition of nitric oxide formation in vivo enhances superoxide release by the perfused liver. *Am. J. Physiol.*, **266**, G783–G788.

Bautista, A.P. and Spitzer, J.J. (1996) Postbinge effects of acute alcohol intoxication on hepatic free radical formation. *Alcoholism Clin. Exp. Res.*, **20**, 502–509.

Bautista, A.P., D'Souza, N.B., Lang, C.H. and Spitzer, J.J. (1992) Modulation of f-met-leu-phe-induced chemotattic activity and superoxide production by neutrophils during chronic ethanol intoxication. *Alcoholism Clin. Exp. Res.*, **16**, 788–794.

Beckman, J.S. and Koppenol, W.H. (1996) Nitric oxide, superoxide and peroxynitrite: the good, the bad and the ugly. *Am. J. Physiol.*, **271**, C1424-C1437.

Bell, H., Bjørneboe, A., Eidsvoll, B., Norum, K.R., Raknerud, N., Try, K., Thomassen, Y. and Drevon, A.C. (1992) Reduced concentration of hepatic α-tocopherol in patients with alcoholic liver cirrhosis. *Alcohol Alcohol.*, **27**, 39–46.

Bjørneboe G.-E.A., Bjørneboe, A., Hagen B.F., Mørland, J. and Drevon C.A. (1987) Reduced hepatic α-tocopherol content after long-term administration of ethanol to rats. *Biochem. Biophys. Res. Communs.*, **918**, 236–241.

Bosch-Morell, F., Martìnez-Soriano, F., Colell, A., Fernandez-Checa, J.C. and Romero, F.J. (1998) Chronic ethanol feeding induces cellular antioxidants decrease and oxidative stress in rat peripheral nerves. Effect of S-adenosyl-L-methionine and N-acetyl-L-cysteine. *Free Rad. Biol. Med.*, **25**, 365–368.

Boveris, A., Llesuy, S., Azzalis, L.A., Giavarotti, L., Simon, K.A., Junqueira, V.B., Porta, E.A., Videla, L.A. and Lissi, E.A. (1997) In situ rat brain and liver spontaneus chemiluminescence after acute ethanol intake. *Toxicol. Lett.*, **93**, 23–28.

Buttke, T.M. and Sandstrom, P.A. (1994) Oxidative stress as a mediator of apoptosis. *Immunol. Today*, **15**, 7–10.

Cahill, A., Wang, X. and Hoek, J.B. (1997) Increased oxidative damage to mitochondrial DNA following chronic ethanol consumption. *Biochim. Biophys. Res. Commun.*, **235**, 286–290.

Cairo, G., Castrusini, E., Minotti, G. and Bernelli-Zazzera A. (1996) Superoxide and hydrogen peroxide-dependent inhibition of iron regulatory protein activity: a protective stratagem against oxidative injury. *FASEB J.*, **10**, 1326–1335.

Cederbaum, A.I. (1989) Oxygen radical generation by microsomes: Role of iron and implications for alcohol metabolism and toxicity, *Free Rad. Biol. Med.*, **7**, 559–562.

Chamulitrat, W. and Spitzer, J.J. (1996) Nitric oxide and liver injury in alcohol-fed rats after lipipolysaccaride administration. *Alcohol. Clin. Exp. Res.*, **20**, 1065–1070.

Chapman, R.W., Morgan, M.J., Bell, R. and Sherlock, S. (1983) Hepatic iron uptake in alcoholic liver disease, *Gastroenterology*, **84**, 143–148.

Chen, S. and Sulik, K.K. (1996) Free radicals and ethanol-induced cytotoxicity in neural crest cells. *Alcohol. Clin. Exp. Ther.*, **20**, 1071–1076.

Chen, S., Schenker, S. and Henderson, G.I. (1997) 4-hydroxynonenal levels are enhanced in fetal liver mitochondria by in utero ethanol exposure. *Hepatology*, **25**, 142–147.

Clarren, S.K. and Smith, D.W. (1978) The fetal alcohol syndrome. *N. Engl. J. Med.* **298**, 1063–1067.

Clot, P., Tabone, M., Aricò, S. and Albano, E. (1994) Monitoring oxidative damage in patients with liver cirrhosis and different daily alcohol intake. *Gut*, **35**, 1637–1643.

Clot, P., Bellomo, G., Tabone, M., Aricò, S. and Albano, E. (1995) Detection of antibodies against proteins modified by hydroxyethyl free radicals in patients with alcoholic cirrhosis. *Gastroenterology*, **108**, 201–207.

Clot, P., Albano, E., Elliasson, E., Tabone, M., Aricò, S., Israel, Y., Moncada, Y. and Ingelman-Sundberg, M. (1996) Cytochrome P4502E1 hydroxyethyl radical adducts as the major antigenic determinant for autoantibody formation among alcoholics. *Gastroenterology*, **111**, 206–216.

Clot, P., Parola, M., Bellomo, G., Dianzani, U., Carini, R., Tabone, M., Aricò, S., Ingelman-Sundberg, M. and Albano, E. (1997) Plasma membrane hydroxyethyl radical adducts cause antibody-dependent cytotoxicity in rat hepatocytes exposed to alcohol. *Gastroenterology*, **113**, 265–276.

Coleman, W.B., Cahill, A., Ivester, P. and Cunningham (1994) Differential effects of ethanol consumption on synthesis of cytoplasmic and mitochondrial encoded subunits of the ATP synthase. *Alcohol. Clin Exp. Res.*, **18**, 947–950.

Colell, A., Garcia-Ruiz, C., Morales, A., Ballesta, A., Ookhtens, M., Rodes, J., Kaplowitz, N. and Fernandez-Checha, J.C. (1997) Transport of reduced glutathione in hepatic mitochondria and mitoplasts from ethanol-treated rats: Effect of membrane physical properties and S-adenosyl-L-methionine. *Hepatology*, **26**, 699–708.

Cottalasso, D., Gazzo, P., Dapino, D., Domenicotti, C., Pronzato, M.A., Traverso, N., Bellocchio, A., Nanni, G. and Marinari, U.M. (1996) Effect of chronic ethanol consumption on glycosylation processes in rat liver microsomes. *Alcohol Alcohol.*, **31**, 51–59.

Devi, B,G., Henderson, G.I., Frosto, T.A. and Shenker, S. (1993) Effect of ethanol on rat fetal hepatocyte: studies on replication, lipid peroxidation and glutatione. *Hepatology*, **18**, 648–659.

Di Luzio, N.R. (1963) Prevention of acute ethanol-induced fatty liver by antioxidants. *Physiologist*, **6**, 169–173.

Dianzani, M.U. (1985) Lipid peroxidation in ethanol poisoning: a critical reconsideration, *Alcohol Alcohol.*, **20**, 161–173.

Dianzani, M.U. (1991) Biochemical aspects of fatty liver. In Meeks, E.G., Harrison, S.D. and Bull, R.J. (eds) *Hepatotoxicology*, Boca Raton: CRC Press, pp. 327–400.

Dicker, E. and Cederbaum, A.I. (1992) Increases NADH-dependent production of reactive oxygen intermediates by microsomes after chronic ethanol consumption: Comparisons with NADPH, *Arch. Biochem. Biophys.*, **293**, 274–280.

Dorio, R.J., Hoek, J.B., Rubin, E. and Forman, H.J. (1988) Ethanol modulation of rat alveolar macrophage superoxide production. *Biochem. Pharmacol.*, **37**, 3528–3533.

Dufour, M.C., Stinson, F.S. and Fe Cases, M. (1993) Trends in cirrhosis morbidity and mortality: United States 1979–1988. *Semin. Liver Dis.*, **13**, 109–125.

Dupont, I., Lucas, D., Clot, P., Ménez, C. and Albano, E. (1998) Cytochrome P4502E1 inducibility and hydroxyethyl radical formation among alcoholics. *J. Hepatol.*, **28**, 564–571.

Ekström, G. and Ingelman-Sundberg, M. (1989) Rat liver microsomal NADPH-supported oxidase activity and lipid peroxidation dependent on ethanol-inducible cytochrome P450, *Biochem. Pharmacol.*, **38**, 1313–1319.

Ekström, G., Von Bahr, C. and Ingelman-Sundberg, M. (1989) Human liver microsomal cytochrome P450IIE1. Immunological evaluation of its contribution to microsomal ethanol oxidation, carbon tetrachloride reduction and NADPH oxidase activity, *Biochem. Pharmacol.*, **38**, 689–693.

Eliasson, E. and Kenna, J.G. (1996) Cytochrome P450 2E1 is a cell surface autoantigen in halothane hepatitis. *Mol. Pharmacol.*, **50**, 573–582.

Fernandez-Checa, J.C., Ookhtens, M. and Kaplowitz N. (1987) Effect of chronic ethanol feeding on rat hepatocytic glutathione compartimentation, efflux and response to incubation with ethanol. *J. Clin Invest.*, **80**, 57–62.

Fernandez-Checa, J.C., Garcia-Ruiz, C., Ookhtens, M. and Kaplowitz, N. (1991) Impaired uptake of glutathione by hepatic mitichondria from ethanol fed rats. *J. Clin. Invest.*, **87**, 397–405.

Fernandez-Checa, J.C., Kaplowitz N., Garcia-Ruiz, C., Collel, A., Miranda, M., Marì, M., Ardite, E. and Morales, A. (1997) GSH transport in the mitichondria: defense against TNF-induced oxidative stress and alcohol-induced defect. *Am. J. Physiol.*, **273**, G7–G17.

Forman, H.J. and Boveris, A. (1988) Superoxide radical and hydrogen peroxide in mitochondria. In Pryor, W.A. (ed.) *Free Radicals in Biology*, Vol. V, New York: Academic Press, pp. 65–82.

Foster, J.R., Idle, J.R., Bars, R., Scott, P. and Braganza, J.M. (1993) Induction of drug-metabolizing enzymes in human pancreatic cancer and pancreatitis. *J. Pathol.*, **169**, 457–463.

French, S.W., Wong, K., Jui, L., Albano, E., Hagbjörk, A.-L. and Ingelman-Sundberg, M. (1993) Effect of ethanol on cytochrome P450 (CYP2E1), lipid peroxidation and serum protein adduct formation in relation to liver pathology pathogenesis, *Exp. Mol. Pathol.*, **58**, 61–75.

Fromenty, B. and Pessayre, D. (1995) Inhibition of mitochondrial beta-oxidation as a mechanism of hepatotoxicity. *Pharmacol. Ther.*, **67**, 101–154.

Fromenty, B., Grimbert, S., Mansouri, A., Beaugrand, M., Erlinger, S., Röting, A. and Pessayre, D. (1995) Hepatic mitochondrial DNA deletion in alcoholics: association with microvesicular steatosis. *Gastroenterology*, **108**, 193–200.

Garcia-Ruitz, C., Morales, A., Ballesta, A., Rhodes, J., Kaplowitz, N. and Fernandez-Checa, J.C. (1994) Effect of chronic ethanol feeding on glutathione and functional integrity of mitochondria in periportal and perivenous rat hepatocytes. *J. Clin. Invest.*, **94**, 193–201.

Gergel, D., Nisik, V., Riesz, P. and Cederbaum, A.I. (1997) Inhibition of rat and human cytochrome P4502E1 catalytic actitity and reactive oxygen radical formation by nitric oxide. *Arch. Biochem. Biophys.*, **337**, 239–250.

Grattagliano, I., Vendemiale, G., Sabbà, Buonamico, P. and Altomare, E. (1996) Oxidation of circulating proteins in alcoholics: role of acetaldehyde and xanthine oxidase. *J. Hepatol.*, **25**, 28–36.

Grattagliano, I., Vendemiale, G., Errico, F., Bolognino, A.E., Lillo, F., Salerno, M.T. and Altomare, E. (1997) Chronic ethanol intake induces alterations in rat testis. *J. Appl. Toxicol.*, **17**, 307–311.

Green, D. and Kroemer, G. (1998) The central executioners of apoptosis: caspases or mitochondria? *Trends Cell Biol.*, **8**, 267–271.

Grossrau, R., Federiks, W.M. and Van Noorden, C.J. (1990) Histochemistry of reactive oxygen-species (ROS) generating oxidases in cutaneous and mucus epithelial of laboratory rodents with special reference to xanthine oxidase, *Histochemistry*, **94**, 539–545.

Guerri, C. and Grisolia, S. (1980) Changes in glutathione in acute and chronic alcohol intoxication. *Pharmacol. Biochem. Behav.*, **13**, 53–61.

Hansson, T., Tindberg, N. and Ingelman-Sundberg, M. (1990) Regional distribution of ethanol inducible cytochrome P450IIE1 in rat central nervous system. *Neuroscience*, **34**, 451–458.

Harbrecht, B.G. and Billiar, T.R. (1995) The role of nitric oxide in Kupffer cell-hepatocyte interactions. *Shock*, **3**, 79–87.

Henderson, G.I., Devi, B.G. and Schenker, S. (1995) In utero ethanol exposure elicits oxidative stress in the rat fetus. *Alcohol. Clin. Exp. Ther.*, **19**, 714–720.

Hirokawa, M., Miura, S., Kurose, I., Shigematzu, T., Hokari, R., Higuchi, H., Watanabe, N., Yokoyama, Y., Kimura, H., Kato, S. and Ishii, H. (1998) Oxidative stress and mitochondrial damage precedes gastric mucosal cell death induced by ethanol. *Alcohol. Clin. Exp. Ther.*, **22**, 111S–114S.

Iimuro, Y., Bradford, B.U., Gao, W., Kadiiska, M., Mason, R.P., Stefanovic, B., Brenner, D.A. and Thurman, R.G. (1996) Detection of α-hydroxyethyl free radical adducts in the pancreas after chronic exposure to alcohol in the rat. *Mol. Pharmacol.*, **50**, 656–661.

Irving, M.G., Halliday, J.W. and Powell, L.W. (1988) Association between alcoholism and increased hepatic iron store, *Alcohol. Clin. Exp. Res.*, **12**, 7–12.

Ishak, K.G., Zimmerman, H.J. and Ray, M.B. (1991) Alcoholic liver disease: pathology, pathogenetic and clinical ascpects. *Alcohol. Clin. Exp. Res.*, **15**, 45–66.

Israel, Y., Hurwitz, E., Niemela, O. and Arnon, R. (1986) Monoclonal and polyclonal antibodies against acetaldehyde-containing epitopes in acetaldehyde-protein adducts. *Proc. Natl Acad. Sci. USA*, **83**, 7923–7927.

Israel, Y., Speisky, H., Lança, A.J., Iwamura, S., Hirai, M. and Vargese, G. (1992) Metabolism of hepatic glutathione and its relevance in alcohol induced liver damage. In Clément, B. and Guillouzo, A. (eds) *Cellular and Molecular Aspects of Cirrhosis*, Colloque INSERM/J Vol. 216, London: John Libbey Eurotest Ldt, pp. 25–37.

Jaatinen, P., Saukko, P. and Hervonen, A. (1993) Chronic ethanol exposure increases lipopigment accumulation in human heart. *Alcohol Alcohol.*, **28**, 559–569.

Jaatinen, P., Saukko, P., Sarviharju, M., Kiianmaa, K. and Hervonen, A. (1994) Effect of lifelong ethanol exposure on the ultrastructure and lipopigmentation of rat heart. *Alcohol Alcohol.*, **29**, 269–282.

Jenner, P. and Olanow, C.W. (1996) Oxidative stress and the pathogenesis of Parkinson's disease. *Neurology*, **47**, S161–170.

Jewell, S.A., Di Monte, D., Gentile, A., Guglielmini, A., Altomare, E. and Albano, O. (1986) Decreased hepatic glutathione levels in chronic alcoholic patients. *J. Hepatol.*, **3**, 1–6.

Kamimura, S. and Tsukamoto, H. (1995) Cytokine gene expression by Kupffer cells in experimental alcoholic liver disease. *Hepatol*ogy **21**, 1304–1309.

Kamimura, S., Gall, K., Britton, S.R., Bacon, B.R., Triadafilopulos, G. and Tsukamoto, H. (1992) Increased 4-hydroxynonenal levels in experimental alcoholic liver disease: Association of lipid peroxidation with liver fibrogenesis, *Hepatology*, **16**, 448–453.

Kato, I. (1996) The extent of the problem and the epidemiological aspects of alcohol drinking. In Preedy, V.R. and Watson, R.R. (eds) *Alcohol and the Gastrointestinal Tract*, Boca Raton: CRC Press, pp. 1–17.

Kato, S., Kavase, T., Alderman, J., Inatomi, N. and Lieber, C.S. (1990) Role of xanthine oxidase in ethanol-induced lipid peroxidation, *Gastroenterology*, **98**, 203–210.

Kawase, T., Kato, S. and Lieber, C.S. (1989) Lipid peroxidation and antioxidant defense systems in rat liver after chronic ethanol feeding. *Hepatology*, **10**, 815–821.

Kera, Y., Ohbora, Y. and Komura, S. (1982) Buthionine sulfoximine inhibition of glutathione biosynthesis enhances hepatic lipid peroxidation in rats during acute ethanol intoxication. *Alcohol Alcohol.*, **24**, 519–524.

Kessova, I.G., De Carli, L.M. and Lieber, C. (1998) Inducibility of cytochromes P4502E1 and P4501A1 in the rat pancreas. *Alcohl. Clin. Exp. Res.*, **22**, 501–504.

Knecht, K.T., Bradfort, B.U., Mason R.P. and Thurman, R.G. (1990) In vivo formation of free radical metabolite of ethanol. *Mol. Pharmacol.*, **38**, 26–30.

Knecht, K.T., Thurman R.G. and Mason, R.P. (1993) Role of superoxide and trace transition metals in the production of α-hydroxyethyl radical from ethanol by microsomes from alcohol dehydrogenase-deficient deermice. *Arch. Biochem. Biophys.*, **303**, 339–348.

Koskinas, J., Kenna, J.G., Bird, G.L., Alexander, G.J.M. and Williams, R. (1992) Immunoglobulin A antibody to a 200-kilodalton cytosolic acetaldehyde adduct in alcoholic hepatitis. *Gastroenterology*, **103**, 1860–1867.

Kroemer, G. (1997) The proto-oncogene Blc-2 and its role in regulating apoptosis. *Nature Genetic.*, **3**, 614–620.

Kukielka, E., Dicker, E. and Cederbaum, A.I. (1994) Increased production of reactive oxygen species by rat liver mitochondria after chronic ethanol treatment, *Arch. Biochem. Biophys.*, **309**, 377–386.

Kukielka, E. and Cederbaum, A.I. (1996) Ferritin stimulation of lipid peroxidation by microsomes after chronic ethanol treatment. Role of cytochrome P4502E1. *Arch. Biochem. Biophys.*, **332**, 121–127.

Kurose, I., Higuchi, H., Kato, S., Miura, S., Watanabe, N., Kamegaya, Y., Tomita, K., Takaishi, M., Horie, Y., Fukuda, M., Mizukami, K. and Ishii, H. (1997a) Oxidative stress on mitochondria and cell membrane of cultured rat hepatocytes and perfused liver exposed to ethanol. *Gastroenterology*, **112**, 1331–1343.

Kurose, I., Higuchi, H., Miura, S., Saito, H., Watanabe, N., Hokari, R., Hirokawa, M., Takaishi, M., Zeki, S., Nakamura, T., Ebinuma, H., Kato, S. and Ishii, H. (1997b) Oxidative stress-mediated apoptosis of hepatocytes exposed to acute ethanol intoxication. *Hepatology*, **25**, 368–378.

Lauterburg, B.H., Davies, S. and Mitchell, J.R. (1984) Ethanol suppresses hepatic glutathione synthesis in rats *in vivo*. *J. Pharmacol. Exp. Ther.*, **230**, 7–11.

Lelbach, W.K. (1975) Quantitative aspects of drinking in alcoholic liver cirrhosis. In Khanna, H.M., Israel, Y., and Kalant, H., (eds) *Alcoholic Liver Pathology*, Toronto, Addiction Research Foundation of Ontario, pp. 1–18.

Lecomte, E., Herberth, B., Pirrolet, P., Chancerelle, Y., Arnaud, J., Musse, N., Paille, F., Siest, G. and Artur Y. (1994) Effect of alcohol consumption on blood antioxidant nutrients and oxidative stress indicators. *Am. J. Clin. Nutr.*, **60**, 255–261.

Letteron, P., Duchettelle, V., Berson, A., Fromenty, B., Fish, C., Degott, C., Benhamou, P.J. and Pessayre, D. (1993) Increased ethane exhalation, an in vivo index of lipid peroxidation, in alcohol abusers, *Gut*, **34**, 409–414.

Lieber, C.S., Casini, A., De Carli, L.M., Kim, C-I., Lowe, N., Sasaki, R. and Leo, M.A. (1990) S-adenosyl-L-methionine attenuates alcohol-induced liver injury in baboon. *Hepatology*, **11**, 165–172.

Lieber, C.S. (1994) Alcohol and the liver: 1994 update. *Gastroenterology*, **106**, 1085–1105.

Lieber, C.S., Leo, M.A., Aleynik, S.I., Aleynik, M.K. and De Carli, L. (1997) Polyenylphosphatidylcholine decreases alcohol-induced oxidative stress in baboon. *Alcohol Clin. Exp. Res.*, **21**, 375–379.

Lin. R.T., Gentry, T.R., Ito, D. and Yokoyama, H. (1994) First pass metabolism of ethanol is predominantly gastric. *Alcohol Clin. Exp. Res.*, **17**, 1228–1232.

Loguercio, C., Romano, M., Di Sapio M., Nardi, G., Taranto, D., Grella, A. and Del Vecchio Blanco, C. (1991) Regional variations in total and nonprotein sulphydryl compounds in the human gastric mucosa and effect of ethanol. *Scan. J. Gastroenterol.*, **26**, 1042–1047.

Manns, M.P. and Obermayer-Straub, P. (1997) Cytochrome P450 and uridine triphosphate-glucuronyl transferases: model autoantigens to study drug-induced, virus-induced and autoimmune liver disease. *Hepatology*, **26**, 1055–1066.

Mansouri, A., Fromenty, B., Berson, A., Robin, M.A., Grimbert, S., Beaugrand, M., Erlinger, S. and Pessayre, D. (1997) Multiple hepatic mitochondrial DNA deletions suggest premature oxidative aging in alcoholics. *J. Hepatol.*, **27**, 96–102.

Marinari, U.M., Pronzato, M.A., Cottalasso, D., Rolla, C., Biasi, F., Nanni, G. and Dianzani, M.U. (1987) Inhibition of liver Golgi glycosylation activities by carbonyl products of lipid peroxidation. *Free Rad. Res. Commun.*, **3**, 319–324.

Mascotti, D.P., Rup, D. and Thach, R.E. (1995) Regulation of iron metabolism. Effect mediated by iron, heme and cytokines. *Annu. Rev. Nutr.*, **15**, 239–261.

Minotti, G., Di Gennaro, M., D'Ugo, D. and Granone, P. (1991) Possible source of iron for lipid peroxidation, *Free Rad. Res. Comms.*, **12**, 99–110.

Mira, L., Maia, I., Barreira, L. and Manso, C.F. (1995) Evidence for free radical generation to NADH oxidation by aldehyde oxidase during ethanol metabolism. *Arch. Biochem. Biophys.* **318**, 53–58.

Mizui, T. and Doteuchi, M. (1986) Lipid peroxidation: A possible role on gastric damage induced by ethanol in rats. *Life Sci.*, **38**, 2163–2167.

Moncada, C., Torres, V., Vargese, E., Albano, E. and Israel, Y. (1994) Ethanol-derived immunoreactive species formed by free radical mechanisms. *Mol. Pharmacol.*, **46**, 786–791.

Montoliu, C., Valles, S., Renau-Piqueras, J. and Guerri, C. (1994). Ethanol-induced oxygen radical formation and lipid peroxidation in rat brain: effect of chronic alcohol consumption. *J. Neurochem.*, **63**, 1855–1862.

Montoliu, C., Sancho-Tello, M., Azorin, I., Burgal, M., Valles, S., Renau-Piqueras, J. and Guerri, C. (1995). Ethanol increases cytochrome P4502E1 and induces oxidative stress in astrocytes. *J. Neurochem.*, **65**, 2561–2570.

Moore, D.R., Reinke, L.A. and McCay, P.B. (1995). Metabolism of ethanol to 1-hydroxyethyl radicals in vivo: Detection with intravenous administration of α-(4-pyridyl-1-oxide)N-t-butylnitrone. *Mol. Pharmacol.*, **47**, 1224–1230.

Morimoto, M., Hagbjvrk, A-L., Wan, Y.J.Y., Fu, P.C., Ingelman-Sundberg, M., Albano, E., Clot, P. and French, S.W. (1995) Modulation of alcoholic liver disease by cytochrome P4502E1 inhibitors. *Hepatology*, **21**, 1610–1617.

Multhaup, G., Ruppert, T., Schicksupp, A., Hesse, L., Beher, D., Masters, C.L. and Beyreuther, K. (1997) reactive oxygen species and Alzheimer's disease. *Biochem. Pharmacol.*, **54**, 533–539.

Mutoh, H., Hiraishi, H., Ota, S., Ivey, K.J., Terano, A. and Sugimoto, T. (1990) Role of oxygen radicals in ethanol-induced damage to cultured gastric mucosal cells, *Am. J. Physiol.*, **258**, G603–G609.

Nakano, M., Kikuyama, M., Hasegawa, T., Ito, T., Sakurai, K., Hiraishi, K., Hashimura, E., Adachi, M. (1995) The first observation of O_2^- generation at real time in vivo from non-Kupffer sinusoidal cells in perfused rat liver during acute ethanol intoxication. *FEBS Lett.*, **372**, 140–143.

Nanji, A.A., Khwaja, S., Tahan, S.R. and Sadrzadeh, H.S.M. (1994a) Plasma levels of a novel noncyclooxygenase-derived prostanoid (8-isoprostane) correlate with severity of liver injury in experimental alcoholic liver disease. *J. Pharmacol. Exp. Ther.*, **269**, 1280–1285.

Nanji, A.A., Zhao, S., Sadrzadeh, S.M.H., Dannenberg, A.J., Tahan, S.R. and Waxman D.J. (1994b) Markedly enhanced cytochrome P4502E1 induction and lipid peroxidation is associated with severe liver injury in fish oil-treated ethanol-fed rats. *Alcohol. Clin. Exp. Res.*, **18**, 1280–1285.

Nanji, A.A., Greenberg, S.S., Tahan, S.R., Fogt, F., Loscalzo, J., Sadrzadeh, S.M.H., Xie, J. and Stamler, J.S. (1995a) Nitric oxide production in experimental alcoholic liver disease in the rat: role in protection from injury. *Gastroenterology*, **109**, 899–907.

Nanji, A.A., Griniuviene, B., Sadrzadeh, S.M.H., Levitsky, S. and McCully, J.D. (1995b) Effect of dietary fat and ethanol on antioxidant enzyme mRNA induction in rat liver. *J. Lipid Res.*, **36**, 736–744.

Nanji, A.A., Yang, E.K., Fogt, F., Sadrzadeh, S.M.H. and Dannenberg, A.J. (1996) Medium chain triglycerides and vitamin E reduce the severity of established experimental alcoholic liver disease. *J. Pharmacol. Exp. Ther.*, **277**, 1694–1700.

Nicholls, R., De Jersey, J., Worrall, S. and Wilce, P. (1992) Modification of proteins and other biological molecules by acetaldehyde: adduct structure and functional significance. *Int. J. Biochem.*, **24**, 1899–1906.

Niemelä, O., Parkkila, S., Ylä-Herttuala, S., Halsted, C., Witztum, J.L., Lanca, A., Israel, Y. (1994) Covalent protein adducts in the liver as a result of ethanol metabolism and lipid peroxidation. *Lab. Invest.*, **70**, 537–546.

Niemelä, O., Parkkila, S., Ylä-Herttuala, S., Villanueva, J., Ruebner, B. and Halsted, C.H. (1995) Sequential acetaldehyde production, lipid peroxidation and fibrogenesis in micropigs model of alcohol-induced liver disease. *Hepatology*, **22**, 1208–1214.

Nordback, I.H., Olson, J.L., Chacko, V.P. and Cameron, J.L. (1995) Detailed characterization of experimental acute alcoholic pancreatitis. *Surgery*, **117**, 41–49.

Nordmann, R., Ribière, C., Rouach, H. (1992) Implication of free radical mechanisms in ethanol induced cellular injury, *Free Rad. Biol. Med.*, **12**, 219–240.

Nordmann, R. (1994) Alcohol and antioxidant systems. *Alcohol Alcohol.*, **29**, 513–522.

Norton, I.D., Apte, M.V., Haber, P.S., McCaughan, G.W., Pirola, R.C. and Wilson, J.S. (1998) Cytochrome P4502E1 is present in rat pancreas and is induced by chronic ethanol administration. *Gut*, **42**, 426–430.

Olaso, E. and Friedman, S.L. (1998) Molecular regulation of hepatic fibrogensis. *J. Hepatol.*, **29**, 836–847.

Panchenko, L.P. Pirozhkov, S.V., Papova, S.V. and Antonenkov, V.D. (1987) Effect of chronic ethanol treatment on peroxisomal acyl CoA oxidase activity and lipid peroxidation in rat liver and heart. *Experientia*, **43**, 580–583.

Parola, M., Robino, G., Marra, F., Pinzani, M., Bellomo, G., Leonarduzzi, G., Chiarugi, P., Camandola, S., Poli, G., Waeg, G., Gentilini, P. and Dianzani, M.U. (1998) HNE interacts directly with JNK isoforms in human hepatic stellate cells. *J. Clin. Invest.*, **102**, 1942–1950.

Paronetto, F. (1993) Immunologic reactions in alcoholic liver disease. *Sem. Liver Dis.*, **13**, 183–195.

Patel, T. and Gores, G.J. (1995) Apoptosis and hepatobiliary disease. *Hepatology*, **21**, 1725–1744.

Persson, J.O., Terelius, Y. and Ingelman-Sundberg, M. (1990) Cytochrome P450-dependent formation of reactive oxygen radicals: Isozyme-specific inhibition of P450-mediates reduction of oxygen and carbon tetrachloride, *Xenobiotica*, **20**, 887–900.

Pihan, G., Regillo, C. and Szabo, S. (1987) Free radicals and lipid peroxidation in the ethanol- or aspirin-induced gastric mucosal injury. *Dig. Dis. Sci.*, **32**, 1395–1400.

Polavarapu, R., Spitz, D.R., Sim, J.E., Follansbee, M.H., Oberley, L.W., Rahemtulla, A. and Nanji, A. (1998) Increased lipid peroxidation and impaired antioxidant enzyme function is associated with pathological liver injury in experimental alcoholic liver disease in rats fed diets high in corn oil and fish oil. *Hepatology*, **27**, 1317–1323.

Poli, G. and Parola, M. (1996) Oxidative damage and fibrogenesis. *Free Rad. Biol. Med.*, **22**, 287–305.

Preedy, V.R., Patel, V.B., Why, H.J.F., Corbett, J.M. Dunn, M.J., Richardson, P.J. (1996) Alcohol and the hearth: biochemical alterations. *Cardiovasc. Res.*, **31**, 139–147.

Puntarulo, S. and Cederbaum, A.I. (1989) Chemiluminescence from acetaldehyde oxidation by xanthine oxidase involves generation of and interactions with hydroxyl radicals, *Alcohol. Clin. Exp. Res.*, **13**, 84–90.

Rabshabe-Step, J. and Cederbaum, A.I. (1994) Generation of reactive oxygen intermediates by human liver microsomes in the presence of NADPH or NADH. *Mol. Pharmacol.*, **45**, 150–157.

Rajagopalan, K.V. and Handler, P. (1964) Hepatic aldehyde oxidase. III. The substrate binding side, *J. Biol. Chem.*, **239**, 2027–2032.

Rao, D.N.R., Yang, M.X., Lasker, J.M. and Cederbaum, A.I. (1996) 1-hydroxyethyl radical formation during NADPH- and NADH dependent oxidation of ethanol by human liver microsomes. *Mol. Pharmacol.*, **49**, 814–821.

Reinke, L.A., Lai, E.K., Du, Bose, C.M. and McCay., P.B. (1987) Reactive free radical generation in vivo in heart and liver of ethanol-fed rats: correlation with radical formation in vitro. *Proc. Natl Acad. Sci. USA*, **84**, 9223–9227.

Reinke, L.A., Rau, J.M. and Mc Cay, P.B. (1990) Possible roles of free radicals in alcohol tissue damage. *Free Rad. Res. Comms.*, **9**, 205–211.

Reinke, L.A., Moore, D.R. and Mc Cay, P.B. (1997) Mechanisms for metabolism of ethanol to 1-hydroxyethyl radicals in rat liver microsomes. *Arch. Biochem, Biophys.*, **348**, 9–14.

Renis, M., Calabrese, V., Russo, A., Calderone, A., Barcellona, M.L. and Rizza, V. (1996) Nuclear DNA strand breaks during ethanol-induced oxidative stress in rat brain. *FEBS Lett.*, **390**, 153–156.

Ribiére, C., Hininger, I., Saffer-Boccara, C., Salbourault, D. and Nordmann, R. (1994) Mithochondrial respiratory activity and superoxide radical generation in the liver, brain and hearth after chronic ethanol intake. *Biochem. Pharmacol.*, **47**, 1827–1833.

Robin, M.-A., Maratrat, M., Le Roy, M., Le Breton, F.-P., Bonierbale, E., Dansett, P., Ballet, F., Mansuy, D. and Pessayre, D. (1996) Antigenic targets in tienilic acid hepatitis. Both cytochrome P450 2C11 and 2C11-tienilic acid adducts are transported to the plasma membrane of rat hepatocytes and recognized by human sera. *J. Clin. Invest.*, **98**, 1471–1480.

Ronis, M.J.J., Lindros, K.O. and Ingelman-Sundberg, M. (1996) The CYP2E family. in Ioannides, C. (ed.) *Cytochromes P450: metabolic and toxicological aspects*, Boca Raton: CRC Press, pp. 211–239.

Rosenblum, E.R., Galaver, J.S. and Van Thiel, D.H. (1989) Lipid peroxidation: a mechanism for alcohol-induced testicular injury. *Free Rad. Biol. Med.*, 7, 569–577.

Rouach, H., Houzè, P., Orfanelli, M.T., Gentil, M., Bourdon, R., and Nordmann, R. (1990) Effect of acute ethanol administration of the subcellular distribution of iron in rat liver and cerebellum, *Biochem. Pharmacol.*, **39**, 1095–1100.

Rouach, H., Houze, P., Gentil, M., Orfanelli, M.-T. and Nordmann, R. (1994) Effect of acute ethanol administration on the uptake of ^{59}Fe-labelled tranferrin by rat liver and cerebellum. *Biochem. Pharmacol.*, 47, 1835–1841.

Rouach, H., Fattaccioli, V., Gentil, M., French, S.W., Morimoto, M. and Nordmann, R. (1997) Effect of chronic ethanol feeding on lipid peroxidation and protein oxidation in relation to liver pathology. *Hepatol.*, 25, 351–355.

Sadrzadeh, S.M.H., Nanji, A.A. and Prince, P.L. (1994a) The oral iron chelator 1,2,-dimethyl-3-hydroxypyrid-4-one reduces hepatic free iron, lipid peroxidation and fat accumulation in chronically ethanol-fed rats. *J. Pharmacol. Exp. Ther.*, **269**, 632–636.

Sadrzadeh, S.M.H., Nanji, A.A. and Meydani M. (1994b) Effect of chronic ethanol feeding on plasma and liver α- and γ-tocophetol levels in normal and vitamin E-deficient rats. *Biochem. Pharmacol.*, **47**, 2005–2010.

Sadrzadeh, S.M.H., Meydani, M., Khettry, U. and Nanji, A.A. (1995) High-dose vitamin E supplementation has no effect on ethanol-induced pathological liver injury. *J. Pharmacol. Exp. Ther.*, **273**, 455–460.

Salaspuro, M. (1997) Microbial metabolism of ethanol and acetaldehyde and clinical consequences. *Addition Biol.*, **2**, 35–46.

Santiard, D., Ribiére, C., Nordmann, R., Houee-Levin, C. (1995) Inactivation of Cu, Zn-superoxide dismutase by free radicals derived from ethanol metabolism: a gamma-radiolysis study. *Free Rad. Biol. Med.*, **19**, 121–127.

Savolainen, V.T., Penttilä, A. and Karhunen, P.J. (1992) Delayed increases in liver cirrhosis mortality and frequency of alcoholic liver cirrhosis following an increment and redistribution of alcohol consumption in Finland: Evidence from mortality statistics and autopsy survey covering 8533 cases in 1968–1988. *Alcohol. Clin. Exp. Res.*, **16**, 661–664.

Schöneich, C., Bonifacic, M. and Asmus, K.D. (1989) Reversible H-atom abstraction from alcohols by thiyl radicals: Determination of absolute rate constants by pulse radiolysis. *Free Rad. Res. Comms.*, **6**, 393–405.

Schuessler, H., Schmerler-Dremel, G., Danzer, J. and Jung-Kvrner, E. (1992) Ethanol radical-induced protein-DNA crosslinking: a radiolysis study. *Int. J. Radiat. Biol.*, **62**, 517–526.

Sergent, O., Griffon, B., Morel, I., Chevanne, M., Dubus, M.P., Cillard, P. and Cillard, J. (1997) Effect of nitric oxide on iron-mediated oxidative stress in primary hepatocyte culture. *Hepatology*, **23**, 122–127.

Shaw, S. (1989) Lipid peroxidation, iron mobilization and radical generation induced by alcohol. *Free Rad. Biol. Med.*, 7, 541–547.

Shaw, S., Jayatilleke, E., Ross, W.A., Gordon, E.R., and Lieber, C.S. (1981) Ethanol induced lipid peroxidation: potentiation by long-term alcohol feeding and attenuation by methionine. *J. Lab. Clin. Med.*, **98**, 417–425.

Shaw, S., and Jayatilleke, E. (1990a) Ethanol-induced iron mobilization: role of acetaldehyde-aldehyde oxidase generated superoxide, *Free Rad. Biol. Med.*, **9**, 11–15.

Shaw, S. and Jayatilleke, E. (1990b) The role of aldehyde oxidase in ethanol-induced hepatic lipid peroxidation in the rat, *Biochem. J.*, **268**, 579–583.

Shaw, S., Rubin, K.P. and Lieber, C.S. (1983) Depressed hepatic glutathione and increased diene conjugates in alcoholic liver disease. evidence of lipid peroxidation, *Dig. Dis. Sci.*, **28**, 585–589.

Shimizu, M., Lasker, J.M., Tsutsumi, M. and Lieber, C.S. (1990) Immunohistochemical locatization of ethanol-inducible P450IIE1 in the rat alimentary tract, *Gastroenterology*, **99**, 1044–1053.

Shiratori, Y., Takada, H., Hikiba, G., Nakata, R., Okano, K., Komatsu, Y., Niwa, Y., Matsumura, M., Shiina, S., Omata, M. and Kamii, K. (1993) Production of chemotactic factor interleukin-8 from hepatocytes exposed to ethanol. *Hepatology*, **18**, 1477–1482.

Shirley, M.A., Reidhead, C.T., and Murphy, R.C. (1992) Chemotactic LTB4 metabolites produced by hepatocytes in the presence of ethanol, *Biochem. Biophys. Res. Commun.*, **185**, 604–609.

Sinaceur, J., Ribière, C., Sarburault, D. and Nordmann, R. (1995) Superoxide formation in liver mitochondria during ethanol intoxication: Possible role in alcohol hepatotoxicity, in Poli, G., Cheeseman, K.H., Dianzani, M.U. and Slater, T.F. (eds) *Free Radicals in Liver Injury*, Oxford: IRL Press, pp. 175–177.

Situnayake, R.D., Crump, B.J., Thurnham, D.I., Davies, J.A., Gearty, J. and Davis, M. (1990) Lipid peroxidation and hepatic antioxidants in alcoholic liver disease, *Gut*, **31**, 1311–1317.

Slater, A.G.F., Nobel, C.S.I. and Orrenius, S. (1995) The role of intracellular oxidant in apoptosis. *Biochim. Biophys. Acta*, **1271**, 59–62.

Smith, S.M., Grisham, M.B., Manci, E.A., Granger, D.N. and Kvietys, P.R. (1987) Gastric mucosal injury in the rat. Role of iron and xanthine oxidase. *Gastroenterology*, **92**, 950–955.

Speisky, H., MacDonald, A., Giles, G., Orrego, H. and Israel, Y. (1985) Increased loss and decreased syntheis of hepatic glutathione after acute ethanol administration. *Biochem. J.*, **225**, 565–572.

Stål, P., Johansson, I., Ingelman-Sundberg, M., Hagen, K. and Hultcrantz, R. (1996) Hepatotoxicity induced by iron overload and alcohol. Studies on the role of chelatable iron, cytochrome P450 2E1 and lipid peroxidation. *J. Hepatol.*, **25**, 538–546.

Stowel, A., Hillbom, M., Salaspuro, M. and Lindros, K. (1980) Low acetaldehyde levels in blood, breath and cerebrospinal fluid of intoxicated humans assayed by improved methods, *Adv. Exp. Med. Biol.*, **132**, 635–642.

Stoyanovsky, D.A., Wu, D. and Cederbaum, A.I. (1998) Interaction of 1-hydroxyethyl radical with glutathione, ascorbic acid and α-tocopherol. *Free Rad. Biol. Med.*, **24**, 132–138.

Strubelt, O., Younes, M. and Pentz, R. (1987) Enhancement by glutathione depletion of ethanol-induced hepatotoxicity in vitro and in vivo. *Toxicology*, **45**, 213–223.

Suematzu, T., Matsumura, T., Sato, N., Miyamoto, T., Ooka, T., Kamada, T. and Abe, H. (1981) Lipid peroxidation in alcoholic disease in humans, *Alchol. Clin. Exp. Res.*, **5**, 427–430.

Sultatos, L.G. (1988) Effect of acute ethanol administration on the hepatic xanthine dehydrogenese/oxidase system in the rat, *J. Pharmacol. Exp. Ther.*, **246**, 946–949.

Szelenyi, I. and Brune, K. (1988) Possible role of oxygen free radicals in ethanol induced gastric mucosal damage. *Dig. Dis. Sci.*, **33**, 865–871.

Takase, S., Tsutsumi, M., Kawahara, H., Takada, N. and Takada, A. (1993) The alcohol-altered liver membrane antibody and hepatitis C virus infecion in the progression of alcoholic liver disease. *Hepatology*, **17**, 9–13.

Takeshi, H., Kaplowitz, N., Kamimura, T., Tsukamoto, H. and Fernandez-Checa, J.C. (1992) Hepatic mitochondrial GSH depletion and progression of experimental alcoholic liver disease in rats. *Hepatology*, **16**, 1423–1428.

Tanner, A.R., Bantock, I., Hinks, L., Lloyd, B., Turner, N.R. and Wright, R. (1986) Depressed selenium and vitamin e levels in an alcoholic population. Possible relationship to hepatic injury through increased lipid peroxidation. *Dig. Dis. Sci.*, **31**, 1307–1312.

Teare, J.P., Carmichael, A.J., Burnett, F.R. and Reke, M.O. (1993) Detection of antibodies to acetaldehyde-albumin conjugates in alcoholic liver disease. *Alcohol Alcohol.*, **28**, 11–16.

Thome, J., Zhang, J., Davids, E., Foley, E., Weijers, H.G., Weisbeck, G.A., Bönin, J., Riederer, P. and Gerlach, M. (1997) Evidence for increased oxidative stress in alcohol-dependent patients provided by quantification of in vivo salicilate hydroxylation products. *Alcohol. Clin. Exp. Res.*, **21**, 82–85.

Tophan, R., Coger, M., Pearce, K., Schultz, P. (1989) The mobilization of ferritin by liver cytosol: A comparison of xanthine and NADH as reducing substrates, *Biochem. J.*, **261**, 137–142.

Tsukamoto, H., French, S.W., Benson, N., Delgado, G., Rao, G.A., Larkin, E.C. and Largman, C. (1985) Severe and progressive steatosis and focal necrosis in rat liver induced by continuous intragastric infusion of ethanol and low fat diet. *Hepatology*, **5**, 224–232.

Tsukamoto, H., Towner, S.J., Ciofalo, L.M. and French, S.W. (1986) Ethanol-induced liver fibrosis in rats fed high fat diet. *Hepatology*, **6**, 814–822.

Tsukamoto, H., Horne, W., Kamimura, S., Niemelä, O., Parkkila, S., Ylä-Herttuala, S., and Brittenham, G.M. (1995) Experimental liver cirrhosis induced by alcohol and iron. *J. Clin. Invest.*, **96**, 620–630.

Van Gossum, A., Closset, P., Noel, E., Cremer, M. and Neve, J. (1996) Deficiency in antioxidant factors in patients with alcohol-related chronic pancreatitis. *Dig. Dis. Sci.* **41**, 1225–1231.

Van Pelt, F.N.A.M., Straub, P. and Manns M.P. (1995) Molecular basis of drug-induced immunological liver injury. *Sem. Liver Dis.* **15**, 283–300.

Vendemiale, G., Altomare, E., Trizio, T., Grazie, I.E., De Padova, C., Salerno, M.T., Carrieri, V. and Albano, O. (1989) Effects of oral S-adenosyl-L-methionine on hepatic glutathione in patients with liver disease. *Scand. J. Gastroenterol.* **24**, 407–415.

Victor, B.E., Schmidt, K.L., Smith, G.S. and Miller, T.A. (1991) Protection against ethanol injury in the canine stomach: role of mucosal glutathione. *Am. J. Physiol.*, **261**, G966–G971.

Videla, I.A. and Valenzuela, A. (1982) Alcohol ingestion, liver glutathione and lipoperoxidation metabolic interrelations and pathological implications. *Life Sci.*, **31**, 2395–2407.

Viitala, K., Israel, Y., Blake, J.E. and Niemela, O. (1997) Serum IgA, IgG and IgM antibodies directed against acetaldehyde-derived epitopes: relationship to liver disease severity and alcohol consumption. *Hepatology*, **25**, 1418–1424.

Wang, J-F. Greenberg, S.S. and Spitzer, J.J. (1995) Chronic alcohol administration stimulates nitric oxide formation in the rat liver with or without pretreatment with lipopolysaccharide. *Alcohol. Clin. Exp. Res.*, **19**, 387–393.

Wang, J-F. and Spitzer, J.J. (1997) Alcohol-induced thymocyte apoptosis is accompained by impaired mitochondrial function. *Alcohol*, **14**, 99–105.

Wieland, P. and Lauterburg, B.H. (1995) Oxidation of mitochondrial proteins and DNA following administration of ethanol. *Biochem. Biophys. Res. Commun.*, **213**, 815–819.

Worrall, S., De Jersey, J., Shanley, B.C. and Wilce, P.A. (1990) Antibodies against acetaldehyde-modified epitopes: presence in alcoholic, non-alcoholic liver disease and control subjects. *Alcohol Alcohol.*, **25**, 509–517.

Wu, D. and Cederbaum, A.L. (1996) Ethanol cytotoxicity to transfected HepG2 cell line expressing human cytochrome P4502E1. *J. Biol. Chem.*, **271**, 23914–23919.

Yacoub, L.K., Fogt, F., Griniuviene, B. and Nanji, A.A. (1995) Apoptosis and bcl-2 protein expression in experimental alcoholic liver disease in the rat. *Alcohol. Clin. Exp. Res.*, **19**, 854–859.

Zhao, X., Jie, O., Xie, J., Giles, T.D. and Greenberg, S.S. (1997) Ethanol inhibits inducible nitric oxide synthethase transcriptional and post-transcriptional processes in vivo. *Alcohol. Clin. Exp. Res.*, **21**, 1246–1256.

13 The role of cytokines in the inflammatory response

Carmen M. Arroyo

13.1 Introduction

This chapter focuses on the biochemical and cellular mechanisms that may be responsible for the development of acute cutaneous sulfur mustard (HD) injury and their exploitation for establishing rational approaches to therapeutic intervention. Cytokines play a major role in both acute and chronic inflammatory processes. Irritants including 2-chloroethyl-ethylsulfide (half-sulfur mustard, H-MG; chemical structure: $ClCH_2CH_2SCH_2CH_3$) and bis(2-chloroethyl)sulfide (sulfur mustard, HD; chemical structure: $ClCH_2CH_2SCH_2CH_2Cl$) are thought to cause the release of primary cytokines such as interleukin-1 (IL-1) and tumor necrosis factor alpha (TNF-α) from epidermal cells. These primary cytokines, in turn, stimulate the production of other cytokines by cells in the dermis (fibroblasts, macrophages, endothelial cells, and mast cells). Only then is the inflammatory process fully developed.

Tumor necrosis factor (TNF) is a monokine produced by monocytes and macrophages in response to different stimuli. To determine whether vesicant agents such as half-mustard may induce the release of TNF-α in human monocytes (THP-1), enzyme-linked immunosorbent assay (ELISA) experiments were conducted at different postexposure times. The results indicate that significant increases in the TNF-α (pg/ml) concentration were observed as a function of time when THP-1 cells were exposed to 100 μl of 2 M H-MG. A specific serine-type protease inhibitor, $N^{\alpha}p$-tosyl-L-lysine chloromethyl ketone (TLCK), led to partial but significant inhibition of TNF activation. Furthermore, this laboratory detected the generation of spin adducts of 2-methyl-2-nitrosopropane (MNP) having a resemblance to MNP-adducts generated from hydrogen atom abstraction of protein constituents (Carmichael and Riesz, 1985). The electron paramagnetic resonance (EPR)/spin-trapping data indicate the trapping of by-products of protein degradation after exposure to H-MG. TNF-α may play a role as a biochemical marker for pathophysiological changes induced by H-MG or related agents.

The levels of interleukin-1 beta (IL-1β) in cultured human epidermal keratinocytes (NHEK) following exposure to HD were determined by immunocytochemistry. Human skin keratinocytes release significant amounts of IL-1 cytokine as determined by the Quantikine™ Interleukin-1β kit, ELISA procedure. Exposure of NHEK (~10^6–10^7 cells), to HD (2 mM) and preincubation for 3 hours at 37°C results in significant changes in IL-1 activation. In neonatal NHEK exposed to HD, IL-1β is decreased. Conversely, in adult breast NHEK exposed to HD, IL-1β is increased. Nitric oxide (•NO) has been implicated as the effector molecule that mediates IL-1β. To confirm the involvement of •NO in the expression of the IL-1β, EPR spectroscopy was employed. EPR detectable iron–nitrosyl complex in NHEK exposed to HD (18 hours post exposure to 1 mM) was measured, and the generation of •NO and this induced complex was blocked by *N*-nitro-L-arginine (L-NOARG), a competitive inhibitor of nitric oxide synthase (NOS). Our results show nitric oxide and IL-1 cytokine release following keratinocytes exposure to HD. Based upon this work, it appears possible that IL-1 could be used as a specific marker for epidermal cytoxicity in mechanistic studies of the toxicity of HD and in defining interventive and therapeutic regimens against HD vesication.

13.2 Activation of TNF-α by THP-1 exposed to H-MG

Tumor necrosis factor (TNF), or cachectin, is a cytokine originally thought to play a role in host surveillance against neoplasms (Vilcek *et al.*, 1986; Beutler *et al.*, 1985a, 1985b; Pennica *et al.*, 1984). TNF-α was initially identified as a factor produced by leukocytes and is responsible for infection-induced cachexia. It has been recognized subsequently that TNF has a broader range of effects on host immune responsiveness such as enhancing neutrophil (PMN)–endothelial interactions and facilitating phagocytosis and bacterial killing (Beutler and Cerami, 1986; Shalaby *et al.*, 1985; Klebanoff *et al.*, 1986). Recently, a role for TNF in the generation of free radicals and the pathophysiological changes during sepsis and septic shock has been proposed (Tsujimoto *et al.*, 1986; Lloyd *et al.*, 1993). In addition, the ability of sulfur mustard to elicit protease activity in cultured normal human epidermal keratinocytes (NHEK) has been previously reported (Smith *et al.*, 1991). The exposure of either epithelial cell type to sulfur mustard (HD) and its monofunctional analogue resulted in significantly increased proteolytic activity. The monofunctional sulfur mustard (H-MG) was as effective as HD in inducing proteolytic activity. In addition, both cytokines and proteases can up regulate cellular receptors (Cowan *et al.*, 1998). In this study we sought to determine whether TNF-α is released by human monocytes primed with the monofunctional alkylating agent H-MG and to ascertain the possible relationship between free radical generation and pathophysiological changes upon TNF-α amplification.

13.2.1 Materials and methods

Predicta™ Tumor Necrosis Factor-α kit was obtained from Genzyme Corporation (Catalog No. 1915-01), Cambridge, Massachusetts, USA. Human monocyte (THP-1) cells were purchased from American Type Culture Collection, Parklawn Drive, Rockville, Maryland, USA (ATCC No: TIB 202). 2-Methyl-2-nitrosopropane (MNP, $[(CH_3)_3CNO]_2$..); 2-chloroethyl ethyl sulfide (H-MG); cycloheximide (CH) and N^{α}-*p*-tosyl-L-lysine chloromethyl ketone (TLCK) were obtained from Aldrich Chemical Company, Milwaukee, Wisconsin, USA and used as received. Roswell Park Memorial Institute Tissue Culture Medium (RPMI 1640) vitamin solution (Sigma Catalog No. R7256); 2-mercaptoethanol ($HSCH_2CH_2OH$); dimethyl sulfoxide (DMSO); fetal bovine serum; antipain were acquired from Sigma Chemical Company, St Louis, Missouri, USA.

13.2.1.1 Cells

The characteristics of THP-1, a human monocytic cell line, have been described previously in detail (Tsuchiya *et al.*, 1980). This cell line was maintained in suspension cultures in RPMI 1640 vitamin solution supplemented with 5×10^{-5} M 2-mercaptoethanol and 10% fetal bovine serum. This special culture condition was required to maintain consistency between the EPR experiments and the TNF-α determinations. Cultures were maintained by replacement of the medium once a week, and the maximum cell density obtained was 5×10^5/ml. Studies with H-MG were performed in submerged cultures to allow uniform cell exposure to the agent.

13.2.1.2 Half-mustard exposure

THP-1 cell suspensions (5 ml) in 50 ml tissue culture tubes were exposed to 100 µl of 2 M H-MG (4×10^{-5} M). The tissue culture tubes were maintained at room temperature in a fume hood for 45 minutes to allow venting of the volatile agent and then transferred to a CO_2 incubator at 37°C for a total incubation time of 1 to 48 hr. For exposure time less than 1 hr, the tissue culture tubes were kept at room temperature in a fume hood.

13.2.1.3 ELISA (enzyme-linked immunosorbent assay)

The Predicta™ TNF-α enzyme immunoassay contained a 96-well microtiter plate pre-coated with monoclonal antibody to TNF-α. A measured volume of cell suspension (100 µl of 5×10^5 cells; control, not exposed; H-MG exposed; and treated cells either exposed or not exposed to H-MG), 100 µl of the TNF-α standards (0, 15, 100, 800 and 1200 pg/ml) and 100 µl of TNF-α high/low control were added to each test well and incubated to allow any TNF-α to be bound by the antibodies on the microtiter plate. The wells were subsequently washed (five times) and a biotin-labeled polyclonal antibody to TNF-α was added which binds to captured TNF-α. After working the plates, the wells were washed (five times) and a peroxidase labeled streptavidin reagent that attaches to biotin in the immune complex on the plate was added. The wells were washed (five times) and a substrate (peroxide) and chromagen (tetramethylbenzidine) were added producing a blue color in the presence of the peroxidase enzyme. The color reaction was then stopped by the addition of acid, which changes the blue color to yellow. The intensity of the yellow color is in direct proportion to the amount of TNF-α present in the samples, standards, or controls. The absorbance of each well was read at 450 nm using endpoint analysis on a Molecular Devices V_{max} kinetic microplate reader attached to an IBM PC/XT for data manipulation. A standard curve was constructed (CA-Cricket Graph, Computer Associates) for each time point by averaging duplicate runs as each control value and subtracting from each value the background absorbance for 0 pg/ml. These corrected values were used to quantitate TNF-α concentrations in the controls (minus H-MG), the H-MG exposed or treated cells. Figure 13.1 illustrates the ELISA principle and the time course of H-MG induced activation of TNF-α by THP-1 cells.

13.2.1.4 EPR/spin-trapping

MNP is a free radical scavenger with a high degree of hydrophobicity used for EPR/spin-trapping experiments. Forty-two milligrams of MNP were dissolved in 20 µl of DMSO and stirred with a magnetic bar for 30 min to 1 hr in a water bath at 35 or 40°C to ensure the complete dissolution of MNP. THP-1 cell suspensions (2.9×10^7 cells/ml) were incubated with the spin-trap MNP (17 mM) at 37°C for 5 min in the

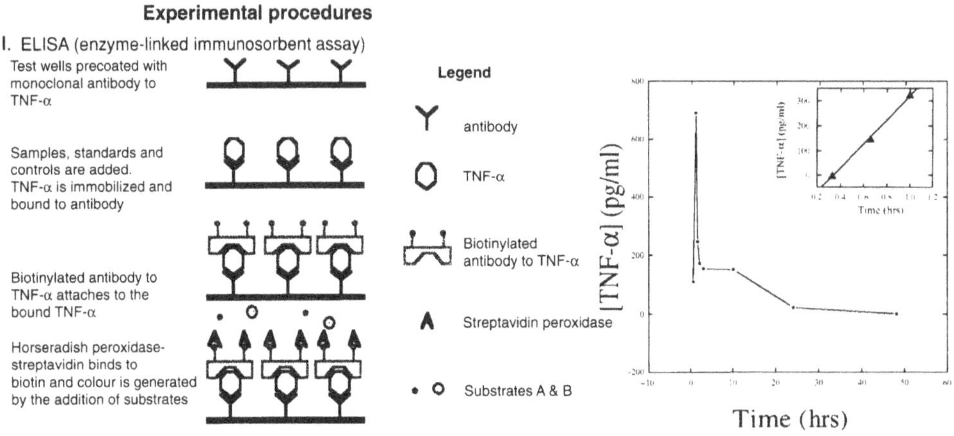

Figure 13.1 Description of the principle of the measurement of tumor necrosis factor (TNF-α) enzyme immunoassay for the quantitative determination of TNF-α in cell culture supernatants. Time course of H-MG induced TNF-α activation by THP-1 cells (right side). THP-1 cells were assayed by ELISA

reaction medium. The MNP/THP-1 mixture was then transferred to an aqueous flat cell (10.5 × 60 × 0.3 mm; Wilmad Glass Company, Inc., Buena, New Jersey, USA) for EPR examination. No signal was detected under the described conditions (see below).

The H-MG exposure experiments were carried out by directly dissolving MNP in 100 µl of 2 M H-MG since it was determined that MNP (42 mg) rapidly dissolves in 100 µl of H-MG and produces a typical blue color solution characteristic of the MNP monomer (Buettner, 1987). The EPR of H-MG/MNP mixture generated a characteristic EPR signal of ditertiary butyl nitroxide (d-tBN), usually formed by MNP decomposition (Makino *et al.*, 1981). To avoid this pitfall, the MNP concentration was kept as low as possible (7 mg/100 µl H-MG). The THP-1 cell suspensions were incubated with the MNP/H-MG viscous mixture for 60 min. The THP-1/MNP/H-MG suspension was then transferred to an aqueous flat cell for EPR analysis. The process was carried out in the dark to prevent photolytic degradation of the spin-trap.

The EPR spectra were recorded and analyzed with a Varian E-109, X-band spectrometer at 100 kHz magnetic field modulation with each experiment being repeated five times. EPR spectra were stored on a COMPAQ DESKPRO 386 computer and the simulated isotropic EPR spectra determined using the multipurpose program EPRDAP written by Dr P. Kuppusamy, USA EPR Inc., Clarksville, Maryland, USA. The hyperfine couplings were measured directly from the spectra and the magnetic field was set at 335.0 mT, microwave power 10 mW, modulation amplitude 0.08 mT and microwave frequency, 9.474 GHz.

13.2.1.5 Data analysis

All data points in the ELISA assays were derived from at least triplicate samples and expressed as the average of the trial minus the absorbance due to the background

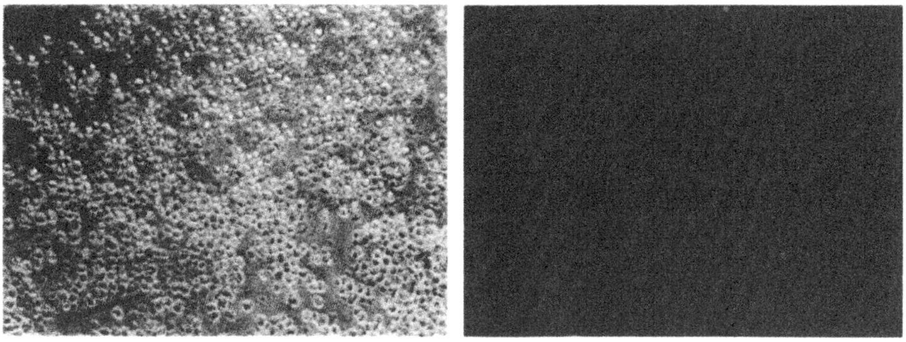

Figure 13.2 Morphological characteristics of THP-1 cells in culture. The cells started to proliferate immediately after the beginning of *in vivo* culturing, and could be expanded to 2 ml cultures in 30 ml test tubes without difficulty. The cells grew in suspension and formed loose clumps (most of the cells remained in suspension with only a few attaching to the plastic surface of the culture dish). The culture reached a saturation density of 1×10^5 to 1×10^5 cells/ml in 14 to 21 days. (A) Control monoblasts and (B) H-MG (2×10^{-4}M) treated cells (24 h postexposure) showing typical changes associated with HD toxicity, including nuclear condensation, paranuclear vacuolation, and necrosis

(0 pg/ml TNF). These absorbance data were fitted against the standard curve to determine the amount of TNF released. The data from the inhibitor studies were corrected for the background absorbance and then analyzed using the standard error analysis within SigmaPlot™ (Jandel Scientific, Corte Madera, California, USA). Results were analyzed by applying a two-way analysis of variance with $p < 0.050$.

13.2.1.6 Photography

Pictures of control monocyctes and H-MG treated cells were obtained using a standard aqueous EPR (flat) sample cell (WG-812; Wilmad Glass Co, Buena, New Jersey, USA) and a Leica Dm-IL inverted microscope equipped for phase contrast (Leica Mikroskopie and Systeme GmbH, Ernst-Leitz-Strabe, Wetzlar, Germany).

13.2.2 Results and discussion

Immediately after the beginning of *in vitro* culturing, the THP-1 cells started to proliferate actively and could be expanded to 2-ml cultures in 30 ml test tubes without difficulty. The cells grew in suspension and formed loose clumps (Figure 13.2A). Most of the cells remained in suspension with only a few attaching to the plastic surface of the culture dish. The culture reached a saturation density of 1 to 2×10^5 cells/ml in 14 to 21 days. The untreated THP-1 cells, with round and ovoid shapes, contained a moderate amount of vacuoles with indented and irregular shape nuclei with prominent nucleoli. After 24 hours treatment with H-MG, cytoplasmic vacuoles of the THP-1 cells were increased (Figure 13.2B), and the nuclei became reniform in shape with distinct nucleoli. This cell line has been recommended as a useful model for

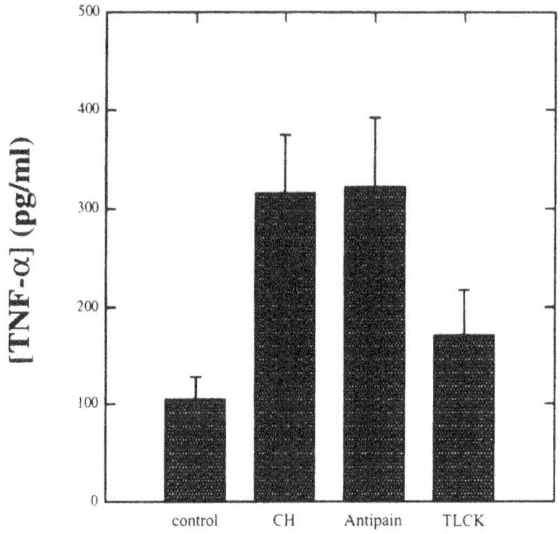

Representative average values obtained:

Control	94.64 ± 0.52
	97.39 ± 0.52
CH	320.90 ± 0.56
	320.22 ± 0.56
Antipain	331.47 ± 0.61
	353.75 ± 0.61
TLCK	162.10 ± 0.28
	152.45 ± 0.28

Figure 13.3 Effect of various inhibitors on activation of TNF-α by THP-1 cells after H-MG exposure. THP-1 cells were pre-treated with 10^{-3} M of cycloheximide (CH), antipain (0.06 mM) and TLCK (0.03 mM) for 30 minutes at 37°C under 5% CO_2 in air followed by H-MG (2×10^{-4} M) exposure for 1 hour. The influence of the protease inhibitors on TNF-α activation was determined by ELISA

the study of the mechanism of maturation or differentiation from monocytes to macrophages and of the biochemical mechanism of phagocytosis (Tsuchiya *et al.*, 1980).

Treatment of human monocytes, THP-1, with H-MG produced copious amounts of tumor necrosis factor (TNF-α) released into the medium (Figure 13.1, right side). The TNF-α activation was a time dependent phenomenon with a linear response (Figure 13.1, right side insert). Maximum TNF-α activation occurred one hour after exposure with a concentration of 0.2 mM H-MG. Subsequently, a reduction in the quantity of released TNF-α was observed for up to 48 hours.

Exposure of a variety of different cell types to sulfur-mustard (HD), or to its monofunctional analog, H-MG, has been shown to produce significant increases in the activity of serine-like proteases (Cowan *et al.*, 1991; Cowan and Broomfield, 1993). TNF-receptors on several human leukemic and cervical carcinoma cell lines are sensitive to hydrolysis by protease (Fiers, 1991). Furthermore, it has been reported that TNF action is not abolished by co-treatment with transcription or translation inhibitors (Suffys *et al.*, 1988). Therefore, the activity of TNF was measured in the presence of cycloheximide, a translation inhibitor, and two identified protease inhibitors, antipain and TLCK. TLCK also specifically inhibits trypsin-like proteases by alkylating the histidine residue in the active site of such enzymes (Suffys *et al.*, 1988). Antipain is a fungus-derived protease inhibitor that inhibits trypsin-type activity. The TNF-α activation was determined in cell suspension medium pre-treated with CH (0.001 M), antipain (0.03 mM) and TLCK (0.03 mM) for 30 min prior to the exposure with H-MG (Figure 13.3). The TNF-α released was increased by 3.0 and 3.1 fold respectively in the presence of CH and antipain (Figure 13.3). The fact that CH and antipain so strongly enhances the cytoxicity of TNF-α is often explained

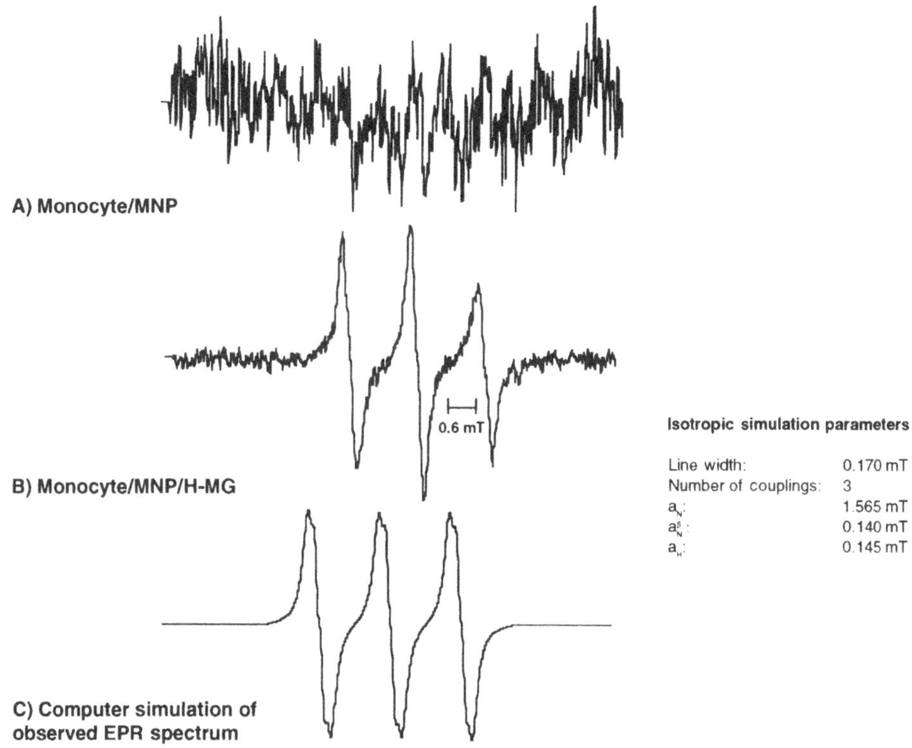

Figure 13.4 Representative EPR spectra obtained from human monocyte cell suspensions. (A) Control consisting of a solution of MNP (1 mg/ml) in phosphate buffer at pH 7.2 and then pre-incubated with THP-1 cell suspensions for 15 min at 37°C. (B) EPR spectrum of MNP-adducts observed when THP-1 suspensions were exposed to H-MG (2×10^{-4} M) in the presence of MNP (7 mg). Spectrum was taken two hours later after exposure. MNP/H-MG mixture. (C) Overall computer simulation of Figure 4B using the following EPR parameters: $a_N = 1.565$ mT; $a_N = 0.140$ mT; $a^{\beta}_{H} = 0.145$ mT and line width of 0.170 mT

by assuming that TNF-α itself induces the synthesis of protective proteins (Suffys *et al.*, 1988), which either interfere with the generation of toxic products or else aid to detoxify these. Figure 13.3 clearly shows that TLCK (0.03 mM) strongly inhibited the release of TNF-α activity.

Human monocytes (THP-1) exposed to H-MG (4×10^{-5} M) in the presence of the spin-trap, MNP (1 mg/ml), were examined for the presence of spin adducts. EPR spectra suggesting the trapping of free radicals were observed in exposed cell suspensions. Representative spectra obtained from control (not-exposed) and exposed cell suspensions are shown in Figures 13.4A and 13.4B. Figure 13.4B shows a signal recorded 4 hours after the H-MG exposure started. This signal consists mainly of a primary triplet ($a_N = 1.565$ mT). A secondary doublet due to a β-proton which further splits into a triplet due to a β-nitrogen was determined. The complexity of the spectrum in Figure 13.4B suggests the trapping of a radical with splitting constants

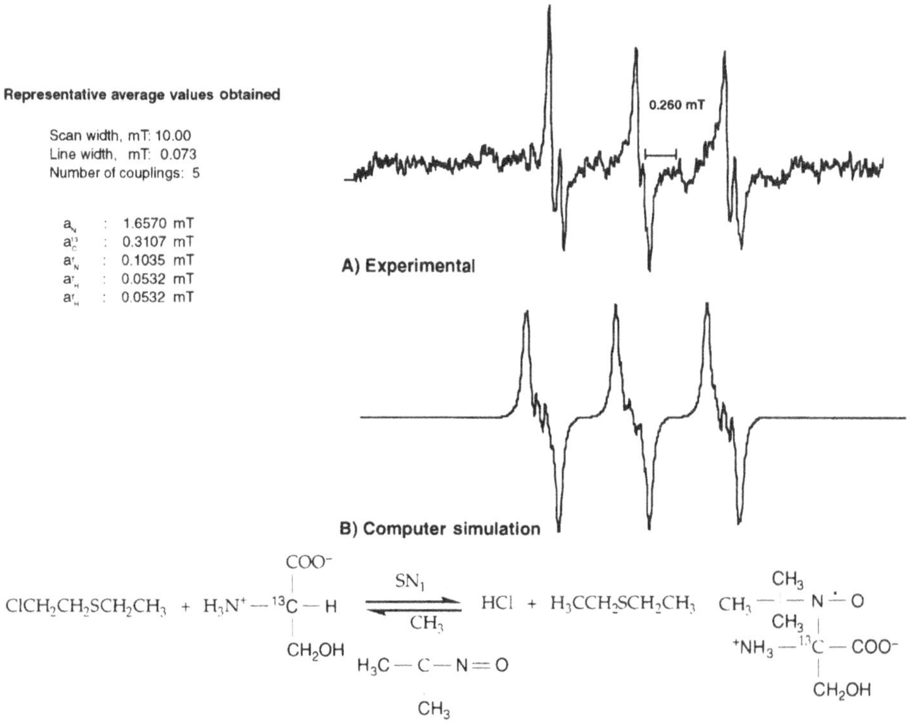

Figure 13.5 Proposed mechanism of the hydrogen atom abstraction by H-MG to ^{13}C-labeled serine in the presence of MNP as the spin-trap. (A) Experimental EPR spectrum; (B) Computer simulation using the EPR parameters given in enclosed table

similar to those of carbon-centered radical adducts with structure R′RH—C$^{\bullet}$—NH_3^+ (a_N = 1.565 mT; a_H^{β} = 0.140 mT and a_N = 0.145 mT). Computer simulation of the observed spectrum was generated using the given set of splitting constants and a line width of 0.170 mT (Figure 13.4C). This spectrum is consistent with the spin-trapping of the H-abstraction radical of polypeptide molecules (Kuwabara, 1986).

Unfortunately, it is very difficult to determine unambiguously the precise structure of the spin-adduct from the EPR signal obtained. Therefore, we conducted isotopic substitution EPR experiments in an attempt to identify the observed adducts. Six ^{13}C-labeled at C_2 representative amino acids of human TNF-α (Kuwabara, 1986) were employed. These included four non-polar, aliphatic amino acids, glycine-$^{13}C_2$, alanine-$^{13}C_2$, valine-$^{13}C_2$ and proline-$^{13}C_2$, an aromatic amino acid, tyrosine $^{13}C_2$, and a polar, uncharged side chain, serine-$^{13}C_2$. The reaction of H-MG with the mentioned ^{13}C-labeled amino acids was studied in the presence of MNP. In a neutral solution (pH ~ 6.5) of serine (0.4 M), only serine MNP-spin adducts were detected. The other ^{13}C-labeled amino acids did not form any MNP-adduct and only the MNP-adduct of di-t-BN was occasionally observed. The serine-MNP-adduct was unstable and decayed in approximately 36 minutes. The resulting spectrum (Figure 13.5A) was shown by simulation to be the sum of five different hyperfine coupling constants (a_N = 1.657 mT; a_C^{13} = 0.311 mT; a_N = 0.1035 mT; $a_{H(2)}^{\gamma}$ = 0.0532 mT (Figure 13.5B)) indicating the presence of MNP-^{13}C-centered adduct. The determined hyperfine coupling of a_C^{13} is in agreement with previously reported MNP-adduct splitting ^{13}C constant (Buettner, 1987).

These studies have shown that a TNF-α message is rapidly synthesized in H-MG stimulated THP-1 cells. This effect peaks at 400- to 600-fold above non-stimulated levels approximately 1 to 2 hrs after stimulation. TNF-α levels fall to a low, but constant, level by 6 hr after H-MG stimulation. The fact that protease or translation inhibitory molecules, i.e. antipain or CH, so strongly enhances the cytotoxicity of TNF has been explained by assuming that TNF itself induces the synthesis of protective proteins (Fiers, 1991), which either interfere with the generation of toxic products or aid to detoxify these materials. The protease apparently essential for TNF action may also be part of a lytic pathway common to cytotoxic natural killer cells (Fiers, 1991). These two phenomena may be connected; TNF-α activation within a cell may result in the synthesis of effectors, which are involved in the induction of proteins promoting cell cycling. But when the mitogenic response is not permitted to proceed, they become toxic and lethal. It seems reasonable to conclude that H-MG-induced TNF-α stimulation of THP-1 cells proceeds by promoting cell cycling due to induction of proteins. Evidence for the presence of serine proteases in HD-toxicity is derived from the work of Higuchi (Higuchi *et al.*, 1988) who showed that culture fluid from mustard-induced inflammatory lesions in rabbit skin show 3 to 6 fold increased levels of proteases.

The spin-trapping technique has been applied to the study of hydrogen atom abstraction by several groups of investigators (Makino *et al.*, 1981; Kuwabara *et al.*, 1986; Rustgi and Riesz, 1978). The EPR and product inhibition results, taken together, indicate that the hydrogen atom abstraction is a product of the action of H-MG and a sensitive amino acid residue. It is interesting to note the presence of Arg–Arg–Ser–Ser tetrapeptides in the first 30 amino acids of the TNF pre-sequence (Pennica *et al.*, 1984), as parts of basic amino acids often serve as cleavage sites for the release of physiologically important peptides from precursor molecules. The evidence outlined so far suggests that not only may damaged proteins accumulate within cells, but some of them may also possess reactive species which can go on to act on previously undamaged protein molecules and other biomolecules. Direct evidence that TNF-α induced events subvert part of the normal electron flow in the mitochondria has been reported (Lancaster *et al.*, 1989). As a result of these reactive intermediate species, oxidation of lipids and proteins occurs, followed by their degradation.

TNF activity has been associated with serine proteases, as inhibitors of both trypsin-like and chymotrypsin-like proteases interfere with TNF action (Fiers, 1991). The inhibitory effect of TLCK on TNF cytotoxicity has been reported (Suffys *et al.*, 1988) and is due to its inhibitory activity on a serine protease(s) (Suffys *et al.*, 1988). We propose that a 'trypsin-likc' serine protease is essentially involved which is proposed to be at the beginning of a cascade of events that may lead eventually to cell lysis (Fiers, 1991; Arroyo *et al.*, 1995). Further aspects of this proposed mechanism can be elucidated only after the protease itself has been identified and characterized.

13.3 Nitrosyl chloride in keratinocytes exposed to HD

The experiments described in this section deal with the effects of sulfur mustard on normal human epidermal keratinocytes (NHEK). Keratinocytes are the cells found in the outer layer of the skin and are a primary target of mustard exposure. Analysis of cultured normal human epidermis using specific ELISA and bioassays for the cytokine interleukin 1 have led to the conclusion that the average adult harbors 20 to 60 μg (0.6 to 1.9 nmol) of IL-1 in his or her epidermis (Kupper *et al.*, 1989). Taking the epidermis as a three-dimensional space with a thickness of 0.1 mm and an area of 1.5 m^2, the concentration of IL-1 in this space is 4 to 12 nmol/l. This concentration exceeds the concentration of IL-1 required to activate certain cells by three orders of magnitude (< 10 pmol/l).

Most keratinocyte interleukin (*in vivo* and *in vitro*) is cell associated; relatively little is released from the cell. Keratinocytes can express higher numbers of IL-1 receptors than can any other cell type studied (Kupper *et al.*, 1989; Blanton *et al.*, 1989). A variety of stimuli can enhance IL-1 receptor expression, *in vitro*. These include phorbol esters, calcium, and UV-radiation (Dinarello, 1991). In this study, we have determined by immunocytochemistry levels of interleukin-1 beta in cultured NHEK following exposure to sulfur mustard.

HD is a known alkylating agent causing, among other things, DNA damage which contributes to cell injury or death. However, the exact mechanism of alkylation remains unclear. What is known is that the action of alkylating agents, such as HD, proceeds through a mechanism involving, as a first step, formation of a cyclic alkylating agent (sulfonium ion) and the release of chloride (equation 13.1) (Ross, 1962):

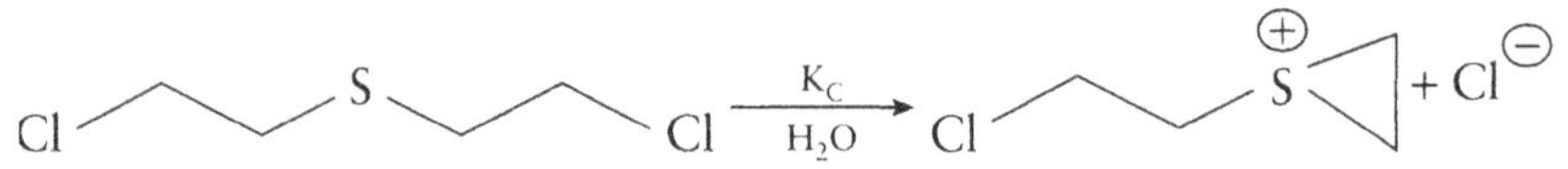

(13.1)

Sulfur mustard — Cyclic ethylene sulfonium ion

The hydrolysis of HD in an aqueous medium at 37°C is rapid, with $t_{1/2}$~2 min (Lawley, 1976; Lieske *et al.*, 1992). Furthermore, the HD may affect or alter various cell pathways prior to reaching the DNA located in the nucleus. Therefore, the action of alkylating agents on target organ cells may proceed via three principal mechanisms: (i) the physical interaction of the alkylating agent with cellular receptors; (ii) the chemical reaction of the alkylating agent with these receptors; or (iii) either type of metabolite–receptor interaction following metabolism of the alkylating agent within the target organ or elsewhere.

Although direct alkylation of DNA and RNA has been widely described, we hypothesize additional alkylation events of potential importance in skin injury. The levels of IL-1β in cultures of NHEK following exposure to HD (3 hours post exposure to 2 mM) were determined by immunocytochemistry. The expression of IL-1β in NHEK was found to be related to cell culture donor age. EPR spectroscopy was used to show the formation of an EPR detectable, g = 2.04, feature characteristic of iron–nitrosyl complex formation, and the generation of this induced complex by NHEK exposed to HD (18 hours post exposure to 1 mM) was blocked by L-NOARG, a competitive inhibitor of NOS. •NO has been implicated as the effector molecule that mediates IL-1β (Corbett *et al.*, 1993). The results show the release of nitric oxide during cytokine IL-1β expression when keratinocytes are exposed to HD. The combination of the nitric oxide with the chloride (Cl^-) in the plasma which is approximately 0.10 M (Faulkner *et al.*, 1968), and the one released from sulfur mustard ($[ClCH_2CH_2]_2S$) upon cyclization to the sulfonium ion may lead to the formation of nitrosyl chloride (NOCl), a known potent alkylating agent. If NOCl is formed, it may play a role in the skin injury.

In addition, EPR and ELISA results show the generation of •NO, therefore, suggesting that NOS plays a role during IL-1 expression when keratinocytes are exposed to HD. Inducible nitric oxide synthase (iNOS) has been associated with inflammatory and autoimmune tissue injury (Kolb *et al.*, 1992). Their findings show the presence of iNOS in human skin that can be localized to keratinocytes in the epidermal layer. This piece of work shows the possible involvement of active nitrogen species in the pathways of cell injury/repair following exposure to sulfur mustards. Active nitrogen species are those derived from nitric oxide generated in cells. The possible involvement of the known alkylating agent NOCl is discussed. The thermodynamic considerations and possible *in vivo* production of NOCl by macrophages and neutrophils have been reported (Koppenol, 1994).

13.3.1 Materials and methods

NHEK (adult and neonatal, Clonetics, San Diego, CA) were cultured to confluency and harvested for experiments. The culture medium used was keratinocyte basal medium, modified Molecular Cellular Developmental Biology 153 (MCDB 153), which was supplemented with bovine pituitary extract (7.5 mg/ml); human recombinant epidermal growth factor (0.1 μg/ml); hydrocortisone (0.5 mg/ml); bovine insulin (5 mg/ml); gentamicin sulfate (50 mg/ml); and amphotericin-B (50 μg/ml). The medium was changed after 2 days of culture. At days 3 and 6, the cells were harvested. After harvesting, the NHEK were resuspended in phosphate medium (5×10^6–5×10^7 cells/0.5 ml). IL-1β and EPR were measured after exposure of the suspension to 1–2 mM HD, 0.1 mM *N*-nitro-L-arginine (L-NOARG; $C_6H_{13}N_5O_4$, Sigma Chemical Co., St Louis, MO), or a combination of these agents (figure legends). IL-1β expression was measured after 3 hr exposure and heme-NO EPR was measured after 18 hr of exposure. The difference in post exposure time between the ELISA and the EPR experiments is due to the sensitivity of the two techniques. The IL-1β expression was measured using a commercially available kit for the ELISA technique (Quantikine™). This kit is specific for IL-1β and shows no cross reactivity with other cytokines (e.g. IL-1α etc.) as stated in the Quantikine™ brochure for IL-1β.

The Quantikine™ human IL-1β Immunoassay (Catalog Number DLB50, R&D Systems, Inc., Minneapolis, MN) was used for the quantitative determination of human IL-1β concentration in the cell cultures. This assay employs the quantitative 'sandwich' enzyme immunoassay technique. A monoclonal antibody specific for IL-1β is coated onto the microtiter plate provided in the kit. Standards and homogenous cell suspensions (100 μl) were pipetted into the wells. Following a wash to remove any unbound antibody–enzyme reagent, a substrate solution was added to the wells and color developed in proportion to the amount of IL-1β bound in the initial step. The color development was stopped and the intensity of the color was measured. The absorbance of each well was read at 450 nm. By comparing the optical density of the samples to the standard curve, the concentration of the IL-1β in the unknown samples was then determined. The assays were run in triplicate, and statistical evaluation was carried out using the paired sample *t* test with significance defined as $p < 0.05$. Representative data are shown in Figure 13.7.

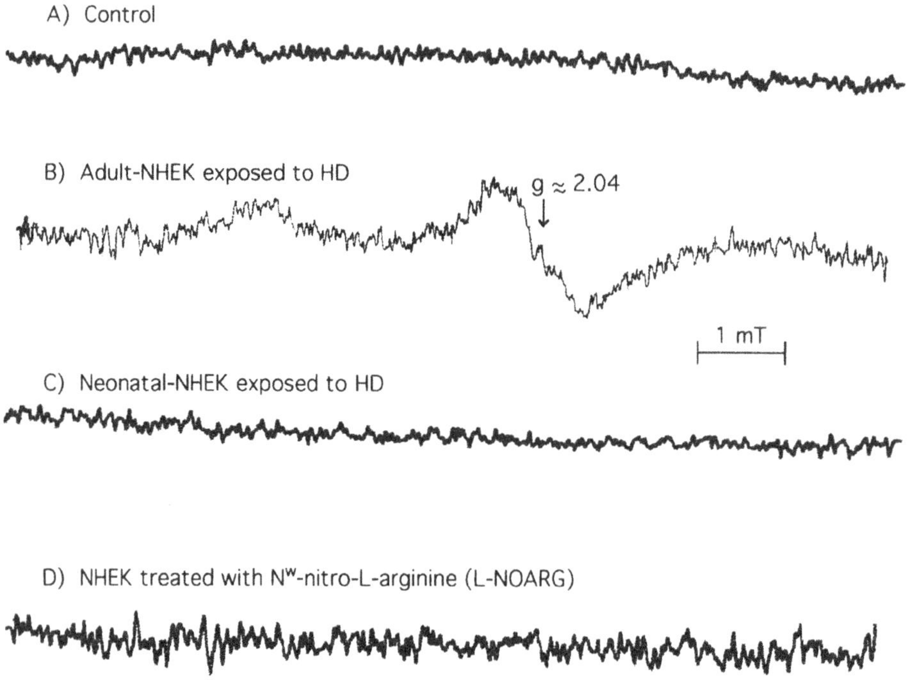

Figure 13.6 Low temperature (liquid nitrogen, 77 K) EPR spectra recorded from NHEK suspensions (3.8×10^6 cells/ml). (A) Control, non-exposed adult-NHEK; (B) adult-NHEK exposed to 1 mM sulfur mustard (HD) collected 18 hours post-HD exposure; (C) neonatal-NHEK exposed to 1 mM HD EPR spectrum obtained 18 hours post exposure; (D) sample treated with N^w-nitro-L-arginine (L-NOARG) to final concentration of 100 μM. EPR conditions were magnetic field, 334.5 mT; modulation frequency, 100 kHz; microwave frequency, 9.475 GHz; microwave power, 10 mW; receiver gain, 5×10^5; modulation amplitude, 0.5 mT and scan rate 0.62 mT/s

Electron paramagnetic resonance spectroscopy experiments were performed using cell suspensions containing at least 5×10^6 keratinocyte cells/0.5 ml suspended in 1 ml of complete MCDB 153 media. The cells were incubated at 37°C for an additional 18 hours in the presence or absence of 1 mM HD or 50–100 μM of L-NOARG at which time the cells were isolated and frozen at −70°C. EPR spectroscopy was performed at 77 K on the cell suspensions using a Varian E-109 spectrometer equipped with an X-band (9.5 GHz) microwave bridge. The instrumental parameters at which the EPR spectra were recorded are given in Figure 13.6 (legend).

13.3.2 Results and discussion

Adult-NHEK cells exposed to HD and incubated for 18 hours at 37°C induced the formation of EPR-detectable signals (Figure 13.6). The EPR spectra in Figure 13.6 were obtained by freezing the samples to 77 K in liquid nitrogen after the incubation at 37°C for 18 hr. Figure 13.6A is the control and represents the EPR spectrum obtained when adult-NHEK were incubated in the absence of HD and then frozen.

However, adult-NHEK exposed to HD and incubated for 18 hours generate the EPR spectrum shown in Figure 13.6B. This EPR spectrum has an approximate g-value of g = 2.04, which is similar to the g-values of known reported iron–nitrosyl complexes, suggesting the formation of •NO (Corbett *et al.*, 1991). The observed EPR spectrum may originate directly from the formation of •NO or one of the reactive nitrogen by-products (NO_x) generated in the biological decomposition of •NO. Neonatal-NHEK incubated in the presence and absence of HD generate the EPR spectrum shown in Figure 13.6C. It is possible that, if •NO is produced by neonatal-NHEK either in the unexposed controls or in the HD-exposed cells, this •NO may originate from another source and have a different function, thus rendering it unavailable to interact with iron to form an iron–nitrosyl type complex as observed in Figure 13.6B. It is also conceivable that low concentrations of •NO are always present in neonatal-NHEK. This concentration may be too low to detect by EPR but high enough to cause other types of biological effects (e.g. cytokine expression).

Since the spectrum in Figure 13.6B suggests that adult-NHEK generate •NO during HD exposure, the cells were incubated with HD in the presence of the specific NOS inhibitor L-NOARG (Fukuto and Chaudhuri, 1995) to confirm the production of •NO. The result from this experiment is shown in Figure 13.6D. It can be seen that the EPR spectrum shown in Figure 13.6B is not generated when the cells are incubated with HD in the presence of L-NOARG. This strongly suggests that •NO is produced when adult-NHEK are exposed to HD.

It is important to establish what role the production of •NO plays during exposure of NHEK to HD. Previous reports have shown that •NO is involved in immune/cytokine regulation (Nathan and Xie, 1994). Therefore, to establish the role •NO plays during HD exposure, the effect that HD exposure has on cytokine regulation must be determined. Initially the effect of HD on the production of IL-1β was chosen because human skin keratinocytes express significant activation of IL-1 determined by the ELISA technique. For this reason, IL-1β was assayed in adult-NHEK and neonatal-NHEK (5×10^6–5×10^7 cells) exposed to HD (2 mM) and incubated for 3 hours at 37°C, using the ELISA technique. This exposure time is sufficient to allow expression of a significant quantity of IL-1β. However, it is not long enough to allow the cells (adult and neonatal) to go through a complete cycle and divide, thus eliminating possible errors due to differences in cell turnover time (~21 hr for neonatal-NHEK and ~24 hr for adult-NHEK). The HD concentration range was chosen because it is generally observed that exposure to approximately 1 mM HD causes formation of microblisters in human skin while full blister formation occurs after exposure to approximately 2 mM HD (Requena *et al.*, 1988; unpublished observation). The results are shown in Figure 13.7. Adult-NHEK exposed to HD (2 mM) show a significant increase in the production of IL-1β (Figure 13.7A). When these cell suspensions are incubated with the specific NOS inhibitor L-NOARG, the production of IL-1β is decreased to the level obtained for the cells (controls) not exposed to HD. In addition, the combined effects of incubation with L-NOARG followed by exposure to HD slightly lower the production of IL-1β when compared to the cell suspensions incubated with L-NOARG alone or to controls. For neonatal-NHEK exposed to HD (2 mM) the production of IL-1β is decreased when compared to the cells (controls) not exposed to HD (Figure 13.7B). Furthermore, incubation of the cell suspensions with L-NOARG alone also decreases the production of IL-1β. The combined effects of incubation with L-NOARG followed by HD exposure are

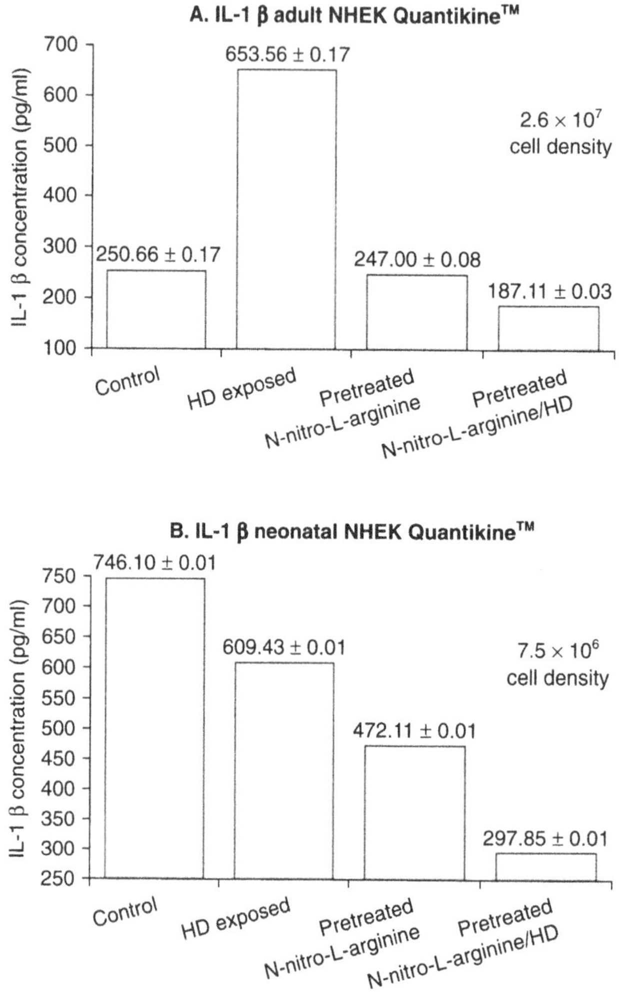

Figure 13.7 IL-1β activation of adult and neonatal NHEK exposed to 2 mM HD. The IL-1β levels in cell suspensions (5×10^6–5×10^7 cells/ml) were measured using IL-1β Quantikine™ ELISA kit. A: 100 μl aliquots of adult-NHEK (2.6×10^7 cells/ml) as a function of IL-1β (pg/ml) produced in HD-exposed and nonexposed controls. B: 100 μl aliquots of neonatal-NHEK (7.5×10^6 cells/ml) as function of IL-1β (pg/ml) produced in HD-exposed and nonexposed controls. The effect of *N*-nitro-L-arginine (L-NOARG) alone and the combined effects of sulfur mustard and L-NOARG on IL-1β expression by NHEK are also included in panel A and panel B

also shown in Figure 13.7B. These show a further decrease in the production of IL-1β when compared to the controls and to the cells exposed to HD alone or cells incubated with L-NOARG alone. The concentration of IL-1β per cell before and after exposure to HD or treatment with L-NOARG is given in Table 13.1. The data show that the neonatal-NHEK contain significantly larger amounts of IL-1β per cell when compared to the adult-NHEK.

Table 13.1 Concentration of interleukin-1β per cell

Cell type	n^1	*Control (pg/cell) × 10^6*	*HD (pg/cell) × 10^6*	*L-NOARG (pg/cell) × 10^6*	*L-NOARG+HD (pg/cell) × 10^6*
Adult	3	9.64	25.14	9.50	7.20
Neonatal	3	99.48	81.26	62.95	39.71

(1) *n = number of experiments. All columns are the average over number of experiments.*

The approximately ten-fold difference in IL-1β concentration per cell in the untreated control adult and neonatal NHEK can be attributed to two factors: (1) a continual higher level of IL-1β in neonatal-NHEK is required for the induction of growth factors for proliferation and other structural proteins required for the normal keratinocyte development; and (2) the fully developed •NO-related immune pathways in adult NHEK differ from the developing •NO-related immune pathways in the neonatal-NHEK (Becherel *et al.*, 1995). For instance, the data (Figure 13.7) show that the production of IL-1β is directly linked to the production of •NO. Therefore, it is possible that the adult-NHEK contain a low concentration of the constitutive form of NOS (cNOS), but are capable of rapidly producing the inducible form of NOS (iNOS) which is activated immediately upon the presence of a foreign toxic substance (Fukuto and Chaudhuri, 1995). This would mean that adult NHEK produce iNOS on demand in the presence of a foreign toxic substance (HD in this case) generating the •NO which triggers the observed increase in IL-1β production. Alternatively, the neonatal-NHEK contain mainly the cNOS and thus require only small concentrations of the iNOS. However, neonatal-NHEK contain a steady concentration of cNOS which is continuously producing low levels of •NO which, in turn, triggers the observed higher level production of IL-1β. Although the adult- and neonatal-NHEK appear to contain different forms of the NOS enzyme, both types of cells are affected in the same manner by the specific NOS inhibitor L-NOARG even though each type of cell (adult and neonatal) reacts in an opposite manner when exposed to HD.

The lower yield of IL-1β (Figure 13.7B) when neonatal-NHEK are exposed to HD as compared to unexposed controls is possibly due to the interaction of HD with the cell surface. It is possible that activation of c-NOS and iNOS originates from different receptors on the cell surface. Therefore, since neonatal-NHEK appear to contain only the cNOS it is likely that the interaction of HD with the cell surface would interfere with the continuous production of •NO and IL-1β (Figure 13.7B). This is consistent with the fact that no EPR signal is observed for HD-exposed neonatal-NHEK when suspensions of these cells were run in the same manner as the adult-NHEK (Figure 13.6). These observations are consistent with previous studies (Hunyadi *et al.*, 1992) that suggest that the surface characteristics of NHEK are continuously changing. These modulations reflect the stage of differentiation and activation of the NHEK. Thus, the NHEK in various stages of differentiation have distinct sets of surface moieties that are expressed in different manners (Figure 13.7).

One thing clear from the results in Figure 13.7 is dependence of the production of IL-1β in adult- and neonatal-NHEK on the generation of •NO. This establishes •NO as an effector molecule in cytokine regulation, at least for IL-1β. This fact is supported by the observation that when the production of •NO is blocked by the specific NOS inhibitor (L-NOARG) the production of IL-1β is also decreased. The results

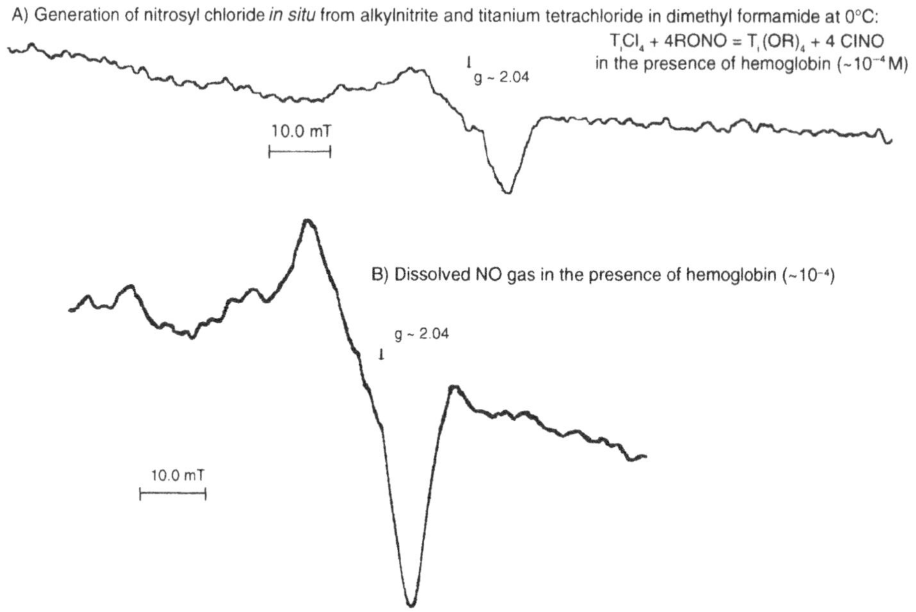

Figure 13.8 Low temperature (77 K) EPR spectrum of hemoglobin after reaction with (A) chemically generated NOCl and (B) dissolved •NO. EPR conditions were magnetic field, 334.5 mT; modulation frequency, 100 kHz; microwave frequency, 9.475 GHz; microwave power, 10 mW; receiver gain, 1.25×10^4; modulation amplitude, 1.0 mT and scan rate 0.833 mT/s

suggest that the production of •NO by the NHEK serves as a direct interleukin-1 converting enzyme activator or that •NO formed triggers a signal at the cell membrane which activates the interleukin-1 converting enzyme.

The fate of the •NO generated by the NHEK after activating the IL-1β system remains to be addressed. Since Cl^- is released upon dissolution of HD in aqueous environments, it is possible that the reactive nitrogen species formed is NOCl. NOCl is a nitrosylating agent that is consistent with the known biological action of HD. In addition, NOCl and •NO react with hemoglobin yielding the EPR spectra shown in Figure 13.8. Figure 13.8A represents the reaction of NOCl with hemoglobin and Figure 13.8B represents the reaction of a solution of dissolved •NO with hemoglobin (1×10^{-4} M). Both these EPR spectra have similar g-values to the one observed in Figure 13.6B. However, in Figure 13.8A the NOCl was generated chemically in a reaction vessel and then carried over with an inert gas (N_2) and bubbled through hemoglobin (1×10^{-4} M) solution. The hemoglobin solution containing the NOCl was rapidly frozen at 77 K in liquid nitrogen prior to obtaining its EPR spectrum. The presence of Cl^- in plasma and cytosol and •NO production or one of its active nitrogen by-products (NO_x) during the expression of IL-1β may feasibly generate the highly reactive alkylating agent NOCl. The Gibbs energy of formation of NOCl is energetically feasible (Williams, 1988). Nitrosyl chloride can nitrosylate organic compounds directly, and therefore its presence poses two dangers: (1) it is a strong oxidant and (2) nitrosylation leads to compounds that are often mutagenic or

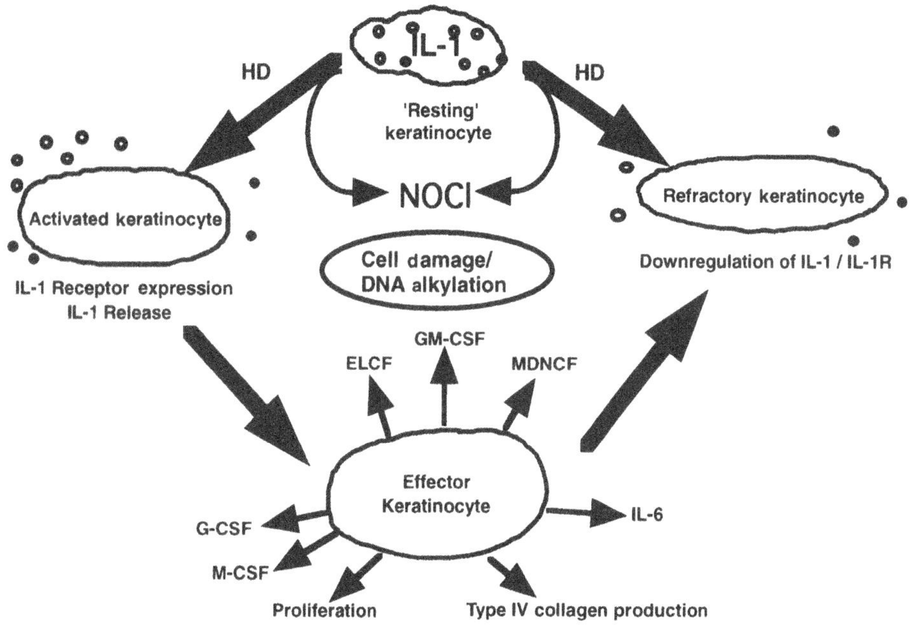

Figure 13.9 Hypothetical model of keratinocyte activation. The cellular activation of keratinocytes that follows the binding of IL-1 to its cell-surface causes gene expression of various products including interleukin-6 (IL-6), granulocyte (G), macrophage (M and GM), colony-stimulating factors (CSFs), epidermal-derived lymphocyte chemotactic factor (ELCF), and mononuclear cell-derived neutrophil chemotactic factor (MDNCF). Other consequences of IL-1 binding to its receptor include proliferation, chemotaxis, and type IV collagen production. Adapted from Kupper (1990)

promutagenic (alkylation of DNA). Therefore, the EPR spectrum in Figure 13.6B could originate from either the direct interaction of •NO with some type of porphyrin-containing molecule or the interaction with NOCl with the porphyrin containing molecule. A hypothetical model consistent with the results is shown in Figure 13.9. Both adult and neonatal NHEK conform to this scheme under normal circumstances.

However, because of the suggested difference in the NOS-type (inducible or constitutive) in these cells, their reaction to HD exposure is quite different. In terms of IL-1β production, keratinocytes in normal skin do not express significant numbers of IL-1 receptors. The normal keratinocyte cells can be in an 'activated' or 'refractory' state. The IL-1β receptor expression is associated with the active state of the keratinocyte and requires an outside stimulus to express the receptor and generate IL-1β. Exposure to HD causes adult NHEK to go into the activated state generating •NO, which in turn activates the IL-1β receptors causing the production of IL-1β. The •NO generated by the exposure of human keratinocytes to HD activates the interleukin-1 converting enzyme producing the observed increase in interleukin-1β (Arroyo *et al.*, 1997). The production of IL-1β triggers a series of other events in the keratinocyte (effector state). These observations are consistent with the results shown in Figure 13.7A. The neonatal-NHEK are continuously producing higher levels of •NO and IL-1β. However, when neonatal-NHEK are exposed to HD the cells go directly into the 'refractory' state either from the active state or the effector state. In the refractory state the IL-1 receptors are downregulated consistent with the results in Figure 13.7B. In both cases NOCl could be produced leading to cell damage and DNA alkylation. These findings may provide guidance to the understanding of sulfur mustard toxicity and also suggest that L-NOARG has significant novel pharmacological effects other than the inhibition of NOS.

In summary, human monocytes can release the pro-inflammatory mediator TNF-α after H-MG exposure as measured in this reported study. As a result of the TNF-α activation, oxidation of lipids and protiens occurs, followed by their degradation; also the fragmentation of DNA would be an expected consequence. In addition, HD induced release of both IL-1β and ·NO from normal human epidermal keratinocytes. The fact that both human monocyte and keratinocyte models implicate these pro-inflammatory mediators in vesicant injury tends to substantiate their role in injury/repair. These results indicate that the human cell model should provide specific aspects to resolving the mechanisms of vesicant-induced injury/repair and determining the efficacy of vesicant antagonists.

Acknowledgements

The author is grateful to Dr Laura Rhoads, NRC fellow at USAMRICD, and Dr Alasdair J. Carmichael for assisting in the inverted microscope phase studies. Dr Arroyo also thanks Dr Clarence A. Broomfield for helping with the vesicant agent exposures and Mr David W. Kahler and Ms Mary Theresa Nipwoda for the cell culture work.

References

Arroyo, C.M., von Tersch, R.L. and Broomfield, C.A. Activation of alpha human tumor necrosis factor (TNF-α) by human monocytes (THP-1) exposed to 2-chloroethyl ethyl sulphide (H-MG). *Human and Experimental Toxicology* 1995; **14**: 547–553.

Arroyo, C.M., Carmichael, A.J. and Broomfield, C.A. Could nitrosyl chloride be produced by human skin keratinocytes and sulfur mustard? A magnetic resonance study. *In Vitro Toxicology* 1997; **10**: 253–61.

Becherel, P.A., Le Goff, L., Ktorza, S., Quaaz, F., Mencia-Huerta, J.M., Dugas, B., Debre, P., Mossalayi, M.D. and Arock, M. Interleukin-10 inhibits IgE-mediated nitric oxide synthase induction and cytokine synthesis in normal human keratinocytes. *Eur. J. Immunol.* 1995; **25**: 2992–5.

Beutler, B., Mahoney, J., Le Trang, N. *et al.* Purification of cachectin, a lipoprotein-lipase suppressing hormone secreted by endotoxin induced RAW 264.7 cells. *J. Exp. Med.* 1985a; **161**: 984–995.

Beutler, B., Greenwald, C., Hulmes, J.D., *et al.* Identity of tumor necrosis factor and the macrophage-secreted factor cachectin. *Nature* 1985b; **316**: 552–4.

Beutler, B. and Cerami, A. Cachectin and tumor necrosis factor as two sides of the same coin. *Nature* 1986; **320**: 584–8.

Blanton, B., Kupper, T.S., McDougall, J. and Dower. Regulation of interleukin-1 and its receptor in human keratinocytes. *Proc. Natl Acad. Sci. USA.* 1989; **86**: 1273–7.

Buettner, G.R. Spin-Trapping: ESR parameters of spin adducts. *Free Radical Biology & Medicine* 1987; **3**: 259–303.

Carmichael, A.J. and Riesz, P. Photoinduced reactions of anthraquinone antitumor agents with peptides and nucleic acid bases: An electron spin resonance and spin-trapping study. *Archives of Biochemistry and Biophysics* 1985; **237**: 433–44.

Corbett, J.A., Lancaster, Jr J.R., Sweetland, M.A. and McDaniel, M.L. Interleukin-1β-induced formation of EPR-detectable iron-nitrosyl complexes in islets of Langerhans. Role of nitric oxide in interleukin-1β-induced inhibition of insulin secretion. *J. Biol. Chem.* 1991; **266**: 21351–4.

Corbett, J.A., Lancaster, J.R., Sweetland, M.A. and McDaniel, M.L. A 1-hour pulse with IL-1β induces formation of nitric oxide and inhibits secretion by rat islets of Langerhans: Evidence for a tyrosine kinase signaling mechanism. *FASEB J.* 1993; **7**: 369–74.

Cowan, F.M., Broomfield, C.A. and Smith, W.J. Effect of sulfur exposure protease activity in human peripheral blood lymphocytes. *Cell Biology and Toxicology* 1991; **7**: 239–248.

Cowan, F.W. and Broomfield, C.A. Putative roles of inflammation in the dermatopathology of sulfur mustard. *Cell Biology and Toxicology* 1993; **9**: 201–213.

Cowan, F.W., Broomfield, C.A. and Smith, W.J. Sulfur mustard exposure enhances Fc receptor expression on human epidermal keratinocytes in cell culture: implications for toxicity and medical countermeasures. *Cell Biol. & Toxicol.* 1998; 14:261–266.

Dinarello, C.A. Interleukin-1 and interleukin-1 antagonism. *Blood* 1991; 77: 1627–52.

Faulkner, W.R., J.W. King and H. Damm, *Handbook of Clinical Laboratory Data*, Chemical Rubber Co., Cleveland, Ohio, 1968.

Fiers, W. Tumor necrosis factor characterization at the molecular, cellular and in vivo level. *Federation of European Biochemical Societies (FEBS) Letters* 1991; **285**: 199–212.

Fukuto, J.M. and Chaudhuri, G. Inhibition of constitutive and inducible nitric oxide synthase: potential selective inhibition. *Annu. Rev. Pharmacol. Toxicol.* 1995; **35**: 165–94.

Higuchi, K., Kajiki, A., Nakamura, M., Harada, S., Pula, P.J., Scott, A.L. and Dannenberg, A.M. Proteases released in organ culture by acute dermal inflammatory lesions produced in vivo in rabbit skin by sulfur mustard: hydrolysis of synthetic peptide substrates for trypsin-like and chymotrypsin-like enzymes. *Inflammation* 1988; **12**: 311–334.

Hunyadi, J., Simon, Jr M. and Dobozy, A. Immune-associated surface markers of human keratinocytes (minireview). *Immunol. Letters* 1992; **31**: 209–16.

Klebanoff, S.J., Vadas, M.A., Harlan, J.M., *et al.* Stimulation of neutrophils by tumor necrosis factor. *J Immunol* 1986; **136**: 4220–5.

Kolb, H. and Kolb-Bachofen, V. Nitric oxide: A pathogenic factor in autoimmunity. *Immunol. Today* 1992; **13**: 157–60.

Koppenol, W.H. Thermodynamic concentrations on the formation of reactive species from hypochlorite, superoxide and nitrogen monoxide. Could nitrosyl chloride be produced by neutrophils and macrophages? *FEBS Lett.* 1994; **347**: 5–8.

Kupper, T.S. Role of epidermal cytokines, in: *Immunophysiology. The role of cells and cytokines in immunity and inflammation.* Oppenheim, J.J. and Shevach, E.M. (eds) Oxford University Press: NY 1990; 285–305.

Kupper, T.S., Lee, F., Bichall, N., Clark, S. and Dower, S.K. Interleukin-1 binds to specific receptors on human keratinocytes and induces granulocyte macrophage colony-stimulating factor mRNA and protein. *J. Clin. Invest.* 1989; **82**: 1787–92.

Kuwabara, M., Inanami, O. and Sato, F. OH-induced free radicals in purine nucleosides and their homopolymers: ESR and Spin-Trapping with 2-methyl-2-nitrosopropane. *International Journal Radiation Biology* 1986; **49**: 829–844.

Lancaster, J.R., Laster, S.M. and Gooding, L.R. Inhibition of target cell mitochondrial electron transfer by tumor necrosis factor. *Federation of European Biochemical Societies (FEBS) Letters* 1989; **248**: 169–174.

Lawley, P.D. Carcinogenesis by alkylating agents. *ACS Monograph* 1976; **173**: 83–234.

Lieske, C.N., Klopcic, R.S., Gross, C.L., Clark, J.H., Dolzine, T.W., Logan, T.P. and Meyer, H.G. Development of an antibody that binds sulfur mustard. *Immunology Letters* 1992; **31**: 117–22.

Lloyd, S.S., Chang, A.K., Taylor, Jr. F.B., Janzen, E.G. and McCay. Free radicals and septic shock in primates: the role of tumor necrosis factor. *Free Radical Biology & Medicine* 1993; **14**: 233–242.

Makino, K., Suzuki, N., Moriya, F., Rohushika, S. and Hatano, H. A fundamental study on aqueous solution of 2-methyl-2-nitrosopropane as spin-trap. *Radiation Research* 1981; **86**: 294–310.

Nathan, C. and Xie, Q. Regulation of biosynthesis of nitric oxide. *J. Biol. Chem.* 1994; **269**: 13725–8.

Pennica, D., Nedwin, G.E., Hayflick, J.S., Seeburg, P.H., Derynck, R., Palladino, M.A., Kohr, W.J., Aggarwal, B.B. and Goeddel, D.V. Human tumor necrosis factor: precursor structure, expression and homology to lymphotoxin. *Nature* 1984; **312**: 724–729.

Requena, L., Requena, C., Sanchez, M., Jaqueti, G., Aguilar, A., Sanchez-Yus, E. and Hernandez-Moro, B. Chemical warfare cutaneous lesions from mustard gas. *Journal of the American Academy of Dermatology* 1988; **19**: 529–536.

Ross, W.C.J. *Biological alkylating agents, fundamental chemistry and design of compounds for selective toxicity.* 1962; Butterworths, London.

Rustgi, S.N. and Riesz, P. An ESR and spin-trapping study of the reactions of $S^{\bullet}O_2^-$ radical with protein and nucleic acid constituents. *International Journal Radiation Biology* 1978; **34**: 301–316.

Shalaby, M.R., Aggarwal, B.B., Rinderknecht, E. *et al.* Activation of human polymorphonuclear neutrophil functions by interferon-gamma and tumor necrosis factors. *J. Immunol.* 1985; **135**: 2069–2073.

Smith, W.J., Cowan, F.M. and Broomfield, C.A. Increased proteolytic activity in human epithelial cells following exposure to sulfur mustard. *FASEB J.* 1991a; **5**: A828.

Suffys, P., Beyaert, R., Van Roy, F. and Fiers, W. Involvement of a serine protease in tumor-necrosis-factor-mediated cytotoxicity. *European Journal of Biochemistry* 1988; **178**: 257–265.

Tsuchiya, S., Yamabe, M., Yamaguchi, Y., Kobayashi, Y., Konno, T. and Tada, K. Establishment and characterization of human acute monocytic leukemia cell line (THP-1). *Int. J. Cancer* 1980; **26**: 171–6.

Tsujimoto, M., Yokota, S., Vilcek, J., *et al.* Tumor necrosis factor provokes superoxide anion generation from neutrophils. *Biochem Biophys Res Commun* 1986; **137**: 1094–1098.

Vilcek, J., Palombella, V.J., Henrikson-Destefano, D. *et al.* Fibroblast growth enhancing activity of tumor necrosis factor and its relationship to other polypeptide growth factors. *J. Exp. Med.* 1986; **163**: 632–8.

Williams, D.L.H. *Nitrosation.* Cambridge University Press: New York 1988; 1–36.

14 Duplicity of thiols and thiyl radicals: protector and foe

Christopher J. Rhodes

Abstract

A spectroscopic technique using positive muons is presented and shown to be extremely effective in the study of thiyl radicals, both in aqueous and non-aqueous (membrane-like) environments; through its agency, structural information and reaction kinetics of these radicals may be obtained. The method is superior to conventional ESR spectroscopy, to which thiyl radicals are invisible in liquid solution, and to pulse-radiolysis which is limited to aqueous media; the rate constants for the reactions of thiyl radical with lipids but also with antioxidants such as β-carotene and (SH) glutathione are found to be enhanced in non-aqueous media. The results from these and other 'artificial' (chemical) studies are placed in the context of the role of thiyl radicals in biological systems.

14.1 Introduction

Thiyl radicals are widely implicated in biological systems (Kalyanaraman, 1995; Halliwell and Gutteridge, 1989) for example as H-atom donors which repair free radical damage on bio-molecules, and in the nitrosylation of protein thiols; it has also been suggested that they are involved in the thiol-induced enhancement of oxidative modification of low-density lipoprotein, although the mechanistic details remain speculative. It has been proposed that thiyl radicals might be detoxified in cells in the presence of glutathione (GSH) and superoxide dismutase. Since thiyl radicals are themselves reactive species, it is of concern to know their subsequent fate, particularly in regard to their reactivity with sensitive cellular constituents; potentially there is a duplicity, in which a thiol might act to repair a free radical damage site, but the resulting thiyl radical causes further damage.

In any mechanistic evaluation of a transient intermediate, it is highly desirable to obtain some details of electronic structure and geometry: while thiyl radicals may be detected indirectly in ESR/spin-trapping experiments (Mile *et al.*, 1992), they have never been studied directly by ESR in solution, although they may be trapped and thereby measured in solid matrices (Nelson *et al.*, 1977; Symons, 1974).

Consideration of the electronic structure of an $RS^•$ radical which, formally, is orbitally degenerate has led to the conclusion that they should be undetectable in solution by ESR in consequence of their highly anisotropic g-tensor (Nelson *et al.*, 1977; Symons, 1974), resulting in extremely broad lines; moreover, the high reactivity of

RS$^{\bullet}$ radicals towards their precursors will favour the detection of product radicals such as RS—SR$_2$ (Symons, 1974) or RSS$^{\bullet}$ (Hadley and Gordy, 1974).

Given these difficulties, which we prefer to avoid, we now introduce an altogether different approach for the study of RS$^{\bullet}$ radicals which uses positive muons as a radioactive probe of magnetic interactions in radicals.

14.2 Background to the muon spectroscopy technique

It is partly the purpose of this chapter to indicate the potential of muon spin rotation (TF-μSR) spectroscopy (Walker, 1983; Roduner, 1988) for studying free radicals related to those extant in biological systems, and since this approach will be novel to the great majority of biochemists, the following overview of the method itself should prove helpful.

Muons, fundamental sub-atomic particles, are a component of cosmic radiation but are formed for research purposes by bombarding a target of suitable material (carbon or beryllium) with medium energy protons.[1] Among the products of the ensuing nuclear reactions that occur are pions (binding components of nuclei); these decay on a nanosecond timescale to muons (*via* $\pi^+ \rightarrow \mu^+ + \nu_\mu$) which may be implanted into matter, for the study of the magnetic properties of solid state, liquid phase and gas phase processes. Depending on the charge of the pion, either positive or negative muons are formed in their decay; negative muons find application, *inter alia*, in the promotion of nuclear fusion, while their positive counterparts apply particularly to chemistry, biology, materials and catalysis research.

While muons are leptons, to use the classification of nuclear physicists, and belong formally to the 'electron' category, as chemical species positive muons behave as protons, forming a bound state with a negative electron which is normally dubbed 'muonium' (μ^+e^-): muonium is equivalent to a normal protium atom (p^+e^-) and indeed shows the chemical properties of a *light* hydrogen atom, undergoing both H-atom abstraction and addition reactions. It serves well, therefore, as a radioactive probe of H-atom chemistry (Rhodes *et al.*, 1995) and of processes involving free radicals since the latter are readily formed by addition of muonium to unsaturated organic substrates, e.g.

$$H_2C{=}CH_2 + Mu^{\bullet} \rightarrow Mu{-}CH_2CH_2^{\bullet}$$

$$R_2C{=}S + Mu^{\bullet} \rightarrow R_2C(Mu){-}S^{\bullet}$$

Thus labelled, these radicals are characterised by a single pair of lines in the TF-μSR spectrum, on the application of a high transverse magnetic field, which represent the $-\frac{1}{2}, +\frac{1}{2}$ m_s spin combination with the muon (m_μ) states; these occur at the precession frequencies from muons which experience the sum of the applied and ($-\frac{1}{2}, +\frac{1}{2}$ m_s) hyperfine magnetic fields. The muon-electron hyperfine coupling constant is obtained from the difference between the high (ν_2) and low (ν_1) frequencies for each radical: $A_\mu = \nu_2 - \nu_1$; as the coupling increases, for a given magnetic field, the frequency ν_2 increases, while concomitantly that ν_1 first decreases, reaches zero and then increases due to a sign change in the transition; in the latter limit the coupling is obtained from the sum of the frequencies: $A_\mu = \nu_2 - (-\nu^1)$. An analysis of the data may be made corresponding to each frequency as described below.

14.3 Experimental section

The μSR experiments were carried out using the μE4 beamline at the Paul Scherrer Institute, Villigen, Switzerland. All reagents were purchased from Fluka or Aldrich in the highest available purity, and were used as received. Solutions (20 wt%) were prepared of thioacetamide, thiobenzamide, *N,N*-dimethylthioformamide, ethylene trithiocarbonate and 3-ethyl-2-thioxo-4-oxazolidone in ethanol, tetrahydrofuran and formamide and were deoxygenated at 10^{-4} mmHg, before being sealed into 35mm o.d. thin-walled pyrex ampoules. In each experiment, the sample was maintained in an applied magnetic field of 0.2T, while being irradiated with positive muons of momentum 85 MeV/c; data of 20–40 million decay events were accumulated and processed as described in the footnote, yielding the muon–electron coupling constants. For each kinetic measurement, the appropriate substrate (linoleic acid, linolenic acid, beta-carotene, (SH) glutathione) was incorporated in the appropriate concentration in the thioacetamide or thiobenzamide solution.

Semi-empirical M.O. calculations were carried out on the John Moores University Vax Cluster, using the PM3 Hamiltonian (Stewart, 1989), as available in the MOPAC6 package (QCPE No. 455). The geometry was optimised using either a UHF or RHF Hamiltonian. In the single point calculations, with the RHF approximation, a configuration interaction was allowed over an active space consisting of the two highest doubly occupied and the singly occupied together with the lowest three unoccupied levels.

14.4 Structural studies of thiyl radicals

In this study, we have investigated the temperature dependences of the muon coupling in thiyl radicals $R_2C(Mu)—S^{\bullet}$, formed by the addition of muonium to the following thiocarbonyl compounds (C═S): thioacetamide, thiobenzamide, *N,N*-dimethylthioformamide, ethylene trithiocarbonate and 3-ethyl-2-thioxo-4-oxazolidinone, as a function of solvent (tetrahydrofuran, ethanol, formamide). Our results are summarised under the following headings.

14.4.1 Temperature dependences

In all cases, other than the radical formed from thiobenzamide, the muon coupling was found to increase with decreasing temperature. This confirms our original conclusion (Rhodes *et al.*, 1988) that the equilibrium conformation of $C(Mu)—S^{\bullet}$ radicals is that in which the C—Mu bond eclipses the density axis of the nominally sulphur centred $3p_z$ orbital. While this is also true for the adduct of ethylene trithiocarbonate, the coupling is in all solvents lower than that of all thiyl radicals other than that from thiobenzamide (*vide infra*), and is due to the increased weighting of conformations in which vacant orbitals on the sulphur atoms interact with the unpaired electron orbital, thus providing stabilisation.

14.4.2 Effect of solvent

This may be illustrated with reference to Table 14.1, which shows muon-electron hyperfine couplings/MHz obtained at 300K.

Table 14.1 Muon hyperfine couplings/MHz for $R_2C(Mu)—S^\bullet$ radicals derived from C=S compounds

Sample	*THF*	*Ethanol*	*Formamide*
Thioacetamide	487	464	435
Thiobenzamide	201	197	181
N,N-dimethylthioformamide	501	500	482
Ethylene trithiocarbonate	345	355	346
3-Ethyl-2-thioxo-4-oxazolidinone	419	420	409

It is clear that, in *all* cases, the coupling is reduced in formamide solution: this, we propose, is caused by specific solvation of the thiyl $S^\bullet$ atom, in the manner $RS^\bullet—C=O$, which will reduce the energy of the $3p_z$ level relative to the C—Mu σ and σ^*, levels, which are responsible for hyperconjugative spin-transfer to the muon. The difference between ethanol and THF is generally small, other than for thioacetamide which may relate to solvation of the NH_2 group by H-bonding in ethanol solution, and a change in conformational 'control' of the radical.

The value is very similar in *all* media for the radical derived from ethylene trithiocarbonate, and the effect of formamide may be offset by additional solvation of the *other* sulphur atoms, thus reducing their C—S σ^* orbital energies concomitantly with that of the $C—S^\bullet$ $3p_z$ level, leading to a reduced net effect.

In the biological context, these results demonstrate a solvent dependence, which reflects changing electronic interactions and overall conformations, suggesting that the properties of $RS^\bullet$ radicals may depend partly on the polarity (philicity) of the environment in which they are formed, i.e. be determined by their location in either hydrophobic or hydrophilic regions. The influence of the medium (aqueous *vs* non-aqueous) on the rate constants for thiyl radical reactions is demonstrated later on.

14.4.3 *The thiobenzamide adduct: $PhC(Mu)(NH_2)S^\bullet$*

We feel that this radical deserves the honour of a separate heading, since its behaviour is quite unlike that of the rest. Clearly, its coupling is reduced to *ca* 40–50% of the other typical values; moreover, in all three solvents the temperature dependence is *negligible*, implying either that the rotation about the $C—S^\bullet$ bond is free or that the conformation is largely *fixed*, by a substantial torsional barrier. Even were the highest value measured – that for DMTF in THF (501 MHz) – to represent the fully eclipsed conformation (i.e. the maximum coupling), free rotation would give *half* this, on the normal $\cos^2\theta$ basis. The appreciable increase in the coupling on cooling, however, tells us that the eclipsed limit has not been reached and so the half-value of *ca* 250 MHz is certainly an underestimate of that for *free* rotation. Since the measured coupling is far lower even than this low estimate, we reject this possibility and propose that the conformation is essentially fixed, and lies in a potential significantly below that of other torsional states.

In order to understand this effect, we have carried out *semi-empirical* M.O. calculations on the radicals $PhC(Mu)(NH_2)S^\bullet$ and $MeC(Mu)(NH_2)S^\bullet$. The results so far, which we are currently trying to refine, predict that the coupling in the former radical should be reduced to about 50% of that in the latter, and that its equilibrium geometry eclipses the phenyl group with the sulphur SOMO, in the manner of an incipient bridging structure. Since this geometry decreases the hyperconjugative overlap between the SOMO and the C—Mu bond, at least qualitatively, the fall in the muon coupling is explained.

14.4.4 Substituent effects

While the influence of phenyl and sulphur substituents, in the adducts of thiobenzamide and ethylene thiocarbonate, may be rationalised respectively in terms of an incipient phenyl bridging structure (Figure 14.1a) and interactions with σ* C—S orbitals, substituent effects in the remaining radicals are more subtle. We feel that the salient common structural feature of the adducts from *N,N,*-dimethylthioformamide, thioacetamide and 3-ethyl-2-thioxo-4-oxazolidinone is the presence of a β-nitrogen atom, the relative electron releasing power of which decreases along this series, belonging to, respectively, Me_2N, H_2N and N—C=O groups.

Considering the competing interaction between the S $3p_z$ SOMO and both the C—Mu and C—N bond σ* orbitals (Figure 14.1b), the energy of the N—C σ* will be raised in the stronger C—N σ-bond as the nitrogen atom becomes more electron releasing: therefore, the S• — C—N σ* interaction will be weaker, and that with the C—Mu bond will dominate further the conformation of the radical and hence raise the coupling. Qualitatively, this would account for the fall in the muon coupling along the series: Me_2N, H_2N, N—C=O.

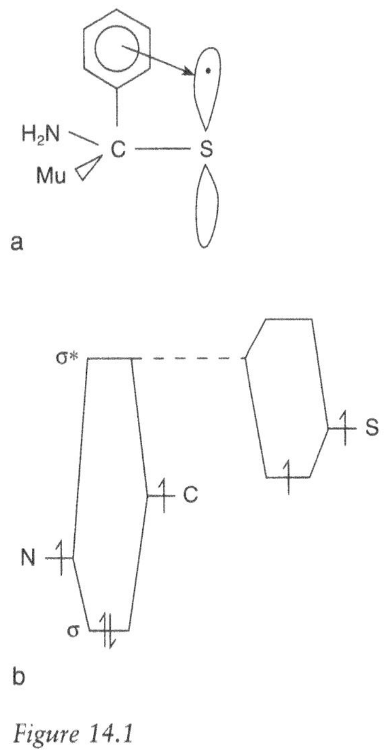

Figure 14.1

14.5 Reactivity of thiyl radicals

Having arrived at some conclusions regarding environmental and substituent effects in thiyl radicals, we now discuss some aspects of reactivity. The importance of the reactivity of thiyl radicals concerns particularly their role in biological systems, in which they are known to undergo a variety of reactions including electron transfer, hydrogen abstraction and addition reactions with biological substrates and xenobiotics (Kalyanaraman, 1995). We emphasise that it is not possible to use ESR spectroscopy to detect thiyl radicals in fluid solution because of extreme relaxation effects which broaden the spectral lines beyond detection; thus the technique is intrinsically inappropriate for the study of the kinetics of reactions involving them.

Of particular interest are the rate constants for the reactions of thiyl radicals with cellular constituents, and there have been a number of such studies made using the technique of pulse-radiolysis. However, pulse-radiolysis necessitates the use of aqueous (or at least partly aqueous) media, since the predominant source of the radicals is provided by the radiolysis products of water produced when a pulse of electrons is deposited therein, i.e. $OH^{\bullet}$ radicals, $H^{\bullet}$ atoms and solvated electrons. While this should provide no difficulty when it is desired to model processes that occur in aqueous regions of cells, it is surely a great limitation in the understanding of radical reactions in cell membranes: these are non-aqueous, and it is known that the rates of radical reactions may be profoundly altered according to the hydrophobic or hydrophilic nature of their medium. Using this method, thiyl radicals may be produced via (S)H-atom abstraction from thiols by carbon centred radicals, which arise from e.g. tert-butyl alcohol from which an H-atom has been abstracted by $OH^{\bullet}$ radicals (Everett *et al.*, 1996).

$$H_2O \rightarrow e^-/solv + OH^{\bullet} + H^{\bullet}$$

$$(CH_3)_3COH + OH^{\bullet}/H^{\bullet} \rightarrow {}^{\bullet}CH_2C(CH_3)_2COH$$

$$RSH + {}^{\bullet}CH_2C(CH_3)_2OH \rightarrow RS^{\bullet} + (CH_3)_3COH$$

Solvated electrons may also be converted to $OH^{\bullet}$ radicals if the system is saturated with N_2O:

$$N_2O + e^-/solv \rightarrow N_2 + O^{-\bullet}$$

$$O^{-\bullet} + H_2O \rightarrow OH^{\bullet} + OH^-$$

Thiyl radicals can be formed via displacement (homolytic bond cleavage) in addition to single electron redox processes involving thiols and disulphides (Asmus, 1990). In the context of toxicology, the classic example of the first kind is the 'repair' process where a carbon centred radical abstracts the (S)H-atom from the thiol:

$$RSH + R_3C^{\bullet} \rightarrow RS^{\bullet} + R_3CH$$

Rate constants for this process run close to 10^8 M^{-1} s^{-1}, which is an order or less below the diffusion controlled limit, and so the process is strongly activation controlled; in contrast, abstraction by $OH^{\bullet}$ radicals requires almost no activation energy. Direct single-electron oxidation of thiols, and particularly their conjugate base (RS^-) with transition metal ions (e.g. Fe^{3+}, Cu^{2+}) and other oxidising species with redox potentials of at least +0.8–+1.0 V is likely for RS^- oxidation and higher (> 1.3 V) for RSH oxidation.

Of similar importance is the displacement and reduction of disulphides, and many organic and inorganic radicals ($R^{\bullet}$) can undergo reactions of the type:

$$R^{\bullet} + RS\text{—}SR \rightarrow RSR + RS^{\bullet}$$

Rate constants for these processes are very strongly activation controlled and in the region 10^6 M^{-1} s^{-1} and lower; therefore, they require relatively high disulphide concentrations in order to compete effectively with other reactions of $R^{\bullet}$ radicals. One further reaction, which has been claimed to be of biological significance – at least in the case of penicillamine disulphide – is the cleavage of the C—S rather than the S—S bond in the manner:

$$R^{\bullet} + RSSR \rightarrow RSS^{\bullet} + \text{products}$$

Energetically, this process is only favourable for strongly stabilised perthiyl radicals, although the process has been implicated in the radiolysis of solid disulphides under cryogenic conditions (Hadley and Gordy, 1974); this is, however, controversial, as Symons has assigned the radiolytic product to radicals of the type $RS\text{—}SR_2^{\bullet}$ (Symons, 1974). We, nonetheless, consider that some alternative reaction, leading to a radical species different from $RS^{\bullet}$, may be of significance in cellular radiation biology.

Thiyl radicals can undergo a variety of reactions, the most obvious being combination to form the disulphide RSSR:

$$RS^{\bullet} + RS^{\bullet} \rightarrow RS\text{—}SR$$

Kinetically, this is a fast process, and is limited only by diffusion; however, it has to compete with many possible radical–molecule reactions, and at the expectedly low concentrations of thiyl radicals present in most biological systems is unlikely to be of particular significance.

Thiyl radicals may act as oxidants, and redox potentials of +0.75 and +1.33 V have been measured for $RS^{\bullet}/RS^-$ and $RS^{\bullet}/RSH$ couples, respectively; electron donors, for example vitamin A (ascorbate), α-tocopherol (vitamin E) and a number of phenothiazine derivatives can reduce thiyl radicals with rate constants of *ca* 10^8 M^{-1} s^{-1}, indicating their great effectiveness in this respect.

The reaction with molecular oxygen now appears to be reversible (Asmus, 1990):

$$RS^{\bullet} + O_2 \rightleftharpoons RSOO^{\bullet}$$

Initially, it was concluded that the reaction occurred by a diffusion controlled addition of oxygen, which was irreversible, but later data indicate that this diffusion control ($k = 10^9$ M^{-1} s^{-1}), in fact, requires reversibility; if it is irreversible, then much lower values, measured in a different set of experiments ($k = 10^7$–10^8 M^{-1} s^{-1}) must be correct. Whatever the final analysis, the reaction between $RS^{\bullet}$ radicals and oxygen is far less efficient than was once believed. Sevilla has correlated the spin-density distribution in peroxyl radicals with their reactivity (Sevilla *et al.*, 1990); RSOO• radicals are found to possess an unusual electronic structure, and hence optical (charge-transfer) properties, and a reduced unpaired electron spin density on the terminal oxygen atom – its reactivity should then be more selective than a normal alkylperoxyl radical.

Several recent studies have also indicated that the 'repair' reaction is reversible:

$$RSH + R_3C^{\bullet} \rightleftharpoons RS^{\bullet} + R_3CH$$

The forward reaction is nearly always the faster process with rate constants in the range 10^7–10^9 M^{-1} s^{-1}; on the other hand, values of 10^3–10^7 M^{-1} s^{-1} have been measured for the reverse reaction, i.e. H-atom abstraction from activated C—H bonds. Asmus has pointed out (Asmus, 1990) that one important consequence of this reversibility is that thiols are not only able to repair a potentially damaging site but indirectly, via their thiyl radicals, can also contribute to biological damage, and that the equilibrium at least provides a rationale for the possible incompleteness of a repair process.

Displacement reactions involving thiyl radicals can also occur with disulphides, and have been shown to proceed through the formation of an intermediate adduct radical:

$$RS^{\bullet} + RSSR \rightleftharpoons [RSS(SR)R^{\bullet}] \rightleftharpoons RSSR + RS^{\bullet}$$

Stability constants of *ca* 100 M^{-1} and rate constants of *ca* 10^6 M^{-1} s^{-1} for the thiyl/disulphide reaction have been measured for the CysS•/cystine couple.

14.5.1 Kinetic studies of thiyl radicals using muon spectroscopy

This method does not require a medium of any particular kind, beyond it containing an appropriate substrate (a C═S compound for thiyl radicals), since the muons may be implanted into any medium and both aqueous and non-aqueous phases could, in principle, be studied. In the present work, we have used non-aqueous solutions ('super-dried' ethanol) since our major interest concerns the potential role of thiyl radicals in membranes, particularly in regard to lipid-peroxidation: the process that destroys the structural integrity of cells and is central to the ageing process. Details of the procedure for measuring radical kinetics have been described previously (Roduner, 1989; Rhodes, 1999).

The derived second order rate constants for reaction of $MeC(Mu)(NH_2)S^{\bullet}$ (from thioacetamide) and $PhC(Mu)(NH_2)S^{\bullet}$ (from thiobenzamide) radicals with selected lipid and antioxidant materials, measured in ethanol solution, are displayed in Table 14.2, along with those for related thiyl radicals measured in partly aqueous solutions using pulse-radiolysis.

The unsaturated acids are chosen to model membrane lipid components, and it is clear that thiyl radicals react with them (probably by H-atom abstraction from allylic groups) at appreciable rates: the rates are faster by a factor of 3–4 than for thiyl radicals with these substrates in aqueous media, and so membrane damage and loss of cellular integrity from the lipid peroxidation that this may initiate is a serious possibility.

What then of potential protective agents? Glutathione is a well established repairer of radical damage in cells (Halliwell and Gutteridge, 1989), and could certainly

Table 14.2 Rate constants ($M^{-1}\ s^{-1} \times 10^{-7}$) for reactions of thiyl radicals with various substrates

	(SH) Glutathione	*β-Carotene*	*Linolenic acid*	*Linoleic acid*
$HO(CH_2)_2S^{\bullet}$	–	250 ± 1[a]	–	3.1[a]
$GS^{\bullet}$	6.62 ± 0.76[d]	22 ± 1[a]	1.9[b]	0.8[b]
$CysS^{\bullet}$	–	26 ± 1[a]	0.9[b]	0.6[b]
$PenS^{\bullet}$	–	–	0.4[b]	0.3[b]
$RS^{\bullet(2+)}$	–	0.42 ± 0.02[a]	–	–
TA	18.7 ± 0.6[c]	57.3 ± 1.6[c]	2.4 ± 0.2[c]	1.4 ± 0.2[c]
TB	16.9 ± 0.8[c]	37.3 ± 1.8[c]	2.0 ± 0.2[c]	1.2 ± 0.1[c]

(References: a, Everett *et al.*, 1996; b, Schoneich and Asmus, 1990; c, this work; d, Mezyk, 1996). Key: $GS^{\bullet} = H_3N^+(^-O_2C)CH(CH_2)_2CONH(HO_2CCH_2NHCO)CHCH_2S^{\bullet}$; $CysS^{\bullet} = H_3N+(^-O_2C)CHCH_2S^{\bullet}$; $PenS^{\bullet} = H_3N^+(^-O_2C)CMe_2S^{\bullet}$; $RS^{\bullet(2+)} = H_3N^+(CH_2)_3H_2N^+(CH_2)_2S^{\bullet}$; TA = $Me(H_2N)(Mu)CS^{\bullet}$; TB = $Ph(H_2N)MuCS^{\bullet}$.

scavenge reactive thiyl radicals (and indeed other radicals that may be produced) given the high rate constants for its reactions with them; again, the rates are faster by a factor of 2–3 than for the equivalent reactions in aqueous media.

β-Carotene has received considerable attention as a dietary supplement supposed to protect against developing cancer, and to alleviate the effects of ageing: the rates are indeed very large for the scavenging of thiyl radicals by this agent, and accords that such a role contributes in these conditions.

Everett and co-workers (Everett *et al.*, 1996) have pointed out that the rate of reaction with β-carotene is accelerated as the lipophilicity of the thiyl radical increases, as demonstrated by the near diffusion controlled limit measured for $HO(CH_2)_2S^{\bullet}$; we view the situation in reverse. That the rate can be as fast as this implies that there is little intrinsic activation barrier to the reaction, and the differences are determined mainly by the rate of diffusion of the thiyl radicals; those which are highly charged will be strongly impeded by solvation in the aqueous media used for pulse-radiolysis measurements, whereas $HO(CH_2)_2S^{\bullet}$ will not. For the radicals $MeC(Mu)(NH_2)S^{\bullet}$ (TA) and $PhC(Mu)NH_2)S^{\bullet}$ (TB), particularly as measured in the more weakly solvating ethanol medium, the rates are higher than for thiyl radicals with a similar degree of steric hindrance around the sulphur atom, measured in aqueous solution.

Support for this argument is given by the fact that both addition and abstraction reactions – with entirely different transition structures – are all accelerated when solvation effects are reduced, both by a non-aqueous medium and low-polarity of the thiyl radical.

It is as though nature has compensated for the potential increase in radical damage in membrane environments by increasing the efficiency of the radical repair processes.

14.6 Evidence for actual biological significance of thiyl radicals

We have outlined the molecular structural nature of thiyl radicals along with their reactivity in aqueous and non-aqueous environments, as determined by muon spectroscopy and pulse-radiolysis; the different behaviour in each likely reflecting differences between membrane and other cell regions. These, however, are 'chemical' studies, in which a system is designed specifically to observe the thiyl radical, and it is only in our imagination that this mimics an actual biological system. So, let us now review the evidence for the involvement of thiyl radicals in real life.

Thiyl radicals result from the single electron oxidation of thiols:

$$RSH \rightarrow RS^{\bullet} + e^{-} + H^{+}$$

They may also result from the 'repair' of a free radical damage site, e.g. at a carbon atom:

$$RSH + RCH_2^{\bullet} \rightarrow RS^{\bullet} + RCH_3$$

For this reason, thiols are efficient radioprotectors, which act mainly by repair of radical damage to essential cellular constituents, such as DNA; the purpose of this section is to consider whether the positive maintenance of cellular function counterbalances the fact that a reactive $RS^{\bullet}$ radical has been produced, since at least under *in vitro* conditions, thiyl radicals are well known to promote cellular damage, such as lipid peroxidation.

Thiyl radicals may, in fact, be formed by a variety of reactions between thiols and cellular constituents (Kalyanaraman, 1995). (S)H-atom abstraction may be accomplished by deleterious radicals (ROI), including superoxide ($O_2^{-\bullet}$) and its conjugate acid, hydroperoxyl ($HO_2^{\bullet}$), along with alkyl peroxyls ($RO_2^{\bullet}$) and hydroxyl ($OH^{\bullet}$) radicals. In addition, enzymes, notably lactoperoxidase and prostaglandin hydroperoxidase and also myoglobin are capable of oxidising thiols in the presence of peroxides (or H_2O_2).

The reaction between peroxynitrite ($^{-}OONO$) and thiols is one of the latest interests regarding the biochemistry of nitric oxide, since NO reacts with $O_2^{-\bullet}$ at close to the diffusion limited rate to form $^{-}OONO$; the latter is a powerful oxidant which has been shown to oxidise thiols to thiyl radicals; other work has shown that $NO_2^{\bullet}$ can also oxidise thiols (Everett *et al.*, 1996).

Once formed there are many possible reactions of thiyl radicals, although their concentration is generally too low for dimerisation (radical combination) to occur; instead, H-atom abstractions, additions to unsaturated moieties and electron transfer reactions occur. Thiyl radicals can also add to molecular oxygen to form $RSOO^{\bullet}$ radicals (Grierson *et al.*, 1992), which, though formal peroxyl radicals, have quite distinct electronic properties (Sevilla *et al.*, 1990), and their spin density distribution is low on the terminal oxygen atom, thus making them more selective than normal (e.g. alkyl) peroxyls in their reactivity. Though the dimerisation of thiyl radicals is disfavoured kinetically by their low concentration, the reaction between $RS^{\bullet}$ and RS^{-}, which in the case of glutathione (GSH) produces $GSSG^{-}$, may well be important in that it eventually generates $O_2^{-\bullet}$ by electron transfer to $O_2^{\bullet}$, since $O_2^{-\bullet}$ and H_2O_2 can be destroyed by superoxide dismutase, catalase and glutathione peroxidase; this particular reaction scheme has been proposed as a major route to detoxification of ROI in cells (Winterbourn, 1993).

14.6.1 *Detection of thiyl radicals in biological processes*

We have already stated that $RS^{\bullet}$ radicals cannot be detected in liquids directly by ESR spectroscopy, although they may be measured in solid matrices, so long as these are composed of molecules that can provide effective hydrogen-bonding to the sulphur atom of the radical, and clearly muon spectroscopy is not feasible for a real system containing actual cells. However, thiyl radicals may be spin-trapped in liquid aqueous solution, which provides the most direct evidence for their formation in various systems, believed to approach the biological environment. There are a number of possible agents that may be used as spin-traps: in all cases there is an increase in the g-factor due to the large spin-orbit coupling from the sulphur atom, and with DMPO and novel phosphorylated variants of it, the hyperfine couplings are also characteristic, being determined by the electronic effects of the sulphur substituent (Kalyanaraman *et al.*, 1996; Karoui *et al.*, 1996).

Actual examples include the use of these traps to catch thiyl radicals derived from haemoglobin and from serum albumins: in these cases, the ESR spectral features show that the thiyl radicals are appreciably immobilised by the large protein molecule to which they are bound. The oxidation of thiols by peroxynitrite and by sulphite ions has similarly been shown to generate thiyl radicals, and by using higher (Q-band) frequencies, greater spectral resolution is obtained among mixtures of spin-adducts, when carbon-centred radicals are produced simultaneously in these systems (Kalyanaraman *et al.*, 1996; Karoui *et al.*, 1996).

Asmus and his coworkers (Schoneich *et al.*, 1989; Schoneich and Asmus, 1990) have made pulse-radiolysis studies of thiyl radicals in homogeneous media containing linoleic and other carboxylic acids containing diene functions in their hydrocarbon chains; efficient H-atom abstraction was found to occur, with the formation of a pentadienyl radical:

$$RS^{\bullet} + CH_2C{=}C{-}C{=}C \rightarrow RSH + CH^{\bullet}{-}C{=}C{-}C{=}C$$

In a biological system, however, the efficiency of most lipid peroxidations is believed to depend on the presence of lipid hydroperoxides and metal ions such as Fe^{2+}; it is at first sight curious that ascorbate/ascorbic acid can both promote or ameliorate oxidation, depending on the relative concentration of these minor species, and this appears to be true for thiols too (Tien *et al.*, 1982). It has been shown that thiol dependent lipid peroxidation is unaffected by the presence of superoxide dismutase, and that the iron-dependent decomposition of linoleic acid hydroperoxides can be promoted by the presence of thiols (Tien *et al.*, 1982).

Of recent interest is the macrophage-dependent oxidation of low-density lipoprotein (Sparrow *et al.*, 1993; Heinecke *et al.*, 1993), for which it is proposed that an initial formation of $RS^{\bullet}$ results in a thiyl peroxyl, $RSOO^{\bullet}$, radical; this then abstracts an H-atom from the lipid in the normal initiation of its peroxidation. The extent of oxidative modification of LDL is generally thought to be determined by the presence of lipid peroxides associated with LDL (Thomas and Jackson, 1991). (Thus, although we noted earlier the reduced reactivity of $RSOO^{\bullet}$ radicals compared with 'normal' peroxyls (Sevilla *et al.*, 1990), they are nonetheless reactive, but perhaps are more selectively so.) It is interesting that, although thiyl radicals are known to react quite rapidly with DMPO, the presence of this spin-trap does not significantly inhibit the macrophage-dependent oxidation of low-density lipoprotein in the presence of thiols (Kalyanaraman, 1995); thus the mechanism cannot be as simple as one is tempted to imagine from the results of simple chemical systems.

It has been suggested that protein thiols might provide the main antioxidant defence against peroxyl radicals formed in human plasma; thiyl radicals are formed in an intermediary role during oxidative stress of blood plasma, and during oxidative modification of LDL, both lipid and apolipoprotein are oxidised. Kalyanaraman and co-workers (Kalyanaraman, 1995) have found that the cysteine residues of apolipoprotein B-100 compete with vitamin E for scavenging peroxyl radicals; a more rapid peroxidation was found for LDL in which the hydrophobic cysteine residues are blocked.

14.6.2 Detoxification of thiyl radicals

It has been proposed recently (Winterbourn, 1993) that thiyl radicals can be deactivated in cells (detoxified) by GSH and superoxide dismutase. The theory goes that an initial oxidation of GSH occurs, by unspecified radicals or by $O_2^{-\bullet}$, whereupon the resulting $GS^\bullet$ radical reacts with GS^- to form the glutathione disulphide radical anion ($GSSG^{-\bullet}$); in the presence of O_2, electron transfer occurs to regenerate $O_2^{-\bullet}$, thus starting 'chain-initiation'; superoxide dismutase then destroys $O_2^{-\bullet}$ in its normal way, producing ultimately $H_2O_2 + O_2$, so acting as a 'radical sink'.

Results by Stadtman and co-workers suggest that thiyl radicals are detoxified by a thiol-specific antioxidant enzyme in bacterial systems, but it is less clear whether a similar mechanism operates to detoxify thiyl radicals in mammalian cells (Yim *et al.*, 1994).

14.7 Conclusions

We have demonstrated that, in contrast to conventional ESR spectroscopy, thiyl radicals may readily be studied in liquid solution using muon spin rotation spectroscopy; moreover, this technique is not limited to aqueous solutions, unlike pulse-radiolysis, and so we have a viable means for obtaining kinetic data more relevant to those processes which occur in the membranes of cells, rather than their aqueous regions. Though this represents a considerable advance, we admit that these studies are still rather idealised, and certainly do not reflect the full impact of the complex interplay which must exist in a real biological system; but rate data, for instance showing the acceleration of reactions between thiyl radicals and membrane constituents in non-aqueous media we feel must be significant, and that thiyl radicals are potentially even more destructive to membrane lipids than previous pulse-radiolysis data would indicate. Significantly, their reaction rates with antioxidants such as glutathione (SH) and β-carotene are also enhanced (and have anyway far greater rate constants), so nature has compensated for this in order to protect the integrity of the vulnerable cellular compartments.

Beyond these purely chemical models, we have, as far as anyone can, surveyed the limited evidence for the role of thiyl radicals, and their study, in actual biological media. In this latter respect, Kalanaraman (1995) has arrived at some salient conclusions and recommendations:

1 The antioxidant mechanism of protection of protein thiols probably involves formation of thiyl radicals.
2 Nitrosylation of thiols and protein thiols by NO derived oxidants is probably mediated by thiyl radicals.
3 Thiol-induced oxidative modification of low-density lipoproteins (LDL) is probably not mediated by thiyl radicals.

4 Spin-traps can be used to inhibit thiyl-radical-mediated reactions in biological systems.
5 Nitrone spin-traps may be used to trap protein-associated thiyl radicals.
6 Although it has been proposed that a thiol-specific detoxification enzyme is needed for detoxification of thiyl radicals in bacterial systems, in mammalian cells thiyl radicals can be detoxified by ascorbate.
7 More sensitive and specific biological markers for thiyl radicals are needed!

Thus, any advance in this area depends partly on the skill of synthetic chemists in making improved spin-traps, but also in identifying (diamagnetic) markers for various kinds of radical damage.

Acknowledgements

I thank Dr Ivan Reid for invaluable experimental support, the Leverhulme Trust, the Paul Scherrer Institute, Liverpool John Moores University and the EPSRC for grants in support of our work on thiyl radicals and membrane-model systems.

Note

1 Muons decay to positrons ($\mu^+ \rightarrow e^+ + \nu_e + \nu_\mu$ on a microsecond timescale, which are detected using scintillation counters, and counted using fast electronics; the decay events are accumulated in 4 data histograms. Generally, the histogram is of the form

$$N(t) = N_0\{B + \exp(-t/t_\mu)[1 + F(t)]\},$$

where N_0 is a normalisation factor, roughly equal to the number of counts in the first channel ($t = 0$), B is the background fraction (usually <1%), t_μ is the muon lifetime and F(t) reflects the time dependence of the muon spin polarisation. For experiments in which a number of frequencies are obtained (as in the present case), F(t) is the sum of contributions of the form

$$F_j(t) = A_j\exp(-\lambda_j t)\cos(\omega_j t + \varnothing_j),$$

corresponding to a muon precession at a specific frequency ν_j, for which A_j is the asymmetry (amplitude), λ_j the relaxation rate, and provides a measure of chemical reactions or physical relaxation processes, and $\varnothing_j$ is the initial phase. The data are analysed by fitting directly the latter expression to the experimental data in Fourier space, which yields the parameters A_j, λ_j, ω_j, $\varnothing_j$ for each frequency.

References

Asmus, K.-D., 1990, Sulfur-centred free radicals, *Methods in Enzymology*, 186, 168–180.

Everett, S.A., Dennis, M.F., Patel, K.B., Maddix, S., Kundu, S.C. and Willson, R.L., 1996, Scavenging of nitrogen dioxide, thiyl, and sulfonyl free radicals by the nutritional antioxidant beta-carotene, *Journal of Biological Chemistry*, 271, 3988–3994.

Grierson, L., Hildenbrand, K. and Bothe, E., 1992, Intramolecular transformation reaction of the glutathione thiyl radical into a non-sulfur-centred radical – a pulse-radiolysis and EPR study, *International Journal of Radiation Biology*, 62, 265–277.

Hadley, J.H. and Gordy, W., 1974, Nuclear coupling of sulfur-33 and the nature of free radicals in irradiated crystals of cystine dihydrochloride, *Proceedings of the National Academy of Sciences of the USA*, 71, 3106–3110.

Halliwell, B. and Gutteridge, J.M.C., 1989, *Free Radicals in Biology and Medicine*, Clarendon Press, Oxford.

Heinicke, J.W., Kawamura, M., Suzuki, L. and Chait, A., 1993, Oxidation of low-density-lipoprotein by thiols – superoxide-dependent and superoxide-independent mechanisms, *Journal of Lipid Research*, 34, 2051–2061.

Kalyanaraman, B., 1995, Thiyl radicals in biological systems: significant or trivial. *Biochemical Society Symposium*, 61, 55–63.

Kalyanaraman, B., Karoui, H., Singh, R.J. and Felix, C.C., 1996, Detection of thiyl radical adducts formed during hydroxyl radical- and peroxynitrite-mediated oxidation of thiols – a high resolution ESR spin-trapping study at Q-band (35 GHz), *Analytical Biochemistry*, 240, 1–7.

Karoui, H., Hogg, N., Frejaville, C., Tordo, P. and Kalyanaraman, B., 1996, Characterization of sulfur-centered radical intermediates formed during the oxidation of thiols and sulfite by peroxynitrite, *Journal of Biological Chemistry*, 271, 6000–6009.

Mezyk, S.P., 1996, Rate constant determination for the reaction of hydroxyl and glutathione thiyl radicals with glutathione in aqueous solution, *Journal of Physical Chemistry*, 100, 8861–8866.

Mile, B., Rowlands, C.C., Sillman, P.D. and Fildes, M., 1992, The EPR spectra of thiyl radical spin adducts produced by photolysis of disulphides in the presence of 2,4,6-tri-tert-butylnitrosobenzene and 5,5-dimethyl-1-pyrroline N-oxide, *Journal of the Chemical Society, Perkin Transactions 2*, 1431–1437.

Nelson, D.J., Petersen, R.L. and Symons, M.C.R., 1977, The structure of intermediates formed in the radiolysis of thiols, *Journal of the Chemical Society, Perkin Transactions 2*, 2005–2015.

Rhodes, C.J., 1999, Radical Behaviour, *Chemistry in Britain*, Vol. 35, No. 2, pp.20–22.

Rhodes, C.J., Roduner, E. and Symons, M.C.R., 1988, Muonium containing thiyl radicals, *Journal of the Chemical Society, Chemical Communications*, 3–4.

Rhodes, C.J., Morris, H. and Reid, I.D., 1995, Patterns of muonium addition to imidazoles: a model of radiation produced hydrogen-atom reactivity with key biological subunits, *Journal of the Chemical Society, Perkin Transactions 2*, 2107–2114.

Roduner, E., 1988, *The Positive Muon as a Probe in Free Radical Chemistry*, Springer, Berlin.

Schoneich, C., Asmus, K.-D., Dillinger, U. and v.Bruchhausen, F., 1989, Thiyl radical attack on polyunsaturated fatty acids: a possible route to lipid peroxidation, *Biochemical and Biophysical Research Communications*, 161, 113–120.

Schoneich, C. and Asmus, K.-D., 1990, Reaction of thiyl radicals with alcohols, ethers and polyunsaturated fatty acids: a possible role of thiyl radicals in thiol mutagenesis, *Radiation and Environmental Biophysics*, 29, 263–271.

Sevilla, M.D., Becker, D. and Yan, M., 1990, Structure and reactivity of peroxyl and sulphoxyl radicals from measurement of oxygen-17 hyperfine couplings: relationship with Taft substituent parameters, *Journal of the Chemical Society, Faraday Transactions*, 86, 3279–3286.

Sparrow, C.P. and Olszewski, J., 1993, Cellular oxidation of low-density-lipoprotein is caused by thiol production in media containing transition-metal ions, *Journal of Lipid Research*, 34, 1219–1228.

Stewart, J.J.P., 1989, Optimisation of parameters for semiempirical methods, *Journal of Computational Chemistry*, 10, 209–220; 221–264.

Symons, M.C.R., 1974, On the electron spin resonance detection of $RS^{\bullet}$ radicals in irradiated solids: radicals of type $RSSR^-$, $RS—SR_2$ and $R_2S—SR_2^+$, *Journal of the Chemical Society, Perkin Transactions 2*, 1618–1620.

Tein, M., Bucher, J.R. and Aust, S.D., 1982, Thiol-dependent lipid-peroxidation, *Biochemical and Biophysical Research Communications*, 107, 279–285.

Thomas, C.E. and Jackson, R.L., 1991, Lipid hydroperoxide involvement in copper-dependent and independent oxidation of low-density lipoproteins, *Journal of Pharmacology and Experimental Therapeutics*, 256, 1182–1188.

Walker, D.C., 1983, *Muon and Muonium Chemistry*, Cambridge University Press, Cambridge.

Winterbourn, C.C., 1993, Superoxide as an intracellular radical sink, *Free Radical Biology and Medicine*, 14, No. 1, 85–90.

Yim, M.B., Chae, H.Z., Rhee, S.G., Chock, P.B. and Stadtman, E.R., 1994, On the protective mechanism of the thiol-specific antioxidant enzyme against the oxidative damage of biomacromolecules, *Journal of Biological Chemistry*, 269, 1621–1626.

15 Synergistic effect of carbon tetrachloride and 1,2-dibromoethane

Giuseppe Poli, Manuela Aragno, Elena Chiarpotto, Simonetta Camandola and Oliviero Danni

15.1 Introduction

Halogenated aliphatic hydrocarbons are among the chemical compounds employed in agriculture and in various industrial fields. As they are quite volatile, their extensive use may seriously contribute to environmental pollution. The well-known hepatotoxin carbon tetrachloride (CCl_4) (Slater, 1972; Slater and Sawyer, 1969) belongs to this family of compounds, and, notwithstanding its recognised toxicity, is still used as fumigant or lipid solvent, often mixed with other haloalkanes. One of these haloalkanes, 1,2-dibromoethane (DBE), also widely employed as a lead scavenger in the gasoline industry, is of particular concern as a likely carcinogen for humans (Alexeeff *et al.*, 1990).

In addition to the individual effects of CCl_4 and DBE is the recognised possibility that concentrations of such chemicals, which on their own are of low toxicity, become highly toxic when in association. A synergistic activation does indeed occur when these two haloalkanes are simultaneously administered to experimental animals or added to suspensions of isolated hepatocytes.

The aim of this chapter is to review the main metabolic mechanisms underlying such synergism and its expression in terms of enhanced oxidative effect on membrane lipid breakdown, irreversible cell damage and modulation of redox-sensitive transcription factors.

15.2 Carbon tetrachloride and 1,2-dibromoethane interactions

15.2.1 Carbon tetrachloride-induced shift of 1,2-dibromoethane metabolism through the inactivation of glutathione transferase

DBE is known to be metabolised in various organs, and particularly in the liver, along two main pathways. A greater amount of the haloalkane undergoes oxidative modification in the endoplasmic reticulum by the cytochrome P450-dependent enzyme system (Tomasi *et al.*, 1983; Khan *et al.*, 1993; Wormhoudt *et al.*, 1996). The most likely product of this metabolic route is bromoacetaldehyde, whose detection is actually quite complicated since it is very reactive (Hill *et al.*, 1978; Van Bladeren *et al.*, 1981). A minor, but still relevant, proportion of DBE is metabolically disposed through an addition reaction with glutathione, catalysed by the cytosolic enzyme glutathione transferase (Van Bladeren *et al.*, 1980; Shih and Hill, 1981). The DBE

Table 15.1 GSH transferase activity in rat isolated hepatocytes treated with CCl_4, DBE, or CCl_4 plus DBE

Groups	*(nmol/mg protein/min)*		
	1h	*2h*	*3h*
Control	268 ± 17	260 ± 19	256 ± 19
CCl_4 172 μM	220 ± 24[a]	224 ± 18[a]	217 ± 14[a]
DBE 50 μM	250 ± 28	276 ± 19	266 ± 13
CCl_4 ± DBE	214 ± 15[a]	209 ± 20[a]	210 ± 14[a]

Values are means ± SD of four experiments.
a Significantly different from the control group ($p \leq 0.05$).

metabolites stemming from the cytochrome P450-dependent pathway appear to be mainly responsible for the acute or subacute toxicity of the compound. The DBE-glutathione adducts are, on the contrary, related to the carcinogenic action (Inskeep *et al.*, 1986; Cmarik *et al.*, 1990).

With regard to DBE acute toxicity, we showed that low concentrations of 1,2-dibromoethane not able *per se* to exert damage in the rat hepatocyte model but became effective when given in combination with CCl_4, at low but sufficient amounts to impair the DBE metabolism. In fact, the metabolic activation of CCl_4 yealds reactive intermediates which are able to inactivate totally liver glutathione transferase (Danni *et al.*, 1991) (Table 15.1). A metabolic shift then operated which led to an enhanced formation of bromoacetaldehyde and/or related reactive products. In fact, following the treatment of rat hepatocyte with CCl_4 plus ^{14}C-DBE, the covalent binding of DBE metabolites to total cell protein was markedly increased in comparison to that of hepatocytes treated only with radiolabeled DBE (Chiarpotto *et al.*, 1993).

15.2.2 Potentiation of CCl_4-induced lipid peroxidation by DBE; in vitro *and* in vivo *data*

Does 1,2-dibromoethane interfere, on the other hand, with CCl_4 metabolism and the consequent production of reactive free radical species? The answer is negative, at least on the basis of experiments carried out with rat hepatocytes poisoned with a CCl_4/DBE mixture in which the carbon tetrachloride was ^{14}C-labeled. CCl_4-dependent protein covalent binding was not modified at all by the association with the other haloalkane (Chiarpotto *et al.*, 1993).

However, DBE was still able to potentiate the effect of CCl_4 while not interfering with its metabolic activation. In fact, as shown in Table 15.2, the reaction of a yet undefined DBE metabolite, most likely bromoacetaldehyde, with cellular glutathione, lowered the level of this primary antioxidant molecule below a threshold value, so facilitating the prooxidant activity of CCl_4. Reported in Table 15.3 is the extent of peroxidation of hepatocyte membrane lipids, monitored in terms of malonaldehyde production, following the treatment of rat isolated hepatocytes with a single dose of CCl_4, DBE and CCl_4 plus DBE. The stimulation of lipid peroxidation due to CCl_4 was markedly potentiated by the simultaneous cell treatment with a dose of DBE which was clearly not prooxidant *per se*.

Table 15.2 Total glutathione content of rat isolated hepatocytes treated with CCl_4, DBE, or CCl_4 plus DBE

Groups	*(nmol/10⁶ cells)*		
	1h	*2h*	*3h*
Control	20 ± 4	20 ± 4	18 ± 3
CCl_4 172 μM	16 ± 4	16 ± 5	13 ± 5
DBE 50 μM	6 ± 3[a]	7 ± 3[a]	7 ± 4[a]
CCl_4 ± DBE	6 ± 2[a]	7 ± 2[a]	5 ± 4[a]

Values are means ± SD of four experiments.
a Significantly different from the control group ($p \leq 0.05$).

Table 15.3 Malonaldehyde production in rat isolated hepatocytes incubated at 37°C in the presence of CCl_4, DBE, or CCl_4 plus DBE

Groups	*(nmol/10⁶ cells)*		
	1h	*3h*	*5h*
Control	0.2 ± 0.1	0.3 ± 0.1	0.3 ± 0.2
CCl_4 172 μM	0.5 ± 0.2[a]	1.1 ± 0.4[a]	1.8 ± 0.3[a]
DBE 50 μM	0.2 ± 0.1	0.4 ± 0.1	0.4 ± 0.1
CCl_4 ± DBE	0.7 ± 0.3	1.9 ± 0.3[b]	4.1 ± 0.4[b]

Values are means ± SD of four experiments.
a Significantly different from the control group ($p \leq 0.05$).
b Significantly different from CCl_4 group ($p \leq 0.05$).

Actually, the synergistic effect of low amounts of DBE and CCl_4 on hepatic lipid peroxidation was first observed by Danni *et al.* (1988) in the whole animal, and then reproduced in the isolated hepatocyte model to better investigate the relative pathomechanisms. The DBE-induced enhancement of lipid peroxidation due to CCl_4 was proved by a statistically significant increase of malonaldehyde steady-state in the rat liver homogenate two hours after dosing with the haloalkane mixture.

15.2.3 The prooxidant synergism of the CCl_4–DBE mixture turns out in an increased cytolytic activity; in vivo *and* in vitro *data*

The second main finding reported in the *in vivo* investigation by Danni and colleagues (1988) was that the prooxidant effect of CCl_4 and of DBE plus CCl_4 was followed by a significant rise of serum sorbitol dehydrogenase, a sensitive marker of irreversible cell damage. Again, while rat treatment with a non-prooxidant, a single dose of 1,2-dibromoethane, did not lead to any cell deletion, the same amount given in association with a low but cytolytic amount of CCl_4 increased the effect of the latter compound by two to three fold. Also, the synergistic effect of the two compounds on liver cytolysis was reproducible in the intact hepatocyte model. In this case DBE-exerted potentiation on CCl_4-dependent cytolysis was shown in terms of increased Trypan Blue cell stainability, as well as release of lactate dehydrogenase

Table 15.4 Lactate dehydrogenase release in the incubation medium of isolated rat hepatocytes in the presence of CCl_4, DBE, or CCl_4 plus DBE

Groups	*(% release)*		
	1h	*3h*	*5h*
Control	3 ± 2	6 ± 2	9 ± 4
CCl_4 172 μM	8 ± 5	12 ± 4	25 ± 8[a]
DBE 50 μM	4 ± 2	8 ± 3	14 ± 7
CCl_4 ± DBE	7 ± 4	16 ± 3[a]	49 ± 9[a,b]

Values are means ± SD of four different experiments. LDH leakage was calculated as percent of total enzyme release as obtained by cell lysis with Triton X-100. The percent release already present before cell incubation (zero time value) has been substracted from all reported data.
a Significantly different from the control group ($p \leq 0.05$).
b Significantly different from CCl_4 group ($p \leq 0.05$).

and GOT and GPT transaminases in the cell incubation medium (Danni *et al.*, 1991) (Table 15.4). Of note was that the increased prooxidant effect of the mixture appeared to significantly precede its enhancing effect on hepatocyte death.

15.2.4 Rat pretreatment with α-tocopherol protects against both prooxidant and cytolytic effects of the two haloalkanes

Of course, only a clear prevention or amelioration of the cytotoxic effect of the CCl_4 and CCl_4 plus DBE treatments would have definitely supported a cause–effect relationship between the enhancement of lipid peroxidation and the observed hepatocyte irreversible damage. Indeed, this hypothesis was proved by the lack of necrogenic effect either when the compounds were administered to α-tocopherol supplemented animals (Danni *et al.*, 1988) or when added to hepatocyte suspensions isolated from vitamin E dosed rats (Danni *et al.*, 1991). In our opinion, the treatment with the vitamin was so successful as it was able to act at the level of the likely crucial biochemical change, that is the CCl_4-induced derangement of the metabolic pathway involving GSH-transferase. In fact, such enzyme activity may well counteract bromoacetaldehyde-dependent toxicity, firstly, by reducing the DBE molecules metabolised through the cytochrome P450 pathway and, secondly, by direct addition reaction with the aldehyde. Indeed, the CCl_4-induced derangement of total GSH-transferase activity was prevented by α-tocopherol supplementation and consequently the potentiation of DBE action was not any more evident (Danni *et al.*, 1991).

15.2.5 CCl_4 effect on DBE metabolism is mimicked by acute treatment with ethanol

When ethanol is introduced in this model of drug interaction, by means of acute pretreatment of the animal with a single dose (2.5 g/kg b.wt.) prior to cell isolation, the DBE dosing previously demonstrated to be insufficient to exert toxicity now shows strong prooxidant and cytolytic activities (Chiarpotto *et al.*, 1995). The ethanol-induced potentiation of DBE hepatotoxicity appears to be dependent on two main effects of acute alcohol intoxication: a GSH depletion further to that induced by animal starvation, and a marked inactivation of hepatocyte total GSH-transferase. A series of events can contribute to the decrease of hepatocyte GSH after a single dose of alcohol, e.g. an addition reaction with acetaldehyde (Vina *et al.*, 1980), lipid peroxidation (Younes and Siegers, 1981; Videla *et al.*, 1987), inhibition of GSH synthesis (Speisky *et al.*, 1985) plus the basal tripeptide consumption in DBE-treated cells for DBE and bromoacetaldehyde detoxification (Khan *et al.*, 1993). Very recently, the acute ethanol treatment of the rat was also proved to decrease total hepatic GSH-transferase activity, presumably through a stimulated efflux of the enzyme into plasma (Yang and Carlson, 1991).

Similar results have been obtained in the whole animal (Aragno *et al.*, 1996). Rat pretreatment with a single dose of ethanol definitely renders cytolytic an otherwise weakly-toxic dose of DBE and this effect is preceded by both inactivation of GSH-transferase and increased production of MDA in the liver.

Even with the ethanol–DBE combination, like in the case of CCl_4–DBE mixture, a crucial role in the synergism is played by a reactive intermediate, an aldehydic compound again, acetaldehyde. In fact, animal pretreatment with methylpyrazole, an inhibitor of alcohol dehydrogenase, prevents ethanol-dependent enhancement of DBE toxicity. On the contrary, disulfiram, which inhibits aldehyde dehydrogenase, worsens the interaction. In both cases, the enzymatic inhibitors interfere with the ethanol-induced change of GSH-transferase activity: methylpyrazole prevents the inactivation due to the alcohol, while disulfiram makes it worse.

Also, CCl_4 toxicity appears potentiated by acute ethanol pretreatment but not as much as DBE. Actually, as regards LDH release and Trypan Blue stainability, the CCl_4-induced cell damage is anticipated rather than increased when ethanol dosing is performed. Moreover, the net rise in MDA production does not differ from the corresponding ethanol-untreated group, because of the endogenous increased level of the aldehyde. An important background mechanism for both rise of basal lipid peroxidation and anticipation of the damage is likely represented by the significant GSH decrease observed in CCl_4-treated hepatocytes only when rat poisoning with ethanol was carried out (Chiarpotto *et al.*, 1995). While chronic ethanol intoxication strongly potentiates carbon tetrachloride toxicity (Lindros *et al.*, 1990; Ikatsu *et al.*, 1991; Hall *et al.*, 1991) essentially because of the induction of cytochrome P450IIE1 (Ingelman-Sundberg and Iornvall, 1984; French *et al.*, 1993), acute poisoning does not have time to exert the same effect. In fact, cytochrome P450 content was found in the normal range both in microsomes and intact hepatocytes from animals dosed with ethanol two hours before sacrifice and subsequent cell isolation (Chiarpotto, unpublished data).

15.2.6 Liver AP-1 activation due to carbon tetrachloride is potentiated by 1,2-dibromoethane

The prooxidant effect of $CCl_4 \pm$ DBE acute poisoning on hepatic lipid membranes determines an actual increase of a variety of lipid peroxidation and breakdown products. It appears likely that especially diffusible molecules among these products are involved in the promotion and amplification of liver injury. Indeed, aldehydes of the hydroxyalkenal class, whose steady-state concentration during CCl_4 poisoning reaches the micromolar range (Poli *et al.*, 1985), can be responsible for part of the damage up to the point of no return. Consistently, the antioxidant treatments able to inhibit the CCl_4-enhanced generations of hydroxyalkenals are very efficiently preventing the haloalkane-induced centrilobular cytolysis (Poli *et al.*, 1989; Parola *et al.*, 1992). The fact that aldehydic compounds, besides those arising from lipid peroxidation, are also involved in the cytotoxic action of DBE and ethanol may be only a coincidence. Certainly this kind of chemical medium-reactive species appears very suitable for amplification of pathophysiological events, not only necrosis. Actually, acute CCl_4 intoxication is consistently followed by liver regeneration. Sustained action of CCl_4 provokes intensive fibrogenesis with excessive deposition of collagen and other extracellular matrix proteins. Hence, it seems of great interest to elucidate whether aldehyde-driven reactions are also involved in the above-mentioned processes. A connection between lipid peroxidation derived aldehydes, for instance produced in the CCl_4 experimental model, and increased collagen synthesis or compensative hepatic hyperplasia could be the modulation of redox-sensitive transcription factors. Typical examples are the nuclear factor kB/rel family (Schreck *et al.*, 1991) and the activator protein-1 family (Angel and Karin, 1991). They are activated by a number of agents, all characterised by an increased production of reactive free radical species, e.g. UV light, hydrogen peroxide, heat shock, and are required for optimal transcription of a great variety of genes involved in cellular proliferation and function (Abate *et al.*, 1990; Schreck *et al.*, 1991) as well as in collagen deposition (Poli and Parola, 1997).

Indeed, a major aldehydic derivative of arachidonate oxidative breakdown, 4-hydroxy-nonenal, has been demonstrated to significantly up-regulate AP-1 nuclear binding in macrophages (Camandola *et al.*, 1997) and hepatic stellate cells (Camandola *et al.*, unpublished data).

Back to rat poisoning with carbon tetrachloride, *in vivo* analyses of the effect of lipid peroxidation as induced by CCl_4, DBE or both on liver AP-1 activity confirmed a significant CCl_4-induced up-regulation of this factor (Zawaski *et al.*, 1993), but, in addition, showed a marked synergistic effect of the mixture. Electrophoretic mobility shift assay was performed on hepatic extracts from animals sacrificed six hours after treatment. Notably, at this time of intoxication the prooxidant effect exerted by CCl_4 and measured in terms of malonaldehyde production was already significantly potentiated by DBE (data not shown) while tissue damage was just appearing in CCl_4 and CCl_4 + DBE groups. As reported in Figure 15.1, while intragastric administration of DBE (20 μl/kg b. wt.) did not influence the DNA binding activity of AP-1, CCl_4 treatment (125 μl/kg b. wt.) led to a 3–5 fold induction of AP-1 binding compared to untreated animals. When the two haloalkanes were administered simultaneously, a dramatic synergistic effect was consistently observed on AP-1 binding which increased by about 20 fold.

15.3 Conclusions

The simultaneous administration to the experimental animal of CCl_4 and DBE in concentrations *per se* exerting low or no toxicity, triggers defined pathomechanisms able to mutually potentiate the toxicity of the single compounds and consequently to markedly amplify the overall resulting damage.

Key events of this synergism appear to be the inactivation of hepatic glutathione

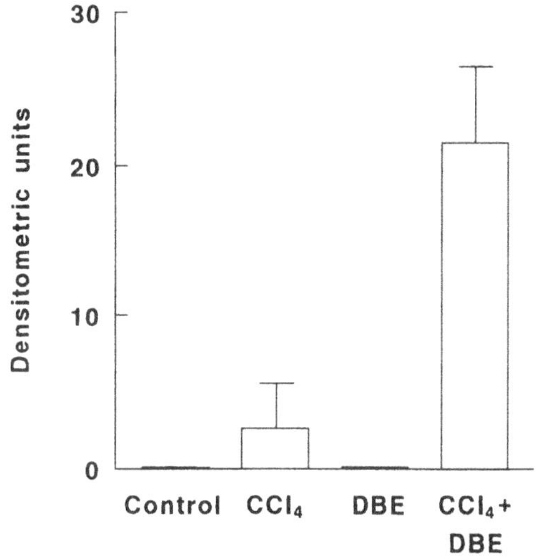

Figure 15.1 Hepatic AP-1 binding activity: synergism between CCl_4 and DBE. Liver extracts were incubated with radiolabeled AP-1 oligonucleotide and protein–DNA complexes were separated from umbound DNA by electrophoresis. Data are the mean changes in AP-1 density bound ± SD of 4 different experiments. Density of bounds was quantitated by Imaging Densitometer (BIO-RAD Laboratories GmbH, Munchen, Germany) and expressed in arbitrary densitometric units

transferase and the increased formation of reactive aldehydes. The latter compounds include lipid peroxidation-derived carbonyls and DBE metabolites of aldehydic nature. Above all, the enhancement of hepatic lipid peroxidation seems to play a primary role in triggering and amplifying cytotoxicity as well as in pivoting tissue reaction to the toxic damage. In fact, its abolition by animal pretreatment with an antioxidant dose of vitamin E provides strong, if not complete, protection against CCl_4–DBE driven tissue perturbation and injury.

While identification of cytotypes involved in hepatic AP-1 up-regulation by CCl_4 plus DBE is in progress, it is already possible to conclude that a remodeling of gene transcription takes place early after acute liver intoxication with the employed haloalkanes and it clearly appears redox-sensitive. In fact, up-regulation of AP-1 activity is preceded by a significant enhancement of membrane lipid peroxidation in the poisoned rat liver and it can be prevented by animal supplementation with α-tocopherol (Camandola *et al*., 1998). If the toxic stimulus, like that represented by CCl_4, is chronically applied, hyperactivation of AP-1 would provide optimal transcription of collagen type I gene, so contributing to the fibrotic differentiation of hepatic connective tissue and to its progression to sclerosis. The demonstrated finding of a synergistic up-regulation of tissue remodeling by a mixture of xenobiotics indicates a mechanism, by which means external factors can actually influence, in either a positive or a negative way, defined processes of pathophysiological interest.

References

Abate, C., Patel, L., Rausher, F.J. and Curran, T. (1990) 'Redox regulation of Fos and Jun DNA-binding activity in vitro' *Science* **249**, 1157–1161.

Alexeef, G.V., Kilgore, W.W. and Li, M-Y. (1990) 'Ethylene dibromide toxicology and risk assessment' in: Reviews of Environmental Contamination and Toxicology (Ware, G.M. ed.) Springer Verlag, New York, **12**, 49–122.

Angel, P. and Karin, M. (1991) 'The role of Jun, Fos and the AP-1 complex in cell-proliferation and transformation' *Biochem. Biophys. Acta*, **1072**, 129–157.

Aragno, M., Tamagno, E., Danni, O., Chiarpotto, E., Biasi, F., Scavazza, F., Albano, E., Poli, G., Dianzani, M.U. (1996) 'In vivo potentiation of 1,2-dibromoethane hepatotoxicity by ethanol through inactivation of glutathione s-transferase', *Chem. Biol. Interact.*, **99**, 277–288.

Camandola, S., Scavazza, A., Leonarduzzi, G., Biasi, F., Chiarpotto, E., Azzi, A. and Poli, G. (1997), 'Biogenic 4-hydroxy-2-nonenal activates transcription factor AP-1 but not NF-kappa B in cells of the macrofage lineage', *Biofactors*, **6**, 173–179.

Camandola, S., Aragno, M., Cutrin, J.C., Tamagno, E., Danni, O., Chiarpotto, E., Parola, M., Leonarduzzi, G., Biasi, F. and Poli, G. (1998) 'Liver AP-1 activation due to carbon tetrachloride is potentiated by 1,2-dibromoethane but is inhibited by alphatocopherol or gadolinium chloride', *Free Rad. Biol. Med.*, accepted, in press.

Chiarpotto, E., Biasi, F., Aragno, M., Scavazza, A., Danni, O., Albano, E. and Poli, G. (1993) 'Change of liver metabolism of 1,2-dibromoethane during simultaneous treatment with carbon tetrachloride' *Cell Biochem. Funct.*, **11**, 71–75.

Chiarpotto, E., Biasi, F., Aragno, M., Scavazza, A., Danni, O., Dianzani, M.U. and Poli, G. (1995) 'Ethanol-induced potentiation of rat hepatocyte damage due to 1,2-dibromoethane' *Alcohol Alcohol*, **30**, 37–45.

Cmarik, J.L., Inskeeo, P.B., Meredith, M.J., Mejer, D.J., Ketterer, B. and Guengerich, F.P. (1990) 'Selectivity of rat and human glutathione s-transferases in activation of ethylene dibromide by glutathione conjugation and DNA binding and induction of unscheduled DNA synthesis in human hepatocytes', *Cancer Res.*, **50**, 2747–2752.

Danni, O., Aragno, M. and Ugazio, G. (1988), 'In vivo studies on halogen compound interactions', *Res. Commun. Chem. Pathol. Pharmacol.*, **61**, 377–390.

Danni, O., Chiarpotto, E., Aragno, M., Biasi, F., Comoglio, A., Belliardo, F., Dianzani, M.U. and Poli, G. (1991), 'Lipid peroxidation and irreversible cell damage: synergism between carbon tetrachloride and 1,2-dibromoethane in isolated rat hepatocytes', *Toxicol. Appl. Pharmacol.*, **110**, 216–222.

French, S.W., Wong, K., Jui, L., Albano, E., Hagbjork, A.L. and Ingelman-Sundberg, M. (1993), 'Effect of ethanol on cytochrome P450 2E1 (CYP2E1), lipid peroxidation, and serum protein adduct formation in ralation to liver pathology pathogenesis', *Exp. Mol. Pathol.*, **58**, 61–75.

Hall, P.D.L.M., Plummer, J.L., Ilsley, A.H. and Cousins, M.J., (1991), 'Hepatic fibrosis and cirrhosis after chronic administration of alcohol and "low dose" carbon tetrachloride vapor in the rat', *Hepatology*, **13**, 815–819.

Hill, D.L., Shih, T.W., Johnston, T.P. and Struck, R.F. (1978), 'Macromolecular binding and metabolism of the carcinogen 1,2-dibromoethane', *Cancer Res.*, **38**, 2438–2442.

Ikasu, H., Okino, T. and Nakajima, T. (1991), 'Ethanol and food deprivation induced enhancement of hepatotoxicity in rats given carbon tetrachloride at low concentration', *Brit. J. Ind. Med.*, **48**, 636–642.

Ingelman-Sundberg, M. and Iornvall, H. (1984), 'Induction of the ethanol-inducible form of rabbit liver microsomal cytochrome P-450 by inhibitors of alcohol dehydrogenase', *Biochem. Biophys. Res. Commun.*, **124**, 375–382.

Inskeep, P.B., Koga, N., Cmarick, J.L. and Guengerich, F.P. (1986) 'Covalent binding of 1,2-dihaloalkanes to DNA and stability of the major DNA adduct, S-2-(N7-guanyl)ethylglutathione', *Cancer Res.*, **46**, 2839–2844.

Khan, S., Sood, C. and O'Brien, P.J. (1993), 'Molecular mechanisms of dibromoalkane cytotoxicity in isolated rat hepatocytes', *Biochem. Pharmacol.*, **45**, 439–447.

Lindros, K.O., Cai, Y. and Penttila, K.E. (1990), 'Role of ethanol-induccible cytochrome P-450 IIE1 in carbon tetrachloride-induced damage to centrilobular hepatocytes from ethanol-treated rats', *Hepatology*, **12**, 1092–1097.

Parola, M., Leonarduzzi, G., Biasi, F., Albano, E., Bocca, M., Poli, G. and Dianzani, M.U. (1992), 'Vitamin E dietary supplementation protects against carbon tetrachloride induced chronic liver damage and cirrhosis', *Hepatology*, **16**, 1014–102.

Poli, G., Dianzani, M.U., Cheeseman, K.H., Slater, T.F., Lang, J. and Esterbauer, H. (1985), 'Separation and characterization of aldehydic products of lipid peroxidation stimulated by carbon tetrachloride or ADP-iron in isolated rat hepatocytes and rat liver microsomal suspensions', *Biochem. J.*, **227**, 629–638.

Poli, G., Cheeseman, K.H., Biasi, F., Chiarpotto, E., Dianzani, M.U., Esterbauer, H. and Slater, T.F. (1989), 'Promethazine inhibits the formation of aldehydic products of lipid peroxidation but not covalent binding resulting from the exposure of rat liver fraction to CCl_4', *Biochem. J.*, **264**, 527–532.

Poli, G. and Parola, M. (1997), 'Oxidative damage and fibrogenesis', *Free Rad. Biol. Med.*, **22**, 287–305.

Schreck R., Rieber, P. and Baeuerle, P.A. (1991), 'Reactive oxygen intermediates as apparently widely used messengers in the activation of the NF-*k*B transcription factor and HIV-1', *EMBO J.*, **10**, 2247–2258.

Slater, T.F. and Sawyer, B.C. (1969), 'The effect of carbon tetrachloride on rat liver microsomes during the first hour of poisoning in vivo and the modifying actions of promethazine', *Biochem. J.*, **111**, 317–324.

Slater, T.F. (1972), 'Free radicals in tissue injury' in: Free Radical Mechanisms in Tissue Injury (Lagnado, J.R. ed.) Pion Ltd, London, pp. 1–283.

Shih, T.W. and Hill, D.L. (1981), 'Metabolic activation of 1,2-dibromoethane by glutathione transferase and by microsomal mixed function oxidase: further evidence for formation of two reactive metabolites', *Res. Commun. Chem. Pathol. Pharmacol.*, **33**, 449–460.

Speisky, H., MacDonald, A., Giles, G., Orrego, H. and Israel, Y. (1985), 'Increased loss and decreased synthesis of hepatic glutathione after acute ethanol administration. Turnover studies', *Biochem. J.*, **225**, 565–572.

Tomasi, A., Albano, E., Dianzani, M.U., Slater, T.F. and Vannini, V. (1983), 'Metabolic activation of 1,2-dibromoethane to a free radical intermediate by rat liver microsomes and isolated hepatocytes', *FEBS Lett.*, **160**, 191–194.

Van Bladeren, P.J., Breimer, D.D., Rotteveel-Smijs, G.M.T., De Jong, R.A.E., Buijs, W., van der Gen, A. and Mohn, G.R. (1980), 'The role of glutathione conjugation in the mutagenicity of 1,2-dibromoethane', *Biochem. Pharmacol.*, **29**, 2975–2982.

Van Bladeren, P.J., Breimer, D.D., Van Huijgevoort, J.A.T.C.M., Vermeulen, N.P.E. and Van der Gen, A. (1981), 'The metabolic formation of N-acetyl-S-hydroxyrthil-L-cysteine from tetradeutero-1,2-dibromoethane. Relative importance of oxidation and glutathione conjugation in vivo', *Biochem. Pharmacol.*, **30**, 2499–2501.

Videla, L.A., Fernandez, V. and Valenzuela, A. (1987), 'Age-dependent changes in rat liver lipid peroxidation and glutathione covalent induced by acute ethanol ingestion', *Cell Biochem. Funct.*, **5**, 273–280.

Vina, J., Estrella, J.M., Guerri, C. and Romer, F.J. (1980), 'Effects of ethanol on glutathione concentrations in isolated hepatocytes', *Biochem. J.*, **188**, 549–552.

Wormhoudt, L.W., Ploemen, J.H.T.M., Commandeur, J.N.M., Van Ommrn, B., Van Bladeren, P.J. and Vermeulen, N.P.E. (1996), 'Cytochrome P450 catalyzed metabolism of 1,2-dibromoethane in liver microsomes of differentially induced rats', *Chem. Biol. Interact.*, **99**, 41–53.

Younes, M. and Siegers, C.P. (1981), 'Mechanistic aspects of enhanced lipid peroxidation following glutathione depletion in vivo', *Chem. Biol. Interact.*, **34**, 257–266.
Yang, C.M. and Carlson, G.P. (1991), 'Effects of ethanol on glutathione conjugation in rat liver and lung', *Biochem. Pharmacol.*, **41**, 923–929.
Zawaski, K., Grueble, A., Kaplan, D., Reddy, S., Mortensen, A. and Novak, R.F. (1993), 'Evidence for enhanced expression of c-fos, c-jung and the CA^{2+}-activated neutral protease in rat liver following carbon tetrachloride administration', *Biochem. Biophys. Res. Commun.*, **197**, 585–590.

16 Reactive oxygen species in physiology and toxicology

Nalini Kaul and Henry Jay Forman

Abstract

Reactive oxygen species produced by cellular metabolic reactions have been implicated in a variety of diseases. Toxic effects of free radicals and related species, such as hydrogen peroxide (H_2O_2), have been recognized for a long time but their role as potential intracellular and extracellular signaling molecules has come to light more recently. The realization that hydrogen peroxide and lipid hydroperoxides result as part of normal metabolism lent support to oxidants being implicated as second messengers. Superoxide ($O_2^{-\bullet}$), hydrogen peroxide, lipid hydroperoxides and nitric oxide are now recognized as important signaling molecules under subtoxic conditions. Reactive oxygen species induce various biological processes including a transient elevation of intracellular free Ca^{2+} concentration, phosphorylation of specific proteins, activation of specific transcription factors, modulation of eicosanoid metabolism and stimulation of cell growth. In this chapter we discuss the production of reactive oxygen species, their biological sources, mechanisms of damage, and how reactive oxygen species can lead to altered signaling mechanisms and the toxic effects they exert. The focus here is on toxicity but it should be kept in mind that that these agents play a significant role in normal physiology.

Any molecule with an unpaired electron is a free radical. In general, this electronic configuration makes free radicals relatively unstable and reactive. While nitric oxide and vitamin E radical, which are of major biological importance, are relatively long lived and fairly stable, most free radicals have a fleeting existence. This makes evaluation of their involvement quite difficult. Often, indirect methods such as measurement of oxidized products, are used. When oxidants have a biological effect at concentrations below those causing oxidation of tissue components, this evaluation becomes even more difficult. Thus, much of what is discussed here relies upon implications stemming from the use of inhibitors and antioxidants.

Naturally occurring free radicals typically have an unpaired electron centered on oxygen or nitrogen. The ubiquitous presence of oxygen in higher species and diatomic oxygen's ability to readily accept electrons has made oxygen centered free radicals the most frequently encountered radical species in biological systems. Oxygen reduction, while permitting versatility in the biosphere, is also life threatening. Incompletely reduced oxygen species, when uncontrolled, can cause nonspecific oxidation and thereby damage biological molecules. The reducing environment of the cell provides ample opportunities for oxygen to undergo univalent reduction. It is well documented that reactive oxygen species such as superoxide, hydrogen peroxide, and hydroxyl radical (•OH) are products of normal metabolism (Chance *et al.*, 1979; Cadenas, 1989). In an aerobic environment these agents appear to be responsible for oxygen toxicity. Singlet oxygen, a product of energy transfer rather than electron transfer, is also produced in cells but plays a minor role except under unusual circumstances, such as in photooxidation.

As part of having evolved as oxygen-dependent metabolism, aerobic cells normally synthesize antioxidant enzymes whose role it is to intercept and inactivate reactive oxygen species. As oxidative stress may vary depending upon the environment, adaptive responses, such as the synthesis of antioxidant and repair enzymes, have evolved to be well-regulated processes. Despite the ability to respond or adapt, oxidative damage remains inescapable during aerobic existence. Oxidative stress has been implicated in a wide range of diseases and various degenerative processes, many of which are age related. These include Parkinson's disease, Alzheimer's disease, cancer, atherosclerosis, myocardial infarction, hypertension, ischemia–reperfusion, cataracts, and chronic inflammatory diseases, such as rheumatoid arthritis and lupus erythematosus. Oxygen and nitrogen radicals along with hydrogen peroxide and hypochlorous acid (derived from hydrogen peroxide) are clearly involved in acute inflammation.

16.1 Production of free radicals

The major pathway accounting for more than 95% of the oxygen consumption of tissues involves the tetravalent reduction of oxygen by mitochondrial cytochrome oxidase; a reaction which is devoid of release of any intermediates. The remaining consumption of oxygen (less than 5%) involves univalent and divalent reduction in which several highly reactive intermediates are produced as follows:

$$O_2 + e^- \rightarrow O_2^{-\bullet} \quad (16.1)$$

$$O_2 + 2e^- + 2H^+ \rightarrow H_2O_2 \quad (16.2)$$

$$2O_2^{-\bullet} + 2H^+ \rightarrow H_2O_2 + O_2 \quad (16.3)$$

$$O_2^{-\bullet} + Fe^{3+} \rightarrow O_2 + Fe^{2+} \quad (16.4a)$$

$$H_2O_2 + Fe^{2+} \rightarrow {}^\bullet OH + {}^-OH + Fe^{3+} \quad (16.4b)$$

Single electron reduction (reaction 16.1) results in the formation of superoxide anion radical. Besides being a nucleophile, superoxide is both a reducing as well as an oxidizing agent. In biological systems the relatively short half-life of superoxide limits its diffusion away from its site of generation. At acidic pH this anion is in its protonated form (hydroperoxyl) ($HO_2^\bullet$) and is more toxic to membranes because of its greater lipophilicity and reactivity (Bielski *et al.*, 1983). The divalent reduction of oxygen yields the non radical species hydrogen peroxide (reaction 16.2). Hydrogen peroxide is fully protonated at physiological pH. Hydrogen peroxide has a relatively long half-life, is membrane permeable, and may cause damage at sites distant from its origin. Hydrogen peroxide is also produced by the dismutation of superoxide. The non-enzymatic reaction is quite rapid (~10^5 mol^{-1} s^{-1}) at physiological pH but is markedly accelerated by superoxide dismutase (> 10^9 mol^{-1} s^{-1}). In the presence of reduced transition metals, such as iron and copper, hydrogen peroxide is reduced to hydroxyl radical ($^{\bullet}OH$) and hydroxyl ion (OH^-) (reaction 16.4). Reduction of the metal may be by superoxide, which may simultaneously cause release of the metal from metalloproteins. Hydroxyl radical, a three electron reduction product of oxygen, is highly reactive, with an extremely short half life and therefore, very limited diffusion capacity. Therefore, it modifies molecules immediately adjacent to its site of generation, usually by hydrogen abstraction resulting in the formation of water and a new free radical. This can result in a chain reaction, such as lipid peroxidation:

$$LH + {}^{\bullet}OH \rightarrow L^{\bullet} + H_2O \tag{16.5}$$

$$L^{\bullet} + O_2 \rightarrow LO_2^{\bullet} \tag{16.6}$$

$$LO_2^{\bullet} + LH \rightarrow LOOH + L^{\bullet} \tag{16.7}$$

where LH is an unsaturated fatty acid.

The first excited state of oxygen is singlet oxygen. This reactive oxygen species can be formed, both enzymatically through a variety of peroxidases including prostaglandin hydroperoxidase (Cadenas *et al.*, 1983) and non-enzymatically through photoxidation or the reaction of hydrogen peroxide with hypochlorous acid. Singlet oxygen is an excellent dienophile and therefore can add to biological dienes such as histidine producing metastable dioxygen bridged products. These can break down to form a large variety of products. Nonetheless, quantitatively singlet oxygen plays only a minor role in biology. The exception is photooxidation in which some agents, which can absorb a photon and be elevated to a triplet state can transfer energy to ground state (triplet) oxygen to form a singlet state.

The reactive oxygen intermediates such as superoxide, $HO_2^{\bullet}$ hydrogen peroxide, hydroxyl radical and $^1O_2$ are called reactive oxygen intermediates or reactive oxygen species. Only superoxide, hydroperoxyl and hydroxyl radical are free radicals. Superoxide can directly inactivate aconitase and reacts with a variety of cellular components. Nonetheless, much of the toxicity of superoxide may be mediated through its reaction with nitric oxide, more properly (but less frequently) called nitrogen monoxide, a relatively stable free radical ($^{\bullet}NO$):

$$O_2^{-\bullet} + {}^{\bullet}NO \rightarrow ONOO^- \tag{16.8}$$

Reaction 16.8 is extraordinarily rapid (nearly diffusion limited, 6.7×10^9 mol^{-1} s^{-1}), which produces peroxynitrite anion. Protonation of peroxynitrite yields peroxynitrous acid, which is unstable and has the potential to react with organic molecules similarly to both nitrogen dioxide ($NO_2^{\bullet}$) and hydroxyl radical. Nitric oxide synthase, forming nitric oxide from arginine is found in neurons (Garthwaite *et al.*, 1988; Bredt and Snyder, 1989) epithelium (Busse and Mulsch, 1990), endothelium (Palmer *et al.*, 1987; Ignarro *et al.*, 1987) and activated macrophages (Stuehr and Marletta, 1985; Hibbs *et al.*, 1987). The physiological actions of nitric oxide in vasodilation and neurotransmission are well established and are the clearest examples of how a free radical can participate in normal physiology. These physiological effects are largely mediated through the activation by nitric oxide of guanylate cyclase. Nitric oxide also participates in physiologically relevant processes through non-enzymatic reactions. For example, activated macrophages apparently use peroxynitrite to destroy harmful organisms. Beckman *et al.* (1990) proposed that chronic inflammation involves peroxynitrite produced as a result of the simultaneous generation of nitric oxide and superoxide by macrophages that have been activated in tissues. More recently Christen *et al.* (1997) suggested that peroxynitrite production by activated macrophages in chronic inflammation may contribute to the etiology of various diseases. By continuously producing peroxynitrite from nitric oxide and superoxide such macrophages can cause extensive damage to proteins, lipids and DNA.

16.2 Biological sources of free radicals

There are several cellular sources of reactive oxygen species. The following is a brief overview of various sources of reactive oxygen species.

16.2.1 Small cytoplasmic molecules

Catecholamines are mainly metabolized by two major pathways involving monoamine oxidase and catechol-O methyl transferase (Kopin, 1985). Non enzymatic autooxidation of catecholamines competes with the enzymatic metabolic pathways and is thus accelerated when the metabolism is impaired (Jewett *et al.*, 1989). It has been suggested that high levels of plasma catecholamines produced under a variety of disease or stressful conditions would have a greater tendency to autooxidize, resulting in production of reactive oxygen species. Thiols, flavins, and tetrahydropterins are other small cytoplasmic molecules which can undergo autooxidation and thus produce reactive oxygen species (Fisher and Kaufman, 1973; Baccanari, 1978; Fridovich, 1983; Proctor and Reynolds, 1984). Generally, autooxidation of catecholamines and these other small molecules involves transition metal catalysis as reaction with oxygen directly is kinetically limited. Oxygen preferentially accepts one electron. Non-metal catalyzed reactions of organic molecules (which are singlet in the ground state) generally involve transfer of pairs of electrons; however, a two electron reaction of oxygen (a triplet) requires spin inversion, which requires energy (Taube, 1965).

16.2.2 Cytoplasmic proteins

Enzymes containing flavins or transition metals can catalyze production of hydrogen peroxide during oxidation of organic substrates; however, this usually involves sequential one electron transfers in which superoxide is a bound intermediate. Some cytoplasmic enzymes can release either superoxide and/or hydrogen peroxide as products of their catalytic process. Xanthine oxidase and aldehyde dehydrogenase are important examples (Freeman and Crapo, 1982; Fridovich, 1983; McCord, 1985; Greene and Paller, 1992). Oxyhemoglobin autooxidation can also produce superoxide (Misra and Fridovich, 1972). Both the α and β chains of oxyhemoglobin can autooxidize (Halliwell and Gutteridge, 1984; Proctor and Reynolds, 1984; Clark *et al.*, 1985).

16.2.3 Plasma membrane enzymes

A main source of reactive oxygen species is the respiratory burst of phagocytic cells. This is characterized by an increase in oxygen consumption and glucose metabolism via the hexose monophosphate shunt (Sbarra and Karnovsky, 1959; Iyer *et al.*, 1961) and generation of reactive oxygen species (Babior *et al.*, 1973). The superoxide generating enzyme responsible for the reactive oxygen species production is NADPH oxidase, which is membrane associated and catalyzes one electron reduction of oxygen to superoxide at the expense of NADPH (Babior *et al.*, 1976; Dewald *et al.*, 1979). Studies have shown that during the process of phagocytosis, neutrophils, monocytes and macrophages release proteases as well as oxygen metabolites into the external tissue environment (Egan *et al.*, 1976). These proteases have been shown to act in a synergistic manner with oxidants to produce tissue damage during inflammation (Varani *et al.*, 1990) in part through the activation of the proteases (Weiss *et al.*, 1985).

16.2.4 Perioxisomes

Perioxisomes have great capacity to form hydrogen peroxide because they contain a high concentration of oxidases. Enzymes such as glycollate oxidase or D-amino acid oxidase can catalyze the divalent reduction of molecular oxygen without releasing superoxide (Reddy *et al.*, 1982; Brunk and Cadenas, 1988; Del Rio *et al.*, 1992).

16.2.5 Mitochondrial electron transport

Although over 95% of the oxygen consumed by the cell is reduced to water by the terminal oxidase of the electron transport chain, cytochrome c oxidase, a small but continuous production of hydrogen peroxide by mitochondria is part of normal physiology. Hydrogen peroxide production in mitochondria is supported by substrates for electron transport (Jensen, 1966; Hinkle *et al.*, 1967; Boveris *et al.*, 1972; Loschen and Azzi, 1975; Chance and Oshino, 1971). During the early 1970s, it was shown that superoxide produced as a leak from the electron transport chain was the precursor of mitochondrial hydrogen peroxide production (Loschen *et al.*, 1974; Forman and Kennedy, 1974; Boveris and Cadenas, 1975; Boveris, 1977). Although there is debate concerning which electron carriers in the respiratory chain actually react with molecular oxgyen, it is agreed that superoxide is produced as a leak of electrons from two complexes of the respiratory chain: complex I (NADH : ubiquinone reductase) and complex III (the cytochrome b–cytochrome c_1–ubiquinone complex). The production of hydrogen peroxide by mitochondria is the result of competition between reactions of superoxide with mitochondrial superoxide dismutase and other rapid reactions of superoxide, such as with nitric oxide and aconitase. The rate of generation of hydrogen peroxide is dependent upon the state of the mitochondria as determined by the concentrations of ADP, substrates and oxygen. The resultant hydrogen peroxide may react within mitochondria or diffuse into the cytosol. Superoxide itself never leaves mitochondria (Forman and Azzi, 1997).

There has been substantial debate concerning the role of mitochondrial superoxide and hydrogen peroxide production in toxicology, aging and in the pathophysiology of age-related diseases. There is no debate about whether hydrogen peroxide is produced by the electron transport chain. What needs to be resolved, however, is whether any changes occur in the rate of this hydrogen peroxide production and if so whether it can be demonstrated to participate in the pathology or whether accumulated damage can result from a more invariant rate of mitochondrial hydrogen peroxide production.

16.2.6 Microsomal electron transport systems

Cytochrome P450 reductase and cytochrome b5 reductase present in the endoplasmic reticulum and nuclear membranes are also sources of hydrogen peroxide and superoxide (Blake, 1987). Although it seems that the majority of toxic compounds exert their toxicity either directly or through their metabolites there are now examples that the production of the excited states of oxygen during metabolism could be responsible for the toxicity of some xenobiotics. For example, the mechanism of paraquat toxicity, a commercial herbicide, has been proposed to result from the reduction of paraquat by the NADPH cytochrome P450 reductase of lung microsomes followed by a cyclic oxidation–reduction process that generates superoxide (Bus *et al.*, 1974; Farrington *et al.*, 1973). Subsequent reactions of superoxide have been cited as the cause of paraquat toxicity although decreased NADPH may contribute (Witschi *et al.*, 1977).

Reactive oxygen species are also formed during the catalytic cycle of cytochrome P450 and cytochrome b5 as well as when they undergo autooxidation (Freeman and Crapo, 1982; Sevanian *et al.*, 1990). Thus, reactive oxygen species can be produced at several sites associated with the endoplasmic reticulum. Under aerobic conditions, cytochrome P450 appears to participate in the one-electron reduction of oxygen, an activity referred to as oxygen reductase activity. During the oxygen reductase activity of cytochrome P450, molecular oxygen is reduced to superoxide anion radicals most likely by autooxidation of a cytochrome P450 ferric–dioxyanion complex. The formation of reactive oxygen species (superoxide, hydrogen peroxide and hydroxyl free radical) presents potential toxicity (Paine, 1978; Goeptar *et al.*, 1995).

Even under anaerobic conditions, cytochrome P450 may also be involved in the reduction of xenobiotics (Goeptar *et al.*, 1995). During the xenobiotic reductase activity of cytochrome P450, xenobiotics are reduced by the ferrous xenobiotic complex. After xenobiotic reduction by cytochrome P450, xenobiotic free radicals are formed that are also capable of reacting directly with tissue macromolecules and leading to irreversible damage (Goeptar *et al.*, 1995). The precise molecular mechanism underlying the xenobiotic reductase activity of cytochrome P450 is not yet fully understood.

16.3 Toxicity of reactive oxygen species

We live in an oxidant-rich, oxygen-dependent environment supplied by the evolution of photosynthesis. However, excess molecular oxygen and its metabolic products can be toxic. All cellular components, lipids, proteins, nucleic acids and carbohydrates may interact with reactive oxygen species and reactive nitrogen species, thereby giving rise to metabolic and cellular disturbances.

16.3.1 Lipids

A primary mechanism proposed to explain cell and tissue damage by reactive oxygen species involves the formation of lipid peroxides within cell membranes and organelles.

It is generally accepted that lipid peroxidation process is initiated after hydrogen abstraction from an unsaturated fatty acid (reactions 16.5–7). Hydroxyl and hydroperoxyl radicals, but neither superoxide nor hydrogen peroxide are able to abstract hydrogen from free unsaturated fatty acids or in phospholipids. The lipid radical formed reacts with molecular oxygen producing a lipid hydroperoxyl radical. This can abstract hydrogen from other unsaturated fatty acids propagating a chain reaction. But lipid hydroperoxides can also react with metals:

$$LOOH + Fe^{2+} \rightarrow LO^{\bullet} + Fe^{3+} + OH^{-} \qquad (16.9)$$

$$LOOH + Fe^{3+} \rightarrow LOO^{\bullet} + Fe^{2+} + H^{+} \qquad (16.10)$$

Some lipid peroxyl radicals form endoperoxides, which give rise to short chain products and new free radicals. The endoperoxyradical on intramolecular rearrangement and reaction with oxygen leads to the formation of a malondialdehyde precursor. In biological systems, lipid peroxidation products that have been identified including malondialdehyde, 4-hydroxynonenal and the alkanes, ethane and pentane (Esterbauer, 1982; Halliwell, 1981; Tappel, 1980). Potentially, the most important events in lipid peroxidation are certainly the radical chain reactions. If not terminated by the reaction of two radicals they proceed concurrently destroying all lipid phases. Fortunately, antioxidant defenses particularly α-tocopherol, generally keep such chains rather short. Thus initiation reactions are probably the more problematic aspect.

Alcohols, aldehydes and volatile hydrocarbons are the final products of lipid peroxidation. Some of the aldehydes, particularly the α,β-unsaturated aldehydes inhibit enzymatic activities and thereby affect protein synthesis and numerous metabolic processes. Interestingly, several of the hydroperoxides and aldehydes also have chemotactic activity for phagocytes and are thus pro-inflammatory (Blake, 1987; Southorn and Powis, 1988a, 1988b). In addition malondialdehyde (MDA), used often as an indicator of lipid peroxidation, causes cross-linking of membrane components and adducts with DNA bases (Nielsen, 1981; Valenzuela, 1991). However 4-HNE induces some specific protein synthesis, heat shock proteins and aldehyde reductase.

Lipid peroxidation causes changes in membrane fluidity, alters semipermeable characteristics of membranes and thus affects their ability to function correctly (Halliwell and Gutteridge, 1984; Horton and Fairhurst, 1987; Webster and Nunn, 1988; Southorn and Powis, 1988b; Niki *et al.*, 1991; Schaich, 1992). The reactions of lipid hydroperoxyl and lipid alkoxyl radicals are not limited to propagation of lipid peroxidation. These reactive species can react with other cellular components by abstraction of hydrogen or addition.

16.3.2 Proteins

Proteins are modified by reactive oxygen species depending on the nature of the oxidant and also the protein moiety that is involved. Most of our knowledge about the modification of proteins comes from pioneering studies (Swallow, 1960; Chio and Tappel, 1969; Schuessler and Schilling, 1984; Garrison, 1987; Stadtman *et al.*, 1981). Results from these studies demonstrate that the oxidation of proteins by reactive oxygen species can lead to oxidation of amino acid residue side chain, cleavage of peptide bonds and formation of covalent protein–protein crosslinked derivatives.

Peptide bond cleavage occurs when radiolysis of water or metal catalyzed cleavage of hydrogen peroxide abstracts hydrogen from the polypeptide backbone forming an alkyl radical. The alkyl radical reacts either with oxygen to form alkyl peroxyl radical or with another alkyl radical to form intra or inter protein cross linkages. The protein peroxyl radical can be converted to alkyl peroxide by a dismutation reaction with free peroxyl radical or reaction with Fe^{2+} or abstraction of an hydrogen atom. Irrespective of how it is formed, the protein alkyl peroxide is then converted to an alkoxyl protein derivative by either dismutation, reaction with a free peroxyl radical or reaction with Fe^{2+}. Finally an alkoxyl radical undergoes conversion to a hydroxyl derivative that will undergo peptide bond scission by an α-amidation pathway. Peptide bond cleavage can also occur as a consequence of attack of hydroxyl radical on the side chains of glutamyl or prolyl residues (Garrison, 1987; Kato *et al.*, 1992).

Amino acid side chain oxidation occurs when amino acid side chains are modified by reactive oxygen species. The amino acids included are arginine, cysteine, histidine, leucine, methionine, phenylalanine, tyrosine, tryptophan, threonine, proline, glutamine and lysine (Stadtman *et al.*, 1981). The oxidation of side chains of lysine, proline, arginine and threonine has been seen to yield carbonyl derivatives. Under physiological conditions, the metal catalyzed protein modification process is believed to be site specific. For example, oxidative modification of glutamine synthetase by various metal catalyzed systems is restricted to the oxidation of amino acid residues located at one or another or both of the two metal binding sites on this enzyme (Farber and R.L., 1986; Rivett and Levine, 1990; Sahakian *et al.*, 1991).

Besides direct oxidation of the protein side chains, carbonyl groups can be introduced into peptides by Michael addition as seen by conjugation of 4-hydroxynonenal with either the ε-amino group of lysine, the imidazole moiety of histidine, or the sulfhydryl group of cysteine residues. These carbonyls can react with a neighbouring moiety to produce more stable compounds (Sayre *et al.*, 1993). There is however some potential for cross-linking (see below). Carbonyl groups can also be introduced into proteins by glycation or glyoxidation (Wolf and Dean, 1987; Mullarkey *et al.*, 1990; Kristal and Yu, 1992). The carbonyl content of proteins has been shown to increase in hepatocytes (Starke-Reed and Oliver, 1989) and lung tissue (Winter and Liehr, 1991) following hypoxia or ischemia–reperfusion, in neutrophils and macrophages during the respiratory burst (Oliver, 1987) and in mice livers following exposure to alcohol (Wieland and Lauterburg, 1995).

Cross linking reactions occur when reactive oxygen species promote protein cross linkages by a variety of mechanisms: (a) oxidation of cysteine sulfhydryl group forming disulfide cross links; (b) interaction of an aldehyde group or a 4-hydroxynonenal protein adduct from one protein with the ε-amino group of a lysine residue in another protein (Cohn *et al.*, 1996); (c) interaction of the carbonyl group of a glycated protein with the ε-amino group of a lysine in another protein; (d) interaction of malondialdehyde with ε-amino groups of lysine residues in two different protein molecules; (e) by formation of carbon–carbon covalent linkages; (f) by the reaction of a carbonyl group produced in the oxidation of an amino acid residue side chain in one protein with the ε-amino group of a lysine residue in another protein. Cross-links can clearly inhibit the function of enzymes or structural proteins. In addition such cross-linked proteins may be undegradable by proteases responsible for protein turnover. The accumulation of such cross-linked proteins is clearly a marker of oxidative injury if not a part of pathology (Friguet *et al.*, 1994; Grune *et al.*, 1995).

Other side chain reactions also occur; for example, peroxynitrite has been shown to nitrate tyrosine (Ischiropoulos and Al-Medi, 1995), and oxidize methionine (Pryor and Squadrito, 1995) and cysteine residues (Gatti *et al.*, 1994; Karoui *et al.*, 1996). This could seriously compromise basic mechanisms of cellular regulation (Hunter, 1995). For example, nitration of tyrosines may interfere with phosphorylation that occurs during signal transduction.

The first consequence of even very limited free radical damage to proteins is inactivation of their function. Changes in protein composition will produce changes in conformation and antigenicity. Oxidative changes in proteins, other than cross-linking renders them more susceptible to proteolysis. For example, it has been shown that as few as two radical modifications per molecule of bovine serum albumin is sufficient to induce both conformational changes and increased susceptibility to proteolysis (Birnboim, 1985).

16.3.3 DNA

Oxidative damage is seen to occur to a large extent in nucleic acids. Reactive oxygen species are mutagenic and may act as promoters of carcinogenesis (Cerutti, 1985; Moody and H.M., 1982). Strand scission has been reported when DNA is exposed to high fluxes of reactive oxygen species but changes in bases and deoxyribose sugars may be more frequent. Hydroxyl radicals are thought to mediate these reactions producing a large number of sugar derived and base derived products in DNA as well as DNA-protein cross links in mammalian chromatin (Dizdaroglu, 1986; Oleinick *et al.*, 1987; von Sonntag, 1987).

Base damage occurs by ionizing radiation, ultraviolet radiation and hydrogen peroxide in combination with metals producing covalent DNA protein cross linkage in living cells and in isolated chromatin (Oleinick *et al.*, 1987; Yamamoto, 1976; Lesko *et al.*, 1982; Fornace and Little, 1979). Reactions with DNA bases of hydroxyl radical, hydride radical, or hydrated electrons, which are created by ionizing radiation, are characterized by addition to the double bonds to give adduct radicals. An abstraction of the H atom by hydroxyl radical from the methyl group of thymine is also seen to occur (von Sonntag, 1987; Hazra and Steenken, 1983; Steenken, 1989). While pyrimidine radicals add oxygen to form peroxyl radical, purine adduct radicals react very slowly with oxygen (Steenken, 1989). Subsequent reactions of base radicals lead to a variety of products from each of the DNA bases (Dizdaroglu, 1986; Teoule, 1987; von Sonntag, 1987).

Reactive nitrogen species also cause nitrosation, nitration and deamination of bases. For example, guanine is converted to xanthine and adenine is converted to hypoxanthine (De Rojas-Walker *et al.*, 1995). 8-Nitroguanine is produced when peroxynitrite attacks DNA (Yermilov *et al.*, 1995). The pattern of DNA base damage can be used to help identify the damaging species.

Sugar damage is suffered when hydroxyl radicals react with DNA, mostly by addition to the double bonds of the bases, although small amounts are also seen to react with sugar moieties by abstracting an H atom, leading to sugar radicals (von Sonntag, 1987). How much of the sugar is attacked depends on whether the DNA is single or double stranded. Reactions of sugar radicals lead to the release of intact bases, sugar moiety alterations and also strand breaks (von Sonntag, 1987; Dizdaroglu *et al.*, 1975, 1977; Beesk, 1979).

16.3.4 Carbohydrates

Carbohydrates are also targets of reactive oxygen species. Glycosylated proteins have been found to be more sensitive to oxidative damage (Freeman and Crapo, 1982; Southhorn and Powis, 1988a, 1988b). Radicals can oxidize monosaccharides and also react with polysaccharides causing depolymerization. In patients with rheumatoid arthritis, the synovial fluid contains large number of neutrophils. Many of these are activated, since the fluid contains increased quantities of products released by activated phagocytic cells, including lysozyme, lactoferrin and prostaglandins. The viscosity of rheumatoid synovial fluid is much lower than normal, because of replacement of glycosoaminoglycan with hylauronic acid. The synovial membrane in rheumatoid patients may synthesize hyaluronic acid of a shorter chain length than usual, but hyaluronic acid can also be subject to radical induced degradation in synovial fluid. Such depolymerization of hyaluronic acid may occur resulting in poor synovial fluid viscosity (Greenwald and Moy Wai, 1980; McCord, 1974).

16.4 Oxidant stress and signaling

For many years oxidative stress was considered synonymous with cell and tissue injury. It is now understood that reactive oxygen species also cause functional modulation without disruption of cellular integrity. Indeed, recent evidence has suggested that endogenous generation of reactive oxygen species is an important component of signal transduction. Nonetheless, many studies have involved exposure of cells to extracellular reactive oxygen species. Ultraviolet radiation has been seen to initiate cellular signaling by activating receptors for growth factors and cytokines such as epidermal growth factor (EGF), tumor necrosis factor-α (TNF-α) and interleukin-1 (IL-1) (Coffer *et al.*, 1995; Sachsenmaier *et al.*, 1994). Ultraviolet radiation produces reactive oxygen species that are required for receptor kinase activation (Huang *et al.*, 1996). Rosette and Karin (1996) recently showed clustering of several receptor types including TNF-α, EGF and IL-1 on exposure of cells to ultraviolet radiation. These receptors did not require ligand binding to activate signaling. Anti-EGF receptor antibodies were seen to cause dimerization and clustering of receptors, which is apparently sufficient for activation of EGF receptor kinase and for activation of ERK-MAP kinase pathway. In addition, cells transfected with a EGF receptor with a mutated inactive kinase domain were unable to function in ultraviolet radiation-induced signal transduction because the inactive kinase did not initiate down stream signaling (Sachsenmaier *et al.*, 1994). This suggests reactive oxygen species may cause effects on the signaling proteins at or near the cell membrane, which will in turn initiate signaling pathways. The observation that hydrogen peroxide and lipid hydroperoxides are produced as part of normal metabolism has led to the proposal that these oxidants function as second messengers. The precedence for free radicals to function as second messengers is already established by nitric oxide. Reactive oxygen species also fit the classic pattern for second messengers, such as cyclic AMP and inositol-polyphosphates, in relatively short lived, small, diffusible and ubiquitous.

Growth factor receptor signaling has also been linked to reactive oxygen species production. For example, platelet derived growth factor (PDGF) binding to the PDGF receptor directly causes a transient increase in intracellular hydrogen peroxide as seen by oxidation of dichlorofluorescin. This transient increase in hydrogen peroxide was required for growth induction by PDGF (Sundaresan *et al.*, 1995). EGF was also seen to induce hydrogen peroxide in A431 cells (Bae *et al.*, 1997). Hydrogen peroxide generation has been linked to the tyrosine kinase activity of the EGF receptor as seen by the observation that EGF failed to induce hydrogen peroxide generation in cells expressing a kinase inactive receptor.

Membrane receptor generated signaling is often coupled to cytosolic signal transduction through adapter proteins linked to the small G-protein, ras, or other members of the ras family. Activated ras can stimulate another small G-protein, rac, that binds and activates the membrane bound NADPH-oxidase complex to produce reactive oxygen species. Rac-dependent reactive oxygen species generation was observed in the pathways of EGF, PGDF, TNF-α and IL-1 (Sundaresan *et al.*, 1996) in NIH-3T3 cells. Through as yet undetermined pathways, oxidants activate the JNK and/or p38 MAP kinase pathways, presumably through activation of upstream kinases (Lin *et al.*, 1995; Derijard *et al.*, 1995; Moriguchi *et al.*, 1996; Raingeaud *et al.*, 1996). Following activation, JNK and/or p38 MAP kinase translocate to the nucleus where they phosphorylate and thereby activate transcription factors. Several studies have shown the JNK and p38 MAP kinase pathways are responsive to reactive oxygen species (Lin *et al.*, 1995; Derijard *et al.*, 1995; Moriguchi *et al.*, 1996; Raingeaud *et al.*, 1996). Thus, reactive oxygen species regulate various signaling cascades and thereby various transcription factors that are under their control. Issues that remain include determining the actual targets of oxidants, the modifications (disulfide bonding, phosphorylation, etc.) of the targets, the oxidant species involved, and the turn-off processes that are essential to any normal signal transduction process.

16.4.1 Protein phosphorylation

Protein phosphorylation plays a key role in signaling. The state of phosphorylation is set by two classes of enzymes: protein kinases and protein phosphatases. Depending upon the cellular context, these two types of enzymes may either antagonize or cooperate with each other during the signal transduction process. An imbalance between these two processes may impair normal cell function including growth and cellular transformation. Many oxidant initiated signaling processes are known to involve PKC. Gopalakrishna and Anderson (1989) first reported direct stimulation of PKC by high concentrations of hydrogen peroxide. Recently, alteration of PKC activity from a calcium and phospholipid dependence to calcium and phospholipid independence by physiologically relevant concentrations of hydrogen peroxide has been shown in macrophages (Kaul and Forman, 1996).

Much evidence has accumulated that suggests protein tyrosine kinases and protein tyrosine phosphatases are modulated by reactive oxygen species (Schultze-Osthoff *et al.*, 1995; Brumell *et al.*, 1996). Oxidant stimulation of phosphorylation can be achieved by either directly activating protein tyrosine kinases or inhibiting protein tyrosine phosphatases. It has been suggested that the mechanism of some oxidant induced receptor activation is the activation of a protein tyrosine phosphatase via modification of a critical sulfhydryl group in the phosphatase (Hecht and Zick, 1992; Caselli *et al.*, 1995). Regardless of the mechanism, the evidence for oxidative activation of tyrosine phosphorylation is accumulating. For example, hydrogen peroxide stimulates tyrosine phosphorylation in various cell lines (Zick and Sagi-Eisenberg,

1990), P72 tyrosine kinase is activated by hydrogen peroxide and ultraviolet radiation and hydrogen peroxide both stimulate the tyrosine kinase, ZAP-70 (Schieven *et al.*, 1994). Inhibition of protein tyrosine phosphatases by oxidants like hydrogen peroxide (Hadari *et al.*, 1993; Sullivan *et al.*, 1994) has been reported. Therefore the stimulation of resultant biological processes may be due to inhibitory activities of oxidants. Thiol oxidation of these phosphatases has been reported to result in inhibition of enzyme activity (Nemani and Lee, 1993; Guy *et al.*, 1993) and it has been proposed that oxidant induced phosphorylation is due to the inactivation of redox-sensitive protein phosphatase.

Oxidative activation of phospholipases and MAP kinases, may be indirectly activated through activation of protein tyrosine kinases. For example, hydrogen peroxide, fatty acid hydroperoxides and 4-hydroxynonenal activate phospholipase D (PLD) in endothelial cells. In normal signaling, PLD may be activated through PKC-α, a calcium-dependent isoform. While some studies suggest oxidative changes in intracellular calcium will use this pathway to enhance PLD activity (see below), other studies suggest that oxidant mediated PLD activation is not calcium dependent nor dependent on PKC. In contrast, tyrosine phosphorylation may be important in oxidant mediated PLD activation as seen in HL-60 cells where peroxides of vanadium activated PLD activity through tyrosine phosphorylation.

16.4.2 Calcium signaling

Oxidants generally have profound effects on the cellular control of calcium homeostasis. At low subtoxic concentrations, hydroperoxides can cause transient changes in intracellular free calcium concentration ($[Ca^{2+}]_i$) that resemble the physiologic changes occurring in receptor-mediated signaling. Lethal concentrations of hydroperoxides produce substantial and toxic elevation of $[Ca^{2+}]_i$. The mechanism of these changes in $[Ca^{2+}]_i$ at high and low hydroperoxide concentrations are as dramatically different as the effect on cell viability. Indeed, low concentrations of hydroperoxides may cause enhancement of cell function and release calcium from intracellular pools while the lethal concentrations of hydroperoxides cause massive influx from the extracellular medium and the initiation of either necrotic or apoptotic cell death. Here however, we will focus on the more subtle effects that alter cell function rather than on concentrations that wreck havoc upon cell integrity.

Some of the signaling mechanisms sensitive to the action of oxidants, which regulate $[Ca^{2+}]_i$ or are regulated by changes in $[Ca^{2+}]_i$ include phospholipases, prostaglandins, cGMP, the constituent form of nitric oxide synthetase, ion channels, classic forms of protein kinase C and calmodulin-dependent kinases. Direct effects by ROOH on calmodulin, the calcium binding protein that acts as a nearly ubiquitous $[Ca^{2+}]_i$ detector and regulator of function have not been demonstrated. Nevertheless, sensitivity to changes in $[Ca^{2+}]_i$ makes calmodulin-dependent activities, such as calmodulin dependent kinases and protein phosphatase 2b potential indirect targets of hydroperoxides through modulation of $[Ca^{2+}]_i$. The activity of PLD appears to be enhanced by low concentrations of hydroperoxide mediated by $[Ca^{2+}]_i$ elevation (Natarajan and Garcia, 1993; Natarajan *et al.*, 1993). At higher concentrations of hydroperoxide, PLD is inhibited (Rice *et al.*, 1992). Thus, activities of a variety of calcium dependent enzymes that are central to signal transduction are altered by oxidative stress.

The steady state $[Ca^{2+}]_i$ is approximately 70–100 nM, which is 10^4 times lower than that of the extracellular environment. Active extrusion by ion pumps, exchange with sodium, and sequestration within organelles maintains this gradient. Membrane receptor mediated signaling involving elevation of $[Ca^{2+}]_i$ starts with activation of a heterotrimeric G protein coupled to the receptor and activation of phospholipase C. An increase in inositol-1,4,5-trisphosphate (IP_3) in the cytosol catalyzed by phospholipase C then stimulates Ca^{2+} release from the endoplasmic reticulum. $[Ca^{2+}]_i$ elevation during signal transduction occurs through both release from the endoplasmic reticulum and opening of channels in the plasma membrane. The latter can occur through changes in membrane potential that signal for opening of channels in the plasma membrane or through G proteins coupled to receptors, which then act directly on channels, or through a response to emptying of the endoplasmic reticular Ca^{2+} pool. During physiological signaling, opening of plasma membrane Ca^{2+} channels results in small transient changes in $[Ca^{2+}]_i$. Influx of calcium through the plasma membrane caused by exposure to hydroperoxides usually occurs only with relatively high concentrations of the oxidant. Once hydroperoxide-induced calcium influx does occur, it can cause a many times greater increase in $[Ca^{2+}]_i$ than that observed under physiologic stimulation. This great elevation of $[Ca^{2+}]_i$ is often associated with cell death.

The calcium pool in the endoplasmic reticulum is maintained through a balance of uptake and release. Calcium is actively imported into the organelle through an ATPase, while release occurs by activation of an IP_3 sensitive receptor. A basal release of Ca^{2+} occurs due to a low steady state production of IP_3. It has been demonstrated that oxidized glutathione can decrease the dissociation constant of the IP_3 receptor (Renard *et al.*, 1992) and this may contribute to some of the increase in $[Ca^{2+}]_i$ observed during oxidative stress. In the alveolar macrophage however, release of an intracellular pool of calcium by hydroperoxides is clearly from another source than the endoplasmic reticulum (Hoyal *et al.*, 1996). Instead, the apparent release of annexin VI, a membrane bound calcium-binding protein, from the plasma membrane, has been associated with the elevation of $[Ca^{2+}]_i$ in hydroperoxide treated alveolar macrophages (Hoyal *et al.*, 1996).

16.4.3 Activation of transcription factors

Oxidants have been shown to modulate the activity of the transcription factors nuclear factor-kappa B (NF-κB), activator protein 1 (AP-1) and the antioxidant response element binding protein (ARE).

NF-κB is a protein complex involved in transcription of genes involved in inflammation, such as tumor necrosis factor-α (TNF-α) (Baeuerle, 1994). NF-κB is activated in response to various stimuli including ultraviolet radiation, viruses, cytokines (including TNF-α itself), phorbol esters, okadaic acid and hydrogen peroxide. Thus, oxidants, which are a component of inflammation, may also exacerbate inflammation through generation of cytokines and chemokines. Many of these diverse stimuli purportedly produce reactive oxygen species. For example, TNF-α and IL-1 cause release of superoxide by fibroblasts (Hennet *et al.*, 1993) while phorbol esters cause increased superoxide production by NADPH oxidase via protein kinase C activation. The signals from these various stimuli appear to converge into a common pathway that results in NF-κB activation. Nonetheless, mounting evidence suggests that separate hydrogen peroxide-dependent and independent pathways for NF-κB can exist within the same cell. Baeuerle (1994) suggested a novel signal transduction pathway for NF-κB activation involving reactive oxygen species as second messengers based upon results with Jurkat and Wurzburg cell lines that responded to exogenous hydrogen peroxide and lipid hydroperoxides but not superoxide. Nonetheless, exogenous hydrogen peroxide does not stimulate NF-κB in all cell lines (Israel *et al.*, 1992).

The predominant form of NF-κB exists in the cytoplasm as a heterodimer of p65 (Rel A) and p50 complexed to the inhibitory subunit I-κB (Baeuerle, 1994). On activation by phorbol esters and cytokines I-κB dissociates allowing NF-κB to migrate to the nucleus where it binds to DNA. A role for reactive oxygen species in NF-κB activation that is stimulated through receptor binding or activation by phorbol esters rather than direct addition of hydroperoxides has been suggested by inhibition of NF-κB activation by antioxidants. Support for this proposal is provided by Schmidt *et al.* (1995) who reported that TNF-α or okadaic acid induced NF-κB activation is diminished in cell lines over expressing catalase. Mouse epidermal cells overexpressing either catalase or superoxide dismutase were compared to control, in their response to TNF-α or okadaic acid. Catalase decreased NF-κB activation while superoxide dismutase increased it. Superoxide dismutase competes with other cell components for superoxide and thereby potentially increases net hydrogen peroxide production.

Thus, the opposite effects of over expressing superoxide dismutase and catalase suggests hydrogen peroxide as the possible activating agent. From our own studies, with NF-κB, we found that the stimulated production of superoxide and/or hydrogen peroxide (the respiratory burst) or exogenous hydrogen peroxide activate NF-κB in both J774A.1 cells and rat alveolar macrophages (Kaul and Forman, 1996). Our results showed that catalase significantly inhibited the activation of NF-κB in both cell types, while superoxide dismutase was ineffective. These results suggested that hydrogen peroxide produced by the respiratory burst was involved in the activation of NF-κB.

A role for reactive oxygen species as activators of AP-1 has also been demonstrated by several groups. Exposure of HeLa cells to ultraviolet radiation and to a lesser extent to hydrogen peroxide has been seen to lead to a rapid enhancement of AP-1 to DNA binding within a few minutes (Devary *et al.*, 1991; Nose, 1991). AP-1 is composed of c-Jun and c-Fos. Expression of c-jun and c-fos is rapidly induced by a variety of different extracellular stimuli including mitogens, phorbol esters, ultraviolet radiation, and reactive oxygen species. The exact mechanisms of the induction of these early response genes is unclear. On another level, oxidants affect AP-1 mediated transcription by inhibiting the binding of AP-1 to DNA. This is a process that is regulated by the protein Ref-1, which apparently catalyzes the reduction of a critical sulfhydryl involved with DNA binding (Xanthoudakis and Curran, 1992).

Antioxidant responsive elements (ARE) were first described in the promoter/enhancer regions of rat GSH-S-transferase Ya subunit and NAD(P)H:quinone reductase genes (Favreau and Pickett, 1993; Rushmore *et al.*, 1991). Although ARE appears to have the ability to regulate transcription in response to oxidative stress, the details are still largely unresolved. Dependent on the species and gene, ARE contains imperfect or perfect AP-1 binding consensus sequences (Rushmore *et al.*, 1991; Jaiswal, 1994). The transcription factor, which specifically binds to the ARE has not been conclusively identified. Whether the proteins that bind to ARE are of the AP-1 family (Fos/Jun) or are unrelated is unresolved as is the question of whether the binding changes with oxidative stress (as with AP-1) or not (as with OxyR) (Jaiswal, 1994;

Wang and Williamson, 1994; Daniel, 1995; Nguyen *et al.*, 1994). It is known that cysteine and serine residues in the same relative positions in the DNA-binding domains of the Fos/Jun proteins are highly conserved. It has been postulated that the cysteine may be reversibly oxidized by changes in the intracellular redox state mediated by a protein, now called Ref-1 (Abate *et al.*, 1990; Bannister *et al.*, 1991; Xanthoudakis and Curran, 1992). Recently, it has been reported that quinone mediated generation of reactive oxygen species induces the ARE mediated expression of the mouse gluthathione S-transferase gene (Pinkus *et al.*, 1995). This correlation remains unknown in the case of AREs from other genes including NQO1 and NQO2 genes. However hydrogen peroxide has been shown to induce the ARE mediated expression of rat GST Ya, rat NQO1 and human NQO1 genes (Li and Jaiswal, 1994; Favreau and Pickett, 1993; Rushmore *et al.*, 1991).

Nitric oxide is a short lived mediator that can be induced in a variety of cell types and produces many physiologic and metabolic changes in target cells. An important observation of the inhibition of action of an endothelium derived relaxing factor by superoxide led to the discovery of the ability of mammalian tissues to synthesize nitric oxide (Ignarro, 1989; Furchgott *et al.*, 1990). Nitric oxide acts primarily through its activation of cyclic GMP synthesis, which mediates its best known physiologic functions.

The interaction of nitric oxide with superoxide results in the production of peroxynitrite with a rate constant of 6.7×10^9 mol^{-1} s^{-1}. Recent studies suggest that peroxynitrite also participates in cellular signaling processes (Wu *et al.*, 1994; Davidson *et al.*, 1995; Mayer *et al.*, 1995). Some actions of peroxynitrite that might originate in signaling like mechanisms include vascular relaxation (Wu *et al.*, 1994), uncoupling of receptor action (Lipton *et al.*, 1993), inhibition of mitochondrial respiration, and the inhibition of glyceraldehyde-3-phosohate dehydrogenase (Mohr *et al.*, 1994). Large concentrations of exogenous peroxynitrite are needed to produce these responses. Thus, a principal question is whether exogenous peroxynitrite production produces these effects at non toxic levels. Recent studies suggest peroxynitrite readily forms in tissues at physiologically relevant levels of nitric oxide (Kooy and Royall, 1994; Davidson *et al.*, 1995). Studies suggest that thiol nitrosation is elicited by peroxynitrite and this process seems to result in signaling-like responses (Wu *et al.*, 1994; Davidson *et al.*, 1995; Mayer *et al.*, 1995). At low concentrations of peroxynitrite it is possible that signaling will occur but at greater concentrations, peroxynitrite produces toxicity. Reactive nitrogen species impair the function of the NMDA receptor for glutamate in nerve cells (Lipton *et al.*, 1993), inhibit cellular energy metabolism (Mohr *et al.*, 1994; Radi *et al.*, 1994), and inhibit catalase and hydrogen peroxide elicited cGMP-mediated relaxation of vascular smooth muscle cells (Mohazzab-H *et al.*, 1995). Some of the potential sites of impairment of energy metabolism by peroxynitrite include the iron–sulfur center of mitochondrial aconitase (Castro *et al.*, 1994; Hausladen and Fridovich, 1994), cellular energy metabolism and glyceraldehyde-3-phosphate dehydrogenase (Mohr *et al.*, 1994).

16.5 Oxidative stress and inflammation

Free radicals are known to play a major role in inflammation (McCord, 1974; Henson and Johnston, 1986; Lunec, 1987; Ward, 1988). The inflammatory process, which is a response to injury, involves enhanced vascular permeability and leukocyte infiltration into the damaged area. Polymorphonuclear leukocytes and the macrophages on encountering a foreign particle, undergo both phagocytosis and stimulation of reactive oxygen species production (the respiratory burst). The foreign material is engulfed and is exposed to reactive oxygen species. Myeloperoxidase, which is released by polymorphonuclear leukocytes and eosinophils into the phagosome, catalyzes the reaction of hydrogen peroxide with halide ions to form the hypohalous acids, HOCl or HOBr, which are strong oxidants.

The inflammatory response is essential, but also potentially hazardous as stimulation of reactive oxygen species production may damage surrounding tissue and cause a change in extracellular fluid viscosity. Examples of inflammatory diseases in which reactive oxygen species damage has been proposed are gout, myasthenia gravis and lupus erythematosus (Southorn and Powis, 1988b). Oxidant processes involving stimulation of the respiratory burst of polymorphonuclear leukocytes along with their arachidonic acid metabolism are now believed to account for many of the deleterious effects of inflammation that lead to cell injury. These processes tend to damage the microcirculatory endothelia of all organs and have therefore been implicated in many inflammation–immune-related conditions, both in animal models (Del Rio *et al.*, 1992) and in humans (Clark *et al.*, 1985).

Inflammatory arthritis, characterized by joint inflammation is regarded as an example that demonstrates reactive oxygen species involvement in pathology (Halliwell and Gutteridge, 1985, 1986; Greenwald, 1985). The rheumatoid joint contains both acute and chronic inflammatory components. A neutrophil leucocytosis is invariably present in the synovial cavity and these cells release superoxide and hydrogen peroxide. Iron, which is capable of catalyzing hydroxyl radical production is also present within the cavity (Gutteridge *et al.*, 1981). Such iron is possibly released from transferrin because of the low pH generated within the cavity and particularly at the synovial–cartilage junction. The extracellular environment contains low levels of superoxide dismutase and catalase and there is considerable evidence of lipid peroxidation, ascorbate oxidation, hylauronic acid degradation and oxidative damage to proteins (Blake *et al.*, 1986). Free radical damage to immunoglobulins (IgG) provides a possible explanation for the presence of rheumatoid factor. Rheumatoid factors are antibodies of various immunoglobins subclasses, directed against self IgG with the consequent formation of complexes. Such complexes interact with rheumatoid factor and have the ability to stimulate normal neutrophils to generate further radical species. It has been proposed that tolerance to self antigen IgG is maintained by T suppressor cells within the synovial membrane which are themselves susceptible to reactive oxygen species damage (Allan *et al.*, 1986).

The protein α-1 protease inhibitor (α-1PI), a normal constituent of extracellular fluids, which suppresses the activity of neutrophil derived elastase, is damaged by reactive oxygen species. Exposure of α-1PI to reactive oxygen species leads to oxidation of methionine residues. Oxidation of this methionine on α-1PI is believed to be the mechanism whereby cigarette smoke produces emphysema, a slow inflammatory process in the lung. Thus, these two inflammatory diseases, rheumatoid arthritis and emphysema share a common component in their mechanism. The cartilage matrix also appears susceptible to free radical damage. The polypeptide core of the proteoglycan appears to be particularly suspectible. Proteins rich in proline such as collagen give rise to many fragments of defined length after exposure to hydroxyl radical due to non random protein damage. The rheumatoid factor in these patients is modified by reactive oxygen species. Complexes formed with synovial fluid and serum also stimulate neutrophils to form reactive oxygen species. Oxidants also activate latent collagenase thus further potentiating damage.

16.6 Oxidative stress in apoptosis, necrosis and cell growth

There is compelling evidence that reactive oxygen species are among the agents that can activate apoptosis or programmed cell death. For example, induction of programmed cell death occurs in some cells treated with low concentrations of hydrogen peroxide (Lennon *et al.*, 1991) or lipid peroxides (Sandstorm *et al.*, 1994). Similarly, treatments that promote intracellular reactive oxygen species formation, i.e. with chemicals that inhibit intracellular antioxidant functions (like depletion of intracellular glutathione) or that promote reactive oxygen species formation (e.g. transition metals and quinones), can also induce apoptosis (Hockenbery *et al.*, 1993; Ratan *et al.*, 1994).

The inflammatory mediator, TNF-α, exerts its effects via the 55kDa TNF receptor (TNF-R1). TNF-α can induce cytokine production in macrophages but can also induce both programmed cell death and necrosis in other cells (Laster *et al.*, 1988). The signaling cascade initiated through binding to TNF-R1 has been proposed to generate reactive oxygen species, possibly through activation of lipoxygenases and poorly understood effects upon mitochondrial metabolism including electron transport (Heller and Kronke, 1994). Both reactive oxygen species-dependent and independent pathways activated by TNF-R1 probably exist. Thus, TNF-α may induce programmed cell death and reactive oxygen species formation by separate pathways, while the reactive oxygen species may induce programmed cell death or necrosis depending on cellular defenses (Schultze-Osthoff *et al.*, 1994).

Fas is a cell surface receptor that upon activation induces programmed cell death (Nagata and Golstein, 1995). An intracellular domain conserved between TNF-R1 and Fas is critical for the signal leading to apoptosis mediated by these receptors. The role of reactive oxygen species in Fas-induced apoptosis is controversial. On the one hand Fas-mediated programmed cell death has been shown to occur in low oxygen, not requiring reactive oxygen species (Jacobson and Raff, 1995). On the other hand, the ability of the Fas antigen to induce apoptosis has been shown to be blocked by thioredoxin and exogenous superoxide dismutase (Schultze-Osthoff *et al.*, 1994; Matsuda *et al.*, 1991; Hirose *et al.*, 1993).

Cytochrome c, is a mitochondrial intermembrane protein normally involved in the electron transport between complexes III and IV of the respiratory chain. Recently, cytochrome c has been reported to be released into the cytosol from mitochondria and this has been proposed as a general prerequisite for the apoptotic process (Liu *et al.*, 1996; Yang *et al.*, 1997; Kharbanda *et al.*, 1997). Along the same lines, the protein, Bcl-2 has been shown to prevent apoptosis, possibly by blocking the release of cytochrome from mitochondria (Yang *et al.*, 1997; Kharbanda, 1997; Kluck, 1997). Accumulated cytochrome c may trigger apoptosis by activating caspases (Kluck *et al.*, 1997). For example, binding of cytochrome c activates the proteolytic activity of Apaf-1, a mammalian CED-4 homolog (Zou *et al.*, 1997).

Reactive oxygen species mediated DNA damage has been shown to elicit the activation of polyADP-ribose transferase and p53 accumulation, both of which have been linked to apoptosis (Maltzman and Czyzyk, 1984; Schwartzman and Cidlowski, 1993). The polymerization of ADP-ribose with proteins causes rapid depletion of

NAD/NADH pools, and loss of ATP stores, which results in cell death. Also reactive oxygen species readily react with polyunsaturated fatty acids and cholesterol present in membranes and this direct oxidation can cause apoptosis (Christ *et al.*, 1993). An excellent example of this are the HPETEs which are potent inducers of apoptosis, especially in TNF-α mediated apoptosis (Larrick and Wright, 1990; Matthews *et al.*, 1987; Chang *et al.*, 1992).

It is evident while low levels of reactive oxygen species can have positive regulatory effects in mammalian cell growth, increasing concentrations can have a negative or down regulating effect on growth. Furthermore, while low levels of reactive oxygen species were seen to be growth stimulatory (Burdon and V., 1993) higher levels of reactive oxygen species have been seen to have both a negative effect on proliferation and a signaling role in apoptosis as seen in BHK-21, 208 F and HeLa cells (Burdon *et al.*, 1989, 1990; Burdon and V., 1993).

High levels of hydrogen peroxide ($\geq$ 1 mM) have been shown to increase cell death due to necrosis rather than apoptosis. There is a question as to whether models using such levels of hydrogen peroxide or other hydroperoxides have any relevance to actual pathophysiology or merely represent test tube chemistry.

16.7 Oxidative stress and carcinogenesis

Carcinogenesis is the malignant transformation of a cell. There is strong evidence implicating reactive oxygen species in both initiation and promotion of multistage carcinogenesis (Sies, 1985; Cerutti, 1985; Imlay and Linn, 1988; Saul *et al.*, 1987; Ames, 1983; Sun, 1990; Willson, 1982). Prooxidant states can be caused by different classes of agents, including hyperbaric oxygen, radiation, xenobiotic metabolites, modulators of cytochrome P450 or the electron transport chain and perioxisome proliferators. Many initiators have been shown to produce or cause production of free radicals. Reactive oxygen species have been demonstrated to cause mutations in mammalian cells (Hsie, 1987). Bleomycin, which promotes production of reactive oxygen species through redox cycling, has been found to be highly mutagenic to mammalian cells (Cunningham *et al.*, 1984). Hydrogen peroxide has been shown to be associated with induction of cancer (Shamberger, 1972; Ito *et al.*, 1981) and also with molecular damage leading to transformation of cells *in vitro* (Kennedy *et al.*, 1984). Most effective initiators of carcinogenesis act on DNA, wherein they induce single strand and/or double strand breaks, crosslinking, and chemical aberrations.

Reactive oxygen species involvement in tumor promotion is suggested by the following observations: (a) many free radical generating compounds are tumor promoters; (b) promoters such as phorbol esters stimulate generation of reactive oxygen species; (c) many antioxidants appear to oppose tumor promotion (Fischer *et al.*, 1988; Kensler, 1986). Many classes of tumor promoters act through a prooxidant mechanism. These include ultraviolet radiation, reactive oxygen species generating compounds and peroxisome proliferators. Benzoyl peroxide, decanoyl peroxide and cumene hydroperoxide also cause initiation and promotion of cancer in mouse epidermis purportedly due to free radical generation (Slaga *et al.*, 1981; Pryor, 1976). With cumene peroxide, alkyl radicals are also produced and these may be involved in tumor promotion (Kensler and Taffe, 1986). Involvement of free radicals produced during benzo(a)pyrene metabolism in tumor promotion has also been proposed (Lesko *et al.*, 1975).

Free radical generators have been seen to mimic the action of tumor promoters. For example, both phorbol esters and xanthine plus xanthine oxidase induce both ornithine decarboxylase activity in primary cultures of murine keratinocytes and anchorage independent growth in the murine JB6 epidermal cell line (Kensler and Taffe, 1986; Friedman and Cerutti, 1983). As the induction of ornithine decarboxylase (which usually occurs in tumor promotion) and anchorage independent growth is one characteristic of a transformed phenotype (Bouek and Mayorca, 1979), reactive oxygen species involvement in tumor promotion is strongly implied.

In summary, the observations that reactive oxygen species participate in and/or alter cell signaling has now been described in many systems. While the direct toxic effects of free radicals in various pathologies is well recognized, understanding the mechanisms whereby free radicals participate in signal transduction is still in the early stages of development. What is clear is that such effects on cell signaling may play as great a role in the toxicity of reactive oxygen species as does wholesale oxidation of cellular constituents.

Acknowledgements

The authors thank Dr Julio Girón-Calle for his comments. Work from this laboratory cited in the text was supported by grants HL37556 and ES05511 from the National Institutes of Health.

References

Abate, C., Patel, L., Rauscher, I. and Curran, T. (1990) *Science*, **249**, 1157–1161.

Allan, I., Lunec, J., Salmon, M. and Bacon, P.A. (1986) *Agents and Actions*, **19**, 351–352.

Ames, B.N. (1983) *Science*, **221(4617)**, 1256–1264.

Babior, B.M., Curnutte, J.T. and McMurrich, B.J. (1976) *Journal of Clinical Investigation*, **58**, 989–996.

Babior, B.M., Kipnes, R.S. and Curnutte, J.T. (1973) *Journal of Clinical Investigation*, **52**, 741.

Baccanari, D.P. (1978) *Archives of Biochemistry and Biophysics*, **191(1)**, 351–357.

Bae, Y.S., Kang, S.W., Seo, M.S., Baines, I.C., Tekle, E., Chock, P.B. and Rhee, S.G. (1997) *Journal of Biological Chemistry*, **272**, 217–221.

Baeuerle, P.A.H., T. (1994) *Annu. Rev. Immunol.*, **12**, 141–179.

Bannister, A.J., Cook, A. and Kouzarides, T. (1991) *Oncogene*, **6**, 1243–1250.

Beckman, J.S., Beckman, T.W., Chen, J. and Marshall, P.A. (1990) *Proceedings National Academy of Sciences, USA*, **87**, 1620–1624.

Beesk, F.D., M. Schulte-Frohlinde, D. von Sonntag, C. (1979) *Int. J. Radiat. Biol. Relat. Stud. Phys. Chem. Med.*, **36(6)**, 565–576.

Bielski, B.H.J., Arudi, R.L. and Sutherland, M.W. (1983) *Journal of Biological Chemistry*, **258**, 4759–4761.

Birnboim, H.C.K.-K. (1985) *Proceedings National Academy of Sciences, USA*, **82**, 6820–6824.

Blake, D.R., Lunec, J., Winyard, P. and Brailsford, S. (1986) *Iron, free radicals and chronic inflammation*, Churchill Livingstone, Edinburgh.

Blake, D.R.A., and Lunec, J. (1987) *British Medical Bulletin*, **43(2)**, 371–385.

Bouek, N. and Mayorca, G.D. (1979) In *Methods in enzymology*, Vol. LVII (Ed. Jakoby, W.H.) Academic Press, New York, pp. 296–302.

Boveris, A. (1977) (Eds, Reivich, M., Coburn, R. and Lahiri, S.) *Tissue Hypoxia and Ischemia.*

Boveris, A. and Cadenas, E. (1975) *FEBS Letters*, **54**, 311.

Boveris, A., Oshino, N. and Chance, B. (1972) *Biochem J*, **128**, 617–630.

Bredt, D.S. and Snyder, S.H. (1989) *Proceedings National Academy of Sciences, USA*, **86(22)**, 9030–9033.

Brumell, J.H., Burkhardt, A.L., Bolen, J.P. and Grinstein, S. (1996) *Journal of Biological Chemistry*, **271(3)**, 1455–1461.

Brunk, U. and Cadenas, E. (1988) *APMIS*, **96(1)**, 3–13.

Burdon, R.H., Gill, V. and Rice-Evans, C. (1989) *Free Radical Research Communication*, **7(3–6)**, 149–159.

Burdon, R.H., Gill, V. and Rice-Evans, C. (1990) *Free Radical Research Communication*, **11(1–3)**, 65–76.

Burdon, R.H. and V., G. (1993) *Free Radical Research Communication*, **19(3)**, 203–213.

Bus, J.S., Aust, S.D. and Gibson, J.E. (1974) *Biochemical and Biophysical Research Communications*, **58**, 749.

Busse, R. and Mulsch, A. (1990) *FEBS Letters*, **265(1–2)**, 133–136.

Cadenas, E. (1989) *Annual Review of Biochemistry*, **58**, 79–110.

Cadenas, E., Stess, H., Nastaincyzk, W. and Ullrich, V. (1983) *Hoppe- Seylers, Z Physiol Chem*, **304**, 519–528.

Caselli, A., Chiarugi, P., Camici, G., Manao, G. and Ramponi, G. (1995) *FEBS Letters*, **374**, 249–252.

Castro, L., Rodriguez, M. and Radi, R. (1994) *Journal of Biological Chemistry*, **269**, 29409–29415.

Cerutti, P.A. (1985) *Science*, **227(4685)**, 375–381.

Chance, B. and Oshino, N. (1971) *Biochem J*, **122**, 225–233.

Chance, B., Sies, H. and Boveris, A. (1979) *Physiological Reviews*, **59**, 527–605.

Chang, D.J., Ringold, G.M. and Heller, R.A. (1992) *Biochemical and Biophysical Research Communications*, **188(2)**, 538–546.

Chio, K.S. and Tappel, A.L. (1969) *Biochemistry*, **8**(7), 2827–2832.

Christ, M., Luu, B., Mejia, J.E., Moosbrugger, I. and Bischoff, P. (1993) *Immunology*, **78**(3), 455–460.

Christen, S., Woodall, A.A., Shigenaga, M.K., Southwell-Keely, P.T., Duncan, M.W. and Ames, B.N. (1997) *Proceedings National Academy of Sciences, USA*, **94(7)**, 3217–22.

Clark, I.A., Cowden, W.B. and Hunt, N.H. (1985) *Med. Res. Rev.*, **5**(3), 297–332.

Coffer, P.J., Burgering, B.M., Peppelenbosch, M.P., Bos, J.L. and Kruijer, W. (1995) *Oncogene*, **11**(3), 561–569.

Cohn, J.A., Tsai, L., Friguet, B. and Szweda, L.I.A.B.B. (1996) **328(1)**, 158–164.

Cunningham, M.L., Ringrose, P.S. and Lokesh, B.R. (1984) *Mutation Research*, **135**(3), 199–202.

Daniel, V. (1995) *Critical Reviews in Biochemistry and Molecular Biology*, **28**, 173–207.

Davidson, C.A., Kaminski, P.M. and Wolin, M.S. (1995) *Circulation*, **92**, 1–391.

De Rojas-Walker, T., Tamir, S., Ji, H., Wishnock, J.S. and Tannenbaum, S.R. (1995) *Chemical Research in Toxicology*, **8**(3), 473–477.

Del Rio, L.A., Sandalino, L.M., Palma, J.M., Bueno, P. and Corpas, F.J. (1992) *Free Radical Biology & Medicine*, **13**(5), 557–580.

Derijard, B., Raingeaud, J. and Barett, T. (1995) *Science*, **267**, 682–5.

Devary, Y., Gottlieb, R.A., Lau, L.F. and Karin, M. (1991) *Molecular and Cellular Biology*, **11**, 2804–2811.

Dewald, B., Baggiolini, M., Curnutte, J.T. and Babior, B.M. (1979) *Journal of Clinical Investigation*, **63**, 21–29.

Dizdaroglu, M. (1986) *Biotechniques*, **4**, 536–546.

Dizdaroglu, M., von Sonntag, C. and Schulte-Frohlinde, D. (1975) *J. Amer. Chem. Soc.*, **97**(8), 2277–2278.

Dizdaroglu, M., Schulte-Frohlinde, D. and von Sonntag, C. (1977) *Z Naturforsch.*, **32(11–12)**, 1021–1022.

Egan, R.W., Paxton, J. and Kuehl, F.A.J. (1976) *Journal of Biological Chemistry*, **251(23)**, 7329–7335.

Esterbauer, H. (1982) *Free radicals, lipid peroxidation and cancer*, Academic Press, New York and London.

Farber, J.M. and R.L., L. (1986) **261**, 4574–4578.

Farrington, J.A., Ebert, M., Land, E.J. and Fletcher, K. (1973) *Biochimica et Biophysica Acta*, **314**, 372–381.

Favreau, L.V. and Pickett, C.B. (1993) *Journal of Biological Chemistry*, **268**, 19875–19881.

Fischer, S.M., Floyd, R.A. and Copeland, E.S. (1988) *Cancer Research*, **48(10)**, 3882–3887.

Fisher, D.B. and Kaufman, S. (1973) *Journal of Biological Chemistry*, **248**, 4300–4304.

Forman, H.J. and Azzi, A. (1997) *Faseb J*, **11**, 374–5.

Forman, H.J. and Kennedy, J.A. (1974) *Biochemical and Biophysical Research Communications*, **60**, 1044–1050.

Fornace, A.J. and Little, J.B. (1979) *Cancer Research*, **39(3)**, 704–710.

Freeman, B.A. and Crapo, J.D. (1982) *Laboratory Investigation*, **47**, 412–426.

Fridovich, I. (1983) *Annual Review of Pharmacology and Toxicology*, **23**, 239–257.

Friedman, J. and Cerutti, P.A. (1983) *Carcinogenesis*, **4(11)**, 1425–1427.

Friguet, B., Stadtman, E.R. and Szweda, L.I. (1994) *Journal of Biological Chemistry*, **269(34)**, 21639–21643.

Furchgott, R.F., Khan, M.T. and Jothianandan, K.D. (1990) In *In: Endothelium-derived relaxing factors* (ed. G.M. Rubanyi and P.M. Vanhoutte) Basel, Karger, pp. 8–21.

Garrison, W.M. (1987) *Chem. Rev.*, **87**, 381–98.

Garthwaite, J., Charles, S.L. and Chess-Williams, R. (1988) *Nature*, **336(6197)**, 385–388.

Gatti, R.M., Radi, R. and Augusto, O. (1994) *FEBS Letters*, **348**, 223–227.

Goeptar, A.R., Scheerens, H. and Vermeulen, N.P.E. (1995) *Critical Reviews in Toxicology*, **25(1)**, 25–65.

Gopalakrishna, R. and Anderson, W.B. (1989) *Proceedings National Academy of Sciences, USA*, **86**, 6758–6762.

Greene, E.L. and Paller, M.S. (1992) *American Journal of Physiology*, **263(2 Pt 2)**, F251–255.

Greenwald, R.A. (1985) *Free Radical Biology & Medicine*, **1(3)**, 173–177.

Greenwald, R.A. and Moy Wai, W. (1980) *Arthritis Rheum*, **23**, 455–463.

Grune, T., Reinheckel, T., Joshi, M. and Davies, K.J.A. (1995) *Journal of Biological Chemistry*, **270**, 2344–2351.

Gutteridge, J.M.C., Rowley, D.A. and Halliwell, B. (1981) *Biochemical Journal*, **199**, 263–265.

Guy, G.R., Cairns, J., Ng, S.B. and Tan, Y.H. (1993) *Journal of Biological Chemistry*, **268**, 2141–2148.

Hadari, Y.R., Geiger, B., Nadiv, O., Sabanay, I., Roberts, C.T.J., LeRoith, D. and Zick, Y. (1993) *Molecular Cellular Endocrinology*, **97**, 9–17.

Halliwell, B. (1981) *Age Pigments*, Elsevier, North-Holland, New York and Amsterdam.

Halliwell, B. and Gutteridge, J.M. (1984) *Biochemical Journal*, **219**, 1–14.

Halliwell, B. and Gutteridge, J.M.C. (1985) *Molec. Aspects Med.*, **8**, 89–193.

Halliwell, B. and Gutteridge, J.M.C. (1986) *Archives of Biochemistry and Biophysics*, **246**, 501–514.

Hausladen, A. and Fridovich, I. (1994) *Journal of Biological Chemistry*, **269**, 29405–29408.

Hazra, D.K. and Steenken, S. (1983) *J. Amer. Chem. Soc.*, **105**, 4380–4386.

Hecht, D. and Zick, Y. (1992) *Biochemical and Biophysical Research Communications*, **188**, 773–779.

Heller, R. and Kronke, M. (1994) *Journal of Cell Biology*, **126**, 5–9.

Hennet, T., Richter, C. and Peterhans, E. (1993) *Biochemical Journal*, **289**, 587–592.

Henson, P.M. and Johnston, R.B. (1986) *Journal of Clinical Investigation*, **79**, 699–674.
Hibbs, J.B.J., Taintor, R.R. and Vavrin, Z. (1987) *Science*, **235**, 473–476.
Hinkle, P.C., Butow, R.A., Racker, E. and Chance, B. (1967) *J Biol Chem*, **242**, 5169–5173.
Hirose, K., Longo, D.L., Oppenheim, J.J. and Matsushima, K. (1993) *FASEB Journal*, **7**, 361–368.
Hockenbery, D.M., Oltvai, Z.N., Yin, X.M., Milliman, C.L. and Korsmeyerm, S.J. (1993) *Cell*, **75**, 241–251.
Horton, A.A. and Fairhurst, S. (1987) *Critical Reviews in Toxicology*, **18**, 27–79.
Hoyal, C.R., Thomas, A.P. and Forman, H.J. (1996) *Journal of Biological Chemistry*, **271**, 29205–29210.
Hsie, A.W. (1987) In *Anticarcinogenesis and radiation protection* (Cerutti, P.A., Nygaard, O.F. and Simic, M.G., eds) Plenum Press, New York, pp. 115–119.
Huang, R.P., Wu, J.X. and Adamson, E. (1996) *Journal of Cell Biology*, **133**, 211–220.
Hunter, T. (1995) *Cell*, **60**, 225–236.
Ignarro, L.J. (1989) *Circulation Research*, **65**, 1–21.
Ignarro, L.J., Buga, G.M., Wood, K.S., Byrns, R.E. and Chaudhuri, G. (1987) *Proceedings National Academy of Sciences, USA*, **84**, 9265–9269.
Imlay, J.A. and Linn, S. (1988) *Science*, **240**, 1302–1309.
Ischiropoulos, H. and Al-Medi, A.-B. (1995) *FEBS Letters*, **364**, 279–282.
Israel, N., Gougerotpocidalo, M.A., Alliet, F. and Virelizier, J.L. (1992) *Journal of Immunology*, **149**, 3386–3393.
Ito, A., Watanabe, H., Naito, M. and Naito, Y. (1981) *Gann*, **72**, 174–175.
Iyer, G.Y.N., Islam, M.F. and Quastel, J.H. (1961) *Nature*, **192**, 535–541.
Jacobson, M.D. and Raff, M.C. (1995) *Nature*, **374**, 814–816.
Jaiswal, A.K. (1994) *Biochemical Pharmacology*, **48**, 439–444.
Jensen, P.K. (1966) *Biochim Biophys Acta*, **122**, 157–166.
Jewett, S.L., Eddy, L.J. and Hochstein, P. (1989) *Free Radical Biology & Medicine*, **6**, 323–326.
Karoui, H., Hogg, N., Fréjaville, C., Tordo, P. and Kalyanaraman, B. (1996) *Journal of Biological Chemistry*, **271**, 6000–6009.
Kato, Y., Uchida, K. and Kawashki, S. (1992) *Journal of Biological Chemistry*, **267**, 23646–23651.
Kaul, N. and Forman, H.J. (1996) *Free Radical Biology & Medicine*, **21**, 401–405.
Kennedy, A.R., Troll, W. and Little, J.B. (1984) *Carcinogenesis*, **5**, 1213–1218.
Kensler, T.M. and Taffe, B.G. (1986) *Adv Free Rad Biol Med*, **2**, 347–387.
Kharbanda, S., Pandey, P., Schofield, L., Israel, S., Roncinske, R., Yoshida, K., Bharti, A., Yuan, Z.M., Saxena, S., Weichselbaum, R., Nalin, C. and Kufe, D. (1997) *Proc Natl Acad Sci USA*, **94**, 6939–6942.
Kluck, R.M., Martin, S.J., Hoffman, B.M., Zhou, J.S., Green, D.R. and Newmeyer, D.D. (1997) *EMBO J*, **16**, 4639–4649.
Kooy, N.W. and Royall, J.A. (1994) *Archives of Biochemistry and Biophysics*, **310**, 352–359.
Kopin, I.J. (1985) *Pharmacological Reviews*, **37**, 333–386.
Kristal, B.S. and Yu, B.P. (1992) *J. Gerontol.*, **47**, B104–107.
Larrick, J.W. and Wright, S.C. (1990) *FASEB Journal*, **4**, 3215–3223.
Laster, S.M., Wood, J.G. and Gooding, L.R. (1988) *Journal of Immunology*, **141**, 2629–2634.
Lennon, S.V., Martin, S.J. and Cotter, T.G. (1991) *Cell Prolif.*, **24**, 203–214.
Lesko, S.A., Caspary, W. and Lorentzen, R. (1975) *Biochemistry*, **14**, 3978–3984.
Lesko, S.A., Drocourt, J.L. and Yang, S.U. (1982) *Biochemistry*, **21**, 5010–5015.
Li, Y. and Jaiswal, A.K. (1994) *European Journal of Biochemistry*, **226**, 31–39.
Lin, A., Minden, A. and Martinetto, H. (1995) *Science*, **268**, 286–290.
Lipton, S.A., Choi, Y.-B., Pan, Z.-H., Lei, S.Z., Chen, H.-S. V., Scher, N.J., Loscalzo, J., Singel, D.J. and Stamler, J.S. (1993) *Nature*, **364**, 626–632.

Liu, X., Kim, C.N., Yang, J., Jemmerson, R. and Wang, X. (1996) *Cell*, **86**, 147–157.
Loschen, G. and Azzi, A. (1975) *Recent Adv Stud Cardiac Struct Metab*, 7, 3–12.
Loschen, G., Azzi, A., Richter, C. and Flohe, L. (1974) *FEBS Letters*, **42**, 68.
Lunec, J., Griffiths, H.R. and Blake, D.R. (1987) *ISI Atlas Sci Pharmco*, 45–48.
Maltzman, W. and Czyzyk, L. (1984) *Mol. Cell. Biol.*, **4**, 1789–1694.
Matsuda, M., Masutani, H. and Nakamura, H. (1991) *Journal of Immunology*, **147**, 3837–3841.
Matthews, N., Neale, M.L., Jackson, S.K. and Stark, J.M. (1987) *Immunology*, **62**, 153–155.
Mayer, B., Schrammel, A., Klatt, P., Koesling, D. and Schmidt, K. (1995) *Journal of Biological Chemistry*, **270**, 17355–17360.
McCord, J.M. (1974) *Science*, **185**, 529–531.
McCord, J.M. (1985) *New England Journal of Medicine*, **312**, 159–163.
Misra, H.P. and Fridovich, I. (1972) *Journal of Biological Chemistry*, **247**, 6960.
Mohazzab-H, K.M., Fayngersh, R.P. and Wolin, M.S. (1995) *Circulation*, **92**, I-701.
Mohr, S., Stamler, J.S. and Brune, B. (1994) *FEBS Letters*, **348**, 223–227.
Moody, C.S. and H.M., H. (1982) *Proceedings National Academy of Sciences, USA*, **79(9)**, 2855–2859.
Moriguchi, T., Kuroyanagi, N. and Yamaguchi, K. (1996) *Journal of Biological Chemistry*, **271**, 13675–9.
Mullarkey, C.J., Edelstein, D. and Brownlee, M. (1990) *Biochemical and Biophysical Research Communications*, **173**, 932–939.
Nagata, S. and Golstein, P. (1995) *Science*, **267**, 1449–1456.
Natarajan, V. and Garcia, J.G. (1993) *Journal of Laboratory and Clinical Medicine*, **121**, 337–347.
Natarajan, V., Taher, M.M., Roehm, B., Parinandi, N.L., Schmid, H.H., Kiss, Z. and Garcia, J.G. (1993) *Journal of Biological Chemistry*, **268**, 930–937.
Nemani, R. and Lee, E.Y.C. (1993) *Archives of Biochemistry and Biophysics*, **300**, 24–29.
Nguyen, T., Rushmore, T.H. and Pickett, C.B. (1994) *Journal of Biological Chemistry*, **269**, 13656–13662.
Nielsen, H. (1981) *Lipids*, **16**, 215–222.
Niki, E., Yamamoto, Y. and Komuro, E. (1991) *American Journal of Clinical Nutrition*, **53**, 2015–2055.
Nose, K., Shibanuma, M., Kikuchi, K., Kageyama, H., Sakiyama, S. and Kuroki, T. (1991) *Eur J Biochem*, **201**, 99–106.
Oleinick, N.L., Chiu, S., Ramakrishnan, N. and Xue, L. (1987) *British Journal of Cancer*, **55 (suppl.V11)**, 1135–40.
Oliver, C.N. (1987) *Archives of Biochemistry and Biophysics*, **253**, 62–72.
Paine, A.J. (1978) *Biochemical Pharmacology*, **27**, 1805–1813.
Palmer, R.M.J., Ferrige, A.G. and Moncada, S. (1987) *Nature*, **327**, 524–526.
Pinkus, R., Weiner, L.M. and Daniel, V. (1995) *Biochemistry*, **34**, 81–88.
Proctor, P.H. and Reynolds, E.S. (1984) *Physiol Chem Phys Med.*, **16**, 175–195.
Pryor, W.A. (1976) In *Free radicals in biology*, Vol. 1 (Pryor, W.A., ed.) Academic Press, New York, pp. 1–50.
Pryor, W.A. and Squadrito, G.L. (1995) *American Journal of Physiology*, **268**, L699–L722.
Radi, R., Rodriguez, M., Castro, L. and Telleri, R. (1994) *Arch Biochem Biophys*, **308**, 89–95.
Raingeaud, J., Whitmarsh, A.J. and Barrett, T. (1996) *Molecular and Cellular Biology*, **16**, 1247–55.
Ratan, R.R., Murphy, T.H. and Barbaran, J.M. (1994) *Journal of Neurochemistry*, **62**, 376–379.
Reddy, J.K., Warren, J.R., Reddy, K. and Lalwani, N.D. (1982) *Annals of the New York Academy of Sciences*, **386**, 81–110.
Renard, D.C., Seitz, M.B. and Thomas, A.P. (1992) *Biochemical Journal*, **284**, 507–512.

Rice, K.L., Duane, P.G., Archer, S.L., Gilboe, D.P. and Niewoehner, D.E. (1992) *American Journal of Physiology*, **263**, L430–L438.
Rivett, A.J. and Levine, R.L. (1990) *Archives of Biochemistry and Biophysics*, **278**, 26–34.
Rosette, C. and Karin, M. (1996) *Science*, **274**, 1194–7.
Rushmore, T.H., Morton, M.R. and Pickett, C.B. (1991) *Journal of Biological Chemistry*, **266**, 11632–11639.
Sachsenmaier, C., Radler-Pohl, A. and Zinck, R. (1994) *Cell*, **78**, 963–72.
Sahakian, J.A., Shames, B.D. and Levine, R.L. (1991) *FASEB Journal*, **5**, A1177.
Sandstorm, P.A., Tebbey, P.W., Van Cleave, S. and Buttke, T.M. (1994) *Journal of Biological Chemistry*, **269**, 798–801.
Saul, R.L., Gee, P. and Ames, B.N. (1987) In *Modern Biologic Theories of Aging* (Ed. Warner, H.R.) Raven Press, New York.
Sayre, L.M., Arora, P.K., Iyer, R.S. and Salomon, R.G. (1993) *Chem Res Toxicol*, **6**, 19–22.
Sbarra, A.J. and Karnovsky, M.L. (1959) *Journal of Biological Chemistry*, **234**, 1355–1362.
Schaich, K.H. (1992) *Lipids*, **27**, 209–218.
Schieven, G.L., Mittler, R.S., Nadler, S.G., Kirihara, J.M., Bolen, J.B., Kanner, S.B. and Ledbetter, J.A. (1994) *Journal of Biological Chemistry*, **269**, 20718–20726.
Schmidt, K.N., Amstad, P., Cerruti, P. and Baeuerle, P.A. (1995) *Chemico-Biological Interactions*, **2**, 13–22.
Schuessler, H. and Schilling, K. (1984) *International Journal of Radiation Biology*, **45**, 267–281.
Schultze-Osthoff, K., Krammer, P.H. and Droge, W. (1994) *EMBO Journal*, **13**, 14587–4596.
Schultze-Osthoff, K., Los, M. and Baeuerle, P.A. (1995) *Biochemical Pharmacology*, **50**, 735–741.
Schwartzman, R.A. and Cidlowski, J.A. (1993) *Endocrine Rev.*, **14**, 133–151.
Sevanian, A., Nordenbrand, K., Kim, E., Ernester, L. and Hochstein, P. (1990) *Free Radical Biology & Medicine*, **8**, 145–152.
Shamberger, R.J.J.N.C.I. (1972) *J. Natl. Cancer Inst.*, **48**, 1491–1497.
Sies, H. (1985) *Oxidative Stress*, Academic Press, New York.
Slaga, T.J., Klein-Szanto, A.J.P., Triplett, L.L., Yotti, L.P. and Trosko, J.E. (1981) *Science*, **213**, 1023–1025.
Southhorn, P.A. and Powis, G. (1988a) *Mayo Clin. Proc.*, **63**, 381–389.
Southorn, P. and Powis, G.M.C.P. (1988b) *Mayo Clin. Proc.*, **63**, 390–408.
Stadtman, E.R., Chock, P.B. and Rhee, S.G. (1981) *Curr. Top. Cell. Regul.*, **18**, 79–83.
Starke-Reed, P.E. and Oliver, C.N. (1989) *Archives of Biochemistry and Biophysics*, **275**, 559–567.
Steenken, S. (1989) *Chemical Reviews*, **89**, 503–520.
Stuehr, D.J. and Marletta, M.A. (1985) *Proceedings National Academy of Sciences, USA*, **82**, 7738–7742.
Sullivan, S.G., Chiu, D.T.-Y., Errasfa, M., Wang, J.M., Qi, J.-S. and Stern, A. (1994) *Free Radical Biology & Medicine*, **16**, 399–403.
Sun, Y. (1990) *Free Radicals in Biology and Medicine*, **8**, 583–599.
Sundaresan, M., Yu, Z.-X. and Ferrans, V.J. (1995) *Science*, **270**, 296–9.
Sundaresan, M., Yu, Z.X. and Ferrans, V.J. (1996) *Biochemical Journal*, **318**, 379–82.
Swallow, A.J. (1960) In *Radiation Chemistry of Organic Compounds* (Swallow, A.J., ed.) Pergamon Press, New York, pp. 211–224.
Tappel, A.L. (1980) Vol. IV (Ed. W.A. Pryor) Academic Press, New York and London, pp. 1–45.
Taube, H. (1965) *J. Gen. Physiol.*, **2**, 29.
Teoule, R. (1987) *International Journal of Radiation Biology*, **51**, 573–589.
Valenzuela, A. (1991) *Life Sciences*, **48 1991**, 301–309.

Varani, J., Phan, S.H., Gibbs, D.F., Ryan, U.S. and Ward, P.A. (1990) *Laboratory Investigation*, **63**, 683–689.

von Sonntag, C. (1987) *The chemical basis of radiation biology*, Taylor and Francis, London.

Wang, B. and Williamson, G. (1994) *Biochimica et Biophysica Acta*, **1219**, 645–652.

Ward, P.A., Warren, J.S. and Johnston, K.J. (1988) *Free Radicals in Biology and Medicine*, **5**, 403–408.

Webster, N.R. and Nunn, J.P. (1988) *British Journal of Anaesthesia*, **60**, 98–108.

Weiss, S.J., Peppin, G., Ortiz, X., Ragsdale, C. and Test, S.T. (1985) *Science*, **227**, 747–749.

Wieland, P. and Lauterburg, B.H. (1995) *Biochemical and Biophysical Research Communications*, **213**, 815–819.

Willson, R.L.I. (1982) In *Free Radicals, Lipid Peroxidation and Cancer* (ed., N.M.D.S.T.) Academic Press, London.

Winter, M.L. and Liehr, J.G. (1991) *Journal of Biological Chemistrty*, **266**, 14446–14450.

Witschi, H., Kacew, S., Hirai, K.-I. and Bellomo, B. (1977) *Chemico-Biological Interactions*, **19**, 143–160.

Wolf, S.P. and Dean, R.T. (1987) *Biochemical Journal*, **245**, 243–250.

Wu, M., Pritchard, K.A., Kaminski, P.M., Fayngersh, R.P., Hintze, T.H. and Wolin, M.S. (1994) *American Journal of Physiology*, **266**, H2108–H2113.

Xanthoudakis, S. and Curran, T. (1992) *EMBO Journal*, **11**, 653–665.

Yamamoto, O. (1976) In *Aging, Carcinogenesis and radiation biology* (ed., S.K.) Plenum Press, New York, pp. 165–192.

Yang, J., Liu, X., Bhalla, K., Kim, C.N., Ibrado, A.M., Cai, J., Peng, T.-I., Johnes, D.P. and Wang, X. (1997) *Science*, **275**, 1129–1132.

Yermilov, V., Rubio, J. and Oshima, H. (1995) *FEBS Letters*, **376**, 207–210.

Zick, Y. and Sagi-Eisenberg, R. (1990) *Biochemistry*, **29**, 10240–10245.

Zou, H., Henzel, W.J., Liu, X., Lutschg, A. and Wang, X. (1997) *Cell*, **90**, 405–413.

Part IV

Free radicals in specific disease states

17 The role of free radicals in chemical carcinogenesis

Joseph R. Landolph

Abstract

The purpose of this chapter is to discuss the prior and recent evidence for the involvement of free radicals in the molecular mechanism of chemical carcinogenesis. This review is focussed on the involvement of free radicals in polycyclic aromatic hydrocarbon carcinogenesis, and on carcinogenesis induced by specific carcinogenic compounds of the metals nickel, chromium, iron, and copper, and also by specific carcinogenic compounds of the metalloid arsenic. In addition, a brief discussion of the role of endogenous oxidants generated in mammalian cells and the possibility of their playing a role in carcinogenesis is also discussed.

There is some evidence for stimulation of arachidonic acid release by polycyclic aromatic hydrocarbons such as benzo(a)pyrene and 3-methylcholanthrene and its subsequent metabolism to prostaglandins and leukotrienes, and the possibility of simultaneous generation of superoxide and hydroxyl radicals from this pathway. There is also documentation for the generation of quinone metabolites during the metabolism of benzo(a)pyrene (BaP) and other polycyclic aromatic hydrocarbons, and some evidence that these quinones could redox cycle and generate superoxide, although much more work needs to be done in this area. There is also some evidence for the generation of radicals of BaP itself during the metabolism of BaP by cytochrome P450. The role of these radicals, if any, in BaP carcinogenesis can only be understood by further research in this area.

In the case of nickel carcinogenesis, there is preliminary data that specific nickel compounds may stimulate arachidonic acid metabolism, but more work is needed to define whether significant oxygen radical generation occurs off this pathway in response to stimulation by nickel compounds and whether radical generation is definitively involved in the molecular mechanisms of nickel compound carcinogenesis. There is accumulating and convincing evidence that nickel ions complex with proteins and amino acids, and that under specific circumstances, these nickel complexes may generate hydroxyl radicals at the site of DNA, leading to site-directed mutagenesis. Nickel compounds are well-known to generate chromosome breaks, and nickel-induced oxygen radical generation leading to site-directed mutagenesis may be one of the molecular mechanisms of nickel carcinogenesis. Specific insoluble chromium compounds are also carcinogens. The exact proximate carcinogen in the process of chromium carcinogenesis is still not known, and has been variously cast as chromium(V), chromium(III), or oxygen radicals generated when chromium functions as a redox-active transition metal. Combinations of these species may also act together to induce carcinogenesis upon treatment of cells with insoluble hexavalent chromium compounds. Further research is needed to clarify the mechanisms of chromium carcinogenesis, whether oxygen radicals are generated during this process, and if so, whether they play a role in chromium carcinogenesis.

During certain situations of copper or iron overload in mammalian organs, particularly liver, there is evidence that carcinogenesis may occur, and that iron compounds may be involved in this process. There is accumulating evidence that copper and iron compounds may participate in redox reactions leading to the generation of superoxide and hydroxyl radicals. Work in progress in a number of laboratories suggests that copper or iron-induced oxygen radical generation may in some cases be responsible for liver cancer in humans during situations of copper or iron overload.

Finally, specific arsenic compounds are believed to be carcinogens or co-carcinogens in humans. The molecular mechanisms of arsenic carcinogenesis remain obscure. While the involvement of arsenic-induced radical generation has been postulated to be involved in the mechanisms of arsenic carcinogenesis, these processes occur at very high concentrations of arsenic compounds, and this mechanism has not been substantiated to date as a participating event in the molecular mechanisms of arsenic carcinogenesis. Further work is required in this area. Therefore, to date, the evidence is strongest for the generation of radicals during polycyclic aromatic hydrocarbon carcinogenesis and during nickel, copper, and iron carcinogenesis. Further work is needed to substantiate the exact biological role of radicals during carcinogenesis by these compounds.

17.1 Introduction

The purpose of this chapter is to summarize the knowledge on the involvement of free radicals in the process of chemical carcinogenesis, to critically evaluate this knowledge, and to propose future directions for research in this area. No attempt is made to be exhaustive in this very large body of literature. Rather, the author chose to focus particularly on polycyclic aromatic hydrocarbons and specific carcinogenic metal salts of the elements nickel and chromium, and specific compounds of the metalloid element, arsenic. These specific compounds were chosen firstly because these compounds are known human carcinogens, and secondly because there is evidence that these compounds may cause generation of oxygen or other radicals during the process of carcinogenesis that they induce.

At the same time, there are situations in which an overload of copper or iron in animals or humans leads to liver carcinogenesis. In these situations, there is belief, based on experimental data in animals or on observations in humans, that overloads of these metals may lead to liver cancer, particularly in humans. Of course, it is widely believed that free radicals, particularly oxygen radicals such as superoxide and hydroxyl radical, and activated oxygen metabolites such as hydrogen peroxide, may be potentially involved in cancer and also may be involved in many other diseases, such as neurological diseases and aging (reviewed in Ames *et al.*, 1994). However, this review is focussed on the role of oxygen and other radicals in cancer.

It is also well-recognized that the area of free radical involvement in disease processes is often very difficult to substantiate conclusively. There are many reasons for this. Firstly, the lifetime of many radicals in biological systems is very short, because there are many radical-quenching systems with which they can react. Secondly, there are often many artefacts to the measurement of free radicals, particularly in biological systems. Thirdly, radicals often do not leave concrete and easily measurable types of damage in their wake, such as the carcinogen–DNA covalent adducts that many chemical carcinogens make. Fourthly, often the data concerning the existence of oxygen or other radicals are circumstantial or inferential, as often occurs when various inhibitors are used to demonstrate the existence of free radicals in biological systems. Fifthly, many radicals that are detectable are very reactive, and their various reactions often lead to other radicals or other reactive species that have not yet been identified, or are not easily identified (reviewed in Dix *et al.*, 1994). For all these reasons, the literature on the roles of free radicals in biological systems must be interpreted critically and cautiously.

Against this background, it is now very clear that chemical carcinogenesis by polycyclic aromatic hydrocarbons (PAH) in whole animals (reviewed in Boutwell, 1964; Pitot and Dragan, 1994) and in cultured murine and human cells (reviewed in Barrett and Ts'o, 1978; Landolph, 1985a,b) is a multi-step process. It is also now considered axiomatic that for carcinogenic polycyclic aromatic hydrocarbons (PAH), such as benzo(a)pyrene, 3-methylcholanthrene, and related members of this family, these compounds are metabolized by specific cytochrome P450 enzymes (reviewed in Miller and Miller, 1977; Sugimura, 1992; Pitot and Dragan, 1994) to reactive intermediates, including bay-region diol-epoxides and hydroxy-epoxides that bind covalently to DNA (reviewed in Miller and Miller, 1977; Sugimura, 1992), leading to mutations in bacteria (McCann *et al.*, 1975) and mammalian cells (Huberman *et al.*, 1976; Barrett and Ts'o, 1978; Landolph and Heidelberger, 1979) and morphological and neoplastic transformation of mammalian cells (Huberman *et al.*, 1976; Barrett and Ts'o, 1978; Landolph and Heidelberger, 1979). When these mutations occur in proto-onocogenes, they may activate these proto-oncogenes to oncogenes (Shih *et al.*, 1979; Cooper *et al.*, 1980; Shih and Weinberg, 1982; Eva and Aaronson, 1983; Sukumar *et al.*, 1983; Balmain *et al.*, 1984). When these mutations occur in tumor suppressor genes, they may mutationally inactivate these tumor suppressor genes or cause chromosome breakage, leading to deletion of these tumor suppressor genes (Weissman *et al.*, 1987; Finlay *et al.*, 1989; reviewed in Klein, 1987; Sager, 1989; Stanbridge, 1990). It is now widely believed that combinations of activation of proto-oncogenes to oncogenes and inactivations of tumor suppressor genes, such that on the order of five to seven events of these types occur, leads to carcinogenesis and the generation of the first tumor cell (Land *et al.*, 1983; reviewed in Bishop, 1987; Weinberg, 1985; Klein, 1987; Sager, 1989; Stanbridge, 1990). These events likely occur one at a time; such a mutation in one cell may lead to the first activation of the first proto-oncogene to an oncogene, and then the cell bearing this oncogene may divide so that a field of cells now bears this activated oncogene.

At some later point in time, in a random manner, a second mutation in either a tumor suppressor gene, inactivating it, or in a second proto-oncogene, activating it to a second oncogene, may occur in one of the group of cells bearing the first mutated oncogene. This process of mutation, expansion of the cells bearing the mutation, and a later mutation in one of the progeny of the prior mutated cells continues until the first tumor cell arises. This first or primordial tumor cell now bears five to seven genes, each with a mutation, which causes the generation of activated oncogenes and the inactivation of tumor suppressor genes (reviewed in Landolph, 1985a,b, 1990a,b, 1994, 1996). It is important to note that this primordial tumor cell is actually a degraded cell that lacks the mechanisms regulating growth and mechanisms coupling the growth of this cell to that of its neighboring cells. This primordial tumor cell then can further divide and then undergo further genetic change, such that metastatic cells arise in this population.

For compounds of metallic elements such as nickel and chromium, there is abundant evidence that these metals bind to DNA; nickel binds more srongly to protein, which then may bind to DNA (reviewed in Costa, 1996). It is very clear that nickel may generate oxygen radicals and cause lipid peroxidation and chromosome breakage (reviewed in Costa, 1996; Kasprzak, 1996). Specific insoluble chromium compounds also cause chromsome breakage (Gasiorek and Bauchinger, 1981; Larramendy *et al.*, 1981; Nishimura and Umeda, 1979; reviewed in Costa, 1996). While both specific insoluble nickel and insoluble chromium compounds cause carcinogenesis in animals and in humans, the exact molecular mechanisms by which they cause carcinogenesis are not known (reviewed in Costa, 1996; Landolph, 1989, 1990a,b, 1994, 1996). For copper and iron compounds, there are specific situations in which an overload of these compounds causes carcinogenesis in animals and is believed linked to human liver cancer (reviewed in Ames *et al.*, 1994), but the molecular mechanisms by which these compounds cause cancer is still not clear and requires further investigation. A discussion of our knowledge in these areas is described in detail below.

17.2 Stimulation of the arachidonic acid cascade, oxygen radical generation, and chemical carcinogenesis

17.2.1 Stimulation of arachidonic acid metabolism by chemical carcinogens and tumor promoters – direct evidence

It has been known for many years, since the original work of Levine *et al.* (1977), that many cytotoxic agents, including chemical carcinogens (Hassid and Levine, 1977; Levine, 1977), tumor promoters (Tragni *et al.*, 1986), and UV radiation (De Leo *et al.*, 1985), when added to cells perturb cell membrane physiology, which activates phospholipase A2, which then causes the enzymatic release of arachidonic acid from its esterified form in cell membranes. This released arachidonic acid is then metabolized enzymatically to prostaglandins, leukotrienes, and thromboxanes (Ali *et al.*, 1980; Zipser and Laffi, 1985; Samuelsson *et al.*, 1987; Vane and Botting, 1987; Robinson *et al.*, 1990). Human PGH synthase or cyclo-oxygenase, the enzyme that metabolizes arachidonic acid to prostaglandins (Hassid and Levine, 1977; Levine, 1977; Robinson *et al.*, 1990; reviewed in Zipser and Laffi, 1985; Vane and Botting, 1987; Higgs *et al.*, 1987; Marnett, 1992), has been molecularly cloned and mapped to a specific chromosomal position (Funk *et al.*, 1991). A novel TPA-inducible cyclo-oxygenase has also been molecularly cloned from Swiss 3T3 mouse fibroblasts (Kunjubu *et al.*, 1991).

During the metabolism of arachidonic acid by cyclo-oxygenase, there is evidence that a certain amount of superoxide may be generated. When this superoxide dismutates via the action of superoxide dismutase, to yield hydrogen peroxide, and the hydrogen peroxide plus superoxide react with iron in the Fenton/Haber–Weiss reactions, hydroxyl radicals are generated (Halliwell, 1987; Marnett, 1992; Kasprzak, 1996). The significance of this process is that hydroxyl radical is extremely reactive and reacts with DNA and other macromolecules so rapidly that these reactions are measured at the rates of diffusion (reviewed in Halliwell, 1987). These reactions may lead to mutations in mammalian cells (Ames, 1994).

Carcinogen-stimulated arachidonic acid release and its consequent metabolism to prostaglandins, leukotrienes, and thromboxanes, with the simultaneous generation of activated oxygen metabolites such as superoxide, and hydrogen peroxide, and oxygen radicals such as hydroxyl radical, may play a role in the molecular mechanisms of chemical carcinogenesis. This role is just beginning to become understood (reviewed in Marnett, 1992; Landolph, 1994). There is evidence for the involvement of the prostaglandin cascade in tumor promotion in mouse skin (reviewed in Slaga *et al.*, 1983; Kensler and Trush, 1984; Troll and Weisner, 1985; Cerutti, 1985; Marnett, 1992). Treatment of mice with the tumor promoter 12-O-tetradecanoyl-phorbol-13-acetate (TPA), during the promotion phase of two-step carcinogenesis, stimulates cell growth and leads to the release of arachidonic acid, the metabolism of arachidonic acid by cyclo-oxygenase and lipoxygenases, and the consequent generation of reactive oxgyen species and oxygen radicals (reviewed in Slaga *et al.*, 1983; Kensler and Trush, 1984; Troll and Weisner, 1985; Cerutti, 1985; Tragni *et al.*, 1986; Kunjubu *et al.*, 1991; Marnett, 1992). Prostaglandin synthesis occurs during tumor promotion and may be necessary for tumor promotion, but its exact role in tumor promotion is not yet understood (Slaga *et al.*, 1983; Kensler and Trush, 1984; Marnett, 1992).

17.2.2 *Inhibition of chemical carcinogenesis: insights into arachidonic acid involvement and consequent involvement in chemical carcinognesis by use of inhibitors*

Many studies of the role of arachidonic acid metabolism in tumor promotion and chemical carcinogenesis have employed inhibitors of arachidonic release and metabolism. For instance, use of dexamethasone, an inhibitor of arachidonic acid release, has been shown to inhibit complete carcinogenesis in animals. Similarly, the use of inhibitors of cyclo-oxygenase mediated metabolism, such as aspirin, indomethacin, piroxicam, sulindac, and ibuprofen also inhibit complete chemical carcinogenesis in animals (reviewed in Slaga *et al.*, 1983; Kensler and Trush, 1994; Marnett, 1992). There are also conflicting and controversial findings on the ability of aspirin, a cyclo-oxygenase inhibitor (Zipser and Laffi, 1985; Vane and Botting, 1987; Higgs *et al.*, 1987; Funk *et al.*, 1991; Weissman, 1991; Marnett, 1992), to inhibit human colon tumor formation (Thun *et al.*, 1991; Marnett, 1992).

In addition, certain anti-oxidants, such as vitamin E and vitamin C also inhibit two-step carcinogenesis (Slaga *et al.*, 1983; Kensler and Trush, 1985; Troll and Weisner, 1985; Cerutti, 1985). This cumulative body of evidence therefore implies that tumor promoter-induced arachidonic acid release and metabolism of arachidonic acid by prostaglandin endoperoxide synthase (PGH synthase, Marnett, 1992), lipoxygenase (Metzger *et al.*, 1984), or possibly both enzyme systems (reviewed in Marnett, 1992), may lead to the generation of oxygen radicals, which may attack DNA and may cause mutations during the process of chemical carcinogenesis (reviewed in Kensler and Trush, 1984).

There is also evidence that in animals or cultured animal cells treated with chemical carcinogens such as benzo(a)pyrene (Frenkel *et al.*, 1988; Frenkel, 1992), tumor promoters such as TPA (Wei and Frenkel, 1991; Frenkel and Gleichauf, 1991), and ionizing radiations (Tofigh and Frenkel, 1989; Patel *et al.*, 1992), there is oxidant generation and oxidative damage to DNA bases (reviewed in Frenkel *et al.*, 1988; Frenkel, 1992). Further, inhibitors of tumor promotion, such as sarcophytol, also inhibit oxidative damage to DNA (Wei and Frenkel, 1992). This suggests that both tumor promotion and complete carcinogenesis (treatment of animals with one carcinogen at high concentrations), may involve carcinogen-induced arachidonic acid release (Hassid and Levine, 1977; Levine, 1977; Tragni *et al.*, 1986), metabolism of arachidonic acid to prostaglandins (Hassid and Levine, 1977; Levine, 1977; Tragni *et al.*, 1986) and leukotrienes (Zipser and Laffi, 1985), and consequent generation of oxygen radicals and oxidative damage to DNA bases (Slaga *et al.*, 1983; Cerutti, 1985; Tofigh and Frenkel, 1989; Wei and Frenkel, 1991; Frenkel *et al.*, 1991; Frenkel and Gleuchauf, 1991; Patel *et al.*, 1992; Wei and Frenkel, 1992; Frenkel, 1992). This could be expected to lead to mutational activation of specific proto-oncogenes to oncogenes and to mutational inactivation or deletion of tumor suppressor genes.

17.2.3 Use of model cell culture systems to study the involvement of arachidonic acid metabolism and resultant active oxygen species and oxygen radicals in chemically induced morphological and neoplastic transformation

Many studies to date have utilized model cell culture systems to study the role of carcinogen-induced arachidonic acid release and its metabolism to prostaglandins and leukotrienes and the resultant generation of active oxygen species and oxygen radicals, such as superoxide and hydroxyl radicals. These systems have also been used to study the ability of radicals generated thereby to play a role in the induction of morphological transformation of cultured cells.

One of these cell culture systems, C3H/10T1/2 Cl 8 (10T1/2) mouse embryo cells, is a contact-inhibited, permanent cell line derived from the embryos of C3H mice by Reznikoff *et al.* many years ago (Reznikoff *et al.*, 1973a). 10T1/2 cells have been widely used to study the cellular and molecular mechanisms of chemically induced morphological and neoplastic transformation (Reznikoff *et al.*, 1973b; Landolph and Heidelberger, 1970; reviewed in Landolph, 1985a,b). In 10T1/2 cells, we showed that chemical carcinogens such as BaP and N-acetoxy-acetylamino-fluorene induced morphological transformation and mutation to ouabain resistance over the same concentration ranges (Landolph and Heidelberger, 1979). I postulated that this indicated that these two mutagenic carcinogens were inducing morphological transformation in part by causing mutations in and therefore activation of, specific cellular proto-oncogenes to oncogenes (reviewed in Landolph, 1985a,b). I further postulate that this indicates that these two mutagenic carcinogens also cause mutations in and therefore inactivation of, specific tumor suppressor genes. Therefore, in total, I postulate that these two mutagenic carcinogens cause both mutation in and activation of proto-oncogenes to oncogenes, and mutation in and therefore inactivation of, tumor suppressor genes, and that a total of approximately five of these events leads to neoplastic transformation of a normal mouse fibroblast into a tumorigenic mouse fibrosarcoma cell.

In addition, we have previously shown in a series of publications that mutation to ouabain resistance arises from chemically induced (Landolph and Heidelberger, 1979), stable (Landolph *et al.*, 1980a), specific (Landolph *et al.*, 1980b), base substitution mutations (Landolph and Jones, 1982) that map to murine chromosome #3 (Landolph and Fournier, 1983) and encode a ouabain-resistant sodium, potassium adenosine triphosphatase enzyme activity (Shibuya *et al.*, 1989). Hence, this is a very well-defined mutational assay (Shibuya *et al.*, 1989; reviewed in Landolph, 1985a,b). Next, we showed that the ratio of chemically induced morphological transformation frequencies to chemically induced frequencies of mutation to ouabain resistance is on the order of 20–100 (Landolph and Heidelberger, 1979). We therefore postulated that 20–100 genes are targets for chemically induced morphological transformation (Landolph and Heidelberger, 1979; reviewed in Landolph, 1985b). This figure of 20–100 is close to the number of known oncogenes and tumor suppressor genes known today, which is rapidly approaching 100. Hence, chemically induced carcinogen-DNA covalent adducts and consequent mutation, leading to resultant activation of proto-oncogenes to oncogenes and mutational inactivation of or deletion of tumor suppressor genes almost certainly is a part of the molecular mechanism of chemically induced neoplastic transformation.

Are there other molecular events besides covalent carcinogen–DNA adducts that contribute to chemically induced neoplastic transformation? We believe that there may be. In particular, what is the evidence for an involvement of oxygen radicals and other radicals in the process of chemical carcinogenesis? Benedict's laboratory many years ago showed that the anti-oxidant ascorbate (vitamin C) inhibited morphological transformation induced by the carcinogenic polycyclic aromatic hydrocarbon, 3-methylcholanthrene (MCA) in cultured 10T1/2 cells (Benedict *et al.*, 1980). In addition, studies by Bertram and co-workers indicated that retinoids, which are also anti-oxidants and cell-differentiation-inducing agents, reversibly inhibited the induction of morphological transformation by MCA in 10T1/2 cells (Merriman and Bertram, 1982). Many years ago, Mondal *et al.* developed an assay for initiation and promotion of cell transformation in 10T1/2 cells (Mondal *et al.*, 1976). Cerutti and co-workers utilized this assay to show that addition of exogenous superoxide dismutase to cultures of 10T1/2 cells inhibited two-step transformation of 10T1/2 cells initiated by the (anti) 7,8-dihydrodiol-9,10-epoxide of benzo(a)pyrene (BaP) and promoted by the tumor promoter TPA (Zimmerman and Cerutti, 1984). Interestingly, Cerutti's laboratory also showed that hydroperoxy eicoastetraenoic acids (HPETES) were clastogenic in 10T1/2 cells (Ochi and Cerutti, 1987). This could suggest that carcinogen-induced arachidonic acid release and metabolism of arachidonic acid to HPETES, which then damage chromosomes, is a part of the mechanism of multi-step chemically induced cell transformation, and hence of chemical carcinogenesis.

Work from our own laboratory has shown that BaP and MCA-induced morphological transformation of 10T1/2 cells can be inhibited by dexamethasone, an inhibitor of arachidonic acid release, can be inhibited by aspirin (acetylsalicylic acid), an inhibitor of both cyclo-oxygenase and lipoxygenase-mediated arachidonic acid metabolism, and can also be inhibited by nordihydroguariaretic acid, an inhibitor of lipoxygenase activity (Landolph *et al.*, 1998; Cerepnalkoski and Landolph, 1998). An conceptual integration of these results suggests that PAH such as BaP and MCA, and tumor promoters, such as TPA, stimulate the release of arachidonic acid and its conversion to prostaglandins and leukotrienes, with consequent generation of superoxide. The resultant superoxide molecules may then dismute to hydrogen peroxide via the action of the enzyme, superoxide dismutase (reviewed in Halliwell, 1987). Finally, superoxide and hydrogen peroxide may then form hydroxyl radicals via Fenton or Haber–Weiss reactions (reviewed in Halliwell, 1987; Kasprzak, 1996). The resultant hydroxyl radicals may then attack DNA to generate 8-hydroxyguanine, a mutagenic base, in cellular DNA. We postulate that this 8-hydroxyguanosine, if it occurs in specific proto-oncogenes, may activate them to oncogenes, or may inactivate tumor suppressor genes if it occurs in these genes, contributing to the induction of chemically induced neoplastic cell transformation. We postulate that these events may be responsible for tumor promotion in part and also for the promotional stages of complete carcinogenesis. I further postulate that for complete carcinogens such as BaP and MCA, the metabolism of these carcinogens and their covalent binding to cellular DNA, coupled to the generation of hydroxyl radicals via the arachidonic acid cascade and the subsequent Fenton/Haber–Weiss reactions, and the reaction of the hydroxyl radicals with cellular DNA to cause mutations in cellular proto-oncogenes, activating them to oncogenes, and mutations in cellular tumor suppressor genes, inactivating these tumor suppressor genes, together leads to neoplastic cell transformation and chemical carcinogenesis by PAH.

In the field of radiation carcinogenesis, Little *et al.* (1983), and Hall and Borek (1983) have also shown that X-ray initiated and TPA-promoted transformation of 10T1/2 cells and Syrian hamster embryo cells could be inhibited by exogenously added catalase and superoxide dismutase. This suggests that superoxide and hydrogen peroxide may play a role in X-ray induced morphological cell transformation. One likely possible explanation of these results is that X-rays promote the production of superoxide and hydrogen peroxide, and that these react to generate hydroxyl radical via Fenton and Haber–Weiss reactions, which then causes mutations in the proto-oncogenes and tumor suppressor genes, again suggesting a role of oxygen radicals in X-ray initiated cell transformation (reviewed in Landolph, 1994). Overall, therefore, there is evidence for carcinogen-induced release of arachidonic acid and its metabolism to prostaglandins and leukotrienes. The possibility of generation of significant amounts of hydroxyl radicals is likely if this pathway is over-stimulated by addition of large amounts of carcinogens to cells in culture or in animals. These hydroxyl radicals may cause generation of 8-hydroxyguanosine and other chemically altered bases in DNA. This may lead to activation of proto-oncogenes to oncogenes and inactivation or deletion of tumor supressor genes if hydroxyl radicals are generated in sufficient amounts to react specifically with tumor suppressor genes and proto-oncogenes.

17.3 Metabolism of polycyclic aromatic hydrocarbons, free radicals, and chemical carcinogenesis

Work from many laboratories over a number of years, particularly those of Lesko and Lorentzen, has provided evidence for the oxidation of BaP diols by free radical mechanisms to yield BaP semiquinone radical anions. During this process, oxygen is reduced to produce superoxide anion radicals, which then dismutates to hydrogen peroxide (Lesko and Lorentzen, 1985). This could indicate that oxygen radicals, such as hydroxyl radical, may be involved in BaP carcinogenesis.

Many years ago, experiments from the laboratory of Ts'o (Lesko *et al.*, 1975; Lorentzen *et al.*, 1977) indicated the formation of a 6-oxobenzo(a)pyrene radical in rat liver homogenates incubated with BaP. Whether this radical can react with DNA has not been pursued, but this possibility remains. In addition, it is well known that the 1,6- and 3,6- and 6,12-quinones of benzo(a)pyrene are metabolites of BaP, and it was found by Lorentzen *et al.* (1979) and Lesko and Lorentzen (1985) that these quinones were cytotoxic to cultured Syrian hamster embyro cells and that the cytotoxicity was dependent upon molecular oxygen. Further evidence accrued that this might be due to the redox cycling of these quinones to generate oxygen radicals, and these authors proposed that this might play a role in the molecular mechanisms of carcinogenesis induced by benzo(a)pyrene. The formation of BaP quinones is facile during the cytochrome P450-mediated metabolism of BaP (Lesko and Lorentzen, 1985). These authors have shown that BaP quinones induce DNA single stranded breaks in phage T7 DNA. These authors hold that this is due to Fenton-like reactions in which BaP quinones are reduced and react with molecular oxygen to generate superoxide anions, which dismute to generate hydrogen peroxide, and then in Fenton-like reactions generates hydroxyl radicals (Lesko and Lorentzen, 1985). They further postulate that this ability of BaP quinones to generate superoxide and thence hydroxyl radicals gives BaP the ability to serve both as an initiator and a promoter, hence as a complete carcinogen. They postulate that the ability of BaP to generate oxygen radicals may endow it with promoting ability (Lesko and Lorentzen, 1985).

This is a very attractive hypothesis, and it has interesting parallels in the cytotoxicity of the compound adriamycin, a well-known anthracycline derivative that is used today as a cancer chemotherapeutic agent against certain solid tumors and some leukemias and is considered a weak carcinogen (discussed in Landolph *et al.*, 1980b). Adriamycin was shown to redox cycle and to generate hydroxyl radicals many years ago, and hydroxyl radical generation, as well as the ability of adriamycin to bind to DNA, are thought to be major determinants of its cytotoxicity, its mutagenicity, and its weak carcinogenicity (discussed in Landolph *et al.*, 1980b). Hence, a similar hypothesis for the ability of the quinones of BaP to redox cycle and generate hydroxyl radicals would also be an attractive hypothesis for part of the mechanism of the carcinogenicity of BaP. However, for many years, little further work has been performed in this area. Hence, to date, this hypothesis still remains speculative. It would be a very attractive hypothesis that the formation of these quinones during the metabolism of BaP by cytochrome P450, their redox cycling to generate hydroxyl radicals, and the attack of the hydroxyl radicals upon DNA to generate mutations in proto-oncogenes, activating them to oncogenes, and attack of these radicals upon DNA to generate inactivating mutations in tumor suppressor genes, could be part of the mechanism of carcinogenicity by BaP. It would be a further hypothesis that the quinone-generated hydroxyl radicals and the mutations that they caused in DNA could be a promotion-like step in the overall process of carcinogenesis by BaP, and that the covalent adducts of BaP to DNA bases could be responsible for the initiating part of BaP carcinogenesis. Further research in this area must occur to advance the elegant hypotheses and prior work of Lesko, Lorentzen, and Ts'o described above to provide support for or against this hypothesis of BaP generated quinones and their redox cycling to generate superoxide and hydroxyl radicals, which could cause mutations in DNA, leading to part of the molecular mechanism of BaP carcinogenesis.

17.4 Metal-induced free radical generation and metal salt carcinogenesis

17.4.1 Molecular mechanisms of nickel-induced carcinogenesis

Epidemiological studies have indicated that there are higher frequencies of nasal and respiratory cancer in workers who mine and refine ores containing nickel compounds (reviewed in Pederson *et al.*, 1973; Hernberg, 1977; Costa, 1996; Landolph, 1989, 1990a,b, 1994, 1996). There are often compounding aspects to this situation, since often many of the nickel workers have been shown to be cigarette smokers. It is likely that in this situation, both cigarette smoke carcinogens, such as BaP, dibenz (c,g) carbazole, and the tobacco specific nitrosamines, NNN and NNK, along with the co-carcinogens and promoters in cigarette smoke, interact in a synergistic or additive manner with the nickel compounds to increase the frequency of respiratory cancer in the nickel workers.

Injection of specific nickel compounds into animals induces fibrosarcomas at the site of injection, so a number of specific nickel compounds are animal carcinogens (reviewed in Costa, 1996; Landolph, 1989, 1990a,b, 1994, 1996). In addition, many nickel compounds, particularly the insoluble nickel compounds, are able to induce morphological and neoplastic transformation of cultured Syrian hamster embryo cells (reviewed in Costa, 1996) and of cultured C3H/10T1/2 Cl 8 mouse embryo cells (Miura *et al.*, 1989). It is likely that the insoluble nickel compounds are the most carcinogenic among the nickel compounds (reviewed in Landolph, 1989, 1990a,b, 1994, 1996). These insoluble nickel compounds that are believed to be the most carcinogenic include nickel subsulfide, crystalline nickel monosulfide, and various forms of nickel oxide (there are over 30 different types of nickel oxide; reviewed in Landolph, 1989, 1990a,b, 1994; Landolph *et al.*, 1996). The soluble nickel compounds, including nickel sulfate and nickel chloride, are not strongly active as cell transforming agents or animal carcinogens, and are inactive in C3H/10T1/2 Cl 8 mouse embryo cells (Miura *et al.*, 1989). It is not clear whether soluble nickel compounds are carcinogenic to humans, since human exposure is to a complex array of nickel compounds, including both soluble and insoluble nickel compounds (reviewed in Costa, 1996; Landolph, 1989, 1990a,b, 1994; Landolph *et al.*, 1996). Further work is needed to answer this question. However, at present, it is believed that specific insoluble nickel compounds are carcinogenic, and that these likely contribute to the carcinogenicity of the process of mining and refining nickel compounds (reviewed in Costa, 1996; Kasprzak, 1996; Landolph, 1989, 1990a,b, 1994; Landolph *et al.*, 1996).

A number of interesting papers from various laboratories, including those of Kasprzak and Costa, have indicated that nickel complexes with certain amino acids are redox active and can generate hydroxyl radicals (Costa, 1996; Kasprzak, 1996). Current hypotheses are that nickel can bind avidly to proteins, and that if these protein- or amino acid–nickel complexes bind to DNA, they may be able to react with molecular oxygen and thereby generate hydroxyl radicals at the site to which the nickel–protein complexes attach to DNA (Costa, 1996; Kasprzak, 1996). This should result in mutations in the DNA. These have not yet been shown to occur despite work in many mammalian mutagenesis assays (reviewed in Landolph, 1989, 1990a,b, 1994; Landolph *et al.*, 1996). However, nickel complex generation of hydroxyl radicals may be an interesting model for nickel-induced chromosome breakage, since nickel compounds have been shown to efficiently cause chromosome breakage in mammalian cells (Nishimura and Umeda, 1979; Saxholm *et al.*, 1981).

We have also shown in our laboratory that nickel compounds are able to induce anchorage independence in diploid human fibroblasts (Biedermann and Landolph, 1987). We have further shown that acetylsalicylic acid, a known inhibitor of cyclo-oxygenase, is able to inhibit nickel-induced anchorage independence in diploid human fibroblasts (Biedermann and Landolph, manuscript in preparation, 1998). This suggests that nickel compounds may be able to induce arachidonic acid release and its metabolism by cyclo-oxygenase, with consequent generation of hydroxyl radicals, and that these may also be able to induce genetic change in mammalian cells, leading to induction of anchorage independence. Further work is needed at the level of studies of arachidonic acid metabolism under these conditions and to trap and identify hydroxyl radicals before concrete conclusions can be drawn from these suggestive inhibitor experiments. However, the intriguing possibility exists that nickel compounds may be able to stimulate arachidonic release and its metabolism via cyclo-oxygenase, leading to generation of hydroxyl radicals and their attack on DNA to cause chromsome breakage and possibly mutations, which may inactivate tumor suppressor genes.

To date, our laboratory and other laboratories have shown that insoluble nickel compounds can induce morphological transformation of cultured C3H/10T1/2 mouse embryo cells (Miura *et al.*, 1989) and Syrian hamster cells (reviewed in Costa, 1996), and we have also shown that nickel compound induce anchorage independence in diploid human fibroblasts (Biedermann and Landolph, 1987). Our laboratory has shown that the nickel compounds do not induce mutation to ouabain resistance in 10T1/2 cells (Miura *et al.*, 1989) and do not induce mutation to either ouabain resistance or to 6-thioguanine resistance in human diploid fibroblasts at concentrations that induce morphological transformation in 10T1/2 cells or anchorage independence in human diploid fibroblasts (Biedermann and Landolph, 1987). Hence, we have postulated either a) that the nickel compounds generate oxygen radicals which are not easily shown to be mutagenic in classical mutagenesis systems, or b) that the nickel compounds cause chromosome breakage, leading to inactivation of tumor suppressor genes, and perhaps to breakage of elements of negative regulatory controlling elements from proto-oncogenes, activating them to oncogenes (reviewed by Landolph, 1989, 1990a,b, 1994; Landolph *et al.*, 1996).

To address these possibilities, we have begun studies of whether any proto-oncogenes are amplified or rearranged in nickel-induced transformed 10T1/2 cell lines. We have found that this is not the case, and we have tentatively ruled out this hypothesis for a number of oncogenes that we have studied (Verma, A. and Landolph, J.R., manuscript in preparation). Secondly, we have asked whether there is over-expression of known proto-oncogenes, and we have also ruled out signficant elevations of mRNAs transcribed from known proto-oncogenes that we have studied (Verma, A. and Landolph, J.R., manuscript in preparation). Hence, utilizing mRNA differential display, we have begun to determine whether specific genes are over-expressed in nickel-induced transformed 10T1/2 cell lines and whether specific genes are under-expressed in the transformed cell lines. In preliminary experiments, we have found that numerous genes are over-expressed and numerous genes are under-expressed in the nickel-induced transformed cell lines. Our initial experimental data, extrapolated from a small number of gels to 100% coverage of the mRNA pool, indicate that there are a total of approximately 300 genes that would be over-expressed or under-expressed in nickel-induced transformed 10T1/2 cell lines. We are now in the presence of isolating, sequencing, identifying, and characterizing these genes (Verma, A., Ramnath, J., Ohshima, S., and Landolph, J.R., manuscript in preparation). Further insight into the molecular mechanisms of nickel-induced cell transformation, and hence nickel carcinogenesis, will be gained by the results of these experiments. We expect to identify secondary genes whose expression is influenced, the primary genes that are targets for nickel-induced genetic damage (approximately 5–10 genes), and the genetic damage that nickel causes to these primary genes. It is our current hypothesis that each of the five primary genes that are altered, whether this an activated oncogene or an inactivated tumor suppressor gene, influences that activity of, on the average, 60 genes by affecting major biochemical pathways involved in the control of cellular growth and differentiation in the cell.

17.4.2 Possible radical involvement in chromium carcinogenesis

Specific chromium compounds are also carcinogenic in animals, and epidemiological studies have indicated that workers who mine chromium ores or work in the pigment industries where specific chromium compounds are components of the pigments, such as lead chromate, show evidence of increased frequencies of lung cancer (reviewed in Landolph, 1989, 1990a,b, 1994; Landolph *et al.*, 1996; Klein, 1996). A number of chromium compounds have been shown to induce morphological and neoplastic transformation of C3H/10T1/2 Cl 8 mouse embryo cells (Patierno *et al.*, 1987) and Syrian hamster cells (reviewed in Klein, 1996; Landolph, 1989, 1990a,b, 1994; Landolph *et al.*, 1996). Current belief is that the insoluble compounds of chromium are the most carcinogenic, such as lead chromate (reviewed in Landolph, 1989, 1990a,b, 1994; Landolph *et al.*, 1996). The insoluble chromium compounds are taken up by phagocytosis, depositing a bolus of insoluble hexavalent chromium compound into the cell (Patierno *et al.*, 1987; reviewed by Landolph *et al.*, 1989, 1990a,b, 1994). In the case of soluble hexavalent chromium compounds, hexavalent chromium ion is taken up into cells by the somewhat non-specific sulfate anion transport system (reviewed in Klein, 1996). It is believed that reduction of hexavalent chromium to lower valence states is responsible for its toxicity, mutagenicity for some chromium compounds, clastogenicity, and likely also its carcinogenicity (reviewed in Landolph, 1989, 1990a,b, 1994; Landolph *et al.*, 1996; Klein, 1996). The exact nature of the species that attacks DNA during the process of carcinogenesis and that is responsible for chromium carcinogenesis is still not clear (reviewed in Landolph, 1989, 1994). Earlier workers held that the production of intracellular chromium(III) was the important ion responsible for carcinogenesis. This view has not disappeared, but a reasonable summation of this field is that production of chromium(III) is one important species involved in chromium carcinogenesis (reviewed in Landolph, 1989, 1990a,b, 1994; Landolph *et al.*, 1996; Klein, 1996). Further work from the laboratory of the late Professor Karen Wetterhahn stressed the importance of the possibility of a transient chromium(V) species, a radical, in chromium carcinogenesis (reviewed in Landolph, 1990a,b, 1994). Evidence had been presented for the production of chromium(V) during treatment of cells with chromium compounds. However, further work is needed in this area to establish whether chromium(V) is important in chromium carcinogenesis. Other workers had previously held that chromium(VI) could participate in redox reactions, and that oxygen radicals could be generated from this pathway (reviewed in Landolph, 1989, 1990a,b, 1994). However, little evidence in recent years has accumulated to substantiate this pathway as having a role in chromium carcinogenesis. While this hypothesis is attractive, given the transition metal properties of chromium, and since chromium can easily be postulated to participate in reactions similar to the Fenton reaction that iron participates in, further research needs to be conducted in this area to determine whether chromium radicals or chromium-induced oxygen radicals contribute to the mechanisms of chromium carcinogenesis.

17.4.3 Iron and copper overload and liver carcinogenesis

Iron has long been known to be an essential element in biological systems. At the same time, due to its activity as a redox-active metal, it has also been postulated to be able to generate oxygen radicals in Fenton reactions (reviewed in Halliwell, 1987). Hence, iron compounds have also been postulated to be potential human and animal carcinogens. In the human disease, hereditary hemochromatosis, iron accumulation is correlated with cirrhosis of the liver and carcinomas of the liver (reviewed in Poirier and Littlefield, 1996). Up until recently, there has been little experimental evidence in animals to support the hypothesis of the carcinogenicity of iron compounds. Iron has been shown to enhance the carcinogenicity of dimethylnitrosamine and diethylnitrosamine in inducing lung and nasal tumors in hamsters when administered topically as iron oxide, and iron oxide delivered by intratracheal instillation along with BaP has enhanced the carcinogenicity of BaP in hamsters (reviewed in Poirier and Littlefield, 1996).

Recent exciting results from the laboratory of Professor Helmut Bartsch and his co-workers have investigated this hypothesis of the potential carcinogenicity of iron compounds and copper compounds very carefully, and also studied the linkage of this hypothesis to the hypothesis that oxidative stress may be responsible for lipid peroxidation and the consequent generation of aldehydes such as *trans*-4-hydroxy-2-nonenal that react with DNA to form exocyclic adducts (Bartsch *et al.*, 1994; Chung *et al.*, 1996). Some of the background for their studies were the previous studies that these exocyclic adducts, termed etheno adducts, were shown to be formed in DNA treated with the epoxide metabolites of the human carcinogens vinyl chloride andurethane (Bartsch *et al.*, 1994). Their laboratory has developed very sensitive methods utilizing immuno-affinity methods to detect these exocyclic DNA adducts (Chung *et al.*, 1996; El-Ghissassi *et al.*, 1995). These authors have noted that nitric oxide is over-produced when conditions of chronic infection and inflammation occur, and that several human cancers arise against a background of chronic viral infection. They have utilized a mouse model system, the SGL mouse, which produces high levels of NO after these mice are injected with lymphoma cells. In this model mouse system, they found that lipid peroxide-induced etheno adducts occur in tissue in which NO is over-produced. This provides an interesting model that inflammation leads to lipid peroxidation and, hence, production of etheno-DNA adducts in animals (Nair *et al.*, 1997).

Secondly, these workers have utilized a second model in which Long Evans Cinnamon rats develop liver tumors due to a Cu-dependent formation of miscoding etheno-DNA adducts (Nair *et al.*, 1996b). In this model animal system, these rats spontaneously develop hepatitis and 100% of the animals develop hepatocellular carcinoma (Nair *et al.*, 1996b). The rats originally have no symptoms, but they progress such that 40% of them develop a fulminant hepatitis and 60% develop a prolonged hepatitis (Nair *et al.*, 1996b). Those rats developing a prolonged hepatitis progress to develop liver cancer in one year. These workers have shown that the copper content of the liver correlates with the etheno adduct levels (Nair *et al.*, 1996b). Hence, it is likely that copper is participating in redox reactions, generating oxygen radicals, and that these cause lipid peroxidation. As a result of this lipid peroxidation, aldehydes are generated, and these react with cellular DNA, leading to formation of etheno adducts in the DNA (Nair *et al.*, 1996a,b). It is further likely that the etheno adducts are formed in critical genes, such as oncogenes and tumor suppressor genes, and that this leads to cancer in the livers of these animals. Further molecular biology studies in this elegant system should lead to insight as to the exact genes which are damaged, the mutations which are formed in this genes, and the biological effectiveness of these mutations. Such studies are ongoing in the laboratory of Professor Bartsch and his collaborators.

Next, these workers have chosen to study the human genetic diseases, Wilson's Disease and primary haemochromatosis, in which there are copper or iron accumulations, respectively, in the patients affected, resulting in increased frequencies of liver damage and liver cancer. They have measured increased levels of 1,*N6*-ethenoadenosine-deoxy (EdA) adducts and 3,*N4*-ethenodeoxy-cytidine (EdC) DNA adducts in the livers of patients with these diseases. In Wilson's disease patients, there is a roughly linear correlation between the amounts of EdA and EdC adducts *vs.* the levels of copper in the livers of these patients (Nair *et al.*, 1996a). Therefore, these authors hold that the elevated etheno adducts in LEC rats and in Wilson's disease patients suggest that there is a role of oxidative stress in causing lipid peroxidation, and that the products of lipid peroxidation elevate the levels of etheno-DNA adducts. They also suggest that these processes are likely part of the process of liver disease and liver cancer in these patients (Nair *et al.*, 1996a).

Finally, Bartsch *et al.* have also studied the role of dietary ω-6 polyunsaturated fatty acids in influencing the frequency of human colon and prostate cancer (Bartsch and Nair, 1997). They have shown that dietary ω-6 polyunsaturated fatty acids (PUFAs) drastically increase the formation of etheno-DNA base adducts in white blood cells from humans (Nair *et al.*, 1997b). They found a positive association between dietary fat and the increase of colon and prostate cancer (Bartsch and Nair, 1997b). They also found that high dietary ω-6 dietary PUFA (linoleic acid, C18:2) and that low amounts of ω-3 PUFA may be linked to colon cancer risk and breast cancer risk (Bartsch and Nair, 1997b). Finally, they also noted that ω-6 PUFAs enhance experimental mammary tumorigenesis, and that ω-3 PUFAs are in high levels in humans. The work of Professor Bartsch, Dr Nair, and their collaborators has been summarized in an abstract and presented by Professor Bartsch at a recent scientific meeting (Bartsch and Nair, 1997).

These authors conclude that exocyclic DNA adducts derived from lipid peroxidation are good biomarkers to determine the effects of dietary fat intake, and oxidative stress on endogenous DNA damage and may be mechanistically linked to elevated risk for specific human cancers. These authors also stress that the etheno-adducts they have identified and measured should be useful for studies on the biological significance of background levels of such adducts, the repair of such adducts in human tissues, their use in molecular epidemiology studies, ways in which to prevent formation of such adducts, and whether such intervention decreases the risk of specific human cancers (reviewed in Bartsch and Nair, 1997). Indeed, other workers have already found that human glycosylase enzymes excise these adducts from DNA (Hang *et al.*, 1996).

17.4.4 Are free radicals involved in arsenic carcinogenesis?

There is today a significant amount of epidemiological evidence that exposure of humans to specific arsenic compounds correlates with an increased frequency of respiratory cancer (reviewed in Landolph, 1989, 1990a,b, 1994; Smith *et al.*, 1992; Wang and Rossman, 1996; Fan, 1996). This is true in the copper smelting industry, where smelting of copper ores contaminated by arsenic leads to evolution of arsenic trioxide, and worker exposure to arsenic trioxide correlates with an increased frequency of respiratory cancer (reviewed in Landolph, 1989, 1990a,b, 1994; Landolph *et al.*, 1996; Wang and Rossman, 1996). In addition, exposure of vintners to arsenic compounds during the spraying of these compounds as pesticides led to an increased frequency of skin cancers and leukemia. Further, in Taiwan, arsenic compounds often naturally contaminate artesian wells that are used as sources of drinking water and as a source to irrigate crops. When people in Taiwan drink arsenic-contaminated water, there has been shown to be an increased frequency of cancer of the lung and other internal organs. In the past, treatment of patients with psoriasis with Fowler's solution, containing arsenic compounds, has been shown to result in an increased frequency of skin cancer. All these cases of increased frequencies of cancers in humans exposed to arsenic compounds have been discussed at length (reviewed in Landolph, 1989, 1990a,b, 1994; Landolph *et al.*, 1996; Wang and Rossman, 1996). In the United States, risk assessment calculations for cancer induction in humans due to various sources of drinking water are often dominated by the risk attributed to arsenic (Smith *et al.*, 1992; Fan, 1996). Interestingly, past studies of animal carcinogenicity had led to largely negative results for the carcinogenicity of arsenic compounds in animals (reviewed in Landolph, 1989, 1990a,b, 1994; Landolph *et al.*, 1996; Wang and Rossman, 1996). More recent studies are showing positive results when arsenic is added together with other substances. Hence, arsenic compounds may be co-carcinogens, promoters, or progressors of cancer. There may also be a number of co-factors affecting the carcinogenicity of arsenic in humans, such as diet, exposure to other carcinogens, and other unknown factors (reviewed in Landolph, 1989, 1990a,b, 1994; Landolph *et al.*, 1996; Wang and Rossman, 1996; Hotta, 1997; Ferreccio *et al.*, 1997).

To date, it can be stated that the molecular mechanisms of arsenic carcinogenicity are not known (reviewed in Landolph, 1989, 1990a,b, 1994; Landolph *et al.*, 1996; Wang and Rossman, 1996). There had been a few previous reports that various species may be able to react with oxygen to generate dimethylarsenic peroxyl radicals (Okada, 1997). However, these reactions occur at extremely high concentrations of arsenic. Therefore, to date, there is no strong evidence that arsenic can generate oxygen radicals or other radicals at concentrations of arsenic to which humans are likely to be exposed. Therefore, at the present time, there is no strong evidence to indicate that arsenic radicals, dimethylarsenic peroxyl radicals, oxygen radicals, or other radicals participate in the process of arsenic carcinogenesis or co-carcinogenesis.

17.5 Endogenous oxygen radical generation and carcinogenesis

There is now a large body of evidence indicating that a great number of oxygen molecules are processed per day during the course of normal oxidative metabolism in normal cellular processes. As discussed in an excellent review article by Dr Bruce Ames (Ames *et al.*, 1994), there are a number of sources of oxidants in mammals, including superoxide, hydrogen peroxide, and hydroxyl radicals that are generated as unwanted by-products during the normal course of respiration in mammals. Ames and co-workers estimate that each cell in a rat contacts 10^{12} molecules of oxygen each day, and consequently generates 2×10^{10} molecules of superoxide and hydrogen peroxide per day (reviewed in Ames *et al.*, 1994). Ames *et al.* also discussed the functioning of phagocytes as another source of reduced oxygen metabolites and oxygen radicals, and that peroxisomes and cytochrome P450 enzymes also generate reduced oxygen metabolites and oxygen radicals. These activated oxygen metabolites, such as peroxide and superoxide, and radicals, such as hydroxyl radical, have the capacity to attack DNA, causing generation of 8-hydroxyguanosine and other oxidized bases, and these may contribute to generating mutations during the lifetime of a cell, particularly during chronic inflammation processes such as hepatitis B virus infection, which is a co-factor for human liver cancer (along with aflatoxin B1 ingestion) (reviewed in Ames *et al.*, 1994).

We note that there is a large amount of oxygen processed in the mitochondria, and that although the resultant superoxide radicals are generated, then dismutated, and the hydrogen peroxide is acted upon and inactivated by catalase, still there will be a constant generation of a small amount of hydroxyl radicals generated that will attack mitochondrial DNA (Ames *et al.*, 1994). The biological consequences of this in the generation of mutations in mitochondrial DNA are unknown, as are the possible contributions of such mutations in mitochondrial DNA for carcinogenesis, but it is intriguing to speculate that generation of oxygen radicals by mitochondria during the course of normal cellular metabolism may play some role carcinogenesis. Clearly, many more critical experiments will need to be done to test the hypothesis that mitochondrially generated oxygen radicals may play some part in the multi-step process of carcinogenesis in humans.

17.6 Conclusions

Firstly, regarding carcinogenesis by polycyclic aromatic hydrocarbons, we can conclude that this process is multi-step, and that it takes at least five to seven steps beginning with a normal cell to generate a tumor cell. These steps involve mutations in proto-oncogenes, activating them to oncogenes, further cell divisions, and mutations in tumor suppressor genes, inactivating them, or deletions of these genes. Once one cell possesses a total of five to seven mutations, such that some of these activate one or more proto-oncogenes to oncogenes, and one or more either cause mutational inactivation or deletion of specific tumor suppressor genes, it has then become a primordial tumor cell. Once this primordial tumor cell divides and grows to a sizable mass that is clinically recognizable as a tumor, the primary tumor comes into existence. Further cell division and genetic alterations via spontaneous mutation and chromsome losses and gains then leads to a metastatic tumor. In the case of polycyclic aromatic hydrocarbons, some of these steps are due to the metabolism of the hydrocarbons to bay-region diol-epoxides, to hydroxy-epoxides, and to cation radicals at the six position of BaP. These electrophilic species then bind covalently to DNA, resulting in hydrocarbon-DNA covalent adducts. A certain fraction of these adducts are repaired, and a certain fraction of them are processed to generate mutations in cells. This hydrocarbon-induced mutation in proto-oncogenes and tumor suppressor genes, activating the former and inactivating the latter, is responsible for a part of the mechanisms of hydrocarbon carcinogenesis. It is also now believed that hydrocarbon-induced release of arachidonic acid from its esterified position in cellular membranes, leading to its metabolism by cyclo-oxygenase to prostaglandins and by various lipoxygenases to leukotrienes, occurs. In addition, consequent leakage of superoxide, used as an attacking species in these metabolisms, leads to hydrogen peroxide, and reaction of superoxide and hydrogen peroxide in the presence of cellular iron may lead to hydroxyl radical. If this occurs in the vicinity of cellular DNA, hydroxyl radicals may react with DNA to generate 8-hydroxyguanosine and other oxidized DNA bases. These may be processed to mutations in cellular proto-oncogenes, activating them, and to mutations in cellular tumor suppressor genes, inactivating them. Hence, this process may lead to part of the mechanisms of carcinogenesis by carcinogenic polycyclic aromatic hydrocarbons. Further work is needed on the generation of hydroxyl radicals, on their reaction with DNA, and on the formation of mutations in proto-oncogenes and tumor suppressor genes to validate whether this hypothesis is true.

Regarding nickel compound carcinogenesis, it is clear that nickel compounds can generate oxygen radicals, lipid peroxidation, and chromosome breakage. Further work is needed to prove whether specific oncogenes are activated and whether specific tumor suppressor genes are inactivated in nickel-induced transformed cell lines. Work is also needed to determine whether nickel-generated oxygen radicals can induce 8-hydroxyguanosine in cells, and whether and what fraction of these oxidized base products lead to mutations in tumor suppressor genes and in proto-oncogenes.

In chromium carcinogenesis, it is clear that specific insoluble chromium compounds cause chromosome breakage and carcinogenesis in animals and neoplastic transformation of cultured murine and human cells. The carcinogenic species generated from hexavalent chromium compounds are likely chromium(III) at the very least. Whether chromium(V) and other reduced species are generated and play a role in chromium carcinogenesis is unclear at present. It is also unclear whether reduced chromium species can lead to generation of oxygen radicals in Fenton-like reactions. Hence, the role, if any, of oxygen or other radicals in chromium carcinogenesis at present is only speculative. Further experimental work is needed in this area.

In iron and copper storage diseases and in experimental animal models utilizing iron or copper overloads, there is evidence that there is generation of the products of lipid peroxidation, likely due to the generation of oxygen radicals and their attack on cellular lipids. The products of lipid peroxidation leads to the formation of etheno adducts, which are mutagenic due to their formation in animals treated with carcinogenic vinyl chloride. Further measurements of the generation of and the identity of these radicals would strengthen the concept that iron or copper-generated oxygen radicals lead to lipid peroxidation, etheno-DNA adduct formation, and carcinogenesis. Further work is also needed to substantiate whether proto-oncogenes are activated to oncogenes and whether tumor suppressor are mutationally inactivated or deleted by this process.

To date, there is little or no evidence that specific arsenic compounds can generate oxygen or arsenic radicals. Dimethylperoxyl radical has been measured by Okada (1997), but usually at high concentratons of arsenic. Further work is required in this area to substantiate a hypothesis that arsenic radicals or other radicals are produced at physiological conditions of arsenic exposure and that arsenic-induced radical species do indeed play some role in arsenic carcinogenesis. Indeed, the mechanism of arsenic carcinogenesis by arsenic compounds is so obscure at present, that the overall mechanisms of arsenic carcinogenesis, or indeed arsenic co-carcinogenesis would be better investigated, and the hypothesis of arsenic-induced radical generation tested later after some mechanistic insight is gained into arsenic carcinogenicity/co-carcinogenicity.

Finally, it is clear that a huge flux of oxygen is processed by cells every day. Some of this is metabolized to superoxide, which together with superoxide dismutase, will generate hydrogen peroxide. A small amount of hydrogen peroxide and superoxide will, in the presence of cellular iron, generate a small amount of hydroxyl radicals. These radicals in the vicinity of DNA may react with DNA, generating 8-hydroxyguanosine and other oxidized bases. Quantitation of the frequency of these oxidized bases per DNA base and critical experiments designed to determine whether these levels of oxidized bases are mutagenic in cells and can play a significant role in endogenous carcinogenesis remain to be done to substantiate the hypothesis of endogenous carcinogenesis due to oxygen-derived hydroxyl radicals, in the opinion of the author.

Dedications and acknowlegements

The author wishes to thank three senior scientists who had an enormous impact on his career via their mentorship, friendship, and colleagueship, and to dedicate this chapter to them.

Firstly, the author thanks the late Dr Melvin Calvin, University Professor in the Dept of Chemistry at the University of California at Berkeley, Member of the National Academy of Sciences of the United States, and Nobel Laureate for discovering the path of carbon in photosynthesis. Professor Calvin and his collaborators over many decades in his chemical carcinogenesis group made many seminal contributions to the field of chemical carcinogenesis. Dr Calvin was my PhD thesis advisor, and taught me the basics of chemical carcinogenesis and the absolute necessity of critical thinking and rigor in science, but also the necessity of keeping an open mind for new advances and directions in science and for giving collaborators and competitors credit for their accomplishments in a fair way.

Secondly, I would like to thank the late Professor Charles Heidleberger, Member of the National Academy of Sciences of the United States, who also did postdoctoral study with Professor Melvin Calvin at the University of California at Berkeley. When I was a postdoctoral fellow in Professor Heidelberger's laboratory at the USC/Norris Comprehensive Cancer Center at the University of Southern California, he taught me the basic considerations of the use of cell transformation systems to study the molecular and cellular mechanisms of chemical carcinogenesis.

Thirdly, I would like to thank Dr Paul Hochstein, Professor Emeritus, Department of Molecular Pharmacology and Toxicology, USC School of Pharmacy. Professor Hochstein was my collaborator and still is my friend, and he graciously taught me some of the rudiments of free radical chemistry and biochemistry and attracted me into the field of free radicals by virtue of his vast knowledge and seminal contributions to this area, his willingness and delight to share that vast knowledge with others, his mentorship of myself and other junior colleagues, and his large and generous spirit, and continual optimism.

References

Ali, A.E., Barrett, J.C., and Eling, T.C., 1980, Prostaglandin and thromboxane production by fibroblasts and vascular endothelial cells, Prostaglandins, 20, 667–688.

Ames, B.N., Shigenaga, M.K., and Hagen, T.M., 1994, Oxidants, antioxidants, and the degenerative diseases of aging, In: Biological Oxidants and Antioxidants, Eds L. Packer and E. Cadenas, Hippokrates Verlag, Stuttgart, Gremany, pp. 193–212.

Balmain, A., Ramsden, M., Bowden, G.T., and Smith, J., 1984, Activation of the mouse cellular Harvey-ras gene in chemically induced benign skin papillomas, Nature, 307, 658–660.

Barrett, J.C. and Ts'o, P.O.P., 1978, Relationship between somatic mutation and neoplastic transformation, Proc. Natl Acad. Sci. USA, 75, 3297–3301.

Bartsch, H., Barbin, A., Marion, M., Nair, J., and Guichard, Y., 1994, Formation, detection, and role in carcinogenesis of ethenobases in DNA, Drug Metab. Rev., 26, 349–371.

Bartsch, H. and Nair, J., 1997, Role of miscoding etheno-DNA adducts in carcinogenesis, In: Cancer: Genesis – Detection – Therapy: 7th Charles Heidelberger Meeting, Abstract Booklet, pages 17–18, Reisensburg Castle, Gunzburg, Germany, July 2–5, 1997.

Benedict, W.F., Wheatley, W.L., and Jones, P.A., 1980, Inhibition of chemically induced morphological transformation and reversion of the transformed phenotype by ascorbic acid in C3H/10T1/2 cells, Cancer Res., 40, 2796–2801.

Biedermann, J.A. and Landolph, J.R., 1987, Induction of anchorage-independence in human diploid foreskin fibroblasts by carcinogenic metal salts, Cancer Research, 47, 3815–3823.

Bishop, M.J., 1987, The molecular genetics of cancer, Science, 235, 305–311.

Boutwell, R.K., 1964, Some bioloigcal aspects of skin carcinogenesis, Progr. Exptl. Tumor Res., 4, 207–250.

Cerepnalkoski, L. and Landolph, J.R., 1998, Inhibition of 3-methylcholanthrene-induced morphological transformaiton by inhibitors of arachidonic release and metabolism, Manuscript in Preparation.

Cerutti, P.A., 1985, Pro-oxidant states and tumor promotion, Science, 227, 375–381.

Chung, G.-L., Chen, J.-J.C., and Nath, R.G., 1996, Lipid peroxidation as a potential endogenous source for the formation of exocyclic DNA adducts, Carcinogenesis, 17, 2105–2111.

Cooper, G.M., Okenquist, S., and Silverman, L., 1980, Transforming activity of DNA of chemically transformed and normal cells, Nature, 284, 418–421.

Costa, M., 1996, Mechanisms of nickel genotoxicity and carcinogenicity, In: Toxicology of Metals, Ed. Louis Chang, pp. 245–252, CRC Lewis Publishers, Boca Raton, Florida.

De Leo, V.A., Hanson, D., Weinstein, I.B., and Harber, C., 1985, Ultraviolet radiation stimulates the release of arachidonic acid from mammalian cells in culture, Photochemistry and Photobiology, 41, 51–56.

Dix, T.A., Hess, K.M., Medina, M.A., Tilly S.L., and Sullivan, R.W., 1994, Mechanistic studies of DNA oxidation and nicking by perhydroxyl, peroxyl, and hydroxyl radicals, biological oxidants and antioxidants, Eds L. Packer and E. Cadenas, Hippokrates Verlag, Stuttgart, Germany, pp. 13–23.

El-Ghissassi, F., Barbin, A., Nair, J., and Bartsch, H., 1995, Formation of 1,N6-ethenoadenine and 3,N4-ethenocytosine by lipid peroxidation products and nucleic acid bases, Chemical Res. in Toxicology, 8, 273–283.

Eva, A. and Aaronson, S., 1983, Frequent activation of c-kis as a transforming gene in fibrosarcomas induced by methylcholanthrene, Science, 220, 955–956.

Fan, A.M., 1996, Assessment of metals in drinking wter with specific references to lead, copper, arsenic, and selenium, In: Toxicology of Metals, Ed. Louis Chang, pp. 39–54, CRC Lewis Publishers, Boca Raton, Florida.

Ferreccio, C., Gonzalez, C., Milosavjlevic, V., Marshall, G., and Sancha, A.M., 1997, Exposure to arsenic in air and drinking water: results of two epidemiology studies in Chile, Abstract #S-14, In: Arsenic: health effects, mechanisms of actions, and research issues, Meeting Abstract Booklet, Meeting sponsored by the U. S. N. C. I., N. I. E. H. S., and the U. S. E. P. A., Sept. 22–24, 1997, Marriott's Hunt Valley Inn, Hunt Valley, Maryland.

Finlay, C.A., Hinds, P.W., and Levine, A.J., 1989, The p53 proto-oncogene can act as a suppressor of transformation, Cell, 57, 1083–1083.

Frenkel, K., 1992, Carcinogen-mediated oxidant formation and oxidative DNA damage, Pharmac. Ther., 53, 127–166.

Frenkel, K. and Gleichauf, C., 1991, Hydrogen peroxide formation by cells treated with a tumor promoter, Free Rad., Res. Comm., 12–13, 783–784.

Frenkel, K., Donahue, J.M., and Banerjee, S., 1988, Benzo(a)pyrene-induced oxidative DNA damage: a possible mechanism for promotion by complete carcinogens, Oxy-Radicals in Molecular Biology and Pathology, 509–524.

Frenkel, K., Zhong, Z., Wei, H., Karkoszka, J., Patel, U., Georgescu, M., and Solomon, J., 1991, Quantitative high-performance liquid chromatography analysis of DNA oxidized in vitro and in vivo, Anal. Biochem., 196, 126–136.

Funk, C.D., Funk, L.B., Kennedy, M.E., Pong, A.S., and Fitzgerald, G.A., 1991, Human platelet/erythroleukemia cell prostaglandin G/H synthase: cDNA cloning, expression, and gene chromsomal assignment, FASEB J., 5, 2304–2312.

Gasiorek, K. and Bauchinger, M., 1981, Chromosomal changes in human lymphocytes after separate and combined treatment with divalent salts of lead, cadmium, and zinc, Environ. Mutagenesis, 3, 513–518.

Hall, F.J. and Borek, C., 1983, SOD protects against oncogenic transformation, In: Radioprotectors and Anticarcinogens, Nygaard, O.F. and Simic, M.G., eds, Academic Press, pp. 515–525.

Halliwell, B., 1987, Oxidants and human disease: some new concepts, FASEV J., 1, 358–364.

Hang, B., Chenna, A., Rao, S., and Singer, B., 1996, 1,N6-ethenoadenine and 3,N4 ethenocytosine are excised by separate human DNA glycosylases, Carcinogenesis, 17, 155–157.

Hassid, A. and Levine, L., 1977, Induction of fatty acid cyclo-oxygenase activity in canine kidney cells (LDCK) by benzo(a)pyrene, J. Biol. Chem., 252, 6591–6593.

Higgs, G.A., Salmon, J.A., Henderson, B., and Vane, J.R., 1987, Pharmacokinetics of aspirin and salicylate in relation to inhibition of arachidonate cyclo-oxygenase and anti-inflammatory activity, Proc. Natl Acad. Sci. USA, 84, 1417–1420.

Hotta, N., 1996, Discrepancies in occurrence of malignant skin changes among different reaces exposed to inorganic arsenic, In: Arsenic: health effects, mechanisms of actions, and research issues, Abstract #S16, Meeting sponsored by the U. S. N. C. I., the U. S. N. I. E. H. S., and the U. S. E. P. A., Sept. 22–24, 1997, Marriott's Hunt Valley Inn, Hunt Valley, Maryland.

Huberman, E., Mager, R., and Sachs, L., 1976, Mutagenesis and transformation of normal cells by chemical carcinogenesis, Nature (London), 264, 360–361.

Kasprzak, K., 1996, Oxidative DNA damage in metal-induced carcinogenesis, In: Toxicology of Metals, Ed. Louis Chang, pp. 299–320, CRC Lewis Publishers, Boca Raton, Florida.

Kensler, T.W. and Trush, M.A., 1984, role of oxygen radicals in tumor promotion, Environmental Mutagenesis, 6, 593–616.

Klein, G., The approaching era of the tumor suppressor genes, 1987, Science, 238, 1539–1545.

Klein, C.B., 1996, Carcinogenicity and genotoxicity of chromium, In: Toxicology of Metals, Ed. Louis Chang, CRC Lewis Publishers, Boca Raton, Florida, pp. 205–220.

Kunjubu, D.A., Fletcher, B.S., Varnum, B.C., Lim, R.W., and Herschman, H.R., 1991, TS10, a phorbol ester tumor promoter-inducible mRNA from Swiss 3T3 cells, encodes a novel prostaglandin synthase/cyclooxygenase homologue, J. Biol. Chem., 266, 12866–12872.

Land, H., Parada, L.F., and Weinberg, R.A., 1983, Cellular oncogenes and multistep carcinogenesis, Science, 222, 771–778.

Landolph, J.R., 1985a, Chemical transformation of C3H/10T1/2 Cl 8 mouse embro cells: Historical background, assessment of the transformation assay, and evolution and optimization of the transformation assay protocol. In IARC Scientific Publication No. 67, pp. 185–198, Eds Kakunaga, T. and Yamasaki, H.

Landolph, J.R., 1985b, Mechanisms of chemicall induced multistep neoplastic transformaiton in C3H/10T1/2 cells, In: Role of Chemicals and Radiation in Etiology of Neoplasia. Ed. E. Huberman, Argonne National Laboratory Symposium, Raven Press, pp. 195–210.

Landolph, J.R., 1989, Molecular and cellular mechanisms of transformation of C3H/10T1/2 Cl 8 and diploid human fibroblasts by unique, carcinogenic, nonmutagenic metal compounds. A review, Biological Trace Element Res., 21, 459–467.

Landolph, J.R., 1990a, Neoplastic transformation of mammalian cells by carcinogenic metal compounds: cellular and molecular mechanisms, In: CRC Critical Reviews in Toxicology: Biological Effects of Heavy Metals, Volume 2, Chapter 5, pp. 1–18.

Landolph, J.R., 1990b, Neoplastic transformation of mammalian cells by carcinogenic metal compounds: Cellular and molecular mechanisms, In: Biological Effects of Heavy Metals, Volume II, Metal Carcinogenesis, Eds Foulkes, E.L., CRC Press, Boca Raton, Florida, pp. 1–14.

Landolph, J.R., 1994a, Molecular mechanisms of transformation of C3H/10T1/2 Cl 8 mouse embryo cells and diploid human fibroblasts by carcinogenic metal compounds, Environ. Health Persp., 102(53), 119–123.

Landolph, J.R., 1994b, Arachidonic acid metabolism, oxygen radicals, and chemical carcinogenesis, In: Biological Oxidants and Antioxidants, Eds Packer L. and Cadenas, E., pp. 133–143. From the series, New Strategies in Prevention and Therapy, Eds Karlkeinz Schmidt, Hippokrates and Verlag, Stuttgart, Germany.

Landolph, J.R. and Cerapnalkoski, L., 1998, Inhibition of benzo(a)pyrene-indiced morphological transformation by inhibitors of arachidonic acid release and metabolism, Manuscript in Preparation.

Landolph, J.R. and Fournier, R.E.K., 1983, Microcell-mediated transfer of carcinogen-induced ouabain resistance from C3H/10T1/2 mouse fibroblasts to human cells, Mutat. Res., 107, 447–463.

Landolph, J.R. and Heidelberger, C., 1979, Chemical carcinogens produce mutations to ouabain resistance in transformable C3H/10T1/2 Cl 8 mouse embryo cells, Proc. Natl Acad. Sci. UAS, 76, 930–934.

Landolph, J.R. and Jones, P.A., 1982, Mutagenicity of 5-azacytidine and related nucleosides in C3H/10T1/2 and V79 Cells, Cancer Res., 41, 1855–1875.

Landolph, J.R., Telfer, N., and Heidelberger, C., 1980a, Further evidence that ouabain-resistant variants induced by chemical carcinogens in transformable C3H/10T1/2 Cl 8 mouse fibroblasts are mutants, Mutation Res., 72, 295–310.

Landolph, J.R., Bhatt, R.S., Telfer, N., and Heidelberger, C., 1980b, Adriamycin and ouabain induced cytotoxicity and inhibition of 86-ribidium transport in wild-type and ouabain-resistant C3H/10T1/2 mouse fibroblasts, Cancer Res., 40, 4581–4588.

Landolph, J.R., Dews, M., Ozbun, L., and Evans, D.P., 1996, Metal-induced gene expression and neoplastic transformation, In: Toxicology of Metals, Eds Chang, L., CRC Lewis Publishers, Boca Raton, Florida, pp. 321–329.

Larramendy, M.L., Popescu, N.C., and DiPaolo, J.A., 1981, Induction by inorganic metal salts of sister chromatid exchanges and chromosome aberrations in human and Syrian hamstaer cell strains, 1981, Environ. Mutagenesis, 3, 597–606.

Lesko, S.A., Caspary, W.J., Lorentzen, R.J., and Ts'o, P.O.P., 1975, Enzymic formation of 6-oxobenzo(a)pyrene radical in rat liver homogenates from carcinogenic benzo(a)pyrene, Biochemistry, 14, 3978–3984.

Lesko, S.A. and Lorentzen, R.J., 1985, Benao(a)pyrene dione-benzo(a)pyrene diol oxidation–reduction couples; involvement in DNA damage, cellular toxicity, and carcinogenesis, J. Toxicology and Environmental Health, 16, 679–691.

Levine, L., 1977, Chemical carcinogens stimulate canine kidney (MDCK) cells to produce prostaglandins, Nature, 268, 447–448.

Little, J.B., Kennedy, A.R., and Nagasawa, H., 1983, Involvement of free radical intermediates in oncogenic transformation and tumor promotion in vitro, In: Radioprotectors and Anticar-cinogens, Nygard, O.F. and Simic, M.G., Eds, Academic Press, New York, pp. 4870–4893.

Lorentzen, R.J. and Ts'o, P.O.P., 1977, Benzo(a)pyrenedione/benzo(a)pyrenediol oxidation–reduction couples and the generation of reactive reduced oxygen, Biochemistry, 16, 1467–1473.

Lorentzen, R.J., Lesko, S.A., McDonald, K., and Ts'o, P.O.P., 1979, Toxicity of metabolic benzo(a)pyrenediones to cultured cells and the dependence upon molecular oxygen, Cancer Res., 39, 3194–3198.

Marnett, L.J., 1992, Aspirin and the potential role of prostaglandins in colon cancer, Cancer Res., 52, 5575–5589.

McCann, J., Choi, E., Yamasaki, E., and Ames, B.N., 1975, Detection of carcinogens as mutagens in the Salmonella/microsome test: Assay of 300 chemicals, Proc. Natl Acad. Sci. USA, 72, 5135–5139.

Merriman, R.L. and Bertram, J.S., 1982, Reversible inhibiton by retinoids of 3-methylcholanthrene-induced neoplastic transformation in C3H/10T1/2 cells, Cancer Res., 39, 1621–1666.

Metzger, U., Meier, J., Uhlschmid, G., and Weihe, H., 1984, Influence of various prostaglandin synthesis inhibitors on DMG-induced rat colon cancer, Disease of the Colon and Rectum, 27, 366–369.

Miller, J.A. and Miller, E.C., 1977, Origins of human cancer., In: Mechanisms of Carcinogenesis, Book B. Eds Hiatt, H.H., *et al.*, Cold Spring Harbor Laboratory, pp. 605–627.

Miura, T., Patierno, S.R., Sakuramoto, T., and Landolph, J.R., 1989, Morphological and neoplastic transformation of C3H/10T1/2 C1 8 mouse embryo cells by insoluble carcinogenic nickel compounds, Environmental and Molecular Muagenesis, 14, 65–78.

Mondal, S., Brankow, D.M., and Heidelberger, C., 1976, Two-stage carcinogenesis in cultures of C3H/10T1/2 cells, Cancer Res., 36, 2254–2260.

Nair, J., Barbin, A., Guichard, Y., and Bartsch, H., 1995, 1,N6-Ethenodeoxyadenosine and 3,N4-ethenodeoxycytidine in liver DNA from humans and untreated rodents detected by immunoaffinity/32P-postlabelling, Carcinogenesis, 16, 613–617.

Nair, J., Carmichael, P.L., Fernando, R.C., Phillips, D.H., and Bartsch, H., 1996a, Elevated 1,N6-ethenodeoxyadenosine (EdA) and 3,N4-ethenodeoxycytidine (EdC) DNA adducts in liver of patients with Wilson's disease and primary hemochromatosis, Proc. American Association of Cancer Reseach, Abstract #813.

Nair, J., Sone, H., Hagao, M., Barbin, A., and Bartsch, H., 1996b, Copper-dependent formation of miscoding etheno-DNA adducts in the liver of Long Evans Cinnamon (LEC) rats developing hereditary hepatitis and hepatocellular carcinoma, Cancer Res., 56, 1267–1271.

Nair, J., Gal, A., Tamir, S., Tannenbaum, Wogan, T.G., and Bartsch, H., 1997, Enhanced formaton of etheno DNA adducts in spleen of RcsW lymphoma cells injected into SJL mice, producing high levels of nitric oxide (NO), Proc. American Assocation for Cancer Research, Abstract #522, 1997.

Nair, J., Vaca, C.E., Velic, I., Mutanen, M., Valsta, L.M., and Bartsch, H., 1997, High dietary ω-6 polyunsaturated fatty acids drastically increase the formation of etheno-DNA base adducts in white blood cells of female subjects, Cancer Epidemiology Biomarkers Prev., in Press.

Nishimura, M. and Umeda, M., 1979, Induction of chromosomal aberrations in cultured mammalian cells by nickel comopunds, Mutation Res., 68, 337–349.

Ochi, T. and Cerutti, P., 1987, Clastogenic action of hydroperoxy-5,8,11,12-icosatetra-enoic acids on the mouse embryo fibroblasts C3H/10T1/2, Proc. Natl Acad. Sci. USA, 84, 990–994

Okada, S., 1997, Involvement of metabolic methylation in arsenic-induced genetic damage and tumorigenesis, In: Arsenic: health effects, mechanisms of actions, and research issues, Abstract #S31, Meeting sponsored by the U. S. E. P. A., the U. S. N. C. I, and the U. S. N. I. E. H. S., Sept. 22–24, 1997, Marriott's Hunt Valley Inn, Hunt Valley, Maryland.

Patel, U., Bhimani, R., and Frenkel, K., 1992, Mechanism of mutagenicity by 5-hydroperoxymethyl–2′-deoxyuridine, an intermediate product of ionizing radiation, in bacteria, Mutat. Res., 283, 145–146.

Pitot, H.C. and Dragan, Y.D., 1994, Facts and theories concerning the mechanisms of carcinonogenesis, FASEB J., 5, 2280–2286.

Poirier, L.A. and Littlefield, N.A., 1996, Metal interactions in chemical carcinogenesis, In: Toxicology of Metals, Ed. Louis Chang, pp. 289–298, CRC Lewis Publishers, Boca Raton, Florida.

Reznikoff, C.A., Brankow, D.W., and Heidelberger, C., 1973a, Establishment and characterization of a cloned line of C3H mouse embryo cells sensitive to postconfluence inhibition of division, Cancer Res., 33, 3231–3238.

Reznikoff, C.A., Bertram, J.S., Brankow, D.W., and Heidelberger, C., 1973b, Qualitative and quantitative studies of chemical transformation of cloned C3H mouse embryo cells sensitive to postconfluence inhibition of cell division, Cancer Res., 33, 3239–3249.

Robinson, T.W., Sevanian, A., and Forman, H.J., 1990, Inhibition of arachidonic acid release by nordihydroguaiaretic acid and its anti-oxidant action in rat alveolar macrophages and Chinese hamster lung fibroblasts, Toxicology and Applied Pharmacology, 105, 113–122.

Sager, R., Tumor suppressor genes: the puzzle and the promise, 1989, Science, 246, 1406–1412.

Samuelsson, B., Dahlen, S.-E., Lindgren, J.A., Rouzer, C.A., and Serhan, C.N., 1987, Leuk-otrienes and lipoxins: structures, biosynthesis and biological effects, Science, 237, 1171–1176.

Saxholm, H.J.K., Reith, A., and Brogger, A., 1981, Oncogenic transformation and cell lysis in C3H/10T1/2 cells and increased sister chromatid exchange in human lymphocytes by nickel subsulfide, Cancer Research, 41, 4136–4139.

Shibuya, M.L., Miura, T., Lillehaug, J.R., Farley, R.A., and Landolph, J.R., 1989, Ouabain-resistant (Na+,K+)-ATPase enzyme activity in chemically induced ouabain-resistant C3H/10T1/2 cells, J. Molec. Toxicol., 2, 75–98

Shih, C. and Weinberg, R., 1982, Isolation of a transforming sequence from a human bladder carcinoma cell line, Cell, 29, 161–169.

Shih, C., Shilo, B.-Z., Goldfarb, M.B., Dannenberg, A., and Weinberg, R.A., 1979, Passage of phenotypes of chemically transformed cells via transfection of DNA and chromatin, Proc. Natl Acad. Sci USA, 76, 5714–5718.

Slaga, T.J., Solanki, V., and Logani, M., 1983, Studies on the mechanism of action of anti-tumor promoting agents: suggestive evidence for the involvement of free radicals in promotion. In Radioprotectors and Anti-carcinogens, Nygaard, O.F., and Simic M.G., Eds, Academic Press, New York, pp. 471–485.

Smith, A.H., Hopenhayn-Rich, C., Betes, M.N., Goeden, H.M., Hertz-Picciotto, I., Duggan, H.M., Wood, R., Kosnett, M.J., and Smith, M.T., 1992, Cancer risks from arsenic in drinking water, Environ. Health Persp., 97, 259–267.

Stanbridge, E.J., 1990, Identifying tumor suppressor genes inhuman colorectal cancer, Science, 247, 12–13.

Sugmiura, T., 1992, Multi-step carcinogenesis: A 1992 Perspective, Science, 258, 603–607.

Sukumar, K., Notario, V., Martin-Zanca, D., and Barbacid, M., 1983, Induction of mammary carcinomas in rats by nitroso-methylurea involves malignant activation of Ha-ras-1 locus by single point mutations, Nature, 306, 658–661.

Thun, M.J., Namboodiri, G.S., and Heath, C.W., 1991, Aspirin use and reduced risk of fatal colon cancer, New England J. of Med., 325, 1593–1596.

Tofigh, S. and Frenkel, K., 1989, Effect of metals on nucleoside hydroperoxide, a product of ionizing radiation in DNA, Free Radical Biology and Medicine, 7, 131–143.

Tragni, E., Corsini, A., Fumagalli, R., Flaminio, L., and Galli, C.L., 1986, Lack of correlation between TPA-induced prostaglandin biosynthesis and ornithine decarboxylase activity in Balb/c mouse 3T3 fibroblasts, Biochem. and Biophys. Res. Comm., 139, 186–185.

Troll, W.W. and Weisner, R., 1985, The role of oxygen radicals as a possible mechanism of tumor promotion, Ann. Rev. Pharmacol. Toxicol., 25, 509–528.

Vane, J. and Botting, R., 1987, Inflammation and the mechanism of action of anti-inflammatory drugs, FASEB Journal, 1, 89–96.

Wang, Z. and Rossman, T., 1996, The carcinogenicity of arsenic, In: Toxicology of Metals, Ed. Louis Chang, pp. 221–229, CRC Lewis Publishers, Boca Raton, Florida.

Wei, H. and Frenkel, K., 1991, In vivo formation of oxidized DNA bases in tumor-promoter-treated mouse skin, Cancer Res., 51, 4443–4449.

Wei, H. and Frenkel, K., 1992, Suppression of tumor promoter-induced oxidative events and DNA damage in vivo by sarcophytol: a possible mechanism of antipromotion, Cancer Res., 52, 2298–2303.

Weinberg, R.A., 1985, The action of oncogenes in the cytoplasm and nucleus, Science, 230, 770–776.

Weissman, B.E., Saxom, P.J., Pasquale, S.R., Jones, G.R., Geiser, A.G., and Stranbridge, E.J., 1987, Introduction of a normal human chromosome 11 into a Wilm's tumor cell lines controls its tumorigenic expression, Science, 236, 175–180.

Weissman, G., 1991, Aspirin, Scientific American, 264, 84–90.

Zimmerman, R. and Cerutti, P.A., 1984, Active oxygen acts as a promoter of transformation in mouse embryo C3H/10T1/2 Cl 8 fibroblasts, Proc. Natl Acad. Sci. USA, 81, 2085–2087.

Zipser, R.D. and Laffi, G., 1985, Prostaglandins, thromboxanes, and leukotrienes in clinical medicine, Western J. Med., 143, 485–497.

18 Free radical involvement in cardiovascular and respiratory diseases

Frank J. Kelly

Abstract

Considerable interest has developed in the idea that oxidative stress is instrumental in the aetiology of numerous human diseases. Overwhelming evidence of the involvement of low-density lipoprotein (LDL) oxidation in the development of atherosclerosis and the generation of radicals in reperfusion injury has firmly placed cardiovascular disorders at the forefront in this respect. Recent findings of oxidant/antioxidant imbalances in a number of respiratory disorders such as cystic fibrosis, adult respiratory distress syndrome (ARDS) and asthma has ensured that the lung is also receiving increased attention from those involved in free radical research. Definitive proof of free radical formation and oxidative cell damage being causative, rather than a result of other underlying pathologies remains elusive. Therefore in this chapter the role of free radicals in the development and importantly, the progression, of a range of diseases of the heart and lungs will be discussed. As oxidative stress develops from an imbalance between free radical production and available antioxidant defences, the protective role of antioxidants will also be highlighted. Increased consumption of fruit and vegetables, which are good sources of antioxidants, is clearly associated with a lower coronary risk. In this respect, there is convincing evidence of a reduced coronary risk in populations with high blood levels of the antioxidant nutrients, vitamins C and E. Similar data are now emerging in respect of respiratory diseases and hence, here too, the evidence is becoming sufficiently compelling to suggest that antioxidants are potential therapeutic agents for these conditions.

18.1 Heart disease

Cardiovascular disease is the leading cause of morbidity and mortality in industrialised countries, moreover its prevalence is increasing in developing countries. A number of risk factors have been associated with coronary heart disease (CHD). These include smoking, diabetes, high blood pressure and specific dietary components (Ulbricht and Southgate, 1991). In particular, high blood levels of low density lipoprotein (LDL) cholesterol (Esterbauer *et al.*, 1989; Steinberg and Witztum, 1990), and low levels of antioxidants (Ulbricht and Southgate, 1991; Gey *et al.*, 1991) are important risk factors.

18.1.1 LDL modification and atherogenesis

LDL consists of a central core of cholesterol esters arranged in a lamellar fashion. Around this central core cholesterol, phospholipids and apoproteins are distributed. Apoproteins play a central role in tissue uptake of LDL. When oxidative stress results in the oxidation of polyunsaturated fatty acids (PUFAs) within the phospholipids and cholesterol esters of LDL, a series of events begin which lead to atherosclerosis. The fatty streak is the earliest macroscopically evident lesion of atherosclerosis (Ross, 1986). It contains macrophages, oxidised lipid and macrophage-derived foam cells. Oxidised LDL may promote development of the plaque through a number of potentially atherogenic properties. It is a chemoattractant for leukocytes and stimulates cytokine production by macrophages. The lipid peroxides which accumulate within artherosclerotic plaques stimulate smooth muscle cell proliferation and may also injure overlying endothelial cells.

LDL contains vitamin E, which helps prevent oxidation of the PUFAs. Vitamin E (in particular α-tocopherol) acts as a chain-breaking antioxidant by donating a hydrogen atom to a lipid peroxyl radical. Vitamin E is unique among antioxidants in this respect. As a consequence, it plays an important role in the prevention of CHD. Following donation of the hydrogen atom it is thought that the resultant α-tocopheroxyl radical may interact with vitamin C (ascorbic acid) or possibly with reduced glutathione (GSH). As the vitamin E content of PUFA is directly related to the dietary intake of vitamin E, individuals with low dietary vitamin E have decreased LDL vitamin E and hence have an increased susceptibility to atherosclerosis.

The Nurses Health Study of 87,245 middle-aged women showed a progressively reduced incidence of CHD with higher intake of vitamin E (Stampfer *et al.*, 1993). Likewise, the Physicians Health Professionals Study of 40,000 healthy male doctors, reported a risk of 0.64 of CHD in those consuming more than 60 IU of vitamin E daily as compared with those on less than 7.5 IU (Rimm *et al.*, 1993). In a controlled trial of vitamin E supplementation, the Cambridge Heart Antioxidant Study (CHAOS) investigators evaluated the effect of vitamin E (400 or 800 IU of natural-source vitamin E per day) or placebo on the risk of myocardial infarction in 2002 patients with angiographic evidence of coronary atherosclerosis (Stephens *et al.*, 1996). Vitamin E supplementation, at both levels, was shown to decrease the risk of cardiovascular death and non-fatal myocardial infarction by 47%. Overall there was a 77% decrease in the risk of non-fatal myocardial infarction, with the effect apparent after the first year of treatment. The findings of CHAOS represent a watershed in antioxidant research, as they are the first from a prospective clinical trial, which supports the concept of lipid oxidation playing a pivotal role in coronary artery disease. As no benefits in terms of cardiovascular death or total mortality were however observed in CHAOS, further longer term studies are required to examine such effects.

18.1.2 Myocardial ischaemia/reperfusion injury

Ischaemic/reperfusion injury of the heart, which often arises as a consequence of atherosclerosis, is also associated with increased oxidative stress. Atherosclerotic coronary arteries can stop blood flow to the myocardium. If the heart, and hence the patient, is to survive, blood flow must be quickly restored. Paradoxically, reperfusion, the return of blood flow to the heart following a period of ischaemia, itself carries severe consequences (Hearse and Yellon, 1984; Braunwald and Kloner, 1985). These include disturbances of electrical rhythm and life-threatening ventricular arrhythmias. The extent of this so-called *oxygen paradox*, the development of further injury on reperfusion, is proportional not only to the duration and severity of the proceeding hypoxia but also to the pO_2 of the reperfusing blood (Hearse *et al.*, 1978). Thus, it is the readmission of oxygen *per se*, rather than the redistribution of flow, that is responsible for reperfusion-induced injury. This pivotal finding led investigators to examine in detail the rate of oxygen-derived free radical production during early reperfusion. Definitive evidence that oxygen free radicals are generated at this time was obtained using electron spin resonance (ESR) techniques (Garlick *et al.*, 1987; Zweier *et al.*, 1987).

Given that oxygen free radicals are involved in the injury process, many investigators have proposed that antioxidant therapy may effectively reduce reperfusion injury. An array of experimental and clinical studies with antioxidants such as superoxide dismutase, catalase, mercaptoproponylglycine and *N*-acetylcysteine have been undertaken. The publication of positive and negative findings (Opie, 1989; Engler and Gilpin, 1989; Qui *et al.*, 1990; Galinanes *et al.*, 1992) however highlight the difficulties that are often encountered in the investigation of oxidative stress.

18.2 Lung diseases

The adult lung has a surface area of approximately 70 m^2, which maximises transfer of oxygen from the 10,000 to 15,000 litres of air entering the airways each day. The oxygen tension in the airways and alveolar spaces is approximately 3 times higher than in most other tissues. This favours oxidative reactions and hence by virtue of its position and function, the lung epithelium is vulnerable to oxidant damage (Cantin and Bebin, 1991). Inhalation of air itself often exposes the lung to oxidative stress. Many air pollutants, such as ozone, nitrogen dioxide and cigarette smoke are free radicals or have powerful oxidant activity.

The lung is also particularly susceptible to a number of ingested toxins. Bleomycin, an important anti-tumour agent, induces lung fibrosis and appears to be related to free radical formation. The lungs appear to be particularly susceptible as they have no, or little, bleomycin hydrolase activity (Dorr, 1992). The herbicide paraquat is known to be toxic to the lung. Type II cells are particularly sensitive because they selectively take up paraquat and subsequently recycle it via a redox reaction.

The role of free radicals in a variety of lung diseases, including adult respiratory distress syndrome, ARDS (Brigham, 1990), and asthma (Owen *et al.*, 1991), has recently attracted much attention. In all these cases it has been suggested that oxygen free radicals, produced in a variety of ways, are implicated in the damage to the pulmonary epithelium (Table 18.1). Evidence is now accumulating that reactive oxygen species, in conjunction with abnormalities in protective antioxidants, may also be important mediators of tissue damage in patients with cystic fibrosis (Brown and Kelly, 1994; Winklhofer-Roob, 1994). A feature common to all these respiratory diseases is the influx of activated inflammatory cells, such as neutrophils, to the lung.

The generation of oxygen free radicals by activated inflammatory cells may be involved in the pathogenesis of these conditions. Neutrophils, eosinophils and macrophages posses a membrane-bound flavoprotein, cytochrome b245 NADPH oxidase, that is induced during cell activation. Using molecular oxygen, it produces

Table 18.1 Free radical involvement in respiratory diseases

Free radicals	*Respiratory disease*
Excessive production of superoxide, H_2O_2 and HOCl by activated inflammatory cells	Asthma Emphysema ARDS Bronchopulmonary dysplasia Cystic fibrosis
Increased formation of reactive oxygen species by drugs and toxins	Paraquat toxicity
Abnormal oxidative substrate or changes in oxygen concentration	Hyperoxia Hypoxia
Inadequate antioxidant defences	Cystic fibrosis Asthma ARDS

superoxide anions which, if removed by superoxide dismutase, result in H_2O_2 generation. Nathan and Root (1977) estimated that activated macrophages produce H_2O_2 at a rate of $2–5 \times 10^{14}$ mol/hr/cell. Owing to the relatively low reactivity of H_2O_2 it can easily pass across cell membranes, where it may activate intracellular signalling pathways, or lead to the generation of other reactive oxygen species. For example, in the presence of transition metals H_2O_2 leads to the production of the more toxic, hydroxyl radical.

18.2.1 Cystic fibrosis

Cystic fibrosis (CF) is a genetic condition that mainly affects the lungs and gastrointestinal tract. Great advances have been made in understanding the underlying genetic defect in CF. The cause of the severe lung damage that arises, and the subsequent pulmonary fibrosis that develops, however, remains unclear. Irregularities in ion transport leading to inspissation of mucous secretions, along with increased adherence of bacteria to epithelia and reduced muco-ciliary clearance, are all thought to contribute to the recurrent, progressive, pulmonary infections characteristic of the disease. As a consequence, CF patients are subjected to regular bouts of oxidative stress due to recurring inflammation and an impaired ability to absorb certain nutrients, including lipid-soluble antioxidants (Brown and Kelly, 1994). As mentioned above, evidence is accumulating that these reactive oxygen species may be important mediators of tissue damage in patients with CF. In recent years reports of elevated concentrations of lipid (Brown and Kelly, 1994a; Brown *et al.*, 1996), protein (Brown and Kelly, 1994a; Brown *et al.*, 1996) and DNA (Brown *et al.*, 1995) oxidation products in CF patients have appeared in the literature.

Importantly, oxidative stress is not present in all CF patients at all times. Oxidative stress, like the recurring infections, is probably cyclic. Importantly, antioxidant status tends to decrease with age in CF (Brown *et al.*, 1996), hence older CF patients are particularly susceptible to renewed cycles of pulmonary inflammation. It is tempting to speculate that it is this oxidant/antioxidant imbalance that is responsible, in part, for their decline in lung function with advancing age. The reason for the fall in antioxidant status in CF is not clear; however, decreased compliance in taking vitamin supplements may play a role. Alternatively, it is conceivable that repeated cycles of pulmonary inflammation, and associated oxidative stress, also contribute to the decline in antioxidant status. Whatever the exact cause, it is probable that the worsening antioxidant status of the CF adolescent contributes to their deteriorating clinical circumstances.

18.2.2 Adult respiratory distress syndrome

Adult respiratory distress syndrome (ARDS) is characterised by diffuse lung inflammation, hypoxaemia and bilateral pulmonary infiltrates. It can occur in response to a variety of unrelated direct or indirect insults to the lung, including bacterial or viral pneumonia, major trauma and haemolytic shock. The acute pulmonary cellular response in ARDS is a neutrophilic alveolitis (Schuster, 1995). Although the incidence of ARDS is relatively low (3.5 cases per 100,000 population per year in Britain) it remains clinically important due to its high mortality rate of greater than 50%.

There is now a considerable body of evidence suggesting that patients who develop ARDS are subjected to considerable oxidative stress. Increased lipid peroxidation, measured as diene conjugates or thiobarbituric acid-reactive substances (TBARS), have been reported in both ARDS patients (Takeda *et al.*, 1984; Richard *et al.*, 1987) and animal models of the condition (Takeda *et al.*, 1986; Seekamp *et al.*, 1988). The alveoli are the primary site of increased neutrophil activity in ARDS patients. Bronchoalveolar lavage fluid recovered from these patients contains high levels of lipid hydroperoxides. The presence of myeloperoxidase and oxidised α-anti-proteinase in bronchoalveolar lavage fluid obtained from critically-ill patients also supports this view (Cochrane *et al.*, 1983; Weiland *et al.*, 1986; Fantone *et al.*, 1987). The critically-ill patient often requires respiratory support, including supplemental oxygen and it is well established that breathing elevated concentration of oxygen will increase the oxidative stress on the lung (Kelly, 1993).

Decreased circulating levels of antioxidants, especially at the onset of ARDS, also supports the concept that free radicals play an important role in ARDS. Plasma concentration of both ascorbate and ubiquinol-10 are significantly lower in ARDS patients than in control subjects (Cross *et al.*, 1990). Appropriate treatment of ARDS is still open to debate. If the condition does involve a component of oxidative stress then it would seem that some form of antioxidant treatment would be appropriate. It has been found that supplementation of ARDS patients with up to 3 g of orally administered vitamin E frequently results in only a slight increase in plasma vitamin E concentration. This may be due to either reduced absorption or increased utilisation of the vitamin E in this patient group.

18.2.3 Asthma

Asthma is a chronic inflammatory disease of the conducting airways in which several inflammatory cell types play a role, particularly mast cells, lymphocytes and eosinophils. Susceptible individuals have symptoms such as widespread, but variable, airway narrowing, airflow obstruction that is reversible either spontaneously or with treatment, and increased airway responsiveness to a variety of stimuli (Holgate, 1993). Epithelial damage, thickening of the reticular basement membrane and infiltration of the airway wall by a mixed population of mononuclear cells and eosinophils were among the earliest histological features of asthma. Bronchoalveolar lavage and bronchial biopsy studies helped to clarify that the asthmatic airways undergo inflammatory responses involving activated T-helper cells, lymphocytes, eosinophils and mast cells (Jeffery, 1992). Moreover, it was found that altered airway function was as a result of an array of preformed and newly generated vaso-bronchoactive mediators.

Several studies have demonstrated that inflammatory and epithelial cells in asthmatic airways produce increased levels of reactive oxygen species (Postma *et al.*, 1988; Bast *et al.*, 1991; Jarjour and Calhoun, 1994; Sedgwick *et al.*, 1992). Increased superoxide anion levels have been detected *in vivo* following allergen challenge. These are released from activated macrophages and hypodense eosinophils (Sedgwick *et al.*, 1990). Resting neutrophils from asthmatics also generate greater amounts of H_2O_2 than neutrophils from control subjects. Elevated levels of H_2O_2 are present in the breath of asthmatics and appear to be associated with the severity of the condition.

Antioxidant status of asthma patients has usually been established by examining peripheral blood. Moderate to severe asthmatics have decreased plasma levels of catalase and red blood cell glutathione levels (Novak *et al.*, 1991). Conversely, Smith and colleagues (1993) found increased levels of antioxidant defences in bronchoalveolar lavage fluid. These investigators proposed that these increased antioxidant levels might be in response to increased oxidative stress in the airways of asthmatic patients. Recently, De Raeve and colleagues (1997) found that, although bronchial epithelium catalase and glutathione peroxidase activities were similar in asthmatic and control subjects, Cu/Zn superoxide dismutase was lower in asthma patients not receiving steroid therapy. Moreover, following the introduction of steroids, Cu/Zn superoxide dismutase activity increased to control levels.

If oxidative stress does play an important role in the pathogenesis of asthma, antioxidants may play a crucial role in the treatment of the condition. The integrity of the bronchial epithelium and its antioxidant status may be an influencing actor in the susceptibility of individuals to the effects of an inflammatory response following exposure to allergens and the subsequent development of asthma.

References

Bast, A.G., Haenen, R.M., Doelman, C.J. (1991). Oxidants and antioxidants: State of the art. Am J Med, 91 Suppl. 3C 2S–13S.

Bernier, M., Hearse, D.J. (1988). Reperfusion-induced arrhythmias: mechanisms of protection by glucose and mannitol. Am J Physiol 254:H862–H870.

Braunwald, E., Kloner, R.A. (1985). Myocardial reperfusion: a double-edged sword? J Clin Invest 76:1713–1719.

Brigham, K.L. (1990). Oxidant stress and adult respiratory distress syndrome. Eur Respir J 11: Suppl, 482s–4s.

Brown, R.K., Kelly, F.J. (1994). Role of free radicals in the pathogenesis of cystic fibrosis. *Thorax* 49:738–742.

Brown, R.K., Kelly, F.J. (1994a). Evidence for increased oxidative damage in patients with cystic fibrosis. Pediatr Res 36 (4):487–493.

Brown, R.B., McBurney, A., Lunec, J., Kelly, F.J. (1995). Oxidative damage to DNA in patients with cystic fibrosis. Free Rad Biol Med 18 (4):801–806.

Brown, R.K., Wyatt, H., Price, J.F., Kelly, F.J. (1996). Pulmonary dysfunction in cystic fibrosis is associated with oxidative stress. Eur Respir J 9:334–339.

Cantin, A.M., Begin, R. (1991). Glutathione and inflammatory disorders of the lung. *Lung* 169:123–138.

Cochrane, C.G., Spragg, R.G., Revak, S.D. (1983). Pathogenesis of the adult respiratory distress syndrome: evidence of oxidant activity in bronchoalveolar lavage fluid. J Clin Invest 71:754–61.

Cross, C.E., Forte, T., Stocker, R., Louie, S., Yamamoto, Y., Ames, B.N., Frei, B. (1990). Oxidative stress and abnormal cholesterol metabolism in patients with adult respiratory distress syndrome. J Lab Clin Med 115:396–404.

De Raeve, H.R., Thunnissen, F.B., Kaneko, F.T., Gui, F.H., Lewis, M., Kavuru, M.S., Secic, M., Thomassen, M.J., Erzurum, S.C. (1997). Decreased Cu,Zn-SOD activity in asthmatic airway epithelium: correction by inhaled corticosteriod in vivo. Am J Physiol 272:L148–L154.

Dorr, R.T. (1992). Bleomycin pharmacology: mechanism of action and resistance, and clinical pharmacokinetics. Semin Oncol 19 (2 Suppl 5):3–8.

Engler, R., Gilpen, E. (1989). Can superoxide dismutase alter myocardial infarct size? Circulation 79:1137–1142.

Esterbauer, H., Waeg, G., Puhl, H., Dieber-Rotheneder, M., Tatzber, F. (1992). Inhibition of LDL oxidation by antioxidants. Free Radicals and Aging:145–157.

Fantone, J.C., Feltner, D.E., Beieland, J.K., Ward, P.A. (1987). Phagocyte cell-derived inflammatory mediators and lung disease. Chest 91:428–35.

Galinanes, M., Qiu, Y., Ezrin, A., Hearse, D.J. (1992). PEG-SOD and myocardial protection: studies in the blood and crystalloid perfused rabbit hearts. *Circulation* 86:672–682.

Garlick, P.B., Davies, M.J., Hearse, D.J., Slater, T.F. (1987). Direct detection of free radicals in the reperfused rat heart using electron spin resonance spectroscopy. Circ Res 61:757–760.

Gey, K.F., Puska, P., Jordan, P., Moser, U.K. (1991). Inverse correlation between plasma vitamin E and mortality from ischemic heart disease in cross-cultural epidemiology. Am J Clin Nutr. 1991; 53 (1 Suppl):326S–334S.

Gey, K.F., Moser, U.K., Jordan, P., Stahelin, H., Eichholzer, M., Ludin, E. (1993). Increased risk of cardiovascular disease at suboptimal plasma concentrations of essential antioxidants; an epidemiological update with special attention to carotene and vitamin C. Am J Clin Nutr 57 (suppl):787s–97s.

Hearse, D., Yellon, D. (1984). Reduction in infarct size; a growing controversy. Cardiovascular Focus. 1984; 12:1–4.

Hearse, D.J., Yellon, D.M. (1988). Therapeutic approaches to myocardial infarct size reduction. Raven Press, London.

Hearse, D.J., Humphrey, S.M., Bullock, G.R. (1978). The oxygen paradox and the calcium paradox: two facets of the same problem? J Mol Cell Cardiol 10:641–668.

Holgate, S.T. (1993). Asthma: past, present and future. E Respir J, 6:1507–1520.

Jarjour, N.N., Calhoun, W.J. (1994). Enhanced production of oxygen radicals in asthma. J Lab Clin Med 123:131–137.

Jeffery, P.K. (1992). Pathology of asthma. Br Med Bull 48:23–39.

Kelly, F.J. (1993). Free radical disorders of preterm infants. Br Med Bull 3:41–56.

Nathan, C.F., Root, R.K. (1977). Hydrogen peroxide release from mouse peritoneal macrophages: dependence on sequential activation and triggering. J Exp Med 146:1648–1662.

Novak, Z., Nemeth, K., Gyurkovitis, S., Varga, I., Matkovics, B. (1991). Examination of the role of oxygen free radicals in bronchial asthma in childhood. Clin Chim Acta 201:247–252.

Opie, L.H. (1989). Reperfusion injury and its pharmacological modification. Circulation 80:1049–1062.

Owen, S., Pearson, D., Suarez-Mendez, V., O'Driscoll, R., Woodcock, A. (1991). Evidence of free radical activity in asthma. N Engl J Med 325:586–7.

Postma, D.S., Renkema, T.E., Noordhooek, J.A., Faber, H., Sluiter, H.J., Kauffman, H. (1988). Association between nonspecific bronchial reactivity and superoxide anion production by polymorphonuclear leukocytes in chronic air-flow obstruction. Am Rev Respir Dis 137:57–61.

Qiu, Y., Bernier, M., Hearse, D.J. (1990). The influence of N-acetylcysteine on cardiac function and rhythm disorders during ischemia and reperfusion. Cardioscience 1:65–74.

Richard, C., Lemonnier, F., Thibaut, M., Couturier, M., Riou, B., Auzepy, P. (1987). Lipoperoxidation and vitamin E consumption during adult respiratory distress syndrome. Am Rev Respir Dis 135:425a.

Rimm, E.B., Stampfer, M.J., Ascherio, A., Giovannucci, E., Colditz, G.A. and Willett, W.C. (1993). Vitamin E consumption and the risk of coronary heart disease in men. N Engl J Med 328:1450–1456.

Ross, R. (1986). The pathologies of atherosclerosis: an update. N Engl J Med 314:488–500.

Schuster, D.P. (1995). What is acute lung injury? What is ARDS? BMJ 107:1721–1726.

Sedgwick, J.B., Calhoun, W.J., Vrtis, R.F., Bates, M.E., McAllister, P.K., Busse, W.W. (1992). Comparison of airway and blood eosinophil function after in vivo antigen challenge. J Immunol 149:3710–3718.

Seekamp, A., LaLonde, C., Zhu, D., Demling, R.H. (1988). Catalase prevents prostanoid release and lung lipid peroxidation after endotoxemia in sheep. J Appl Physiol 65:1210–6.

Smith, L., Houston, M., Anderson, J. (1993). Increased levels of glutathione in bronchoalveolar lavage fluid from patients with asthma. Am Rev Respir Dis 147:1461–1464.

Stampfer, M.J. and Rimm, E.B. (1993) A review of the epidemiology of dietary antioixdants and risk of coronary artery disease. Can J Cardiol 9:14B–18B.

Steinberg, D., Witztum, J.L. (1990). Lipoproteins and atherogenesis. Current concepts. JAMA. 1990; 264 (23):3047–52.

Steinberg, D., Parthasarathy, S., Carew, T.E., Khoo, J.C., Witztum, J.L. (1989). Modifications of low-density lipoprotein that increase its atherogenicity. N Engl J Med 320 (14):915–924.

Stephens, N.G., Parsons, A., Schofield, P.M., Kelly, F.J., Cheeseman, K., Mitchinson, M.J., Brown, M.J. (1996) A randomised controlled trial of vitamin E in patients with coronary disease: the Cambridge heart antioxidant study (CHAOS). The Lancet 347:781–786.

Takeda, K., Shimada, Y., Amano, M. (1984). Plasma lipid peroxides and alpha-tocopherol in critically ill patients. Crit Care Med 12:957–9.

Takeda, K., Shimada, Y., Okada, Y., Amano, M., Sakai, T., Yoshiya, I. (1986). Lipid peroxidation in experimental septic rats. Crit Care Med 14:719–23.

Ulbricht, T.L., Southgate, D.A. (1991). Coronary heart disease: seven dietary factors. Lancet 338:985–992.

Weiland, J.E., Davis, W.B., Holter, J.F., Mohamme, J.R., Dorinsky, P.M., Gadek, J.E. (1986). Lung neutrophils in the adult respiratory distress syndrome. Am Rev Respir Dis 133:218–25.

Winklhofer-Roob, B.M. (1994). Oxygen free radicals in cystic fibrosis: the concept of an oxidant-antioxidant imbalance. Acta Paediatr Scand 395:49–57.

Zweier, J.L., Flaherty, J.T., Weisfeldt, M.L. (1987). Direct measurement of free radical generation following reperfusion of ischemic myocardium. Proc Natl Acad Sci USA 84 (5):1404–7.

Part V

Free radicals in the diet

19 NMR evaluation of thermally-induced peroxidation in culinary oils

Martin Grootveld, Christopher J.L. Silwood and Andrew W.D. Claxson

19.1 Introduction

A high percentage of humans are continually exposed to oxidised oils and fats in the diet, which arise from either shallow or deep fat frying processes, and the possibility that regular consumption of such materials may be deleterious to human health has evoked much interest (reviewed in Smith, 1988). The most important reaction involved in the oxidative degradation of lipids is the autoxidation of polyunsaturated fatty acids (PUFAs), a process which occurs during the heating of culinary oils and fats. PUFAs are particularly susceptible to oxidative damage by virtue of the facile abstraction of one of their *bis*-allyic methylene group hydrogen atoms on exposure to light or radical species of sufficient reactivity, a process facilitated by the low bond dissociation energy of the methylene group C—H bonds. Subsequently, one major reaction pathway for the resulting resonance-stabilised carbon-centred pentadienyl lipid radical generated in this manner involves its interaction with molecular oxygen to produce a peroxyl radical which in turn can abstract a hydrogen atom from an adjacent PUFA to form a conjugated hydroperoxydiene (CHPD) and a further pentadienyl lipid radial species. In the absence of sufficient quantities of chain-terminating, lipid-soluble antioxidants such as vitamin E (α-tocopherol, α-TOH), the process is repeated many times and represents an autocatalytic, self-perpetuating chain reaction. Conjugated diene lipid hydroperoxides are particularly unstable at standard frying temperatures (*ca.* 180°C) and are degraded to a wide variety of secondary peroxidation products which include saturated and unsaturated aldehydes, di and epoxyladehydes, lactones, furans, ketones, oxo and hydroxy acids, and saturated and unsaturated hydrocarbons. This lipid peroxidation process has also been implicated in the pathogenesis of many human diseases, e.g. atherosclerosis.

Thermal stressing of culinary oils according to routine standard frying/cooking practices (domestic or otherwise) gives rise to and/or perpetuates the radical-dependent peroxidation of PUFAs therein. A wide range of aldehydes arise from the thermally-induced decomposition of CHPDs via several processes, including the β-scission of pre-formed alkoxyl radicals. Such aldehydic fragments (*n*-alkanals, *trans*-2-alkenals, *trans,trans*- and *cis,trans*-alka-2,4-dienals, 4-hydroxy-*trans*-2-alkenals, and malondialdehyde) have the capacity to exert a variety of toxicological effects in view of their extremely high reactivity with critical biomolecules (DNA base adducts, proteins such the apoliprotein B (apo B) moiety of low-density-lipoprotein (LDL), peptides, free amino acids, endogenous thiols such as glutathione, etc.) (Addis and Park, 1989).

Interaction of these aldehydes with DNA can give rise to genotoxic events and possibly cancer.

Toxic effects which putatively arise from short-term feeding of heated and/or oxidised oils and fats to experimental animals include loss of appetite, diarrhoea, growth retardation, cardiomyopathy, hepatomegaly, haemolytic anaemia, and accumulation of peroxides in adipose tissue (highly oxidised cod liver oil) (Sanders, 1983); elevated kidney and liver weights, cellular damage in various organs and a modified fatty acid composition of tissue lipids (oils and fats subjected to the heat and oxidation associated with normal usage) (Yoshida and Kujimoto, 1989; Alexander *et al.*, 1987). Moreover, in one long-term investigation Kaunitz (1978) reported that the consumption of mildly oxidised oils by rats throughout their lifespan gave rise to an increase in the frequency of hepatic bile duct and cardiac fibrotic lesions.

Recent studies indicate that dietary-derived lipid peroxidation products may contribute significantly to the pathogenesis of atherosclerosis (Addis, 1990), a hypothesis supported by reports that such species can accelerate all three stages of the disease process, i.e. endothelial injury, accumulation of plaque and thrombosis (Yagi, 1998). The observation that concentrations of lipid peroxides in serum obtained from humans and animals with atherosclerosis are significantly higher than those in corresponding samples collected from relevant controls has provided further evidence consistent with this proposal, and consumption of peroxidised oils has been suggested to enhance the accumulation of oxidised lipids in macrophages and monocytes (Smith and Kummerow, 1987), a critical event in the disease process. Furthermore, animal studies have shown that diets containing thermally-stressed, PUFA-laden oils exhibit a greater atherogenicity than those containing unheated oils (Kritchevsky and Tepper, 1967). Indeed, Staprans *et al.* (1996) recently examined the ability of oxidised dietary lipids to accelerate the development of atherosclerosis in New Zealand White rabbits and found that feeding with an oxidised lipid-rich diet gave rise to a 100% increase in fatty streak lesions in the aorta.

Our laboratory has recently reported that typical *trans*-2-alkenal compounds generated from the thermally-induced peroxidation of PUFAs are readily absorbed from the gut into the systemic circulation, metabolised (primarily via the addition of glutathione across their electrophilic carbon–carbon double bonds), and excreted in the urine as C-3 mercapturate conjugates in rats (Grootveld *et al.*, 1998). Hence, such aldehydes have the ability to covalently derivatise lysine residues of the apo B moiety of LDL, rendering it susceptible to uptake by macrophages, a critical stage in the production of foam cells by macrophages (Witzum and Steinberg, 1991).

Additional toxicological investigations concerning heated oils have focused on their mutagenic properties. Indeed, the mutagenic potential of repeatedly used deep-frying fats has been previously assessed using the Ames test (Hageman *et al.*, 1988). Fat samples were fractionated into polar and non-polar fractions by column chromatography. Polar fractions increased the number of revertants without S-9 mix in various strains, but strain TA97 was the most sensitive. Non-polar fractions showed no mutagenicity. Mutagenic activity of polar fractions was positively correlated with thiobarbituric acid-reactive substances (TBARS) suggesting the involvement of lipid oxidation products in mutagenicity (Hageman *et al.*, 1988). Polar fractions of lipid extracted from pre-fried potatoes as well as from fried potatoes were also investigated and the frying process was found to marginally increase the number of revertants in strain TA97 without S-9 mix (Hageman and Hermans, 1990). The mutagenicity of the lipid fractions of fried potatoes was not related to the heating time of the fat. Methanol extracts of fat-free residues of fried potatoes increased numbers of revertants in strain TA97 after metabolic activation, indicating that different classes of mutagens had been isolated. Urine samples from a small number (six) of healthy, non-smoking volunteers collected during the 24 hours following consumption of portions of potatoes fried in repeatedly used fat showed no increase in mutagenicity compared with control samples. However, the precise molecular nature of mutagens formed during deep frying as well as their metabolic fate in humans remains unclear (Hageman and Hermans, 1990), and hence further investigations are required.

Despite the availability of much epidemiological and experimental evidence relating the dietary consumption of saturated or polyunsaturated fatty acids to the development and progression of human disease, the precise autoxidation status of culinary oils and fats ingested (i.e. the molecular nature and levels of CHPDs, aldehydes, etc. therein) has not hitherto been sufficiently considered. Indeed, the toxicological hazards associated with the regular ingestion of unheated and/or thermally-stressed culinary oils and fats may, at least in part, be ascribable to aldehydic products generated from the autoxidation of PUFAs.

In this communication we describe the detection and quantification of PUFA-derived peroxidation products (notably aldehydes and their conjugated hydroperoxy/hydroxydiene precursors) in culinary oils and fats by very high field proton (^1H) nuclear magnetic resonance (NMR) spectroscopy (Grootveld and Rhodes, 1994), a virtually non-invasive multicomponent analytical technique. The influence of episodes of thermal stressing (i.e. those which simulate their domestic or commercial usage) on the generation, nature and concentrations of such products is also reported in detail. We also present data concerning: (1) the abilities of selected lipid-soluble chain-breaking antioxidants to modulate the thermally-induced production of lipid peroxidation products in culinary oils; and (2) the influence of thermal stressing episodes on the stability, and oxidative consumption level, of such dietary antioxidants. The dietary, physiological, biochemical and toxicological significance of the results obtained are discussed.

19.2 Materials and methods

19.2.1 Thermal stressing of commercially-available culinary oils and fats

Sunflower seed oil [containing 64% (w/w) polyunsaturates, 22% (w/w) monounsaturates and 14% (w/w) saturates], corn (maize) oil [57% (w/w) polyunsaturates, 30% (w/w) monounsaturates and 13% (w/w) saturates], groundnut (peanut) oil [34% (w/w) polyunsaturates, 45% (w/w) monounsaturates and 21% (w/w) saturates] and olive oil [12% (w/w) polyunsaturates, 72% (w/w) monounsaturates and 12% (w/w) saturates] were commercially-available samples obtained from local retail outlets. Coconut oil [containing 94% (w/w) saturates and an unspecified content of mono- and polyunsaturates] was purchased from Sigma Chemical Co. (Poole, Dorset, UK).

Samples of control (unheated) and repeatedly-utilised culinary frying oils were kindly supplied by a fast-food establishment. Electronic integration of the *bis*-allylic-CH_2, *w*-CH_3 and highly unsaturated fatty acid acyl chain terminal-CH_3 group proton resonances ($\delta = 2.76$, 0.90 and 0.95 ppm, respectively) in 400 MHz ^1H NMR spectra of the above control (unheated) sample revealed that the polyunsaturate content of this material was 38 molar %, of which approximately one-quarter comprised highly unsaturated fatty acids, i.e. those with ≥ 3 unconjugated double bonds (predominantly linolenoylglycerol species).

Pentanal, hexanal, *trans*-2-pentenal, *trans*-2-hexenal, *trans*-2-heptenal, *trans*-2-octenal, *trans*-2-nonenal, *trans,trans*-nona-2,4-dienal and *trans,trans*-deca-2,4-dienal) were purchased from Aldrich Chemical Co. (Gillingham, Dorset, UK). 2-Thiobarbituric acid, tetramethylsilane (TMS), α-tocopherol (α-TOH) and 2,[6]-di-*tert*-butyl-*p*-cresol [butylated hydroxytoluene (BHT)] were obtained from Sigma, and deuterated chloroform (C^2HCl_3) was purchased from Gross Scientific Ltd (Great Baddow, Essex, UK).

Samples of each of the above culinary oils (18.48, 20.00 or 25.00 g) and fats (10.00 g) were placed in glass vessels (details of which are given below) and heated at a temperature of 180°C on an electronically controlled hot-plate in the presence of atmospheric O_2 for periods of up to 90 min. Aliquots (1.0 ml) of the oils and fats were removed at time-points of 30, 60 and 90 min, and then cooled to ambient temperature prior to storage (in the manner described below) and ^{1}H NMR measurements. Selected culinary oils were subjected to a further 30 min heating episode at an elevated temperature (250°C). The temperature of these samples was continuously maintained at 180 ± 3°C (or 250 ± 5°C, where appropriate) throughout the heating process using a calibrated thermometer.

Since the nature and size (capacity) of the glassware employed to heat the above culinary oils was found to exert a major influence on the concentrations of thermally-induced autoxidation products generated (a consequence of differing effective surface areas of the culinary oils or fats, i.e. the amount of these materials exposed to atmospheric O_2 during periods of thermal stressing), the same size and type of vessel was utilised to heat the same class of culinary oil for the purpose of comparative, quantitative ^{1}H NMR investigations. The glass vessels employed for each series of heating episodes consisted of 25, 50, 100 and 250 ml volume beakers (corn oil: sample surface areas 7.75, 12.20, 18.41 and 34.64 cm^2 respectively); 100 ml volume beakers (sunflower seed oil, olive oil, coconut oil, lard, and beef and lamb fat: sample surface area 18.41 cm^2); and 100 ml volume conical flasks (groundnut oil: sample surface area 24.64 cm^2).

Samples were stored in 2 ml capacity glass sample tubes in the dark at ambient temperature for durations of 12 hr–334 days. As expected, the storage time under these conditions was also found to influence the concentrations of ^{1}H NMR-detectable, PUFA-derived peroxidation products in culinary oils and fats, and therefore each class of samples examined were stored for exactly the same time periods prior to ^{1}H NMR analysis. For each group of samples, the duration of storage was recorded and is specified in the Figure legends (section 19.3).

19.2.2 Proton NMR measurements

^{1}H NMR measurements on the above samples were conducted on Bruker AMX-400 [University of London Intercollegiate Research Sciences (ULIRS), King's College facility, University of London, UK] or Bruker AMX-600 [ULIRS, Queen Mary and Westfield College facility, University of London, UK] spectrometers operating at frequencies of 400.13 and 600.13MHz respectively and a probe temperature of 298 K. Typically, a 0.20 ml aliquot of each culinary oil was diluted to a volume of 0.60 ml with deuterated chloroform (C^2HCl_3) which provided a field frequency lock, the samples treated with 10 μl of a 5.00×10^{-3} mol dm^{-3} solution of 1,3,5-trichlorobenzene in C^2HCl_3 (internal quantitative ^1H NMR standard: $\delta = 7.227$ ppm), rotamixed and then placed in 5-mm diameter NMR tubes. For culinary fats and coconut oil, accurately weighed samples (*ca.* 60 mg) were directly dissolved in 0.60 ml of C^2HCl_3, the solutions thoroughly rotamixed and treated with 10 μl of the above 1,3,5-trichlorobenzene solution, and then transferred to 5-mm diameter NMR tubes. Pulsing conditions were: 64–256 free induction decays (FIDs), 72° pulses, a relaxation delay of 2.00 s and an acquisition time of 1.28 s. Chemical shifts were referenced to tetramethylsilane ($\delta = 0.00$ ppm, internal) and/or residual chloroform ($\delta = 7.262$ ppm). Resonances present in spectra were assigned by a consideration of chemical shift values, coupling patterns and coupling constants. Exponential line-broadening functions of 0.20 Hz were routinely employed for purposes of processing.

Two-dimensional shift-correlated ^1H NMR (COSY) spectra were acquired using the standard sequence of Aue *et al.* (Aue *et al.*, 1976), with 2,048 data points in the t_2 dimension, 512 increments of t_1, a relaxation delay of 2.00 s, and 64 transients.

^1H—^1H J-resolved experiments employed the standard sequence $90°–t_1/2–180°–t_1/2$-ACQ[12], 90° on the AMX-600 spectrometer being equivalent to 7.9 μs. Acquisition parameters were: 128 t_1 increments, each of magnitude 2,048 data points; spectral width 8,621 Hz in f_2 and 100 Hz in f_1; 64 transients in each case; 4 dummy scans; relaxation delay 2.0 s; acquisition time 0.24 s. Unshifted sine-bell window functions were applied in each dimension before transformation of the matrix of eventual size 2,048 × 256 data points.

Resonances present in spectra of culinary oils and fats were assigned by a consideration of chemical shift values, coupling patterns and coupling constants. The molecular nature of particular classes of aldehydes detectable in spectra of thermally-stressed culinary oils was confirmed by standard additions of authentic, commercially-available compounds (*n*-alkanals, *trans*-2-alkenals and alka-2-4-dienals). The relative intensities of selected signals were determined by electronic integration, and the concentrations of aldehydes present were computed by comparing their resonance areas with that of the added 1,3,5-trichlorobenzene (final concentration 8.20×10^{-5} mol dm^{-3}).

19.2.3 *Selective estimation of the bi-functional aldehyde malondialdehyde (MDA) in culinary oil samples*

A 0.75 ml aliquot of a 1.00 g dm^{-3} solution of 2-thiobarbituric acid in doubly-distilled water was added to 2.00 ml volumes of control and thermally-stressed culinary oil samples, the mixture thoroughly rotamixed, centrifuged (20 min, 500 × g), the lower (aqueous) phase removed and then heated at a temperature of 96°C for 20 min. A 0.30 ml aliquot of the resulting pink-coloured solution was then diluted to a final volume of 4.00 ml with doubly-distilled water prior to spectrophotometric analysis. Zero-order and corresponding second-derivative electronic absorption spectra of these samples were obtained on a Kontron Uvikon 860 spectrophotometer (scan rate 120 nm/min), and MDA concentrations were determined using an ε_{532} value of 1.56×10^5 M^{-1} cm^{-1} for the 2:1 TBA:MDA adduct (Willis, 1969).

19.3 Results

19.3.1 Multicomponent analysis of PUFA-derived autoxidation products in culinary oils and fats

Heating of the PUFA-rich, vegetable-derived culinary oils according to standard frying procedures generated a complex series of aldehydic group proton (—C<u>H</u>O) resonances in their ^{1}H NMR spectra characteristic of *trans*-2-alkenals (doublet located at 9.48 ppm, *j* = 8.2 Hz), *trans,trans*-alka-2,4-dienals (doublet at 9.52 ppm, *j* = 8.2 Hz) and *n*-alkanals (prominent triplet at 9.74 ppm, *j* = 1.7 Hz) (Pouchert and Behnke, 1992). Also detectable was an aldehydic group proton doublet (δ = 9.63 ppm, *j* = 8.2 Hz) of relatively weak intensity, a resonance assignable to *cis,trans*-alka-2,4-dienal species.

Subjection of such culinary oils to more rigorous episodes of thermal stressing yielded a series of further —C<u>H</u>O resonances of relatively low intensity [specifically those located at 9.56 (*d*, *j* = 7.4 Hz), 9.58 (*d*, *j* = 8.3 Hz), 9.79 (*t*, ^3j = 1.2 Hz), 9.82 (*t*, ^3j = 1.0 Hz) and 9.86 ppm (*t*, ^3j = 1.4 Hz)]. The expanded aldehydic (8.00–10.50 ppm) regions of typical 400 MHz ^{1}H NMR spectra acquired on control (unheated) and thermally-stressed samples of a commercially-available sunflower seed oil preparation are shown in Figures 19.1(a)–(c).

Two-dimensional COSY ^{1}H NMR spectroscopy greatly facilitated distinction between the four classes of aldehyde detectable in thermally/oxidatively-stressed culinary oils [illustrated for a heated sample of corn oil in Figure 19.1(d)]. Indeed, spectra acquired demonstrated clear connectivities between (1) the 9.48 ppm doublet and olefinic proton resonances located at 6.10 (*dd*, *j* 15.5, 8.2 Hz) and 6.85 ppm (*dt*, *j* 6.9, 15.5 Hz), (2) the 9.52 ppm doublet and olefinic proton multiplet signals at 6.04 and 7.07 ppm (*dd*, *j* 15.5, 10.1 Hz), (3) the 9.74 ppm triplet and an α-methylene group multiplet at 2.44 ppm (*dt*), and (4) the 9.63 ppm doublet and olefinic proton multiplets centred at 6.15 and 7.39 ppm (*dd*), confirming the above resonance assignments. Connectivities between the *trans*-2-alkenal olefinic proton multiplet resonances (δ = 6.10 and 6.85 ppm) and their corresponding 4-position methylene group signal (*dt*, δ = 2.335 ppm) were also observable.

Figure 19.1 (*opposite*) Expanded aldehydic proton regions of 400 MHz ^{1}H NMR spectra of a commercially-available sample of sunflower seed oil (a) before, (b) after heating at a temperature of 180°C in the presence of atmospheric O_2, and (c), as (b), but after a further 30 min period of heating at an elevated temperature (250°C). A 25g quantity of this culinary oil was heated in a 100 ml volume beaker (oil surface area 34.64 cm^2) on an electronically-controlled hot-plate and samples were placed in 2 ml capacity glass sample tubes and stored in the dark at ambient temperature for a duration of 18 hr prior to ^{1}H NMR analysis. (d) Partial 400 MHz two-dimensional COSY ^{1}H NMR spectrum of a commercially-available sample of corn (maize) oil subjected to a 90 min episode of heating at 180°C in the presence of atmospheric O_2 (this material was heated and stored as described above prior to analysis). Typical spectra are shown. Abbreviations: 1,2,3 and 4, aldehydic group (—C<u>H</u>O) protons of *trans*-2-alkenals, *trans,trans*-alka-2,4-dienals, *cis,trans*-alka-2,4-dienals and *n*-alkanals; 5, aldehydic group proton resonance (*t*, ^3j = 1.0 Hz) of a further aldehyde autoxidation product containing a methylene (—CH_2—) group in the α(2)-position; $CHCl_3$, residual chloroform; TCB, 1,3,5-trichorobenzene (internal quantitative ^{1}H NMR standard).

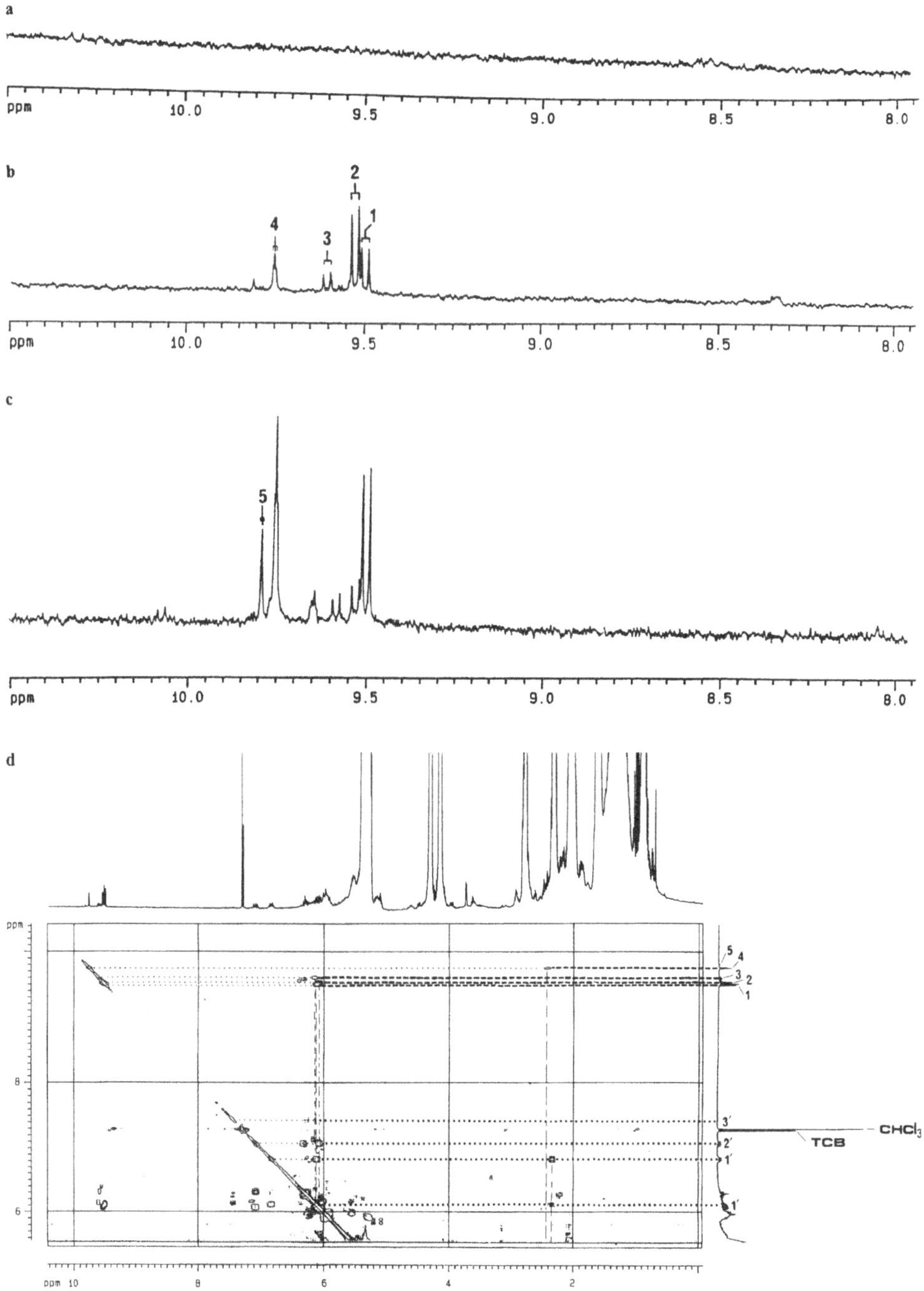
a
ppm
10.0
9.5
9.0
8.5
8.0
b
2
1
3
4
c
5
d
5
4
3
2
1
3'
2'
1'
CHCl3
TCB
8
6
4
2
ppm 10

The molecular nature of the aldehydic species giving rise to the above resonances was further confirmed by making standard additions of aldehydes known to be generated from the autoxidation of PUFAs (pentanal, hexanal, *trans*-2-pentenal, *trans*-2-hexenal, *trans*-2-heptenal, *trans*-2-octenal, *trans*-2-nonenal, *trans,trans*-nona-2,4-dienal and *trans,trans*-deca-2,4-dienal) to thermally-stressed culinary oil samples. However, the aldehydic group proton resonances present in the ^{1}H NMR spectra of thermally/oxidatively-stressed oils are predominantly ascribable to a mixture of *trans*-2-heptenal and -octenal (δ = 9.48 ppm), *trans,trans*-nona- and/or *trans,trans*-deca-2,4-dienals (δ = 9.52 ppm) and hexanal (δ = 9.74 ppm) since these adducts are the predominant *trans*-2-alkenal, *trans,trans*-alka-2,4-dienal and *n*-alkanal aldehydic components arising from the peroxidation of glycerol-bound linoleate (Esterbauer, 1982).

The above aldehydes arise from the fragmentation of CHPDs which are particularly unstable at standard frying temperatures ($\geq$ 180°C) and high field ^{1}H NMR spectra of thermally/oxidatively-stressed culinary oils also contained signals characteristic of these primary autoxidation products, i.e. conjugated diene system olefinic proton multiplets located in the 5.4–6.7 ppm chemical shift range and broad hydroperoxide —OO<u>H</u> group singlets (δ = 8.3–8.9 ppm) attributable to *cis,trans*- and *trans,trans*-isomers of 9- and 13-hydroperoxy-octadecadienoylglycerols (Gardner and Weisleder, 1972; Chan and Levett, 1977; Neff *et al.*, 1990) (typical examples of the ^{1}H NMR detection of these species in culinary oils are shown in Figures 19.2 and 19.3). As expected, culinary oil spectra containing *cis,trans*-CHPD resonances also contained a multiplet (*dt*) centred at 4.35 ppm assignable to the hydroperoxy group-bearing carbon methine proton [—C<u>H</u>(OOH)—] of these isomers (Neff *et al.*, 1990). Two-dimensional COSY ^{1}H NMR spectra of such samples showed clear linkages between multiplet resonances arising from the coupled CHPD conjugated diene system

Figure 19.2 (*opposite* and p. 382) Partial (5.00–10.50 ppm regions of) 400 MHz ^{1}H NMR spectra of (a) control (unheated) and (b) repeatedly-utilised samples of culinary frying oils obtained from a fast-food restaurant. Typical spectra are shown. After collection, the above samples were stored in the dark at ambient temperature for a period of 12 hr prior to analysis. (c) and (d) As (a) and (b), respectively, but after prolonged storage in the dark at ambient temperature (334 days). Abbreviations: as Figure 19.1, with 1′, 2′ and 3′, vinylic proton resonances of *trans*-2-alkenals, *trans,trans*-alka-2,4-dienals and *cis,trans*-alka-2,4-dienals respectively; a, b, c, d and e, vinylic proton resonances of the conjugated diene systems of 13- and/or 9-hydroperoxy-substituted octadecadienoylglycerol adducts (the *cis*-9,*trans*-11 and *trans*-10,*cis*-12 isomers, respectively) as denoted for the latter in (b); p and q, vinylic proton resonances of the conjugated diene systems of 13- and/or 9-hydroperoxy-substituted octadecadienoylglycerol adducts in the *trans,trans*-configuration (the *trans*-9,*trans*-11 and *trans*-10,*trans*-12 isomers, respectively) as specified for the latter in (d); —OO<u>H</u>, broad hydroperoxy group proton signals of conjugated hydroperoxydienes; Phth, aromatic ring protons of phthalic acid (a trace contaminant arising from the plasticware in which the samples were obtained). Two multiplet resonances ascribable to conjugated diene system protons of *cis,trans*-conjugated hydroxydienes (with the hydroxy-substituent in the 9- and/or 13-positions) have chemical shift values of 5.64 and 6.28 ppm (Gardner and Weisleder, 1972) and hence are obscured by those arising from *trans,trans*-hydroperoxydiene species (p and q). The arrow in spectrum (d) denotes a singlet resonance (δ = 8.02 ppm) generated in spectra during repeated frying episodes.

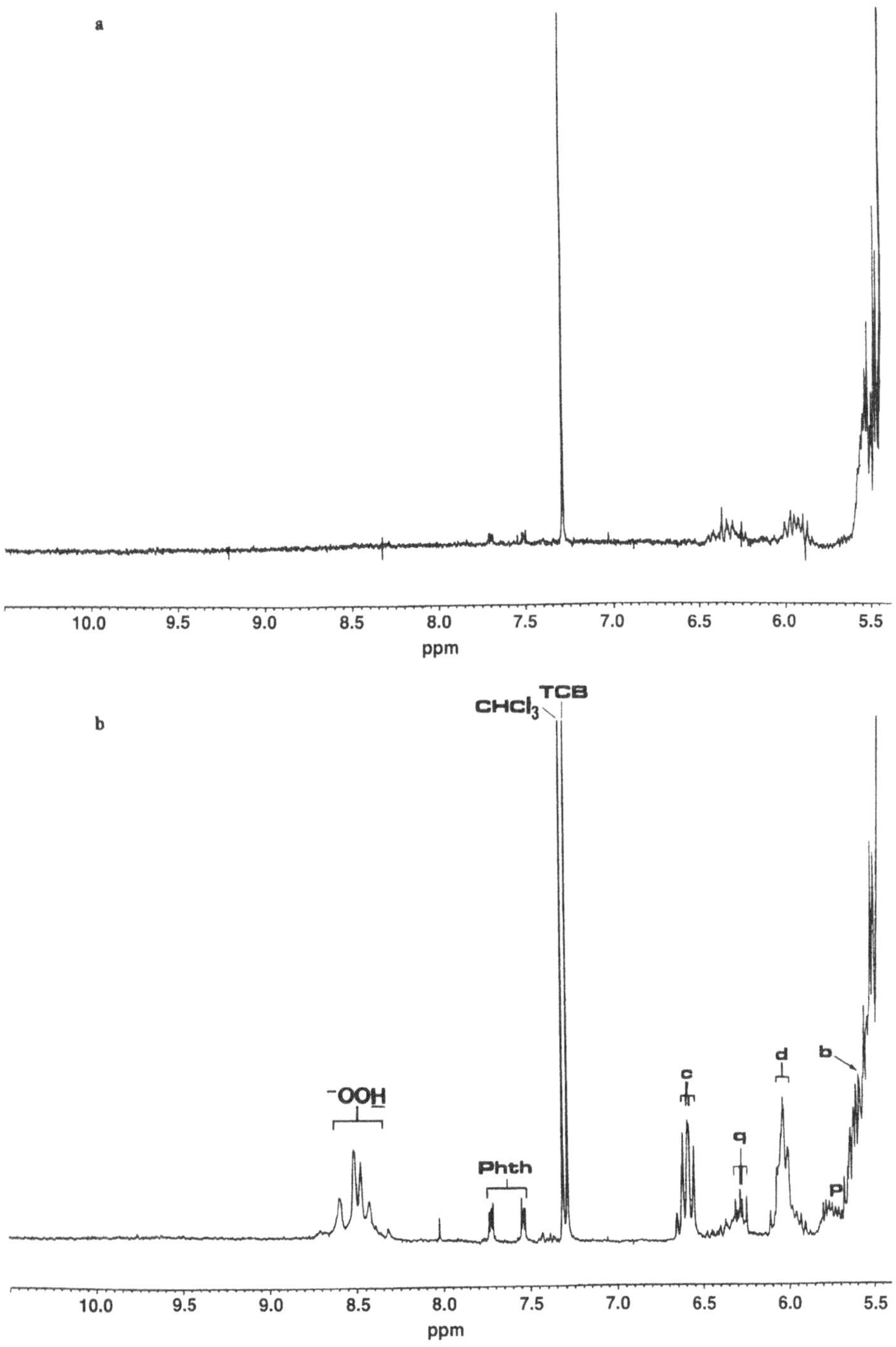
a
10.0
9.5
9.0
8.5
8.0
7.5
7.0
6.5
6.0
5.5
ppm
b
CHCl3
TCB
-OOH
Phth
c
q
d
b
p
10.0
9.5
9.0
8.5
8.0
7.5
7.0
6.5
6.0
5.5
ppm

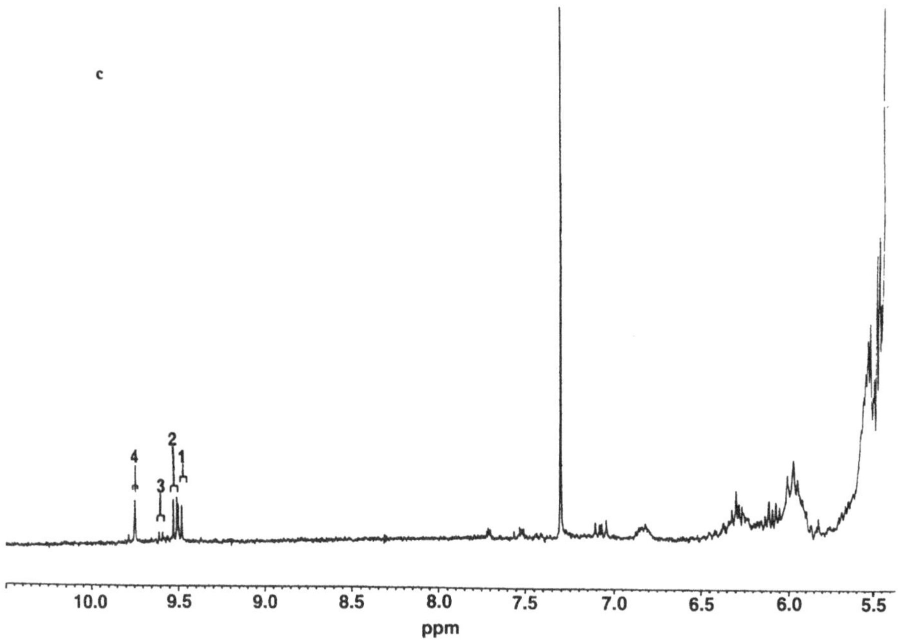
c
4
3
2
1
10.0
9.5
9.0
8.5
8.0
7.5
7.0
6.5
6.0
5.5
ppm

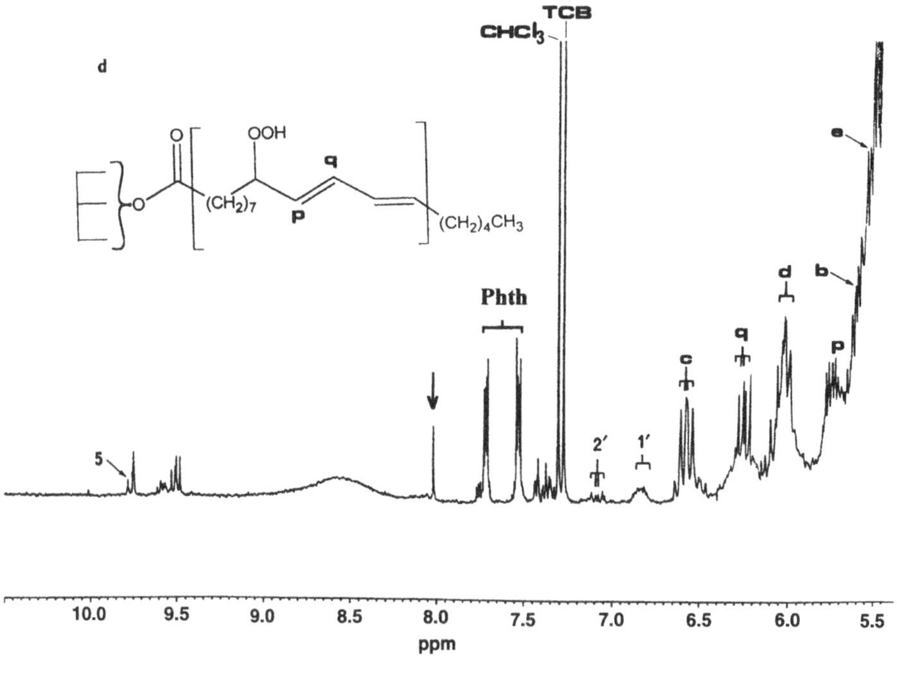
d
TCB
CHCl3
OOH
O
q
p
(CH2)7
(CH2)4CH3
Phth
5
2′
1′
c
q
d
b
e
p
10.0
9.5
9.0
8.5
8.0
7.5
7.0
6.5
6.0
5.5
ppm

protons [e.g. the 9-, 10-, 11-, 12- and 13-position protons of *cis,trans*-9-hydroperoxy-octadecadienoylglycerol adducts (protons a, b, c, d and e respectively)]. Comparisons of these two-dimensional shift-correlated spectra with those acquired on relevant chemical model systems (autoxidised samples of methyl, ethyl and propyl linoleates) confirmed the above resonance assignments.

The thermally-induced, autoxidative consumption of glycerol-bound PUFAs was also detectable by ^1H NMR analysis, i.e. selective decreases in the intensities of the *mono*- and *bis*-allylic-CH_2 group resonances (δ = 2.06 and 2.76 ppm respectively), and that of the vinylic protons located at 5.38 ppm. Moreover, the low field-shifted acyl chain terminal-CH_3 group triplet resonance (δ = 0.95 ppm) detectable in spectra of vegetable-derived culinary oils with a relatively high content of highly unsaturated fatty acids (i.e. those with ≥ 3 unconjugated double bonds, predominantly linolenoylglycerols) markedly diminished in intensity when such samples were heated according to common frying practices. For example, the highly unsaturated fatty acid content of a typical culinary oil was reduced from 9.7 to 6.4 molar % after being utilised for repeated frying episodes in a restaurant, data consistent with the increased susceptibility of linolenoylglycerols to autoxidation when expressed relative to that of linoleoylglycerols (de Man, 1976).

High field ^1H NMR analysis also revealed the presence of high levels of potentially toxic aldehydes and their CHPD precursors in samples of repeatedly-used frying oils obtained from fast-food take-away establishments (Figure 19.2). Indeed, 400 MHz spectra of such repeatedly-utilised oils contained intense aldehydic proton resonances, whereas those of unheated control samples analysed within 18 hours of collection contained little or no NMR-detectable, autoxidised PUFA-derived aldehydes [Figures 19.2(a) and (b)]. However, prolonged incubation (334 days) in the dark at ambient temperature (23 ± 2°C) yielded high concentrations of CHPDs in both unheated and repeatedly-used samples, and also modified the pattern and intensities of the aldehydic proton resonances (Figures 19.2(c) and (d)], observations which clearly have important ramifications regarding the autoxidation of PUFAs in culinary oils during periods of storage by manufacturers, retail outlets and consumers.

Interestingly, the conjugated diene regions of ^1H—^1H J-resolved spectra of samples corresponding to the one-dimensional NMR profiles shown in Figures 19.2(b) and (d) also contained resonances ascribable to conjugated hydroxydiene species (Gardner and Weisleder, 1972; Tallent *et al.*, 1966) [i.e. multiplets centred at 6.28 (*dd*, *j* = 13.9, 10.1 Hz), 5.93 (*dd*, *j* = 11.0, 10.1 Hz), 5.64 (*dd*) and 5.30 ppm (*dt*), corresponding to the 11-, 10-, 12- and 9-position vinylic protons, respectively, of 13-hydroxy-*cis*-*9,trans-11*-octadecadienoylglycerol and/or the 11-, 12-, 10- and 13-position vinylic protons, respectively, of 9-hydroxy-*trans*-10,*cis*-12-octadecadienoylglycerol]. Such conjugated hydroxydienes arise from the thermally-induced fragmentation of their corresponding *cis,trans*-hydroperoxydienes (equations 19.1 and 19.2).

$$LOOH \rightarrow LO^\bullet + {}^\bullet OH \qquad (19.1)$$

$$LO^\bullet + L'H \rightarrow LOH + L'^\bullet \qquad (19.2)$$

The expanded 5.00–10.00 ppm regions of the 400 MHz ^1H NMR spectra of control and heated samples of a typical commercially-available groundnut oil preparation are shown in Figures 19.3(a) and (b) respectively. Interestingly, the spectrum of the unheated sample contains prominent resonances attributable to CHPDs (predominantly

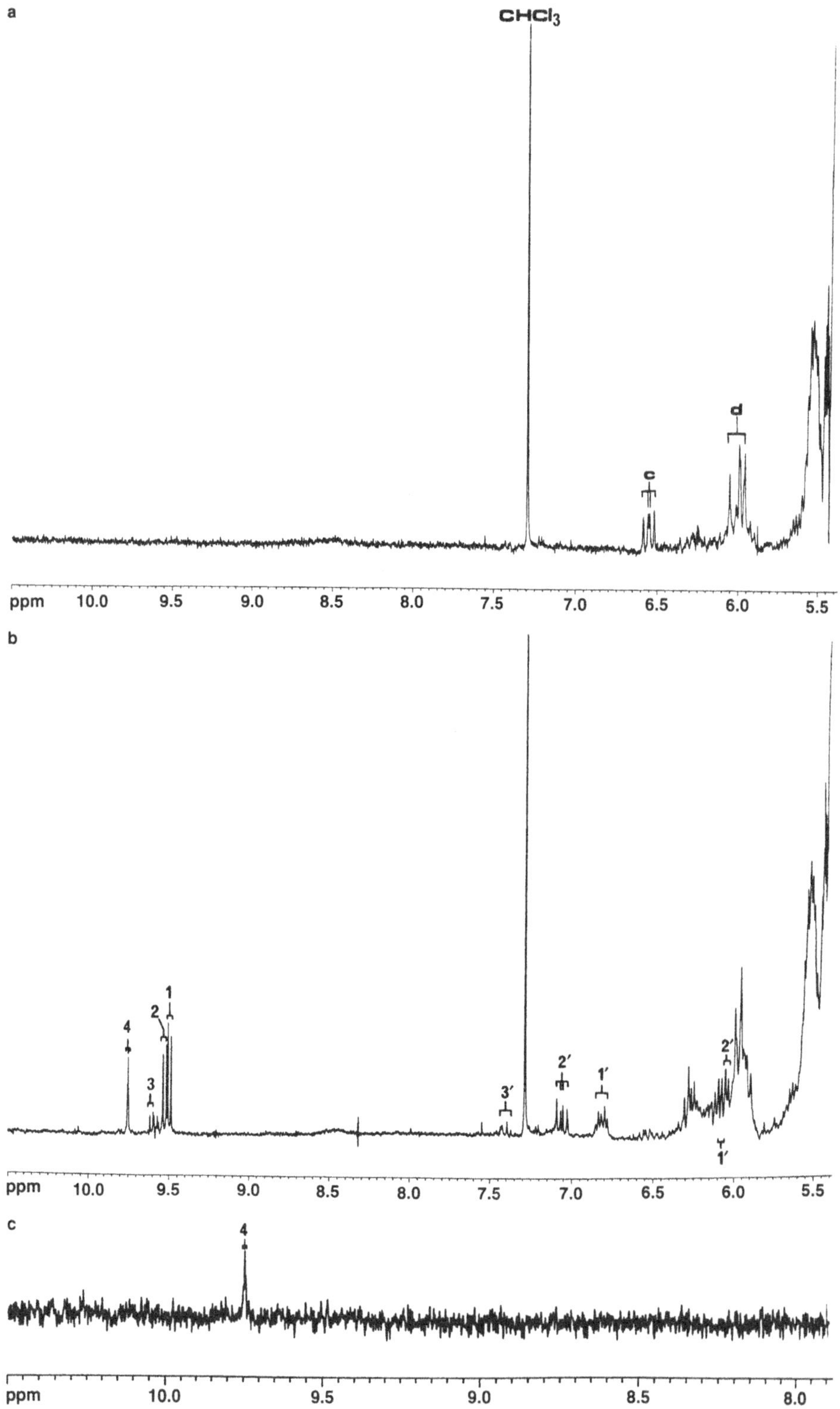

a
CHCl3
c
d
ppm 10.0 9.5 9.0 8.5 8.0 7.5 7.0 6.5 6.0 5.5
b
4
3
2
1
3′
2′
1′
2′
1′
ppm 10.0 9.5 9.0 8.5 8.0 7.5 7.0 6.5 6.0 5.5
c
4
ppm 10.0 9.5 9.0 8.5 8.0

cis,trans-isomers) which presumably arise during isolation, preparation and storage of this material by the manufacturers, or storage by the retail shop from which it was purchased. Indeed, such CHPD signals were present in many of the unheated (control) PUFA-rich culinary oil spectra examined, together with those arising from relatively low levels (i.e. 2×10^{-4} mol dm^{-3}) of *n*-alkanals and *trans*-2-alkenals. Figure 19.3(c) shows the facile ^{1}H NMR detection of these aldehydes in a further unheated commercially-available groundnut oil sample.

Heating of groundnut oil at a temperature of 180°C for periods of up to 90 min resulted in time-dependent reductions in the intensities of the above CHPD resonances which were accompanied by the production, or substantial increases in the intensities of, signals attributable to *n*-alkanals, *trans*-2-alkenals, *trans,trans*-alka-2,4-dienals and *cis,trans*-alka-2,4-dienals, data fully consistent with the thermally-induced decomposition of PUFA-derived hydroperoxides. In view of the reported pro-atherogenic actions of this culinary oil (Kritchevsky, 1991), these data further confirm the importance of determining the precise autoxidation status of such materials prior to the instigation of associated *in vivo* studies.

Spectrophotometric analysis of aqueous extracts of unheated and thermally-stressed culinary oil samples afforded a selective estimation of the water-soluble *bi*-functional aldehyde MDA via its reaction with TBA at 95°C to produce a pink-coloured chromophore (the 2:1 TBA:MDA adduct). The product generated from the reactions of such aqueous extracts with TBA had electronic absorption spectra that were very similar to that of the adduct arising from the reaction of an authentic, pure MDA sample with this reagent (data not shown), and second-derivative spectrophotometric analysis confirmed the selectivity of this method, i.e. no interfering absorption bands ascribable to the chromophoric TBA adducts of alternative aldehydes were detectable. Reaction of an aqueous extract obtained from a typical commercially-available soyabean oil sample pre-heated at 180°C for a 90 min period had an A_{532} value corresponding to an MDA concentration of 1.21×10^{-5} mol kg^{-1}, assuming a 100% recovery of this aldehyde into the lower, aqueous phase.

Thermal stressing of olive oil preparations yielded very low levels of selected classes of aldehyde (*trans*-2-alkenals and *n*-alkanals), which only became NMR-detectable after rigorous episodes of heating ($\geq$ 90 min at 180°C), data presumably reflecting the low levels of PUFAs present in this culinary oil [*ca.* 12% (w/w)]. Similarly, subjection of coconut oil [1–3% (w/w) PUFAs], lard [4–11% (w/w) PUFAs] and beef and lamb fat (dripping) to episodes of heating at 180°C in the manner described above generated little or no NMR-detectable aldehydes. Figure 19.4 shows the expanded

Figure 19.3 (*opposite*) (a) and (b) Expanded 5.00–10.50 ppm regions of 400 MHz ^{1}H NMR spectra acquired on a commercially-available sample of groundnut (peanut) oil before and after subjecting to an episode of thermal stressing (90 min at 180°C in the presence of atmospheric O_2), respectively. Typical spectra are shown. A 25g quantity of this culinary oil was heated in a 100 ml volume conical flask (oil surface area 24.64 cm^2) and samples removed at the 90 min time-point were stored in the dark at ambient temperature for a 912 hr period prior to ^{1}H NMR analysis [the unheated (control) sample was stored in the same manner for an equivalent length of time]. (c) Expanded aldehydic proton region of the 400 MHz ^{1}H NMR spectrum of a further commercially-available groundnut oil sample subjected to ^{1}H NMR analysis within 4 hr after purchase from a local retail outlet). Abbreviations: as Figures 19.1 and 19.2

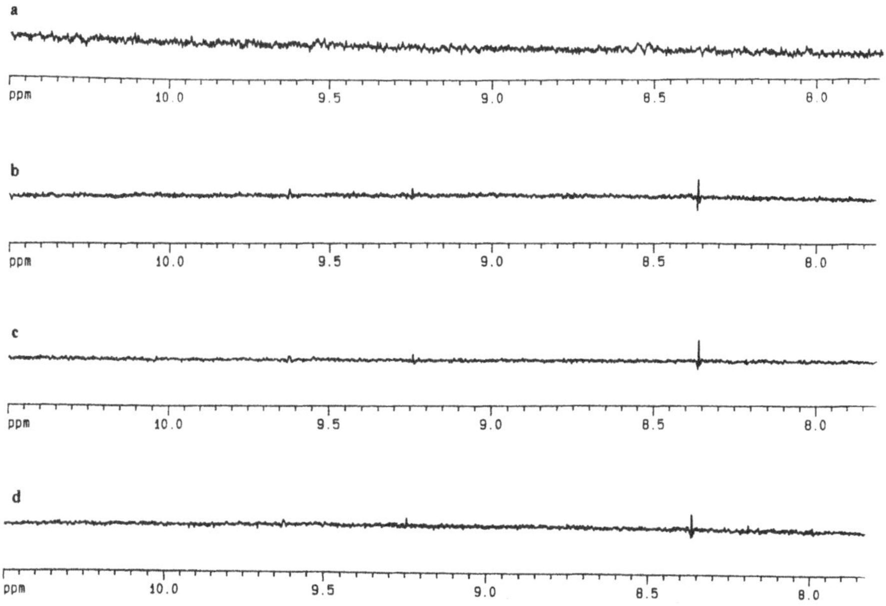

Figure 19.4 Expanded 5.00–10.50 ppm regions of the 400 MHz ^{1}H NMR spectra of a commercially-available sample of coconut oil obtained (a) before, and at (b) 30, (c) 60 and (d) 90 min after heating at a temperature of 180°C in the presence of atmospheric O_2 (100 ml volume beaker heating vessel). Typical spectra are shown. Samples removed at the above time-points were stored in the dark at ambient temperature for a period of 106 days prior to ^{1}H NMR analysis [the unheated (control) sample was stored in the same manner for an equivalent length of time]. Abbreviations: as Figure 19.1

5.00–10.50 ppm regions of the 400 MHz ^{1}H NMR spectra of control and thermally-stressed samples of a commercially-available coconut oil preparation. However, certain of the spectra acquired on heated dripping samples contained resonances assignable to conjugated hydroxydienes [multiplets located at 5.93 (*dd*, *j* 11.0, 10.1 Hz), 6.28 (*ad*, *j* 13.9, 10.1 Hz) and 5.64 ppm (*dd*) corresponding to the 10-, 11- and 12-position conjugated diene system olefinic protons, respectively, of 13-hydroxy-*cis*-9,*trans*-11-octadecadienoylglycerol (Gardner and Weisleder, 1972)].

19.3.2 Influence of heating vessel capacity on aldehyde generation

The nature and size (capacity) of the glassware utilised for laboratory experiments involving the heating of culinary oils was found to exert a substantial influence on the concentrations of thermally-induced autoxidation products generated therein (a consequence of differing effective surface areas of the materials investigated, i.e. the quantity of oil exposed to atmospheric O_2 during episodes of thermal stressing). Hence, fixed volumes (20.0 ml) of a commercial corn oil preparation were subjected to 30, 60 and 90 min periods of heating at 180°C in glass vessels of increasing size in order to further evaluate this. Table 19.1 gives the concentrations of saturated and α,β-unsaturated aldehydes produced during heating of this culinary oil in the above

Table 19.1 Concentrations of α,β-unsaturated and saturated aldehydes generated with increasing time of heating at 180°C for 18.48 g quantities of a commercially-available corn oil sample heated in 25, 50, 100 and 250 ml volume glass beakers (each sample was subjected to ^{1}H NMR analysis immediately after each period of thermal stressing). The concentrations of aldehydes present at a heating time of 0 min represent those generated during the elevation of the oil temperature from 22 (ambient) to 180°C. Typical data are shown

		Concentration of aldehyde (10^{-3} mol kg^{-1}) in corn (maize) oil						
Capacity (ml) and surface area (cm^2) of heating vessel	*Period of heating at 180°C (min)*	trans-2-*alkenals*	trans,trans-*alka-2,4-dienals*	cis,trans-*alka-2,4-dienals*	*further α,β-unsaturated aldehydes[a] (total)*	n-*alkanals*	*aldehyde giving rise to δ = 9.82 ppm signal* (t)	*further aldehydes with—CH_2—group in 2(α)-position[b] (total)*
25 ml : 7.75 cm^2	0	0.42	0.46	n.d.	n.d.	0.60	n.d.	n.d.
	30	0.62	0.62	0.05	n.d.	0.82	n.d.	n.d.
	60	1.37	1.46	0.17	n.d.	1.72	n.d.	n.d.
	90	1.43	1.63	0.20	n.d.	1.88	0.16	0.02
50 ml : 12.20 cm^2	0	0.59	0.74	n.d.	n.d.	0.74	n.d.	n.d.
	30	1.37	1.43	0.27	0.23	1.65	0.12	n.d.
	60	2.59	3.10	0.72	0.48	3.32	0.24	n.d.
	90	3.82	4.55	0.91	0.85	4.09	0.39	0.11
100 ml : 18.41 cm^2	0	0.87	0.90	n.d.	n.d.	0.92	0.14	n.d.
	30	2.59	3.01	0.62	0.28	3.23	0.27	n.d.
	60	3.92	3.67	0.78	0.83	5.35	0.37	0.24
	90	9.57	7.65	1.23	1.20	10.11	0.92	0.62
250 ml : 34.64 cm^2	0	1.29	1.53	0.53	0.34	1.95	0.27	0.02
	30	4.54	4.50	0.72	0.56	5.27	0.46	0.26
	60	9.03	7.20	1.23	1.32	10.11	0.92	0.49
	90	12.57	11.31	1.63	2.42	14.87	1.88	1.19

a α,β-unsaturated aldehydes responsible for the —C<u>H</u>O doublet resonances located at 9.56 and 9.58 ppm.
b Aldehydes giving rise to the —C<u>H</u>O triplet signals at 9.79 and 9.86 ppm.
n.d., none detectable.

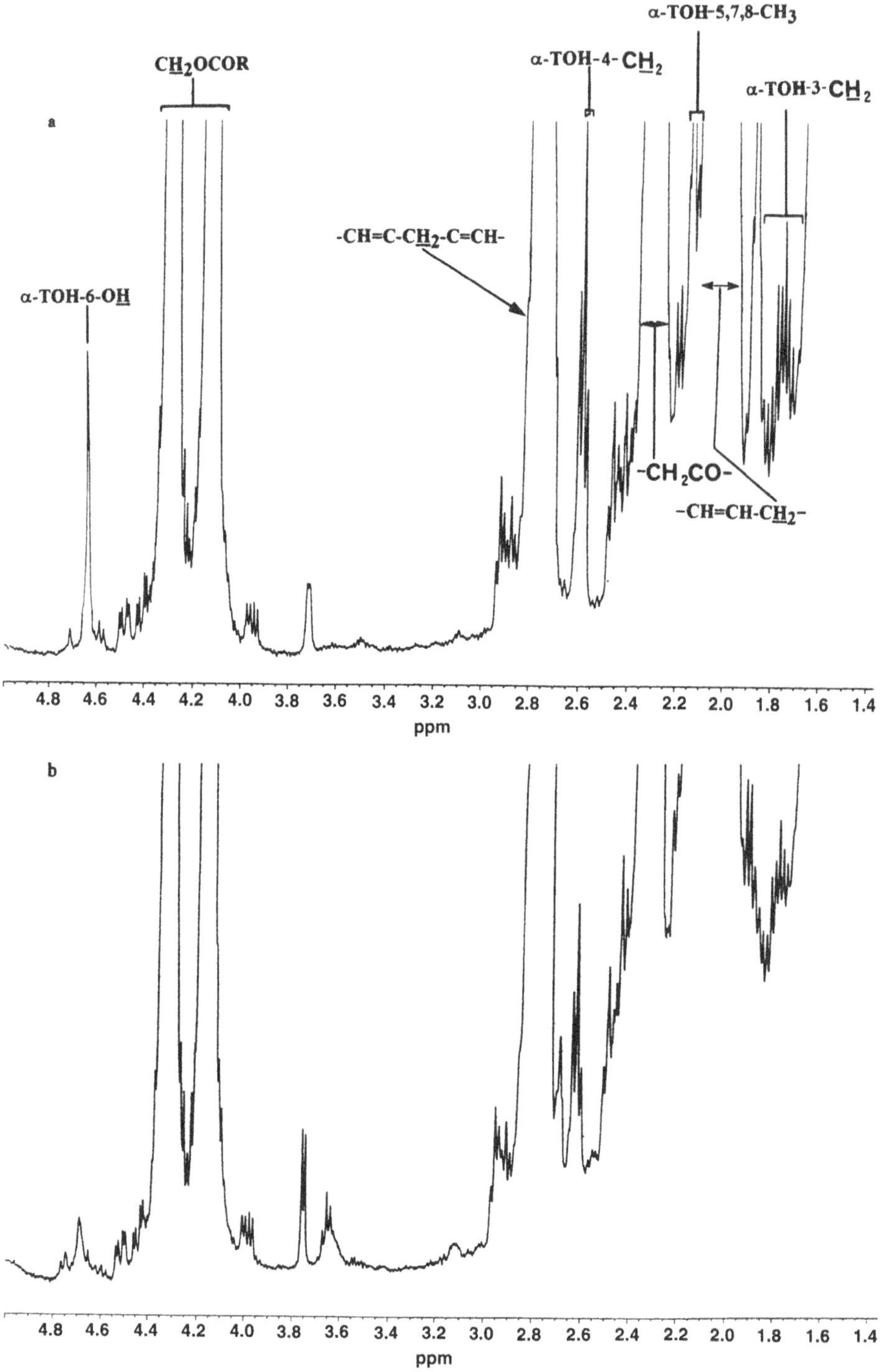

Figure 19.5 Expanded 1.00–5.00 ppm regions of the 400 MHz ^{1}H NMR spectra of a commercially-available sample of sunflower seed oil containing added α-tocopherol (5.80×10^{-2} mol kg^{-1}) (a) before and (b) after heating for a period of 90 min at a temperature of 180°C followed by a further 30 min at 250°C, in the presence of atmospheric O_2. Typical spectra are shown. The control (unheated) and thermally-stressed samples were stored

manner. Clearly, the levels of each class of aldehyde generated are enhanced with increasing vessel size, those present in the sample collected subsequent to heating in a 250 ml beaker for 90 min being extremely high (total concentrations of *n*-alkanals and α,β-unsaturated aldehydes 1.49 and 2.79×10^{-2} mol kg^{-1} respectively), an observation of much significance in view of the prevalence of domestic shallow frying procedures employing PUFA-rich culinary oils.

In view of their high volatilities (e.g. pentanal, hexanal, *trans*-2-heptenal and *trans,trans*-hepta-2,4-dienal have boiling points of 103, 131, 91 and 84°C respectively (Pouchert and Behnke, 1992)) the concentration of each class of aldehyde reported in Table 19.1 represents that remaining in the sample after specified thermal stressing episodes (i.e. that potentially available for human ingestion) and hence is considered to be an extreme lower limit for the total amount generated. Previous investigations involving the gas chromatographic analysis of lipid oxidation products in vegetable oils have focused on the detection and quantification of selected aldehydes (together with ketones, hydrocarbons and pyrazine derivatives) in the headspace gas of such samples (reviewed in Grob and Habich, 1985).

19.3.3 Consumption of α-tocopherol (α-TOH) and synthetic dietary antioxidants in culinary oils during episodes of the thermal stressing

In view of their powerful abilities to break peroxidative chain reactions (Burton and Ingold, 1986), lipid-soluble dietary antioxidants such as α-TOH (vitamin E) should, in principle, block the thermally-induced generated of aldehydes from glycerol-bound PUFAs present in culinary oils, and high resolution ^1H NMR spectroscopy was employed to investigate this. However, both α-TOH and the synthetic antioxidant 2,6-di-*tert*-butyl-4-methylphenol (BHT) were found to offer only a limited level of protection against aldehyde production when pre-added to culinary frying oils at high concentrations. Indeed, in a typical experiment, added α-TOH at a concentration of 25 mg/ml (5.80×10^{-2} mol kg^{-1}) reduced the levels of *n*-alkanals and *trans*-2-alkenals generated by only 30 and 46% respectively in sunflower seed oil heated for 90 min at 180°C followed by a further 30 min at 250°C [added concentrations of 10 mg/ml (2.32×10^{-2} mol kg^{-1}) were virtually ineffective]. Similarly, BHT at an added level of 10 mg/ml (4.54×10^{-2} mol kg^{-1}) suppressed the concentrations of *n*-alkanals and *trans*-2-alkenals produced in the same experimental system by only 17 and 22% respectively.

The above experiments provided further useful information regarding the oxidative and/or thermally-induced consumption of the above chain-breaking antioxidants in culinary oils subjected to episodes of heating. Indeed, the α-TOH phenolic-O<u>H</u> group proton resonance (*s*, δ = 4.635 ppm) detectable in ^1H NMR spectra of samples of sunflower seed oil pre-treated with this antioxidant (25 mg/ml) was markedly diminished in intensity after thermal stressing according to standard frying practices [Figures 19.5(a) and (b)]. Furthermore, the intensity of its heterocyclic ring 4-position

in the dark at ambient temperature for an 18 hr period prior to analysis. Abbreviations: α-TOH 6-O<u>H</u> and 5,7,8-CH_3, α-tocopherol 6-position-O<u>H</u>, and 5, 7 and 8-position-CH_3 aromatic substituent groups, respectively; α-TOH-3 and –4, α-tocopherol flexible heterocyclic ring 3- and 4-position —CH_2— groups respectively. Appropriate triacylglycerol resonances are also indicated

methylene group proton resonance (t, $\delta = 2.59$ ppm) was also reduced subsequent to heating. Although the decrease in the intensities of α-TOH resonances observed may be partially attributable to the volatilisation of this antioxidant [boiling point (b.pt.) 210°C], these ^{1}H NMR-detectable modifications were accompanied by the generation of an intense brown colouration (λ_{max} 270 nm in hexane), an observation concordant with the production of α-TOH oxidation products such as α-tocopherylquinone (Hess *et al.*, 1977); the intensity of this colouration was much greater than that observed on heating untreated sunflower seed oil in the same manner. Therefore, these data are consistent with the consumption of α-TOH via its suppression of peroxidative chain reactions and/or its direct thermally-mediated oxidation.

Similarly, ^{1}H NMR spectra acquired on sunflower seed oil pre-treated with 10 mg/ml BHT revealed that the characteristic *tert*-butyl, methyl and aromatic ring proton singlet resonances of this lipid-soluble antioxidant ($\delta = 1.42$, 2.27 and 6.97 ppm respectively) were substantially diminished in intensity after subjecting these samples to episodes of thermal stressing, a phenomenon attributable to its consumption by PUFA-derived peroxyl and/or alkoxyl radicals, thermally-induced oxidation (O_2^- mediated) and/or its volatilisation (b.pt. 265°C), especially at the higher temperature applied (250°C).

^{1}H NMR analysis also demonstrated that thermal stressing of an authentic sample of α-TOH in the manner described above gave rise to the consumption of this naturally-derived chain-breaking antioxidant (data not shown). Indeed, minor time-dependent decreases in its phenolic-O$\underline{H}$ group resonance were observed throughout a 90 min period of heating at 180°C, and an additional episode of thermal stressing at 250°C for 30 min markedly reduced the intensity of this signal to 71% of its initial, pre-heating value (when normalised to that of its 4-position —CH_2— and aromatic substituent —CH_3 groups), a process synchronous with the development of an intense brown colouration in the heated material. The thermally-induced consumption of α-TOH was further accompanied by the generation of a complex series of product resonances in its ^{1}H NMR spectrum, the most prominent being clearly linked doublets located at 5.57 and 6.48 ppm (data not shown).

These data indicate that thermal stressing of vegetable-derived culinary oils may severely deplete their α-TOH content, a phenomenon of much importance in view of the common and regular use of these materials for frying purposes, and recommended dietary requirements for vitamin E.

19.4 Discussion

The results obtained in this investigation demonstrate that PUFA-rich culinary oils produce much higher levels of cytotoxic aldehydes than those with a low PUFA content, or predominantly saturated fats (lard, dripping, etc.) when subjected to episodes of thermal/oxidative stressing. Such aldehydic products have the ability to exert a range of toxicological effects and previous investigations have shown that they interfere with the growth of cultured animal and bacterial cells, block macrophage action, express chemotactic actions upon neutrophils in biofluids, stimulate thrombin production *in vivo* and are mutagenic in bacterial test systems (Gutteridge *et al.*, 1976; Tappel, 1975; Turner *et al.*, 1975; Barrowcliffe *et al.*, 1984; Schauenstein *et al.*, 1977; Van Hinsbergh, 1984). Moreover, these adducts inhibit protein synthesis and inactivate enzymes (Addis, 1986; Kristal and Yu, 1992).

With the exception of direct damage to the gastrointestinal epithelium, the toxicological hazards putatively posed by aldehydes generated from the autoxidation of PUFA-rich culinary oils are, of course, critically dependent on the rate and extent of their *in vivo* absorption from the gut into the systemic circulation. Our recent investigations have shown that typical autoxidised PUFA-derived *trans*-2-alkenals, are readily absorbed *in vivo*, metabolised and excreted in the urine as water-soluble mercapturate conjugates (Grootveld *et al.*, 1998), data further supporting the hypothesis that dietary-derived aldehydes play an important role in the pathogenesis of a variety of human clinical conditions (e.g. ischaemic heart disease, inflammatory joint diseases, etc.).

The multicomponent analysis of control (unheated) and thermally-stressed culinary oils and fats by high resolution ^{1}H NMR spectroscopy provides valuable information regarding the molecular nature and concentrations of PUFA-derived autoxidation products therein (specifically four or more classes of aldehyde and their isomeric conjugated hydroperoxy/hydroxydiene precursors). Such information is rapidly obtained (spectral acquisition time *ca.* 12 min per sample) and the materials investigated here required only a minimal level of sample preparation prior to analysis. Previous investigations of this nature have involved labour-intensive, onerous laboratory methods which supply only a limited amount of chemical information concerning the autoxidation status of such materials. Indeed, many of these methods are unable to distinguish between different classes of conjugated diene species or aldehydes (e.g. spectrophotometric determination of thiobarbituric acid-reactive substances [TBARS] (Addis, 1986)). The wealth of specific structural (i.e. qualitative) and quantitative analytical data provided by the technique employed in our study is, of course, a critical primary requirement for future investigations of the toxicological/pro-atherogenic effects associated with the regular dietary consumption of culinary oils and fats. Such data should provide valuable information regarding the average daily intake of PUFA-derived oxidation products, information which will eventually be of much use to those undertaking epidemiological research.

Moreover, high resolution ^{1}H NMR analysis will permit future studies involving evaluations of the efficacy of further procedures aimed at diminishing the levels of PUFA autoxidation products present in repeatedly-utilised culinary oils, e.g. frying oil filtration systems incorporating filter aids such as highly refined cellulose fibres and/or activated charcoal.

The application of appropriate two-dimensional NMR techniques to the analysis of frying media may also serve to provide much valuable information regarding the molecular nature and concentrations of alternative oil and fat triacylglycerol deterioration products, e.g. thermally-induced cyclic fatty acid monomers and dimers together with adducts arising from hydrolysis (diacylglycerols, monoacylglycerols, free fatty acids and glycerol), and experiments to evaluate this are currently in progress.

Acknowledgements

We are very grateful to the Commission of the European Communities (Agriculture and Agro-industry, including Fisheries: DG-XII) for financial support, the University of London Inter-collegiate Research Services for the provision of NMR facilities, Jane Hawkes (Department of Chemistry, King's College, London, UK) for excellent technical assistance, and Melissa Boucher and Sheila Sinclair for help with typing the manuscript.

References

Addis, P.B. (1986). *Food Chem. Toxicol.*, **24**, 1021–1030.

Addis, P.B. (1990). *Food Nutr. News*, **62**, 1–4.

Addis, P.B. and Park, S.W. (1989). In Taylor, S.L. and Scanlan, R.A. (eds), *Food Toxicology. A Perspective on the Relative Risks*, New York: Marcel Dekker, pp. 247–330.

Alexander, J.C., Valli, V.E. and Chanin, B.E. (1987). *J. Toxicol. Environ. Health*, **21**, 295–309.

Aue, W.P., Bartholdi, E. and Ernst, R.R. (1976). *J. Chem. Phys.*, **64**, 2229–2246.

Barrowcliffe, T.W., Gray, E., Kerry, P.J. and Gutteridge, J.M.C. (1984). *Thromb. and Haemostas.*, **52**, 7–10.

Burton, G.W. and Ingold, K.U. (1986). *Acc. Chem. Res.*, **19**, 194–201.

Chan, H.W. and Levett, G. (1977). *Lipids*, **12(1)**, 99–104.

de Man, J.M. (1976). *Principles of Food Chemistry*, Westport, CT: AVI, p. 60.

Esterbauer, H. (1982). In McBrien, D.C.H. and Slater, T.F. (eds), *Aldehydic Products of Lipid Peroxidation*, London and New York: Academic Press, pp. 101–112.

Gardner, H.W. (1972). *Lipids*, **7(3)**, 191–193.

Gardner, H.W. and Weisleder, D. (1972). *Lipids*, **7(3)**, 191–193.

Grob, K. and Habich, A. (1985). *J. Chromatogr.*, **321**, 45–58.

Grootveld, M. and Rhodes, C.J. (1994). In Winyard, P.G. and Blake, D.R. (eds), *The Immunopharmacology of Free Radicals*, London: Academic Press, pp. 1–21.

Grootveld, M., Atherton, M.D., Sheerin, A.N., Hawkes, J., Blake, D.R., Richens, T.E. *et al.* (1998). *J. Clin. Invest.*, **101(6)**, 1210–1218.

Gutteridge, J.M.C., Lamport, P. and Dormandy, T.L. (1976). *J. Med. Microbiol.*, **9**, 105–110.

Hageman, G., Hermans, R., Ten Hoor, F. and Kleinjans, J. (1988). *J. Mutat. Res.*, **204**, 593–604.

Hageman, G., Hermans, R., Ten Hoor, F. and Kleinjans, J. (1990). *J. Food Chem. Toxicol.*, **28**, 75–80.

Hess, J.L., Pallansch, M.A., Harich, K. and Bunce, G.E. (1977). *Anal. Biochem.*, **83**, 401–407.

Kaunitz, H. (1978). In Kabara, J.J. (ed.) *The Pharmacological Effect of Lipid*, Champaign, IL: The American Oil Chemists Society, pp. 203–210.

Kristal, B.S. and Yu, B.P. (1992). *J. Gerontol.*, **47(4)**, B107–B114.

Kritchevsky, D. (1991). *Int. J. Tissue Reactions*, **13(2)**, 59–65.

Kritchevsky, D. and Tepper, S.A. (1967). *J. Atheroscler. Res.*, 7, 647–651.

Neff, W.E., Frankel, E.N. and Miyashita, K. (1990). *Lipids*, **25(1)**, 33–39.

Pouchert, C.J. and Behnke, J. (1992). *The Aldrich Library of ^{13}C and ^{1}H FT-NMR Spectra*, Aldrich Chemical Co. Inc.

Sanders, T.A.B. (1983). In Allen, J.C. and Hamilton, R.J. (eds), *Rancidity in Foods*, Amsterdam: Elsevier, pp. 59–66.

Schauenstein, E., Esterbauer, H. and Zollner, H. (1977). *Aldehydes in Biological Systems*, Pion Press.

Smith, R.L. (1988). *Diet, Blood Cholesterol and Coronary Heart Disease: A Critical Review of the Literature*, Santa Monica, USA: Vector Enterprises.

Smith, T. and Kummerow, F.A. (1987). In Watson, R.R. (ed.), *Nutrition and Heart Disease Vol. I*, CRC Press, pp. 45–64.

Strapans, I., Rapp, J.H., Pan, X.M., Hardman, D.A. and Feingold, K.R. (1996). *Arterioscler. Thromb. Vasc. Biol.*, **16**, 533–538.

Tallent, W.H., Harris, J., Wold, I.A. and Lundin, R.E. (1996). *Tetrahedron Lett.*, 4329.

Tappel, A.L. (1975). In Trump, B.J. and Arstila, A.V. (eds), *Pathology of Cell Membranes Vol. 1*, London and New York: Academic Press, pp. 145–170.

Turner, S.R., Campbell, J.A. and Lynn, W.S. (1975). *J. Exp. Med.*, **141**, 1437–1441.

Van Hinsbergh, V.W.H. (1984). *Atherosclerosis*, **53**, 113–118.
Willis, E.D. (1969). *Biochem. J.*, **113**, 315–324.
Witzum, J.L. and Steinberg, D. (1991). *J. Clin. Invest.*, **88**, 1785–1792.
Yagi, K. (1988). In Ando, W. and Morooka, Y. (eds), *The Role of Oxygen in Chemistry and Biology*, Amsterdam: Elsevier, pp. 383–390.
Yoshida, H. and Kujimoto, G. (1989). *G. Ann. Nutr. Metab.*, **33**, 153–161.

Part VI

Chemical models

20 The structure and electronic properties of oxy intermediates in the enzymatic cycle of cytochrome P450s

Dan Harris

20.1 Introduction

Cytochrome P450s are a class of heme proteins involved in the metabolism of a wide variety of xenobiotic substances in many species in nature. In mammalian species, these enzymes are found in a number of organs, most notably the liver, and are the enzymes primarily responsible for the transformation of a number of exogenous compounds into products which are soluble, benign forms and may be excreted. The products of such P450 metabolism, however, can be toxic or even carcinogenic and hence the importance of fundamental studies of determinants of both enzymatic efficacy and product specificity.

This enzymatic class, with over 300 isozymes, is believed to perform monoxygenase reactions largely via a common enzymatic cycle involving a number of oxy radical intermediates. The site of enzymatic chemistry always contains a cysteinate ligated to a ferric heme. Most discussions of this class of enzymes enzymatic function center around a depiction of this cycle, similar to that shown in Figure 20.1, involving transformation of a ferric heme species to an active oxidative form, which the preponderance of evidence suggests is analogous to the oxyferryl species in peroxidases. While common chemical intuition makes plausible many of the features of this enzymatic mechanism and associated chemistry, in fact, little definitive information exists as to the precise structural and electronic nature of the intermediate radical forms in this enzymatic cycle.

Much of what has been presumed about the nature of intermediate species in this mechanism has been indirect, deduced from either model compound studies or other classes of enzymes with some commonalities, e.g. peroxidases. It is difficult, given the extremely transient nature of the oxygen intermediates in this cycle, to firmly establish a direct association of a set of measured intermediate properties with a definitive chemical species. While this facet has by no means inhibited progress in many areas of P450 research, one may determine the geometric and electronic structures of these transient intermediate radical species employing techniques in computational chemistry. Indeed, additional knowledge of the structure and properties of intermediate oxy intermediates may facilitate a broader comprehension of the role of protein and water in their stabilization and/or conversion to either the oxyferryl species or decoupled products in advance of solution of unknown structures of the elusive binding site architectures of many mammalian P450s. Furthermore, while the oxyferryl species is thought to be the enzymatically active species in the preponderance of cytochrome P450 catalysis, other radical intermediates, such as the ferric peroxide

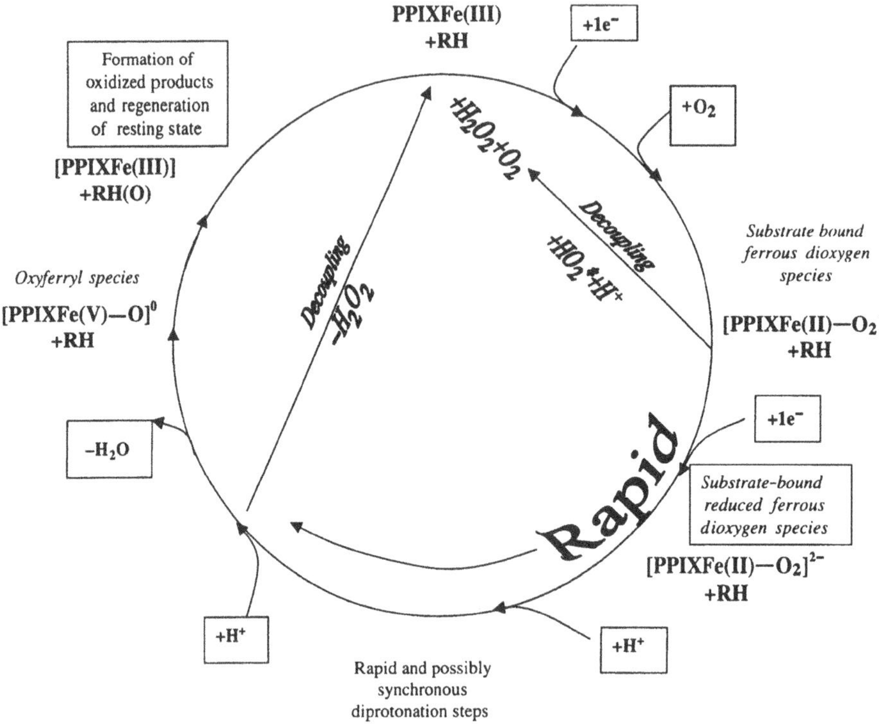

Figure 20.1 The enzymatic cycle of cytochrome P450s

species, may themselves play a role in some enzymatic chemistry (Matsunaga *et al.*, 1993; Vaz *et al.*, 1991; Roberts *et al.*, 1991).

The enzymatic cycle in Figure 20.1, depicting the cysteinate-ferric heme as PPIX (for protoporphyrin IX), indicates that the cycle begins with the binding of substrate. In cases where this substrate binding results in displacement water as a ligand of the heme there is often observed a heme spin state change from low to high spin. Occasionally this spin state change has functional significance in that it alters the redox potential of the heme, facilitating the first one-electron reduction. This reduction of the heme is essential to the next step, which is the binding of triplet molecular oxygen. This substrate bound oxyferrous species (denoted $[PPIXFe(II)—O_2]^{1-}$+RH in Figure 20.1) is the last metastable species in this enzymatic cycle, i.e. it is sufficiently stable for some spectroscopic characterization. While the steps leading from the resting state to formation of the oxyferrous species have been amenable to experimental characterization (Wanatabe and Groves, 1992; Groves and Han, 1995), the steps following formation of the oxyferrous species proceed rapidly through a succession of postulated intermediates. A reduction of the oxyferrous species is believed to be virtually commensurate with the addition of two protons to form either the oxyferryl species, which performs the monoxygenase chemistry, or generation of hydrogen peroxide and regeneration of the resting state. The latter case, which invoves proton additions to the reduced oxyferrous species to form hydrogen peroxide, is but one form of 'decoupling' of the monoxygenation P450 cycle, the other two being the result of a dismutation of the superoxide form resulting from the oxyferrous species or a decoupling step stemming from a $2e^-$ addition to the oxyferryl species (an overall $4H^+/4e^-$) process (not shown in Figure 20.1).

The structures of the ferric substrate-free and substrate bound resting states of several P450 isozymes have now been characterized. Numerous substrate bound crystal structures of the first crystallized P450*cam* as well as the resting state have been solved (Raag and Poulos, 1992). In addition, the substrate bound forms of both cytochrome P450*eryF* (Cupp-Vickery and Poulos, 1995) and P450-BM3 (Li and Poulos, 1996) have been determined. Finally the substrate free crystal structures of P450-BM3 (Ravichandran *et al.*, 1993; Li and Poulos, 1995) and P450*terp* (Haseman *et al.*, 1994) are known. While the structures of the stable ferric resting and substrate (including product bound) forms are known, detailed structural features of the less stable intermediates in the reaction cycle are more elusive.

Of the three oxygen containing heme intermediates in the P450 enzymatic cycle, most is known about the oxyferrous species because it is stable enough to have been characterized by a number of spectroscopic methods, particularly that of EXAFS. An EXAFS study of oxyferrous intermediate in P450*cam* has been reported by Dawson and co-workers and indicates an iron–sulfur distance of 2.37 Å, an iron-oxygen distance of 1.78 Å, and an iron-nitrogen distance of 2.0 Å (Dawson *et al.*, 1986; Dawson and Sono, 1987). Additional evidence for the structure of the P450 oxy ferrous species comes from an X-ray structure of a model dioxygen P450 heme complex (Schappacher *et al.*, 1987). This model contains a phenyl mercaptide substituted for the cysteine ligand and a picket fence porphyrin. In the crystal structure of this model compound, the oxygen is bound in an 'end-on' fashion to the heme iron similar to the carbon monoxide bound structure of cytochrome P450*cam* (Raag and Poulos, 1989). In this P450*cam* crystal structure, CO also binds end on to the heme iron with the C as the proximal ligand atom, and CO occupies what is believed to also be the oxygen binding pocket. In addition to structural information, a number of spectroscopic properties have been measured for both the oxyferrous P450*cam* species and the oxyferrous P450 heme model compound. The oxyferrous P450*cam* species lacks an ESR signal (Davdov *et al.*, 1991; Kobayshi *et al.*, 1994) and has had its Mossbaüer spectra characterized (Peterson *et al.*, 1972; Sharrock *et al.*, 1976). Resonance Raman studies, in conjunction with isotopic labeling, have associated the observed frequency of 1140 cm^{-1} with the O—O stretching frequency of the bound dioxygen (Bangcharoenpaurpong *et al.*, 1986). The measured spectroscopic properties of the model compound include its vis–UV, ESR, and Mossbaüer spectra (Schappacher *et al.*, 1987). These are all similar to the oxyferrous species of cytochrome P450*cam*. A recent report of the spectrum of the vis–UV spectra of the oxyferrous has been reported in conjunction with the spectrum of a species resulting from the reduction of the oxyferrous species in a D251N mutant of the wt P450*cam* isozyme (Benson *et al.*, 1997). A computational study of the oxyferrous and reduced oxyferrous species for a model methyl mercapto porphine system indicates the plausibility of that experimental assignment (Harris *et al.*, 1997).

Kinetic isotope effect measurements have demonstrated the requirement of two protons in the proton assisted pathway from the reduced oxyferrous species to formation of the compound I (oxyferryl species) (Aikens and Sligar, 1994). While the two proton additions to the dioxygen via pathway A or B in Figure 20.2 may be synchronous/simultaneous, the rapidity of the reaction following the second reduction and of proton addition reactions make it unlikely that this facet will be understood in

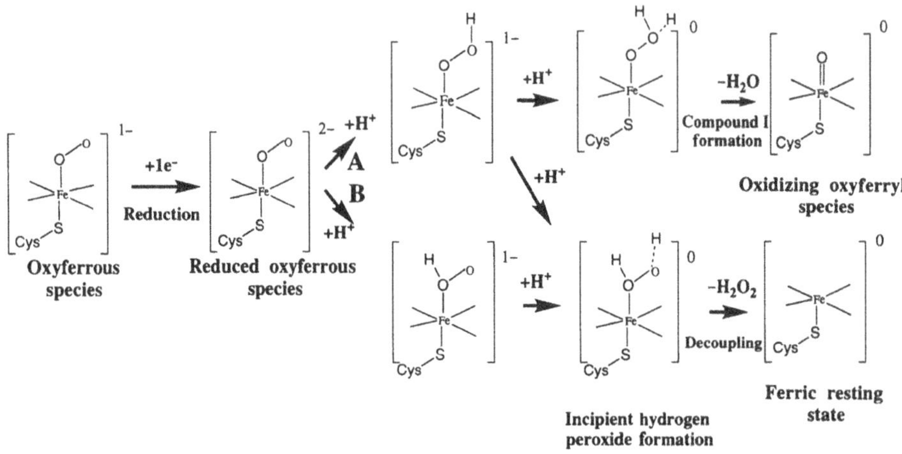

Figure 20.2 The reduction and protonation steps in formation of the oxyferryl or hydrogen peroxide products from the initial oxyferrous species

the near future. Aside from the role of hydrogen bond networks themselves biasing the dioxygen atom(s) to which protons are delivered, the order and synchronicity of proton additions to the proximal vs. distal oxygen may also play a role in the pathway determination.

Quantum chemical methods are well suited for providing a direct link between structure, properties, and chemistry for many systems including species which are metastable/unstable due to their enhanced reactivity. One may, in principle, calculate the geometric structure, electronic distribution, and spectral properties of such species using a judicious choice of quantum chemical methods.

Several significant problems, however, underlie the accurate calculation of the electronic structure of such a quantum mechanical system in any credible calculation of intermediates of the cytochrome P450 enzymatic cycle. The system is an open-shell electronic system, contains a transition metal iron requiring the inclusion of d-orbitals in the calculation, and is a large computational task for *ab initio* quantum mechanical methods. Given the open-shell electronic description of hemes containing iron, prediction of the properties of the lowest energy species requires the calculations of several possible species with different electronic configurations in order to deduce the ground state configuration. Further complicating the work on such systems, the energy differences between electronic states with differing spin multiplicities are rather small, sometimes as small as a few kcal/mol, making the accurate prediction of the prediction of the correct spin state of the ground (lowest) electronic state difficult.

Many electron systems require the numerical solution of equations describing the energetic interaction of each electron with nuclei as well each electrons interaction with all remaining (n–1) electrons, all within the framework of quantum mechanics. This must be done in Hartree–Fock based methods in an approximate manner by solving equations in which each electron in essence travels in the 'mean-field' of the remaining electrons. Such a calculation is referred to a self-consistent field (SCF) calculation. While such an approximate calculation embodied in the standard Hartree–Fock method recovers 99% of the total (nonrelativistic) energy of an atom or molecule, calculation of small energy differences as embodied in the calculation of electron or proton affinities, electronic excitation energies (spectra), potential surfaces or reaction energy barriers requires the correction of this mean-field approximation to include a facet referred to as electron correlation.

Quantum chemical calculations aiming to adequately describe the electronic structure, geometries, and chemistry of the heme system in cytochrome P450s requires the inclusion of electron correlation. Electron correlation in molecular systems embodies the principle that electrons in atomic/molecular orbitals have 'instantaneous' coulombic interactions which tend to keep them far apart, i.e. the electron motions are correlated. Accurate treatment of the geometric and electronic structure of such systems of transition metal systems as well as any system involving bond breaking and formation requires the inclusion of electron-correlation, which accounts for the quantum effects of pairs of electrons in orbitals avoidance of one another. Many of the quantum chemical methods which include electron correlation are either computationally too expensive or are virtually impossible given the computational work necessary for large systems such as hemes.

One of the methods potentially applicable to heme studies is that of density functional theory (DFT), pioneered by the seminal work of Hohenberg, Kohn, and Sham (Hohenberg and Kohn, 1964; Kohn and Sham, 1965). This method is a promising alternative to the traditional *ab initio* methods for characterizing heme systems because it includes electron correlation in the guise of an exchange–correlation functional and are not as computationally intensive as traditional correlated electron SCF *ab initio* methods, such as multiconfigurational selfconsistent field–configuration interaction (MCSCF–CI) methods. Density functional theory states that the ground state properties of a system of interacting particles (electrons and nuclei of atoms) are solely determined by its electron density. The formalism underlying density functional theory rigorously relates the energy of the system to the electron density. The only unknown in such a formalism is the exact form of the density functional which may be used to calculate either the contributions due to electronic exchange interactions and electron correlation in pure DFT, or merely the correlation contribution in hybrid Hartree–Fock/DFT approaches. Fortunately, much progress has been made in the search for improved functionals in recent years such that it is now realistic to begin to apply DFT to problems of biological relevance. Nonlocal density functional theory improves on the originally cast form of this theory by including 'nonlocal' terms which include the gradient of the density. The use of nonlocal corrections has been demonstrated to improve the accuracy of transition metal–ligand and metal–metal bond distances (Fan and Ziegler, 1991), as well as the structure and thermochemistry of iron chlorides and their positive and negative ions (Bach *et al.*, 1996). Nonlocal DFT has also been applied to a range of systems of biochemical interest including optimized geometries for a five coordinate model of the oxyferryl heme complex (Ghosh *et al.*, 1993) and iron sulfur clusters relevant to these centers in proteins (Mouesca *et al.*, 1994). It has most recently been applied to the calculation of the properties of compound I in peroxidases at experimentally determined geometries (Kuramochi *et al.*, 1997). The Parinello method, a valence electronic (frozen core) DFT theory, has been used for the calculation of the equilibrium geometries and electronic structure of a number of iron porphyrin five coordinate complexes including CO, O_2, imidazole, and NO, as well as six coordinate porphyrin complexes with imidazole as the proximal iron ligand and either CO or O_2 as the distal ligands (Rovira *et al.*, 1997). While work remains in refining the accuracy of density functional methods as applied to certain types of problems, this work is largely in the domain of improving the quality of the functionals used to calculate the energy and properties of the system.

In the work described here we report the computed geometries, electronic distributions, electrostatic properties of the elusive oxygen intermediates in the pathway from the oxyferrous cytochrome P450 species through formation of the oxyferryl species using a model consisting of a methylmercaptate bound to a iron–porphine as an approximation to the protoporphyrin IX–heme found in all cytochrome P450s. The species considered in this work are those species in the enzymatic steps shown in the steps in Figure 20.2. Initially, the DFT results of the computed model oxyferrous structure are compared with EXAFS data for the oxyferrous P450*cam* and the crystal structure of the oxyferrous Weiss model compound. The effects of reduction on the structure of the oxyferrous species are presented. Finally, the structures of the oxyferrous, the reduced oxyferrous and each of the possible forms resulting from distal and proximal protonation of the reduced oxyferrous species are calculated using the nonlocal DFT method. The ground state spin multiplicities and excess spin distributions on atoms of the model P450 system for each of the species are reported. Electrostatic potential surfaces are calculated from the ground state wavefunctions and the electrostatic potential of the reduced oxyferrous species compared with DFT energy minimized complexes of the reduced oxyferrous species with water, hydrogen bonded either to the distal or proximal oxygen. Features of the electrostatic potential and stability of the protonated species are compared and contrasted to elucidate possible features relevant to the enzymatic branching between the oxyferryl (pathway A) versus decoupling (pathway B) shown in Figure 20.2.

20.2 Theoretical and technical aspects

20.2.1 Model used and choice of initial geometries

The model of the P450 heme site used in all calculations is an iron–porphine complex with no porphyrin substituents and a methyl mercaptate (SCH_3—) axial ligand for the heme iron. The initial geometry chosen for the oxyferrous species was based on the Weiss model compound structure (Schappacher *et al.*, 1987), thus, an initial end-on dioxygen geometry was used for this species. The initial Fe—S, Fe—O, Fe—N and O—O distances used were: 2.3 Å, 1.8 Å, 2.0 Å, and 1.8 Å respectively. All models were constructed in the UNICHEM 4 interface (DGauss 4.0/Unichem Oxford Molecular, Beaverton, Oregon).

For the initial structure of the reduced oxyferrous species, the optimized geometry for the end-on structure of the oxyferrous species was used. A previously postulated bridged dioxygen bound reduced oxyferrous form has been shown by DFT calculations to not be a thermodynamically relevant form of the reduced oxyferrous species in that it is *ca* 28 kcal/mol higher in energy than the end-on dioxygen binding geometry (Harris *et al.*, 1997).

Initial structures of the distally and proximally protonated reduced oxyferrous species were generated starting with the DFT energy optimized reduced oxyferrous mercapto porphine system and adding the proton at *ca* 1 Å from the incipient protonated oxygen and then performing energy optimization. Two distinct initial structures of water complexed with the reduced oxyferrous species were constructed by adding a single water in an orientation such that one of the two water hydrogens was initially within 2.0 Å of either the distal or proximal oxygen of the previously optimized structure of the reduced oxyferrous species. The reduced oxyferrous–water complexes were then geometry optimized in an additional set of DFT calculations.

20.2.2 Density functional DFT calculations

Density functional calculations were performed using DGauss version 4.0 in conjunction with the UNICHEM interface (DGauss 4.0/Unichem, Oxford Molecular, Beaverton, Oregon). All DFT calculations were performed using Becke's 1988 functional, which includes the Slater exchange along with corrections involving the gradient in the density (nonlocal corrections) (Becke, 1988). The gradient corrected Perdew–Wang–'91 (BPW91) (Perdew and Wang, 1992) correlation functional was used, given its superior performance in past studies (Harris *et al.*, 1997) over the Lee–Yang–Parr (BLYP) (Lee *et al.*, 1988) gradient-corrected correlation functionals. A double zeta valence polarization basis (DZVP) set was used in all calculations. These include diffuse d-functions on all atoms except hydrogen. The atomic basis sets in DGauss employed atomic centered Gaussian basis functions optimized for use with DFT (Godbout *et al.*, 1992; Sosa *et al.*, 1992).

This program employs an additional auxiliary approximation in the use of the fitting basis (Dunlap *et al.*, 1979) to express the charge density in a series expansion in Gaussian basis functions. The energy expression, employing this fitting basis, may be computed in order N^3 in contrast to the exact energy expression which scales as N^4. Assessment of the use of this approximation in reproducing the results for small compounds containing iron previously examined by Bauchlicher using couple cluster methods were found to be well reproduced.

In all calculations, a fine density grid was chosen (*ca* 107844 grid points). All the SCF iterations were converged to 2×10^{-6} in the density matrix and 2×10^{-8} in the energy. All optimizations converged within the chosen criterion of 0.0008 hartrees/angstrom. All calculations were made using 2-CPUs of a J90 provided by Oxford Molecular, Beaverton, Oregon.

Iron porphines have closely spaced, low lying electronic states of different spin multiplicity, differing in the spin pairing in open shell iron orbitals. While density functional theory was originally cast for the ground state (Hohenberg and Kohn, 1964) it is equally valid to apply it to the lowest excited states with different spin multiplicities and symmetries (Gunnarsson and Lundqvist, 1976). The spin state species, which is the ground state of oxyferrous and reduced oxyferrous species was determined by optimizing the complexes as all plausible multiplets consistent with the total number of electrons. In principle, singlet and triplet states are plausible candidates for low lying electronic states of the oxyferrous P450 heme species and doublet and quartet states for the reduced oxyferrous P450 heme species. Consequently, geometry optimizations were performed on both of these spin state surfaces. Specifically, an initial SCF was converged for each of these possible spin states of these oxy intermediates. The structures were subsequently optimized until a stationary point was located on each spin state surface located using the unrestricted density functional code embodied in DGauss. The structure and electronic description of the ground state was determined as the lower energy of the two spin state alternatives for each species.

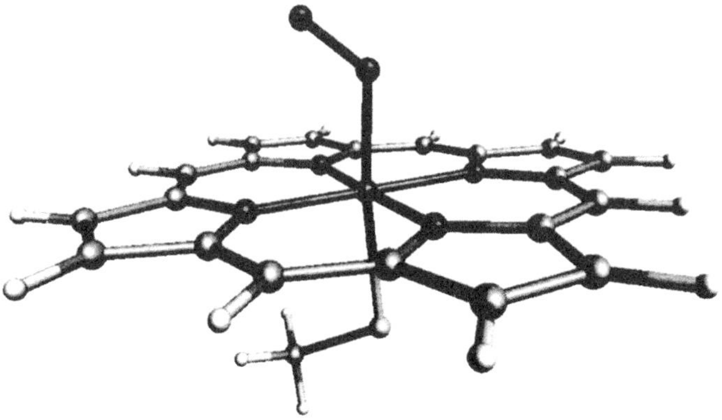

Figure 20.3 Starting structure of model oxyferrous P450: an oxyferrous porphine modeled on the Weiss model compounds

20.3 Results and discussion

20.3.1 Structure of the oxy intermediates of cytochrome P450 deduced from density functional theoretical energy optimizations

20.3.1.1 Comparison of calculated and experimental geometries of the oxyferrous species

Figure 20.3 shows the structure of the methylmercapto iron porphine used as a computationally tractable model of the heme in cytochrome P450s. Previous work at both the *ab initio* and semiempirical level indicate the elimination of the prosthetic side chains of the protoporphyrin IX has only minor consequences on geometric and electronic structure, the largest of which are small perturbations on the energies of nearly degenerate molecular orbitals. The model starting structure was chosen to preserve the salient features of iron–ligand interactions present in the heme system, which are major determinants of the electronic structure of cytochrome P450 intermediates. It is this structural approximation, which has been used to examine the effects of reduction and protonation of intermediates in the enzymatic cycle of cytochrome P450s (cf. Figure 20.2) throughout this study.

The starting point of DFT energy–structure optimization of the oxyferrous form was one approximately that of the Weiss model compound (Schappacher *et al.*, 1987), as discussed in methods above. This model compound structure was of sufficient resolution to discern the mode of oxygen binding and to determine with accuracy all of the geometric features of the system except the O—O bond distance, a consequence of multiple occupancy of the distal oxygen locations in the asymmetric unit of the unit cell. The model compound used by these workers was a picket fence porphyrin with a phenyl-mercaptate linked to the iron. This model compound crystal structure along with the EXAFS structural information (Dawson *et al.*, 1987) serves as structural benchmarks to assess the adequacy of DFT for predictions of the structural intermediates in the P450 enzymatic cycle.

The oxyferrous species in cytochrome P450*cam* as well as the Weiss model compound results from the binding of triplet molecular oxygen to a ferrous porphyrin species, which is a quintet electronic state with four unpaired electrons. The resultant oxyferrous species has been determined to be a diamagnetic singlet state (Davdov *et al.*, 1991; Kobayshi *et al.*, 1994) with no unpaired electrons. The DFT calculation of the two plausible candidate oxyferrous species with differing spin pairing, a triplet and a singlet species, indicates the oxyferrous species to be a singlet ground state in agreement with experiment (Harris *et al.*, 1997).

Table 20.1 presents the calculated structural parameters of the singlet electronic state oxyferrous model species in this study with the geometric parameters determined from the crystal structure of the oxyferrous Weiss model compound and EXAFS bond lengths for cytochrome P450*cam*. The results in this table indicate good agreement between calculated values of the Fe—S, Fe—O, and Fe—N bond length and experimental values. The most notable deviations for calculated and experimental bond lengths are the O—O and C—S bond lengths. In the case of the O—O bond length, the deviation is a principally a manifestation of disorder in the crystal structure result as noted by the authors of this study itself, who note that more credible estimates of O—O bond lengths in such a valence lie in the range of 1.28–1.31 Å (Schappacher *et al.*, 1987). The C—S deviations between computed and the model compound experimental study are largely a reflection of the differences in the model systems used to emulate the cytochrome P450–heme system, with a phenyl mercapto group used in the Weiss study and a methyl mercapto functional group approximating the cysteinate heme in our own. No conclusions may be drawn from the noted deviation in the S—Fe—O1—O2 torsion angle given the high variability in the Weiss model structure and the isolated nature of the model structure in the calculation. The remainder of the calculated angles reported are in fair agreement given the differences in the model systems in which they are present.

Both the calculated and experimentally determined value of the Fe—O1—O2 (iron—proximal oxygen—distal oxygen) angle indicate that the binding motif of the dioxygen molecule in the oxyferrous species is 'end-on'. This end-on binding geometry therefore results in two oxygens, one proximal and one distal to the heme, in two distinct (chemical) bonding environments. How this inequivalence manifests itself in calculated electronic distributions and electrostatic properties is not easy to predict *a priori* but is examined in properties/calculations discussed below.

20.3.1.2 Structure of the reduced oxyferrous species

The addition of a single electron to the oxyferrous species results in a stable (to dissociation) doublet electronic state. This calculated electronic state is consistent with one report of the ESR signature of a species arising from reduction of the oxyferrous species via irradiation (Davdov *et al.*, 1991). Moreover, the calculated spectra of this species is predicted to be red shifted with respect to the spectra of the oxyferrous species (Harris *et al.*, 1997), confirming the plausibility of the assignment recently reported for a species resulting from the reduction of the oxyferrous species in cytochrome P450*cam* (Benson *et al.*, 1997).

The result of adding an electron to the oxyferrous species (the 2nd electronic reduction in the enyzmatic cycle of cytochrome P450s) is shown in Table 20.1. The principal effects of reduction of the oxyferrous species are the elongation of the Fe—S and Fe—O bonds by 0.14 Å and 0.13 Å respectively. The underlying reason for

Table 20.1 Comparison of the calculated nonlocal DFT optimized geometries[a] of oxyferrous and reduced oxyferrous porphine with experimental geometric data for the oxyferrous forms of a model compound and P450*cam*

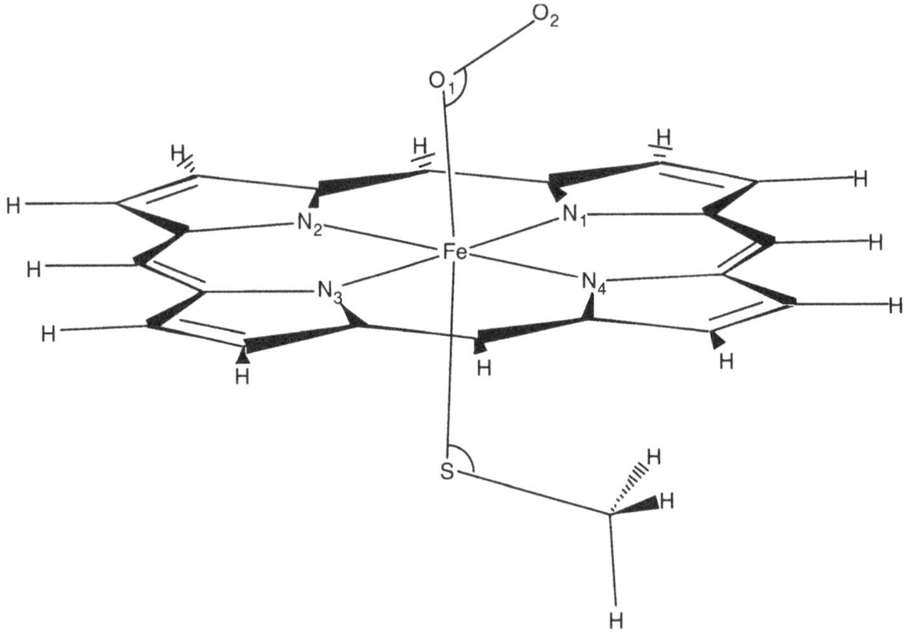

	Oxyferrous species		*Reduced oxyferrous species*
Geometric quantity	Experiment	Calculated	Calculated
Fe—O1	1.82 (1.78)[c]	1.82	1.95
O1—O2	1.14[b]	1.31	1.33
Fe—N1	1.98 (2.0)[c]	2.01	2.01
Fe—N2	1.99	2.01	2.01
Fe—N3	2.01	2.04	2.03
Fe—N4	1.99	2.04	2.03
Fe—S	2.37 (2.37)[c]	2.32	2.46
C2—S	1.72	1.84	1.84
<Fe—O1—O2	128°	122°	120°
<S—Fe—O1	177°	173°	175°
<C—S1—Fe		110°	109°
<N4—Fe—O1—O2	23°–45°	45°	45°
<C—S—Fe—N4		45°	45°

a Distances in angstroms and angles in degrees.
b Position of O_2 disordered in model compound crystal structure resulting in O1—O2 inaccuracy.
c Except for this footnote all values are from (Schappacher *et al.*, 1987) – Weiss Model. EXAFS cytochrome P450*cam* and chloroperoxidase: (Dawson *et al.*, 1986).

this facet may be seen by examining the nature of the lowest unoccupied molecular orbital (LUMO) of the oxyferrous species. As shown in Figure 20.4, this species has significant antibonding character in the Fe—O and Fe—S interactions given the mismatch in signs (color coded in the figure) of the wavefunctions of the atomic

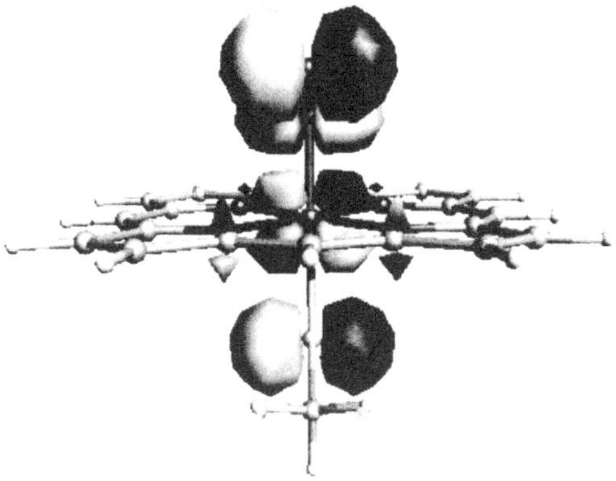

Figure 20.4 The oxyferrous species showing the antibonding character of the LUMO (lowest unoccupied molecular orbital) distributed on the Fe—S—O atoms

orbitals on those centers in this LUMO orbital. This mismatch results in decreasing the bonding electron density in the region between the Fe, S and oxygen nuclei in the reduced oxyferrous species compared to the oxyferrous species, resulting in elongation of the Fe—S and Fe—O bonds. Note that the reduction has only small effects on the remainder of the structure and bonding. In particular, the O—O bond length does not change significantly. While the reduction process results in a slight destabilization of the oxy intermediate form (*ca* +37 kcal mol^{-1}) relative to the oxyferrous form it is, nevertheless, a bound state (nondissociative).

20.3.1.3 Structure of the distal and protonated reduced oxyferrous species

Table 20.2 shows a summary of structural features of both the oxyferrous species and reduced oxyferrous species which change significantly by reduction of the oxyferrous or protonation of the reduced oxyferrous species. The table incorporates the results from Table 20.1 that illustrated that reduction of the oxyferrous species results in elongation of both the Fe—O (0.13 Å) and Fe—S (0.14 Å) bonds. In addition, this table shows that single protonation of either the distal or the proximal oxygen atoms of the reduced oxyferrous species results in significant changes in Fe—O and O—O and Fe—S lengths but in differing manners for the distal and proximal protonated reduced oxyferrous species.

Protonation of the proximal oxygen results in weakening the Fe—O bond (lengthening Fe—O by *ca* 0.09 Å), but makes virtually no change in the (O—O) oxygen—oxygen bond. In contrast, protonation of the distal oxygen (labeled O2 in the diagram above Table 20.2) strengthens the Fe—O bond and weakens the O—O bond as manifested by a decrease in the Fe—O bond length by −0.06 Å and increase in the O—O bond by 0.13 Å. The Fe—S bond is strengthened as evident by its reduction in length by 0.14 Å for distal protonation and 0.19 Å for proximal protonation.

Also shown in Table 20.2 are results from the DFT energy optimization of a single water associated with either the distal or the proximal oxygen of the reduced oxyferrous species. The structural results cited illustrate that while protonation demonstrably effects the structure of the Fe—O, O—O, and Fe—S bonding, mere

Table 20.2 Structural summary of Fe—O, O—O and Fe—S bond length changes with reduction and protonation

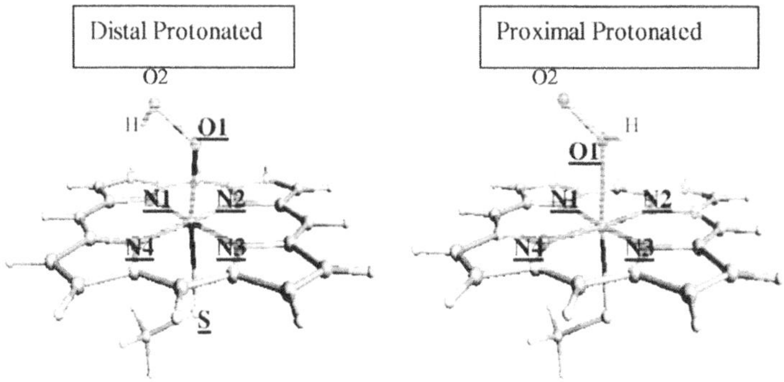

Species	*Structure*[a,b]			*Relative energetics*		
	$R_{Fe—O}$	$R_{O—O}$	$R_{Fe—S}$	*H_2O interaction energy*	*Reduction*	*'Proton-affinity'*
Oxyferrous	1.82	1.31	2.32	–	0.0	–
Reduced oxyferrous	1.95	1.33	2.46	0.0	37.0	0.0
Distal protonated oxyferrous	1.89	1.46	2.32	–	–	–422.6
Proximal protonated oxyferrous	2.06	1.41	2.27	–	–	–404.2
Distal H_2O hydrogen bonded reduced oxyferrous	1.95	1.35	2.46	–11.0	–	–
Proximal H_2O hydrogen bonded reduced oxyferrous	1.98	1.34	2.46	–11.2	–	–

a All bond distances in angstroms. All energies are relative energies in kcal mol^{-1}.
b Changes in bond distances from the reduced oxyferrous species are given in parentheses.
c The reference species for relative energetics is apparent as the zero of energy for that column. Note that all of the species listed are 'stable' to dissociation in that they have total absolute (self consistent field energies) in the order of –2800 hartrees.

hydrogen bonding does not significantly alter the structure. Figure 20.5 summarizes the Fe—O (Panel A) and O—O (Panel B) bond lengths determined for the oxy intermediates beginning with the oxyferrous form and ending with the singly protonated reduced oxyferrous species.

20.3.1.4 Additional protonation of the distal and proximal protonated reduced oxyferrous species: structure of the oxyferryl species

Figure 20.6 shows, in addition to all of the calculated structures of the fully optimized intermediates discussed above, the results after partial DFT optimization of the species resulting from adding a second proton to the distally and proximally protonated species. Conforming to chemical expectations, geometry optimization of the twice distal protonated oxygen results in stable products of the oxyferryl species plus

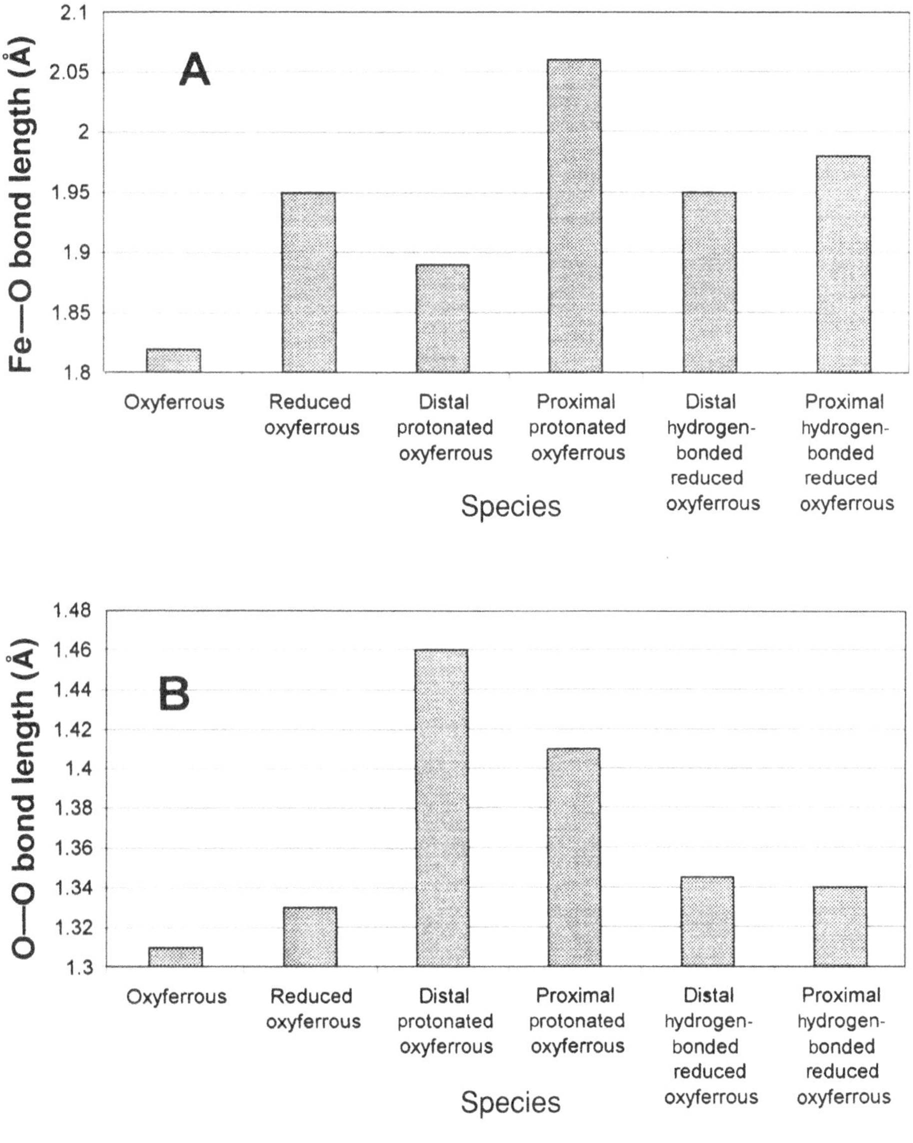

Figure 20.5 The DFT calculated A) Fe—O and B) O—O bond lengths for the oxyferrous, reduced oxyferrous, distal and proximal protonated, and distal and proximal hydrogen bonded reduced oxyferrous species

water. The results clearly indicate the formation of hydrogen peroxide with a long 2.4 Å Fe—O bond length for a combination of distal and proximal oxygen protonation. The results of complete optimization of the oxyferrous, reduced oxyferrous and singly protonated species to a local minima resulted in optimized species with structure and bonding perturbed by the chemical transformation (reduction or proton addition), but nonetheless qualitatively similar in structure and bonding to the starting structure. In contrast, after a few DFT cycles of energy–structure optimization of the diprotonated species, one obtains species with structure and bonding significantly

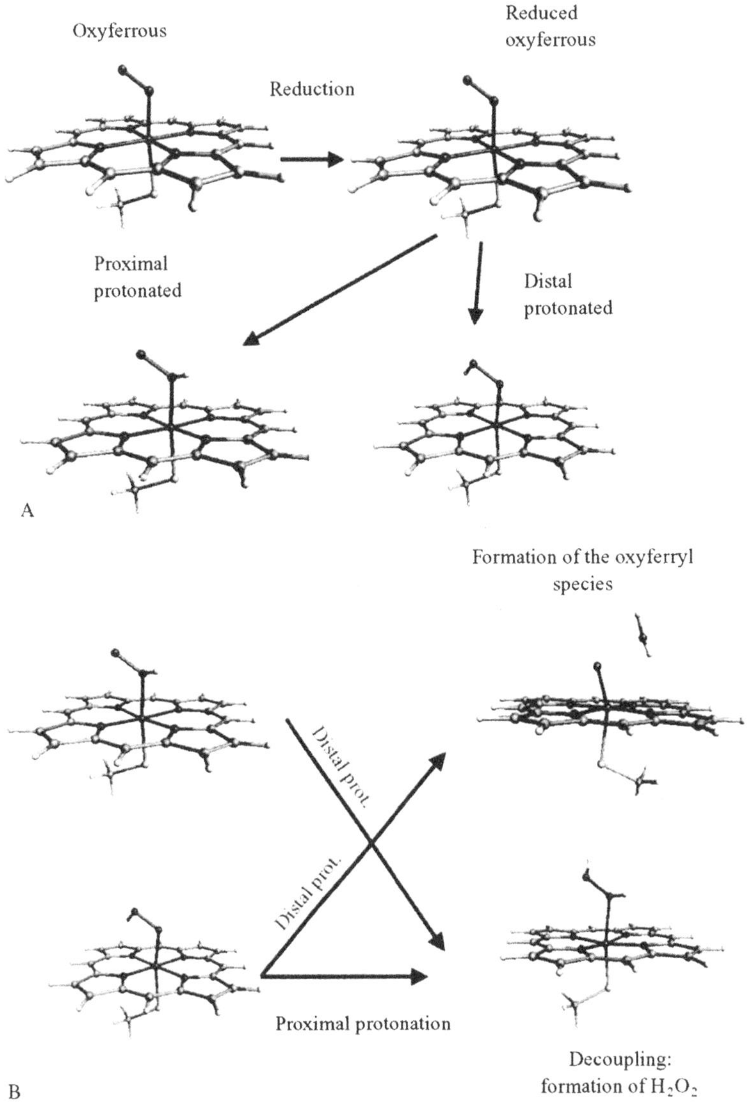

Figure 20.6 A) The structures of the geometry optimized oxyferrous, reduced oxyferrous, and the distal and proximal protonated reduced oxyferrous species. B) The structures of the species resulting from partial optimization of the diprotonated reduced oxyferrous species (oxyferryl + H_2O and hydrogen peroxide + resting state porphine)

and radically different from the starting structures in that, in essence, one obtains product species where the O—O bond is no longer intact.

The results obtained for diprotonation are not remarkable or surprising, in that one expects such an optimization procedure to result in such 'dissociated' products in a case where the species resulting from the proton additions are more stable compared to the reactant species and no significant activation barrier exists between reactants and products. The latter study was performed merely to test the adequacy of density functional theory to give the expected results using the same basis sets and exchange–correlation functional used to predict the oxyferrous, reduced oxyferrous as well as each of the single protonated reduced oxyferrous species. Rigorous examination of the nature of the energy surface between reactants and products, e.g. the presence of transition states and determination of activation energy barriers, require a different sort of calculation.

The diprotonated results are worth noting, however, because no activation energy barrier is expected in such proton donation schemes, and consequently one expects to obtain the noted products. Contrasting these 'diprotonated' structural results with those for the other species emphasizes the facet that: in addition to the oxyferrous and reduced oxyferrous species, which we expect to be stable intermediates in the P450 enzymatic cycle, the distal and proximal singly protonated species are viable intermediates, i.e. while there are distinct structural changes in the Fe—S, Fe—O and O—O bonds, the protonated dioxygen (i.e. Fe—OOH or Fe—(HO)O) moiety remains bound to the iron porphine.

20.3.2 Relative energies of the oxyferrous, reduced oxyferrous, water complexed reduced oxyferrous and singly protonated reduced oxyferrous species

Table 20.2 reports the relative energies of several plausible intermediates in the enzymatic cycle of cytochrome P450. The absolute quantum mechanically derived energies are not reported here for conciseness, but are large negative quantities indicating that these are putative molecular intermediates which do not spontaneously dissociate to compound I.

The relative energies in Table 20.2 of the relative energy of the reduced oxyferrous species relative to the oxyferrous form indicate that the reduction is destabilizing in the reaction pathway in that the reduced oxyferrous species is 37 kcal mol^{-1} higher in energy than the unreduced oxyferrous form. The relative energies for a single water hydrogen bonded to the reduced oxyferrous species indicate that such interactions stabilize the reduced oxyferrous species by approximately 11 kcal mol^{-1} for both the distal and proximal hydrogen bonded forms. This demonstrates the extent to which interactions with a single water or a polar amino acid in the binding site, short of complete proton transfer, stabilize this intermediate. It also indicates that the distal and proximal sites are likely to be equivalent as electrostatic interaction sites, since hydrogen bonding, as probed in the calculations of the water complexed reduced oxyferrous species, has large contributions due to electrostatics. These results concur with conclusions based on mutational studies (Martinis *et al.*, 1989; Imai *et al.*, 1989) as well as molecular dynamics (Harris and Loew, 1994) which indicated the possible import of hydrogen bonding in stabilization of oxy intermediates in the P450 enzymatic cycle.

Table 20.2 also shows that the distal single protonated species is, in fact, more stable than the proximal protonated reduced oxyferrous form. Thermodynamically, distal protonation appears to be favored over the proximal protonation. Therefore, while hydrogen bonding to the distal and proximal oxygen are equivalent, a largely electrostatic phenomenon, these results indicate actual proton additions, where actual bond formation occurs, are not. The values listed for the approximate proton affinities of the distal and proximal oxygens are large negative quantities, but are not atypical for protonation of a doubly charged negative species. Clearly the changes in energies with such protonations, however, are not to be viewed as stabilization of the reduced oxyferrous species but merely part of an overall chemical process requiring diprotonation.

20.3.3 Excess spin distributions of the reduced oxyferrous and protonated reduced oxyferrous and oxyferryl intermediates

Figure 20.7 shows the nature of the radical species by illustrating the excess spin distributions each of the species examined, while Table 20.3 summarizes the excess spin populations on the S, Fe, O, and porphine atoms. In Figure 20.7, the distribution of the excess electron spin density of the radical species is indicated by contours about those atoms having an excess of 'up' versus 'down' electronic spin population. As mentioned above, the ground state of the oxyferrous species was a singlet electronic state, indicating no unpaired electrons and is therefore not included in this figure. As clearly shown in this figure, as well as Table 20.3, the reduced oxyferrous and the two singly protonated forms have similar descriptions as regards the location of the unpaired electron in the doublet ground state, which is distributed over the oxygen atoms. In addition, the protonated reduced oxyferrous species have a small amount of unpaired spin density on the iron and sulfur as well.

The electronic structure of the oxyferryl species of this model P450 system currently under investigation, without the protoporphyrin IX (heme) prosthetic groups, indicates the lowest energy quartet of this model P450 system to be quartet A_{2u} state

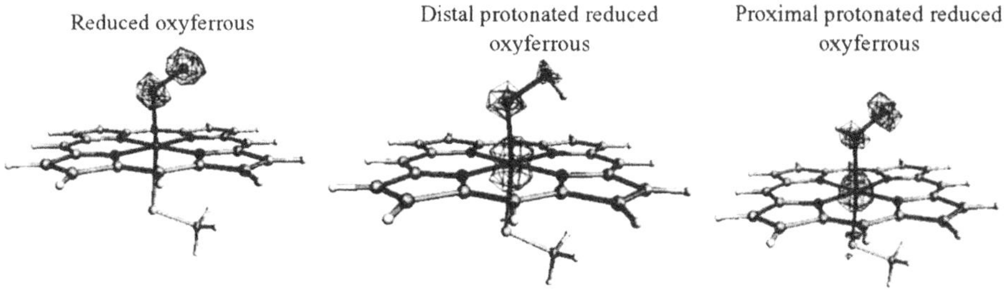

Figure 20.7 Excess spin densities of the reduced oxyferrous and protonated reduced oxyferrous species

Table 20.3 Excess spin densities($\rho\uparrow$-$\rho\downarrow$) for the ground state doublet species of the reduced oxyferrous and protonated reduced oxyferrous species

	Reduced oxyferrous	*Distal protonated reduced oxyferrous*	*Proximal protonated reduced oxyferrous*
Spin population on atoms[a]			
Fe	−0.22	0.59	0.39
O1[b]	0.42	0.24	0.12
O2[b]	0.51	0.06	0.36
S	−0.02	0.18	0.19
Porphine π	+0.32		

a Mulliken open shell (unpaired spin) populations.
b O1 is the proximal oxygen and O2 the distal oxygen.

with a total of two electrons on the Fe═O unit (total spin S = 1) and additional electron in the a_{2u} porphyrin π orbital (S = ½). These investigations, however, also indicate the presence of a low lying doublet state with a similar unpaired spin distribution on the Fe, O, S, and porphine orbitals to the $^4A_{2u}$ electronic state but with the electron spins in the porphine π and mercapto sulfur orbitals to the unpaired electron spins on the Fe and O atoms. Investigations are underway to ascertain the relative energy ordering of these two closely spaced electronic states in the oxyferryl species as a function of basis set and density functionals.

20.3.4 Molecular electrostatic potential surfaces of the oxy intermediates

Having calculated the structures of the oxy intermediates using DFT, one may calculate the molecular electrostatic potential (MEP) for grid points outside the van der Waals surface of the oxy intermediates using the ground state wavefunction of each of these species. The electrostatic potential calculated in this way indicates the manner in which potential polar groups in the binding site and in particular hydrogen bonding amino acid sidechains and water will interact with each of the species. This is particularly true of the prediction of hydrogen bond interactions, since a large portion of the total energetics of hydrogen bonding is due to electrostatic components.

Figure 20.8 shows the molecular electrostatic potentials of three of the putative intermediate oxyradical species in the P450 enzymatic cycle. The first panel shows

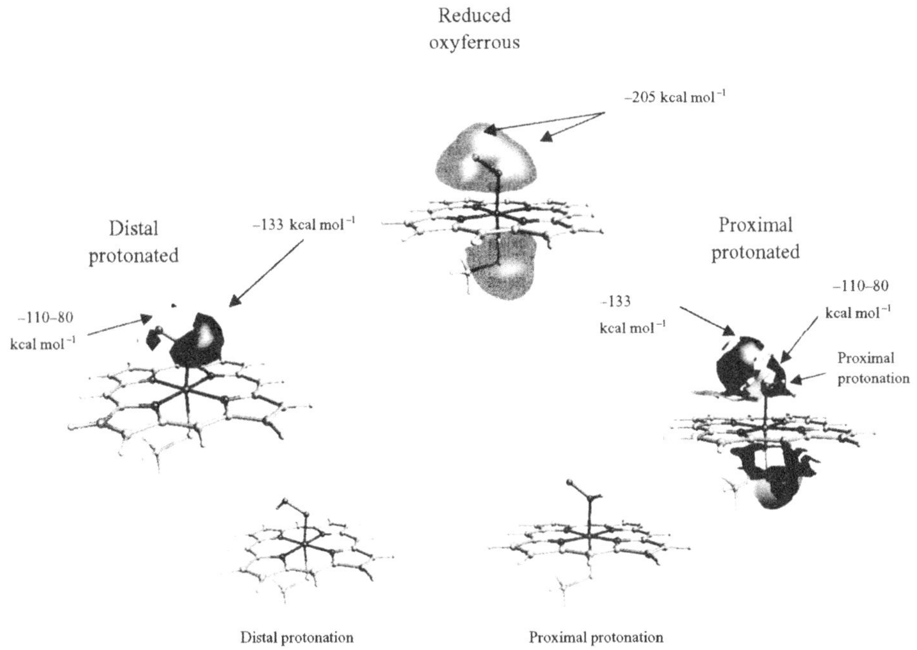

Figure 20.8 The molecular electrostatic potential surfaces of the reduced oxyferrous (top), distal protonated reduced oxyferrous (bottom, left two figures) and proximal reduced oxyferrous (bottom, right two figures) species. The range of the molecular electrostatic potential near the proximal and distal oxygens is indicated next to each representation

the MEP of the reduced oxyferrous species. It indicates that there is a pronounced minimum (blue) near the bound dioxygen atoms, and that the minima near the distal and proximal oxygen atoms are equivalent. Note this is consistent with the results in Table 20.2, which indicated that the interaction energies of a single water hydrogen bonded to the distal and proximal oxygen atoms were equivalent. These results suggest that *a priori* there should be no preference for a hydrogen bonding network terminating on the distal oxygen than on the proximal oxygen. Clearly, however, substrate and/or binding site features can alter this preference, so that the nature of the substrate and P450 binding site is crucial to determining the hydrogen bond networks surrounding the bound dioxygen in the P450 oxy intermediate forms. Previous molecular dynamics studies of the reduced oxyferrous form of P450eryF indicate that wild type (efficacious) production of the oxyferryl species may be correlated with the formation of hydrogen bonding patterns in which two stable hydrogen bond networks terminate on the distal (terminal) oxygen atom of the reduced oxyferrous species (Harris and Loew, 1996), thus permitting simultaneous delivery of two protons from some ultimate proton source. This hypothesis is supported by observations that mutations perturbing hydrogen bond networks promote decoupling of the P450eryF enzymatic cycle (Cupp-Vickery *et al.*, 1996).

The bottom of Figure 20.8 shows the MEPS of the distal and proximal protonated reduced oxyferrous species potential surfaces. Both pairs of figures for the distal and proximal protonated species illustrate a facet, that while in hindsight obvious, is nonetheless of interest mechanistically. Examining the MEPS of the pair of figures for the distal protonated species, one notes, in either the expanded scale MEP to the far left, or the figure adjacent, that the protonation of the distal oxygen results in a significant difference in the electrostatic potential of 10–25 kcal mol^{-1} between the distally protonated oxygen and the proximal oxygen. Such an inequivalence would indicate that if there is sufficient time between the first and second proton additions for adjustment of hydrogen bonding (e.g. of waters or a polar amino acid proximate to the dioxygen atoms), the hydrogen bonding pattern and perhaps even the pathway for delivery of the second proton will favor the unprotonated oxygen from the standpoint of electrostatics alone. The two figures (MEPS) to the bottom right of Figure 20.8, showing two different views of the MEPS of the proximal protonated reduced oxyferrous species, indicate a similar phenomenon for the proximal as for the distal protonated species. Thus, asynchronous diprotonation itself could favor changes in hydrogen bonding and/or proton delivery, which would enhance delivery of the two protons to different oxygens, i.e. favoring decoupling. Mutations of P450 enzymes or substrate modifications that enhance complete asynchronous addition of the protons may alter the electrostatic features favoring decoupling in the absence of other influences.

20.4 Conclusions

Density functional theoretical methods have been used to probe the nature of the oxy intermediates in the ubiquitous enzymatic cycle of cytochrome P450s using a model approximation to the complete heme as methyl mercaptate bound to the iron porphine. Employing this correlated electron method, we have reported the energy optimized structures of the plausible putative intermediate forms leading from the oxyferrous species through the oxyferryl species, determined the geometric and electronic structures, as well as determined the ground state spin multiplicities and electrostatic potential surfaces of these species. The relative energies of the putative species have been determined from their ground state DFT energies. The geometric changes due to: 1) reduction of the oxyferrous species, 2) distal oxygen protonation, 3) proximal oxygen protonation, as well as 4) hydrogen bonding of water to the distal and, 5) proximal oxygens have been determined.

The computed geometry of the oxyferrous model P450 species employed in this study indicates an end-on binding geometry, analogous to a model compound crystal structure of this oxyferrous species (Schappacher *et al.*, 1987). The geometric (bond lengths and angles) structure of the DFT calculated oxyferrous form is in good agreement with both EXAFS data for the oxyferrous species in cytochrome P450*cam* (Dawson *et al.*, 1986; Dawson and Sono, 1987) as well as the Weiss model compound crystal structure (Schappacher *et al.*, 1987). The comparison of the DFT computed structure of the oxyferrous species with these experimental determinations of the structure of this enzymatic intermediate provides validation of the ability of this method to compute the structures of intermediates in the P450 enzymatic cycle.

The DFT results for the oxyferrous species indicate it to be a singlet ground state, with no unpaired spin density. DFT optimizations of the oxyferrous species with addition of a single electron indicate the ground state of the reduced oxyferrous species to be a doublet ground state with one unpaired electron distributed over both the distal and proximal oxygen atoms. The computed ground state spin multiplicities of the oxyferrous and reduced oxyferrous species are consistent with the reported ESR data for these species (Davdov *et al.*, 1991; Kobayshi *et al.*, 1994). The proximal and protonated reduced oxyferrous species are both doublet ground states with unpaired (excess) electron spin density distributed over the both bound dioxygen atoms, but with some excess spin density on the Fe and S atoms. The nature of the oxyferryl species as probed by these methods is currently under investigation.

The computed structures of the putative intermediates (cf. Figure 20.2) indicate:

1. the reduction of the oxyferrous species results in primarily elongation of the Fe—O and Fe—S bonds,
2. hydrogen bonding of a single water to the distal oxygen or proximal oxygens of the reduced oxyferrous species makes very minor changes in the structure: principally a slight elongation of the O—O bond,
3. the principal effect of proximal protonation of the reduced oxyferrous species is Fe—O bond elongation,
4. in contrast, the principal effect of distal protonation of the reduced oxyferrous form is O—O bond lengthening.

The computation of the ground state structures of the above mentioned oxy forms of the model porphine system indicate them to all be stable intermediates in the sense that they are electronic bound states and as such are not spontaneously dissociative. They are, of course, highly reactive and in that sense metastable (oxyferrous) or unstable species difficult to trap, isolate, and rigorously characterize in the wt P450 enzymes. Both these computations determining that these species are 'intermediates' that do not spontaneously generate compound I (the oxyferryl species), as well as experimental kinetic isotope measurements (Aikens and Sligar, 1994), indicate the requirement of two protons in the conversion of the reduced oxyferrous species to compound I.

Examination of the relative energies of the oxy intermediates in this work indicates that, while reduction results in a destabilization of the reduced oxyferrous species relative to the oxyferrous form, hydrogen bonding to the reduced oxyferrous form stabilizes the reduced oxyferrous–water complexed form by up to 11 kcal mol^{-1} over the isolated species. This indicates the dual role that hydrogen bonding by water/ and or hydrogen bonding amino acid sidechains may play in the enzymatic cycle of cytochrome P450s by both stabilization of intermediate oxy intermediates as well as being the pathway by which protons are ultimately delivered to the distal and proximal oxygens in both the efficacious and dysfunctional (decoupled) pathways. While distal or proximal single protonation significantly reduces the energy of the reduced oxyferrous species, the DFT calculations in this study confirm that neither the reduction step alone nor single protonation result in a species which spontaneously converts to the oxyferryl species. Rather, diprotonation is confirmed in the DFT calculations in this study to be the principal prerequisite for conversion to either the oxyferryl form in the functional pathway or to hydrogen peroxide in the decoupling of the enzymatic pathway.

Examination of the computed molecular electrostatic potential surfaces of the reduced oxyferrous species indicates equivalent minima near the distal and proximal oxygens for interaction with electropositive entities in the P450 enzymes, e.g. hydrogen bonding hydrogens. This is consistent with the observation of equal computed energies for the reduced oxyferrous species hydrogen bonded to water at the distal proximal oxygens. Whether one views the proton transfer events in the enzymatic cycle of P450s as involving the transfer of a protons from a remote ultimate source of protons or merely in the replacement of a proton donated locally, the role and import of hydrogen bond networks in providing a pathway for this transfer process is evident. The calculations in this study indicate there to be no *a priori* basis, in the absence of substrate and the protein environment, for such hydrogen bond networks terminating the distal versus the proximal oxygen atom of the reduced oxyferrous species.

Given the possibility that the addition of protons to the reduced oxyferrous species of cytochrome P450s may be, in some cases, sequential rather than synchronous, the molecular electrostatic potential surfaces of the distal and proximal protonated reduced oxyferrous forms have also been computed. Examination of these surfaces makes apparent the facet that the single protonation, resulting in nonequivalence of the two oxygen centers, results in a difference of 10–20 kcal mol^{-1} in the electrostatic potential minima about the two oxygens. The already protonated center is thus disfavored compared to the unprotonated oxygen center for interaction with electropositive/hydrogen-bonding species. While speculative, this asynchronicity in the protonation steps could facilitate the decoupling of the P450 enzymatic cycle if sufficient time transpires for rearrangement of hydrogen bonding patterns to favor the network terminating on the atomic center associated with the lower electrostatic minima. Consequently, in such an instance, the alternate (unprotonated) oxygen center would likely be the site of the second protonation, favoring formation of hydrogen peroxide. Even without such a hydrogen bond network rearrangement, the gradient in the electrostatic potential indicates the pathway for proton transfer to be biased toward the alternate (unprotonated) oxygen, and the same outcome: decoupling (formation of H_2O_2 rather than the oxyferryl species) as a result of diprotonation. Mutations, which affect the nature and topology of hydrogen bond networks terminating on the reduced oxyferrous species oxygens atoms might determine extent of decoupling by either providing pathways for proton transport to both the distal and proximal oxygens or by enabling asynchronous transfer of the two protons required.

The results of this study illustrate that density functional theoretical methods including electron correlation are adequate for addressing questions of structural and chemical import and may be used to frame significant questions regarding the structure and properties of radical intermediate species involved in the chemistry of enzymes mediating the metabolism of xenobiotics in living systems.

Acknowledgements

The author would like to thank Dr Gilda H. Loew for critical comments and support (NIH-GM 56125) during the course of this research. Additionally, the author would like to thank Oxford Molecular for both providing access to DGauss/Unichem 4.0 and for computer time on the Oxford Molecular Cray-J90.

References

Aikens, J. and Sligar, S. 1994, Kinetic Solvent Isotope Effects during Oxygen Activation by Cytochrome P450*cam* J. Am. Chem. Soc., 116, 1143–1144.

Bach, R.D., Schobe, and D.S. Schlegel, H.B. 1996, J. Phys. Chem., 100, 8770.

Bangcharoenpaurpong, O., Rizos, A.K., Champion, P.M., Jollie, D. and Sligar, S.G. 1986, Resonance Raman Detection of Bound Dioxygen in Cytochrome P450*cam* J. Biol. Chem., 261, 8089.

Becke, A.D. 1988, Density functional exchange energy contribution with correct asymptotic behavior. Phys. Rev. A, 38, 3098–3100.

Benson, D.E., Suslick, K.S. and Sligar, S.G. 1997, Reduced Oxy Intermediate Observed in D251N Cytochrome P450*cam* Biochemistry, 36, 5104.

Cupp-Vickery, J.R. and Poulos, T.L. 1995, Structure of cytochrome P450eryF involved in erythromyocin biosynthesis, Structural Biology, 2, 144–153.

Cupp-Vickery, J.R., Han, O., Hutchinson, C.R. and Poulos, T.L. 1996, Nature Struct. Biol. 3, 632–637.

Davdov, R., Kappl, R., Huttermann, and J. Peterson, J.A. 1991, EPR spectroscopy of reduced oxyferrous cytochrome P450*cam* FEBS, 295, 113.

Dawson, J.H., Kau, L.-S., Penner-Hahn, J.E., Sono, M., Eble, K.S., Bruce, G.S., Hager, L.P. and Hodgson, K.O. 1986, Oxygenated Cytochrome P450-CAM and Choroperoxidase: Direct Evidence for Sulfur Donor Ligation Trans to Dioxygen and Structural Characterization using EXAFS Spectroscopy. J. Am. Chem. Soc., 108, 8114.

Dawson, J.H. and Sono, M. 1987, Cytochrome P-450 and Choroperoxidase: Thiolate-Ligated Heme Enzymes. Spectroscopic Determination of Their Active Site Structures and Mechanistic Implications of Thiolate Ligation. Chem. Rev., 87, 1255.

Dgauss 4.0/Unichem Oxford Molecular, Beaverton Oregon.

Dunlap, B.I., Connolly, J.W.D. and Sabin, J.R. 1979, On some approximations in applications of X (theory). J. Chem. Phys., 71, 3396–402.

Dunlap, B.I., Connolly, J.W.D. and Sabin, J.R. 1979, On first-row diatomic molecules and local density models, 71, 4993–9.

Fan, L. and Ziegler, T. 1991, Optimization of molecular structures by self-consistent and nonlocal density-functional theory. J. Chem. Phys., 95, 7401–8.

Ghosh, A., Almlof, J. and Que, L. Jr. 1993, Density Functional Study of Oxo(Porphyrinato)Iron(IV) Complexes, Models of Peroxidase Compounds I and II. J. Am. Chem. Soc., 98, 5576.

Godbout, N., Salahhub, D.R., Andzelm, J. and Wimmer, E. 1992, Optimization of Gaussian-type basis sets for local spin density functional calculations. I. Boron through neon, optimization technique and validation. Can. J. Chem., 70, 560–571.

Groves, J.T. and Han, Y.-H. 1995, Models and Mechanisms of Cytochrome P450 Action., in Cytochrome P450: Structure, Mechanism, and Biochemistry. 2nd ed. (Paul. R. Ortiz de Montellano, ed.), Plenum Press, New York, pp 1–27.

Gunnarsson, O. and Lundqvist, B.I. 1976, Exchange and correlation in atoms, molecules, and solids by the spin-density formalism. Phys. Rev. B, 13, 4274.

Harris, D. and Loew, G.H. 1994, A Role of Thr 252 in Cytochrome P450*cam* Oxygen Activation. J. Am. Chem. Soc. 116:11671–11674.

Harris, D.L. and Loew, G.H. 1996, Investigation of the Proton Assisted Pathway to Formation of the Catalytically Active, Ferryl Species of P450s by Molecular Dynamics Studies of P450eryF. J. Am. Chem. Soc. 118; 6277–6387.

Harris, D., Loew, G.H. and Waskell, L. 1997, Structure and Spectra of Ferrous Dioxygen and Reduced Ferrous Dioxygen Model Cytochrome P450, JACS submitted.

Haseman, C.A., Ravichandran, K.G., Peterson, J.A. and Deisenhofer, J. 1994, Crystal structure and refinement of cytochrome P450-terp at 2.3 Å resolution. 236, 1169–1185.

Hohenberg, P. and Kohn, L.J. 1964, Inhomogeneous electron gas. Phys. Rev. B, 136, 864–871.

Imai, T., Schimada, H., Watanabe, Y., Matsushima-Hibiya, Y., Makino, R., Koga, H., Horiushi, T. and Ishimura, Y. 1989, Uncoupling of the Cytochrome P-450*cam* monooxygenase reaction by a single mutation, threonine-252 to alanine or valine: A possible role of the hydroxy amino acid in oxygen activation. Proc. Nat. Acad. Sci., USA, 86:7823.

Kobayshi, K., Iwamoto, T. and Honda, K. 1994, Spectra Intermediate in the Reaction of Ferrous Cytochrome P450*cam* with Superoxide Anion. Biochem. and Biophys. Res. Comm., 201, 1348.

Kohn, W. and Sham, L.J. 1965, Self consistent equations including exchange and correlation effects. Phys. Rev. A, 140, 1133–1138.

Kuramochi, H. Noodleman, and Case, D.A. 1997, Density Functional Study on the Electronic Structures of Peroxidase Comounds I and II. J. Am. Chem. Soc., 119, 11442–11451.

Lee, C., Yang, W. and Parr, R.G. 1988, Development of the Cole–Salvetti correlation energy formula into a functional of the electron density. Phys. Rev. B, 37, 785–789.

Li, H. and Poulos, T.L. 1995, Modeling protein-substrate interactions in the heme domain of cytochrome P450-BM-3. Acta Crystallogr. D, 51, 21–32.

Li, H. and Poulos, T.L. 1996, Conformational dynamics in cytochrome P450-substrate interactions. Biochimie, 78, 695–9.

Martinis, S.A., Atkins, W.M., Slayton, P.S. and Sligar, S.G. 1989, A Conserved Residue of Cytochrome P-450 Is Involved in Heme Dioxygen Stability and Activation. J. Am. Chem. Soc., **111**:9252.

Matsunaga, K., Iwawaki, Y., Wanatabe, K., Narimatsu, S., Yamamoto, I., Imaoka, S., Funae, Y. and Yoshimura, H. 1993, Cytcochrome P450 isozymes catalyzing the hepatic microsomal oxidation of 9-anthraldehyde to 9-anthracene carboxylic acid in adult male rats. Biol. Pharm. Bull., 16, 866–869.

Mouesca, J.-M., Chen, J.L., Noodleman, L., Bashford, D. and Case, D.A. 1994, Density Functional/Poisson-Boltzmann Calculations of the Redox Potentials of Iron-Sulfur Clusters. J. Am. Chem. Soc., 116, 11898.

Perdew, J.P. and Wang, Y. 1992, Accurate and simple analytic representation of the electron-gas correlation energy. Phys. Rev., 45, 13244.

Peterson, J.A., Ishimura, Y. and Griffin, B.W. 1972, Pseudomonas putida cytochrome P-450: characterization of an oxygenated form of the hemoprotein. Arch. Biochem. Biophys., 149, 197–208.

Raag, R. and Poulos, P. 1989, Crystal structure of the carbon monoxide-substrate-cytochrome P-450*cam* ternary complex. Biochemistry, 28, 7586–7592.

Raag, R. and Poulos, T.L. 1992, X-Ray Crystallographic Structural Studies of Cytochrome P450*cam*: Factors Controlling Substrate Metabolism. Frontiers in Biotransformations Ruckpahl, R., Ed. Akademie-Verslag: Berlin, 1992 Chapter 1., p1 Taylor and Francis Publ. London.

Ravichandran, K.G., Boodupalli, S.S., Hasemann, C.A. and Peterson, J.A. 1993, Crystal structure of hemoprotein domain of P450BM-3, a prototype for microsomal P450's. Science, 261, 731–736.

Roberts, E.S., Vaz, A.D.N. and Coon, M.J. 1991, Catalysis by cytochrome P450 of an oxidative reaction in xenobiotic aldehyde metabolism: Deformylation with olefin formation. Proc. Nat. Acad. Sci., 88, 9063–8966.

Rovira, C., Kunc, K., Hutter, J., Ballone, P. and Parrinello, M. 1997, Equilibrium Geometries and Electronic Structure of Iron-Porphyrin Complexes: A Density Functional Study. J. Phys. Chem., 101, 8914–8925.

Schappacher, M., Ricard, L., Fischer, J., Weiss, R., Bill, E., Montiel-Montoya, R., Winkler, H. and Trautwein, A.X. 1987, Synthesis, structure and spectroscopic properties of two models for the active site of the oxygenated state of cytochrome P450. Eur J. Biochem. 168(2):419–430.

Sharrock, M., Debrunner, P.G., Schulz, C., Lipscomb, J.D., Marshall, V. and Gunsalus, I.C. 1976, Cytochrome P450*cam* and its complexes. Mossbauer parameters of the heme iron. Biochim Biophys Acta, 420, 8–26.

Sosa, C., Andzelm, J., Elkin, B.C., Wimmer, E., Dobbs, K.D. and Dixon D.A. 1992, A local density functional study of the structure and vibrational frequencies of molecular transition-metal compounds. J. Phys. Chem., 96, 6630–6.

Vaz, A.D.N., Roberts, E.S. and Coon, M.J. 1991, Olefin formation in the oxidative deformylation of aldehydes by cytochrome P450. Mechanistic implications for catalysis by oxygen derived peroxide. J. Am. Chem. Soc., 113, 5886–5887.

Wanatabe, Y. and Groves, J.T. 1992, Molecular Mechanism of Oxygen Activation by Cytochrome P450, in: The Enzymes, 3rd ed., Vol. XX. (D. Sigaman, ed.), Academic Press, New York, pp 406–453.

21 Spin-trapping – applications to photochemical and photobiological problems

Colin F. Chignell

21.1 Introduction

The interaction of light (natural or artificial) with chemical agents present in the skin or eyes often results in photosensitization (Urbach, 1992). The chemical photosensitizer may be endogenous (protoporphyrin in erythropoietic porphyria), a drug (chlorpromazine, declomycin), a topical agent (halogenated salicylanilides in soap, musk ambrette in perfume), or an environmental agent (polycyclic aromatic hydrocarbons in coal tar). Photosensitivity may manifest itself as phototoxicity and/or photoallergy. Phototoxicity generally occurs during a subject's first exposure to light and chemical agent and often results in an exaggerated erythemal response ('sunburn'). Phototoxicity will occur in all individuals provided that they are exposed to sufficient amounts of light and the offending chemicals. Photoallergy may be preceded by a phototoxic response. As this subsides delayed abnormal responses may occur including papular, eczematous and urticarial reactions. The latter may persist for months to years even after avoidance of the photoallergen. As with non-light dependent allergic responses, photoallergy is not observed in all subjects exposed to the photoallergen.

21.2 Photochemistry of photosensitizers

The initial step in all forms of photosensitization is the absorption of radiation (visible light or UVA) by the chemical or its metabolites. When a molecule in the ground state (S_0) absorbs a photon of light an electron moves rapidly (within 10^{-15} s) to an excited singlet energy level (^1S, Figure 21.1). Once there, the electron can return to the S_0 state by emitting energy (fluorescence) or by passing its energy to the environment (internal conversion). The ^1S state is fairly short-lived ($\sim10^{-9}$ s) and so photochemistry does not generally occur from this state. The electron may also move ('intersystem crossing') to a triplet energy level (^3S) where it can return to the ground state by emitting energy (phosphorescence) or non-radiatively via intersystem crossing. However, because the ^3S state is generally long-lived (10^{-3} s to seconds) it is from this state that most photochemistry occurs.

The reactions that occur from the triplet (or singlet) state may follow a number of different pathways depending on the chemical nature of the sensitizer, the solvent and other molecules (substrates) present in the system. Two major processes may occur which have been termed Type I and Type II. In Type I (free radical or redox)

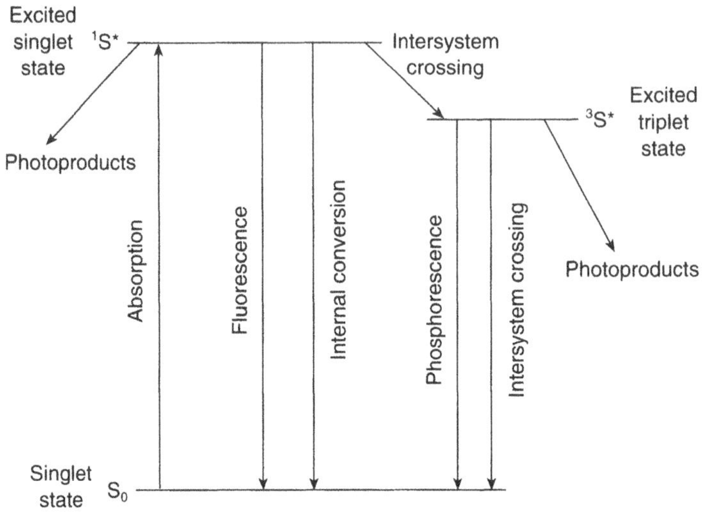

Figure 21.1 Energy diagram

reactions the triplet sensitizer may abstract an electron (or hydrogen atom) from a substrate molecule (D) (Equations 21.1 and 21.2).

$$S_O \xrightarrow{h\nu} {}^1S^* \longrightarrow {}^3S^* \qquad (21.1)$$

$$^3S^* + D \rightarrow S^{\bullet -} + D^{+\bullet} \qquad (21.2)$$

Alternatively, the sensitizer may act as an electron donor and the substrate molecule may function as an acceptor (A) (Equation 21.3).

$$^3S^* + A \rightarrow S^{+\bullet} + A^{\bullet -} \qquad (21.3)$$

If oxygen is the electron acceptor then the product is superoxide (Equation 21.4).

$$^3S^* + O_2 \rightarrow S^{+\bullet} + O_2^{-\bullet} \qquad (21.4)$$

Superoxide may also be formed by reaction of the reduced form of the substrate (Equation 21.2) with oxygen (Equation 21.5).

$$S^{\bullet -} + O_2 \rightarrow S + O_2^{-\bullet} \qquad (21.5)$$

In all of these cases the initial photoproducts are free radicals.

Free radicals may also be generated if the excited molecule undergoes homolytic fission. This often occurs with halogenated photosensitizers, particularly those containing chlorine or bromine atoms (X = Cl, Br) attached to an aromatic ring (Equation 21.6).

$$ArX \xrightarrow{h\nu} Ar^{\bullet} + X^{\bullet} \tag{21.6}$$

Photolysis of peroxides gives rise to highly reactive alkoxyl and hydroxyl radicals (Equation 21.7)

$$ROOH \xrightarrow{h\nu} RO^{\bullet} + {}^{\bullet}OH \tag{21.7}$$

In Type II (energy transfer) reactions, electronic excitation energy is transferred from $^*S^3$ to ground state (triplet) oxygen to give singlet oxygen (Equation 21.8).

$$^3S^* + {}^3O_2 \rightarrow S_o + {}^1O_2 \tag{21.8}$$

The latter is a highly oxidizing form of oxygen that reacts with many different biomolecules much more rapidly than ground state oxygen (Frimer, 1985). For example, singlet oxygen reacts with unsaturated fatty acids ('ene addition') to give a peroxide (Equation 21.9) (Frimer and Stephenson, 1985). In the presence of a reducing agent the peroxide is converted to a highly reactive alkoxyl radical (Equation 21.10) which can initiate lipid peroxidation (Girotti, 1990).

$$R^1{-}CH_2{-}CH{=}CH{-}R^2 \xrightarrow{^1O_2} R^1{-}CH{=}CH{-}\underset{\displaystyle OOH}{\underset{|}{CH}}{-}R^2 \tag{21.9}$$

$$ROOH \xrightarrow{[Red]} RO^{\bullet} + OH^- \tag{21.10}$$

21.3 Free radicals and EPR

Free radicals from photochemical reactions may be detected by a number of different techniques which are summarized in Table 21.1 along with their respective advantages and disadvantages. Electron paramagnetic (spin) resonance, EPR (ESR) is the only spectroscopic technique that directly detects free radicals and can determine their structure. For example, Figure 21.2 shows the EPR spectrum observed when benzanthrone (BA), a photosensitizing dye intermediate, is irradiated in alkaline ethanol. The spectrum was attributed (Dabestani *et al.*, 1992) to the benzanthrone anion radical, $BA^{\bullet-}$, formed by the reaction of triple benzanthrone with the solvent (Equation 21.11).

$$^3BA^* + EtO^- \rightarrow BA^{\bullet-} + EtO^{\bullet} \tag{21.11}$$

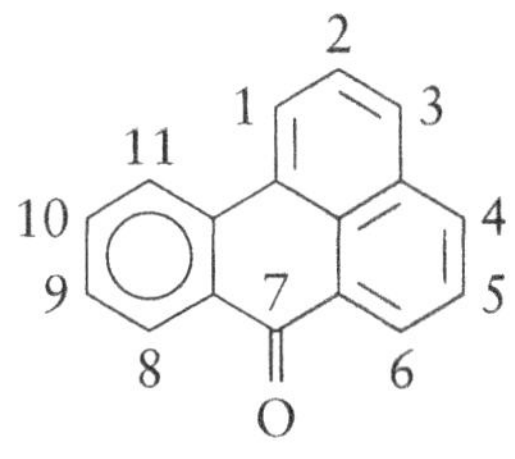

BA

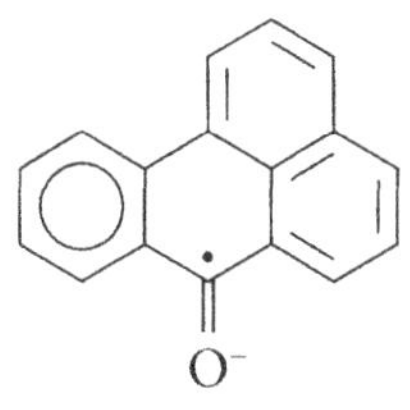

$BA^{\bullet-}$

Table 21.1 Methods for detecting and identifying free radicals

Method	*Advantages*	*Disadvantages*
Product analysis	Employs well-established analytical techniques	Does not provide unequivocal evidence for presence of radicals
Inhibition by free radical scavengers (e.g. anti-oxidants)	Simple technique	Little information about structure of radical
Inhibition by superoxide dismutase (SOD)	Simple, highly specific technique	Only applicable to superoxide
Pulse radiolysis/flash photolysis	Optical detection Lifetime measurement	May not identify radical
Electron paramagnetic (spin) resonance		
A. Direct	Yields structural information	Radical may not be long-lived enough for direct detection
B. Indirect (spin-trapping)	Wide application	Structural information may be lost (e.g. DMPO/$^{\bullet}$C)

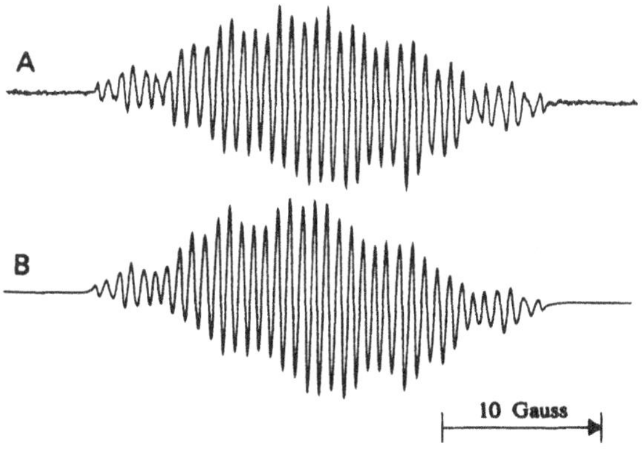

Figure 21.2 EPR spectrum of the benzanthrone anion radical obtained by irradiating (394 nm) BA (4.3 mM) in ethanol containing 0.1 N NaOH. Sample was nitrogen gassed before irradiation. (B) Computer simulation of (A) using the coupling constants in Table 21.2

The coupling constants for the ten $BA^{\bullet -}$ protons shown in Table 21.2 were assigned on the basis of spin density calculations and analogy to a structurally related radical anion from phenalenone.

However, since most free radicals generated photolytically are chemically reactive, they cannot be observed by direct EPR. For such radicals the technique of spin-trapping may be employed (Chignell, 1990). Spin-trapping is a technique in which a short-lived reactive free radical ($R^{\bullet}$) combines with a diamagnetic molecule ('spin-trap', ST) to form a more stable free radical ('spin adduct', $SA^{\bullet}$) which can be

Table 21.2 Hyperfine coupling constants (a_H) and spin densities for the benzanthrone anion radicals ($BA^{\bullet -}$)

Position	*Calculated spin density*[a]	$a_H(G)$
1	0.107	3.8
2	0.022	1.7
3	0.155	6.3
4	0.124	6.2
5	0.021	1.0
6	0.090	3.7
8	0.0014	0.7
9	0.045	2.3
10	0.024	1.7
11	0.016	0.8

a Restricted Hartree–Fock with configuration interaction (6 levels).

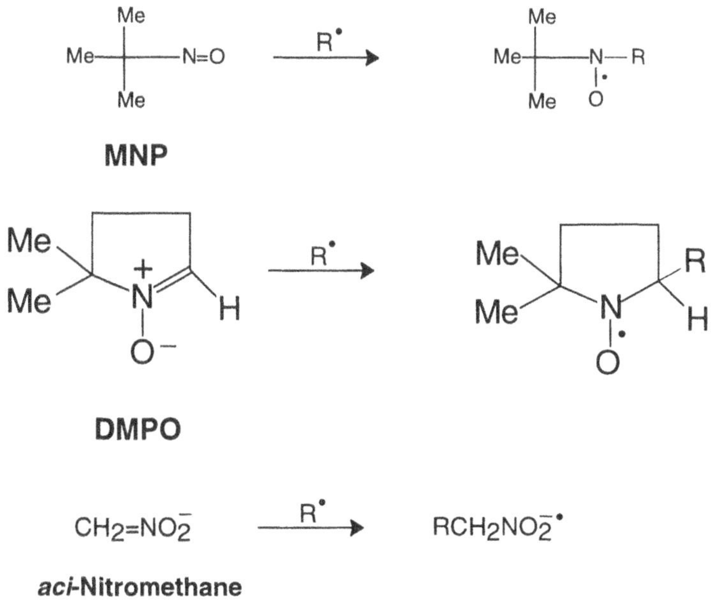

Figure 21.3 Spin-trapping with MNP, DMPO and aci-nitromethane

detected by EPR (Janzen, 1971). Spin-traps are generally nitroso compounds (MNP, DBNBS), nitrones (DMPO, DEPMPO, PBN, POBN) or *aci*-anions of nitroalkanes. In the case of the nitroso and nitrone spin-traps, the product is a nitroxide while the *aci*-anions of nitroalkanes form nitroanion radicals (Figure 21.3).

$$R^{\bullet} + ST \rightarrow SA^{\bullet} \quad (21.12)$$

The structures of the most useful nitrone spin-traps are shown in Figure 21.4. The advantages of the nitrone spin-traps for photochemical studies include no visible

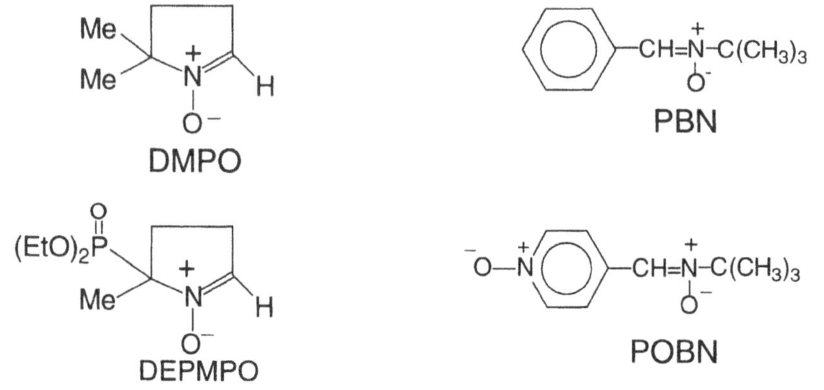

Figure 21.4 Structures of nitrone spin-traps

absorption bonds (so no spin-trap photochemistry takes place), they are monomeric (cf. nitroso spin-traps) and they are excellent for detecting oxygen-centered radicals (*vide infra*). DEPMPO, a new nitrone spin-trap recently introduced by Tordo and coworkers (Frejaville *et al.*, 1995), is particularly useful for detecting superoxide in aqueous solutions as the resultant adduct has a long half-life. One disadvantage of nitrone spin-traps is that they react with nucleophiles to form hydroxylamines which upon oxidation generate so-called 'phantom' adducts (Figure 21.5). An additional problem is that it is often difficult to determine the structure of the radical as adducts often have similar EPR spectra (*vide infra*). This is particularly true for carbon-centered radicals. The EPR spectrum of the promazinyl ($P^{\bullet}$) adduct of DMPO, generated by the UVA irradiation of chlorpromazine (Equation 21.13) is shown in Figure 21.6A. The hyperfine coupling constants are a_N = 15.9 G and a_H = 24.3 G which give rise to a 6-line spectrum (Motten *et al.*, 1985). However, many carbon-centered radical adducts of DMPO give very similar spectra (Li *et al.*, 1988) so that uneqivocal identification of the parent radical is difficult.

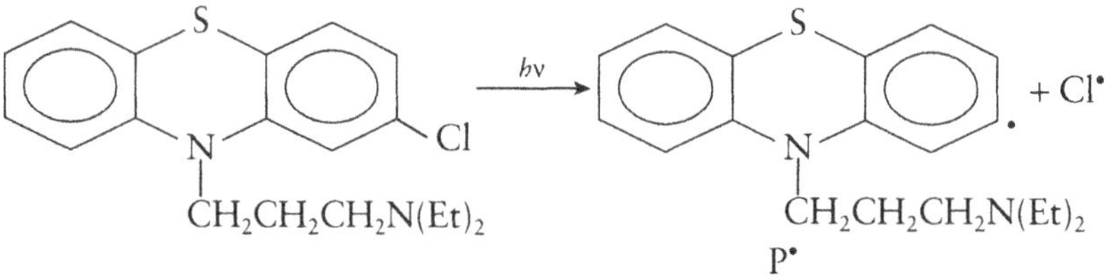

(21.13)

The structures of the some nitroso spin-traps are shown in Figure 21.7. The most commonly used nitroso spin-trap is MNP. Aromatic nitroso spin-traps have also been employed, however, they suffer from the disadvantage that couplings involving the aromatic ring hydrogen atoms produce very complex spin adduct EPR spectra. This problem may be alleviated by placing *t*-butyl substituents in the *o*- and *p*-positions (Figure 21.7). Water solubility of these traps can be increased by incorporating a sulfonic acid group (e.g. DBNBS).

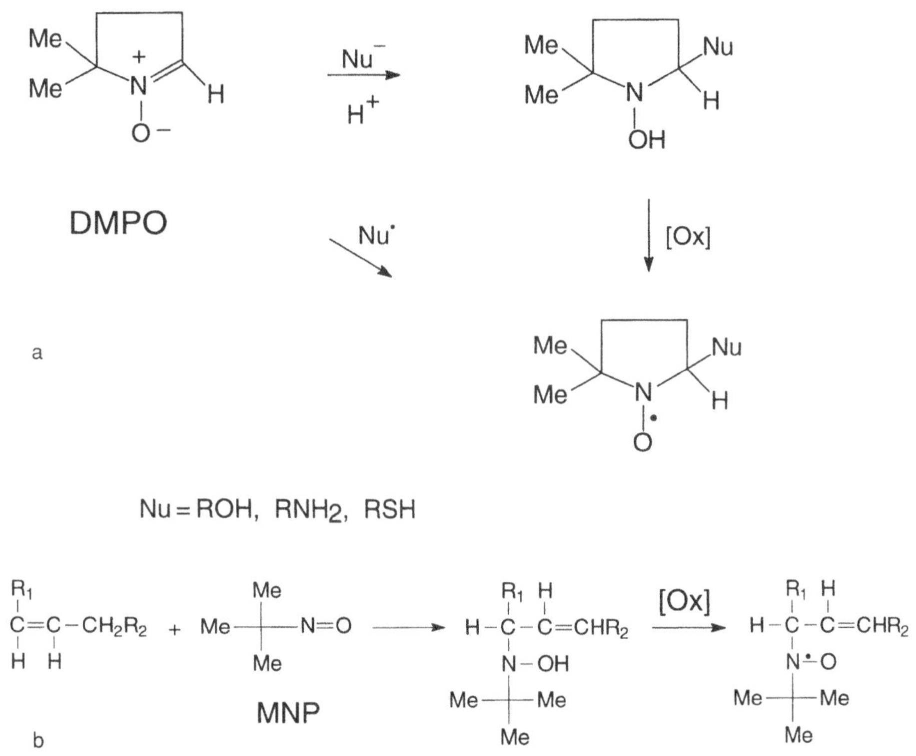

Figure 21.5 Phantom adducts from DMPO and MNP. (a) Reactions of nitrones with nucleophiles; (b) 'ene' reaction of nitroso spin-traps

The main advantage of the nitroso spin-traps is that they generate rich EPR spectra upon reaction with a radical which makes identification much easier. The EPR spectrum of the promazinyl adduct of MNP is shown in Figure 21.6. The three ring protons of the promazine adduct (at positions 1, 3 and 4) have different couplings (1.99 G, 1.95 G and 0.92 G) which provides good evidence for the assigned structure. However, nitroso spin-traps suffer from a number of disadvantages including thermal and photochemical instability of the adducts and the presence of a low energy visible spectrum (λ_{max} ~650 nm). In photochemical experiments the latter may result in photolysis of the spin-trap. For example irradiation of MNP gives a strong signal form di-*tert*-butyl nitroxide which may obscure any signal from an adduct that is present (*vide infra*). Nitroso spin-traps also form dimers which are not active as trapping agents. They are also unreliable for oxygen-centered radicals at least in aqueous systems. Finally, nitroso spin-traps undergo addition reactions ('ene') with carbon–carbon double bonds. Although the resultant product is an EPR silent hydroxylamine, it is readily oxidized by oxygen and chemical oxidizers to give a 'phantom' adduct (Figure 21.5).

When more than one radical is generated in a system, the relative concentrations of the spin adducts (SA) observed may not accurately reflect the amounts of free radicals generated since the rates of reaction between the trap (ST) and the radical ($R^•$) can differ widely depending both on the trap itself and the radical or radicals under

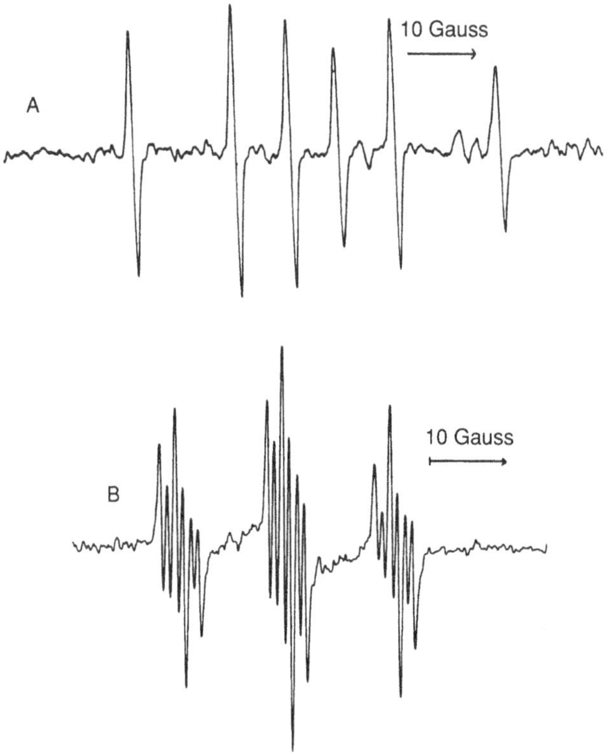

Figure 21.6 The EPR spectra of MNP/P• and DMPO/P• adducts generated by the irradiation of chlorpromazine

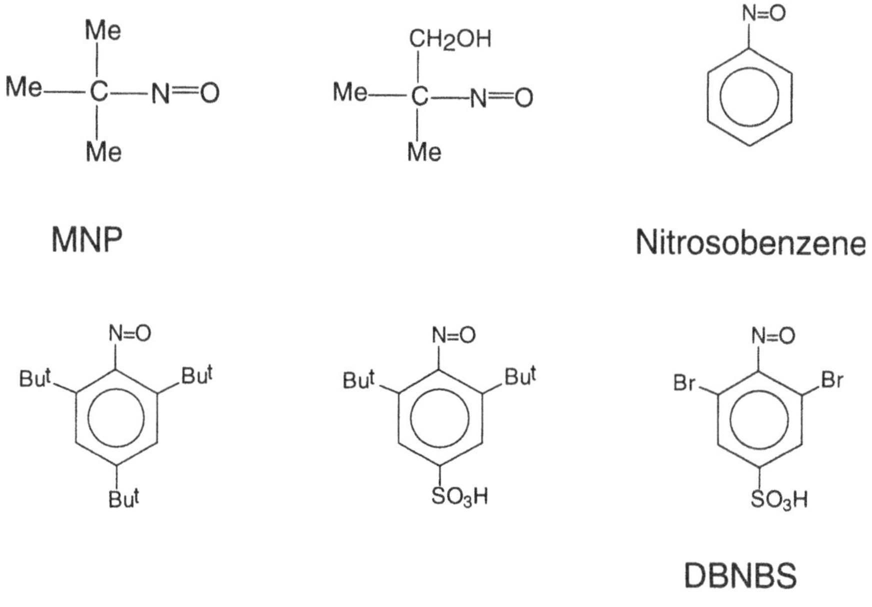

Figure 21.7 Structures of nitroso spin-traps

Table 21.3 Rate constants for the reaction of spin-traps with radicals[a]

Spin-trap	*Radical*	*Solvent*	*Rate constant* ($M^{-1}s^{-1}$)
MNP	$^{\bullet}C(CH_3)_3$	C_6H_6	$1–9 \times 10^6$
MNP	$^{\bullet}OC(CH_3)_3$	C_6H_6	1.5×10^6
MNP	$^{\bullet}SC_6H_6$—P—Br	C_6H_6	2.2×10^8
MNP	$^{\bullet}CH_2R$	C_6H_6	$1–9 \times 10^6$
MNP	$^{\bullet}OCH_3$	MeOH	1.6×10^6
PBN	$^{\bullet}C(OH_3)_3$	C_6H_6	1×10^4
PBN	$^{\bullet}C_6H_5$	C_6H_6	1.2×10^7
PBN	$^{\bullet}CHR$	C_6H_6	1×10^5
PBN	$^{\bullet}COC_6H_5$	C_6H_6	1×10^{10}
PBN	$^{\bullet}CH_2OH$	H_2O	4.3×10^7
PBN	$CO_2^{-\bullet}$	H_2O	1.5×10^7
DMPO	$^{\bullet}CH_2R$	C_6H_6	4×10^5
DMPO	$^{\bullet}O(CH_3)_3$	C_6H_6	5.1×10^8
DMPO	$O_2^{-\bullet}$	H_2O	10
DMPO	$^{\bullet}O_2H$	H_2O	6.6×10^3
DMPO	$^{\bullet}OH$	H_2O	$2–4 \times 10^9$
DMPO	e^-	H_2O	1×10^{10}

a Data from reference (Li *et al.*, 1988).

investigation. Table 21.3 lists some rate constants for the reaction of radicals with MNP, PBN, and DMPO. In organic solvents DMPO reacts with carbon-centered radicals more slowly than oxygen-centered radicals. In aqueous systems the hydrated electron reacts with DMPO at a rate that is about nine orders of magnitude faster than the superoxide anion. Thus in a system where both species are being generated at the same rate the DMPO/H$^{\bullet}$ adduct (Equation 21.14) will predominate.

$$DMPO + e_{aq}^- \longrightarrow DMPO^- \xrightarrow{H^+} DMPO/H^{\bullet} \qquad (21.14)$$

Since spin adducts of nitrones are nitroxides bearing an α-hydrogen atom, they are not as stable as nitroxide spin labels in which the nitroxide moiety is flanked by quaternary carbon atoms. Spin adducts may decay by several mechanisms including disproportionation, reduction and oxidation (Figure 21.8). If the product is a hydroxylamine it may be possible to regenerate the spin adduct simply by bubbling oxygen through the solution or by adding a chemical oxidizing agent.

Some radicals, notably the π-radicals, do not add to spin-traps. In a π-radical the electron is delocalized, usually over an aromatic ring system, and so its reactivity is decreased. The phenoxyl radical is a good example of this type of radical. While MNP is a poor trap for oxygen-centered radicals in aqueous solution such radicals react readily with nitrones such as DMPO. Thus it is important to use more than one spin-trap, particularly when negative results are initially obtained.

21.4 Spin-trapping in photochemical systems

In photochemical systems the photochemistry of the spin-traps becomes important if the trap absorbs radiation from the light source. For example, the nitroso spin-traps

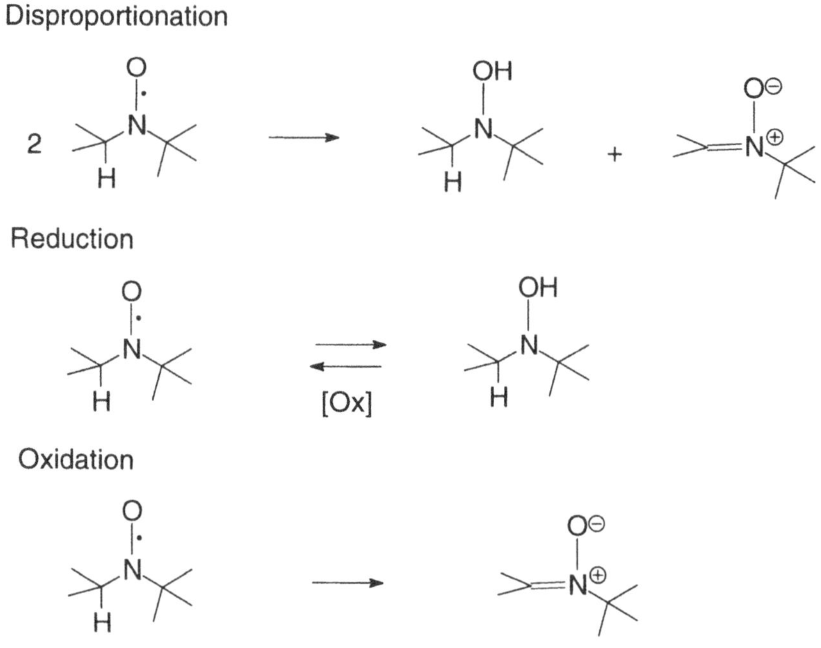

Figure 21.8 Decay mechanisms for spin adducts

have absorption maxima in the visible region (Timpe *et al.*, 1991). Irradiation of MNP with visible light generates a strong EPR signal from di-*tert*-butyl nitroxide (Equation 21.15) which can obscure the spectrum from any adduct that is formed. This problem is more acute in organic solvents than aqueous systems and is exacerbated when oxygen is removed. In contrast, nitrone spin-traps have very weak absorption spectra above 300 nm (Timpe *et al.*, 1991) and so trap photochemistry is not generally a problem above this wavelength. However, it should be emphasized that spin-traps are often used at fairly high concentrations which means that even weak absorption above 300 nm may be sufficient to produce spin-trap photochemistry (Chignell *et al.*, 1994b).

$$\underset{\text{MNP}}{(CH_3)_3CNO} \xrightarrow{h\nu} (CH_3)_3C^\bullet + NO^\bullet \xrightarrow{\text{MNP}} (CH_3)_3C\text{—}\overset{\overset{\text{O}}{|\bullet}}{\text{N}}\text{—}C(CH_3)_3 \qquad (21.15)$$

In water, UVB irradiation of DMPO results in photoionization of the trap generating both the hydrogen atom and hydroxyl radical adducts (Equations 21.14, 21.16, 21.17). In cyclohexane solvent radicals are formed (Chignell *et al.*, 1994b). Irradiation of DBNBS may result in loss of the sulfonate goup as $SO_3^{\bullet-}$ which adds to the trap to give the DBNBS/$SO_3^{\bullet-}$ adduct.

$$DMPO \xrightarrow{h\nu} DMPO^{+\bullet} + e^-_{aq} \qquad (21.16)$$

$$DMPO^{+\bullet} + H_2O \rightarrow DMPO/^\bullet OH + H^+ \qquad (21.17)$$

Another problem that may occur in photochemical systems is that the spin-trap can act as a quencher of the excited state of the photosensitizer under investigation (Timpe *et al.*, 1991). If quenching is physical then the excited state photosensitizer is returned to the ground state and no chemistry takes place. However, if chemical quenching occurs then photoproducts will be formed that would otherwise not be observed in the absence of the spin-trap. For example, the reaction of 6-mercaptopurine (PSH) with MNP results in electron (or hydrogen atom) transfer and the generation of the thiyl radical (Moore *et al.*, 1994) which is then trapped by MNP (Equation 21.19).

$$^{3}(PSH)^{*} + MNP \rightarrow PS^{\bullet} + MNP/H^{\bullet} \qquad (21.18)$$

$$PS^{\bullet} + MNP \rightarrow MNP/PS^{\bullet} \qquad (21.19)$$

Photochemical systems frequently generate non-radical species that may react with the spin-trap. For example, visible irradiation of an aerated aqueous solution containing the photosensitizer Rose Bengal and DMPO generates the DMPO/$^{\bullet}$OH adduct (Bilski *et al.*, 1996). The mechanism probably involves addition of $^{1}O_2$ to the nitrone double bond to form a biradical which decomposes to give DMPO/$^{\bullet}$OH and free $^{\bullet}$OH. Nitro and nitroso compounds are also formed as a result of opening of the DMPO pyrroline ring (Bilski *et al.*, 1996).

21.5 Spin-trapping of oxygen-centered radicals

Oxygen-centered radicals are thought to play an important role in many pathological processes including photosensitization (Simic *et al.*, 1985). Nitrones, particularly DMPO, are good spin-traps for oxygen-centered radicals. The EPR spectra of DMPO adducts of $^{\bullet}$OH, *t*-BuO$^{\bullet}$, EtOO$^{\bullet}$, and $O_2^{-\bullet}$ in aqueous solution are shown in Figure 21.9A. These spectra have been simulated using well established EPR parameters (Li *et al.*, 1988). The DMPO/$^{\bullet}$OH and DMPO/*t*-BuO$^{\bullet}$ adducts are easily recognized and can be differentiated from each other. However, DMPO/EtOO$^{\bullet}$ and DMPO/$O_2^{-\bullet}$ adducts exhibit a small coupling due to the γ-hydrogen atom and are quite similar, although they can be readily distinguished from the DMPO/$^{\bullet}$OH and DMPO/*t*-BuO$^{\bullet}$ adducts. Figure 21.10 shows the EPR spectrum generated by irradiation of an aerobic aqueous solution of Rose Bengal and DMPO in the presence of a reducing agent (NADH). To determine whether this spectrum is due to DMPO/$O_2^{-\bullet}$ the enzyme superoxide dismutase is added and the experiment repeated. Complete inhibition of the DMPO/$O_2^{-\bullet}$ EPR signal confirms that superoxide has been trapped.

While the DMPO/$O_2^{-\bullet}$ signal in Figure 21.10 is quite strong, frequently the EPR spectrum of this adduct is much weaker and quickly disappears when the light source is turned off. The yield of DMPO/$O_2^{-\bullet}$ can sometimes be improved by removing Fe^{2+} or other redox metals by treating the buffer with Chelex resin. Alternatively, a chelating agent (e.g. DETAPAC) can be added directly to the buffer (Buettner and Oberley, 1978). Trapping superoxide at slightly acid pH may also improve its detectability because the hydroperoxyl radical, $^{\bullet}$OOH ($pK_a \simeq 4.6$), adds much faster to DMPO than superoxide (Table 21.3). A more recent approach involves the use of a new spin-trap, DEPMPO (Frejaville *et al.*, 1995). As may be seen from Table 21.4, DEPMPO reacts a little faster than DMPO with superoxide. More importantly the

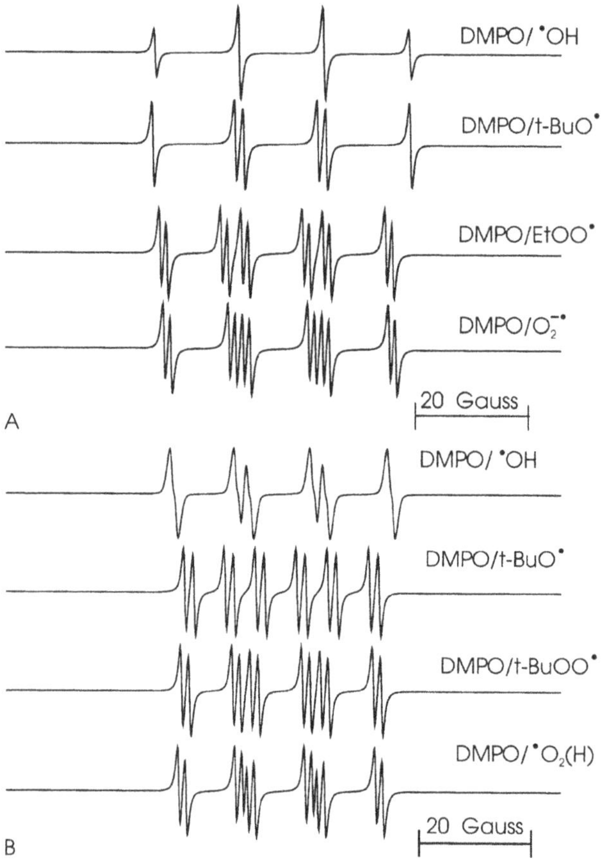

Figure 21.9 The EPR spectra of oxygen-centered adducts of DMPO in water (A) or benzene/toluene (B)

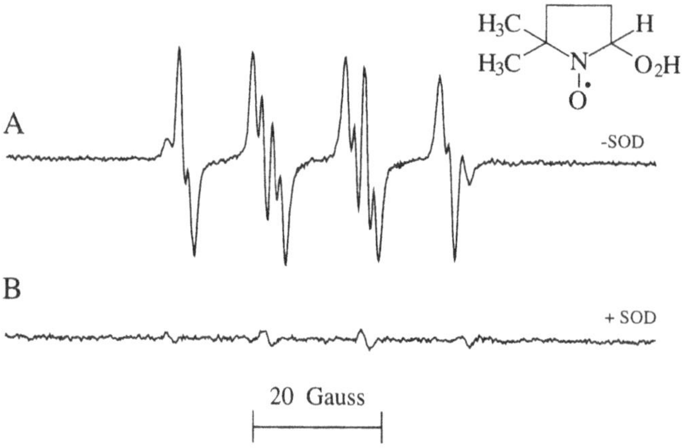

Figure 21.10 EPR spectrum of DMPO/$O_2^{-•}$ and the effect of superoxide dismutase. Rose Bengal (0.1 mM), NADH (2 mM), DMPO (100 mM) in aerated phosphate buffer (50 mM, pH7, chelexed) illuminated with monochromate light at 550 nm

Table 21.4 Reaction of DMPO and DEPMPO with superoxide[a]

Spin-trap	*Rate of reaction with superoxide*	*Decay rate of adduct*	*Adduct half life*
DMPO	~10 M^{-1} s^{-1}	1.4×10^{-2} s^{-1}	50s
DEPMPO	~15 M^{-1} s^{-1}	7.8×10^{-4} s^{-1}	890s

a Data from reference (Frejaville *et al.*, 1995).

resultant adduct has a much longer half life. Thus this spin-trap is finding increasing use for the detection of superoxide, particularly in biological systems. The only major drawback with DEPMPO is that the phosphorus atom doubles the number of EPR adduct lines as compared to the corresponding DMPO adduct (Frejaville *et al.*, 1995). This may make radical identification difficult if more than one species is present.

As mentioned above the DMPO/$O_2^{-\bullet}$ adduct is not very stable. In particular this adduct may decompose to give the hydroxyl radical which can react with DMPO to give DMPO/$^\bullet$OH (Finkelstein *et al.*, 1982). Thus DMPO/$^\bullet$OH observed in any system that contains superoxide must be viewed with suspicion. The addition of SOD will abolish both the DMPO/$O_2^{-\bullet}$ and DMPO/$^\bullet$OH signals if the latter is derived from the former. As already mentioned (*vide supra*), any photochemical reaction that generates singlet oxygen will also give rise to artefactual DMPO/$^\bullet$OH signals.

The EPR signals of DMPO adducts from oxygen-centered radicals in an aprotic solvent (benzene or toluene) are shown in Figure 21.9B. Again the DMPO/$^\bullet$OH and DMPO/*t*-BuO$^\bullet$ adducts are readily identified while the DMPO/*t*-BuOO$^\bullet$ and DMPO/$^\bullet O_2H$ adducts are not distinguishable based on their EPR parameters. SOD of course cannot be used in a manner analogous to that already described for the same adducts in aqueous systems so that unequivocal identification of (hydro)peroxyl adducts may not be possible.

A knowledge of the rate constant for the reaction of a given radical with a spin-trap can be used to determine the rate constant for the reaction of a second substrate with the same radical. This approach has been recently applied to the measurement of the rate of reaction of the hydroxyl radical with melatonin and related indoles (Matuszak *et al.*, 1997).

21.6 Spin-trapping of radicals from photosensitizing chemicals

A listing of some of the photosensitizing chemicals studied in the author's laboratory is shown in Table 21.5. In all cases the radicals generated in Type I reactions were identified by EPR using either direct detection, where the radical was sufficiently stable, or spin-trapping where reactive radicals were generated. The generation of singlet oxygen in a Type II reaction was determined by its luminescence at 1280 nm. From the data in Table 21.5 several conclusions can be drawn. Photoallergic compounds tend to generate carbon-centered radicals. This is reasonable in that photoallergy requires the formation of a photoallergen and carbon-centered radicals would be expected to react with proteins to give covalent adducts with allergenic properties. In contrast phototoxic chemicals generate active forms of oxygen, superoxide and/or singlet oxygen. Those compounds that are most phototoxic appear to generate both forms of active oxygen.

Table 21.5 Photosensitivity response and photochemistry of photosensitizing chemicals

Chemical agent	*Photosensitivity response*		*Photochemistry*		*Reference*
	Phototoxicity	*Photoallergy*	*Type I*	*Type II*	
Topical					
4-aminobenzoic acid and esters	+	+	$Ar^{\bullet}$, $H^{\bullet}$	+	Chignell *et al.*, 1980; Chignell *et al.*, 1981
halogenated salicylanilides	+	+	$Ar^{\bullet}$	–	Chignell and Sik, 1989
bithionol, fentichlor	–	+	$Ar^{\bullet}$	–	Li and Chignell, 1987; Li and Chignell, 1989
musk abrette	–	+	$ArNO_2^{-\bullet}$, $O_2^{-\bullet}$	–	Motten *et al.*, 1983
gentian violet	+	–	$O_2^{-\bullet}$	–	Fischer *et al.*, 1984
anthralin	+	–	$O_2^{-\bullet}$	+	Dabestani *et al.*, 1990
curcumin	+	–	$O_2^{-\bullet}$	+	Chignell *et al.*, 1994
Environmental					
disperse blue 35	+	–	$O_2^{-\bullet}$	+	Dabestani *et al.*, 1991
benzanthrone	+	–	$O_2^{-\bullet}$	+	Dabestani *et al.*, 1992
coal tar (anthracene)	+	–	–	+	Sigman *et al.*, 1991
Drugs					
antibacterial sulfonamides	+	+	$Ar^{\bullet}$, SO_2NH_2, $SO_3^{\bullet-}$, $^{\bullet}NH_2$		Chignell *et al.*, 1980; Chignell *et al.*, 1981
benoxaprofen	+	–	$O_2^{-\bullet}$	+	Reszka and Chignell, 1983
chlorpromazine	+	–	$Ar^{\bullet}$, $CPZ^{+\bullet}$, e^-_{aq}	+	Motten *et al.*, 1985; Hall *et al.*, 1987
chlorpromazine sulfoxide	+	?	$^{\bullet}OH$, $CPZ^{+\bullet}$		Buettner *et al.*, 1986
amiodarone	+	–	$Ar^{\bullet}$, e^-_{aq}, $O_2^{-\bullet}$		Li and Chignell, 1987
tetracyclines	+	–	$O_2^{-\bullet}$	+	Li *et al.*, 1987

References

Bilski, P., Reszka, K., Bilski, M., and Chignell, C.F., 1996, Oxidation of the spin-trap 5,5-dimethyl-1-pyrroline N-oxide by singlet oxygen in aqueous solution, J. Amer. Chem. Soc., 118, 1330–8.

Buettner, G.R. and Oberley, L.W., 1978, Considerations in the spin-trapping of superoxide and hydroxyl radical in aqueous systems using 5,5-Dimethyl-1-pyrroline-1-oxide, Biochem. Biophys. Res. Commun., 83, 69–74.

Buettner, G.R., Motten, A.G., Hall, R.D., and Chignell, C.F., 1986, Free radical production by chlorpromazine sulfoxide. An ESR spin-trapping and flash photolysis study, Photochem. Photobiol., 33, 5–10.

Chignell, C.F., 1990, Spin-trapping studies of photochemical reactions, Pure & Appl. Chem., 62, 301–305.

Chignell, C.F. and Sik, R.H., 1989, Spectroscopic studies of cutaneous photosensitizing agents. XIV. The spin-trapping of free radicals formed during the photolysis of halogenated salicylanilide antibacterial agents, Photochem. Photobiol., 50, 287–295.

Chignell, C.F., Kalyanaraman, B., Mason, R.P., and Sik, R.H., 1980, Spectroscopic studies of cutaneous photosensitizing agents. 1. Spin-trapping of photolysis products from sulfanilamide, 4-aminobenzoic acid and related compounds, Photochem. Photobiol., 32, 563–571.

Chignell, C.F., Kalyanaraman, B., Sik, R.H., and Mason, R.P., 1981, Spectroscopic studies of cutaneous photosensitizing agents. II. Spin-trapping of photolysis products from sulfanilamide and 4-aminobenzoic acid using 5,5-dimethyl-1-pyrroline-1-oxide (DMPO), Photochem. Photobiol., 34, 147–156.

Chignell, C.F., Motten, A.G., Sik, R.H., Parker, C.E., and Reszka, K., 1994b, A spin-trapping study of the photochemistry of 5,5-dimethyl-1-pyrroline N-oxide (DMPO), Photochem. Photobiol., 59, 5–11.

Chignell, C.F., Bilski, P., Reszka, K., Motten, A.H., Sik, R.H., and Dahl, T.A., 1994a, Spectral and photochemical properties of curcumin, Photochem. Photobiol., 59, 295–302.

Dabestani, R., Hall, R.D., Sik, R.H., and Chignell, C.F., 1990, Spectroscopic studies of cutaneous photosensitizing agents. XV. Anthralin and its oxidation product 1,8-dihydroxyanthraquinone, Photochem. Photobiol., 52, 961–971.

Dabestani, R., Reszka, K., Davis, D.G., Sik, R.H., and Chignell, C.F., 1991, Spectroscopic studies of cutaneous photosensitizing agents. XVI. Disperse blue 35, Photochem. Photobiol., 54, 37–42.

Dabestani, R., Sik, R.H., Motten, A.G., and Chignell, C.F., 1992, Spectroscopic studies of cutaneous photosensitizing agents. XVII, Benzanthrone. Photochem. Photobiol., 55, 533–539.

Finkelstein, E., Rosen, G.M., and Rauckmann, J.R., 1980, Production of hydroxyl radical by decomposition of superoxide spin-trapped adducts, Mol. Pharmacol., 21, 262–265.

Fischer, V., Harrelson, W.G., Chignell, C.F., and Mason, R.P., 1984, Spectroscopic studies of cutaneous photosensitizing agents. V. Spin-trapping and direct electron spin resonance investigations of the photoreduction of gentian (crystal) violet, Photobiochem. Photobiophys., 7, 111–119.

Frejaville, C., Karoui, H., Tuccio, B., Le Moigne, F., Culcasi, M., Pietri, S., Lauricella, R., and Tordo, F., 1995, 5-(Diethoxyphosphoryl)-5-methyl-1-pyrroline N-oxide: A new efficient phosphorylated nitrone for the *in vitro* and *in vivo* spin-trapping of oxygen-centered radicals, J. Med. Chem., 38, 258–265.

Frimer, A.A. (ed.), 1985, Singlet Oxygen Volume II Reaction Modes and Products Part 1. CRC Press, Boca Raton, FL.

Frimer, A.A. and Stephenson, L.M., 1985, The singlet oxygen ene reaction, Singlet Oxygen Volume II Reaction Modes and Products Part 1. CRC Press, Boca Raton, FL, pp. 67–92.

Girotti, A.W., 1990, Photodynamic lipid peroxidation in biological systems, Photochem. Photobiol., 51, 497–509.

Hall, R.D., Buettner, G.R., Motten, A.G., and Chignell, C.F., 1987, Near-infrared detection of singlet molecular oxygen produced by photosensitization with promazine and chlorpromazine, Photochem. Photobiol., 46, 295–300.

Janzen, E.G., 1971, Spin-trapping, Accts. Chem. Res., 4, 31–40.

Li, A.S.W. and Chignell, C.F., 1987a, Spectroscopic studies of cutaneous photosensitizing agents. IX. A spin-trapping study of the photolysis of amiodarone and desthylamiodarone, Photochem. Photobiol., 45, 191–197.

Li, A.S.W. and Chignell, C.F., 1987b, Spectroscopic studies of cutaneous photosensitizing agents. XII. Spin-trapping study of the free radicals generated during the photolysis of photoallergens bithionol and fentichlor, Photochem. Photobiol., 46, 445–452.

Li, A.S.W. and Chignell, C.F., 1989, Spectroscopic studies of cutaneous photosensitizing agents. XIII. pH dependence of the photochemistry of photoallergens bithionol and fentichlor: An electron spin resonance study of the free radical photoproducts, Photochem. Photobiol., 49, 25–32.

Li, A.S.W., Roethling, H.P., Cummings, K.B., and Chignell, C.F., 1987, $O_2^{-\bullet}$ Photogenerated from aqueous solutions of tetracycline antibiotics (pH 7.3) as evidenced by DMPO spin-trapping and cytochrome c reduction, Biochem. Biophys. Res. Comm., 146, 1191–1195.

Li, A.S.W., Cummings, K.B., Roethling, H.P., Buettner, G.R., and Chignell, C.F., 1988, A spin-trapping database implemented on the IBM PC/AT, J. Magn. Reson., 79, 140–142.

Matuszak, Z., Reszka, K.J., and Chignell, C.F., 1997, Reaction of melatonin and related indoles with hydroxyl radical. EPR and spin-trapping investigations, Free Rad. Biol. Med., 23, 367–372.

Moore, D.E., Sik, R.H., Bilski, P., Chignell, C.F., and Reszka, K.J., 1994, Photochemical sensitization by azathioprine and its metabolites. Part 3. A direct EPR and spin-trapping study of light-induced free radicals from 6-mercaptopurine and its oxidation products, Photochem. Photobiol., 60, 574–581.

Motten, A.G., Buettner, G.R., and Chignell, C.F., 1985, Spectroscopic studies of cutaneous photosensitizing agents – VIII. A spin-trapping study of light induced free radicals from chlorpromazine and promazine, Photochem. Photobiol., 42, 9–15.

Motten, A., Chignell, C.F., and Mason, R.P., 1983, Spectroscopic studies of cutaneous photosensitizing agents. VI. Identification of the free radicals generated during the photolysis of musk ambrette, musk xylene and musk ketone, Photochem. Photobiol., 38, 671–678.

Reszka, K. and Chignell, C.F., 1983, Spectroscopic studies of cutaneous photosensitizing agents. IV. The photolysis of benoxaprofen, an anti-inflammatory drug with photoxic properties, Photochem. Photobiol., 38, 281–291.

Sigman, M.E., Zingg, S.P., Pagni, R.M., and Burns, J.H., 1991, Photochemistry of anthracene in water, Tetrahedron Lett., 32, 5737–5740.

Simic, M.G., Taylor, K.A., Ward, J.F., and Sonntag, C., 1985, Oxygen Radicals in Biology and Medicine, Plenum Press.

Timpe, H.-J., Ulrich, S., and Ali, S., 1991, Spin-trapping as a tool for quantum yield determinations in radical forming processes, J. Photochem. Photobiol. A: Chem., 61, 77–89.

Urbach, F. (ed.), 1992, Biological Responses to Ultraviolet A Radiation, Valdenmar Publishing Co, KS.

22 Thermodynamic considerations of free radical reactions

John Butler

22.1 Introduction

Reduction and oxidation reactions often occur via one-electron processes wherein reactive free radicals are formed as intermediates. Generally, the overall direction of a redox reaction is controlled by the feasibility of forming such radical intermediates and in many instances, this feasibility also determines the kinetics of the reactions. Thermodynamically, the feasibility of a reaction is related to the one-electron reduction potentials of the couples involved.

The object of this review is to discuss some biological applications of reduction potentials with some emphasis on drug activation, radicals in proteins and redox cycling. This is by no means a comprehensive review on reduction potentials and the interested reader is advised to consult more specialised texts (e.g. Koppenol and Butler, 1985; Wardman, 1989; Butler and Hoey, 1993a). Similarly, the implications of redox reactions in antitumour drug design have also been omitted as these are discussed in a recent review (Butler, 1998).

22.2 Reduction

The ease of one- or two-electron reduction of a compound (D) is related to the reduction potentials of the corresponding couples:

$$D + e^- \rightleftharpoons D^{\bullet-} \qquad E[D/D^{\bullet-}] \qquad (22.1)$$

$$D + 2e^- \rightleftharpoons D^{2-} \qquad E[D/D^{2-}] \qquad (22.2)$$

These potentials essentially represent one half of an electron transfer equilibrium. For one-electron reduction potentials, equilibrium can be in the form:

$$D^{\bullet-} + S \rightleftharpoons D + S^{\bullet-} \qquad (22.3)$$

The free energy change, ΔG, of this equilibrium is given by:

$$\Delta G = -nF\Delta E = -RT \ln K_3 \qquad (22.4)$$

Where $n = 1$, F is the Faraday constant, R is the gas constant, T is the absolute temperature and K_3 is the equilibrium constant. In reality, the equilibrium constant represents the products of the activities of the components. However, at low ionic strengths, these can be considered as being equivalent to concentrations. ΔE is defined as the difference in the reduction potentials of the two couples:

$$\Delta E = E[S/S^{\bullet -}] - E[D/D^{\bullet -}] \qquad (22.5)$$

Thus, if the reduction potentials of the two couples are known, the position of the equilibrium can be calculated. It can also be seen from the above equations that if the reduction potential of the $D/D^{\bullet -}$ couple is more negative than that of the $S/S^{\bullet -}$ couple, ΔE becomes a positive number and ΔG is negative. Hence, the equilibrium constant, K_3 is positive (i.e. the equilibrium goes in the direction of the right-hand side). Thus for example, if the above equilibrium represents the reaction of durosemiquinone radicals with oxygen, as the one-electron reduction potential of the duroquinone couple, $E[Q/Q^{\bullet -}]$ is –244 mV and the corresponding value for the oxygen/superoxide radical couple, $E[O_2/O_2^{\bullet -}]$ is –155 mV, durosemiquinone radicals can reduce oxygen to produce superoxide radicals.

22.3 Oxidation

The reduction potentials for the oxidation of a radical species, can be derived and treated in a similar manner as the above couples:

$$D^{\bullet +} + e^- \rightleftharpoons D \qquad E[D^{\bullet +}/D] \qquad (22.6)$$

However, as by convention, the oxidised species are always on the right hand side, the sign of the reduction potentials may appear to have a different meaning. Strong oxidising radicals have high positive $E[D^{\bullet +}/D]$ values. This can be illustrated by the reactions of tryptophan radicals with tyrosine. The tryptophan couple, $[Trp^{\bullet}, H^+/TrpH]$ has a reduction potential of +1050 mV at pH 7 whereas the corresponding potential for the tyrosine couple, $[TyrO^{\bullet}, H^+/TyrOH]$ is about +950 mV (Prutz *et al.*, 1989; Harriman, 1987). As the potential for the tryptophan couple is more positive at pH 7 than that for the tyrosine couple, tryptophan radicals will oxidise tyrosine to tyrosine radicals at pH 7. These radicals are discussed in more detail below.

22.4 pH dependence of reduction potentials

As stated above, reduction potentials can be considered as representing one half of an electron transfer equilibrium. If this equilibrium involves protons, then the equilibrium constant and hence the reduction potential, has to be dependent on pH.

The *standard reduction potential*, E_o, is defined as the potential of a couple at unit activity (or unit fugacity if a gas is involved). These potentials refer to the specific condition wherein the hydrogen ion concentration is 1 mol dm^{-3} and therefore are independent of pH.

The relationship between the standard reduction potential and the potential at any pH (or indeed, under any other condition including ionic strength and different fugacities), is given by the Nernst equation:

$$E_i = E_o + \frac{RT}{nF} \ln \frac{(\textit{Product of activites of all oxidants})}{(\textit{Product of activites of all reductants})} \qquad (22.7)$$

It can be seen from this simple equation that if an oxidant or a reductant is in equilibrium with different protonated forms, then the activities of these forms have to be included in the equation. Superoxide radicals for example, can exist in two forms:

$$O_2 + e^- \rightleftharpoons O_2^{-\bullet} \qquad E[O_2/O_2^{-\bullet}] \tag{22.8}$$

$$O_2 + H^+ + e^- \rightleftharpoons HO_2^\bullet \qquad E[O_2,H^+/HO_2^\bullet] \tag{22.9}$$

The pK of superoxide radicals is 4.7 and the reduction potential at any pH is a composite of the two potentials. However, the radical exists mainly in the form of $O_2^{-\bullet}$ above the pK and hence the potential is essentially independent of pH. Below the pK, the $HO_2^\bullet$ radical is the predominant species and so there will be a pH dependence on the reduction of oxygen. The overall dependence of pH on the reduction of oxygen to form superoxide radicals ($O_2^{-\bullet}$ and $HO_2^\bullet$) is related by the term, RT/F ln (K + H^+). As a general rule, the reduction potential value of any couple involving protons decreases as the pH increases. Hence, reducing radicals become better reducing agents ($E[D,H^+/DH^\bullet]$ becomes more negative) and oxidising radicals become weaker oxidants ($E[D^\bullet,H^+/DH]$ also becomes more negative).

An important consequence of these pH dependencies is that the direction of an electron transfer can be reversed at different pHs. It was mentioned above that the reduction potentials of the couples [$TyrO^\bullet$, H^+/TyrOH] and [$Trp^\bullet$, H^+/TrpH] are about +950 mV and +1050 mV, respectively at pH 7.

Tryptophan (TrpH) is oxidised by a variety of radicals, including hydroxyl radicals, to form $TrpH^+$ radicals. These radicals have a pK of 4.3:

$$TrpH^{\bullet+} + e^- \rightleftharpoons TrpH \tag{22.10}$$

$$TrpH^{\bullet+} \rightleftharpoons Trp^\bullet + H^+ \tag{22.11}$$

From substituting the relative concentrations of the oxidised forms ($TrpH^{\bullet+}$ and $Trp^\bullet$) and the reduced form (TrpH) into the Nernst equation, it can easily be shown (Wardman, 1989; Clark, 1960) that the potential for the oxidation of tryptophan at any pH (E_i) is given by:

$$E_i = E_o + \frac{RT}{F} \ln \frac{[H^+]}{[H^+ + K_{11}]} \tag{22.12}$$

The pH dependence of the reduction potential of tryptophan is shown in Figure 22.1.

Tyrosine is also oxidised to form a radical but in contrast to tryptophan, it is the pK of the phenoxyl group of the parent molecule (pK = 10.1) which affects the pH dependence of the reduction potential:

$$TyrO^\bullet + e^- \rightleftharpoons TyrO^- \tag{22.13}$$

$$TyrOH \rightleftharpoons TyrO^- + H^+ \tag{22.14}$$

In this case, there is only one oxidised form ($TyrO^\bullet$) whereas there are two reduced forms ($TyrO^-$ and TyrOH). From the Nernst equation:

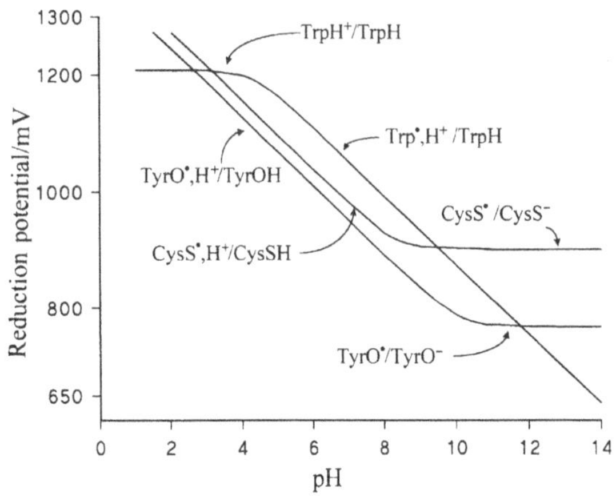

Figure 22.1 pH dependence of the reduction potentials of some redox active amino acids (adapted from Prutz *et al.*, 1989)

$$E_i = E_o + \frac{RT}{F} \ln[H^+ + K_{14}] \tag{22.15}$$

The pH dependence on the reduction potential is included in Figure 22.1. It can be seen from this figure that tryptophan radicals should be capable of oxidising tyrosine to tyrosine radicals at physiologically relevant pH values. This reaction has been shown to occur in a variety of different peptides and proteins (Prutz *et al.*, 1982, 1986; Bobrowski *et al.*, 1997). However, this figure also shows that there are two crossover points where the direction of the reactions can change. It has been demonstrated that oxidised tyrosine radicals do indeed oxidise tryptophans to radicals in strong acid and alkali conditions (Butler *et al.*, 1986).

Several studies (Prutz *et al.*, 1986, 1989) have now shown that other radical transfer reactions can occur in proteins. The order of transfer follows the reduction potentials of the couples, $E[D^•,H^+/DH]$

$$\text{Methionine} \rightarrow \text{Tryptophan} \rightarrow \text{Cysteine} \leftrightharpoons \text{Tyrosine} \tag{22.16}$$

Hence in theory, an oxidised methionine radical, formed by the reaction of a hydroxyl radical on a methione residue in a protein, can oxidise a tryptophan residue. The tryptophan radical can oxidise cysteine and the resulting cysteine radical can be in equilibrium with a tyrosine radical (Prutz *et al.*, 1989). The radical transfer can be terminated by intermolecular/intramolecular radical–radical interactions (Prutz *et al.*, 1983), which can result in the formation of protein dimers and aggregates (Jollow *et al.*, 1995; Stadtman, 1993; Schuessler *et al.*, 1986). The pH dependence of the reduction potential of cysteine radicals is also included in Figure 22.1.

The presence of a cysteine radical in the amino acid radical cascade is interesting as this radical can also form an equilibrium with other cysteines:

$$\text{CysS}^• + \text{CysSH} \leftrightharpoons (\text{CysS—SCys})^{•-} + H^+ \tag{22.17}$$

Disulfide radicals $(CysS—SCys)^{\bullet-}$ are very powerful *reducing* agents, $E[(CysS—SCys)/(CysS—SCys)^{\bullet-}]$ being approximately –1600 mV (Surdhar and Armstrong, 1986), and can readily reduce oxygen to superoxide radicals or any of the oxidised amino acids.

Studies of this type are important as naturally occurring amino acid radicals are detected in several enzymes. Tryptophan, tyrosine and modified tyrosine radicals exist as redox centres in several enzymes including cytochrome c peroxidase, ribonucleotide reductase, photosystem II, prostaglandin H synthase, amine oxidase and galactose oxidase (Miller *et al.*, 1994; Babcock *et al.*, 1997; Pedersen and Finazzi-Agrò, 1993). In view of the above radical cascade, it is surprising that these radicals can have relatively long lifetimes in the presence of the other amino acids. Why does the tryptophan radical in cytochrome c peroxidase not transfer to tyrosine or cysteines? Why do natural tyrosine radicals not come into equilibrium with cysteine residues or not react rapidly with thiols outside the protein, including glutathione?

It is evident that these natural amino acid radicals must be in a different environment to the radicals that are formed in other proteins by radical attack. It is interesting to note however, that all of the previous pulse radiolysis studies showed that the radical transfers could be observed in a variety of different proteins (e.g. Prutz *et al.*, 1982, 1983) and it has been proposed that the these transfers may occur even when the residues cannot come into physical contact (Bobrowski *et al.*, 1992; Weinstein *et al.*, 1991).

Recent studies have concentrated on mutations of *E. coli* ribonucleotide reductase which has a tyrosine radical involved in electron transfer. These studies have shown that the reactivities of the tyrosine radicals with added phenols, hydrazines, hydroxyurea, dithionite and ascorbate are dependent on the structures of some of the amino acids near the radical. This has been explained by changes in the hydrophobic environment of the radical (Gerez *et al.*, 1997). Significantly, several EPR studies on these naturally occurring tyrosine radicals have shown that the different protein environments dramatically affect the line shapes (Barry, 1993; Hoganson and Babcock, 1992).

The hydrophobic environment can affect hydrogen bonding and even the geometry of the radical (Himo *et al.*, 1997; Davydov *et al.*, 1996). However, another consequence of the environment will be a change in the apparent pKs of the tyrosine and cysteines. It is obvious from Figure 22.1 that a relatively small decrease in the pK of tyrosines or an increase in the pKs of cysteines in proteins can separate the values of the $(D^{\bullet},H^{+}/DH)$ potentials such that the transfer equilibrium will not occur.

22.5 Measurement of reduction potentials

Reduction potentials can be determined by many different techniques which involve oxidation and reduction reactions at electrode surfaces. Some of these conventional electrochemical techniques have been used to determine the one-electron reduction potential of some radicals in aqueous solution (Harriman, 1987; Tocher and Edwards, 1995). However, free radicals are usually very short-lived in polar solvents due to radical–radical reactions and it can be difficult to establish true reversible equilibria. Fast kinetic methods are therefore generally more reliable.

Pulse radiolysis is a technique wherein a short pulse of ionising radiation (typically, 5 ns–5 μs) is used to excite and ionise the water molecules in an aqueous solution

(Butler and Land, 1996). As a consequence of these reactions, known amounts of hydrated electrons, hydrogen atom and hydroxyl radicals are produced from the water within about 1 ns. These radicals can rapidly react with different solutes to form other radicals which can come into equilibria:

$$D^{\bullet-} + S \rightleftharpoons D + S^{\bullet-} \qquad (22.3)$$

The reactions of the radicals are usually measured as a function of time using optical spectroscopy. The equilibrium constant can thus be accurately measured and if the reduction potential of the standard, $E[S/S^{\bullet-}]$ is known, the reduction potential of the unknown couple, $E[D/D^{\bullet-}]$ can be determined using equation 22.4.

The main advantage of this technique is that the radicals are observed directly. Thus, if the radicals behave unexpectedly such that, for example, they react with each other (e.g. $2\ D^{\bullet-} \rightarrow$ products), rather than form an equilibrium, this can be observed by the nature of the kinetics. Pulse radiolysis has been used to determine hundreds of reduction potentials involving both oxidised and reduced radicals. The majority of these values have been compiled (Wardman, 1989).

22.6 Redox enzymes

Many different types of compounds can be reduced by mammalian reductases either by one- or two-electron steps. The one-electron reducing enzymes include NADPH:cytochrome P450 reductase (E.C. 1.6.2.3), NADPH:cytochrome b_5 reductase (E.C. 1.6.2.2), NADPH:ubiquinone oxidoreductase (E.C. 1.6.5.3) and xanthine oxidase (E.C. 1.2.3.2). The two-electron reducing enzymes include DT-diaphorase (E.C. 1.6.99.2), xanthine dehydrogenase and carbonyl reductase (Butler and Hoey, 1993a; Ross *et al.*, 1996), although it would appear that the main two-electron reducing enzyme in mammalian cells is DT-diaphorase (Ross *et al.*, 1994).

The different modes of reduction by these enzymes depend on the nature of the reducing moieties in the enzyme (FAD, FMN and/or metals) and the structure/ease of reducibility of the compound. Thus, for example, DT-diaphorase will bind compounds and hold them at a specific electron transfer site while the two electrons (or rather, a hydride) are donated (Tedeschi *et al.*, 1995) In this manner, the enzyme avoids radical production *in vivo*. However, several *in vitro* studies have shown that the reduction of some quinones by DT-diaphorase can result in radical production. These radicals are certainly not formed by the enzyme but are produced as a consequence of a subsequent equilibrium:

$$D + DH_2 \rightleftharpoons D^{\bullet-} + D^{\bullet-} + 2H^+ \qquad (22.18)$$

The pKs of hydroquinones are typically >11 whereas the pKs of semiquinones are usually between 4 and 5 (Swallow, 1982). Hence, at physiologically relevant pH values, DH_2 is a protonated hydroquinone and $D^{\bullet-}$ is an unprotonated semiquinone.

This equilibrium can be considered to be an electron transfer process wherein one radical is reduced by another. The one- and two-electron potentials of any compound are therefore related by:

$$2E[D/DH_2] = E[D^{\bullet-}/DH_2] + E[D/D^{\bullet-}] \qquad (22.19)$$

The value of the equilibrium constant, K_{18}, is essentially determined by the stability of the radical and in the majority of cases, the radicals will rapidly react with each other ($K_{18} \ll 1$). In some cases, internal hydrogen bonding can stabilise the radical. Thus, the semiquinones from duroquinone (Baxendale and Hardy, 1953), doxorubicin (Mukherjee *et al.*, 1989) and EO9 (Butler *et al.*, 1996) have all been shown to be in equilibrium with their hydroquinones at physiological pH. Indeed, it has been demonstrated that when EO9 is reduced by DT-diaphorase, semiquinone radicals can be detected by EPR (Bailey *et al.*, 1998).

The equilibrium constant, K_{18} also determines the stability of DH_2 in the presence of oxygen:

$$DH_2 + O_2 \rightarrow D + H_2O_2 \tag{22.20}$$

The direct two-electron oxidation by oxygen is spin forbidden and hence in the absence of metal ions (see below), the autoxidation probably occurs via one-electron steps:

$$D + DH_2 \rightleftharpoons D^{\bullet-} + D^{\bullet-} + 2H^+ \tag{22.18}$$

$$D^{\bullet-} + O_2 \rightleftharpoons D + O_2^{-\bullet} \tag{22.21}$$

$$O_2^{-\bullet} + DH_2 \rightleftharpoons H_2O_2 + D^{\bullet-} \tag{22.22}$$

and,

$$O_2^{-\bullet} + O_2^{-\bullet} + 2H^+ \rightarrow H_2O_2 + O_2 \tag{22.23}$$

The evidence for the involvement of superoxide radicals in the autoxidation of hydroquinones has come from studies wherein it was shown that superoxide dismutase dramatically alters the rate of autoxidation (Cadenas *et al.*, 1995; Jarabak and Jarabak, 1995).

The reduction of a compound by the one-electron reducing enzymes can be considered as a simple electron transfer equilibrium:

$$Enz^{\bullet-} + D \rightleftharpoons Enz + D^{\bullet-} \tag{22.24}$$

The free energy change, ΔG for the equilibrium is given by:

$$\Delta G = -nF\ \Delta E = -RT\ \ln K_{24} \tag{22.25}$$

where, K_{24} is the equilibrium constant and $\Delta E = E(D/D^{\bullet-}) - E(Enz/Enz^{\bullet-})$. According to Marcus theory, the rates of outer-sphere electron transfer reactions are dependent on ΔG. Thus, for any one-electron reducing enzyme, the rates of reduction of different substrates will depend on their relative $E[D/D^{\bullet-}]$ values. The structures of the compounds will also affect the apparent rates due to Michaelis–Menten binding but well below the V_{max} values, the rates should have a dependency on their reduction potentials.

This has been shown to be true for a variety of enzymes and different substrates including the reductions of quinones and nitrocompounds by cytochrome P450 reductase (Livertoux *et al.*, 1996; Butler and Hoey, 1993b; Cenas *et al.*, 1994a), the reduction of nitroimidazoles by xanthine oxidase (Clarke *et al.*, 1982), the reduction of 2,4-dinitrobenzamide mustards by bacterial nitroreductases (Atwell *et al.*, 1996), the reduction of quinones by a trypanothione reductase (Cenas *et al.*, 1994b) and the reduction of nitrofurans, 5-nitroimidazoles and nitrobenzenes by NADPH:adrenodoxin reductase (Marcinkeviciene *et al.*, 1990). Similarly, the rates of oxidation of hydroquinones by cytochrome f (Rich, 1982) and the oxidation of some phenols and indole-3-acetic acids by horseradish peroxidase (Candeias *et al.*, 1996) are dependent on the $E[D^{\bullet},H^{+}/DH]$ values.

If these enzyme rates can affect a biological function of a cell then the reduction potentials of the compounds can also be related to a biological effect. Hence for example, the cytotoxicities of some nitroaromatics (Adams *et al.*, 1979, 1980), benzotriazine di-*N*-oxides (Zeman *et al.*, 1989), and nitroheterocyclics (Cowan *et al.*, 1994; Smith and Edwards, 1995) can be related to the one-electron reduction potentials of the compounds. Similarly, the reduction potentials of some nitroheterocycles (Olive, 1981), naphthoquinones (Hakura *et al.*, 1994) and nitropolycyclic aromatic hydrocarbons (Jung *et al.*, 1991) have been related to their abilities to act as mutagens. More recent studies have shown that the inhibition of Epstein–Barr virus activation by anthraquinones can be correlated with reduction potentials (Koyama *et al.*, 1997).

It is now widely acknowledged that redox cycling and oxidative stress play an important role in the viability of cells and modulate many diverse events such as DNA synthesis, enzyme activation, selective gene expression, the regulation of the cell cycle and the triggering of cell death (e.g. Butler, 1998; Sun and Oberley, 1996; Powis *et al.*, 1997). These processes are mediated by redox signalling pathways which can be affected by compounds which change the oxidation state of cysteines (Kirkpatrick *et al.*, 1997; Oblong *et al.*, 1995).

22.7 Drugs, oxygen and metals

The reduction potential range for radical production in mammalian enzymes is approximately between about −450 mV for cytochrome P450 reductase ($[D/D^{\bullet-}]$) and +1000 mV for the peroxidase enzymes ($[D^{\bullet},H^{+}/DH]$). Under normal biological conditions, the radicals that can be produced from these reduction/oxidation reactions are not toxic to the cell. This is essentially because the crucial biological targets (DNA bases, lipids, aliphatic amino acids etc.) are not readily oxidised or reduced. Furthermore, the cell has a myriad of controls which prevent or repair most types of free radical damage including superoxide dismutase, catalase, glutathione peroxidase, vitamins and antioxidants, glycosylases and exonucleases (Halliwell, 1994; Breimer, 1991; Wood, 1996).

The functional reduction potential range can be extended considerably if metals and oxygen are involved:

$$D \rightarrow D^{\bullet-} \tag{22.26}$$

$$D^{\bullet-} + O_2 \rightleftharpoons D + O_2^{-\bullet} \tag{22.21}$$

$$D^{\bullet-} + M^{(n+1)+} \rightarrow D + M^{n+} \tag{22.27}$$

$$O_2^{-\bullet} + M^{(n+1)+} \rightarrow O_2 + M^{n+} \quad (22.28)$$

$$2O_2^{-\bullet} + 2H^+ \rightarrow H_2O_2 + O_2 \quad (22.23)$$

$$M^{n+} + H_2O_2 \rightarrow M^{(n+1)+} + OH^- + {}^\bullet OH \quad (22.29)$$

These reactions represent the main mechanisms whereby simple redox active compounds can damage cells. The possible production of hydroxyl radicals ($^\bullet OH$) can extend the redox range as they are extremely reactive and can undergo —H abstraction reactions, —OH additions and oxidations, ($E_o[^\bullet OH,H^+/H_2O]$ = 2730 mV (Wardman, 1989)). Hence, these radicals can react with all of the DNA bases and sugars, and will oxidise or undergo —H abstraction reactions with amino acids, lipids and vitamins. The damaging processes can continue as the radicals which are initially produced from the hydroxyl radical reactions can react with oxygen:

$$^\bullet OH + RH \rightarrow R^\bullet + H_2O \quad (22.30)$$

$$R^\bullet + O_2 \rightarrow ROO^\bullet \quad (22.31)$$

The resulting peroxy radicals can also react with metal ions and produce more damage:

$$ROO^\bullet + M^{n+} + H^+ \rightarrow ROOH + M^{(n+1)+} \quad (22.32)$$

$$ROOH + M^{n+} \rightarrow RO^\bullet + OH^- + M^{(n+1)+} \quad (22.33)$$

If RH is a lipid in a membrane, the alkoxyl radicals ($RO^\bullet$) can undergo further —H abstraction reactions, resulting in a chain reaction.

Reduction potentials can only predict the value of an equilibrium constant. The feasibility of a reaction occurring therefore also depend on the concentrations of the reactants and the stability of the products. Hence, a reaction with a large positive ΔG can go to completion if one of the products is a radical which reacts with solutes at very high rates. Examples of this are reactions which produce $^\bullet OH$ radicals. Similarly, a reaction with a large negative ΔG may occur slowly, or not at all, due to kinetic constraints (for example, reaction 22.23). Nonetheless, although these provisos have to considered, it is possible to use reduction potentials to assess the feasibility of biological damage occurring as a consequence of the above metal catalysed redox reactions.

As mentioned previously, the mammalian one-electron reducing enzymes function at potentials of about −450 mV. Hence the initial step of the redox cycling pathway (reaction 22.26), can occur for compounds which have $E[D/D^{\bullet -}]$ values more positive than −450 mV. The reduction potential for the oxygen couple, $E[O_2/O_2^{-\bullet}]$ is −155 mV under conditions of 1 M oxygen and so equilibrium (22.21) will only be in favour of superoxide production with compounds which have reduction potential less than this value. Thus, in order for compounds to be reduced by the one-electron reducing enzymes and be capable of forming superoxide radicals, they need to have reduction potentials ($E[D/D^{\bullet -}]$) of between −450 and −155 mV. There are numerous types of compounds which can have one-electron reduction potentials in this range including, quinones, nitroimidazoles, di-nitrobenzenes, nitroquinolines and azobenzenes (Wardman, 1989).

The reduction of the metal by superoxide radicals (or the drug radical), can occur if the reduction potential of the metal ($E[M^{(n+1)+}/Mn^{+}]$) is more positive than −155 mV. This is a very low potential for redox active metal complexes and indeed, this means that the majority of the biologically relevant complexes of iron and copper should be reduced by superoxide radicals. These include those in mammalian heme proteins (cytochromes, heme- and myoglobins, oxidases, peroxidases etc.). Similarly, the iron complexes formed from cyanide, sulphate, EDTA and DETAPAC can also be reduced by superoxide radicals. The role of iron in redox cycling is discussed in more detail in Chapter 9.

The dismutation of superoxide radicals (reaction 22.23) is usually catalysed in aerobic cells by superoxide dismutases. This reaction can be considered as being composed of two coupled reactions involving superoxide radicals. One reaction is the reduction of $O_2^{-\bullet}$ to hydrogen peroxide ($E[O_2^{-\bullet},2H^+/H_2O_2]$ = +940 mV at pH 7 (Koppenol and Butler, 1985) and the other is the oxidation of $O_2^{-\bullet}$ to oxygen ($E[O_2/O_2^{-\bullet}]$ = −155 mV). Thermodynamically, the dismutation reaction is very favourable although in the absence of superoxide dismutase, it is relatively slow. This is because in aqueous solution, the reaction has to occur via the protonated, $HO_2^{\bullet}$ form of the superoxide radical (Koppenol and Butler, 1985, 1987).

The reactions between the metal complexes and hydrogen peroxide (reaction 22.29) are Fenton reactions and depend on the reduction potentials of hydrogen peroxide ($E[H_2O_2, H^+/{}^{\bullet}OH]$ = 320 mV at pH 7 (Koppenol, 1989) and the metal complex, $E[M^{(n+1)+}/M^{n+}]$. Any metal complex with a reduction potential ($E[M^{(n+1)+}/M^{n+}]$) of less than 320 mV at pH 7 should be able to produce hydroxyl radicals from hydrogen peroxide. If this is to be coupled with the back reduction of the metal by superoxide radicals (reaction 27), the reduction potential of the metal complex has to be between −155 mV and +320 mV.

Interestingly, several metal complexes including those in superoxide dismutase, cytochromes and haemoglobins should be able to undergo reaction 22.29, but it appears that hydroxyl radicals are not formed. Similarly, many different types of reduced radicals including semiquinones and superoxide radicals should be capable of reducing hydrogen peroxide to ${}^{\bullet}OH$, but the reaction also does not occur. It is obvious that the reaction is not initiated by a simple one-electron reduction step and a $H_2O_2^-$ species is not formed. Numerous studies have proposed that the hydrogen peroxide has to interact with a metal to initially produce a ferryl-type intermediate. In the case of simple metal complexes, the hydrogen peroxide initially replaces a ligand (usually water). However, the ligands in some metalloproteins and in particular, the ligands in cytochromes, are not so readily replaced and hence the ferryl type reactions with hydro-gen peroxide cannot take place except, perhaps, at very high hydrogen peroxide concentrations.

Hydrogen peroxide reacts stoichiometrically with human methaemoglobin to form a stable ferryl–oxo species and a protein based radical (Patel *et al.*, 1996; Gunther *et al.*, 1995). These products can oxidise other additives to radical cations (Kelder *et al.*, 1991) and initiate lipid peroxidation (Davies, 1990). Similar ferryl intermediates have been proposed to occur when hydrogen peroxide reacts with simple metal complexes. This has led to a great debate in the literature as to the nature of the main oxidising species produced in the Fenton reaction (e.g. Zhao and Jung, 1995; Rush and Koppenol, 1987; Rush and Koppenol, 1986; Tomita *et al.*, 1994; Wink *et al.*, 1994). However, the most likely explanation is that as some form of a metal–peroxide complex is always initially produced, this complex can react directly with the substrates, can decompose to produce a higher oxidation state of the metal or release $^{\bullet}OH$ radicals (Goldstein *et al.*, 1993). Thus, the nature of the oxidising species is very much dependent on the structure of the metal complex and the relative concentrations of all the reactants.

Similar arguments can be applied to the reactions of the metal complexes with organic peroxides and alkoxides (reactions 22.32 and 22.33). Theoretical values at pH 7 of $E[ROO^{\bullet},H^+/ROOH] = 1000$ mV and $E[ROOH/RO^{\bullet}] = 1900$ mV have been derived from considerations of bond energies (Koppenol, 1990). Hence reactions 22.32 and 22.33 should readily occur with many transition metal complexes.

Acknowledgments

I am indebted to the many colleagues in different institutes and for helpful discussions and numerous preprints. In particular, I would like to thank Ted Land and Wim Koppenol for their invaluable comments.

This work was supported by the Cancer Research Campaign.

References

Adams, G.E., Clarke, E.D., Flockhart, I.R., Jacobs, R.S., Sehmi, D.S., Stratford, I.J., Wardman, P., Watts, M.E., Parrick, J., Wallace, R.G. and Smithen, C.E., 1979, Structure-activity relationships in the development of hypoxic cell radiosensitizers. I. Sensitization efficiency. *Int. J. Radiat. Biol. Phys. Chem. Med.*, **35**, 133–150.

Adams, G.E., Stratford, I.J., Wallace, R.G., Wardman, P. and Watts, M.E., 1980, Toxicity of nitro compounds toward hypoxic mammalian cells in vitro: dependence on reduction potential. *J. Natl. Cancer Inst.*, **64**, 555–560.

Atwell, G.J., Boyd, M., Palmer, B.D., Anderson, R.F., Pullen, S.M., Wilson, W.R. and Denny, W.A., 1996, Synthesis and evaluation of 4-substituted analogues of 5-[N,N-bis (2-chloroethyl)amino]-2-nitrobenzamide as bioreductively activated prodrugs using an Escherichia coli nitroreductase. *Anticancer Drug Des.*, **11**, 553–567.

Babcock, G.T., Espe, M., Hoganson, C., Lydakis Simantiris, N., McCracken, J., Shi, W., Styring, S., Tommos, C. and Warncke, K., 1997, Tyrosyl radicals in enzyme catalysis: some properties and a focus on photosynthetic water oxidation. *Acta Chem. Scand.*, **51**, 533–540.

Bailey, S.B., Lewis, A.D., Knox, R.J., Patterson, L.H., Fisher, G.P. and Workman, P., 1998, Reduction of EO9 by DT-diaphorase. A detailed kinetic study and analysis of metabolites. *Biochem. Pharmacol.*, **56**, 613–621.

Barry, B.A., 1993, The role of redox-active amino acids in the photosynthetic water-oxidizing complex. *Photochem. Photobiol.*, **57**, 179–188.

Baxendale, J.H. and Hardy, H.R., 1953, The formation constant of durosemiquinone, *Trans. Faraday Soc.*, **49**, 1433–1437.

Bobrowski, K., Wierzchowski, K.L., Holcman, J. and Ciurak, M., 1992, Pulse radiolysis studies of intramolecular electron transfer in model peptides and proteins. IV. Met/S:.Br→Tyr/O. radical transformation in aqueous solution of H-Tyr-$(Pro)_n$-Met-OH peptides. *Int. J. Radiat., Biol.*, **62**, 507–516.

Bobrowski, K., Holcman, J., Poznanski, J. and Wierzchowski, K.L., 1997, Pulse radiolysis studies of intramolecular electron transfer in model peptides and proteins. 7. Trp→TyrO radical transformation in hen egg-white lysozyme. Effects of pH, temperature, Trp62 oxidation and inhibitor binding. *Biophys. Chem.*, **63**, 153–166.

Breimer, L.H., 1991, Repair of DNA damage induced by reactive oxygen species. *Free Rad. Res. Comm.*, **14**, 159–171.

Butler, J., 1998, Redox Cycling Drugs. In: *DNA Damage by Free Radicals* (eds B. Halliwell and O.I. Aruoma). OICA Press, St Lucia pp. 131–157.

Butler, J. and Hoey, B.M., 1993a, Redox Cycling drugs and DNA damage. In *DNA and Free Radicals* (eds B. Halliwell and O.I. Aruoma). Ellis Horwood, London pp. 243–265.

Butler, J. and Hoey, B.M., 1993b, The one electron reduction potential of several substrates can be related to their reduction rates by cytochrome P450 reductase. *Biochim. Biophys. Acta*, **1161**, 73–76.

Butler, J. and Land, E.J., 1996, Pulse-Radiolysis. In: *Free Radicals. A Practical Approach* (eds N.A. Punchard and F.J. Kelly), Oxford University Press, London pp. 47–61.

Butler, J., Land, E.J., Prutz, W.A. and Swallow, A.J., 1986, Reversibility of charge transfer between tryptophan and tyrosine. *J.C.S. Chem. Comm.*, **4**, 348–349.

Butler, J., Spanswick. V. and Cummings, J., 1996, The autoxidation of the reduced forms of EO9. *Free Rad. Res.*, **25**, 141–146.

Cadenas, E., 1995, Antioxidant and prooxidant functions of DT-diaphorase in quinone metabolism. *Biochem. Pharmacol.*, **49**, 127–140.

Candeias, L.P., Folkes, L.K., Porssa, M., Parrick, J. and Wardman, P., 1996, Rates of reaction of indoleacetic acids with horseradish peroxidase compound I and their dependence on the redox potentials. *Biochemistry*, **35**, 102–108.

Cenas, N., Anusevicius, Z., Bironaite, D., Bachmanova, G.I., Archakov, A.I. and Ollinger, K., 1994a, The electron transfer reactions of NADPH: cytochrome P450 reductase with non-physiological oxidants. *Arch. Biochem. Biophys.*, **315**, 400–406.

Cenas, N.K., Arscott, D., Williams, C.H. and Blanchard, J.S., 1994b, Mechanism of reduction of quinones by Trypanosoma congolense trypanothione reductase. *Biochemistry*, **33**, 2509–2515.

Clark, W.M., 1960, *Oxidation-Reduction Potentials of Organic Systems*. Williams and Wilkins, Baltimore.

Clarke, E.D., Golding, K.H. and Wardman, P., 1982, Nitroimidazoles as anaerobic electron acceptors for xanthine oxidase. *Biochem. Pharmacol.*, **31**, 3237–3242.

Cowan, D.S., Matejovic, J.F., Wardman, P., McClelland, R.A. and Rauth, A.M., 1994, Radiosensitizing and cytotoxic properties of DNA targeted phenanthridine-linked nitroheterocycles of varying electron affinities. *Int. J. Radiat. Biol.*, **66**, 729–738.

Davies, M.J., 1990, Detection of myoglobin-derived radicals on reaction of metmyoglobin with hydrogen peroxide and other peroxidic compounds. *Free Rad. Res. Comm.*, **10**, 361–370.

Davydov, R., Sahlin, M., Kuprin, S., Gräslund, A. and Ehrenberg, A., 1996, Effect of the tyrosyl radical on the reduction and structure of the Escherichia coli ribonucleotide reductase protein R2 diferric site as probed by EPR on the mixed-valent state. *Biochemistry*, **35**, 5571–5576.

Gerez, C., Elleingand, E., Kauppi, B., Eklund, H. and Fontecave, M., 1997, Reactivity of the tyrosyl radical of Escherichia coli ribonucleotide reductase – control by the protein. *Eur. J. Biochem.*, **249**, 401–407.

Goldstein, S., Meyerstein, D. and Czapski, G., 1993, The Fenton reagents. *Free Rad. Biol. Med.*, **15**, 435–445.

Gunther, M.R., Kelman, D.J., Corbett, J.T. and Mason, R.P., 1995, Self-peroxidation of metmyoglobin results in formation of an oxygen-reactive tryptophan-centered radical. *J. Biol. Chem.*, **270**, 16075–16081.

Hakura, A., Mochida, H., Tsutsui, Y. and Yamatsu, K., 1994, Mutagenicity and cytotoxicity of naphthoquinones for Ames Salmonella tester strains. *Chem. Res. Toxicol.*, **7**, 559–567.

Halliwell, B., 1994, Free radicals and antioxidants: a personal view. *Nutr. Rev.*, **52**, 253–265.

Harriman, A., 1987, Further comment on the redox potentials of tryptophan and tyrosine. *J. Phys. Chem.*, **91**, 6102–6104.

Himo, F., Gräslund, A. and Eriksson, L.A., 1997, Density functional calculations on model tyrosyl radicals. *Biophys. J.*, 72, 1556–1567.

Hoganson, C.W. and Babcock, G.T., 1992, Protein-tyrosyl radical interactions in photosystem II studied by electron spin resonance and electron nuclear double resonance spectroscopy: comparison with ribonucleotide reductase and in vitro tyrosine. *Biochemistry*, **31**, 11874–11880.

Jarabak, R. and Jarabak, J., 1995, Effect of ascorbate on the DT-diaphorase-mediated redox cycling of 2-methyl-1,4-naphthoquinone. *Arch. Biochem. Biophys.*, **318**, 418–423.

Jollow, D.J., Bradshaw, T.P. and McMillan, D.C., 1995, Dapsone induced hemolytic anemia. *Drug Metab. Rev.*, **27**, 107–124.

Jung, H., Shaikh, A.U., Heflich, R.H. and Fu, P.P., 1991, Nitro group orientation, reduction potential, and direct-acting mutagenicity of nitro-polycyclic aromatic hydrocarbons. *Environ. Mol. Mutagen.*, **17**, 169–180.

Kelder, P.P., de Mol, N.J. and Janssen, L.H., 1991. Mechanistic aspects of the oxidation of phenothiazine derivatives by methemoglobin in the presence of hydrogen peroxide. *Biochem. Pharmacol.*, **42**, 1551–1559.

Kirkpatrick, D.L., Ehrmantraut, G., Stettner, S., Kunkel, M. and Powis, G., 1997, Redox active disulfides: the thioredoxin system as a drug target. *Oncol. Res.*, **9**, 351–356.

Koppenol, W.H. and Butler. J., 1985, The energetics of the interconversion reactions of oxyradicals. *Free Rad. Biol. Med.*, **1**, 91–131.

Koppenol, W.H. and Butler, J., 1987, Mechanisms of reactions involving singlet oxygen and the Superoxide Anion. *FEBS Lett.*, **83**, 1–5.

Koppenol, W.H., 1989, Generation and thermodynamic properties of oxyradicals. In: *Focus on membrane lipid peroxidation*, *Vol 1*. (ed. C. Vigo-Pelfrey). CRC Press, Boca Raton pp. 1–13.

Koppenol, W.H., 1990, Oxyradical reactions: from bond-dissociation energies to reduction potentials. *FEBS Lett.*, **264**, 165–167.

Koyama, J., Tagahara, K., Osakai, T., Tsujino, Y., Tsurumi, S., Nishino, H. and Tokuda, H., 1997, Inhibitory effects on Epstein-Barr virus activation of anthraquinones: correlation with redox potentials. *Cancer Lett.*, **115**, 179–183.

Livertoux, M.H., Lagrange, P. and Minn, A., 1996, The superoxide production mediated by the redox cycling of xenobiotics in rat brain microsomes is dependent on their reduction potential. *Brain Res.*, **725**, 207–216.

Marcinkeviciene, J., Cenas, N., Kulys, J., Usanov, S.A., Sukhova, N.M., Selezneva, I.S. and Gryazev, V.F., 1990, Nitroreductase reactions of the NADPH: adrenodoxin reductase and the adrenodoxin complex. *Biomed. Biochim. Acta*, **49**, 167–172.

Miller, M.A., Han, G.W. and Kraut, J., 1994, A cation binding motif stabilizes the compound I radical of cytochrome c peroxidase. *Proc. Natl Acad. Sci. USA*, **91**, 11118–11122.

Mukherjee, T., Land, E.J., Swallow, A.J. and Bruce, J.M., 1989, One-electron reduction of adriamycin and daunomycin: short-term stability of the semiquinones. *Arch. Biochem. Biophys*, **272**, 450–458.

Oblong, J.E., Chantler, E.L., Gallegos, A., Kirkpatrick, D.L., Chen, T., Marshall, N. and Powis, G., 1995, Reversible inhibition of human thioredoxin reductase activity by cytotoxic alkyl 2-imidazolyl disulfide analogues. *Cancer Chemother. Pharmacol.*, **34**, 434–438.

Olive, P.L., 1981, Correlation between the half-wave reduction potentials of nitroheterocycles and their mutagenicity in Chinese hamster V79 spheroids. *Mutat. Res.*, **82**, 137–145.

Patel. R.P., Svistunenko, D.A., Darley Usmar, V.M., Symons, M.C. and Wilson, M.T., 1996, Redox cycling of human methaemoglobin by H_2O_2 yields persistent ferryl iron and protein based radicals. *Free Rad. Res.*, **25**, 117–123.

Pedersen, J.Z. and Finazzi-Agrò, 1993, Protein-radical enzymes. *FEBS Lett*, **325**, 53–58.

Powis, G., Gasdaska, J.R. and Baker, A., 1997, Redox signaling and the control of cell growth and death. *Adv. Pharmacol.*, **38**, 329–359.

Prutz, W.A., Butler, J., Land, E.J. and Swallow, A.J., 1982, Charge transfer between tryptophan and tyrosine in proteins. *Biochim. Biophys Acta*, **705**, 150–162.

Prutz, W.A., Butler, J. and Land, E.J., 1983, Phenol coupling initiated by one-electron oxidation of tyrosine units in peptides and proteins. *Int. J. Radiat. Biol.*, **44**, 183–196.

Prutz, W.A., Butler, J., Land, E.J. and Swallow, A.J, 1986, Unpaired electron migration between aromatic and sulfur peptide units. *Free Rad. Res. Comms.*, **2**, 69–73.

Prutz, W.A., Butler, J., Land, E.J. and Swallow, A.J., 1989, The role of sulphur peptide functions in free radical transfer. A pulse radiolysis study. *Int. J. Radiat. Biol.*, **55**, 539–556.

Rich, P.R., 1982, A physicochemical model of quinone-cytochrome b-c complex electron transfers. In: *Function of quinones in energy conserving systems* (ed. B.L. Trumpower). Academic Press, London pp. 73–83.

Ross, D., Beall, H.D., Traver, R.D., Siegel, D., Phillips, R.M. and Gibson, N.W., 1994, Bioactivation of quinones by DT-diaphorase. Molecular, biochemical and chemical studies. *Oncol., Res.*, **6**, 493–500.

Ross, D., Beall, H.D., Siegel, D., Traver, R.D. and Gustafson, D.L., 1996, Enzymology of bioreductive drug activation. *Br. J. Cancer*, **74** (Suppl. XXVIII), S1–S8.

Rush, J.D. and Koppenol, W.H., 1986, Oxidizing intermediates in the reaction of ferrous EDTA with hydrogen peroxide. Reactions with organic molecules and ferrocytochrome c. *J. Biol. Chem.*, **261**, 6730–6733.

Rush, J.D. and Koppenol, W.H., 1987, The reaction between ferrous polyaminocarboxylate complexes and hydrogen peroxide: an investigation of the reaction intermediates by stopped flow spectrophotometry. *J. Inorg. Biochem.*, **29**, 199–215.

Schuessler, H., Davies, J.V., Scherbaum, W. and Jung, E., 1986, Reactions of reducing radicals with ribonuclease. *Int. J. Radiat. Biol.*, **50**, 825–839.

Smith, M.A. and Edwards, D.I., 1995, Redox potential and oxygen concentration as factors in the susceptibility of Helicobacter pylori to nitroheterocyclic drugs. *J. Antimicrob. Chemother.*, **35**, 751–764.

Stadtman, E.R., 1993, Oxidation of free amino acids and amino acid residues in proteins by radiolysis and by metal-catalyzed reactions. *Annu. Rev. Biochem.*, **62**, 797–821.

Sun, Y. and Oberley, L.W., 1996, Redox regulation of transcriptional activators. *Free Rad. Biol. Med.*, **21**, 335–348,

Surdhar, P.S. and Armstrong, D.A., 1986, Redox potentials of some sulfur-containing radicals. *J. Phys. Chem.*, **90**, 5919–5917.

Swallow, A.J., 1982, Physical chemistry of semiquinones. In: *Function of quinones in energy conserving systems* (ed. B.L. Trumpower). Academic Press, London pp. 59–72.

Tedeschi, G., Chen, S. and Massey, V., 1995, DT-diaphorase. Redox potential, steady-state, and rapid reaction studies. *J. Biol. Chem.*, **270**, 1198–1204.

Tocher, J.H. and Edwards, D.I., 1995, The interaction of nitroaromatic drugs with aminothiols. *Biochem. Pharmacol.*, **50**, 1367–1371.

Tomita, M., Okuyama, T., Watanabe, S. and Watanabe, H., 1994, Quantitation of the hydroxyl radical adducts of salicylic acid by micellar electrokinetic capillary chromatography: oxidizing species formed by a Fenton reaction. *Arch. Toxicol.*, **68**, 428–433.

Wardman, P., 1989, Reduction potentials of one-electron couples involving free radicals in aqueous solution, *J. Phys. Chem. Ref. Data*, **18**, 1637–1755.

Weinstein, M., Alfassi, Z.B., DeFelippis, M.R., Klapper, M.H. and Faraggi, M., 1991, Long range electron transfer between tyrosine and tryptophan in hen egg-white lysozyme. *Biochim. Biophys Acta*, **1076**, 173–178.

Wink, D.A., Wink, C.B., Nims, R.W. and Ford, P.C., 1994, Oxidizing intermediates generated in the Fenton reagent: kinetic arguments against the intermediacy of the hydroxyl radical. *Environ. Health Perspect.*, **102** (Suppl 3), 11–15.

Wood, R.D., 1996, DNA repair in eukaryotes. *Ann. Rev. Biochem.*, **65**, 135–167.

Zeman, E.M., Baker, M.A., Lemmon, M.J., Pearson, C.I., Adams, J.A., Brown, J.M., Lee, W.W. and Tracy, M., 1989, Structure-activity relationship for benzotrazine di-N-oxides. *Int. J. Rad. Oncol. Biol. Phys.*, **16**, 977–981.

Zhao, M.J. and Jung, L., 1995, Kinetics of the competitive degradation of deoxyribose and other molecules by hydroxyl radicals produced by the Fenton reaction in the presence of ascorbic acid. *Free Rad. Res.*, **23**, 229–243.

23 The use of QSAR for the prediction of free radical toxicity

Mark T.D. Cronin and Thuy T. Tran

23.1 Introduction

There is an increased emphasis to identify potentially harmful compounds, with regard both to human exposure to new pharmaceutical, detergent or perfume compounds and to environmental exposure. Importance is placed on the development of 'safe' new compounds for commercial exploitation and risk assessment of existing compounds. One of the approaches to identify compounds with associated hazard is the use of quantitative structure–activity relationships (QSARs). A QSAR aims to relate statistically the biological activity of a chemical to its physico-chemical structure. It is increasingly being viewed as a viable alternative to whole-animal toxicity testing and as a technique to prioritise existing chemicals for testing (Richner and Weidenhaupt, 1997). The reader is referred elsewhere for detailed descriptions of the use of QSARs to predict toxicity (Cronin and Dearden, 1995a–d).

Ideally a QSAR study to predict toxicity should be performed using chemicals having the same mechanism of toxic action (Cronin and Dearden, 1995a). This was achieved historically by the utilisation of congeneric series of compounds, making the assumption that simple changes in the structure, while affecting physico-chemical properties such as hydrophobicity, or molecular size, did not affect the mechanism of action. More recently effort has been placed into the elucidation and identification of mechanism of action from molecular structure alone. For the prediction of acute environmental toxicity for instance, the structural determinants of the narcoses modes of action, and compounds that may cause toxicity via electrophilic reactions, are becoming well established (Schultz *et al.*, 1997a). At the far end of the current spectrum of knowledge are the identification, and the prediction of toxicity, of compounds with the potential to form free radicals, or reactive oxygen species (ROS) (Bradbury *et al.*, 1995). This chapter will give an account of the efforts in this important area and make suggestions for likely progress.

It should be apparent that there are a number of important issues and distinctions to be recognised when the prediction of toxicity of free radicals is considered. Firstly, the prediction of toxicity is based not on the free radical itself, but on the compounds that have the tendency to propagate free radicals. This is because the secondary radicals of the free radical chain reactions may react with compounds of biological importance and consequently cause injurous pathologies to the cells. Secondly, this implies that there will be two general areas for the prediction of toxicity via free radical mechanisms. Initially, compounds that have the capability to produce free radicals must be identified qualitatively and indeed this is the sole purpose of some of the predictive methods described herein. Once these have been identified there is the subsequent quantitative prediction of relative potency *in vitro* or *in vivo*. Thus QSAR analysis is performed not on the relative toxicity of a range of different free radicals, but on the relative propensity of a compound to produce the free radical.

QSAR analysis of toxicity has a number of advantages. At its most basic level it is a predictive technique to estimate the toxicity of chemicals. However, the analysis may also provide more subtle information regarding the mechanism of action of compounds. Unfortunately, QSAR is seen too often as the cure for all ills and often disappointment and disillusionment follow when significant numbers of poor predictions are made. This raises a number of issues that must be recognised by the reader. Firstly, a QSAR is only as good as the data that go into it. Poor quality biological data (i.e. those with a large amount of error or uncertainty associated with them) will only provide only a poor quality model and any claim to the contrary should be viewed with the utmost suspicion! Secondly, a QSAR is applicable for predictive purposes only for compounds that reside within the area of knowledge of the biological activity of the training set. Thus, for example, a QSAR based on a series of non-polar narcotics will not provide an accurate estimate of the toxicity of a compound that is able to form free radicals.

Finally, for a QSAR to be truly predictive it should be capable of making an estimate of toxicity from chemical structure alone (so there is no necessity for the chemical to be available, or even to exist). As part of the remit of this review a frank assessment of the applicability of the individual predictive methods will be attempted. This should not be seen, however, as undue criticism of any one methodology, as it is felt that each study should be viewed on individual merits and by the advancement made to the science.

23.2 Identification of compounds likely to form free radicals

Free radical reactions are an essential component of the normal metabolic processes. The production of reactive oxygen radicals (ROR) is natural in biological systems and usually does not result in harmful effects. However, production is induced also by external factors such as drugs, food additives, environmental pollutants and radiation. These may form potentially toxic free radicals in the biological systems. Uncontrolled ROR production is implicated in the processes of aging and various other pathological conditions (Halliwell and Gutteridge, 1990; Aruoma, 1994).

Generally organic radicals are very short lived and react rapidly with each other or with neighbouring molecules. Mechanistically, this usually involves hydrogen abstraction, addition to unsaturated bonds or electron transfer. In this way free radicals become cytotoxic when important biological macromolecules are affected and injury is made worse by the presence of compounds capable of propagating free radicals. Biological systems are notoriously complex so that *in vivo* reactions of radicals may follow numerous pathways, enzymic or non-enzymic redox processes. Therefore, the degree of their toxicity is affected by the concentration at each biological component, the likelihood of the latter encountering the radical and their reaction rate constant indicating the reactivity of the radical with that particular macromolecule. Once this is established, the importance of each pathway may be judged and the toxicity of the compound determined.

The studies of biologically important radicals are usually carried out in aqueous solutions. This is to emulate the conditions under which reactions in biological systems occur; namely, the aqueous environment or at the interface between aqueous and lipid compartments. Such studies aim to comprehend the biological processes whereby enzymic methods of radical production are used in conjunction with physicochemical methods of radical production. In addition, most studies examine factors that modulate radical production to assist in the prediction of toxicity and for the elucidation of mechanisms of action. Other studies have been based upon the identification of compounds that have the ability to produce free radicals. These characteristics will be reviewed here.

A gross method for the identification of compounds able to produce free radicals is given in the section on skin sensitisation. This is no more than the recognition that certain chemical features of compounds can give rise to free radicals. Since this is a gross assessment of the ability to produce free radicals, it may be rather overpredictive in nature.

An accurate parameter used to indicate compounds that are able to induce oxidative stress is the redox potential of the compound. There are definite 'windows' of redox potentials associated with free radical propagators. For instance, quinones are capable of one-electron reduction only if their redox potentials fall within the ranges −240 mV to −170 mV and −50 mV to 25 mV (Powis and Appel, 1980; Powis *et al.*, 1981; O'Brien, 1991). Thus the redox potential of a quinone could be used as a general indication of the ability of that compound to redox cycle.

In cases where wide-scale identification of compounds susceptible to redox cycling is needed, the redox potential may be viewed as an appropriate parameter. However, this is hampered by the availability of a good database. While databases of reduction potentials are available (Wardman, 1989), in comparison to those for other commonly utilised parameters (such as the uptake and distribution of a compound which is related to its hydrophobicity and is quantified by the logarithm of the n-octanol/water partition coefficient, log P, for which tens of thousands of measurements have been made (Leo, 1993)), this is comparatively small. Thus a method to calculate redox potential would be advantageous as an aid towards a viable QSAR for toxicity prediction.

To this end Bradbury *et al.* (1995) investigated the role of quantum chemical parameters to model redox potentials. Their study found that within structural classes the one-electron reduction potentials of benzoquinones, naphthoquinones, phenols and nitrobenzenes, could be estimated from global and local electronic parameters that are considered to be related to delocalisation. This is illustrated by the relation of the one-electron reduction potentials (mV) of phenols at pH 7.0 (E^1_7) to E_{HOMO}, the energy of the highest occupied molecular orbital:

$$E^1_7 = -700E_{HOMO} - 5600 \qquad (23.1)$$

$n = 18 \quad r = 0.890 \quad$ s not given $\quad F = 60.4$

where: n = number of compounds tested, r = regression coefficient, s = standard error of estimate and F = Fisher's test.[1]

In addition, amongst other good correlations, it was observed that the one and two-electron reduction potentials for nitrobenzenes and aziridinylbenzoquinones at pH 7.0 ($E^{1/2}_7$) were well modelled by the energy of the lowest unoccupied molecular orbital (E_{LUMO}):

$$E_7^{1/2} = -200E_{LUMO} - 710 \quad (23.2)$$

$n = 47 \quad r = 0.872 \quad$ s not given $\quad F = 134.5$

Thus, these equations, while not being able to predict the redox potential for all compounds, do clearly provide a starting point for the prediction of the reduction potential of benzoquinones, naphthoquinones, phenols and nitrobenzenes. This is an appropriate starting point for the further work required in this area.

23.3 Acute toxicity

Toxicity data may be obtained from environmentally important species for ecotoxicological purposes, or from mammals, such as those in the testing of new pharmaceutical compounds. Such data, ascribed to single species, are reported often as a measure of hazard for a compound. At present, there is a desire to replace such tests, particularly those of the LD_{50} type, and develop alternatives.

An overriding requirement for the prediction of toxicity using QSAR techniques is reliable toxicity data. It is less likely that a useful predictive model will be produced when the data on which it is based may be subject to intra- and inter-laboratory variability. In the area of acute toxicity it is surprising how few reliable toxicity data are available openly for QSAR modelling. Data on the acute toxicity of free radicals are particularly scarce, and hence a limited number of studies have been performed.

It is perhaps appropriate here to consider the mechanistic approach taken to predict the toxicity of xenobiotics. To model such toxicity MacFarland (1970) proposed the following generic QSAR:

$$\log [\text{toxicity}]^{-1} = a(\log \text{ of penetration}) + b(\log \text{ of interaction}) + c \quad (23.3)$$

This is only true if the toxicity is given as a concentration to produce a defined effect. Toxicity is considered to be a two stage process involving firstly the uptake of a chemical into either the cellular membrane or cytoplasm (penetration). Following uptake there may be some interaction of the chemical with cellular contents. The xenobiotic's reactions with macromolecules by electrophilic reactions with nucleophiles (the stereo-electronic interaction term) is easily quantified by the energy of the lowest unoccupied molecular orbital (E_{LUMO}). Yet the interaction term of the compounds of interest to this paper (i.e. the capability to produce the free radical) has been more difficult to quantify as described. In equation 23.3, a and b are the coefficients for each of the terms and c is the intercept.

As for mechanisms of action where there is no reactivity associated with toxicity (e.g. mechanisms associated with the narcoses modes of action), toxicity is often reported to be related to hydrophobicity (as quantified by log P) (van Wezel and Opperhuizen, 1995). Even for some reactive mechanisms of action, providing the mechanism of action is similar for all compounds, the relative variation in toxicity may be related solely to the hydrophobicity of the compound (Cronin *et al.*, 1998; Cronin and Schultz, 1998). In these instances it is assumed that reactivity remains constant. They hypothesised that this may be legitimate with regard to free radical propagators, i.e. once it is at the site of action the compound's intrinsic ability to form free radicals may be relatively constant. If this were true, the toxicity of such compounds would be related solely to their ability to reach the active site and therefore QSARs would be based upon log P alone. However, a review of the literature suggests that this is not the case. Despite the limited number of QSARs available for the prediction of acute toxicity of compounds that may generate free radicals, those that are accessible tend to suggest that hydrophobicity is of little importance. Although the issue is not so definite, a number of important issues raised are outlined below.

Ideally, a QSAR to predict toxicity should be based upon a single mechanism of action (Cronin and Dearden, 1995a). Some of the most commonly studied compounds that generate free radicals are the quinones and napthoquinones. It is apparent that these compounds may produce toxicity by at least two mechanisms of action (Henry and Wallace, 1995, 1996). Quinones and naphthoquinones essentially are made up of carbonyl groups conjugated with aromatic double bonds. As such they are able to act as Michael-type acceptors and arylate nucleophilic sites on macromolecules *in vivo*. In addition to such electrophilic reactions, quinones may be reduced enzymatically by one-electron or two-electron processes. This may propagate a range of secondary free radicals or enter futile redox cycling (termination). It is likely that at least some quinones and naphthoquinones are capable of both mechanisms of action, so that both will occur to some extent.

Free radical reactions of quinone, one of the compounds that participate in electron transport *in vivo*, follows one electron transfer with reducing or oxidising radicals. The oxidising alkyl radical normally has low reactivity with organic compounds (RH) (equation 23.4) but has high reactivity towards oxygen (equation 23.5) and quinones (equations 23.6, 23.7).

$$^{\bullet}CH_3 + RH \rightarrow CH_4 + R^{\bullet} \text{ (adduct radical)} \qquad (23.4)$$

$$^{\bullet}CH_3 + O_2 \rightarrow CH_3O_2^{\bullet} \qquad (23.5)$$

Quinones react with non-reducing radicals (methyl or phenyl) by addition reactions or are rapidly oxidised to another quinone.

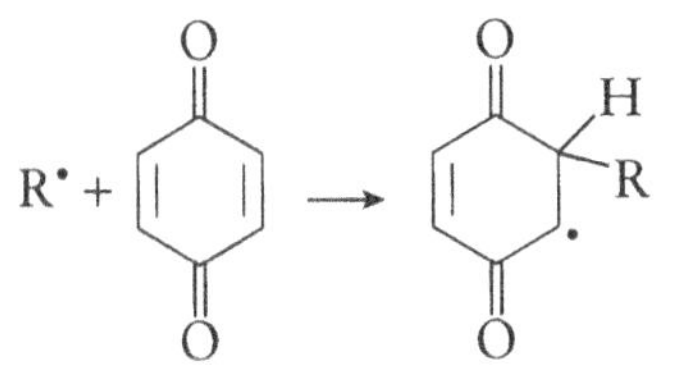

Q ADDUCT (23.6)

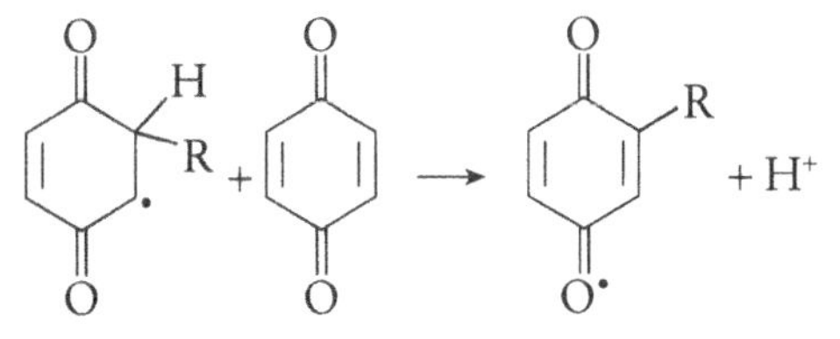

Q Q (23.7)

The reducing radicals are produced by hydrogen abstraction by α-hydroxy and α-aminoalkyl radicals from the α position of amines and amino acids or by one-electron reduction and protonation of carbonyl compounds. Quinones can also react very rapidly with reducing radicals to produce semiquinone radicals:

$$(CH_3)_2C^{\bullet}OH + Q \rightarrow (CH_3)CO + H^+ + Q^{\bullet-} \quad (23.8)$$

The semiquinone radical decays by disproportionation to the original quinone and dihydro derivative:

$$2Q^{\bullet-} + 2H^+ \rightarrow Q + QH_2 \quad (23.9)$$

In the case of semiquinones, these reactions are often assumed to be an outer sphere electron transfer. The rate of decay of this radical is dependent upon its nature where $Q^{\bullet-} + Q^{\bullet-}$ is unreactive but rapid reaction is observed with $Q^{\bullet-} + Q^{\bullet}H$ and $Q^{\bullet}H + Q^{\bullet}H$ semiquinone states. The environmental pH also affects the rate of decay where at high pH, the rate is increased according to equation 23.9 (Adams and Michael, 1967; Rao and Hayon, 1973; Anderson, 1983).

The toxicity of a quinone radical adduct depends upon the substitution pattern. Substitution at positions other than at the α-carbon ($C^{\bullet}$) has only minor effects depending on the electron directing effect of the groups. For example, the strong electron withdrawing substituent on the α-carbon of the alkyl radical alone increases the oxidising power of this radical. α-Substitution of carboxyl or haloalkyl substituents confers only a slightly modified behaviour compared to non-substituted alkyl radicals. In contrast, substitution with OH, OR, NH_2, or NR_2 results in a strongly reducing radical and carbonyl substitution yields an oxidising radical.

Despite the overwhelming evidence of semiquinone radical toxicity, biological systems have protective antioxidant enzymes that scavenge this radical and render it harmless. During normal metabolic processes, the flavoprotein NADPH-cytochrome P_{450} reductase participates as an initial electron donor in the activation of different xenobiotics, with the formation of an initial semiquinone radical form and subsequent production of reactive oxygen species. The endoplasmic reticulum and enzymes such as NADH-cytochrome b_5 reductase, NADH-dehydrogenase, xanthine dehydrogenase and aldehyde oxidase can catalyse this initial one-electron reduction step:

$$O_2 \xrightarrow{e^-} O_2^{-\bullet} \xrightarrow{e^-} H_2O_2 \xrightarrow{e^-} OH^{\bullet} \xrightarrow{e^-} H_2O \quad (23.10)$$

$$2H^+ \qquad OH^- \qquad H$$

The metabolism of quinones by flavoenzymes can occur by either one- or two-electron reduction routes. The one-electron route often results in cytotoxicity by the formation of deleterious semiquinone radicals which can rapidly reduce dioxygen, forming superoxide anion radical $O_2^{-\bullet}$ and regenerating the quinone. Subsequent removal of this radical and other ROR by dismutation of superoxide dismutase rapidly leads to conditions of oxidative stress. Toxicity ensues as redox cycling of the quinone continues (see Figure 23.1).

However, the two-electron route occurs in competition with the one-electron route forming hydroquinones without production of the free radical semiquinone

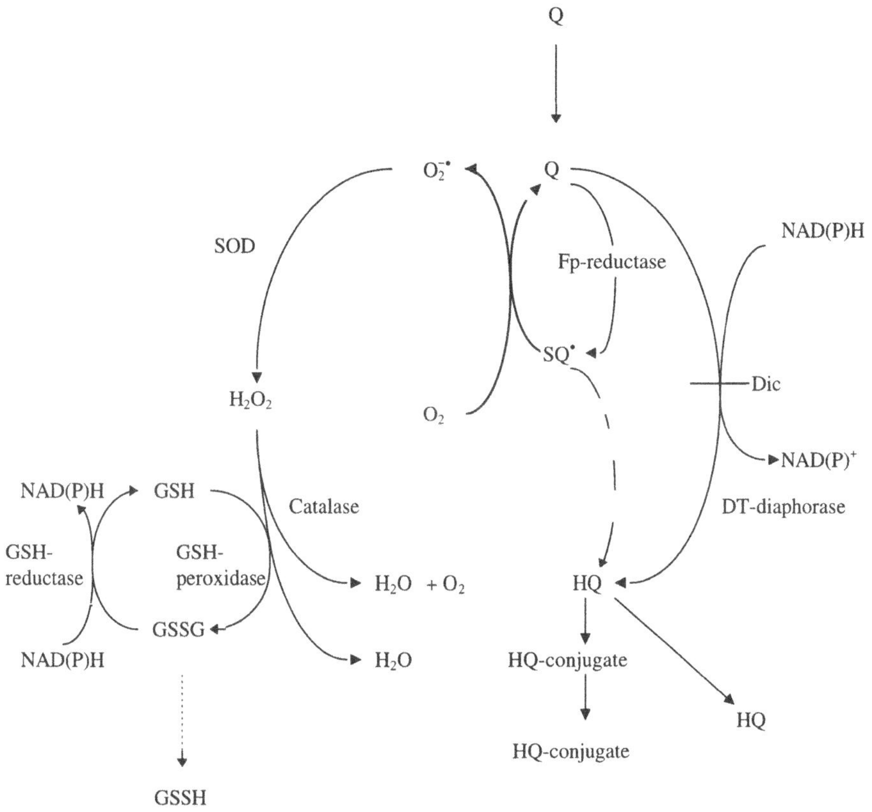

Figure 23.1 Scheme showing the metabolism of quinone *in vivo*

intermediates (see Figure 23.1). This DT-diaphorase-catalysed reaction serves to protect the cells as hydroquinones are often less reactive and more easily excreted from the cell than are semiquinone radicals.

Confusion inevitably ensues over the mechanism of toxic action for the quinones and naphthoquinones. These problems are exacerbated by the paucity of toxicity data available for QSAR analysis. With the exception of the limited data sets that are mentioned herein there appear to be few, if any, reliable data sets are available for analysis. There are also a number of experimental problems that must be overcome when testing quinones and naphthoquinones. These include such problems as poor aqueous solubility and solvent effects such as hydrolysis in aqueous solution that will alter the test solution, thus adding to uncertainty in the quantification of the concentration in the test solutions.

As a rule of thumb it is recommended that the absolute minimum number of compounds required for a QSAR analysis is five to minimise the risk of chance correlations. Furthermore, only one physico-chemical parameter should be included in a QSAR per five compounds; thus if two parameters are to be included in an equation the number of compounds required rises to ten (Topliss and Costello, 1972). This means that for some of the consistent toxicity values that are published, there are insufficient data to perform a meaningful QSAR analysis.

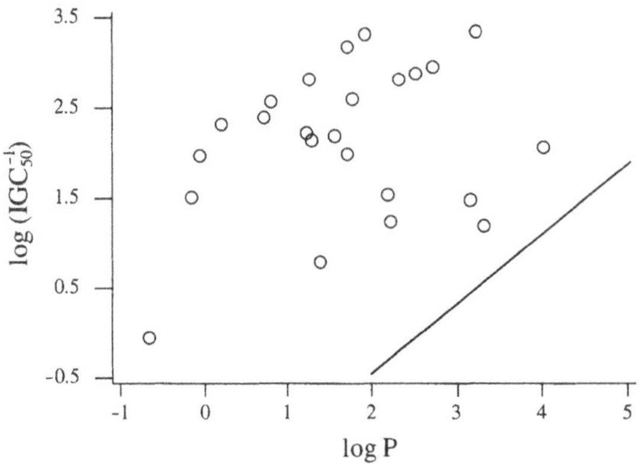

Figure 23.2 Plot of the toxicity of quinones and naphthoquinones to *Tetrahymena pyriformis* against hydrophobicity (data from Schultz *et al.*, 1997 and Schultz and Bearden, 1998). Also shown is the QSAR for non-polar narcosis (equation 23.11)

Despite these problems, reliable toxicity data for quinones and naphthoquinones are provided by Schultz *et al.* (1997b) and Schultz and Bearden (1998) respectively. These toxicity data are measured in *Tetrahymena pyriformis*, a unicellular ciliate, over forty hours (Schultz, 1996). Figure 23.2 shows the relationship of the toxicity data for these compounds with log P. Also shown on Figure 23.2 is the QSAR for the toxicity of non-polar narcotics to *Tetrahymena pyriformis*, the so-called baseline toxicity QSAR (Schultz, 1996):

$$\log (IGC_{50}^{-1}) = 0.8 \log P - 2.0 \qquad (23.11)$$

A number of conclusions can be drawn from Figure 23.2. Firstly, the toxicity of quinones and naphthoquinones is elevated above that of non-polar narcosis. This confirms that the mechanism of action of these compounds is 'bioreactive'. Secondly, while general trends are discernable, hydrophobicity is clearly not the main physico-chemical property affecting toxicity.

In their analysis of the eighteen quinones for which the toxicity could be measured, Schultz *et al.* (1997b) noted a number of trends, but were unable to establish a QSAR to describe the toxicity of all eighteen compounds. The reason for this undoubtedly was the 'mixture' of mechanisms of action within this class of compounds. Despite this a number of conclusions could be drawn from the analysis of the toxicity data. *ortho*-Substituted quinones were found to be the most toxic. In addition, the substitution of the 1,4-benzoquinone moeity affected toxicity. The presence of electron-donating substituents (such as methoxy or hydroxy groups) also decreased toxicity. In contrast, the fully substituted tetramethyl- and tetrachloro-derivatives were significantly less toxic than 1,4-benzoquinone (despite a noticeable increase in hydrophobicity of the latter compound). It was hypothesised that the *mono*- and *di*-substituted benzoquinones were able to arylate and the electrophilic mechanism of toxic action predominates. However, the fully substituted benzoquinones were not able to undergo electrophilic reactions and excess toxicity was caused by redox cycling.

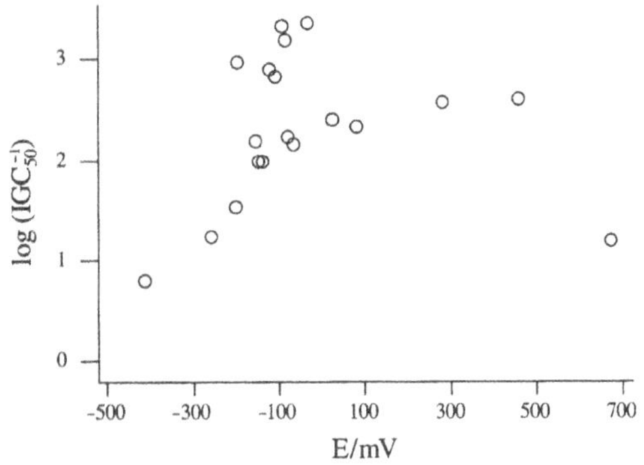

Figure 23.3 Plot of the toxicity of quinones and naphthoquinones to *Tetrahymena pyriformis* against redox potential (E/mV) (data from Schultz *et al.*, 1997 and Schultz and Bearden, 1998)

Within the data set, there are five methyl-substituted 1,4-benzoquinone derivatives that have measured redox potentials within the known 'window' of reductive activity. For these potential redox cyclers toxicity is found to be independent of hydrophobicity and electrophilicity, but is strongly correlated with the redox potential:

$$\log(IGC_{50}^{-1}) = 0.004\ (E/mV) + 2.45 \qquad (23.12)$$

$$n = 5 \quad r = 0.938 \quad s = 0.18 \quad F = 21$$

Thus it may be concluded that toxicity is related directly to the ability of compounds to redox cycle and the extent of toxicity is dependent upon the magnitude of this ability.

This point is further illustrated by the analysis of the toxicity to *Tetrahymena pyriformis* of eight naphthoquinones by Schultz and Bearden (1998). Yet again there is no observable relationship with hydrophobicity (see Figure 23.2). However, for those compounds with redox potentials within the 'window' of redox cycling (i.e. –240 to 25 mV) a strong correlation is observed with the redox potential:

$$\log(IGC_{50}^{-1}) = 0.0127\ (E/mV) + 4.14 \qquad (23.13)$$

$$n = 7 \quad r = 0.940 \quad s = 0.27 \quad F = 38$$

One compound, 2-hydroxy-1,4-naphthoquinone, has a redox potential of –415 E/mV and so falls outside the 'window'. It is, however, more toxic than predicted by the above equation and has toxicity considerably greater than baseline non-polar narcosis. It must be assumed therefore that it elicits its toxicity through an electrophilic mechanism of action (Schultz and Bearden, 1998). It is of interest in the comparison of the relative effects to different test systems that Munday *et al.* (1995) reported that 2-hydroxy-1,4-naphthoquinone is less toxic than 2-methyl-1,4-naphthoquinone to fibroblasts. This may indicate that the *Tetrahymena* test system is more susceptible than fibroblasts to electrophilic toxicants.

The toxicity data for both the naphthoquinone and quinone data sets are plotted against redox potential in Figure 23.3. This shows that despite the closely related chemical structures there is no universal relationship of toxitity for these compounds to redox cycle.

Further evidence for the importance of redox cycling in the production of toxic effects was provided by Moret *et al.* (1996). These workers studied a range of antitumour, *in vivo* toxicities, and other biological activities for a group of quinones and naphthoquinones substituted with a range of functional groups including alkyl and halogen groups and most notably aziridinyl. Despite the range of exotic substituents, cytotoxicity (as expressed by one of the principal components following principal component analysis on the whole range of biological activities) was found to be related to the half-wave potential of the adduct, although no formal QSAR representing this fact was presented. This strengthens the theory that even for a structurally complex set of quinones toxicity is largely governed by the ability to redox cycle, and by formation of the associated free radicals. Thus, if toxicity is governed by this ability, then predictions must utilise this physico-chemical property.

The role of redox cycling in the production of free radical-related toxicity is exemplified further by Munday *et al.* (1994). In this study the authors measured a number of *in vivo* and *in vitro* toxicological parameters for seven 2-alkyl substituted 1,4-napthoquinones. The alkyl substituents were methyl, ethyl, propyl, butyl, pentyl, hexyl and decyl. There were good correlations between the responses *in vivo* and *in vitro*, with 2-methyl-1,4-naphthoquinone being the most active in terms of toxicity, which was found to decrease with increasing alkyl chain length. This may be explained by the increase in the reducing power of alkyl substitution on the α-carbon ($C^{\bullet}$), whereas alkyl substitution on the O or N attatched to that carbon lowers it. The reducing power of these radicals may be deduced from the rate constants of their reactions or from measurements of redox potential or the polarographic half wave potential for the adducts (Neta, 1989).

Also measured by Munday *et al.* (1994) was the ability of each compound to produce superoxide. This was studied as the abundance of oxygen *in vivo* aids the rapid formation of peroxyl radicals by electron transfer from carbon-centred radicals to O_2.

$$^{\bullet}CH_3 + O_2 \rightarrow CH_3O_2^{\bullet} \quad k_4 = \sim 5 \times 10^9\ M^{-1}\ s^{-1} \qquad (23.14)$$

Peroxyl radicals ($CH_3O_2^{\bullet}$) produced this way are more labile and may either react as an oxidant or undergo non-catalysed or base catalysed decomposition to form $O_2^{-\bullet}$ which is less reactive and therefore may reduce or oxidise.

Once again, the 2-methyl substituted 1,4-napthoquinones produced the highest levels of superoxide, with free radical production decreasing as alkyl chain lengthened. Munday *et al.* (1994) postulated that the increase in alkyl chain length sterically hinders free radical formation. Unfortunately the authors did not report the redox potential for the compounds tested so it is not possible to ascertain whether or not they all fall within the 'window' for redox cycling. Wisely, the authors did not attempt a QSAR study on these data. It should be noted that whilst these data all fall within the criterion of having been measured within one laboratory, their use for QSAR is rather inappropriate. For instance, had log P been measured, or calculated, inverse linear relationships with the toxicological parameters (i.e. decreasing toxicity with increasing log P) would have been observed. This could lead to confusion regarding mechanisms of action.

This study also demonstrates that with a more judicious choice of chemicals to test (i.e. those with a more varied set of substituents such as the inclusion of oxygen or nitrogen), considerably more toxicological and structural information could have been gained from these analyses. A number of other similar datasets, based broadly on a naphthoquinone-type structures, are available. These were reviewed by Ahn and Sok (1996) who noted that, generally, the toxicity or anticancer effect data of compounds with the basic naphthoquinone-type structures show decreased efficacy and ability of the compound to undergo redox cycling with increasing hydrophobic content and steric hindrance respectively.

Intracellular redox cycling is a diversion from the normal electron transport whereby the xenobiotic competes with the endogenous cytochromes for electrons. Therefore, to summarise, the redox potential of xenobiotics is the overwhelming controlling factor for the prediction of the acute toxicity of compounds able to generate reactive free radical adducts. Despite this, to date, no one QSAR has been established satisfactorily for the prediction of acute toxicity. However, it is now understood that the toxic effects of quinone type compounds may be species specific and triggered by a mixture of two toxic mechanisms of action depending firstly on the initiating radical and the radical adduct and secondly on electrophilic reactivity.

23.4 Skin sensitisation

Reactive oxygen species (ROS) and other free radicals are responsible for a large number of deleterious dermatological responses (Maccarrone *et al.*, 1997). Skin sensitisation results from a compound penetrating the skin, reacting with a skin protein and hence initiating the immunological response. Initial contact does not result in the allergic response but subsequent challenge will do so (Payne and Walsh, 1994). It is an important concern for any industry whose products may come into contact with the skin (e.g. cosmetics, detergents, etc.). There are many types of functional groups with the potential to cause skin allergy which is due predominantly to electrophilic reactions between the hapten (compounds with the potential to cause allergy) and the skin protein. Also implicated in causing allergy are compounds that can produce free radicals (Cronin and Basketter, 1994).

The identification of free radical generators that may cause skin sensitisation has been attempted. Structural features recognised for these particular free radical generators have been established. Examples of these include quinones (Barratt *et al.*, 1994) and di- or polyhydroxy compounds, haloamides, diacyl or picroyl peroxides, and hydroperoxides (Payne and Walsh, 1994). This information forms the basis of rules that exist within the DEREK knowledge-based expert system for predicting toxicity (Ridings *et al.*, 1996). The role of expert systems for predicting toxicity is reviewed elsewhere (Cronin and Dearden, 1995d; Benfenati and Gini, 1997; Dearden *et al.*, 1997; Cronin, 1998). Once a compound is inputted into DEREK the structure is 'searched' for the presence of these fragments and such fragments are highlighted. This is a relatively simplistic approach to hazard identification and while it may be thorough (i.e. it identifies all skin sensitisers correctly in a test set) it is dogged by the problem of falsely predicting non-sensitisers as sensitisers (Barratt and Basketter, 1994). This is undoubtedly as a result of the prediction failing to take into account modulating factors of substituents on the structural fragments being identified. Other problems have arisen from prediction of ionised compounds as false positives (Barratt and Basketter, 1994).

More detailed investigations have been performed on sub-sets of compounds causing skin sensitisation. Whilst Basketter *et al.* (1992) found that hydrophobicity is clearly important, other workers have studied quinone-type compounds. Cremer *et al.* (1987) demonstrated that molecular orbital parameters could account for the nucleophilic reactivity of some skin sensitisers; however, a data set including only one compound known not to be a skin sensitiser was chosen, so true quantification was not possible.

Bertrand *et al.* (1997) hypothesised that eugenol-type compounds could be metabolised to the *ortho*-quinone, or quinone methide intermediate. However, neither of these studies discussed in detail the possible role or influence of radicals upon the skin sensitisation process.

The DEREK expert system for toxicity prediction has been proposed as an approach to identify any compound that has the capability to generate a free radical. Although the method has global application, it does not contain sufficient chemical knowledge to differentiate between compounds that may contain other modulating features that would prevent redox cycling. This approach is of interest however and it should be remembered that compounds capable of undergoing free radical reactions should not be considered solely to be indicative of skin sensitisation. However, this ability may contribute to any of the toxicities described in this chapter.

23.5 Teratogenicity

That free radicals have undoubtedly been implicated as one of the causes of birth defects, miscarriage, or reduction in live birth weight is of utmost concern. Therefore, the prediction of such toxic effects via a QSAR approach would be advantageous. However, despite teratogenicity being a vital endpoint to consider, there are few toxicity data available and so little QSAR analysis has been performed (Schultz and Dawson, 1990; Cronin and Dearden, 1995c).

From those teratogenicity data available, Hansch and Zhang (1995) evaluated a number of studies. Three QSARs were derived from the data of Oglesby *et al.* (1992). These are for the toxicity of phenols to rat embryos *in vitro*. The re-analysis reveals the following three surprisingly similar QSARs (note that intercepts are not reported, neither are the original data in order to verify the equations):

Tail Defect Test:

$$\log 1/C = -0.56\ \sigma^{+} \tag{23.15}$$

$n = 9 \quad r = 0.913$

where σ^{+} is the Hammett constant for electron-accepting substituents.

Tail Defect with Hepatocytes:

$$\log 1/C = -0.58\ \sigma^{+} \tag{23.16}$$

$n = 10 \quad r = 0.912$

Somite Test:

$$\log 1/C = -0.65\ \sigma^+ \qquad (23.17)$$

$n = 11 \quad r = 0.910$

The above QSARs are similiar to the following QSAR for the $(CH_3)_3CO^{\bullet}$ radical reaction with phenol reported below:

$$\log k = -0.71\ \sigma^+ + 0.73 \qquad (23.18)$$

$n = 7 \quad r = 0.990$

Hansch and Zhang (1995) noted that hydrophobicity was not found to be important to induce toxicity in these QSARs, but rather the high electron density on the aromatic ring. Indeed, the interaction of π and *p*-electrons of substituents with the π-electron system of the aromatic ring results in a resonance substituent effect, i.e. the Hammett characteristic, and causes the polarisation of the ring system. Electron-donating substituents will direct electron density to the ring system resulting in a partial positive charge at the substituent. This explains the increasing reactivity of aromatic compounds with electrophilic reactants. In contrast, electron-withdrawing substituents will shift π-electrons in the opposite direction and a partial negative charge will arise on the substituent. Thus, electron-withdrawing substituents accelerate the reaction of aromatic compounds with nucleophilic reactants (increasing nucleophilic radicals) and in contrast, decelerate their reaction with electrophiles.

Moreover, the interaction of substituents with π-electrons in the aromatic ring directly affects the dissociation energy of the C—H bond that must be broken to yield a free radical. Clearly, for the phenols, the ability to form the phenoxyl radical is important since its reactivity is directly related to the electron withdrawing properties of the substituents as modelled by the Hammett constant σ^+ (equations 23.15–23.18). Therefore, this resonance effect is the most important component of total substituent effect used to model reactivity of phenolic compounds. The authors suggest this is evidence for the reactive free radical toxicity of phenols to produce birth defects. Nevertheless, it is important to note that the use of Hammett constant to predict the relative toxicity of xenobiotics depends upon the nature of the aromatic compound and the mechanism of the reaction studied.

The ability of aromatic compounds to form and/or react with free radicals has been modelled by Hammett constants as described (equations 23.15–23.18). However, polar (inductive) effects may also play an important role for free radical reactions. There are two main arguments; the transition states of free radical reactions are stabilised by polar structures I, II, III below

$$R^{\bullet} + H{-}Ar \rightarrow [\ \underset{\text{I}}{R...H...Ar} \leftrightarrow \underset{\text{II}}{\overset{\delta^-}{R}...H...\overset{\delta^+}{Ar}} \leftrightarrow \underset{\text{III}}{\overset{\delta^+}{R}...H...\overset{\delta^-}{Ar}}\ ] \rightarrow RH + Ar^{\bullet} \qquad (23.19)$$

and polar effects arise through 'space transmission of electrons'. In this way, the reactivity of a compound to generate a free radical is a measure of its ability to allow hydrogen abstraction by the free radical. Indeed, this is valid for the formation of secondary aromatic free radicals by hydrogen abstraction reactions (Afanas'ev, 1989).

The QSAR models (equations 23.15 to 23.18) above demonstrate the ability of QSAR to elucidate mechanisms of action described by Hammett constant σ^+. However their use for predictive purposes is restricted to a very limited number of compounds because of the scarce availability of biological activity data.

Richard and Hunter (1996) also showed that QSAR analysis is possible on a limited congeneric series. The following model was developed for the developmental toxicity (log $1/BC_m$ – the μM concentration required to produce a 5% increase in the number of mouse embryos with neural tube defects) of ten haloacetic acids:

$$\log 1/BC_m = 1.41\ pKa - 42.8\ E_{LUMO} + 5.77 \qquad (23.20)$$

$n = 10 \quad r = 0.960 \quad s = 0.38 \quad$ F not given

The authors demonstrate that all ten haloacetic acids are active via the same mechanism of action described by E_{LUMO} and pKa. They postulated that the mechanism of action is based upon the effect of the formation of a pH gradient between the maternal and embryonic compartments which favours the accumulation of weak acids in the foetal tissue. Once in the embryo the acids undergo some form of redox cycling. This is confirmed by equation 23.20, with E_{LUMO} being related to both electrophilicity and redox cycling (see above for details on the latter). Unfortunately, the redox potentials of the acids themselves were not used as parameters in this analysis. It should be noted however that this analysis was restricted solely to the ten haloacetic acids; as the E_{LUMO} values of the ten compounds are all within 0.1 of each other, any large structural variation would take the values outside the range of the training set.

The excellent review of Schultz and Dawson (1990) demonstrates that whilst congeneric series of data are available, few QSARs can be developed. Those that have been developed are class-based models. They cannot be used to extrapolate outside this class. Furthermore, one is not able to infer a great deal concerning the potential mechanisms of action, perhaps because of multiple mechanisms of action, including those involving free radicals, coming into play.

23.6 Mutagenicity

Mutagenesis is considered to be one of the precursors of carcinogenesis. The QSAR studies of mutagenic activity have emphasised the potential role of compounds that can generate, by pro-carcinogen activation, deleterious secondary free radicals *in vivo*. Free radicals have been implicated in the mutagenesis of cells resulting from the disruption of DNA or chromosomal aberration. The application of QSAR techniques to the study of mutagenic effects has been widespread and well reviewed (Cronin and Dearden, 1995c; Benigni and Giuliani, 1996; Benefenati and Gini, 1997). The reasons for this are undoubtedly the importance of the endpoint for risk assessment purposes, that a large number of reproducible toxicity data have been established in short-term *in vitro* tests such as the Ames test, and that there is a strong mechanistic basis to this toxicity.

Hakura *et al.* (1994, 1995) demonstrated that the mutagenic activity of naphthoquinones and benzoquinones, respectively, was a result of a mixture of toxic mechanisms of action, namely electrophilic reactivity and the ability to be reduced to secondary free radicals. The cytotoxicity of naphthoquinones and benzoquinones was related to the oxidative damage produced by their free radicals when they were metabolically activated by reactive oxygen species such as hydroxyl radical ($OH^{\bullet}$) and superoxide anion radical ($O_2^{-\bullet}$). The nature of substituents on the quinones affected the reduction potential. While the authors did not attempt to quantify the relationships, their studies demonstrate that similar mechanisms of action are important in both acute toxicity and mutagenicity.

Electrophilic activity and the ability to be reduced to free radicals mechanisms in relation to mutagenicity are demonstrated more quantitatively by Mekenyan *et al.* (1996). These authors studied chromosomal aberrations and spindle disturbances in mammalian liver cells for eight regioisomers of pyrene, benzo(a)pyrene and phenanthrene quinones. Reactivity descriptors were calculated to account for both alkylating effects and the production of the respective intermediate anion-radical. After analysis of the quantum chemical information it was concluded that the genotoxic activity of the studied quinones was an integrated effect of the two mechanisms.

The value of quantum chemical calculations to establish reaction mechanism was further illustrated by Tsai *et al.* (1994) on a study of the genotoxicity of a series of alkylbenzenes. While a metabolic pathway involving a radical transition was proposed, quantum chemical calculations demonstrated that this is not in fact the underlying mechanism of genotoxicity for alkylbenzenes. Calculations on various aspects of the metabolic route indicated that it is the stability of the carbonium ion (calculated as the differences in the heats of formation of the carbonium ions and an intermediate alcohol) that controls whether or not the compound is genotoxic. Compounds forming stable carbonium ions were deemed likely to be genotoxic.

Crebelli *et al.* (1995) identified halogenated aliphatic hydrocarbons that were able to produce free radical species. A test set of fifty-five compounds was tested for the induction of mitotic chromosome malsegregation, mitotic arrest and lethality in the mould *Aspergillus nidulans*. As part of this study it was observed that for a subset of twenty-seven compounds there was a partial correlation between the ability to induce lipid peroxidation (a crude assessment of the ability to form free radicals) and chromosome segregration disturbance. These parameters provided a discriminant model to identify free radical producers. The ability to induce lipid peroxidation was found to dependent on electrophilicity (as measured by E_{LUMO}) and the ease of homolytic cleavage (as measured by the greatest bond length between a carbon and a halogen). Clearly this has application as a predictive tool for the halogenated aliphatic hydrocarbons, but will not be applicable outside this group. The use of the distance parameter, however, may suggest that this, or other quantum chemical parameters, could be applicable to other groups of compounds undergoing homolytic cleavage.

Another means of identifying compounds that may form reactive species has been proposed by Lewis *et al.* (1994). They have developed the Computer-Optimised Molecular Parametric Analysis of Chemical Toxicity (COMPACT) method to predict whether a chemical has the potential to act as a substrate for one or more of the cytochromes P450 (see Dearden *et al.* (1997) for a review of this technique). The COMPACT methodology proposes that the structural characteristics of a molecule determine its ability to fit into the binding site on an enzyme (Lewis, 1997). Amongst the family of cytochromes P_{450}, substrate activation of CYP2E results in the generation of reactive oxygen species. Lewis *et al.* (1993) proposed that CYP2E substrates are characterised by a molecular diameter of less than 6.5 Å. Such measures of molecular diameter are easily obtained by molecular modelling techniques. Lewis *et al.* (1994) demonstrated the applicability of this technique to the identification of CYP2E substrates by determining the molecular diameters of 19 acyclic terpenes commonly used as food additives. In each case the molecular diameter was greater than 6.5 Å, implying that none of the acyclic terpenes considered could form reactive species. The use of such an easily calculated parameter as molecular diameter, which requires neither the compound to be available, nor complex molecular orbital calculations, could have widespread applicability.

Smith *et al.* (1996) have demonstrated the use of EVA (Eigen-VAlue) descriptors to model the mutagenicity of 39 compounds with published Ames data. EVA is the vector format containing the summary of the molecule's infrared vibrational spectrum in the frequency range of 0–400 cm^{-1}. This model was developed with EVA descriptors that had a 81.0% goodness of fit for the prediction of mutagenicity of a further 19 compounds indicated by the Ames test. However, more effort is required to find the association between the EVA descriptors and the mechanism of action for mutagenicity. The EVA descriptor may be an alternative parameter for the modelling of the free radical propagation, and hence mutagencity, of compounds.

Mutagenicity is a mechanistically complex toxicity endpoint that is a precursor to many cancers. Free radicals are implicated in the mutagenic process. While the identification of compounds capable of forming free radicals has potential benefit (e.g. from a knowledge-based expert system), this is by no means the whole story. Redox potential is an important indicator of the ability of compounds to become free radicals, but more subtle effects are highlighted by the use of quantum chemical parameters such as E_{LUMO}, electron density, bond energy and resonance energy. Such parameters could be used to give an indication of the ease of metabolic activation of xenobiotics to secondary free radical species.

23.7 Conclusions

QSAR is a useful technique to predict the toxicity of compounds either for risk assessment purposes of new compounds, or to prioritise the testing of existing compounds. Efforts to use QSAR to predict the toxicity associated with free radicals are hampered by a number of fundamental issues. Reliable toxicity data for modelling are scarce and difficult to come by due to, for example, the problems of the short lives of free radicals. Such experimental problems are exacerbated by the relatively long duration of toxicity tests compared to the stability of free radicals. QSAR modelling of the toxicity of xenobiotics is also made more complex by the inevitable mixture of mechanisms, electrophilic or free radical generating in nature, which may contribute to the biological activity. For this reason, the redox potential of xenobiotics may be the best descriptor for the prediction of the acute toxicity of compounds able to generate free radical adducts.

The identification of compounds likely to generate free radicals is possible also from knowledge-based expert systems. While local models based on very specific classes of compounds or mechanisms of toxic are available, these as yet seem only poorly developed for combination with the knowledge-based systems. The prediction of a range of toxicological endpoints resulting from free radicals is thus, as yet, relatively poorly developed. More reliable data, and more mechanistically based modelling, are required.

Acknowledgment

The authors gratefully acknowledge the input provided by Dr Terry Schultz and the critical assessment of this manuscript by Prof. John Dearden.

Note

1 The correlation coefficient r is being used as a basic indicator of the goodness of fit of the correlations where ± 1 is the best fit and 0 is the worst fit. Many QSARs are expressed in a linear or multiple regression form with n = number of data points, s = standard error of estimate which is a measure of the predictive ability of the QSAR for compounds of similiar structure to the ones in the training set, and F = Fisher's test which gives an indication of the significance of the QSAR model. A credible QSAR should have 5 compounds per parameter, small s value and a large F value.

References

Adams, G.E., Michael, B.D. (1967) Pulse radiolysis of benzoquinone and hydroquinone. Semiquinone formation by water elimination from trihydroxycyclohexadienyl radicals. *Transaction of the Faraday Society* 63:1171–1175.

Afanas'ev, I.B. (1989) Radical reactions of aromatic compounds. In *Chemical kinetics of small organic radicals: Vol IV. Reactions in special systems.* Alfassi, Z.B. (ed.) CRC Press, Boca Raton, Florida. pp. 97–159.

Ahn, B.-Z., Sok, D.E. (1996) Michael acceptors as a tool for anticancer drug design. *Current Pharmaceutical Design* 2:247–262.

Anderson, R.F. (1983) The biomolecular decay rates of the flavosemiquinones of riboflavin, FMN, and FAD. *Biochimica et Biophysica Acta* 723:78–82.

Aruoma, O.I. (1994) Nutrition and health aspects of free radicals and antioxidants. *Food and Chemical Toxicology* 32:671–683.

Barratt, M.D., Basketter, D.A. (1994) Structure-activity relationships for skin sensitization: an expert system. In: Rougier, A., Goldberg, A.M., Maibach, H.I. (Eds) *In vitro skin toxicology: irritation, phototoxicity, sensitization.* Mary Ann Liebert, New York, pp. 293–301.

Barratt, M.D., Basketter, D.A., Chamberlain, M., Admans, G.D., Langowski, J.J. (1994) An expert system rulebase for identifying contact allergens. *Toxicology in Vitro* 8:1053–1060.

Basketter, D.A., Roberts, D.W., Cronin, M., Scholes, E.W. (1992) The value of the local lymph node assay in quantitative structure-activity investigations. *Contact Dermatitis* 27:137–142.

Benfenati, E., Gini, G. (1997) Computational predictive programs (expert systems) in toxicology. *Toxicology* 119:213–225.

Benigni R., Giuliani, A. (1996) Quantitative structure-activity relationship (QSAR) studies of mutagens and carcinogens. *Medical Research Reviews* 16:267–284.

Bertrand, F., Basketter, D.A., Roberts, D.W., Lepoittevin, J.P. (1997) Skin sensitization to eugenol and isoeugenol in mice: possible metabolic pathways involving *ortho*-quinone and quinone methide intermediates. *Chemical Research in Toxicology* 10:335–343.

Bradbury, S.P., Mekenyen, O., Veith, G.D., Zaharieva, N. (1995) SAR models for futile metabolism: one-electron reduction of quinones, phenols and nitrobenzenes. *SAR and QSAR in Environmental Research* 4:109–124.

Crebelli, R., Andreoli, C., Carere, A., Conti, L., Crochi, B., Cotta-Ramusino, M., Benigni, R. (1995) Toxicology of halogenated aliphatic hydrocarbons: structural and molecular determinants for the disturbance of chromosome segregation and the induction of lipid peroxidation. *Chemico-Biological Interactions* 98:113–129.

Cremer, D., Hausen, B.M., Schmalle, H.W. (1987) Toward a rationalization of the sensitizing potency of substituted *p*-benzoquinones: reaction of nucleophiles with *p*-benzoquinones. *Journal of Medicinal Chemistry* 30:1678–1681.

Cronin, M.T.D. (1998) Computer-aided prediction of drug toxicity in high throughput screening. *Pharmacy and Pharmacology Communications* 4: 157–163.

Cronin, M.T.D., Basketter, D.A. (1994) Multivariate QSAR analysis of a skin sensitization database. *SAR and QSAR in Environmental Research* 2:159–179.

Cronin, M.T.D., Dearden, J.C. (1995a) QSAR in Toxicology 1. Prediction of Aquatic Toxicity. *Quantitative Structure-Activity Relationships* 14:1–5.

Cronin, M.T.D., Dearden, J.C. (1995b) QSAR in Toxicology 2. Prediction of Acute Mammalian Toxicity and Interspecies Relationships. *Quantitative Structure-Activity Relationships* 14:117–120.

Cronin, M.T.D., Dearden, J.C. (1995c) QSAR in Toxicology 3. Prediction of Chronic Toxicities. *Quantitative Structure-Activity Relationships* 14:329–334.

Cronin, M.T.D., Dearden, J.C. (1995d) QSAR in Toxicology 4. Prediction of Non-lethal mammalian toxicological endpoints, and expert systems for toxicity prediction. *Quantitative Structure-Activity Relationships* 14:518–523.

Cronin, M.T.D., Schultz, T.W. (1998) Structure-toxicity relationships for three mechanisms of toxic action to *Vibrio fischeri*. *Ecotoxicology and Environmental Safety* 39:65–69.

Cronin, M.T.D., Gregory, B.W., Schultz, T.W. (1998) Response surface-based analyses of nitrobenzene toxicity to *Tetrahymena pyriformis*. *Chemical Research in Toxicology* 11:902–908.

Dearden, J.C., Barratt, M.D., Benigni, R., Bristol, D.W., Combes, R.D., Cronin, M.T.D., Judson, P.N., Payne, M.P., Richard, A.M., Tichy, M., Worth, A.P., Yourick, J.J. (1997) The development and validation of expert systems for predicting toxicity. *ATLA* 25:223–252.

Hakura, A., Mochida, H., Tsutsui, Y., Yamatsu, K. (1994) Mutagenicity and cytotoxicity of naphthoquinones for Ames *Salmonella* tester strains. *Chemical Research in Toxicology* 7:559–567.

Hakura, A., Mochida, H., Tsutsui, Y., Yamatsu, K. (1995) Mutagenicity of benzoquinones for Ames *Salmonella* tester strains. *Mutation Research* 347:37–43.

Halliwell, B., Gutteridge, J.M.C. (1990) Role of free radicals and catalytic metal ions in human disease: an overview. *Methods in Enzymology* 186:1–85.

Hansch, C., Zhang, L. (1995) Comparative QSAR: radical toxicity and scavenging. Two different sides of the same coin. *SAR and QSAR in Environmental Research* 4:73–82.

Henry, T.R., Wallace, K.B. (1995) The role of redox cycling versus arylation in quinone-induced mitochondrial dysfunction: a mechanistic approach in classifying reactive toxicants. *SAR and QSAR in Environmental Research* 4:97–108.

Henry, T.R., Wallace, K.B. (1996) Differential mechanisms of cell killing by redox cycling and arylating quinones. *Archives of Toxicology* 70:482–489.

Leo, A.J. (1993) Calculating log P(oct) from structures. *Chemical Reviews* 93:1281–1306.

Lewis, D.F.V. (1997) Quantitative structure-activity relationships in substrates, inducers and inhibitors of cytochrome P4501 (CYP1). *Drug Metabolism Reviews* 29:589–650.

Lewis, D.F.V., Ioannides, C., Parke, D.V. (1993) Validation of a novel molecular orbital approach (COMPACT) to the safety evaluation of chemicals by comparison with *Salmonella* mutagenicity and rodent carcinogenicity data evaluated by the U.S. NCI/NTP. *Mutation Research* 291:61–77.

Lewis, D.F.V., Ioannides, C., Walker, R., Parke, D.V. (1994) Safety evaluation of food chemicals by 'COMPACT'. 1. A study of some acyclic terpenes. *Food and Chemistry Toxicology* 32:1053–1059.

Maccarrone, M., Catani, M.V., Iraci, S., Melino, G., Agrò, A.F. (1997) A survey of reactive oxygen species and their role in dermatology. *Journal of the European Academy of Dermatology and Venereology.* 8:185–202.

McFarland, J.W. (1970) On the parabolic relationship between drug potency and hydrophobicity. *Journal of Medicinal Chemistry* 13:1192–1196.

Mekenyan, O., Sbrana, I., Turchi, G. (1996) QSAR for clastogenic effects induced by regioisomers of PAH quinones. *Polycyclic Aromatic Compounds* 11:253–260.

Moret, E.E., de Boer, M., Hilbers, H.W., Tollenaere, J.P., Janssen, L.H.M., Holthuis, J.J.M., Driebergen, R.J., Verboom, W., Reinhoudt, D.N. (1996) *In vivo* activity and hydrophobicity of cytostatic aziridinyl quinones. *Journal of Medicinal Chemistry* 39:720–728.

Munday, R., Fowke, E.A., Smith, B.L., Munday, C.M. (1994) Comparative toxicity of alkyl-1,4-naphthoquinones in rats: relationship to free radical production *in vitro*. *Free Radical Biology and Medicine* 16:725–731.

Munday, R., Smith, B.L., Munday, C.M. (1995) Comparative toxicity of 2-hydroxy-3-alkyl-1,4-naphthoquinones in rats. *Chemico-Biological Interactions* 98:185–192.

Neta, P. (1989) Reactions of radicals in biologically important molecules. In *Chemical kinetics of small organic radicals: Vol IV. Reactions in special systems.* Alfassi, Z.B. (ed.) CRC Press, Boca Raton, Florida. pp. 161–186.

O'Brien, P.J. (1991) Molecular mechanisms of quinone cytotoxicity. *Chemico-Biological Interactions* 80:1–41.

Oglesby, L.A., Eborn-McCoy, M.T., Logsdon, T.R., Copeland, F., Beyer, P.E., Kavlock, R.J. (1992) *In vitro* embryotoxicity of a series of para-substituted phenols: structure, activity and correlation with *in vivo* data. *Teratology* 45:11–33.

Payne, M.P., Walsh, P.T. (1994) Structure-activity relationships for skin sensitization potential: development of structural alerts for use in knowledge-based toxicity prediction systems. *Journal of Chemical Information and Computer Sciences* 34:154–161.

Powis, G., Appel, P.L. (1980) Relationship of the singlet-electron reduction potential of quinones to their reduction by flavoproteins. *Biochemical Pharmacology* 29:2567–2572.

Powis, G., Svingen, B.A., Appel, P. (1981) Quinone-stimulated superoxide formation by subcellular fractions, isolated hepatocytes and other cells. *Molecular Pharmacology* 20:387–394.

Rao, P.S. and Hayon, E. (1973) Ionization constants and spectral characteristics of some semiquinone radicals in aqueous solution. *Journal of Physical Chemistry* 77:2274–2280.

Richard, A.M., Hunter, E.S. (1996) Quantitative structure-activity relationships for the developmental toxicity of haloacetic acids in mammalian whole embryo culture. *Teratology* 53:352–360.

Richner, P., Weidenhaupt, A. (1997) Environmental risk assessment of chemical substances. *Chimia* 51:222–227.

Ridings, J.E., Barratt, M.D., Cary, R., Earnshaw, C.G., Eggington, C.E., Ellis, M.K., Judson, P.N., Langowski, J.J., Marchant, C.A., Payne, M.P., Watson, W.P., Yih, T.D. (1996) Computer prediction of possible toxic action from chemical structure: an update on the DEREK system. *Toxicology* 106:267–279.

Schultz, T.W. (1996) *Tetrahymena* in aquatic toxicology: QSARs and ecological hazard assessment. In, *Proceedings of the International Workshop on a Protozoan Test Protocol with Tetrahymena in Aquatic Toxicity Testing* (W. Pauli and S. Berger, Eds). German Federal Environmental Agency, Mauerstr 45–52, 14191 Berlin, pp. 31–66.

Schultz, T.W., Bearden, A.P. (1998) Structure-toxicity relationships for selected naphthoquinones to *Tetrahymena pyriformis*. *Bulletin of Environmental Contamination and Toxicology.* 61:405–410.

Schultz, T.W., Dawson, D.A. (1990) Structure-activity relationships for teratogenicity and developmental toxicity. In: *Practical Applications of Quantitative Structure-Activity Relationships (QSAR) in Environmental Chemistry and Toxicology.* Karcher W. and Devillers J. (eds.). Kluwer Academic Publishers, Dordrecht, pp. 389–409.

Schultz, T.W., Sinks, G.D., Cronin, M.T.D. (1997a) Identification of mechanisms of toxic action of phenols to *Tetrahymena pyiformis* from molecular descriptors. In, *Quantitative Structure-Activity Relationships in Environmental Sciences – VII* (F. Chen and G. Schuurmann Eds). SETAC Press, Pensacola, USA, pp. 329–342.

Schultz, T.W., Sinks, G.D., Cronin, M.T.D. (1997b) Quinone-induced toxicity to *Tetrahymena*: structure-activity relationships. *Aquatic Toxicology* 39:267–278.

Smith, M.D., Greenwood, R., Rees, R.W., Livingstone, D.J. (1996) A QSAR study of mutagenicity. 11th European Symposium on Quantitative Structure-Activity Relationships: Computer-assisted lead finding and optimization. September 1996, Lausanne. Poster presentation.

Topliss, J.G., Costello, R.J. (1972) Chance correlations in structure-activity studies using multiple regression analysis. *Journal of Medicinal Chemistry* 15:1066–1068.

Tsai, R.-S., Carrupt, P.-A., Testa, B., Caldwell, J. (1994) Structure-genotoxicity relationships of allylbenzenes and propenylbenzenes: a quantum chemical study. *Chemical Research in Toxicology* 7:73–76.

van Wezel, A.P., Opperhuizen, A. (1995) Narcosis due to environmental pollutants in aquatic organisms: Residue-based toxicity, mechanisms, and membrane burdens. *Critical Reviews in Toxicology* 25:255–279.

Wardman, P. (1989) Reduction potentials of one-electron couples involving free radicals in aqueous solution. *Journal of Physical Chemistry Reference Data* 18:1637–1755.

Index

Note: page numbers in *italic* type denote figures and tables

Printed and bound by CPI Group (UK) Ltd, Croydon, CR0 4YY

01/11/2024

01782599-0020